Lecture Notes in Computer Science   11361

Commenced Publication in 1973
Founding and Former Series Editors:
Gerhard Goos, Juris Hartmanis, and Jan van Leeuwen

More information about this series at http://www.springer.com/series/7412

C. V. Jawahar · Hongdong Li ·
Greg Mori · Konrad Schindler (Eds.)

Computer Vision – ACCV 2018

14th Asian Conference on Computer Vision
Perth, Australia, December 2–6, 2018
Revised Selected Papers, Part I

 Springer

Editors
C. V. Jawahar
IIIT Hyderabad
Hyderabad, India

Greg Mori
Simon Fraser University
Burnaby, BC, Canada

Hongdong Li
ANU
Canberra, ACT, Australia

Konrad Schindler (iD)
ETH Zurich
Zurich, Zürich, Switzerland

ISSN 0302-9743 ISSN 1611-3349 (electronic)
Lecture Notes in Computer Science
ISBN 978-3-030-20886-8 ISBN 978-3-030-20887-5 (eBook)
https://doi.org/10.1007/978-3-030-20887-5

LNCS Sublibrary: SL6 – Image Processing, Computer Vision, Pattern Recognition, and Graphics

This Springer imprint is published by the registered company Springer Nature Switzerland AG
The registered company address is: Gewerbestrasse 11, 6330 Cham, Switzerland

Preface

The Asian Conference on Computer Vision (ACCV) 2018 took place in Perth, Australia, during December 2–6, 2018. The conference featured novel research contributions from almost all sub-areas of computer vision.

This year we received a record number of conference submissions. After removing the desk rejects, 979 valid, complete manuscripts were submitted for review. A pool of 34 area chairs and 1,063 reviewers was recruited to conduct paper reviews. Like previous editions of ACCV, we adopted a double-blind review process to determine which of these papers to accept. Identities of authors were not visible to reviewers and area chairs; nor were the identities of the assigned reviewers and area chairs visible to authors. The program chairs did not submit papers to the conference.

Each paper was reviewed by at least three reviewers. Authors were permitted to respond to the initial reviews during a rebuttal period. After this, the area chairs led discussions among reviewers. Finally, a physical area chairs was held in Singapore, during which panels of three area chairs deliberated to decide on acceptance decisions for each paper. At the end of this process, 274 papers were accepted for publication in the ACCV 2018 conference proceedings, of which five were later withdrawn by their authors.

In addition to the main conference, ACCV 2018 featured 11 workshops and six tutorials.

We would like to thank all the organizers, sponsors, area chairs, reviewers, and authors. Special thanks go to Prof. Guosheng Lin from Nanyang Technological University, Singapore, for hosting the area chair meeting. We acknowledge the support of Microsoft's Conference Management Toolkit (CMT) team for providing the software used to manage the review process.

We greatly appreciate the efforts of all those who contributed to making the conference a success.

December 2018

C. V. Jawahar
Hongdong Li
Greg Mori
Konrad Schindler

Organization

General Chairs

Kyoung-mu Lee	Seoul National University, South Korea
Ajmal Mian	University of Western Australia, Australia
Ian Reid	University of Adelaide, Australia
Yoichi Sato	University of Tokyo, Japan

Program Chairs

C. V. Jawahar	IIIT Hyderabad, India
Hongdong Li	Australian National University, Australia
Greg Mori	Simon Fraser University and Borealis AI, Canada
Konrad Schindler	ETH Zurich, Switzerland

Advisor

Richard Hartley	Australian National University, Australia

Publication Chair

Hamid Rezatofighi	University of Adelaide, Australia

Local Arrangements Chairs

Guosheng Lin	Nanyang Technological University, Singapore
Ajmal Mian	University of Western Australia, Australia

Area Chairs

Lourdes Agapito	University College London, UK
Xiang Bai	Huazhong University of Science and Technology, China
Vineeth N. Balasubramanian	IIT Hyderabad, India
Gustavo Carneiro	University of Adelaide, Australia
Tat-Jun Chin	University of Adelaide, Australia
Minsu Cho	POSTECH, South Korea
Bohyung Han	Seoul National University, South Korea
Junwei Han	Northwestern Polytechnical University, China
Mehrtash Harandi	Monash University, Australia
Gang Hua	Microsoft Research, Asia

Rei Kawakami	University of Tokyo, Japan
Tae-Kyun Kim	Imperial College London, UK
Junseok Kwon	Chung-Ang University, South Korea
Florent Lafarge	Inria, France
Laura Leal-Taixé	TU Munich, Germany
Zhouchen Lin	Peking University, China
Yanxi Liu	Penn State University, USA
Oisin Mac Aodha	Caltech, USA
Anurag Mittal	IIT Madras, India
Vinay Namboodiri	IIT Kanpur, India
P. J. Narayanan	IIIT Hyderabad, India
Carl Olsson	Lund University, Sweden
Imari Sato	National Institute of Informatics
Shiguang Shan	Chinese Academy of Sciences, China
Chunhua Shen	University of Adelaide, Australia
Boxin Shi	Peking University, China
Terence Sim	National University of Singapore, Singapore
Yusuke Sugano	Osaka University, Japan
Min Sun	National Tsing Hua University, Taiwan
Robby Tan	Yale-NUS College, USA
Siyu Tang	MPI for Intelligent Systems
Radu Timofte	ETH Zurich, Switzerland
Jingyi Yu	University of Delaware, USA
Junsong Yuan	State University of New York at Buffalo, USA

Additional Reviewers

Ehsan Abbasnejad	Ognjen Arandjelovic	Nick Barnes
Akash Abdu Jyothi	Anil Armagan	Peter Barnum
Abrar Abdulnabi	Chetan Arora	Joe Bartels
Nagesh Adluru	Mathieu Aubry	Paul Beardsley
Antonio Agudo	Hossein Azizpour	Sima Behpour
Unaiza Ahsan	Seung-Hwan Baek	Vasileios Belagiannis
Hai-zhou Ai	Aijun Bai	Boulbaba Ben Amor
Alexandre Alahi	Peter Bajcsy	Archith Bency
Xavier Alameda-Pineda	Amr Bakry	Ryad Benosman
Andrea Albarelli	Vassileios Balntas	Gedas Bertasius
Mohsen Ali	Yutong Ban	Ross Beveridge
Saad Ali	Arunava Banerjee	Binod Bhattarai
Mitsuru Ambai	Monami Banerjee	Arnav Bhavsar
Cosmin Ancuti	Atsuhiko Banno	Simone Bianco
Vijay Rengarajan Angarai Pichaikuppan	Aayush Bansal	Oliver Bimber
	Dániel Baráth	Tolga Birdal
Michel Antunes	Lorenzo Baraldi	Horst Bischof
Djamila Aouada	Adrian Barbu	Arijit Biswas

Soma Biswas	Jiacheng Chen	James Crowley
Henryk Blasinski	Jianhui Chen	Jinshi Cui
Vishnu Boddeti	Jiansheng Chen	Zhaopeng Cui
Federica Bogo	Jiaxin Chen	Bo Dai
Tolga Bolukbasi	Jie Chen	Hang Dai
Terrance Boult	Kan Chen	Xiyang Dai
Thierry Bouwmans	Longbin Chen	Yuchao Dai
Abdesselam Bouzerdoum	Ting Chen	Carlo Dal Mutto
Ernesto Brau	Tseng-Hung Chen	Zachary Daniels
Mathieu Bredif	Wei Chen	Mohamed Daoudi
Stefan Breuers	Xi'ai Chen	Abir Das
Marcus Brubaker	Xiaozhi Chen	Raoul De Charette
Anders Buch	Xilin Chen	Teofilo Decampos
Shyamal Buch	Xinlei Chen	Koichiro Deguchi
Pradeep Buddharaju	Yunjin Chen	Stefanie Demirci
Adrian Bulat	Erkang Cheng	Girum Demisse
Darius Burschka	Hong Cheng	Patrick Dendorfer
Andrei Bursuc	Hui Cheng	Zhiwei Deng
Zoya Bylinskii	Jingchun Cheng	Joachim Denzler
Weidong Cai	Ming-Ming Cheng	Aditya Deshpande
Necati Cihan Camgoz	Wen-Huang Cheng	Frédéric Devernay
Shaun Canavan	Yuan Cheng	Abhinav Dhall
Joao Carreira	Zhi-Qi Cheng	Anthony Dick
Dan Casas	Loong Fah Cheong	Zhengming Ding
M. Emre Celebi	Anoop Cherian	Cosimo Distante
Hakan Cevikalp	Liang-Tien Chia	Ajay Divakaran
François Chadebecq	Chao-Kai Chiang	Mandar Dixit
Menglei Chai	Shao-Yi Chien	Thanh-Toan Do
Rudrasis Chakraborty	Han-Pang Chiu	Jose Dolz
Tat-Jen Cham	Wei-Chen Chiu	Bo Dong
Kwok-Ping Chan	Donghyeon Cho	Chao Dong
Sharat Chandran	Nam Ik Cho	Jingming Dong
Chehan Chang	Sunghyun Cho	Ming Dong
Hyun Sung Chang	Yeong-Jun Cho	Weisheng Dong
Yi Chang	Gyeongmin Choe	Simon Donne
Wei-Lun Chao	Chiho Choi	Gianfranco Doretto
Visesh Chari	Jonghyun Choi	Bruce Draper
Gaurav Chaurasia	Jongmoo Choi	Bertram Drost
Rama Chellappa	Jongwon Choi	Liang Du
Chen Chen	Hisham Cholakkal	Shichuan Du
Chu-Song Chen	Biswarup Choudhury	Jean-Luc Dugelay
Dongdong Chen	Xiao Chu	Enrique Dunn
Guangyong Chen	Yung-Yu Chuang	Thibaut Durand
Hsin-I Chen	Andrea Cohen	Zoran Duric
Huaijin Chen	Toby Collins	Ionut Cosmin Duta
Hwann-Tzong Chen	Marco Cristani	Samyak Dutta

Pinar Duygulu
Ady Ecker
Hazim Ekenel
Sabu Emmanuel
Ian Endres
Ertunc Erdil
Hugo Jair Escalante
Sergio Escalera
Francisco Escolano Ruiz
Bin Fan
Shaojing Fan
Yi Fang
Aly Farag
Giovanni Farinella
Rafael Felix
Michele Fenzi
Bob Fisher
David Fofi
Gian Luca Foresti
Victor Fragoso
Bernd Freisleben
Jason Fritts
Cheng-Yang Fu
Chi-Wing Fu
Huazhu Fu
Jianlong Fu
Xueyang Fu
Ying Fu
Yun Fu
Olac Fuentes
Jan Funke
Ryo Furukawa
Yasutaka Furukawa
Manuel Günther
Raghudeep Gadde
Matheus Gadelha
Jürgen Gall
Silvano Galliani
Chuang Gan
Zhe Gan
Vineet Gandhi
Arvind Ganesh
Bin-Bin Gao
Jin Gao
Jiyang Gao
Junbin Gao

Ravi Garg
Jochen Gast
Utkarsh Gaur
Xin Geng
David Geronimno
Michael Gharbi
Amir Ghodrati
Behnam Gholami
Andrew Gilbert
Rohit Girdhar
Ioannis Gkioulekas
Guy Godin
Nuno Goncalves
Yu Gong
Stephen Gould
Venu Govindu
Oleg Grinchuk
Jiuxiang Gu
Shuhang Gu
Paul Guerrero
Anupam Guha
Guodong Guo
Yanwen Guo
Ankit Gupta
Mithun Gupta
Saurabh Gupta
Hossein Hajimirsadeghi
Maciej Halber
Xiaoguang Han
Yahong Han
Zhi Han
Kenji Hara
Tatsuya Harada
Ali Harakeh
Adam Harley
Ben Harwood
Mahmudul Hasan
Kenji Hata
Michal Havlena
Munawar Hayat
Zeeshan Hayder
Jiawei He
Kun He
Lei He
Lifang He
Pan He

Yang He
Zhenliang He
Zhihai He
Felix Heide
Samitha Herath
Luis Herranz
Anders Heyden
Je Hyeong Hong
Seunghoon Hong
Wei Hong
Le Hou
Chiou-Ting Hsu
Kuang-Jui Hsu
Di Hu
Hexiang Hu
Ping Hu
Xu Hu
Yinlin Hu
Zhiting Hu
De-An Huang
Gao Huang
Gary Huang
Haibin Huang
Haifei Huang
Haozhi Huang
Jia-Bin Huang
Shaoli Huang
Sheng Huang
Xinyu Huang
Xun Huang
Yan Huang
Yawen Huang
Yinghao Huang
Yizhen Huang
Wei-Chih Hung
Junhwa Hur
Mohamed Hussein
Jyh-Jing Hwang
Ichiro Ide
Satoshi Ikehata
Radu Tudor Ionescu
Go Irie
Ahmet Iscen
Vamsi Ithapu
Daisuke Iwai
Won-Dong Jang

Dinesh Jayaraman
Sadeep Jayasumana
Suren Jayasuriya
Hueihan Jhuang
Dinghuang Ji
Mengqi Ji
Hongjun Jia
Jiayan Jiang
Qing-Yuan Jiang
Tingting Jiang
Xiaoyi Jiang
Zhuolin Jiang
Zequn Jie
Xiaojie Jin
Younghyun Jo
Ole Johannsen
Hanbyul Joo
Jungseock Joo
Kyungdon Joo
Shantanu Joshi
Amin Jourabloo
Deunsol Jung
Anis Kacem
Ioannis Kakadiaris
Zdenek Kalal
Nima Kalantari
Mahdi Kalayeh
Sinan Kalkan
Vicky Kalogeiton
Joni-Kristian Kamarainen
Martin Kampel
Meina Kan
Kenichi Kanatani
Atsushi Kanehira
Takuhiro Kaneko
Zhuoliang Kang
Mohan Kankanhalli
Vadim Kantorov
Nikolaos Karianakis
Leonid Karlinsky
Zoltan Kato
Hiroshi Kawasaki
Wei Ke
Wadim Kehl
Sameh Khamis
Naeemullah Khan

Salman Khan
Rawal Khirodkar
Mehran Khodabandeh
Anna Khoreva
Parmeshwar Khurd
Hadi Kiapour
Joe Kileel
Edward Kim
Gunhee Kim
Hansung Kim
Hyunwoo Kim
Junsik Kim
Seon Joo Kim
Vladimir Kim
Akisato Kimura
Ravi Kiran
Roman Klokov
Takumi Kobayashi
Amir Kolaman
Naejin Kong
Piotr Koniusz
Hyung Il Koo
Dimitrios Kosmopoulos
Gregory Kramida
Praveen Krishnan
Ravi Krishnan
Hiroyuki Kubo
Hilde Kuehne
Jason Kuen
Arjan Kuijper
Kuldeep Kulkarni
Shiro Kumano
Avinash Kumar
Soumava Roy Kumar
Kaustav Kundu
Sebastian Kurtek
Yevhen Kuznietsov
Heeseung Kwon
Alexander Ladikos
Kevin Lai
Wei-Sheng Lai
Shang-Hong Lai
Michael Lam
Zhenzhong Lan
Dong Lao
Katrin Lasinger

Yasir Latif
Huu Le
Herve Le Borgne
Chun-Yi Lee
Gim Hee Lee
Seungyong Lee
Teng-Yok Lee
Seungkyu Lee
Andreas Lehrmann
Na Lei
Spyridon Leonardos
Marius Leordeanu
Matt Leotta
Gil Levi
Evgeny Levinkov
Jose Lezama
Ang Li
Chen Li
Chunyuan Li
Dangwei Li
Dingzeyu Li
Dong Li
Hai Li
Jianguo Li
Stan Li
Wanqing Li
Wei Li
Xi Li
Xirong Li
Xiu Li
Xuelong Li
Yanghao Li
Yin Li
Yingwei Li
Yongjie Li
Yu Li
Yuncheng Li
Zechao Li
Zhengqi Li
Zhengqin Li
Zhuwen Li
Zhouhui Lian
Jie Liang
Zicheng Liao
Jongwoo Lim
Ser-Nam Lim

Kaimo Lin
Shih-Yao Lin
Tsung-Yi Lin
Weiyao Lin
Yuewei Lin
Venice Liong
Giuseppe Lisanti
Roee Litman
Jim Little
Anan Liu
Chao Liu
Chen Liu
Eryun Liu
Fayao Liu
Huaping Liu
Jingen Liu
Lingqiao Liu
Miaomiao Liu
Qingshan Liu
Risheng Liu
Sifei Liu
Tyng-Luh Liu
Weiyang Liu
Xialei Liu
Xianglong Liu
Xiao Liu
Yebin Liu
Yi Liu
Yu Liu
Yun Liu
Ziwei Liu
Stephan Liwicki
Liliana Lo Presti
Fotios Logothetis
Javier Lorenzo
Manolis Lourakis
Brian Lovell
Chen Change Loy
Chaochao Lu
Feng Lu
Huchuan Lu
Jiajun Lu
Kaiyue Lu
Xin Lu
Yijuan Lu
Yongxi Lu

Fujun Luan
Jian-Hao Luo
Jiebo Luo
Weixin Luo
Khoa Luu
Chao Ma
Huimin Ma
Kede Ma
Lin Ma
Shugao Ma
Wei-Chiu Ma
Will Maddern
Ludovic Magerand
Luca Magri
Behrooz Mahasseni
Tahmida Mahmud
Robert Maier
Subhransu Maji
Yasushi Makihara
Clement Mallet
Abed Malti
Devraj Mandal
Fabian Manhardt
Gian Luca Marcialis
Julio Marco
Diego Marcos
Ricardo Martin
Tanya Marwah
Marc Masana
Jonathan Masci
Takeshi Masuda
Yusuke Matsui
Tetsu Matsukawa
Gellert Mattyus
Thomas Mauthner
Bruce Maxwell
Steve Maybank
Amir Mazaheri
Scott Mccloskey
Mason Mcgill
Nazanin Mehrasa
Ishit Mehta
Xue Mei
Heydi Mendez-Vazquez
Gaofeng Meng
Bjoern Menze

Domingo Mery
Pascal Mettes
Jan Hendrik Metzen
Gregor Miller
Cai Minjie
Ikuhisa Mitsugami
Daisuke Miyazaki
Davide Modolo
Pritish Mohapatra
Pascal Monasse
Sandino Morales
Pietro Morerio
Saeid Motiian
Arsalan Mousavian
Mikhail Mozerov
Yasuhiro Mukaigawa
Yusuke Mukuta
Mario Munich
Srikanth Muralidharan
Ana Murillo
Vittorio Murino
Armin Mustafa
Hajime Nagahara
Shruti Nagpal
Mahyar Najibi
Katsuyuki Nakamura
Seonghyeon Nam
Loris Nanni
Manjunath Narayana
Lakshmanan Nataraj
Neda Nategh
Lukáš Neumann
Shawn Newsam
Joe Yue-Hei Ng
Thuyen Ngo
David Nilsson
Ji-feng Ning
Mark Nixon
Shohei Nobuhara
Hyeonwoo Noh
Mehdi Noroozi
Erfan Noury
Eyal Ofek
Seong Joon Oh
Seoung Wug Oh
Katsunori Ohnishi

Iason Oikonomidis
Takeshi Oishi
Takahiro Okabe
Takayuki Okatani
Gustavo Olague
Kyle Olszewski
Mohamed Omran
Roy Or-El
Ivan Oseledets
Martin R. Oswald
Tomas Pajdla
Dipan Pal
Kalman Palagyi
Manohar Paluri
Gang Pan
Jinshan Pan
Yannis Panagakis
Rameswar Panda
Hsing-Kuo Pao
Dim Papadopoulos
Konstantinos Papoutsakis
Shaifali Parashar
Hyun Soo Park
Jinsun Park
Taesung Park
Wonpyo Park
Alvaro Parra Bustos
Geoffrey Pascoe
Ioannis Patras
Genevieve Patterson
Georgios Pavlakos
Ioannis Pavlidis
Nick Pears
Pieter Peers
Selen Pehlivan
Xi Peng
Xingchao Peng
Janez Perš
Talita Perciano
Adrian Peter
Lars Petersson
Stavros Petridis
Patrick Peursum
Trung Pham
Sang Phan
Marco Piccirilli

Sudeep Pillai
Wong Ya Ping
Lerrel Pinto
Fiora Pirri
Matteo Poggi
Georg Poier
Marius Popescu
Ronald Poppe
Dilip Prasad
Andrea Prati
Maria Priisalu
Véronique Prinet
Victor Prisacariu
Hugo Proenca
Jan Prokaj
Daniel Prusa
Yunchen Pu
Guo-Jun Qi
Xiaojuan Qi
Zhen Qian
Yu Qiao
Jie Qin
Lei Qin
Chao Qu
Faisal Qureshi
Petia Radeva
Venkatesh Babu
    Radhakrishnan
Ilija Radosavovic
Bogdan Raducanu
Hossein Rahmani
Swaminathan Rahul
Ajit Rajwade
Kandan Ramakrishnan
Visvanathan Ramesh
Yongming Rao
Sathya Ravi
Michael Reale
Adria Recasens
Konstantinos Rematas
Haibing Ren
Jimmy Ren
Wenqi Ren
Zhile Ren
Edel Garcia Reyes
Hamid Rezatofighi

Hamed Rezazadegan
    Tavakoli
Rafael Rezende
Helge Rhodin
Alexander Richard
Stephan Richter
Gernot Riegler
Christian Riess
Ergys Ristani
Tobias Ritschel
Mariano Rivera
Antonio Robles-Kelly
Emanuele Rodola
Andres Rodriguez
Mikel Rodriguez
Matteo Ruggero Ronchi
Xuejian Rong
Bodo Rosenhahn
Arun Ross
Peter Roth
Michel Roux
Ryusuke Sagawa
Hideo Saito
Shunsuke Saito
Parikshit Sakurikar
Albert Ali Salah
Jorge Sanchez
Conrad Sanderson
Aswin Sankaranarayanan
Swami Sankaranarayanan
Archana Sapkota
Michele Sasdelli
Jun Sato
Shin'ichi Satoh
Torsten Sattler
Manolis Savva
Tanner Schmidt
Dirk Schnieders
Samuel Schulter
Rajvi Shah
Shishir Shah
Sohil Shah
Moein Shakeri
Nataliya Shapovalova
Aidean Sharghi
Gaurav Sharma

Baoyuan Wang
Chaohui Wang
Chaoyang Wang
Chunyu Wang
De Wang
Dong Wang
Fang Wang
Faqiang Wang
Hongsong Wang
Hongxing Wang
Hua Wang
Jialei Wang
Jianyu Wang
Jinglu Wang
Jinqiao Wang
Keze Wang
Le Wang
Lei Wang
Lezi Wang
Lijun Wang
Limin Wang
Linwei Wang
Pichao Wang
Qi Wang
Qian Wang
Qilong Wang
Qing Wang
Ruiping Wang
Shangfei Wang
Shuhui Wang
Song Wang
Tao Wang
Tsun-Hsuang Wang
Weiyue Wang
Wenguan Wang
Xiaoyu Wang
Xinchao Wang
Xinggang Wang
Yang Wang
Yin Wang
Yu-Chiang Frank Wang
Yufei Wang
Yunhong Wang
Zhangyang Wang
Zilei Wang
Jan Dirk Wegner

Ping Wei
Shih-En Wei
Wei Wei
Xiu-Shen Wei
Zijun Wei
Bihan Wen
Longyin Wen
Xinshuo Weng
Tom Whelan
Patrick Wieschollek
Maggie Wigness
Jerome Williams
Kwan-Yee Wong
Chao-Yuan Wu
Chunpeng Wu
Dijia Wu
Jiajun Wu
Jianxin Wu
Xiao Wu
Xiaohe Wu
Xiaomeng Wu
Xinxiao Wu
Yi Wu
Ying Nian Wu
Yue Wu
Zheng Wu
Zhirong Wu
Jonas Wulff
Yin Xia
Yongqin Xian
Yu Xiang
Fanyi Xiao
Yang Xiao
Dan Xie
Jianwen Xie
Jin Xie
Fuyong Xing
Jun Xing
Junliang Xing
Xuehan Xiong
Yuanjun Xiong
Changsheng Xu
Chenliang Xu
Haotian Xu
Huazhe Xu
Huijuan Xu

Jun Xu
Ning Xu
Tao Xu
Weipeng Xu
Xiangmin Xu
Xiangyu Xu
Yong Xu
Yuanlu Xu
Jia Xue
Xiangyang Xue
Toshihiko Yamasaki
Junchi Yan
Luxin Yan
Wang Yan
Keiji Yanai
Bin Yang
Chih-Yuan Yang
Dong Yang
Herb Yang
Jianwei Yang
Jie Yang
Jin-feng Yang
Jufeng Yang
Meng Yang
Ming Yang
Ming-Hsuan Yang
Tien-Ju Yang
Wei Yang
Wenhan Yang
Yanchao Yang
Yingzhen Yang
Yongxin Yang
Zhenheng Yang
Angela Yao
Bangpeng Yao
Cong Yao
Jian Yao
Jiawen Yao
Yasushi Yagi
Mang Ye
Mao Ye
Qixiang Ye
Mei-Chen Yeh
Sai-Kit Yeung
Kwang Moo Yi
Alper Yilmaz

Xi Yin

Zhaozheng Yin

Xianghua Ying

Ryo Yonetani

Donghyun Yoo

Jae Shin Yoon

Ryota Yoshihashi

Gang Yu

Hongkai Yu

Ruichi Yu

Shiqi Yu

Xiang Yu

Yang Yu

Youngjae Yu

Chunfeng Yuan

Jing Yuan

Junsong Yuan

Shanxin Yuan

Zejian Yuan

Xenophon Zabulis

Mihai Zanfir

Pablo Zegers

Jiabei Zeng

Kuo-Hao Zeng

Baochang Zhang

Cha Zhang

Chao Zhang

Dingwen Zhang

Dong Zhang

Guofeng Zhang

Hanwang Zhang

He Zhang

Hong Zhang

Honggang Zhang

Hua Zhang

Jian Zhang

Jiawei Zhang

Jing Zhang

Kaipeng Zhang

Ke Zhang

Liang Zhang

Linguang Zhang

Liqing Zhang

Peng Zhang

Pingping Zhang

Quanshi Zhang

Runze Zhang

Shanghang Zhang

Shu Zhang

Tianzhu Zhang

Tong Zhang

Wen Zhang

Xiaofan Zhang

Xiaoqin Zhang

Xikang Zhang

Xu Zhang

Ya Zhang

Yinda Zhang

Yongqiang Zhang

Zhang Zhang

Zhen Zhang

Zhoutong Zhang

Ziyu Zhang

Bin Zhao

Bo Zhao

Chen Zhao

Hengshuang Zhao

Qijun Zhao

Rui Zhao

Heliang Zheng

Shuai Zheng

Stephan Zheng

Yinqiang Zheng

Yuanjie Zheng

Zhonglong Zheng

Guangyu Zhong

Huiyu Zhou

Jiahuan Zhou

Jun Zhou

Luping Zhou

Mo Zhou

Pan Zhou

Yang Zhou

Zihan Zhou

Fan Zhu

Guangming Zhu

Hao Zhu

Hongyuan Zhu

Lei Zhu

Menglong Zhu

Pengfei Zhu

Shizhan Zhu

Siyu Zhu

Xiangxin Zhu

Yi Zhu

Yizhe Zhu

Yuke Zhu

Zhigang Zhu

Bohan Zhuang

Liansheng Zhuang

Karel Zimmermann

Maria Zontak

Danping Zou

Qi Zou

Wangmeng Zuo

Xinxin Zuo

Contents – Part I

Oral Session O1: Learning

Dual Generator Generative Adversarial Networks for Multi-domain Image-to-Image Translation

Hao Tang[1]([⊠]), Dan Xu[2], Wei Wang[3], Yan Yan[4], and Nicu Sebe[1]

[1] University of Trento, Trento, Italy
hao.tang@unitn.it
[2] University of Oxford, Oxford, UK
[3] École Polytechnique Fédérale de Lausanne, Lausanne, Switzerland
[4] Texas State University, San Marcos, USA

Abstract. State-of-the-art methods for image-to-image translation with Generative Adversarial Networks (GANs) can learn a mapping from one domain to another domain using unpaired image data. However, these methods require the training of one specific model for every pair of image domains, which limits the scalability in dealing with more than two image domains. In addition, the training stage of these methods has the common problem of model collapse that degrades the quality of the generated images. To tackle these issues, we propose a Dual Generator Generative Adversarial Network (G^2GAN), which is a robust and scalable approach allowing to perform unpaired image-to-image translation for multiple domains using only dual generators within a single model. Moreover, we explore different optimization losses for better training of G^2GAN, and thus make unpaired image-to-image translation with higher consistency and better stability. Extensive experiments on six publicly available datasets with different scenarios, *i.e.*, architectural buildings, seasons, landscape and human faces, demonstrate that the proposed G^2GAN achieves superior model capacity and better generation performance comparing with existing image-to-image translation GAN models.

Keywords: Generative Adversarial Network ·
Image-to-image translation · Unpaired data · Multi-domain

1 Introduction

Generative Adversarial Networks (GANs) [6] have recently received considerable attention in various communities, *e.g.*, computer vision, natural language processing and medical analysis. GANs are generative models which are particularly designed for image generation tasks. Recent works have been able

This work was supported by NIST 60NANB17D191. We gratefully acknowledge the gift donations of Cisco, Inc. and the support of NVIDIA Corporation with the donation of the GPUs used for this research. This article solely reflects the opinions and conclusions of its authors and neither NIST, Cisco, nor NVIDIA.

C. V. Jawahar et al. (Eds.): ACCV 2018, LNCS 11361, pp. 3–21, 2019.
https://doi.org/10.1007/978-3-030-20887-5_1

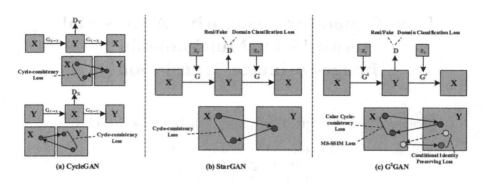

Fig. 1. A motivation illustration of the proposed G^2GAN (c) compared with CycleGAN (a) [49] and StarGAN [5] (b). For the multi-domain image-to-image generation task w.r.t. m image domains, CycleGAN needs to train $m(m-1)$ generator/discriminator pairs, while the proposed G^2GAN only needs to train dual generators and one discriminator. StarGAN [5] shares the same generator for both the translation and reconstruction tasks, while G^2GAN employs task-specific generators (G^t and G^r) which allow for different network designs and different levels of parameter sharing.

to yield promising image-to-image translation performance (*e.g.*, pix2pix [8] and BicycleGAN [50]) in a supervised setting given carefully annotated image pairs. However, pairing the training data is usually difficult and costly. The situation becomes even worse when dealing with tasks such as artistic stylization, since the desired output is very complex, typically requiring artistic authoring. To tackle this problem, several GAN approaches, such as Cycle-GAN [49], DualGAN [47], DiscoGAN [10], ComboGAN [1] and DistanceGAN [3], aim to effectively learn a hidden mapping from one image domain to another image domain with unpaired image data. However, these cross-modal translation frameworks are not efficient for multi-domain image-to-image translation. For instance, given m image domains, pix2pix and BicycleGAN require the training of $A_m^2 = m(m-1) = \Theta(m^2)$ models; CycleGAN, DiscoGAN, DistanceGAN and DualGAN need $C_m^2 = \frac{m(m-1)}{2} = \Theta(m^2)$ models or $m(m-1)$ generator/discriminator pairs; ComboGAN requires $\Theta(m)$ models.

To overcome the aforementioned limitation, Choi *et al.* propose StarGAN [5] (Fig. 1(b)), which can perform multi-domain image-to-image translation using only one generator/discriminator pair with the aid of an auxiliary classifier [25]. More formally, let X and Y represent training sets of two different domains, and $x \in X$ and $y \in Y$ denote training images in domain X and domain Y, respectively; let z_y and z_x indicate category labels of domain Y and X, respectively. StarGAN utilizes the same generator G twice for translating from X to Y with the labels z_y, *i.e.*, $G(x, z_y) \approx y$, and reconstructs the input x from the translated output $G(x, z_y)$ and the label z_x, *i.e.*, $G(G(x, z_y), z_x) \approx x$. In doing so, the same generator shares a common mapping and data structures for two different tasks, *i.e.*, translation and reconstruction. However, since each task has unique information and distinct targets, it is harder to optimize the generator and to make it gain good generalization ability on both tasks, which usually leads to blurred generation results.

In this paper, we propose a novel Dual Generator Generative Adversarial Network (G²GAN) (Fig. 1(c)). Unlike StarGAN, G²GAN consists of two generators and one discriminator, the translation generator G^t transforms images from X to Y, and the reconstruction generator G^r uses the generated images from G^t and the original domain label z_x to reconstruct the original x. Generators G_t and G_r cope with different tasks, and the input data distribution for them is different. The input of G_t is a real image and a target domain label. The goal of G_t is to generate the target domain image. While G_r accepts a generated image and an original domain label as input, the goal of G_r is to generate an original image. For G_t and G_r, the input images are a real image and a generated image, respectively. Therefore, it is intuitive to design different network structures for the two generators. The two generators are allowed to use different network designs and different levels of parameter sharing according to the diverse difficulty of the tasks. In this way, each generator can have its own network parts which usually helps to learn better each task-specific mapping in a multi-task setting [31].

To overcome the model collapse issue in training G²GAN for the multi-domain translation, we further explore different objective losses for better optimization. The proposed losses include (i) a color cycle-consistency loss which targets solving the "channel pollution" problem [39] by generating red, green, blue channels separately instead of generating all three at one time, (ii) a multi-scale SSIM loss, which preserves the information of luminance, contrast and structure between generated images and input images across different scales, and (iii) a conditional identity preserving loss, which helps retaining the identity information of the input images. These losses are jointly embedded in G²GAN for training and help generating results with higher consistency and better stability. In summary, the contributions of this paper are as follows:

- We propose a novel Dual Generator Generative Adversarial Network (G²GAN), which can perform unpaired image-to-image translation among multiple image domains. The dual generators, allowing different network structures and different-level parameter sharing, are designed to specifically cope with the translation and the reconstruction tasks, which facilitates obtaining a better generalization ability of the model to improve the generation quality.
- We explore jointly utilizing different objectives for a better optimization of the proposed G²GAN, and thus obtaining unpaired multi-modality translation with higher consistency and better stability.
- We extensively evaluate G²GAN on six different datasets in different scenarios, such as architectural buildings, seasons, landscape and human faces, demonstrating its superiority in model capacity and its better generation performance compared with state-of-the-art methods on the multi-domain image-to-image translation task.

2 Related Work

Generative Adversarial Networks (GANs) [6] are powerful generative models, which have achieved impressive results on different computer vision tasks, *e.g.*, image generation [26,35], editing [4,36] and inpainting [15,46]. However, GANs are difficult to train, since it is hard to keep the balance between the generator and the discriminator, which makes the optimization oscillate and thus leading to a collapse of the generator. To address this, several solutions have been proposed recently, such as Wasserstein GAN [2] and Loss-Sensitive GAN [28]. To generate more meaningful images, CGAN [23] has been proposed to employ conditioned information to guide the image generation. Extra information can also be used such as discrete category labels [16,27], text descriptions [20,29], object/face keypoints [30,41], human skeleton [37,39] and referenced images [8,14]. CGAN models have been successfully used in various applications, such as image editing [27], text-to-image translation [20] and image-to-image translation [8] tasks.

Image-to-Image Translation. CGAN models learn a translation between image inputs and image outputs using neutral networks. Isola *et al.* [8] design the pix2pix framework which is a conditional framework using a CGAN to learn the mapping function. Based on pix2pix, Zhu *et al.* [50] further present Bicycle-GAN which achieves multi-modal image-to-image translation using paired data. Similar ideas have also been applied to many other tasks, *e.g.* generating photographs from sketches [33]. However, most of the models require paired training data, which are usually costly to obtain.

Unpaired Image-to-Image Translation. To alleviate the issue of pairing training data, Zhu *et al.* [49] introduce CycleGAN, which learns the mappings between two unpaired image domains without supervision with the aid of a cycle-consistency loss. Apart from CycleGAN, there are other variants proposed to tackle the problem. For instance, CoupledGAN [18] uses a weight-sharing strategy to learn common representations across domains. Taigman *et al.* [38] propose a Domain Transfer Network (DTN) which learns a generative function between one domain and another domain. Liu *et al.* [17] extend the basic structure of GANs via combining the Variational Autoencoders (VAEs) and GANs. A novel DualGAN mechanism is demonstrated in [47], in which image translators are trained from two unlabeled image sets each representing an image domain. Kim *et al.* [10] propose a method based on GANs that learns to discover relations between different domains. However, these models are only suitable in cross-domain translation problems.

Multi-domain Unpaired Image-to-Image Translation. There are only very few recent methods attempting to implement multi-modal image-to-image translation in an efficient way. Anoosheh *et al.* propose a ComboGAN model [1], which only needs to train m generator/discriminator pairs for m different image domains. To further reduce the model complexity, Choi *et al.* introduce StarGAN [5], which has a single generator/discriminator pair and is able to perform the task with a complexity of $\Theta(1)$. Although the model complexity is low, jointly learning both the translation and reconstruction tasks with the same generator requires the sharing of all parameters, which increases the optimization complexity and reduces the generalization ability, thus leading to unsatisfactory generation performance. The proposed approach aims at obtaining a good balance between the model capacity and the generation quality. Along this research line, we propose a Dual Generator Generative Adversarial Network (G^{2}GAN), which achieves this target via using two task-specific generators and one discriminator. We also explore various optimization objectives to train better the model to produce more consistent and more stable results.

3 G^{2}GAN: Dual Generator Generative Adversary Networks

We first start with the model formulation of G^{2}GAN, and then introduce the proposed objectives for better optimization of the model, and finally present the implementation details of the whole model including network architecture and training procedure.

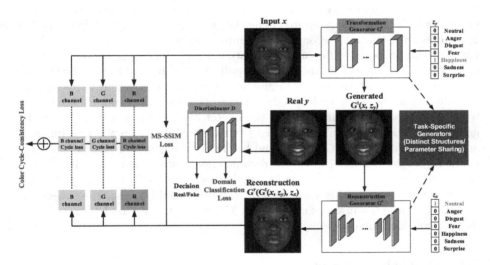

Fig. 2. The framework of G^{2}GAN. z_x and z_y indicate the category labels of domain X and Y, respectively. G^t and G^r are task-specific generators. The generator G^t converts images from domain X into domain Y and the generator G^r inputs the generated image $G^t(x, z_y)$ and the original domain label z_x and attempts to reconstruct the original image x during the optimization with the proposed different objective losses.

3.1   Model Formulation

In this work, we focus on the multi-domain image-to-image translation task with unpaired training data. The overview of the proposed G^2GAN is depicted in Fig. 2. The proposed G^2GAN model is specifically designed for tackling the multi-domain translation problem with significant advantages in the model complexity and in the training overhead compared with the cross-domain generation models, such as CycleGAN [49], DiscoGAN [10] and DualGAN [47], which need to separately train $C_m^2 = \frac{m(m-1)}{2}$ models for m different image domains, while ours only needs to train a single model. To directly compare with StarGAN [5], which simply employs the same generator for the different reconstruction and translation tasks. However, the training of a single generator model for multiple domains is a challenging problem (refer to Sect. 4), we proposed a more effective dual generator network structure and more robust optimization objectives to stabilize the training process. Our work focuses on exploring different strategies to improve the optimization of the multi-domain model aiming to give useful insights in the design of more effective multi-domain generators.

Our goal is to learn all the mappings among multiple domains using dual generators and one discriminator. To achieve this target, we train a translation generator G^t to convert an input image x into an output image y which is conditioned on the target domain label z_y, *i.e.* $G^t(x, z_y) \rightarrow y$. Then the reconstruction generator G^r accepts the generated image $G^t(x, z_y)$ and the original domain label z_x as input, and learns to reconstruct the input image x, *i.e.* $G^r(G^t(x, z_y), z_x) \rightarrow x$ through the proposed different optimization losses, including the color cycle-consistency loss for solving the "channel pollution" issue and the MS-SSIM loss for preserving the information of luminance, contrast and structure across scales. The dual generators are task-specific generators which allows for different network designs and different levels of parameter sharing for learning better the generators. The discriminator D tries to distinguish between the real image y and the generated image $G^t(x, z_y)$, and to classify the generated image $G^t(x, z_y)$ to the target domain label z_y via the domain classification loss. We further investigate how the distinct network designs and different network sharing schemes for the dual generators dealing with different sub-tasks could balance the generation performance and the network complexity. The multi-domain model StarGAN [5] did not consider these aspects.

3.2   Model Optimization

The optimization objective of the proposed G^2GAN contains five different losses, *i.e.*, color cycle-consistency loss, multi-scale SSIM loss, conditional least square loss, domain classification loss and conditional identity preserving loss. These optimization losses are jointly embedded into the model during training. We present the details of these loss functions in the following.

Color Cycle-Consistency Loss. It is worth noting that CycleGAN [49] is different from the pix2pix framework [8] as the training data in CycleGAN are unpaired, and thus CycleGAN introduces a cycle-consistency loss to enforce forward-backward consistency. The core idea of "cycle consistency" is that if we translate from one domain to the other and translate back again we should arrive at where we started. This loss can be regarded as "pseudo" pairs in training data even though we do not have corresponding samples in the target domain for the input data in the source domain. Thus, the loss function of cycle-consistency is defined as:

$$\mathcal{L}_{cyc}(G^t, G^r, x, z_x, z_y) = \mathbb{E}_{x \sim p_{\mathrm{data}}(x)}[\|G^r(G^t(x, z_y), z_x) - x\|_1]. \tag{1}$$

The optimization objective is to make the reconstructed images $G^r(G^t(x, z_y), z_x)$ as close as possible to the input images x, and the L_1 norm is adopted for the reconstruction loss. However, the "channel pollution" issue [39] exists in this loss, which is because the generation of a whole image at one time makes the different channels influence each other, thus leading to artifacts in the generation results. To solve this issue, we propose to construct the consistence loss for each channel separately, and introduce the color cycle-consistency loss as follows:

$$\mathcal{L}_{colorcyc}(G^t, G^r, x, z_x, z_y) = \sum_{i \in \{r,g,b\}} \mathcal{L}_{cyc}^i(G^t, G^r, x^i, z_x, z_y), \tag{2}$$

where, x^b, x^g, x^r are three color channels of image x. Note that we did not feed each channel of the image into the generator separately. Instead, we feed the whole image into the generator. We calculate the pixel loss for the red, green, blue channels separately between the reconstructed image and the input image, and then sum up the three distance losses as the final loss. By doing so, the generator can be enforced to generate each channel independently to avoid the "channel pollution" issue.

Multi-scale SSIM Loss. The structural similarity index (SSIM) has been originally used in [43] to measure the similarity of two images. We introduce it into the proposed G^{2}GAN to help preserving the information of luminance, contrast and structure across scales. For the recovered image $\widehat{x} = G^r(G^t(x, z_y), z_x)$ and the input image x, the SSIM loss is written as:

$$\mathcal{L}_{\mathrm{SSIM}}(\widehat{x}, x) = [l(\widehat{x}, x)]^\alpha [c(\widehat{x}, x)]^\beta [s(\widehat{x}, x)]^\gamma, \tag{3}$$

where

$$l(\widehat{x}, x) = \frac{2\mu_{\widehat{x}}\mu_x + C_1}{\mu_{\widehat{x}}^2 + \mu_x^2 + C_1}, \quad c(\widehat{x}, x) = \frac{2\sigma_{\widehat{x}}\sigma_x + C_2}{\sigma_{\widehat{x}}^2 + \sigma_x^2 + C_2}, \quad s(\widehat{x}, x) = \frac{\sigma_{\widehat{x}x} + C_3}{\sigma_{\widehat{x}}\sigma_x + C_3}. \tag{4}$$

These three terms compare the luminance, contrast and structure information between $\widehat{x}$ and x respectively. The parameters $\alpha > 0$, $\beta > 0$ and $\gamma > 0$ control the relative importance of the $l(\widehat{x}, x)$, $c(\widehat{x}, x)$ and $s(\widehat{x}, x)$, respectively; $\mu_{\widehat{x}}$ and μ_x are the means of $\widehat{x}$ and x; $\sigma_{\widehat{x}}$ and σ_x are the standard deviations of $\widehat{x}$ and x; $\sigma_{\widehat{x}x}$ is the covariance of $\widehat{x}$ and x; C_1, C_2 and C_3 are predefined parameters. To make the model benefit from multi-scale deep information, we refer to a multi-scale implementation of SSIM [44] which constrains SSIM over scales. We write the Multi-Scale SSIM (MS-SSIM) as:

$$\mathcal{L}_{\text{MS-SSIM}}(\widehat{x}, x) = [l_M(\widehat{x}, x)]^{\alpha_M} \prod_{j=1}^{M} [c_j(\widehat{x}, x)]^{\beta_j} [s_j(\widehat{x}, x)]^{\gamma_j}. \tag{5}$$

Through using the MS-SSIM loss, the luminance, contrast and structure information of the input images is expected to be preserved.

Conditional Least Square Loss. We apply a least square loss [21,49] to stabilize our model during training. The least square loss is more stable than the negative log likelihood objective $\mathcal{L}_{CGAN}(G^t, D_s, z_y) = \mathbb{E}_{y \sim p_{\text{data}}(y)}[\log D_s(y)] + \mathbb{E}_{x \sim p_{\text{data}}(x)}[\log(1 - D_s(G^t(x, z_y)))]$, and is converging faster than the Wasserstein GAN (WGAN) [2]. The loss can be expressed as:

$$\mathcal{L}_{LSGAN}(G^t, D_s, z_y) = \mathbb{E}_{y \sim p_{\text{data}}(y)}[(D_s(y) - 1)^2] + \mathbb{E}_{x \sim p_{\text{data}}(x)}[D_s(G^t(x, z_y))^2], \tag{6}$$

where z_y are the category labels of domain y, D_s is the probability distribution over sources produced by discriminator D. The target of G^t is to generate an image $G^t(x, z_y)$ that is expected to be similar to the images from domain Y, while D aims to distinguish between the generated images $G^t(x, z_y)$ and the real images y.

Domain Classification Loss. To perform multi-domain image translation with a single discriminator, previous works employ an auxiliary classifier [5,25] on the top of discriminator, and impose the domain classification loss when updating both the generator and discriminator. We also consider this loss in our optimization:

$$\mathcal{L}_{classification}(G^t, D_c, z_x, z_y) = \mathbb{E}_{x \sim p_{\text{data}}(x)}\{-[\log D_c(z_x|x) + \log D_c(z_y|G^t(x, z_y))]\}, \tag{7}$$

where $D_c(z_x|x)$ represents the probability distribution over the domain labels given by discriminator D. D learns to classify x to its corresponding domain z_x. $D_c(z_y|G^t(x, z_y))$ denotes the domain classification for fake images. We minimize this objective function to generate images $G^t(x, z_y)$ that can be classified as the target labels z_y.

Conditional Identity Preserving Loss. To reinforce the identity of the input image during conversion, a conditional identity preserving loss [38,49] is used. This loss can encourage the mapping to preserve color composition between the input and the output, and can regularize the generator to be near an identity mapping when real images of the target domain are provided as the input to the generator.

$$\mathcal{L}_{identity}(G^t, G^r, z_x) = \mathbb{E}_{x \sim p_{\text{data}}(x)}[\|G^r(x, z_x) - x\|_1]. \tag{8}$$

In this way, the generator also takes into account the identity preserving via the back-propagation of the identity loss. Without this loss, the generators are free to change the tint of input images when there is no need to.

Full G^{2}GAN Objective. Given the losses presented above, the complete optimization objective of the proposed G^{2}GAN can be written as:

$$\mathcal{L} = \mathcal{L}_{LSGAN} + \lambda_1 \mathcal{L}_{classification} + \lambda_2 \mathcal{L}_{colorcyc} + \lambda_3 \mathcal{L}_{\text{MS-SSIM}} + \lambda_4 \mathcal{L}_{identity},$$

$$(9)$$

where λ_1, λ_2, λ_3 and λ_4 are parameters controlling the relative importance of the corresponding objectives terms. All objectives are jointly optimized in an end-to-end fashion.

3.3   Implementation Details

G^{2}GAN Architecture. The network consists of a dual generator and a discriminator. The dual generator is designed to specifically deal with different tasks in GANs, *i.e.* the translation and the reconstruction tasks, which has different targets for training the network. We can design different network structures for the different generators to make them learn better task-specific objectives. This also allows us to share parameters between the generators to further reduce the model capacity, since the shallow image representations are sharable for both generators. The parameter sharing facilitates the achievement of good balance between the model complexity and the generation quality. Our model generalizes the model of StarGAN [5]. When the parameters are fully shared with the usage of the same network structure for both generators, our basic structure becomes a StarGAN. For the discriminator, we employ PatchGAN [5,8,13,49]. After the discriminator, a convolution layer is applied to produce a final one-dimensional output which indicates whether local image patches are real or fake.

Network Training. For reducing model oscillation, we adopt the strategy in [35] which uses a cache of generated images to update the discriminator. In the experiments, we set the number of image buffer to 50. We employ the Adam optimizer [11] to optimize the whole model. We sequentially update the translation generator and the reconstruction generator after the discriminator updates at each iteration. The batch size is set to 1 for all the experiments and all the models were trained with 200 epochs. We keep the same learning rate for the first 100 epochs and linearly decay the rate to zero during the next 100 epochs. Weights were initialized from a Gaussian distribution with mean 0 and standard deviation 0.02.

4   Experiments

In this section, we first introduce the experimental setup, and then show detailed qualitative and quantitative results and model analysis.

Table 1. Description of the datasets used in our experiments.

Dataset	Type	# Domain	# Translation	Resolution	Unpaired/Paired	# Training	# Testing	# Total
Facades [40]	Architectures	2	2	256 × 256	Paired	800	212	1,012
AR [22]	Faces	4	12	768 × 576	Paired	920	100	1,020
Bu3dfe [48]	Faces	7	42	512 × 512	Paired	2,520	280	2,800
Alps [1]	Natural seasons	4	12	–	Unpaired	6,053	400	6,453
RaFD [12]	Faces	8	56	1024 × 681	Unpaired	5,360	2,680	8,040
Collection [49]	Painting style	5	20	256 × 256	Unpaired	7,837	1,593	9,430

4.1 Experimental Setup

Datasets. We employ six publicly available datasets to validate our G^{2}GAN. A detailed comparison of these datasets is shown in Table 1, including Facades, AR Face, Alps Season, Bu3dfe, RaFD and Collection style datasets.

Parameter Setting. The initial learning rate for Adam optimizer is 0.0002, and β_1 and β_2 of Adam are set to 0.5 and 0.999. The parameters $\lambda_1, \lambda_2, \lambda_3, \lambda_4$ in Eq. 9 are set to 1, 10, 1, 0.5, respectively. The parameters C_1 and C_2 in Eq. 4 are set to 0.01^2 and 0.03^3. The proposed G^{2}GAN is implemented using deep learning framework PyTorch. Experiments are conducted on an NVIDIA TITAN Xp GPU.

Baseline Models. We consider several state-of-the-art cross-domain image generation models, *i.e.* CycleGAN [49], DistanceGAN [3], Dist. + Cycle [3], Self Dist. [3], DualGAN [47], ComboGAN [1], BicycleGAN [50], pix2pix [8] as our baselines. For comparison, we train these models multiple times for every pair of two different image domains except for ComboGAN [1], which needs to train m models for m different domains. We also employ StarGAN [5] as a baseline which can perform multi-domain image translation using one generator/discriminator pair. Note that the fully supervised pix2pix and BicycleGAN are trained on paired data, the other baselines and G^{2}GAN are trained with unpaired data. Since BicycleGAN can generate several different outputs with one single input image, and we randomly select one output from them for comparison. For a fair comparison, we re-implement baselines using the same training strategy as our approach.

4.2 Comparison with the State-of-the-Art on Different Tasks

We evaluate the proposed G^{2}GAN on four different tasks, *i.e.*, label↔photo translation, facial expression synthesis, season translation and painting style transfer. The comparison with the state-of-the-arts are described in the following.

Task 1: Label↔Photo Translation. We employ Facades dataset for the label↔photo translation. The results on Facades were only meant to show that the proposed model is also applicable on translation on two domains only and could produce competitive performance. The qualitative comparison is shown in Fig. 3. We can obverse that ComboGAN, Dist. + Cycle, Self Dist. fail to generate reasonable results on the photo to label translation task. For the opposite mapping, *i.e.* (labels→photos), DualGAN, Dist. + Cycle, Self Dist., StarGAN and pix2pix suffer from the model collapse problem, which leads to reasonable but blurry generation results. The proposed G^{2}GAN achieves compelling results on both tasks compared with the other baselines.

Task 2: Facial Expression Synthesis. We adopt three face datasets (*i.e.* AR, Bu3dfe and RaFD) for the facial expression synthesis task with similar settings as in StarGAN. Note that for AR dataset, we not only show the translation results of the neutral expression to other non-neutral expressions as in [5], but also present the opposite mappings, *i.e.* from non-neutral expressions to neutral expression. For Bu3dfe dataset, we only show the translation results from neutral expression to other non-neutral expressions as in [5] because of the space limitation. As can be seen in Fig. 4, Dist. + Cycle and Self Dist. fail to produce faces similar to the target domain. DualGAN generates reasonable but blurry faces. DistanceGAN, StarGAN, pix2pix and BicycleGAN produce much sharper results, but still contain some artifacts in the generated faces, *e.g.*, twisted mouths of StarGAN, pix2pix and BicycleGAN on "neutral2fear" task. CycleGAN, ComboGAN and G^{2}GAN work better than other baselines on this dataset. We can also observe similar results on the Bu3dfe dataset as shown in Fig. 5 (Left). Finally, we present results on the RaFD dataset in Fig. 5 (Right). We can observe that our method achieves visually better results than CycleGAN and StarGAN.

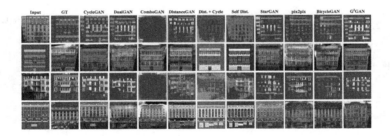

Fig. 3. Comparison with different models for mapping label↔photo on Facades.

Fig. 4. Comparison with different models for facial expression translation on AR.

Task 3: Season Translation. We also validate G^{2}GAN on the season translation task. The qualitative results are illustrated in Fig. 6. Note that we did not show pix2pix and BicycleGAN results on Alps dataset since this dataset does not contain ground-truth images to train these two models. Obviously Dual-GAN, DistanceGAN, Dist. + Cycle, Self Dist. fail to produce reasonable results. StarGAN produces reasonable but blurry results, and there are some artifacts in the generated images. CycleGAN, ComboGAN and the proposed G^{2}GAN are able to generate better results than other baselines. However, ComboGAN yields some artifacts in some cases, such as the "summer2autumn" sub-task. We also present one failure case of our method on this dataset in the last row of Fig. 6. Our method produces images similar to the input domain, while Cycle-GAN and DualGAN generate visually better results compared with G^{2}GAN on "winter2spring" sub-task. It is worth noting that CycleGAN and DualGAN need to train twelve generators on this dataset, while G^{2}GAN only requires two generators, and thus our model complexity is significantly lower.

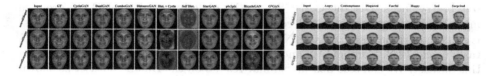

Fig. 5. Comparison with different models for facial expression translation on Bu3dfe (Left) and RaFD (Right) datasets

Table 2. Results on RaFD.

Model	AMT	IS	FID
CycleGAN [49]	19.5	1.6942	52.8230
StarGAN [5]	24.7	1.6695	51.6929
G^{2}GAN (Ours)	**29.1**	**1.7187**	**51.2765**

Table 3. Results on collection style set.

Model	AMT	FID	CA
CycleGAN [49] (ICCV 2017)	16.8% ± 1.9%	47.4823	73.72%
StarGAN [5] (CVPR 2018)	13.9% ± 1.4%	58.1562	44.63%
G^{2}GAN (Ours)	**19.8% ± 2.4%**	**43.7473**	**78.84%**
Real data	–	–	91.34%

Task 4: Painting Style Transfer. Figure 7 shows the comparison results on the painting style dataset with CycleGAN and StarGAN. We observe that Star-GAN produces less diverse generations crossing different styles compared with CycleGAN and G^{2}GAN. G^{2}GAN has comparable performance with CycleGAN, requiring only one single model for all the styles, and thus the network complexity is remarkably lower compared with CycleGAN which trains an individual model for each pair of styles.

Quantitative Comparison on All Tasks. We also provide quantitative results on the four tasks. Different metrics are considered including: (i) AMT perceptual studies [8,49], (ii) Inception Score (IS) [32], (iii) Fréchet Inception Distance

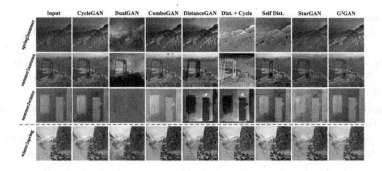

Fig. 6. Comparison with different models for season translation on Alps.

Table 4. AMT "real vs fake" study on Facades, AR, Alps, Bu3dfe datasets.

% Turkers label *real*	label→photo	photo→label	AR	Alps	Bu3dfe
CycleGAN [49] (ICCV 2017)	8.8% ± 1.5%	4.8% ± 0.8%	**24.3% ± 1.7%**	39.6% ± 1.4%	16.9% ± 1.2%
DualGAN [47] (ICCV 2017)	0.6% ± 0.2%	0.8% ± 0.3%	1.9% ± 0.6%	18.2% ± 1.8%	3.2% ± 0.4%
ComboGAN [1] (CVPR 2018)	4.1% ± 0.5%	0.2% ± 0.1%	4.7% ± 0.9%	34.3% ± 2.2%	**25.3% ± 1.6%**
DistanceGAN [3] (NIPS 2017)	5.7% ± 1.1%	1.2% ± 0.5%	2.7% ± 0.7%	4.4% ± 0.3%	6.5% ± 0.7%
Dist. + Cycle [3] (NIPS 2017)	0.3% ± 0.2%	0.2% ± 0.1%	1.3% ± 0.5%	3.8% ± 0.6%	0.3% ± 0.1%
Self Dist. [3] (NIPS 2017)	0.3% ± 0.1%	0.1% ± 0.1%	0.1% ± 0.1%	5.7% ± 0.5%	1.1% ± 0.3%
StarGAN [5] (CVPR 2018)	3.5% ± 0.7%	1.3% ± 0.3%	4.1% ± 1.3%	8.6% ± 0.7%	9.3% ± 0.9%
pix2pix [8] (CVPR 2017)	4.6% ± 0.5%	1.5% ± 0.4%	2.8% ± 0.6%	–	3.6% ± 0.5%
BicycleGAN [50] (NIPS 2017)	5.4% ± 0.6%	1.1% ± 0.3%	2.1% ± 0.5%	–	2.7% ± 0.4%
G²GAN (Ours, fully-sharing)	4.6% ± 0.9%	2.4% ± 0.4%	6.8% ± 0.6%	15.4% ± 1.9%	13.1% ± 1.3%
G²GAN (Ours, partially-sharing)	8.2% ± 1.2%	3.6% ± 0.7%	16.8% ± 1.2%	36.7% ± 2.3%	18.9% ± 1.1%
G²GAN (Ours, no-sharing)	**10.3% ± 1.6%**	**5.6% ± 0.9%**	22.8% ± 1.9%	**47.7% ± 2.8%**	23.6% ± 1.7%

(FID) [7] and (iv) Classification Accuracy (CA) [5]. We follow the same perceptual study protocol from CycleGAN and StarGAN. Tables 2, 3 and 4 report the performance of the AMT perceptual test, which is a "real vs fake" perceptual metric assessing the realism from a holistic aspect. For Facades dataset, we split it into two subtasks as in [49], label→photo and photo→label. For the other datasets, we report the average performance of all mappings. Note that from Tables 2, 3 and 4, the proposed G^2GAN achieves very competitive results compared with the other baselines. Note that G^2GAN significantly outperforms StarGAN trained using one generator on most of the metrics and on all the datasets. Note that paired pix2pix shows worse results than unpaired methods in Table 4, which can be also observed in DualGAN [47].

We also use the Inception Score (IS) [32] to measure the quality of generated images. Tables 2 and 5 report the results. As discussed before, the proposed G^2GAN generates sharper, more photo-realistic and reasonable results than Dist. + Cycle, Self Dist. and StarGAN, while the latter models present slightly higher IS. However, higher IS does not necessarily mean higher image quality. High quality images may have small IS as demonstrated in other image generation [19] and super-resolution works [9,34]. Moreover, we employ FID [7] to measure the performance on RaFD and painting style datasets. Results are shown in Tables 2 and 3, we observe that G^2GAN achieves the best results compared with StarGAN and CycleGAN.

Finally, we compute the Classification Accuracy (CA) on the synthesized images as in [5]. We train classifiers on the AR, Alps, Bu3dfe, Collection datasets respectively. For each dataset, we take the real image as training data and the generated images of different models as testing data. The intuition behind this setting is that if the generated images are realistic and follow the distribution of the images in the target domain, the classifiers trained on real images will be able to classify the generated image correctly. For AR, Alps and Collection datasets we list top 1 accuracy, while for Bu3dfe we report top 1 and top 5 accuracy. Tables 3 and 5 show the results. Note that G^2GAN outperforms the baselines on AR, Bu3dfe and Collection datasets. On the Alps dataset, StarGAN achieves slightly better performance than ours but the generated images by our model contains less artifacts than StarGAN as shown in Fig. 6.

4.3   Model Analysis

Model Component Analysis. We conduct an ablation study of the proposed G^2GAN on Facades, AR and Bu3dfe datasets. We show the results without the conditional identity preserving loss (I), multi-scale SSIM loss (S), color cycle-consistency loss (C) and double discriminators strategy (D), respectively. We also consider using two different discriminators as in [24,39,45] to further boost our performance. To study the parameter sharing for the dual generator, we perform experiments on different schemes including: fully-sharing, *i.e.* the two generators share the same parameters, partially-sharing, *i.e.* only the encoder part shares the same parameters, no-sharing, *i.e.* two independent generators. The basic generator structure follows [5]. Quantitative results of the AMT score and the classification accuracy are reported in Table 6. Without using double discriminators slightly degrades performance, meaning that the proposed G^2GAN can achieve good results trained using the dual generator and one discriminator. However, removing the conditional identity preserving loss, multi-scale SSIM loss

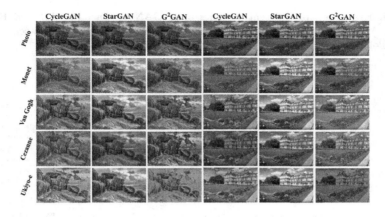

Fig. 7. Comparison on the multi-domain painting style transfer task.

and color cycle-consistency loss substantially degrades the performance, meaning that the proposed joint optimization objectives are particularly important to stabilize the training and thus produce better generation results. For the parameter sharing, as shown in Tables 4, 5 and 7, we observe that different-level parameter sharing influences both the generation performance and the model capacity, demonstrating our initial motivation.

Table 5. Results of Inception Score (IS) and Classification Accuracy (CA).

Model	Facades	AR		Alps		Bu3dfe	
	IS	IS	CA	IS	CA	IS	CA
CycleGAN [49] (ICCV 2017)	3.6098	2.8321	@1:27.333%	4.1734	@1:42.250%	1.8173	@1:48.292%, @5:94.167%
DualGAN [47] (ICCV 2017)	3.7495	1.9148	@1:28.667%	4.2661	@1:53.488%	1.7176	@1:40.000%, @5:90.833%
ComboGAN [1] (CVPR 2018)	3.1289	2.4750	@1:28.250%	4.2438	@1:62.750%	1.7887	@1:40.459%, @5:90.714%
DistanceGAN [3] (NIPS 2017)	3.9988	2.3455	@1:26.000%	4.8047	@1:31.083%	1.8974	@1:46.458%, @5:90.000%
Dist. + Cycle [3] (NIPS 2017)	2.6897	**3.5554**	@1:14.667%	**5.9531**	@1:29.000%	3.4618	@1:26.042%, @5:79.167%
Self Dist. [3] (NIPS 2017)	3.8155	2.1350	@1:21.333%	5.0584	@1:34.917%	**3.4620**	@1:10.625%, @5:74.167%
StarGAN [5] (CVPR 2018)	**4.3182**	2.0290	@1:26.250%	3.3670	**@1:65.375%**	1.5640	@1:52.704%, @5:94.898%
pix2pix [8] (CVPR 2017)	3.6664	2.2849	@1:22.667%	-	-	1.4575	@1:44.667%, @5:91.750%
BicycleGAN [50] (NIPS 2017)	3.2217	2.0859	@1:28.000%	-	-	1.7373	@1:45.125%, @5:93.125%
G²GAN (Ours, fully-sharing)	4.2615	2.3875	@1:28.396%	3.6597	@1:61.125%	1.9728	@1:52.985%, @5:95.165%
G²GAN (Ours, partially-sharing)	4.1689	2.4846	@1:28.835%	4.0158	@1:62.325%	1.5896	@1:53.456%, @5:95.846%
G²GAN (Ours, no-sharing)	4.0819	2.6522	@1:**29.667%**	4.3773	@1:63.667%	1.8714	@1:**55.625%**, @5:**96.250%**

Table 6. Evaluation of different variants of G²GAN on Facades, AR and Bu3dfe datasets. All: full version of G²GAN, I: Identity preserving loss, S: multi-scale SSIM loss, C: Color cycle-consistency loss, D: Double discriminators strategy.

Model	label→photo	photo→label	AR		Bu3dfe	
	% Turkers label real	% Turkers label real	% Turkers label real	CA	% Turkers label real	CA
All	10.3% ± 1.6%	5.6% ± 0.9%	22.8% ± 1.9%	@1:29.667%	23.6% ± 1.7%	@1:55.625%, @5:96.250%
All - I	2.6% ± 0.4%	4.2% ± 1.1%	4.7% ± 0.8%	@1:29.333%	16.3% ± 1.1%	@1:53.739%, @5:95.625%
All - S - C	4.4% ± 0.6%	4.8% ± 1.3%	8.7% ± 0.6%	@1:28.000%	14.4% ± 1.2%	@1:42.500%, @5:95.417%
All - S - C - I	2.2% ± 0.3%	3.9% ± 0.8%	2.1% ± 0.4%	@1:24.667%	13.6% ± 1.2%	@1:41.458%, @5:95.208%
All - D	9.0% ± 1.5%	5.3% ± 1.1%	21.7% ± 1.7%	@1:28.367%	22.3% ± 1.6%	@1:53.375%, @5:95.292%
All - D - S	3.3% ± 0.7%	4.5% ± 1.1%	14.7% ±1.7%	@1:27.333%	20.1% ± 1.4%	@1:42.917%, @5:91.250%
All - D - C	8.7% ± 1.3%	5.1% ± 0.9%	19.4% ± 1.5%	@1:28.000%	21.6% ± 1.4%	@1:45.833%, @5:93.875%

Table 7. Comparison of the overall model capacity with different models.

Method	# Models	# Parameters with $m = 7$
pix2pix [8] (CVPR 2017)	$A_m^2 = m(m - 1)$	57.2M × 42
BicycleGAN [50] (NIPS 2017)		64.3M × 42
CycleGAN [49] (ICCV 2017)	$C_m^2 = \frac{m(m-1)}{2}$	52.6M × 21
DiscoGAN [10] (ICML 2017)		16.6M × 21
DualGAN [47] (ICCV 2017)		178.7M × 21
DistanceGAN [3] (NIPS 2017)		52.6M × 21
ComboGAN [1] (CVPR 2018)	m	14.4M × 7
StarGAN [5] (CVPR 2018)	1	53.2M × 1
G²GAN (Ours, fully-sharing)	1	53.2M × 1
G²GAN (Ours, partial-sharing)	1	53.8M × 1
G²GAN (Ours, no-sharing)	1	61.6M × 1

Overall Model Capacity Analysis. We compare the overall model capacity with other baselines. The number of models and the number of model parameters on Bu3dfe dataset for different m image domains are shown in Table 7. BicycleGAN and pix2pix are supervised models so that they need to train A_m^2 models for m image domains. CycleGAN, DiscoGAN, DualGAN, DistanceGAN are unsupervised methods, and they require C_m^2 models to learn m image domains, but each of them contains two generators and two discriminators. ComboGAN requires only m models to learn all the mappings of m domains, while StarGAN and G²GAN only need to train one model to learn all the mappings of m domains. We also report the number of parameters on Bu3dfe dataset, this dataset contains

7 different expressions, which means $m = 7$. Note that DualGAN uses fully connected layers in the generators, which brings significantly larger number of parameters. CycleGAN and DistanceGAN have the same architectures, which means they have the same number of parameters. Moreover, G^{2}GAN uses less parameters compared with the other baselines except StarGAN, but we achieve significantly better generation performance in most metrics as shown in Tables 2, 3, 4 and 5. When we employ the parameter sharing scheme, our performance is only slightly lower (still outperforming StarGAN) while the number of parameters is comparable with StarGAN.

5   Conclusion

We propose a novel Dual Generator Generative Adversarial Network (G^{2}GAN), a robust and scalable generative model that allows performing unpaired image-to-image translation for multiple domains using only dual generators within a single model. The dual generators, allowing for different network structures and different-level parameter sharing, are designed for the translation and the reconstruction tasks. Moreover, we explore jointly using different loss functions to optimize the proposed G^{2}GAN, and thus generating images with high quality. Extensive experiments on different scenarios demonstrate that the proposed G^{2}GAN achieves more photo-realistic results and less model capacity than other baselines. In the future, we will focus on the face aging task [42], which aims to generate facial image with different ages in a continuum.

References

1. Anoosheh, A., Agustsson, E., Timofte, R., Van Gool, L.: Combogan: unrestrained scalability for image domain translation. In: CVPR Workshop (2018)
2. Arjovsky, M., Chintala, S., Bottou, L.: Wasserstein GAN. In: ICML (2017)
3. Benaim, S., Wolf, L.: One-sided unsupervised domain mapping. In: NIPS (2017)
4. Brock, A., Lim, T., Ritchie, J.M., Weston, N.: Neural photo editing with introspective adversarial networks. In: ICLR (2017)
5. Choi, Y., Choi, M., Kim, M., Ha, J.W., Kim, S., Choo, J.: StarGAN: unified generative adversarial networks for multi-domain image-to-image translation. In: CVPR (2018)
6. Goodfellow, I., et al.: Generative adversarial nets. In: NIPS (2014)
7. Heusel, M., Ramsauer, H., Unterthiner, T., Nessler, B., Hochreiter, S.: GANs trained by a two time-scale update rule converge to a local Nash equilibrium. In: NIPS (2017)
8. Isola, P., Zhu, J.Y., Zhou, T., Efros, A.A.: Image-to-image translation with conditional adversarial networks. In: CVPR (2017)
9. Johnson, J., Alahi, A., Fei-Fei, L.: Perceptual losses for real-time style transfer and super-resolution. In: Leibe, B., Matas, J., Sebe, N., Welling, M. (eds.) ECCV 2016. LNCS, vol. 9906, pp. 694–711. Springer, Cham (2016). https://doi.org/10.1007/978-3-319-46475-6_43
10. Kim, T., Cha, M., Kim, H., Lee, J., Kim, J.: Learning to discover cross-domain relations with generative adversarial networks. In: ICML (2017)

11. Kingma, D., Ba, J.: Adam: a method for stochastic optimization. In: ICLR (2015)
12. Langner, O., Dotsch, R., Bijlstra, G., Wigboldus, D.H., Hawk, S.T., Van Knippenberg, A.: Presentation and validation of the radboud faces database. Cogn. Emot. **24**(8), 1377–1388 (2010)
13. Li, C., Wand, M.: Precomputed real-time texture synthesis with markovian generative adversarial networks. In: Leibe, B., Matas, J., Sebe, N., Welling, M. (eds.) ECCV 2016. LNCS, vol. 9907, pp. 702–716. Springer, Cham (2016). https://doi.org/10.1007/978-3-319-46487-9_43
14. Li, T., et al.: BeautyGAN: instance-level facial makeup transfer with deep generative adversarial network. In: ACM MM (2018)
15. Li, Y., Liu, S., Yang, J., Yang, M.H.: Generative face completion. In: CVPR (2017)
16. Liang, X., Zhang, H., Xing, E.P.: Generative semantic manipulation with contrasting GAN. In: ECCV (2018)
17. Liu, M.Y., Breuel, T., Kautz, J.: Unsupervised image-to-image translation networks. In: NIPS (2017)
18. Liu, M.Y., Tuzel, O.: Coupled generative adversarial networks. In: NIPS (2016)
19. Ma, L., Jia, X., Sun, Q., Schiele, B., Tuytelaars, T., Van Gool, L.: Pose guided person image generation. In: NIPS (2017)
20. Mansimov, E., Parisotto, E., Ba, J.L., Salakhutdinov, R.: Generating images from captions with attention. In: ICLR (2015)
21. Mao, X., Li, Q., Xie, H., Lau, R.Y., Wang, Z., Smolley, S.P.: Least squares generative adversarial networks. In: ICCV (2017)
22. Martinez, A.M.: The AR face database. CVC TR (1998)
23. Mirza, M., Osindero, S.: Conditional generative adversarial nets. arXiv preprint arXiv:1411.1784 (2014)
24. Nguyen, T., Le, T., Vu, H., Phung, D.: Dual discriminator generative adversarial nets. In: NIPS (2017)
25. Odena, A., Olah, C., Shlens, J.: Conditional image synthesis with auxiliary classifier GANs. In: ICML (2017)
26. Park, E., Yang, J., Yumer, E., Ceylan, D., Berg, A.C.: Transformation-grounded image generation network for novel 3D view synthesis. In: CVPR (2017)
27. Perarnau, G., van de Weijer, J., Raducanu, B., Álvarez, J.M.: Invertible conditional GANs for image editing. In: NIPS Workshop (2016)
28. Qi, G.J.: Loss-sensitive generative adversarial networks on Lipschitz densities. arXiv preprint arXiv:1701.06264 (2017)
29. Reed, S., Akata, Z., Yan, X., Logeswaran, L., Schiele, B., Lee, H.: Generative adversarial text-to-image synthesis. In: ICML (2016)
30. Reed, S.E., Akata, Z., Mohan, S., Tenka, S., Schiele, B., Lee, H.: Learning what and where to draw. In: NIPS (2016)
31. Ruder, S.: An overview of multi-task learning in deep neural networks. arXiv preprint arXiv:1706.05098 (2017)
32. Salimans, T., Goodfellow, I., Zaremba, W., Cheung, V., Radford, A., Chen, X.: Improved techniques for training GANs. In: NIPS (2016)
33. Sangkloy, P., Lu, J., Fang, C., Yu, F., Hays, J.: Scribbler: controlling deep image synthesis with sketch and color. In: CVPR (2017)
34. Shi, W., et al.: Real-time single image and video super-resolution using an efficient sub-pixel convolutional neural network. In: CVPR (2016)
35. Shrivastava, A., Pfister, T., Tuzel, O., Susskind, J., Wang, W., Webb, R.: Learning from simulated and unsupervised images through adversarial training. In: CVPR (2017)

36. Shu, Z., Yumer, E., Hadap, S., Sunkavalli, K., Shechtman, E., Samaras, D.: Neural face editing with intrinsic image disentangling. In: CVPR (2017)
37. Siarohin, A., Sangineto, E., Lathuilière, S., Sebe, N.: Deformable GANs for pose-based human image generation. In: CVPR (2018)
38. Taigman, Y., Polyak, A., Wolf, L.: Unsupervised cross-domain image generation. In: ICLR (2017)
39. Tang, H., Wang, W., Xu, D., Yan, Y., Sebe, N.: GestureGAN for hand gesture-to-gesture translation in the wild. In: ACM MM (2018)
40. Tyleček, R., Šára, R.: Spatial pattern templates for recognition of objects with regular structure. In: Weickert, J., Hein, M., Schiele, B. (eds.) GCPR 2013. LNCS, vol. 8142, pp. 364–374. Springer, Heidelberg (2013). https://doi.org/10.1007/978-3-642-40602-7_39
41. Wang, W., Alameda-Pineda, X., Xu, D., Fua, P., Ricci, E., Sebe, N.: Every smile is unique: Landmark-guided diverse smile generation. In: CVPR (2018)
42. Wang, W., Yan, Y., Cui, Z., Feng, J., Yan, S., Sebe, N.: Recurrent face aging with hierarchical autoregressive memory. In: IEEE TPAMI (2018)
43. Wang, Z., Bovik, A.C., Sheikh, H.R., Simoncelli, E.P.: Image quality assessment: from error visibility to structural similarity. IEEE TIP **13**(4), 600–612 (2004)
44. Wang, Z., Simoncelli, E.P., Bovik, A.C.: Multiscale structural similarity for image quality assessment. In: Asilomar Conference on Signals, Systems and Computers (2003)
45. Xu, R., Zhou, Z., Zhang, W., Yu, Y.: Face transfer with generative adversarial network. arXiv preprint arXiv:1710.06090 (2017)
46. Yeh, R., Chen, C., Lim, T.Y., Hasegawa-Johnson, M., Do, M.N.: Semantic image inpainting with perceptual and contextual losses. In: CVPR (2017)
47. Yi, Z., Zhang, H., Gong, P.T., et al.: DualGAN: unsupervised dual learning for image-to-image translation. In: ICCV (2017)
48. Yin, L., Wei, X., Sun, Y., Wang, J., Rosato, M.J.: A 3D facial expression database for facial behavior research. In: FGR (2006)
49. Zhu, J.Y., Park, T., Isola, P., Efros, A.A.: Unpaired image-to-image translation using cycle-consistent adversarial networks. In: ICCV (2017)
50. Zhu, J.Y., et al.: Toward multimodal image-to-image translation. In: NIPS (2017)

Pioneer Networks: Progressively Growing Generative Autoencoder

Ari Heljakka[1,2]($\boxtimes$) (iD), Arno Solin[1] (iD), and Juho Kannala[1] (iD)

[1] Department of Computer Science, Aalto University, Espoo, Finland
{ari.heljakka,arno.solin,juho.kannala}@aalto.fi
[2] GenMind Ltd., Helsinki, Finland

Abstract. We introduce a novel generative autoencoder network model that learns to encode and reconstruct images with high quality and resolution, and supports smooth random sampling from the latent space of the encoder. Generative adversarial networks (GANs) are known for their ability to simulate random high-quality images, but they cannot reconstruct existing images. Previous works have attempted to extend GANs to support such inference but, so far, have not delivered satisfactory high-quality results. Instead, we propose the Progressively Growing Generative Autoencoder (PIONEER) network which achieves high-quality reconstruction with 128×128 images without requiring a GAN discriminator. We merge recent techniques for progressively building up the parts of the network with the recently introduced adversarial encoder-generator network. The ability to reconstruct input images is crucial in many real-world applications, and allows for precise intelligent manipulation of existing images. We show promising results in image synthesis and inference, with state-of-the-art results in CELEBA inference tasks.

Keywords: Computer vision · Autoencoder · Generative models

1 Introduction

Recent progress in generative image modelling and synthesis using generative adversarial networks (GANs, [5]) has taken us closer to robust high-quality image generation. In particular, progressively growing GANs (ProgGAN, [9]) can synthesize realistic high-resolution images with unprecedented quality. For example, given a training dataset of real face images, the models learnt by ProgGAN are capable of synthesizing face images that are visually indistinguishable from face images of real people.

However, GANs have no inference capability. While useful for understanding representations and generating content for training other models, the capability for realistic image synthesis alone is not sufficient for most applications. Indeed,

Electronic supplementary material The online version of this chapter (https://doi.org/10.1007/978-3-030-20887-5_2) contains supplementary material, which is available to authorized users.

© Springer Nature Switzerland AG 2019
C. V. Jawahar et al. (Eds.): ACCV 2018, LNCS 11361, pp. 22–38, 2019.
https://doi.org/10.1007/978-3-030-20887-5_2

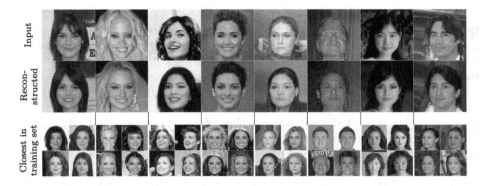

Input

Recon-
structed

Closest in
training set

Fig. 1. Examples of PIONEER network reconstruction quality in 128 × 128 resolution (randomly chosen images from the CELEBA test set). Here, images are encoded into 512-dimensional latent feature vector and simply decoded back to the original dimensionality. Below each image pair, we show the four closest face matches to the input image in the training set (with respect to structural similarity [32] of the face, cropped as in [9]).

in most computer vision tasks, the learnt models are used for feature extraction from existing real images. This motivates generative autoencoder models that allow both generation and reconstruction so that the mapping between the latent feature space and image space is bi-directional. For example, image enhancement and editing would benefit from generation and inference capabilities [2]. In addition, unsupervised learning of generative autoencoder models would be widely useful in semi-supervised recognition tasks. Yet, typically the models such as variational autoencoders (VAEs, [8,11]) generate samples not as realistic nor rich with fine details as those generated by GANs. Thus, there have been many efforts to combine GANs with autoencoder models [2–4,12,16,24,29], but none of them has reached results comparable to ProgGAN in quality.

In this paper, we propose the **ProgressIvely grOwiNg gEnerative autoEncodeR** (PIONEER) network that extends the principle of progressive growing from purely generative GAN models to autoencoder models that allow both generation and inference. That is, we introduce a novel generative autoencoder network model that learns to encode and reconstruct images with high quality and resolution as well as to produce new high-quality random samples from the smooth latent space of the encoder. Our approach formulates its loss objective following [29], and we utilize spectral normalization [19] to stabilize training—to gain the same effect as the 'improved' Wasserstein loss [6] used in [9].

Similarly to [29], our approach contains only two networks, an encoder and a generator. The encoder learns a mapping from the image space to the latent space, while the generator learns the reciprocal mapping. Examples of reconstructions obtained by mapping a real input face image to the latent space and back using our learnt encoder and generator networks at 128 × 128 resolution are shown in Fig. 1. Examples of synthetic face images generated from randomly sampled latent features by the generator are shown in Fig. 3. In these examples, the model is trained using the CELEBA [14] and CELEBA-HQ [9] datasets in a completely unsupervised manner. We also demonstrate very smooth interpolation between tuples of test images that the network has never seen before, a task that is difficult and tedious to carry out with GANs.

In summary, the key contributions and results of this paper are: *(i)* We propose a generative image autoencoder model whose architecture is built up progressively, with a balanced combination of reconstruction and adversarial losses, but without a separate GAN-like discriminator; *(ii)* We show that at least up to 128×128 resolution, this model can carry out inference on input images with sharp output, and up to 256×256 resolution, it can generate sharp images, while having a simpler architecture than the state-of-the-art of purely generative models; *(iii)* Our model gives improved image reconstruction results with larger image resolutions than previous state-of-the-art on CELEBA. The PyTorch source code of our implementation is available at https://aaltovision.github.io/pioneer.

2   Related Work

PIONEER networks belong to the family of generative models, with variational autoencoders (VAEs), autoregressive models, GAN variants, and other GAN-like models. The core idea of a GAN is to jointly train so-called generator and discriminator networks so that the generator learns to output samples from the same distribution as the training set [5], when given random input vectors from a low-dimensional latent space, and the discriminator simultaneously learns to distinguish between the synthetic and real training samples. The generator and discriminator are differentiable, jointly learnt via backpropagation using alternating optimization of an adversarial loss, where the discriminator is updated to maximize the probability of correctly classifying real and synthetic samples and the generator is updated to maximize the probability of discriminator making a mistake. Upon convergence, the generator learns to produce samples that are indistinguishable from the training samples (within the limits of the discriminator network's capacity).

Making the aforementioned training process stable has been a challenge, but the Wasserstein GAN [1] improved the situation by adopting a smooth metric for the distance between the two probability distributions [6]. In Karras *et al.* [9], the Wasserstein GAN loss from [6] is combined with the idea of progressively growing the layers and image resolution of the generator and discriminator during training, yielding excellent image synthesis results. Progressive growing has been used successfully also, for example, by [31]. There is also a line of work on other regularizers that stabilize the training (*e.g.* [19,20,25]).

However, it is well understood that the capability for realistic image synthesis alone is not sufficient for applications and there is a need for better unsupervised feature learning methods that are able to capture the semantically relevant dependencies of input data into a compact latent representation [3]. In their basic form, GANs are not suitable for this purpose as they do not provide means of learning the inverse mapping that projects the data back to latent space.

Nevertheless, there have been many recent efforts which utilize adversarial training for learning bi-directional generative models that would allow both image synthesis and reconstruction in a manner similar to autoencoders. For example, the recent works [3] and [4] simultaneously proposed an approach that employs three deep neural networks (generator, encoder, and discriminator) for learning bi-directional mappings between the data space and latent space. Instead of just samples, the discriminator is trained to discriminate tuples of samples with their latent codes, and it is shown that at the global optimum the generator and encoder learn to invert each other. Further, several others have proposed 3-network approaches that add some form of reconstruction loss and combine ideas of GAN and VAE: [12] extends VAE with a GAN-like discriminator for sample space (also used by [2]), [16,18] do the same with a GAN-like discriminator for the latent space, and [24] adds yet another discriminator (for the VAE likelihood term). While the previous methods have advanced the field, they still have not been able to simultaneously provide high quality results for both synthesis and reconstruction of high resolution images. Most of these methods struggle with even 64×64 images.

Recently, Ulyanov *et al.* [29] presented an autoencoder architecture that simply consists of two deep networks, a generator θ and encoder ϕ, representing mappings between the latent space and data space, and trained with a combination of adversarial optimization and reconstruction losses. That is, given the data distribution X and a simple prior distribution Z in the latent space, the updates for the generator aim to minimize the divergence between Z and $\phi(\theta(Z))$, whereas the updates for the encoder aim to minimize the divergence between Z and $\phi(X)$ and simultaneously maximize the divergence between $\phi(\theta(Z))$ and $\phi(X)$. In addition, the adversarial loss is supplemented with reconstruction losses both in the latent space and image space to ensure that the mappings are reciprocal (*i.e.* $\phi(\theta(\mathbf{z})) \simeq \mathbf{z}$ and $\theta(\phi(\mathbf{x})) \simeq \mathbf{x}$). The results of [29] are promising regarding both synthesis and reconstruction but the images still have low resolution. Scaling to higher resolutions requires a larger network which makes adversarial training less stable.

We combine the idea of progressive network growing [9] with the adversarial generator–encoder (AGE) networks of [29]. However, the combination is not straightforward, and we needed to identify a proper set of techniques to stabilize the training. In summary, our contributions result in a model that is simpler than many previous ones (*e.g.* having a large discriminator network just for the purpose of training the generator is wasteful and can be avoided), provides better results than [29] already in small (64×64) resolutions, and enables training and good results with larger image resolutions than previously possible. The differences to [3,24], and, for example, [27] are substantial enough to perceive by quick visual comparison.

3 PIONEER Networks

Our generative model achieves three key goals that define a good encoder–decoder model: *(i)* faithful reconstruction of the input sample, *(ii)* high sample quality (whether random samples or reconstructions), and *(iii)* rich representations. The final item can be reformulated as a 'well-behaved' latent space that lends itself to high-quality interpolations between given test samples and captures the diversity of features present in the training set. Critically, these requirements are strictly parametrized by our target resolution—there are several models that achieve many of the said goals up to 32×32 image resolution, but very few that have shown good results beyond 64×64 resolution.

PIONEER networks achieve the reconstruction and representation goals up to 128×128 resolution and the random sample generation up to 256×256 resolution, while using a combination of simple principles. A conceptual description in the next subsection is followed by some theory (Sect. 3.2) and more practical implementation details (Sect. 3.3).

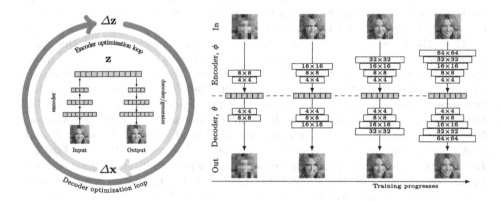

Fig. 2. The network grows in phases during which the image resolution doubles. The adversarial/reconstructive training criterion is continuously applied at each step, adapting to the present input–output resolution. The circular arrows illustrate the two modes of learning: *(i)* reconstruction of real training samples, and *(ii)* reconstruction of the latent representation of randomly generated samples.

3.1 Intuition

The defining training and architecture principles of PIONEER networks are shown in Fig. 2; on the left hand side, the competing objectives are presented in the double loop, and on the right, the progressively growing structure of the network is shown stepping up through $4 \times 4, 8 \times 8, 16 \times 16, \ldots$, doubling the resolution in each phase. The input $\mathbf{x}$ is squeezed through the encoder into a latent representation $\mathbf{z}$, which on the other hand is again decoded back to an image $\hat{\mathbf{x}}$.

The motivation behind the progressively growing setup is to encourage the network to catch the fundamental structure and variation in the inputs at lower resolutions to help the additional layers specialize in fine-tuning and adding details and nuances when reaching the higher resolutions.

The network has encoder–decoder structure with no *ad hoc* components (such as separate discriminators as in [2,12,16,18,24]). Similar to GANs, the encoder and decoder are not trained as one, but instead as if they were two competing networks. This requires the encoder to become sensitive to the difference between training samples and generated (decoded) samples, and the decoder to keep making the difference smaller and smaller. While GANs achieve this with the complexity cost of a separate discriminator network, we choose to just learn to encode the samples in a source-dependent manner. This encoding could be, then, followed by a classification layer, but instead we train the encoder so that the distribution of latent codes of training samples *approach* a certain reference distribution, while the distribution of codes of generated samples *diverges* from it (see AGE [29]).

3.2 Encoder–Decoder Losses

As in variational autoencoders, we choose the Kullback–Leibler (KL) divergence as the metric in latent space. Our reference distribution is unit Gaussian with a diagonal covariance matrix. Each sample $\mathbf{x} \in X$ is encoded into a latent vector $\mathbf{z} \in Z$, giving rise to the posterior distribution $q_\phi(\mathbf{z} \mid \mathbf{x})$ on a d-dimensional sphere. The KL-divergence between such a distribution and a d-dimensional unit Gaussian is (see the reasoning in [29], but with the following corrections):

$$\mathrm{KL}[q_\phi(\mathbf{z} \mid \mathbf{x}) \,\|\, \mathcal{N}(\mathbf{0}, \mathbf{I})] = -\frac{d}{2} + \sum_{j=1}^{d} \left[\frac{\sigma_j^2 + \mu_j^2}{2} - \log(\sigma_j) \right], \tag{1}$$

where μ_j and σ_j are the empirical sample mean and standard deviation of the encoded samples in the latent vector space with respect to dimension $j = 1, 2, \ldots, d$, and $\mathcal{N}(\mathbf{0}, \mathbf{I})$ denotes the unit Gaussian.

The encoder ϕ and decoder θ are connected via two reconstruction error terms. We measure reconstruction error $L_\mathcal{X}$ with L1 distance in sample space $\mathcal{X}$ for the encoder, and code reconstruction error $L_\mathcal{Z}$ with cosine distance in latent code space $\mathcal{Z}$ for the decoder, as follows:

$$L_\mathcal{X}(\boldsymbol{\theta}, \boldsymbol{\phi}) = \mathbb{E}_{\mathbf{x} \sim X} \|\mathbf{x} - \boldsymbol{\theta}(\boldsymbol{\phi}(\mathbf{x}))\|_1, \tag{2}$$

$$L_\mathcal{Z}(\boldsymbol{\theta}, \boldsymbol{\phi}) = \mathbb{E}_{\mathbf{z} \sim Z}[1 - \mathbf{z}^\mathsf{T} \boldsymbol{\phi}(\boldsymbol{\theta}(\mathbf{z}))], \tag{3}$$

where X are the training samples and Z random latent vectors, with $\mathbf{z}$ and $\boldsymbol{\phi}(\mathbf{x})$ normalized to unity.

In other words, a training sample is encoded into the latent space and then decoded back into a generated sample. A random latent vector is decoded into a

random generated sample that is then fed back to the encoder (Fig. 2). This provides an elegant solution to forcing the network to learn to reconstruct training
images. The total loss function of the encoder L_ϕ and decoder L_θ are, then:

$$L_\phi = \quad \mathrm{KL}[q_\phi(\mathbf{z} \mid \mathbf{x}) \,\|\, \mathcal{N}(\mathbf{0}, \mathbf{I})] - \mathrm{KL}[q_\phi(\mathbf{z} \mid \hat{\mathbf{x}}) \,\|\, \mathcal{N}(\mathbf{0}, \mathbf{I})] + \lambda_\mathcal{X} L_\mathcal{X}, \qquad (4)$$

$$L_\theta = -\mathrm{KL}[q_\phi(\mathbf{z} \mid \mathbf{x}) \,\|\, \mathcal{N}(\mathbf{0}, \mathbf{I})] + \mathrm{KL}[q_\phi(\mathbf{z} \mid \hat{\mathbf{x}}) \,\|\, \mathcal{N}(\mathbf{0}, \mathbf{I})] + \lambda_\mathcal{Z} L_\mathcal{Z}, \qquad (5)$$

where $\mathbf{x} \sim X$ and $\hat{\mathbf{x}} = \boldsymbol{\theta}(\mathbf{z})$ with $\mathbf{z} \sim \mathcal{N}(\mathbf{0}, \mathbf{I})$. We fix the hyper-parameters $\lambda_\mathcal{X}$
and $\lambda_\mathcal{Z}$ so they can be read as scaling constants. In practical implementation,
we can simplify the decoder loss to only account for

$$L_\theta = \mathrm{KL}[q_\phi(\mathbf{z} \mid \hat{\mathbf{x}}) \,\|\, \mathcal{N}(\mathbf{0}, \mathbf{I})] + \lambda_\mathcal{Z} L_\mathcal{Z}. \qquad (6)$$

The training is adversarial in the sense that we use each loss function in turn,
first freezing the decoder weights and training only with the loss (4), and then
freezing the encoder weights and training only with the loss (6).

However, in Ulyanov *et al.* [29], this approach was only shown to work with
AGE on images up to 64×64 resolution. Beyond that, we need a larger network
architecture, which is unlikely to work with AGE alone. We confirmed this by
trying out a straightforward extension of AGE to 128×128 resolution (by visual
examination and via results in Table 1). In contrast, to stabilize training, our
model will increase the size of the network progressively, following [9], and utilize
the following techniques.

3.3   Model and Training

The training uses a convolution–deconvolution architecture typically used in generative models, but here, the model is built up progressively during training, as in
[9]. We start training on low resolution images (4×4), bypassing most of the network layers. We train each intermediate phase with the same number of samples.
In the first half of each consecutive phase, we start by adding a trivial down-
sampling (encoder) and upsampling (decoder) layer, which we gradually replace
by fading in the next convolutional–deconvolutional layers simultaneously in the
encoder and the decoder, in lockstep with the input resolution which is also
faded in gradually from the previous to the new doubled resolution (8×8 etc.).
During the second half of each phase, the architecture remains unchanged. After
the first half of the target resolution phase, we no longer change the architecture.

We train the encoder and the generator with loss (4) and (6) in turn, utilizing
various stabilizing factors as follows. The architecture of the convolutional layers
in PIONEER networks largely follows yet simplifies the symmetric structure in
ProgGAN (see Table 2 of [9]), with the provision of replacing its discriminator
with an encoder. This requires removing the binary classifier, allowing us to
connect the encoder and decoder directly via the 512-dimensional latent vector.
We also remove the minibatch standard deviation layer, as it is sensitive to
batch-level statistics useful for a GAN discriminator but not for an encoder.

For stabilizing the training, we employ equalized learning rate and pixelwise feature vector normalization in the generator [9], buffer of images created by previous generators [26], and encoder spectral normalization [19]. We use ADAM [10] with $\beta_1 = 0, \beta_2 = 0.99, \epsilon = 10^{-8}$ and learning rate 0.001. We use 2 generator updates per 1 encoder update. For result visualization (but not training), we use an exponential running average for the weights of the generator over training steps as in [9]. Of these techniques, spectral normalization warrants some elaboration.

To stabilize the training of generative models, it is important to consider the function space within which the discriminator must fit in general, and, specifically, controlling its Lipschitz constant. ProgGAN uses improved Wasserstein loss [6] to keep the Lipschitz constant close to unity. However, this loss formulation is not immediately applicable to the slightly more complex AGE-style loss formulation, so instead, we adopted GAN spectral normalization [19] to serve the same purpose. In this spectral normalization approach, the spectral norm of each layer of the encoder (discriminator) network is constrained directly at each computation pass, allowing the network to keep the Lipschitz constant under control. Crucially, spectral normalization does not regularize the network via learnable parameters, but affects the scaling of network weights in a data-dependent manner.

In our experiments, it was evident that without such a stabilizing factor, the progressive training would not remain stable beyond 64×64 resolution. Spectral normalization solved this problem unambiguously: without it, the training of the network was consistently failing, while with it, the training almost consistently converged. Other strong stabilization methods, such as the penalty on the weighted gradient norm [25], might have worked here as well.

Fig. 3. Randomly generated face image samples with PIONEER networks using CELEBA for training at resolutions 64×64 (top) and 128×128 (middle), and using CELEBA-HQ for 256×256 (bottom).

4   Experiments

PIONEER networks are more most immediately applicable to learning image datasets with non-trivial resolutions, such as CELEBA [14], LSUN [30], and IMA-GENET. Here, we run experiments on CELEBA and CELEBA-HQ [9] (with training/testing split 27000/3000) and LSUN bedrooms. For comparing with previous works, we also include CIFAR-10, although its low-resolution images (32 × 32) were not expected to be most relevant for the present work.

Training with high resolutions is relatively slow in both ProgGAN and our method, but we believe that significant speed optimization is possible in future work. In fact, it is noteworthy that you *can* train these models for a long time without running into typical GAN problems, such as 'mode collapse' or ending up oscillating around a clearly suboptimal point. We trained the PIONEER model on CELEBA with one Titan V GPU for 5 days up to 64 × 64 resolution (172 epochs), and another 8 days for 128 × 128 resolution. We separately trained on CELEBA-HQ up to 256 × 256 resolution with four Tesla P100 GPUs for 10 days (1600 epochs), and on LSUN with two Tesla P100 GPUs for 9 days.

Throughout the training, we kept the hyper-parameters fixed at $\lambda_{\mathcal{Z}} = 1000\,d$ and $\lambda_{\mathcal{X}} = 10\,d$, where d is the dimensionality of the latent space (512), taking advantage of the hyper-parameter search done by [29]. After the progressive growth phase of the training, we switched to $\lambda_{\mathcal{X}} = 15\,d$ to emphasize sample reconstruction [28].

4.1   CELEBA and CELEBA-HQ

The CELEBA dataset [14] contains over 200k images with various resolutions that can be square-cropped to 128 × 128. CELEBA-HQ [9] is a subset of 30k of those images that have been improved and upscaled to 1024 × 1024 resolution. We train with CELEBA up to 128 × 128 resolution, and with CELEBA-HQ up to 256 × 256. In order to compare with previous works, we also trained our network for 64 × 64 images from CELEBA.

We ran our experiments as follows. Following the approach described in Sect. 3.3, we trained the network progressively through each intermediate resolution until we reach the target resolution (64 × 64, 128 × 128, or 256 × 256), for the same number of steps in each stage. For the final stage with the target resolution, we would continue training for as long as the Fréchet Inception Distance (FID, [7]) measures of the randomly generated samples showed improvements. During the progression of the input resolution, we adapted minibatch size to accommodate for the available memory.

For random sampling metrics, we use FID and Sliced Wasserstein Distance (SWD, [21]) between the training distribution and the generated distribution. FID measures the sample quality and diversity, while SWD measures the similarity in terms of Wasserstein distance (earth mover's distance). Batch size is 10000 for FID and 16384 for SWD. For reconstruction metrics, we use the root-mean-square error (RMSE) between the original and the reconstructed image.

We present our results in three ways. First, the model must be able to reconstruct random test set images and retain both sufficient quality and faithfulness of the reconstruction. Often, there is a trade-off between the two [23]. Previous models have often seemed to excel with respect to the quality of the reconstruction image, but in fact, the reconstruction turns out to be very different from the original (such as a different person's face). Second, we must be able to randomly sample images from the latent space of the model, and achieve sufficient quality and diversity of the images. Third, due to its inference capability, PIONEER networks can show interpolated results between input images without any additional tricks, such as first solving a regression optimization problem for the image, as often done with GANs (*e.g.* [22]).

Table 1. Comparison of Fréchet Inception Distance (FID) against 10,000 training samples, Sliced Wasserstein Distance (SWD) against 16384 samples, and root-mean-square error (RMSE) on test set, in the 64 × 64 and 128 × 128 CELEBA dataset on inference-capable networks. ProgGAN with L1 regression has the best overall sample quality (in FID/SWD), but not best reconstruction capability (in RMSE). A pretrained model by the author of [29] was used for AGE for 64 × 64. For 128 × 128, we enlarged the AGE network to account for the larger inputs and a 512-dimensional latent vector, trained until the training became unstable. ALI was trained on default CELEBA settings following [4] for 123 epochs. The error indicates one standard deviation for separate sampling batches of a single (best) trained model. For all numbers, **smaller is better**.

	64 × 64			128 × 128		
	FID	SWD	RMSE	FID	SWD	RMSE
ALI	58.88 ± 0.19	25.58 ± 0.35	18.00 ± 0.21	–	–	–
AGE	26.53 ± 0.08	17.87 ± 0.11	4.97 ± 0.06	154.79 ± 0.43	22.33 ± 0.74	9.09 ± 0.07
ProgGAN/L1	**7.98 ± 0.06**	**3.54 ± 0.40**	2.78 ± 0.05	–	–	–
PIONEER	8.09 ± 0.05	5.18 ± 0.19	**1.82 ± 0.02**	**23.15 ± 0.15**	**10.99 ± 0.44**	**8.24 ± 0.15**

Reconstruction. Given an unseen test image, the model should be able to encode the relevant information (such as hair color, facial expression, *etc.*) and decode it into a natural-looking face image expressing the features. Unlike in image compression, the model does not aim to replicate the input image *per se*, but capture the essentials. In Fig. 1, we show PIONEER reconstructions for CELEBA 128 × 128 test images, coupled with the four closest samples in the training set (in terms of structural similarity [32] of the face as cropped in [9]).

We compare reconstructions against inference-capable models: AGE [29] and ALI [4]. We also train ProgGAN for reconstruction as follows (compare to *e.g.*, [13,15,22]). We train the network normally until convergence, and then use the latent vector of the discriminator also as the latent input for the generator (properly normalized). Finally, we re-train the discriminator–generator network as an autoencoder that attempts to reconstruct input images, with L1 reconstruction loss. When re-training, we only modify the discriminator subnetwork, since allowing the generator to change would inevitably lead to lower-quality generated images. (We also tried training a fully connected layer on top of the existing hidden layer, but training became almost prohibitively slow without improved results.) Like most of the previous results, we find that the network (ProgGAN/L1) can fairly well reconstruct samples that it has generated itself, but performs much worse when given new real input images.

For networks that support both inference and generation, we can feed input images and evaluate the output image. In Fig. 4, we show the output of each network for the given random CELEBA test set images. As seen from the figure, at 64 × 64 resolution, PIONEER outperforms the baseline networks in terms of the combined output quality and faithfulness of the reconstruction. At 64 × 64 resolution, PIONEER's FID score of 8.09 in Table 1 outperforms AGE and ALI, the relevant inference baselines. ProgGAN/L1 outperforms the rest in sample quality (FID/SWD), but is worse in faithfulness (RMSE).

Without modifications, ALI and AGE have not thus far been shown to work with 128 × 128 resolution. We managed to run AGE for 128 × 128 resolution by enlarging the network to account for the larger inputs and a 512-dimensional latent vector, and trained until the training became unstable. For ALI, enlarging the network for 128 × 128 was not tried. ProgGAN excels in sample generation for higher resolutions, but as discussed, it is not designed for reconstruction or inference. Therefore we ran it only for 64 × 64, already showing this difference.

Fig. 4. Comparison of reconstruction quality between PIONEER, ALI, and AGE in 64 × 64. The first row in each set shows examples from the test set of CELEBA (not cherry-picked). The reproduced images of PIONEER are much closer to the original than those of AGE or ALI. Note the differences in handling of the 5th image from the left. ALI and AGE were trained as in Table 1. (For more examples, see Supplementary)

Dreaming up Random Samples. A model that focuses on reconstruction is unlikely to match the quality of the models that only focus on random sample generation. Even though our focus is on excelling in the former category, we do not fall far behind the state-of-the-art provided by ProgGAN in generating new samples. Figure 3 shows samples generated by PIONEER at 64 × 64, 128 × 128, and 256 × 256 resolutions. The ProgGAN SWD results in [9] were based on a more aggressive cropping of the dataset, so the values are not comparable. For AGE and ALI, the FID and SWD scores are clearly worse (see Table 1) even at low resolutions, and the methods do not generalize well to higher resolutions.

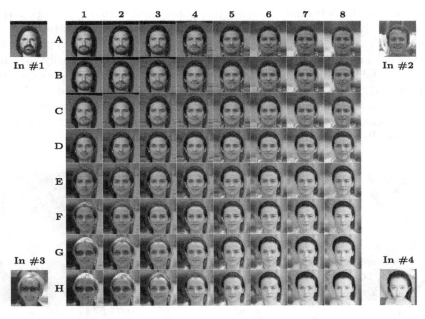

(Best viewed in high resolution / zoomed-in.)

Fig. 5. Interpolation study on test set input images at 128 × 128 resolution. Unlike many works, we interpolate between the (reconstructions of) unseen test images given as input—not between images the network has generated on its own.

Inference Capabilities. Finally, we provide an example of input-based interpolation between different (unseen) test images. In Fig. 5 we have four different test images, one in each corner of the tile figure. Thus image A1 corresponds to the reconstruction of Input #1, A8 to Input #2, H1 to Input #3, and H8 to Input #4. The rest of the images are produced by interpolating between the reconstructions in the latent space—for example, between A1 and A8. As can be seen in the figure, the latent space is well-behaved and even the glasses in Input #3 do not cause problems. We emphasize that compared to many GAN methods, the interpolations in PIONEER can be done elegantly and without separate optimization stage needed for each input sample.

4.2 LSUN Bedrooms

The LSUN dataset [30] contains images of various categories in 256×256 resolution or higher. We choose the category of bedrooms, often used for testing generative models. For humans, comparing randomly generated samples is more difficult on this dataset than with faces, so quantitative metrics are important to separate between the subtle differences in quality and diversity of captured features.

We ran the LSUN training similarly to CELEBA, but with only a single target resolution of 128×128. We present randomly generated samples from LSUN bedrooms (Fig. 6) at 128×128 resolution. Comparing to the non-progressive GANs of [6] and [17], we see that PIONEER output quality visually matches them, while falling slightly behind the fully generative ProgGAN, as expected. The FID of 37.50 was reached with no hyper-parameter tuning specific to LSUN.

For networks that support both inference and generation, we would not expect to achieve the same quality metrics as with purely generative models, so these results are not directly comparable.

Fig. 6. Generated images of LSUN bedrooms at 128×128 resolution. Results can be compared to the image arrays in [6,9,17].

4.3 CIFAR-10

The CIFAR-10 dataset contains 60,000 labeled images at 32×32 resolution, spanning 10 classes. As our method is fully supervised, we do not utilize the label information. During the training, we found that progressive growing seemed to provide no benefits. Therefore, we trained the PIONEER model otherwise as normal, but started at 32×32 resolution and did not use progressive growing.

We used the same architecture, losses and algorithm as for the other datasets, instead of trying to optimize our approach to get the best results in CIFAR-10. We confirmed that the approach works, but it is not particularly suitable for this kind of a dataset without further modifications. Generated samples are provided in Fig. 7. We believe that with some natural modifications, the model will be able to compete with GAN-based methods, but we leave this for our future work.

Fig. 7. Generated images of CIFAR-10 at 32×32 resolution.

5   Discussion and Conclusion

In this paper, we proposed a generative image autoencoder model that is trained with the combination of adversarial and reconstruction losses in sample and latent space, using progressive growing of generator and encoder networks, but without a separate GAN-like discriminator. We showed that this model can both generate sharp images—at least up to 256×256 resolution—and carry out inference on input images at least up to 128×128 resolution with sharp output, while having a simpler architecture than the state-of-the-art of purely generative models [9]. We demonstrated the inference via sample reconstruction and smooth interpolation in the latent space, and showed the overall generative capability by generating new random samples from the latent space and measuring the quality and diversity of the generated distribution against baselines.

We emphasize that evaluation of generative models is heavily dependent on the resolution, and there is a multitude of models that have been shown to work on 64×64 resolution, but not on 128×128 or above. Reaching higher resolutions is not only a matter of raw compute, but the model needs to be able to cope with the increasing information and be regularised suitably in order not to loose the representative power or become instable.

We found that training is more stable using spectral normalization, which also suits our non-GAN loss architecture and loss. The model provides image reconstruction results with larger image resolutions than previous state-of-the-art. Importantly, our model has only few hyper-parameters and is robust to train.

The only hyper-parameter that typically needs to be tuned between datasets is the number of epochs spent on intermediate resolutions. Our results indicate that the GAN paradigm of a separate discriminator network may not be necessary for learning to infer and generate image data sets. GANs do currently remain the best option if one is only interested in generating random samples. Like GANs, our model is heavily based on the general idea of 'adversarial' training, construed as setting the generator–encoder pair up with opposite gradients to each other with respect to the source of the data (that is, simulated vs. observed).

As Karras *et al.* [9] point out for GANs, the principle of growing the network progressively may be more important than the specific loss function formulation. Likewise, even though the AGE formulation for the latent space loss metrics is relatively simple, we believe that there are many ways in which the encoder can be set up to achieve and exceed the results we have demonstrated here.

In future work, we will also continue training the network to carry out faithful reconstructions at 256×256, 512×512, and 1024×1024 resolutions, omitted from this paper primarily due to the extensive amount of computation (or preferably, further optimization) required. We will also further investigate whether the CELEBA-HQ dataset is sufficiently diverse for this purpose.

Acknowledgments. We thank Tero Karras, Dmitry Ulyanov, and Jaakko Lehtinen for fruitful discussions. We acknowledge the computational resources provided by the Aalto Science-IT project. Authors acknowledge funding from the Academy of Finland (grant numbers 308640 and 277685) and GenMind Ltd.

References

1. Arjovsky, M., Chintala, S., Bottou, L.: Wasserstein generative adversarial networks. In: Proceedings of the 34th International Conference on Machine Learning, pp. 214–223 (2017)
2. Brock, A., Lim, T., Ritchie, J.M., Weston, N.: Neural photo editing with introspective adversarial networks. In: Proceedings of the International Conference on Learning Representations (ICLR) (2017)
3. Donahue, J., Krähenbühl, P., Darrell, T.: Adversarial feature learning. In: Proceedings of the International Conference on Learning Representations (ICLR) (2017)
4. Dumoulin, V., et al.: Adversarially learned inference. In: Proceedings of the International Conference on Learning Representations (ICLR) (2017)
5. Goodfellow, I.J., et al.: Generative adversarial networks. In: Advances in Neural Information Processing Systems (NIPS), pp. 2672–2680 (2014)
6. Gulrajani, I., Ahmed, F., Arjovsky, M., Dumoulin, V., Courville, A.C.: Improved training of Wasserstein GANs. In: Advances in Neural Information Processing Systems (NIPS), pp. 5767–5777 (2017)
7. Heusel, M., Ramsauer, H., Unterthiner, T., Nessler, B., Hochreiter, S.: GANs trained by a two time-scale update rule converge to a local Nash equilibrium. In: Advances in Neural Information Processing Systems (NIPS), pp. 6626–6637 (2017)
8. Jimenez Rezende, D., Mohamed, S., Wierstra, D.: Stochastic backpropagation and approximate inference in deep generative models. In: Proceedings of the 31st International Conference on Machine Learning, pp. 1278–1286 (2014)

9. Karras, T., Aila, T., Laine, S., Lehtinen, J.: Progressive growing of GANs for improved quality, stability, and variation. In: Proceedings of the International Conference on Learning Representations (ICLR) (2018)
10. Kingma, D.P., Ba, J.: Adam: a method for stochastic optimization. In: Proceedings of the International Conference on Learning Representations (ICLR) (2015)
11. Kingma, D., Welling, M.: Auto-encoding variational Bayes. In: Proceedings of the International Conference on Learning Representations (ICLR) (2014). https://arxiv.org/abs/1312.6114
12. Larsen, A., Kaae Sønderby, S., Larochelle, H., Winther, O.: Autoencoding beyond pixels using a learned similarity metric. In: Proceedings of the 33rd International Conference on Machine Learning, pp. 1558–1566 (2016)
13. Lipton, Z.C., Tripathi, S.: Precise recovery of latent vectors from generative adversarial networks. In: Proceedings of the International Conference on Learning Representations (ICLR) (2017)
14. Liu, Z., Luo, P., Wang, X., Tang, X.: Deep learning face attributes in the wild. In: Proceedings of the 2015 IEEE International Conference on Computer Vision (ICCV), pp. 3730–3738 (2015)
15. Luo, J., Xu, Y., Tang, C., Lv, J.: Learning inverse mapping by autoencoder based generative adversarial nets. In: Liu, D., Xie, S., Li, Y., Zhao, D., El-Alfy, E.M. (eds.) Neural Information Processing (ICONIP) 2017. LNCS, vol. 10635, pp. 207–216. Springer, Heidelberg (2017). https://doi.org/10.1007/978-3-319-70096-0_22
16. Makhzani, A., Shlens, J., Jaitly, N., Goodfellow, I., Frey, B.: Adversarial autoencoders. arXiv preprint arXiv:1511.05644 (2015)
17. Mao, X., Li, Q., Xie, H., Lau, R.Y.K., Wang, Z., Smolley, S.P.: Least squares generative adversarial networks. In: Proceedings of the 2017 IEEE International Conference on Computer Vision (ICCV), pp. 2813–2821 (2017)
18. Mescheder, L., Nowozin, S., Geiger, A.: Adversarial variational Bayes: unifying variational autoencoders and generative adversarial networks. In: Proceedings of the 34th International Conference on Machine Learning, pp. 2391–2400 (2017)
19. Miyato, T., Kataoka, T., Koyama, M., Yoshida, Y.: Spectral normalization for generative adversarial networks. In: Proceedings of the International Conference on Learning Representations (ICLR) (2018)
20. Qi, G.J.: Loss-sensitive generative adversarial networks on Lipschitz densities. arXiv preprint arXiv:1701.06264 (2017)
21. Rabin, J., Peyré, G., Delon, J., Bernot, M.: Wasserstein barycenter and its application to texture mixing. In: Bruckstein, A.M., ter Haar Romeny, B.M., Bronstein, A.M., Bronstein, M.M. (eds.) SSVM 2011. LNCS, vol. 6667, pp. 435–446. Springer, Heidelberg (2012). https://doi.org/10.1007/978-3-642-24785-9_37
22. Radford, A., Metz, L., Chintala, S.: Unsupervised representation learning with deep convolutional generative adversarial networks. In: Proceedings of the International Conference on Learning Representations (ICLR) (2016)
23. Rosca, M., Lakshminarayanan, B., Mohamed, S.: Distribution matching in variational inference. arXiv preprint arXiv:1802.06847 (2018)
24. Rosca, M., Lakshminarayanan, B., Warde-Farley, D., Mohamed, S.: Variational approaches for auto-encoding generative adversarial networks. arXiv preprint arXiv:1706.04987 (2017)
25. Roth, K., Lucchi, A., Nowozin, S., Hofmann, T.: Stabilizing training of generative adversarial networks through regularization. In: Advances in Neural Information Processing Systems, pp. 2018–2028 (2017)

26. Shrivastava, A., Pfister, T., Tuzel, O., Susskind, J., Wang, W., Webb, R.: Learning from simulated and unsupervised images through adversarial training. In: Proceedings of the IEEE Conference on Computer Vision and Pattern Recognition (2017)
27. Tabor, J., Knop, S., Spurek, P., Podolak, I., Mazur, M., Jastrzębski, S.: Cramer-wold autoencoder. arXiv preprint arXiv:1805.09235 (2018)
28. Ulyanov, D., Vedaldi, A., Lempitsky, V.: Adversarial generator-encoder networks (2018). https://github.com/DmitryUlyanov/AGE. gitHub repository
29. Ulyanov, D., Vedaldi, A., Lempitsky, V.: It takes (only) two: adversarial generator-encoder networks. In: Proceedings of the Thirty-Second AAAI Conference on Artificial Intelligence (AAAI-2018), pp. 1250–1257 (2018)
30. Yu, F., Seff, A., Zhang, Y., Song, S., Funkhouser, T., Xiao, J.: LSUN: construction of a large-scale image dataset using deep learning with humans in the loop. arXiv preprint arXiv:1506.03365 (2015)
31. Zhang, H., et al.: StackGAN: text to photo-realistic image synthesis with stacked generative adversarial networks. In: Proceedings of the 2017 IEEE International Conference on Computer Vision (ICCV) (2017)
32. Zhou, W., Bovik, A.C., Sheikh, H.R., Simoncelli, E.P.: Image qualifty assessment: from error visibility to structural similarity. IEEE Trans. Image Process. **13**(4), 600–612 (2008)

Editable Generative Adversarial Networks: Generating and Editing Faces Simultaneously

Kyungjune Baek, Duhyeon Bang, and Hyunjung Shim[✉]

School of Integrated Technology, Yonsei University, Songdogwahak-ro 85, Yeonsu-gu, Incheon, South Korea
{bkjbkj12,duhyeonbang,kateshim}@yonsei.ac.kr

Abstract. We propose a novel framework for simultaneously generating and manipulating the face images with desired attributes. While the state-of-the-art attribute editing techniques have achieved the impressive performance for creating realistic attribute effects, they only address the image editing problem, using the input image as the condition of model. Recently, several studies attempt to tackle both novel face generation and attribute editing problem using a single model. However, their image quality is still unsatisfactory. Our goal is to develop a single unified model that can simultaneously create and edit high quality face images with desired attributes. A key idea of our work is that we decompose the image into the latent and attribute vector in low dimensional representation, and then utilize the GANs framework for mapping the low dimensional representation to the image. In this way, we can address both the generation and editing problem by training the proposed GANs, namely Editable GAN. For qualitative and quantitative evaluations, the proposed GANs outperform recent algorithms addressing the same problem. Also, we show that our model can achieve the competitive performance with the state-of-the-art attribute editing technique in terms of attribute editing quality.

Keywords: Generative adversarial networks ·
Attribute editing and generation

1 Introduction

Facial attribute manipulation has been an important problem in computer vision and graphics field in past several decades. Traditional techniques [3, 4, 20] develop geometric and radiometric facial models, and reparameterize their model parameters for representing various facial attributes such as expression, race, gender, hair and eye colors, etc. Specifically, they derive the compact representation of attribute parameters by a linear combination of facial parameters. Such a linear model allows real time face rendering and manipulation. However, due to the

© Springer Nature Switzerland AG 2019
C. V. Jawahar et al. (Eds.): ACCV 2018, LNCS 11361, pp. 39–55, 2019.
https://doi.org/10.1007/978-3-030-20887-5_3

simplicity of model, it is limited to represent the diversity of attributes from individuals.

Recently, there are interesting approaches to directly generating faces of realistic attributes without constructing facial models. The most representative models include [5,12,26], which utilize non-linear deep generative adversarial networks (GANs) for face generation. Owing to the impressive visual quality, GANs receive increasing attentions from various researchers and practitioners. GANs aim to reproduce complex data distribution based on adversarial learning between two networks, a generator and a discriminator. The original GANs and its variants are fully unsupervised in that they do not require labeled data to learn the data generation process. As a drawback, they are incapable of controlling various attributes of data during generation. To address this issue, the conditional information such as image labels [18,19] is utilized for controlling the generation process of GANs, specifying attributes of generated image in a supervised fashion. Likewise, He et al., Choi et al. and Sun et al. [5,12,26] achieve the facial attribute editing by adopting the conditional information.

For the successful facial attribute editing, existing approaches pursue two objectives; the facial identity after manipulation should be retained, and the effect of desired attribute should be clearly observable in the resultant face. In order to preserve the facial identity, the encoder-decoder architecture is commonly adopted, allowing to reconstruct the input face. For controlling the effect of attributes, attribute classifiers are popularly employed, where they provide the condition of generation process. As a result, existing techniques maintain the quality of input faces while generating realistic attributes.

Although the encoder-decoder architecture is an effective tool for the identity preservation, this always requires the input face for attribute editing, thus incapable of generating new faces. VAE/GAN [16] addresses this problem by combining the VAE with GANs architecture, and achieves both the data generation and reconstruction/editing simultaneously. Unfortunately, their visual quality and reconstruction accuracy are much worse than other facial editing approaches because the variational inference degrades the quality of reconstruction/editing as well as that of generation [9]. Later, to reconstruct the image for editing, Perarnau et al. [21] proposes a new framework, namely IcGAN, that combines the conditional GANs and two encoders without variational inference. The first encoder maps the image generated by the generator into the latent vector. The second encoder is used to map the image to the condition vector. To train the IcGAN, the conditional GANs is first trained and then two encoders are later updated by fixing the conditional GANs for stabilizing the training process.

Our goal is similar to VAE/GAN and IcGAN in that we aim to modify facial attributes of input faces as well as to generate new faces with desirable attributes. We focus on improving the quality of generation and reconstruction/editing compared to VAE/GAN and IcGAN. Moreover, we formulate an end-to-end network model as opposed to IcGAN. To this end, our model is developed based on the standard GANs architecture, where the generator learns an unidirectional mapping from the input prior distribution P_z to the data distribution P_{data}. This architecture is advantageous for creating arbitrary faces because data generation can be simply done by sampling z. Yet, we need to pay the complexity for the face manipulation because we should find z corresponding to any face for the reconstruction/editing. Finding z of arbitrary image is equivalent to conducting the inverse generation process, which is extremely complex and challenging. Although several studies [7,8] have addressed this problem by learning a bidirectional mapping between P_z and P_{data}, their generation and reconstruction quality is less attractive than standard GANs. Most recently, high quality bidirectional GANs [2] has been developed for learning the fast and accurate inverse generation process. Authors improve the generation and reconstruction quality of existing bidirectional GANs using a connection network. This network transforms the feature of discriminator to the latent variable z. We employ this connection network into our framework for finding z, thus reconstruct the input facial image by generating it from its latent vector z. In this way, we can bypass the use of encoder-decoder architecture while keeping the ability of recovering the latent vector of any input image. Given the latent vector of input facial image or that of arbitrary facial image, we manipulate its attribute using the feedback from attribute classifier.

The main contributions of our study can be summarized as follows. (1) Our algorithm can generate realistic arbitrary faces as well as input faces with desirable multi-attributes. (2) Owing to the attractive nature of GANs latent space, we can easily identify a novel attribute subspace by analyzing the GANs latent space. As a result, our model can be used, without re-training, for manipulating new attributes, which are not used for training the attribute classifier. (3) Our model is more flexible to handle structural variations in attributes such as poses because image level information is not transferred to the output. (e.g., skip connections) (4) We can control the degree of attribute effects without additional training or information. The code is available at https://github.com/FriedRonaldo/GANs.

2 Related Work

2.1 Conditional GANs

While most of generative models have developed the explicit model distribution for learning the data distribution [9], generative adversarial networks (GANs) proposes an implicit approach to learning the data distribution without model assumption or variational bounds; training the image generation process until the distribution of generated images (i.e., model distribution) can approach to that of real images (i.e., data distribution) in the Jensen-Shannon distance sense. However, training GANs is notoriously unstable because the formulation requires finding Nash equilibrium that is tricky to find with gradient descent method [24]. Lately, the training instability has been addressed by various algorithms by developing the stable network architecture [22] or introducing robust metrics [1,15]. To achieve the stability and high quality of GANs, various studies suggest additional information for better posing the generation problem. In fact, from the aspects of application scenario, it is important to control the generation process as we intend, instead of generating the random images.

To this end, conditional GANs [18] suggests a new framework to control the semantics of generated samples; they formulate the problem as reproducing the conditional data distribution by training the conditional model distribution. Specifically, authors utilize a semantic vector as a condition of both the generator and discriminator. Unfortunately, upon the complexity of semantic information, the complexity of data distribution is also increased, thus the discriminator is overloaded [6]. To alleviate this additional burden, Chongxuan et al. and Odena et al. [6,19] propose the separate classification module for handling the semantic information so that the discriminator is focused on evaluating the model distribution. Their architectures impose a cross entropy loss from the classifier to the generator, that guides the conditional generation task. In this work, we adopt a multi-label classifier that decomposes the attribute learning from the discriminator.

2.2   Facial Attribute and Generation

The work aiming to edit facial attribute targeted to modify one attribute for one training [14,25]. Shen et al. [25] conducts editing the facial attribute by learn the region where the target attribute exists on the face images. He et al. [12] adopts encoder-decoder and classifier. By using skip connection between encoder and decoder, they enhance the reconstruction quality. Choi et al. [5] adds cycle consistency loss to generator's loss to reconstruct images. The entire architecture is similar to multi-cycle-gan with classifier. He et al. [12] and Choi et al. [5] succeed to edit the attributes of facial images but they can not generate images because of meaningless latent vectors and requiring the input images. SLGAN [27] introduces themselves as the first work aiming both the semantic controlled generation and attribute editing. Although they can do both of them, the quality of results is not so good to use as an application. Nevertheless the limitation on quality, the significance is on conducting both generation and editing attributes with one training procedure. In this work, we focus on both facial image generation having desired attributes and editing facial attributes with one training procedure and without degradation on quality of generated or modified samples and stability of training.

3   Background

3.1   Connection Network

Connection network [2] serves as transferring the discriminative feature vector to its corresponding latent vector. The discriminative feature is extracted by GANs discriminator and originally used to distinguish the real and fake. By utilizing this connection network, it is possible to estimate the latent vector of an arbitrary image, either fake or real. Then, the estimated latent vector can be input to a generator to restore the corresponding image. As a result, we establish the bidirectional mapping between the latent vector and image; the generator maps from the latent to the image and the connection network maps from the image to the latent vector.

Unlike the traditional approaches to developing bidirectional GANs, connection network defines the mapping relationship in low dimensional spaces (i.e., low dimensional feature vector and low dimensional latent vector), thus efficient for training. Moreover, because the generator is not involved in estimating the latent vector, the modeling power of generator is solely used for improving the image generation quality.

3.2 Selective Learning for Classification

To govern the effect of attributes in image generation, it is necessary to deliver attribute information to the generator network. Existing attribute editing techniques achieve this objective by adapting the attribute classifier [12,19].

Similar to existing techniques, we also aim to govern facial attributes of the resultant faces using the attribute classifier. To develop the highly accurate classifier, it is important to have a balanced database. Unfortunately, real-world database including CelebA [17] shows the significant bias among different attributes. Several studies [10,23] have developed to alleviate bias in data distribution. For example, Hand et al. [10] suggest balancing the number of attribute samples within each batch. Their approach did not require additional network or external database. Instead, they first set the occurrence ratio of each attribute, and select training samples for each batch according to this ratio. This scheme balances the training data by (1) relatively ignoring the attribute samples that exceed the target ratio, and (2) assigning the larger weight for the insufficient amount of attribute samples.

Although this idea of selective learning is simple and effective, data selection inherently prevents from fully utilizing the entire dataset; it intentionally drops samples from the category with many training data. In this work, we fully utilize the entire training dataset by introducing the weight regularization for both the large and small amount of attribute samples, instead of data selection.

4 Editable Generative Adversarial Networks

The major architectural difference of our model compared to previous models is that we do not use the encoder to transform the image into the latent vector for reconstruction. Instead, we directly generate the image from the latent vector, which is estimated by the separate module, the connection network. For this reason, our networks are capable of generating images and editing attributes simultaneously.

4.1 Formulation

Our network model consists of four networks including a generator G, a discriminator D, an attribute classifier C, and a connection network C_n. First, the generator maps the randomly sampled latent vector into the image. The discriminator and classifier have the sample from the generator as the input, and then output the probability of being real or fake and that of presenting attributes, respectively. Meanwhile, the connection network transforms the feature vector extracted by both the discriminator and the classifier to the latent vector. This latent vector rather inputs to the generator for reconstruction/editing. We adopt DCGAN [22] architecture for GANs.

The connection network plays a key role in editing attributes of the input image. In our framework, the input image is directly applied to the discriminator and classifier, and results in the feature vector. Then, the connection network translates it to the latent vector corresponding to the input. After that, the generator has the latent vector and a user-specific attribute vector as the input, and generates the image, which presents specified attributes assigned by the attribute vector.

The novel face generation is identical to the generation process of the standard GANs. That is, we sample a random latent vector from the uniform distribution and set its binary attribute vector. Then, the generator has the latent vector with the attribute vector for producing the novel face image with specified attributes. By changing the attribute vector, it is possible to control the attribute of generated faces. Figure 1 visualizes the architecture of Editable GAN.

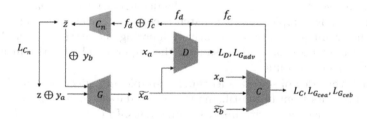

Fig. 1. Network architecture of proposed model. $\oplus$ means concatenation. $\tilde{x}$: generated image, $\tilde{z}$: reconstructed latent vector, y_a, y_b: original and new attribute, x_a, x_b: image with original, new attribute. Descriptions for losses are given in equation 1 through 5.

Adversarial Loss. First, a discriminator provides the critical feedback to the generator for improving the image generation quality. Because the discriminator aims to distinguish real images from dataset and fake images from the generator, this evaluation by the discriminator encourages the generator reproducing the original data distribution. We modify a conditional GANs [18] to reflect the attribute information in a way that the discriminator becomes irrelevant to the attribute, and formulate the objective for training the discriminator as follows.

$$L_D = \mathbb{E}[\log(D(x))] + \mathbb{E}[\log(1 - D(G(z, y_a)))]. \qquad (1)$$

Note x is an image, z is a latent vector randomly sampled from uniform distribution $U[-1, 1]$, and y_a is the attribute vector of an image (x). Same notations will be used in following sections.

Attribute Loss. Although the discriminator assesses the general quality of images, it is incapable of evaluating the presence of the desirable attributes in the output image. It is because either the presence or absence of attributes is equally probable, thus does not contribute for the image quality. To ensure the presence of desirable attributes in the resultant image, the attribute classifier is employed. Similar to existing techniques, the attribute vector is a binary multi-dimensional vector, where positive label is 1, meaning that the attribute appears in the image, and negative label is 0, meaning that the attribute does not appear in the image. As discussed in Sect. 3.2 celebA dataset shows significant bias among different attribute, the attribute classification should deal with the inbalance of dataset. To alleviate the bias on dataset, we adopt the similar method proposed in [10]. Hand et al. [10] balances the gradients from positive samples and negative samples for backpropagating at each batch. Specifically, they ignore the gradients from the samples of which attributes exceed a target ratio. Meanwhile, they increase a weight (i.e., greater than 1) for the gradients from the samples of which attributes are smaller than the target ratio. Rather ignoring the the samples, we provide weights to both of exceeding samples and insufficient samples. Let w_p be the weight of positive samples and w_n be that of negative ones. For each attribute, w_p is calculated by dividing the batch size by $2 \times N$. Likewise, w_n for each attribute is computed by dividing the batch size by $2 \times (M - N)$, where N is the total number of labels and M is the batch size.

$$L_C = \sum_i -w_{pi} y_i \log(S(C(x))) - w_{ni}(1 - y_i) \log(1 - S(C(x))) \qquad (2)$$

Note that S means sigmoid function and y_i means the i_{th} attribute of the input image x. w_p and w_n are selective weight of positive label and negative label, respectively.

Training the Generator. A generator aims (1) to deceive the discriminator by generating realistic faces from the latent vector, and (2) to fool the attribute classifier by producing the desirable attributes in the output image. To this end, both adversarial loss and an attribute constraint are imposed to train the generator.

$$\begin{aligned} L_{adv} &= -\mathbb{E}[\log(1 - D(G(z, y_a)))] \\ L_{G_{ce_a}} &= L_C(G(z, y_a)) \\ L_{G_{ce_{\tilde{a}}}} &= L_C(G(\tilde{z}, y_{\tilde{a}})) \end{aligned} \qquad (3)$$

$y_{\tilde{a}}$ is randomly sampled attribute. $\tilde{z}$ is reconstructed latent vector by connection network.

Connection Network. A connection network plays a role in mapping the image features to the latent vector, which effectively performs the inverse generation process. Because this connection network is trained independently of the generator, it does not introduce additional constraints to the generator training.

In this way, we can bypass the disadvantage of encoder-decoder architecture, which overloads the generator training as the range of latent space is enlarged after induced by the encoder.

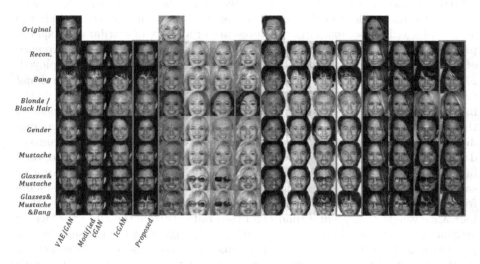

Fig. 2. Comparisons of facial attribute editing. The blue box highlights our results. The first three columns are VAE/GAN, modified cGAN, and IcGAN. For each row, the specified attribute(s) is added to the input image (Color figure online)

While original connection network is built upon the standard GANs architecture with a single generator and a single discriminator, our network model includes the attribute classifier for imposing the desirable attributes in the output image. To correctly reflect those of attribute information in estimating the latent vector, we aggregate the feature vector of the attribute classifier and that of the discriminator by concatenation. More specifically, we derive two feature vectors; f_d from the discriminator and $f_c(G(z, y_a))$ from the classifier, and concatenate them to form the final feature vector. Based on empirical study, we study that the output feature vector of the last fully connected layer is sufficient for mapping to the latent vector. By investigating the role of each feature vector, $f_c(G(z, y_a))$ provides the features related to attribute information to the latent vector, thus the reconstructed image from the latent vector is enforced to hold the style of attributes. Meanwhile, f_d maps various elements other than attributes to a unique latent vector. For example, we observe that f_d is associated with identity or structural information. $\oplus$ means concatenation.

$$\tilde{z} = C_n(\ f_d(G(z, y_a)) \oplus f_c(G(z, y_a)), y_a\ ) \tag{4}$$

$$L_{C_n} = \mathbb{E}[|z - \tilde{z}|] \tag{5}$$

5   Experiments

To evaluate the performance of our Editable GAN, we conduct several tasks on the CelebA dataset [17]. CelebA contains 202,599 facial images for training, validation, and testing with the binary labels of 40 attributes. During the experimental study, we choose 10 attributes whose visual features are distinctive from others. We use 192,599 images for training and 10,000 images for test. Under the same configuration, we train three existing algorithms; (1) IcGAN, (2) VAE/GAN and (3) conditional GANs combined with the connection network [2], namely the modified cGAN. To implement the modified cGAN, we modify the conditional GANs [18] by employing the connection network for establishing both face reconstruction and editing. For all experimental studies, the size of images for all experiments is fixed as 64 × 64.

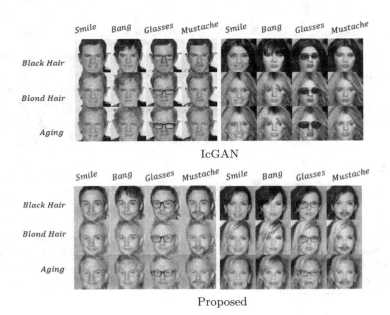

Fig. 3. Novel image generation with diverse attributes using proposed model and IcGAN. For each algorithm, a young male and young female are drawn by sampling random latent vectors. Then, various attributes are added for visualizing their effects

5.1   Qualitative Evaluation

In this paper, our goal is to generate facial images with desired attributes and to edit the attribute of input images. Our evaluation is divided into three folds. (1) To assess the attribute editing of the proposed model and three baseline algorithms (i.e., IcGAN, VAE/GAN and conditional GANs with the connection network), we qualitatively compare the editing results. (2) The quality of novel facial images with desired attributes is evaluated using the proposed model and IcGAN, which produces the most realistic results among three baselines. (3) To show whether the latent space induced by each model is semantically meaningful or now, we perform the latent space interpolation between two latent vectors with interpolated attribute vectors of two images. (4) To show the flexibility of our model to handling structural attribute, we compare our model to AttGAN [12]. Note that AttGAN focuses on editing attribute, thus their results should be better than any algorithms that perform both image generation and editing. Although their work is optimized for image editing, we show that their framework has the limitation in structural editing, such as poses.

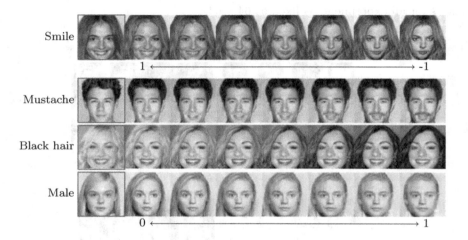

Fig. 4. Controlling the strength of attribute effect. The first row indicates smile attribute between 1 and −1. From the second to forth row, each attribute changes from 0 to 1 when their attributes are mustache, black hair and male. For the third row, blond hair varies from 1 to 0 at the same time. The input images are marked by the red box (Color figure online)

Facial Attribute Editing. We choose the input images from a test set, excluding training. Figure 2 shows the results of attribute editing from the VAE/GAN, the modified cGAN, IcGAN, and the proposed algorithms with the input image. Focusing on the reconstruction results, it is clear that our reconstruction better copes the identity and style much better than others. Analyzing the results of attribute editing, we observe that modified cGAN occasionally misses the attribute. (e.g., hair color) This is expected because the generator was not penalized due to wrong attribute generation during training. Also, VAE/GAN and IcGAN are incapable of generating the mustache on the female faces. These results indicates that VAE/GAN and IcGAN do not properly disentangle the attribute space. Among three baseline algorithms, IcGAN generally performs better. However, even IcGAN sometimes introduces undesirable changes; for example, the hairline of forehead from 'Gender' editing is different from that from input faces. Among all algorithms, our model is the most faithful to reflect the facial attributes than others. Additionally, our attribute editing does not modify other unique attributes of the input face. Also, our results are more natural and realistic under multiple attribute editing than others.

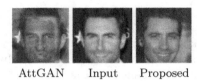

AttGAN   Input   Proposed

Fig. 5. Editing the orientation of the face. We estimate the latent vector of orientation by subtracting profile face from the frontal face, which is computed by averaging the latent vectors of the same person

Generating Facial Image with Desired Attributes. Our Editable GAN can conduct the image generation task with desirable attributes by adopting the attribute classifier. Our model is compared with IcGAN that shows best performance among three competitors. In this experiment, multiple attributes are combined for image generation, where attribute information is noted in each column and row. As shown in Fig. 3, our model can reflect the effect of each attribute and also successfully combine multiple attributes. Moreover, upon varying attributes, our results preserve the identity. Similar to the results of Fig. 2, IcGAN cannot generate the female face with mustache because IcGAN is vulnerable to decoupling highly correlated attributes.

Latent Interpolation of Two Faces and Attributes. By performing the latent interpolation between two faces, we show that (1) our results are not accomplished by data memorization, and (2) our latent and attribute spaces are semantically meaningful. During this empirical study, we find that the smooth transition between the attribute vectors can induce the semantically natural and meaningful changes in the images across the attribute. While the semantic relationship between adjacent images is varied smoothly, each generated image is still sharp and realistic.

Because our latent and attribute spaces are semantically meaningful, our model can represent the strength of attribute effect by controlling the intensity of attribute vector. This is interesting because the attribute vector is a binary vector during training. Although our network never observe the degree of attribute effect during training, it automatically interprets the strength of attributes from the intensity of attribute vector. For example, in the first row of Fig. 4, we apply 'Smile' by interpolating it from 1 to −1, and observe that the changes are smoothly reflected, from a big smile to a cold face. In addition, our model interprets the negative value in the attribute vector, producing a semantically

Table 1. Comparison of image generation quality. We compare the quality of generated images by FID score (mean and standard deviation) with recent algorithms achieving the generation and attribute editing simultaneously. Note that modified cGAN is implemented by a conditional DCGAN combined with a connection network. The lower the value, the better the quality.

Metric	VAE/GAN	cGAN	IcGAN	Proposed
FID	24.05 ± 0.33	23.27 ± 0.46	22.86 ± 0.80	19.92 ± 0.73

opposite attribute. Again, the negative value is never supervised during training, but negative 'Smile' is translated to the cold face. For 'Mustache' attribute, the amount of hairs increases as it changes to a higher value. 'Black hair' attribute also shows the interesting behavior when manipulating its intensity. In the third row of Fig. 4, the input face has the blond hair. Then, we gradually increase 'Black hair' attribute from 0 to 1. Focusing on the middle of the interpolation, we can observe that the generated hair color is neither black nor blond, but a blend of two colors. Utilizing the semantic interpretation of our model, we can create diverse variants of a single face.

Pose Editing by Latent Space Walking. Analogous to the latent space walking, we effectively induce the structural modification such as poses. To do this, we interpolate between one image and its vertically flipped version with a fixed attribute on the latent vector space. Although AttGAN [12] is not our competitor because they only focus on editing, we compare AttGAN with ours in terms of pose editing by latent space walking. In the general scenario of attribute editing, AttGAN should always perform better than other algorithms to generate and edit faces simultaneously. However, when handling the structural changes, we observe that our model can perform clearly better than AttGAN as shown in Fig. 5. Although the pose variation has never taught during training, our model can disentangle the pose variation from the facial identity, successfully conducting the pose changes by latent space arithmetic. On the other hand, AttGAN fails to disentangle the pose attribute in latent space, thus incapable of producing the sharp images with smooth pose transition. This failure with AttGAN is anticipated by their encoder-decoder architecture and skip connections implemented in their network. Unlike GANs architecture, the latent distribution P_z in the encoder-decoder based models tends to be sparse; its range is typically greater than the range of GANs. Hence, the latent space induced by the encoder-decoder architecture rarely possesses semantic interpretation. Also, while skip connections are quite effective to accurately reconstruct the input face, they weaken the semantic power of latent space, as discussed in [28]. As a result, the operation in the latent space does not yield the meaningful interpretation. Figure 6 shows the result of latent interpolation between one image and its flip. In this experiment, we can see the gradual change of the same type of the experiment shown in Fig. 5. As visualized in Fig. 6, the results from AttGAN is akin to the result of an image interpolation, shown in first row of the same figure.

Although each image is produced by the generator after the latent interpolation, the generated image is no longer realistic as if it is an overlap of two images.

5.2   Quantitative Evaluation

For quantitative evaluation, we divide our task into the novel image generation, faithful image reconstruction, and modification. For evaluating the image generation quality, we employ Fréchet Inception Distance (FID) [13]. For measuring the image reconstruction accuracy, we use the structural similarity index (SSIM), the peak signal-to-noise ratio (PSNR). Note that the accurate image reconstruction is critical for achieving successful attribute editing because our attribute editing is based on the image reconstruction framework. Finally, we evaluate the attribute editing by measuring the attribute classification accuracy for the edited images.

Generation Performance. For assessing the image generation quality, we compare our model with VAE/GAN, modified cGAN, and IcGAN using FID score. For the fair comparison, we repeatedly conduct the experiment 10 times, and report the average and standard deviation of FID score for each model as summarized in Table 1. Considering that the smaller value for FID stands for the higher quality, our Editable GAN outperforms other competitors; the performance gap is greater than three standard deviations.

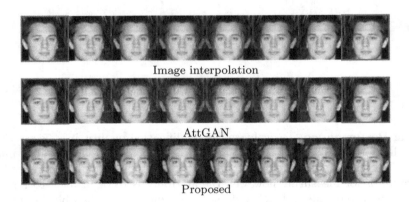

Image interpolation

AttGAN

Proposed

Fig. 6. Interpolation of the face and its vertical flipped on latent vector space. First row shows the interpolation on the image space, weighted summation of rightmost and leftmost. The second and bottom row are done by AttGAN and our model, respectively. The rightmost and leftmost image are input images (*highlighted by red box*) (Color figure online)

Reconstruction Performance. For successful attribute editing, it is important to confirm whether the input face can be faithfully reconstructed using each model. For that, we evaluate our reconstruction accuracy by SSIM and PSNR

Table 2. Comparison of reconstruction performance in terms of SSIM and PSNR. (mean and standard deviation) The higher the value, the better the quality.

Metric	VAE/GAN	cGAN	IcGAN	Proposed
SSIM	0.39 ± 0.0200	0.50 ± 0.0081	0.46 ± 0.0092	0.52 ± 0.0088
PSNR	12.98 ± 0.19	16.15 ± 0.20	15.19 ± 0.19	16.54 ± 0.22

using test images of CelebA dataset, and report the average and standard deviation for VAE/GAN, Modified cGAN, IcGAN, and our Editable GAN. Table 2 summarizes the accuracy of each model. Among all, our model outperforms all others in terms of SSIM and PSNR. It is important to stress that the performance gap between ours and any existing model is greater than two standard deviations. From this result, we can conclude that our model possesses the greater representation power than other competitors.

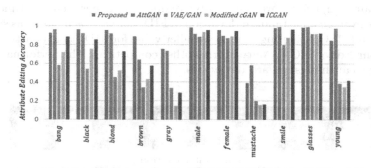

Fig. 7. Facial attribute editing accuracy of each model

Editing Performance. To evaluate the modification quality, it is necessary to check whether the effect of desirable attributes is correctly reflected after image editing. To this end, we develop the pretrained attribute classifier, and use it as the evaluator for measuring how well the attribute is recognizable. To improve the performance of the evaluation classifier, we include residual blocks [11] and apply selective learning method proposed in [10]. We test this evaluation classifier and confirm that its average accuracy is 93.22% over 10 attributes. More specifically, training dataset of CelebA for 10 attributes is applied for developing the evaluation classifier. During this experiment, we exam a single attribute modification case for all models, and the attribute classification accuracy is visualized in Fig. 7. For the comparison, we evaluate the attribute classification accuracy of VAE/GAN, modified cGAN, IcGAN, AttGAN, and our Editable GAN. Based on these experimental results, we observe that the proposed model outperforms our competitors; VAE/GAN, modified cGAN, and IcGAN. Compared to AttGAN [12], we expect that AttGAN should always show the higher performance because the model focuses on attribute editing, specialized for improving the editing performance. Interestingly, Editable GAN is compatible with AttGAN in attribute editing. For some attributes (e.g., black, blond, brown, gray and gender), the proposed model even outperforms AttGAN. Based on three quantitative comparisons, we can conclude that our Editable GAN is effective for simultaneously generating and editing faces with desired attributes.

6 Conclusion

This paper introduces an Editable Generative Adversarial Network (Editable GAN), which establishes (1) the multiple attribute editing of the input face and (2) the novel face generation by controlling semantic attributes. For editing, our model utilizes the attribute classifier to modify multiple attributes of the input face. For the face reconstruction and generation perspective, the proposed model adopts the connection network [2] that estimates the latent vector of the input. We separate the training of connection network from GANs training, thus stabilize the entire training process and retain the quality of image generation. Furthermore, because the image level information is not utilized during training, our model can edit the structural variations such as poses. More importantly, the proposed model is flexible to model the strength of attribute effects, and to handle new semantic attributes, unlabeled during training. Owing to the semantic interpretation in the latent space, we successfully disentangle the attributes of unlabeled categories, and utilize them for image editing.

Acknowledgement. This research was supported by the MSIT (Ministry of Science and ICT), Korea, under the ICT Consilience Creative Program (IITP-2018-2017-0-01015) supervised by the IITP (Institute for Information & communications Technology Promotion), the Ministry of Science and ICT, Korea (2018-0-00207, Immersive Media Research Laboratory), the Basic Science Research Program through the National Research Foundation of Korea (NRF) funded by the MSIP (NRF-2016R1A2B4016236), and ICT R&D program of MSIP/IITP. [R7124-16-0004, Development of Intelligent Interaction Technology Based on Context Awareness and Human Intention Understanding].

References

1. Arjovsky, M., Chintala, S., Bottou, L.: Wasserstein GAN. arXiv preprint arXiv:1701.07875 (2017)
2. Bang, D., Shim, H.: High quality bidirectional generative adversarial networks. arXiv preprint arXiv:1805.10717 (2018)
3. Blanz, V., Vetter, T.: A morphable model for the synthesis of 3D faces. In: Proceedings of the 26th Annual Conference on Computer Graphics and Interactive Techniques, pp. 187–194. ACM Press/Addison-Wesley Publishing Co. (1999)
4. Blanz, V., Vetter, T.: Face recognition based on fitting a 3D morphable model. IEEE Trans. Pattern Anal. Mach. Intell. **25**(9), 1063–1074 (2003)

5. Choi, Y., Choi, M., Kim, M., Ha, J.W., Kim, S., Choo, J.: StarGAN: unified generative adversarial networks for multi-domain image-to-image translation. arXiv preprint arXiv:1711.09020 (2017)
6. Chongxuan, L., Xu, T., Zhu, J., Zhang, B.: Triple generative adversarial nets. In: Advances in Neural Information Processing Systems, pp. 4091–4101 (2017)
7. Donahue, J., Krähenbühl, P., Darrell, T.: Adversarial feature learning. arXiv preprint arXiv:1605.09782 (2016)
8. Dumoulin, V., et al.: Adversarially learned inference. arXiv preprint arXiv:1606.00704 (2016)
9. Goodfellow, I.: NIPS 2016 tutorial: generative adversarial networks. arXiv preprint arXiv:1701.00160 (2016)
10. Hand, E.M., Castillo, C., Chellappa, R.: Doing the best we can with what we have: multi-label balancing with selective learning for attribute prediction. In: AAAI (2018)
11. He, K., Zhang, X., Ren, S., Sun, J.: Deep residual learning for image recognition. In: Proceedings of the IEEE Conference on Computer Vision and Pattern Recognition, pp. 770–778 (2016)
12. He, Z., Zuo, W., Kan, M., Shan, S., Chen, X.: Arbitrary facial attribute editing: only change what you want. arXiv preprint arXiv:1711.10678 (2017)
13. Heusel, M., Ramsauer, H., Unterthiner, T., Nessler, B., Hochreiter, S.: GANs trained by a two time-scale update rule converge to a local Nash equilibrium. In: Advances in Neural Information Processing Systems, pp. 6626–6637 (2017)
14. Kaneko, T., Hiramatsu, K., Kashino, K.: Generative attribute controller with conditional filtered generative adversarial networks. In: IEEE Conference on Computer Vision and Pattern Recognition (CVPR), vol. 2 (2017)
15. Kodali, N., Hays, J., Abernethy, J., Kira, Z.: On convergence and stability of GANs (2018)
16. Larsen, A.B.L., Sønderby, S.K., Larochelle, H., Winther, O.: Autoencoding beyond pixels using a learned similarity metric. arXiv preprint arXiv:1512.09300 (2015)
17. Liu, Z., Luo, P., Wang, X., Tang, X.: Deep learning face attributes in the wild. In: Proceedings of International Conference on Computer Vision (ICCV) (2015)
18. Mirza, M., Osindero, S.: Conditional generative adversarial nets. arXiv preprint arXiv:1411.1784 (2014)
19. Odena, A., Olah, C., Shlens, J.: Conditional image synthesis with auxiliary classifier GANs. arXiv preprint arXiv:1610.09585 (2016)
20. Paysan, P., Knothe, R., Amberg, B., Romdhani, S., Vetter, T.: A 3D face model for pose and illumination invariant face recognition. In: Sixth IEEE International Conference on Advanced Video and Signal Based Surveillance, AVSS 2009, pp. 296–301. IEEE (2009)
21. Perarnau, G., van de Weijer, J., Raducanu, B., Álvarez, J.M.: Invertible conditional GANs for image editing. arXiv preprint arXiv:1611.06355 (2016)
22. Radford, A., Metz, L., Chintala, S.: Unsupervised representation learning with deep convolutional generative adversarial networks. arXiv preprint arXiv:1511.06434 (2015)
23. Rudd, E.M., Günther, M., Boult, T.E.: MOON: a mixed objective optimization network for the recognition of facial attributes. In: Leibe, B., Matas, J., Sebe, N., Welling, M. (eds.) ECCV 2016. LNCS, vol. 9909, pp. 19–35. Springer, Cham (2016). https://doi.org/10.1007/978-3-319-46454-1_2
24. Salimans, T., Goodfellow, I., Zaremba, W., Cheung, V., Radford, A., Chen, X.: Improved techniques for training GANs. In: Advances in Neural Information Processing Systems, pp. 2234–2242 (2016)

25. Shen, W., Liu, R.: Learning residual images for face attribute manipulation. In: IEEE Conference on Computer Vision and Pattern Recognition (CVPR), pp. 1225–1233. IEEE (2017)
26. Sun, R., Huang, C., Shi, J., Ma, L.: Mask-aware photorealistic face attribute manipulation. arXiv preprint arXiv:1804.08882 (2018)
27. Yin, W., Fu, Y., Sigal, L., Xue, X.: Semi-latent GAN: learning to generate and modify facial images from attributes. arXiv preprint arXiv:1704.02166 (2017)
28. Zhang, L., Ji, Y., Lin, X.: Style transfer for anime sketches with enhanced residual U-net and auxiliary classifier GAN. arXiv preprint arXiv:1706.03319 (2017)

Cross Connected Network for Efficient Image Recognition

Lu Yang[1]([✉])(iD), Qing Song[1](iD), Zuoxin Li[2](iD), Yingqi Wu[1](iD), Xiaojie Li[2](iD),
and Mengjie Hu[1](iD)

[1] Beijing University of Posts and Telecommunications, Beijing 100876, China
{soeaver,songqing512,wuyqq,mengjie.hu}@bupt.edu.cn
[2] Beihang University, Beijing 100191, China
{lizuoxin,xiaojieli}@buaa.edu.cn

Abstract. In this work, we describe a novel and highly efficient convolutional neural network for image recognition, which we term the "Cross Connected Network" (CrossNet). We have creatively introduced the *Pod* structure, where the feature map of depthwise convolutions can be reused within the same *Pod* by cross connection. Such a design can make the CrossNet have high performance while less computing resource, especially suitable for mobile devices with very limited computing power. Additionally, we find that depthwise convolutions with large receptive field has better accuracy/computation trade-offs, and further improves CrossNet performance. Our experiments on ImageNet classification and MSCOCO object detection demonstrate that CrossNet can improve the state-of-the-art performance of lightweight networks (such as MobileNets-V1/-V2, ShuffleNets and CondenseNet). We have tested the actual inference time on an ARM-based mobile device. The CrossNet still gets the best performance. Code and models are public available (https://github.com/soeaver/CrossNet-PyTorch).

Keywords: Convolutional Neural Network · Cross connection · Mobile devices

1 Introduction

Convolutional Neural Networks [20,27] have made remarkable progress in image recognition, enabling a series of breakthroughs for challenging visual tasks [9,19,26]. The accuracy of network is improved significantly by going deeper or wider, but it also requires computation at billions of FLOPs. More and more intelligent applications need to be deployed on platforms with limited computational resources, *e.g.* real-time object detection [25,30,37], image matting [41], style transfer [7,21] and so on. Therefore, designing a fast and accurate convolutional neural network is very urgent.

Recent work shows that group convolutions [19,36] and depthwise convolutions [35] are effective and efficient in lightweight networks, *e.g. MobileNet* [15,33]

© Springer Nature Switzerland AG 2019
C. V. Jawahar et al. (Eds.): ACCV 2018, LNCS 11361, pp. 56–71, 2019.
https://doi.org/10.1007/978-3-030-20887-5_4

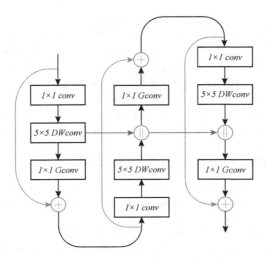

Fig. 1. The *Pod* structure with cross connection. Green lines represent residual connections with '+' symbol, blue lines represent densely connections with '‖' symbol. *DWconv* is depthwise convolution layer and *Gconv* is group convolution layer. (Color figure online)

and *Xecption* [5]. In addition, some recent studies focus on the optimization of model structure. *ShuffleNet* [39] adopts *ResNet*-like [12,13,36] structure and channel shuffle operation to overcome the side effects brought by 1×1 group convolutions. *MobileNet-V2* proposes the inverted residuals and linear bottleneck, which has achieved the state-of-the-art results across challenging datasets. Generally, *DenseNet* provides higher parameter efficiency compared with the *ResNet* due to feature reuse [4]. However, few computer vision works have studied dense connection for lightweight network design, except Huang *et al.* [16]. *CondenseNet* [16] learns the structure of *DenseNet* [17], and propose a novel method to prune redundant connections between layers. But the process of model training is too cumbersome and even affects network generalization. There is still a lack of work about more concise using dense connection in lightweight network design.

This paper introduces a novel and highly efficient convolutional neural network called *CrossNet*. For a given computation complexity budget, we propose the cross connection and *Pod* structure, as show in Fig. 1. In the same *Pod*, the feature maps of depthwise convolutions can be reused by dense connection to reduce computational cost while maintaining accuracy. Moreover, we also use residual connection in the model design to make the network more convenient to optimize. And we give a more reasonable model parameter configuration scheme by discussing the trade-offs between model performance and receptive fields of depthwise convolutions. Our *CrossNet* has better semantic representation but less computational cost, which is very suitable for deploying on mobile devices.

We present extensive experiments on ImageNet [32] classification, and also evaluate the actual inference time on an ARM-based mobile device. A series of experiments shows that our *CrossNet* gains state-of-the-art results among lightweight networks with less inference time. On the ImageNet dataset, a *CrossNet* with 304 MFLOPs and 3.2 MParams[1] achieved a 72.9% top-1 accuracy, which outperforms *MobileNet-V2* by an absolute value of 1.2% with similar computational cost and less parameters. We further demonstrate that *CrossNet* has good ability for transfer learning, and measure the performance on MSCOCO [23] object detection.

The main contributions of this paper are as follows:

- We present a novel and highly efficient convolutional neural network (*CrossNet*) for image recognition, which is specially designed for extremely limited computational resources.
- We creatively introduce the cross connection and *Pod* structure to reuse the feature maps of depthwise convolutions and reduce computational cost while maintaining accuracy. The *CrossNet* demonstrates its superior efficiency and performance over the state-of-the-art lightweight architectures on ImageNet classification. And we further demonstrate that *CrossNet* has good ability for transfer learning on MSCOCO object detection.
- The trade-off between model performance and receptive fields of depthwise convolutions is discussed. And we have analyzed several specific convolution combinations through experiments, further improving the performance of the *CrossNet*.

2   Related Work

Efficient Network Architectures. *MobileNet* [15] uses simpler network structure design and depthwise convolutions, which outperforms *GoogLeNet* with only 1/3 computation. *ShuffleNet* [39] adopts ResNet-like [12,13,36] structure and channel shuffle operation to overcome the side effects brought by 1 × 1 group convolutions, further improves the performance of lightweight network (73.7% top-1 acc. @ 524 MFLOPs). And the *CondenseNet* [16] utilizes DenseNet-like [17] structure with learned group convolution, achieves a similar performance (73.8% top-1 acc. @ 529 MFLOPs). Neural Architecture Search [42] framework uses a policy gradient algorithm to optimize architecture configurations, and proposes the *NasNet − A* with comparable performance (74.0% top-1 acc. @ 564 MFLOPs). Additionally, we notice that a very excellent network structure appeared recently, which is an upgraded version of *MobileNet* and called *MobileNet-V2* [33]. It proposes the inverted residuals and linear bottleneck, which has achieved the state-of-the-art results across challenging datasets and become a new benchmark for lightweight networks (71.7% top-1 acc. @ 300 MFLOPs).

---

[1] In this paper, MFLOPs refers to the number of million multiplication-addition operations, MParams refers to the number of million parameters.

Group and Depthwise Convolutions. Group convolutions and depthwise convolutions are very common and important operations for many efficient neural network architectures. Group convolutions were first proposed by *AlexNet* [19] for spreading the net across two GPUs, which is used to solve the shortage of memory. *ResNeXt* further demonstrates the effectiveness of the group convolutions. Depthwise convolutions are the limit form of group convolutions, which apply a single filter to each input channel. Standard convolutions have the computational cost of:

$$H * W * C_{in} * C_{out} * k * k. \tag{1}$$

Here $H*W$ is the resolution of input feature maps, C_{in} and C_{out} are the input and output channel of the layers, k is the kernel size of convolutions. For depthwise convolutions, there must be $C_{in} = C_{out}$, thus the computational cost is:

$$H * W * C_{in} * k * k. \tag{2}$$

It will bring a great reduction in computation, due to hundreds or even thousands of input channels. *Factorized CNN* [35], *Xception* [5] and *PVANet* [14] are examples of successful use of depthwise convolutions, strike an excellent trade-off between representation capability and computational cost.

Residual and Dense Block. *ResNet* [12,13,36] is one of the most exciting researches in the field of deep learning in recent years, which is constructed by stacking multiple residual bottleneck blocks. Residual block and its varieties [4, 38] have also been widely applied in lightweight network design, *e.g. Xception* [5], *ShuffleNet* [39] and *MobileNet-V2* [33]. *DenseNet* [17] is another successful network structure design, and it has been demonstrated to be more efficient. *CondenseNet* [16] consists of multiple dense blocks, each of which consists of learned group convolutions and depthwise convolutions. Such a structure design can make the *CondenseNet* reuse the feature maps of each dense block, to greatly reduce computational cost while maintaining accuracy. Most lightweight networks can be regarded as variants of *ResNet* or *DenseNet*.

In this work, we presents the "Cross Connected Network", which learns from the feature reuse structure of *DenseNet*, and introduces residual learning to ensure that the network can converge better.

3 Cross Connected Network

In this section, we will introduce the cross connection and *Pod* structure proposed by this work. Additionally, we will explore the trade-off between model performance and receptive fields of depthwise convolutions.

3.1 Cross Connection

The success of *ResNet* [12] and *DenseNet* [17] in the field of computer vision has influenced many recent network structure designs, which means that residual learning and feature reuse has become the most important ideas to design better network architectures. These popular ideas are widely used in lightweight network as depthwise convolutions for a high capability. However, it seems to be difficult to further improve the performance of network by discretizing $N \times N$ convolutions into depthwise convolutions in this structure [15,33]. Obviously, 1×1 convolutions are the next goal of improvement [16,39].

In terms of lightweight networks, 1×1 convolutions occupy most of the computation due to 3×3 depthwise convolutions are very lightweight. Taking $MobileNet$ as an example, the computation time of 1×1 convolutions in this network occupies 94.86% of the total time, and the number of parameters of 1×1 convolution accounts for 74.59% [15]. More generally, we present the computational cost of a residual bottleneck with depthwise convolutions as following:

$$H * W * (e * C) * C + H * W * C * k * k + H * W * C * (e * C). \qquad (3)$$

Where the three terms represent the computational cost of the three convolution layers in a bottleneck, C is the output channel dimension of middle convolution layer, k is the kernel size and e is expansion factor. In general, there are $k = 3$ and $e = 4$. So, the above operation can be:

$$H * W * (4 * C * C + 9 * C + 4 * C * C). \qquad (4)$$

When C is large, the computational cost is dominated by 1×1 convolutions. To reduce the computational cost, a very simple approach is to cut down the input channel C, but it will obviously destroy the network capacity. Different from $ShuffleNet$ [39] and $CondenseNet$ [16] using channel shuffle operation and learned group convolutions to alleviate the side effect, we adopt a concise and effective way without complicated operations. Our goal in this work is to reduce computational cost while maintaining accuracy.

Figure 2 shows part of variations in the residual module. All the variations can be used as the basic unit of the lightweight network. We take the original as the basic for this research. The identity mapping by shortcuts can be seen as a computational unit and defined as:

$$y = \mathcal{F}(x, \{W_i\}) + x. \qquad (5)$$

We denote $e = \mathcal{C}_{in}/\mathcal{C}$ as an expansion factor, where $\mathcal{C}_{in}$ is the input channel dimension of the residual module. The computational cost of the whole module can be given by Eq. (3). Reducing the value of e can effectively reduce the computation cost of the residual module. But this will cause loss of network capacity, due to too small output channel dimension of modules. Correspondingly increasing the value of $\mathcal{C}$ will bring more computational cost. Therefore, in this paper, we adopt a new way of network connection mode, which reuses the feature maps of depthwise convolutions via the densely connection between two adjacent residual modules, as shown in Figs. 1 and 3. This connection mode can implicitly increase the output channel dimension of depthwise convolutions in each residual module. On the contrary, it only increases a small quantity of computational cost in the 2nd 1×1 convolutions.

Alternating residual connections and densely connections, we call this mode **Cross Connection**.

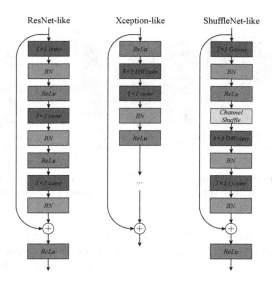

Fig. 2. Variations of the residual module.

3.2 *Pod* Structure

Theoretically, cross connection can reduce computational cost. But each residual module will reuse the feature maps of the last module, which will lead to the rapid increase of the input channel dimension of the 2nd 1×1 convolutions, resulting in the degradation of performance. To solve this problem, we propose the *Pod* structure based on cross connection.

We consider a new operations combination structure between "module" and "stage", called *Pod*. Each *Pod* is composed of several residual modules, and the adjacent residual modules are cross connected. We pass the feature maps of the earlier depthwise convolutional layer as inputs into the later layer. Consequently, each module receives the feature maps of all preceding modules in the same *Pod*. There are only normal feed-forward connections between different *Pods*, and no feature is reused. We refer to the number of modules in one *Pod* as the depth, which is different from the definition of network depth. Identical *Pod* but different weights consists of one "stage". The *Pod* structure is preponderant that it effectively avoids channel dimensional explosion of feature maps caused by cross connection between all the modules by limiting the scope of feature reuse. Moreover, small-range feature reuse can still decrease the computation and promote the accuracy. Additionally, we also find that only replaces the 2nd 1×1 convolutions with group convolutions ($g = 2$) in each module, which can effectively reduce the computational cost and parameters without bring obvious side effect. Channel shuffle operation and learned group convolutions should also be able to further improve the accuracy of *CrossNet*, This is beyond the focus of this paper, but it is suggestive for future research.

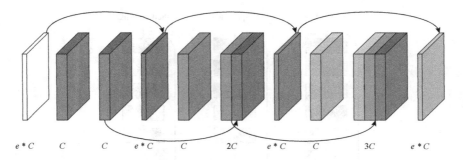

Fig. 3. A *Pod* structure with *depth* = 3, different colors represent different residual modules. The below number represents the output channel dimension of each layer.

Figures 1 and 3 show a *Pod* with depth of 3, and Fig. 3 also indicates the output channel dimension of each operation in the *Pod*. According to Eq. (3), we can calculate the computational cost of a *Pod* with depth of n:

$$\sum_{i=1}^{n} H * W * (e * C * C + k^2 * C + i * e * C * C/2) \tag{6}$$

Here, i denotes the ith module.

3.3   Receptive Fields of Depthwise Convolutions

Receptive fields plays a very important role in the research of the convolutional neural networks [1, 6, 24, 34]. Considering efficiency, the depth of lightweight networks should not be too large (*e.g.* the depth of *MobileNet* [15] is 28, *ShuffleNet* [39] is 50, *MobileNet-V2* [33] is about 50). The receptive fields is sufficient for small resolution input visual tasks, yet when network is transferred to some high resolution input visual tasks (such as object detection [8, 9, 25, 31] and semantic segmentation [1–3, 26, 40]), it may cause bottlenecks.

A simple way to settle this dispute is to increase the depth of the network, which obviously brings the loss of efficiency. Increasing the receptive fields of every 3 × 3 convolution, namely increasing the convolution kernel size is another efficacious method. According to Eq. (3), we can see that it will not make excessive extra computational cost and parameters on depthwise convolutions. In fact, if we replace 3 × 3 depthwise convolutions with 5 × 5, it will only increases 8% of computational cost in *MobileNet* [15].

There are a few previous works which developed their researches on the depthwise convolutions with large receptive fields, such as *Factorized CNN* [35] and *NasNet* [29, 42] literature. In particular, our *CrossNet* will discuss the relationship between receptive fields of depthwise convolutions and model performance through detailed experiments in the next section.

Table 1. *CrossNet* architecture, complexity of computation and parameters. *CrossNet*58 has 4 stages, each stage has {1, 1, 2, 2} *Pods* respective and there are 58 layers of convolution. The expansion factor is 0.4.

Type	Output size	Kernel size	Stride	Pod	DWConv channels				
					$b = 40$	$b = 50$	$b = 60$	$b = 70$	$b = 80$
Image	224×224				3	3	3	3	3
Conv	112×112	3×3	2		24	24	24	24	32
DWConv	112×112	1×1	1		24	24	24	24	32
	56×56	5×5	2		24	24	24	24	32
Stage1	56×56		1	1	40	50	60	70	80
Stage2	28×28		2	1	80	100	120	140	160
Stage3	14×14		2	1	160	200	240	280	320
	14×14		1	1	160	200	240	280	320
Stage4	7×7		2	1	320	400	480	560	640
	7×7		1	1	320	400	480	560	640
Conv	7×7	1×1	1		1024	1024	1024	1024	1536
GlobalPool	1×1	7×7	1						
FC					1000	1000	1000	1000	1000
FLOPs (M)					124	174	234	304	400
Params (M)					1.8	2.2	2.7	3.2	4.5

3.4 *CrossNet* Architecture

Built on cross connection and *Pod* structure, we present the overall *CrossNet* architecture in Table 1. The proposed network is mainly composed of a stack of *Pod* structure grouped into four stages. Before these four stages, we employ three convolution layers to perform two down-sampling operations. The first stage is applied with $stride = 1$, which is composed of one *Pod*. The other three stages are applied with $stride = 2$ and composed of 1, 2, 2 *Pod(s)* respectively. We use the *Pod* structure with a depth of 3, so the depth of the whole network is 58. The 5×5 kernels are used for all the depthwise convolutions. Hyper-parameters within a stage stay the same, and for the next stage the output channel dimension of depthwise convolutions are doubled. We find that when 1/10 expansion ratio of *ResNet* ($e = 4/10 = 0.4$) is used, *CrossNet* has better efficiency.

Similar to *MobileNet-V2* [33], we adopt linear bottlenecks which removes the last ReLU [28] activation function to prevent nonlinearities from destroying too much information. All the activation functions in the network utilize ReLU. Different basic width (denoted as b) can be used to construct *CrossNet* with different computational costs. In Table 1, we enumerate the computational costs and parameters of *CrossNet* at $b = \{40, 50, 60, 70, 80\}$. Some contrast experiments of hyper-parameter will be given in the next section.

Table 2. Single-crop top-1 accuracy rates (%) on the ImageNet validation set of different depths of *Pod*.

Model	Width	Stages	Top-1 Acc. (%)	FLOPs (M)	Params (M)
CrossNet ($d = 1$)	70	{3, 4, 8, 8}	71.9	306	**2.9**
CrossNet ($d = 2$)	70	{2, 2, 3, 3}	72.4	306	3.0
CrossNet ($d = 3$)	70	{1, 1, 2, 2}	**72.9**	**304**	3.2
CrossNet ($d = 4$)	60	{1, 1, 2, 2}	72.8	326	3.9

4   Implementation

In this section, we discuss the accuracy/computation trade-offs of *CrossNet* for image classification. We will analyze the design guideline of the *CrossNet* from two aspects of the depth of *Pod* and depthwise convolutions receptive fields (kernel size of depthwise convolutions).

The Depth of *Pod*. The depth of *Pod* is a very important parameter of *CrossNet*. We set the depth of *Pod* from 1 to 4, and adjust basic width and stages to get the almost same computational cost of each model, as shown in Table 2. We can observe that when the *Pod* depth is 1 (means without cross connection), the top-1 accuracy of model is 71.9% which is the worst in Table 2. But the model without cross connection still gets comparable top-1 accuracy with *MobileNet-V2*, this can be seen as a strong baseline. When the depth of *Pod* is 3, the performance is better than this strong baseline by an absolute values of 1.0%. When the depth of *Pod* is 4, the performance of model is as good as $d = 3$, but the former has more computational cost than the latter (326 MFLOPs *vs* 304 MFLOPs). Therefore, we can consider that $d = 3$ is the best choice.

Kernel Size of Depthwise Convolutions. According to the definition of receptive fields in convolutional neural networks, it is a simple and natural way to apply large kernels or stack multiple small kernels to realize receptive fields of large sizes. The Inception series employ multiple 3×3 or $1 \times n$ ($n \times 1$) convolution layers to clearly demonstrate the effectiveness of large scale receptive fields in vision recognition.

In order to increase the receptive fields of *CrossNet*, we adopt two different ways. The first is the use of 5×5 convolution, and the second is stacking two 3×3 convolutions. The receptive fields of the two methods are the same. Experimental results on the ImageNet classification are shown in Table 3. We adopt $b = 40$ and $b = 70$, the $k = 3$ as a baseline. $k = 3(\times 2)$ represents a stack of two 3×3 depthwise convolutions, *type*1 means there is no batch normalization and activation function between two 3×3 depthwise convolution layers according to [5], and *type*2 is opposite. Through the two groups of experiments in Table 3, we find that $k = 3(\times 2, type1)$ will degrade the performance of the *CrossNet* about 0.1–0.3%, even increasing the computational cost. The *type*2 with batch normalization and activation function can only get the similar accuracy as baseline. Although more

Table 3. The relationship between the receptive fields of depthwise convolutions and the top-1 accuracy of $CrossNet$. $k = 3$ as a baseline, and $k = 3(\times 2)$ represents a stack of two 3×3 depthwise convolutions.

Model	Acc. (%)	FLOPs	Params	Δ Acc. (%)	Δ FLOPs
$(b = 40)$, $k = 3$	68.9	109	1.8		
$(b = 40)$, $k = 3(\times 2, type1)$	68.8	116	1.8	-0.1	$+7$
$(b = 40)$, $k = 3(\times 2, type2)$	69.0	116	1.8	$+0.1$	$+7$
$(b = 40)$, $k = 5$	69.6	124	1.8	$+0.7$	$+15$
$(b = 40)$, $k = 7$	**69.8**	146	1.9	**+0.8**	$+37$
$(b = 70)$, $k = 3$	72.3	279	3.1		
$(b = 70)$, $k = 3(\times 2, type1)$	72.0	292	3.2	-0.3	$+15$
$(b = 70)$, $k = 3(\times 2, type2)$	72.2	292	3.2	-0.1	$+15$
$(b = 70)$, $k = 5$	72.9	304	3.2	$+0.6$	$+25$
$(b = 70)$, $k = 7$	**73.0**	341	3.4	**+0.7**	$+52$

3×3 depthwise convolutions increase the receptive fields and the capacity of network, it also brings problems to optimization. For $k = 5$, the $CrossNet(s)$ achieve 69.6% ($b = 40$) and 72.9% ($b = 70$) top-1 accuracy, passes the baseline by absolute values of 0.7% and 0.6% respectively. This indicates utilizing 5×5 depthwise convolutions to increase the receptive fields of the network is more effective.

In order to further explore the influence of increasing the receptive fields of depthwise convolutions, we also report the results of $k = 7$ with $CrossNet58$ ($b = 40$) and $CrossNet58$ ($b = 70$) in Table 3. Comparing with $k = 5$, the larger receptive fields brings negligible improvement in accuracy (0.2% and 0.1%), but adds considerable computational cost (22 MFLOPs and 27 MFLOPs). We conjecture this is because in the general ImageNet classification task, the feature maps of last stage needs to be padded from 7×7 to 13×13 to fit the depthwise convolutions with $k = 7$ [34]. But padding feature maps with 0-value will introduce too much useless information, resulting in degradation of the performance of the model. From this perspective, a larger receptive fields of depthwise convolutions ($k >= 7$) is more inefficient.

5   Experiments

In this section, we empirically demonstrate the performance and efficiency of $CrossNet$ over the state-of-the-art lightweight networks on competitive benchmark [32]. And we prove the transfer learning ability of $CrossNet$ on MSCOCO [23] object detection.

5.1   ImageNet-1k Dataset

The ImageNet-1k dataset is comprised of 1.28 million training images and 50,000 validation images from 1000 categories, which is widely used to verify the classification ability of convolutional neural networks. The evaluation is measured on the non-blacklist images of the ImageNet2012 validation set.

Training Setup. We implement the $CrossNet$ using PyTorch[2] on a server with 4 NVIDIA Titan X GPUs. Basically following [10,39], we adopt standard data augmentation methods and train the networks using SGD with a mini-batch size of 40 for each GPU (so effective mini-batch size is 160). All models are trained for 240 epochs from scratch, with a cosine shape learning rate [16] which starts from 0.045 and gradually reduces to 0. We set the weight decay to 3e–5. We report the top-1 accuracy using center crop evaluations on the ImageNet validation set, where 224×224 pixels are cropped from each image whose shorter edge is first resized to 256, the results of $CrossNet$ with $b = \{40, 50, 60, 70, 80, 90\}$ are shown in Table 4.

Table 4. Single-crop top-1 accuracy rates (%) and computational complexity on the ImageNet validation set of $CrossNet$ with different basic width. The resolution of input image is 224×224.

Model	Top-1 Acc. (%)	FLOPs (M)	Params (M)
CrossNet58 ($b = 40$)	69.6	124	1.8
CrossNet58 ($b = 50$)	70.5	174	2.2
CrossNet58 ($b = 60$)	71.7	234	2.7
CrossNet58 ($b = 70$)	72.9	304	3.2
CrossNet58 ($b = 80$)	73.8	400	4.5
CrossNet58 ($b = 90$)	74.5	517	5.2

Comparison with State-of-the-Art Lightweight Networks. Table 5 shows the results of $CrossNet$ and several state-of-the-art lightweight networks on the ImageNet-1k dataset. We observe that $CrossNet58$-$b70$ obtains 72.9% top-1 accuracy, which has 304 MFLOPs computational cost and 3.2 million parameters. Compared with the similar computational complexity networks, such as $MobileNet$-$V1$ (0.75), $MobileNet$-$V2$ (1.0), $ShuffleNet$ 1.5× ($g = 3$) and $CondenseNet$ ($G = C = 8$), our $CrossNet$ increases the top-1 accuracy rate by absolute values of 4.5%, 1.2%, 1.4% and 1.9% respectively. Figure 4 (left) shows the comparisons of test error rates $vs.$ computational cost. Among the recent state-of-the-art efficient networks, $CrossNet$ achieves the best accuracy/computation trade-offs and the accuracy improvement is very obvious.

Comparison with $MobileNet$-$V2$. $MobileNet$-$V2$ is a very excellent lightweight network. Our research also draws on the linear bottleneck proposed by it. It should be noted that the article of $MobileNet$-$V2$ only provides the top-1 accuracy of models at 300 MFLOPs and 585 MFLOPs. The accuracy and computational cost of other models are quoted by the official repository.

---

[2] https://pytorch.org/.

Table 5. Single-crop top-1 accuracy rates (%) on the ImageNet validation set and complexity comparisons with other state-of-the-art lightweight networks.

Model	Top-1 Acc.(%)	Top-5 Acc.(%)	FLOPs (M)	Params (M)
AlexNet [19]	57.2	–	720	60
SqueezeNet [18]	57.5	80.3	1700	1.25
Inception-V1 [34]	69.8	89.9	1550	6.8
MobileNet-V1 (1.0) [15]	70.6	–	569	4.2
MobileNet-V2 (1.4) [33]	**74.7**	–	585	6.9
MobileNet-V2 (1.4, our impl.) [33]	74.3	91.9	585	6.9
ShuffleNet 2× ($g = 3$) [39]	73.7	–	524	–
CondenseNet ($G = C = 4$) [16]	73.8	91.7	529	**4.8**
CrossNet58 ($b = 90$)	74.5	**92.1**	**517**	5.2
MobileNet-V1 (0.75) [15]	68.4	88.2	317	**2.6**
MobileNet-V2 (1.0) [33]	71.7	–	300	3.4
MobileNet-V2 (1.0, our impl.)	71.9	90.5	300	3.4
ShuffleNet 1.5× ($g = 3$) [39]	71.5	–	292	–
CondenseNet ($G = C = 8$) [16]	71.0	90.0	**274**	2.9
CrossNet58 ($b = 70$)	**72.9**	**90.9**	304	3.2
MobileNet-V2 (0.75) [33]	69.8	–	209	2.6
MobileNet-V2 (0.75, our impl.) [33]	70.0	89.3	209	2.6
CrossNet58 ($b = 50$)	**70.5**	**89.5**	**174**	**2.2**

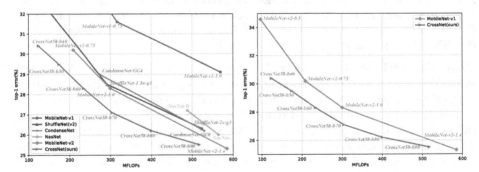

Fig. 4. Top-1 error rate versus computational cost across top performing published convolutional neural network architectures on the ImageNet validation.

Table 6. Actual CPU inference time on mobile device for an input image with resolution 224×224 (smaller number represents better performance). The platform is based on Qualcomm Snapdragon 820 processor. All results are evaluated on ncnn with multithreading.

Model	Top-1 Acc. (%)	FLOPs (M)	Time (ms)
MobileNet-V1 (1.0)	70.6	569	102
MobileNet-V2 (1.0)	71.7	300	77
CrossNet49 ($b = 70$)	72.1	**265**	**60**
CrossNet58 ($b = 70$)	**72.9**	304	79

Figure 4 (right) shows *CrossNet* is better than *MobileNet-V2* at similar computational cost. *CrossNet* is more efficient in the number of parameters which achieves 72.9% top-1 accuracy with only 3.2 million parameters which is superior to *MobileNet-V2* in both performance and number of parameters in Table 5. And when the computational complexity is relatively small, the advantage of *CrossNet* is more significant. This indicates that *CrossNet* has better application prospects in real-time computer vision tasks on mobile device.

Actual Inference Time. Table 6 shows the actual inference time on Qualcomm Snapdragon 820 processor for different models. All results are evaluated on ncnn[3] with multithreading. We find that the accuracy and speed of *CrossNet*58-*b*70 are obviously better than *MobileNet-V1* (1.0). And when the computational complexity is similar, the actual inference time of *CrossNet*58-*b*70 and *MobileNet-V2* (1.0) is also very close (77 ms *vs.* 79 ms). But the accuracy of *CrossNet* is better. In addition, we also show the result of *CrossNet*49-*b*70, which has comparable accuracy but much faster than *MobileNet-V2* (1.0). The comparison of Table 6 demonstrates that *CrossNet* has high practical value mobile device.

5.2   MSCOCO Dataset

Image classification networks provide generic image features that may be transferred to other computer vision tasks. To validate the general applicability, we test our *CrossNet* on the task of MSCOCO object detection. We pre-train *CrossNet* on ImageNet-1k dataset, plug in the models into SSD [25] and Mask R-CNN [11] object detection pipeline with default settings.

CrossNet with SSD. Following the pipeline of [15], we use *CrossNet*58-*b*80 as backbone for feature extraction and perform experiments on the 80 category MSCOCO dataset. We train the union of 80k train images and a 35k subset of val images (called trainval35k), and report the results on a 5k subset of val images (minival). The input image is resized to 300×300, all models were trained using 2 GPU data parallel sync SGD. For a fair comparison, we re-reimplement the baseline that *SSD-MobileNet-V1* with the exactly the same setting. The results are shown in Table 7. *CrossNet*58-*b*80 obtains 19.7% mAP, which outperforms the *MobileNet-V1/MobileNet-V1* (our impl.) and provides less computational cost (400 MFLOPs *vs.* 569 MFLOPs). The results shown in this experiment demonstrate that *CrossNet* is capable of learning better feature representations for transfer learning and benefiting the single stage object detector. We conjecture that this significant gain is partly due to efficient design of *CrossNet* and succinct implementation.

CrossNet with Mask R-CNN. Following the pipeline of [10] and x, we use *CrossNet*58-*b*80-FPN and *MobileNet-V2*-1.0-FPN [22] as backbones for feature extraction and perform experiments on the 80 category MSCOCO dataset. All models were trained using 8 GPU data parallel sync SGD with a mini batch-size of 16 images. The learning rate is 0.02 for the first 60k iterations and finally the

---

[3] https://github.com/Tencent/ncnn.

Table 7. Object detection results on MSCOCO minival set for an input image with resolution 300 × 300 (larger number represents better performance).

Model	mAP [.5, .95]
MobileNet-V1 (1.0) [15]	19.2
MobileNet-V1 (1.0, our impl.)	18.8
CrossNet58 ($b = 80$)	**19.7**

Table 8. Object Detection and **Instance Segmentation** *single-model* results using **Mask R-CNN** on MSCOCO minival set. Models are trained on the trainval35k set with same hyper-parameters and training schedules.

Backbones	AP^{box}	AP^{box}_{50}	AP^{box}_{75}	AP^{seg}	AP^{seg}_{50}	AP^{seg}_{75}
CrossNet58-b70-FPN	31.6	53.1	33.0	28.8	49.3	29.5
MobileNet-V2-1.0-FPN	30.2	51.0	31.7	27.8	47.6	28.7

training process terminates at 90k iterations with a linear learning rate warm up. The input image is resized such that its shorter side has 800 pixels, and only horizontal flipping data augmentation is used. We test the models with no test-time augmentations, the results of object detection and instance segmentation are shown in Table 8. Mask R-CNN using *CrossNet58-b70*-FPN outperforms the *MobileNet-V2*-1.0-FPN by 1.4 points box AP and 1.0 points mask AP, whose complexity are comparable. We conjecture that this significant gain is partly due to *CrossNet*'s simple design and larger receptive fields.

6 Conclusions

In this paper, we proposed the novel "Cross Connected Network" (*CrossNet*), which is able to reuse the feature maps of depthwise convolutions with cross connection in the same *Pod* structure. Alternating residual connections and densely connections makes the network more efficient. Next, we have discussed the trade-off between model performance and depthwise convolutions receptive fields, and have explored how to find more efficient hyper-parameters of *CrossNet*. Extensive experiments on ImageNet-1k and MSCOCO demonstrate that *CrossNet* improves the state-of-the-art performance of lightweight networks and has high practical value on mobile devices.

References

1. Chen, L., Papandreou, G., Kokkinos, I., Murphy, K., Yuille, A.: DeepLab: semantic image segmentation with deep convolutional Nets, Atrous convolution, and fully connected CRFs. arXiv:1606.00915 (2016)
2. Chen, L., Papandreou, G., Schroff, F., Adam, H.: Rethinking Atrous convolution for semantic image segmentation. arXiv:1706.05587 (2017)

3. Chen, L., Zhu, Y., Papandreou, G., Schroff, F., Adam, H.: Encoder-decoder with Atrous separable convolution for semantic image segmentation. arXiv:1802.02611 (2018)
4. Chen, Y., Li, J., Xiao, H., Jin, X., Yan, S., Feng, J.: Dual path networks. In: NIPS (2017)
5. Chollet, F.: Xception: deep learning with depthwise separable convolutions. In: CVPR (2017)
6. Dai, J., et al.: Deformable convolutional networks. In: ICCV (2017)
7. Gatys, L., Ecker, A., Bethge, M.: Image style transfer using convolutional neural networks. In: CVPR (2016)
8. Girshick, R.: Fast R-CNN. In: ICCV (2015)
9. Girshick, R., Donahue, J., Darrell, T., Malik, J.: Rich feature hierarchies for accurate object detection and semantic segmentation. In: CVPR (2014)
10. Goyal, P., et al.: Accurate, large minibatch SGD: training imagenet in 1 hour. arXiv:1706.02677 (2017)
11. He, K., Gkioxari, G., Dollár, P., Girshick, R.: Mask R-CNN. In: ICCV (2017)
12. He, K., Zhang, X., Ren, S., Sun, J.: Deep residual learning for image recognition. In: CVPR (2016)
13. He, K., Zhang, X., Ren, S., Sun, J.: Identity mappings in deep residual networks. In: Leibe, B., Matas, J., Sebe, N., Welling, M. (eds.) ECCV 2016. LNCS, vol. 9908, pp. 630–645. Springer, Cham (2016). https://doi.org/10.1007/978-3-319-46493-0_38. arXiv:1603.05027
14. Hong, S., Roh, B., Kim, K., Cheon, Y., Park, M.: PVANet: lightweight deep neural networks for real-time object detection. arXiv:1611.08588 (2016)
15. Howard, A., et al.: Mobilenets: efficient convolutional neural networks for mobile vision applications. In: CVPR (2017)
16. Huang, G., Liu, S., Maaten, L., Weinberger, K.: Condensenet: an efficient densenet using learned group convolutions. arXiv:1711.09224 (2017)
17. Huang, G., Liu, Z., Weinberger, K.: Densely connected convolutional networks. In: CVPR (2017)
18. Iandola, F., Han, S., Moskewicz, M., Ashraf, K., Dally, W., Keutzer, K.: Squeezenet: alexnet-level accuracy with 50x fewer parameters and ¡0.5MB model size. arXiv:1602.07360 (2016)
19. Krizhevsky, A., Sutskever, I., Hinton, G.: Imagenet classification with deep convolutional neural networks. In: NIPS (2012)
20. LeCun, Y., et al.: Backpropagation applied to handwritten zip code recognition. Neural Comput. 1, 541–551 (1989)
21. Li, S., Xu, X., Nie, L., Chua, T.: Laplacian-steered neural style transfer. In: ACM MM (2017)
22. Lin, T., Dollár, P., Girshick, R., He, K., Hariharan, B., Belongie, S.: Feature pyramid networks for object detection. In: CVPR (2017)
23. Lin, T.-Y., et al.: Microsoft COCO: common objects in context. In: Fleet, D., Pajdla, T., Schiele, B., Tuytelaars, T. (eds.) ECCV 2014. LNCS, vol. 8693, pp. 740–755. Springer, Cham (2014). https://doi.org/10.1007/978-3-319-10602-1_48
24. Liu, S., Huang, D., Wang, Y.: Receptive field block net for accurate and fast object detection. arXiv:1711.07767 (2017)
25. Liu, W., et al.: SSD: single shot multibox detector. In: Leibe, B., Matas, J., Sebe, N., Welling, M. (eds.) ECCV 2016. LNCS, vol. 9905, pp. 21–37. Springer, Cham (2016). https://doi.org/10.1007/978-3-319-46448-0_2
26. Long, J., Shelhamer, E., Darrell, T.: Fully convolutional networks for semantic segmentation. In: CVPR (2015)

27. Matan, O., Burges, C., LeCun, Y., Denker, J.: Multi-digit recognition using a space displacement neural network. In: NIPS (1991)
28. Nair, V., Hinton, G.: Rectified linear units improve restricted Boltzmann machines. In: ICML (2010)
29. Pham, H., Guan, M., Zoph, B., Le, Q., Dean, J.: Efficient neural architecture search via parameter sharing. arXiv:1802.03268 (2018)
30. Redmon, J., Divvala, S., Girshick, R., Farhadi, A.: You only look once: unified, real-time object detection. In: CVPR (2016)
31. Ren, S., He, K., Girshick, R., Sun, J.: Faster R-CNN: towards real-time object detection with region proposal networks. In: NIPS (2015)
32. Russakovsky, O., et al.: Imagenet large scale visual recognition challenge. IJCV (2015)
33. Sandler, M., Howard, A., Zhu, M., Zhmoginov, A., Chen, L.: Inverted residuals and linear bottlenecks: mobile networks for classification, detection and segmentation. arXiv:1801.04381 (2018)
34. Szegedy, C., et al.: Going deeper with convolutions. In: CVPR (2015)
35. Wang, M., Liu, B., Foroosh, H.: Design of efficient convolutional layers using single intra-channel convolution, topological subdivisioning and spatial bottleneck structure. arXiv:1608.04337 (2016)
36. Xie, S., Girshick, R., Dollár, P., Tu, Z., He, K.: Aggregated residual transformations for deep neural networks. In: CVPR (2017)
37. Yu, J., Jiang, Y., Wang, Z., Cao, Z., Huang, T.: UnitBox: an advanced object detection network. In: ACM MM (2016)
38. Zagoruyko, S., Komodakis, N.: Wide residual networks. In: BMVC (2016)
39. Zhang, X., Zhou, X., Lin, M., Sun, J.: ShuffleNet: an extremely efficient convolutional neural network for mobile devices. arXiv:1707.01083 (2017)
40. Zhao, H., Shi, J., Qi, X., Wang, X., Jia, J.: Pyramid scene parsing network. In: CVPR (2017)
41. Zhu, B., Chen, Y., Wang, J., Liu, S., Zhang, B., Tang, M.: Fast deep matting for portrait animation on mobile phone. In: ACM MM (2017)
42. Zoph, B., Vasudevan, V., Shlens, J., Le, Q.: Learning transferable architectures for scalable image recognition. arXiv:1707.07012 (2017)

Answer Distillation for Visual Question Answering

Zhiwei Fang[1,2(✉)], Jing Liu[1,2], Qu Tang[1], Yong Li[3], and Hanqing Lu[1,2]

[1] National Lab of Pattern Recognition, Institute of Automation,
Chinese Academy of Sciences, Beijing, China
{zhiwei.fang,jliu,luhq}@nlpr.ia.ac.cn
[2] University of Chinese Academy of Sciences, Beijing, China
[3] Business Growth BU, JD.com, Beijing, China
liyong5@jd.com

Abstract. Answering open-ended questions in Visual Question Answering (VQA) is a challenging task. As the answers are totally free-form, the answer space for open-ended questions is infinite in theory. This increases the difficulty for algorithms to predict the correct answers. In this paper, we propose a method named answer distillation to decrease the scale of answer space and limit the correct result into a small set of answer candidates. Specifically, we design a two-stage architecture to answer a question: First, we develop an answer distillation network to distill the answers, converting an open-ended question to a multiple-choice one with a short list of answer candidates. Then, we make full use of the knowledge from the answer candidates to guide the visual attention and refine the prediction results. Extensive experiments are conducted to validate the effectiveness of our answer distillation architecture. The results show that our method can effectively compress the answer space and improve the accuracy on open-ended task, providing a new state-of-the-art performance on COCO-VQA dataset.

Keywords: Answer distillation · Visual question answering

1 Introduction

Recent years, Visual Question Answering (VQA) has gained wide attention in deep learning research as it is an important form of artificial intelligence-related tasks. Generally speaking, VQA models require two modalities of information: text and images. The inputs are images and natural language questions about the image and the goal is to generate a natural language answer for the given inputs. Automatically answering questions about visual images is a challenging task and the best performances are still far weaker than human's [3].

As a primary form in VQA, the open-ended task is a research hotspot in this area. Since the answers for open-ended questions are totally free-form, the answer space is usually very large. This causes much difficulty for algorithms to

C. V. Jawahar et al. (Eds.): ACCV 2018, LNCS 11361, pp. 72–87, 2019.
https://doi.org/10.1007/978-3-030-20887-5_5

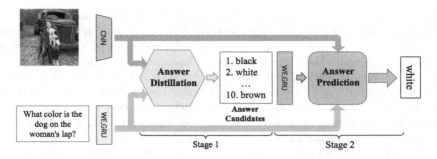

Fig. 1. A two-stage pipeline for visual question answering. Stage 1: Generate answer candidates for open-ended question by answer distillation. Step 2: Pick the correct answer from the answer candidates. WE indicates word embedding module. (Color figure online)

answer a question, because directly seeking the correct answer in a large space is not an easy task. What's more, for a given question-image pair, the number of potential answer candidates is usually very small and most of the answers are irrelevant which should not be reserved. For example, in Fig. 1, since the question is asked about the color of the dog and there are only a few kinds of colors in the image, the possible correct answers can only be "black", "white", "brown", and so on. Some other answers like "sky", "smile" are impossible to become a correct result and should be excluded. Such a process to generate a set of answer candidates for questions is referred to as answer distillation in this paper.

There are two advantages to conduct answer distillation for open-ended questions. First, it can reduce the scale of answer space for the decoder to predict the correct answers. Namely, answer distillation converts an open-ended question to a multiple-choice one. Second, new knowledge from the answer candidates can be introduced and used in answer reasoning. Based on the two advantages, we propose a two-stage architecture for open-ended task in VQA. In the first stage, we distill answers and predict the answer candidates. Then in the second stage, we introduce the candidates to guide the final answer prediction.

The feasibility for answer distillation is built on the two observations in VQA dataset. The first one is that similar questions can be asked about different images and get different answers. For example, the answers for the question in Fig. 1 can be "white", "black", "black and white", "brown", etc., when it is asked about different images. If we ignore the image information, then there is a strong relationship between the question "What color is the dog?" and the answer set {'white", 'black", "black and white", "brown", ...}. We call this kind of relationship Common Sense and it is also learnable. The second observation is that each question-image pair is annotated with multiple answers in common VQA datasets [3,7]. Although the answers may be different, they are all supposed to be answer candidates because these answers are the most reasonable ones from the views of the annotators. Based on the above two observations, we design a multi-task answer distillation network to predict the answer candidates. In the first task, the network takes questions as input to learn the common sense answer candidates. While in the second task, we jointly make use of knowledge from questions, images and the Common Sense to learn multiple-answer candidates.

After answer distillation, an open-ended question is converted to a multiple-choice question with a list of answer candidates. Usually, for open-ended questions, we learn knowledge from questions and images and then reason the answer, in which the *question guided visual attention* mechanism is often employed. But for multiple-choice questions, we can further utilize the knowledge from answer candidates to reason which candidate is more correct. In this study, we further develop an *answer guided visual attention* mechanism to introduce the knowledge of answer candidates to refine answer prediction.

In summary, the main contributions of this study are three points: First, we develop an answer distillation network to predict answer candidates for open-ended questions. Second, an answer guided visual attention mechanism is developed to introduce the knowledge of answer candidates to refine the spatial attention for visual information, providing better answer prediction. Finally, we conduct extensive experiments and the results demonstrate that our methods can effectively predict answer candidates and achieve a new state-of-the-art performance on COCO-VQA dataset.

2   Related Work

Open-Ended Questions in VQA. Open-ended question is a primary form in many VQA datasets [3,7,14,21]. Many methods usually adopt the encoder-decoder architecture: a convolutional neural network (CNN) [15,24,25] to encode visual information; a recurrent neural network (RNN) [5,9,11] to encode question information and a decoder to produce the answer. In order to handle large answer space of open-ended questions, early approaches usually formulate the answer decoder as a generation model and use RNNs to generate a sequence as answer in arbitrary form [18,27]. However, in practice, it is hard to jointly train the encoder and decoder for VQA if the decoder is in form of RNNs [10]. Then [30] proposes to model the decoder as a multiple-label classifier rather than RNNs. They collect a set of the most frequent answers in dataset and assume that the answers of all the questions must be from this answer set. This formulation is simple but effective, and is followed by many approaches such as SAN [28], MCB [6], MLB [12], and so on. Collecting answers by their frequencies is also something like of answer distillation and has been demonstrated effective. However, this kind of answer distillation is still too coarse. In this paper, we further conduct more strict answer distillation to reduce the number of candidate answers to ten or less.

Multimodal Interactions in VQA. Multimodal interactions play an important role in VQA. One of the most popular interactions is the attention mechanism between questions and images. For example, [31] describes how to introduce spatial attention to standard LSTM model for convolutional feature map of image. They prove that image spatial attention mechanism can improve the performance successfully. [28] further develops this scheme with their "stacked attention networks" (SAN) which infers the answer iteratively. Meanwhile, some studies focus on the bilinear pooling for feature fusion in attention, such as Multimodal Compact Bilinear Pooling (MCB) [6], Low-rank Bilinear Polling (MLB) [12], Multimodal Tucker Fusion (MUTAN) [4]. Different from the methods above which only employ spatial attention for feature map of image, [17] introduces "hierarchical co-attention model" (HieCoAtt) which jointly learns attention for both image and question. However, no matter the single image attention or co-attention, the attention mechanisms mentioned above are all from the interactions between questions and images, ignoring the attending of answers. Using knowledge from answers is first been tried by [10]. Since they don't limit the number of answer candidates, the model needs careful sampling of input image-question-answer triplet. Recent work [23] develops high-order attention mechanisms which can handle more than 2 modalities. However, since there is no answer distillation process in their work, the answer candidates must be provided manually. Different from those methods, with our answer distillation architecture, an open-ended question can be first converted to a multiple-choice one, and then the knowledge of answer candidates can be introduced to help the final decision.

3   Answer Distillation for Visual Question Answering

3.1   Overview

In VQA, the goal is to generate an answer for a given question $Q \in \mathcal{Q}$ asked about an image $I \in \mathcal{I}$. In practice, it is usually assumed that the correct answer is from a pre-collected answer set $\mathcal{A}$ (i.e., the answer vocabulary), and formulate answer prediction as a classification problem:

$$\tilde{a} = \underset{a \in \mathcal{A}}{argmax}\, p_\theta(a|I, Q) \tag{1}$$

where $\tilde{a}$ is the predicted answer and θ are the parameters of the model.

Figure 2 illustrates the overall architecture of our method, which includes two stages: answer distillation and answer prediction. In the first stage, the model takes questions and images as inputs and distills the answer set $\mathcal{A}$ to a rather smaller one A. Here A is a set of answer candidates. Then in the second stage, we jointly take account of questions and images as well as the answer candidates to make a decision of which one in A is correct:

$$\tilde{a} = \underset{a \in A}{argmax}\, p_{\theta_2}(a|I, Q, A) \tag{2}$$

where θ_2 are the parameters in the second step.

Usually, in VQA, the inputs are first embedded into feature representations. For image, its representation is typically a feature map $\boldsymbol{v} \in R^{n_v \times d_v}$ extracted

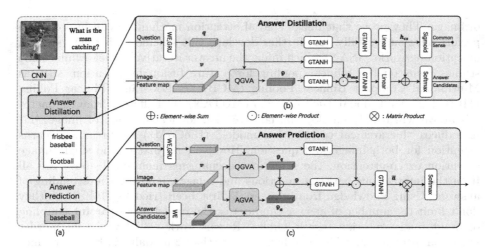

Fig. 2. The overall architecture of our answer distillation method for VQA. WE indicates word embedding module. GTANH indicates non-linear transformation layer of gated-tanh. QGVA and AGVA indicate question guided visual attention and answer guided visual attention, respectively.

from convolutional neural networks (e.g. ResNet152 [8] or FasterRcnn [22]). For question, it can be encoded by GRU recurrent neural networks with a representation of $q \in R^{d_q}$. In the second stage, the representation of answer candidates is also a feature map: $a \in R^{n_a \times d_a}$, where $n_a \equiv |A|$ and the details will be discussed in Sect. 3.3.

3.2 Answer Distillation

Since answer candidates must be learned from both common sense and multiple-answers, we model the training of answer distillation network as a multi-task learning problem.

Learning Common Sense. Firstly, before training, we must build the common sense answers for each question. Let $\mathcal{D} = \{(Q, I, a)\}$ denote training dataset, where (Q, I, a) is a training sample which means the $a \in \mathcal{A}$ is the ground-truth answer for the question $Q \in \mathcal{Q}$ about the image $I \in \mathcal{I}$. Then the common sense answers for a question Q^* can be written as:

$$CSA(Q^*) = \{a | (Q, \cdot, a) \in \mathcal{D}, Q = Q^*\} \tag{3}$$

Then we use the following equation to predict the probabilities p of common sense answers:

$$h_{cs} = F_{cs}(q) \tag{4}$$

$$p = \sigma(W_{cs} h_{cs}) \tag{5}$$

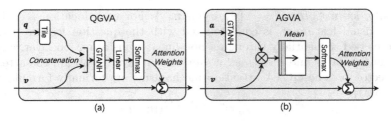

Fig. 3. Illustration of the mechanisms of question guided visual attention (a) and answer guided visual attention (b).

where q is the GRU feature representation of question sequence, F_{cs} is non-linear transformation function, $h_{cs} \in R^{d_h}$ is the hidden state, $W_{cs} \in R^{d_h \times |\mathcal{A}|}$ are the learned weights and σ is sigmoid function. The non-linear transformation can be ReLU or tanh. Here we use the gated tanh (GTANH) non-linear function $f^{m \to n} : x \in R^m \to y \in R^n$ proposed by [1]. It is defined as:

$$y = f^{m \to n}(x) = \tanh(Wx + b) \odot \sigma(W'x + b') \qquad (6)$$

where $W, W' \in R^{n \times m}$ are the learned weights, $b, b' \in R^n$ are the learned bias and $\odot$ is the Hadamard (element-wise) product. In our model, we use two GTANH layers to transform the question feature q to a hidden state h_{cs}:

$$h_{cs} = F_{cs}(q) = f^{512 \to 2048}(f^{d_q \to 512}(q)) \qquad (7)$$

As is shown in Eq. 5, the sigmoid activation function normalizes the final score to $(0, 1)$ which is followed by a modified binary cross-entropy loss:

$$L_{cs} = \gamma[-\sum_{i}^{|\mathcal{A}|} t_i \log(p_i)] + (1 - \gamma)||p||_1 \qquad (8)$$

where $p \in R^{|\mathcal{A}|}$ is the probability array computed by Eq. 5, p_i is the ith element in p, $t_i = 1$ only if the ith candidate $a_i \in \mathcal{A}$ is a common sense answer, otherwise $t_i = 0$, and γ is a scale factor to balance the two terms. The first term is a part of a standard binary cross-entropy formulation which only penalizes the positive positions whose target probability t_i is 1. The reason why we don't penalize negative positions ($t_i = 0$) is that the positions with $t_i = 0$ may also belong to common sense answers. Besides, only penalizing the positive positions also benefits the increase of recall for ground-truth answers. On the other hand, for a given question, we argue that most answers in $\mathcal{A}$ do not belong to common sense answers, i.e., p should be sparse. Thus we add the second term in Eq. 8 to guarantee the sparsity of p.

Learning Multiple-Answers. Multiple-answers are attached to a given question-image pair, thus this task needs both question and visual information. We employ Question Guided Visual Attention (QGVA) for image features.

Specifically, let $v \in R^{n_v \times d_v} = \{v_1, v_2, ..., v_{n_v}\}$. For each location $i \in [1, n_v]$, the local visual feature v_i is concatenated with the question feature q. Then they are passed sequentially through a non-linear transformation layer, a linear transformation layer and a softmax normalization layer to obtain the attention weight vector α. The QGVA attention mechanisms are formally formulated as (see Fig. 3(a)):

$$\lambda_i = W_\alpha f^{(d_v+d_q) \to 512}([v_i, q]) \tag{9}$$

$$\alpha = softmax(\lambda) \tag{10}$$

$$\hat{v} = \sum_{i=1}^{d_v} \alpha_i v_i \tag{11}$$

where W_α are the learned parameters and $\lambda = \{\lambda_1, \lambda_2, ..., \lambda_{n_v}\}$ is unnormalized weight vector. All the visual features are weighted by attention weights α and summed to generate the final single d_v-dim attended feature representation $\hat{v}$.

After obtaining the representations of question (q) and of image ($\hat{v}$), we compute the multimodal fusion representation by a Hadamard product (element-wise product):

$$h_{ma} = f^{d_q \to 512}(q) \odot f^{d_v \to 512}(\hat{v}) \tag{12}$$

The vector h_{ma} is a joint embedding of the question and image.

Since our goal is to predict the answer candidates rather than a single answer with the highest evaluation score, we treat this as a task to learn the probability distribution of occurrence of answers. Besides, introducing the output of common sense task above can also provide some guidance for learning multiple-answers. We formulate the predicted distribution as:

$$p' = softmax(W_{ma} f^{512 \to |\mathcal{A}|}(h_{ma}) + h_{cs}) \tag{13}$$

where W_{ma} are the learned parameters and h_{cs} is the unnormalized output of the common sense task from Eq. 7. We use softmax activation to make sure that the result is a standard probability distribution whose summation is 1. Each location in $p' \in R^{\mathcal{A}}$ represents the possibility of the corresponding answer to be an answer candidate. The target probability t'_i for any answer $a_i \in \mathcal{A}$ is its occurrence frequency in the multiple-answers:

$$t'_i = \frac{N(a_i)}{\sum_{a_j \in \mathcal{A}}(N(a_j))} \tag{14}$$

where $N(a_i)$ is the number of a_i in ground truth answers. Then we use binary cross-entropy loss to measure the distance of predicted and target distributions:

$$L_{ma} = -\sum_{i}^{|\mathcal{A}|}[t'_i \log p'_i + (1 - t'_i) \log (1 - p'_i)] \tag{15}$$

where t'_i and p'_i are the ith elements t' and p', respectively.

Obtaining Answer Candidates. During training, the network is supervised by both common sense loss L_{cs} and multiple-answer loss L_{ma}. During testing, since the multiple-answer branch has contained the knowledge of common sense branch in training, we use the top K answers in p' as the final answer candidates. Here K is a hyper parameter to determine the number of candidates, i.e., $|A| \equiv K$.

3.3   Answer Prediction

Figure 2(c) shows the architecture of the answer prediction network. Let $A = \{a_1, a_2, ..., a_K\}$ denote answer candidates, then the inputs of the network are (Q, I, A), whose feature representations are $(\boldsymbol{q}, \boldsymbol{v}, \boldsymbol{a})$.

Answer Embedding. We use a word embedding module to convert answer candidates to features. The vocabulary of the word embedding is exactly the whole answer vocabulary $\mathcal{A}$, and there is no LSTM or GRU module followed. After the word embedding, we obtain the feature representations for answer candidates: $\boldsymbol{a} \in R^{|A| \times d_a} = \{\boldsymbol{a}_1, \boldsymbol{a}_2, ..., \boldsymbol{a}_{|A|}\}$.

Answer Guided Visual Attention. Figure 3(b) shows the process of answer guided visual attention. The answer representations are first fed into a non-linear layer and transformed to the same dimension d_v of visual features. Then a correlation matrix $C \in R^{n_v \times n_a}$ is obtained by conducting matrix product between image and answer features:

$$C = (f^{d_a \to d_v}(\boldsymbol{a})\boldsymbol{v}^T)^T \tag{16}$$

The $(C)_{i,j}$ represents the correlation of the jth answer candidate and the ith visual patch in image. Then the visual attention weights $\boldsymbol{\alpha}$ are computed by averaging the correlation matrix C across all answer candidates:

$$\boldsymbol{\alpha} = softmax(\mathbb{1}C^T) \tag{17}$$

where $\mathbb{1} \in R^{n_a}$ is an all-one vector. Finally, the attention feature $\hat{\boldsymbol{v}}_a$ is obtained by the weighted sum of visual features:

$$\hat{\boldsymbol{v}}_a = \sum_{i=1}^{d_v} \alpha_i \boldsymbol{v}_i \tag{18}$$

Retrieving the Correct Answer. Let $\hat{\boldsymbol{v}}_q$ denotes the output of the question guided visual attention. The final feature representation of image is formulated as the fusion of $\hat{\boldsymbol{v}}_q$ and $\hat{\boldsymbol{v}}_a$:

$$\hat{\boldsymbol{v}} = \hat{\boldsymbol{v}}_q + \hat{\boldsymbol{v}}_a \tag{19}$$

Then we use question and visual knowledge to reason the feature representation of correct answer $\bar{\boldsymbol{a}}$:

$$\bar{\boldsymbol{a}} = f^{512 \to d_a}(f^{d_v \to 512}(\hat{\boldsymbol{v}}) \odot f^{d_q \to 512}(\boldsymbol{q})) \tag{20}$$

The similarity between $\bar{a}$ and each of the answer candidates is measured by their normalized inner-product:

$$s = softmax(\bar{a}a^T) \tag{21}$$

In training, the objective fusion is softmax loss as in [6]. In testing, we choose the answer who has the maximal similarity in s as the final result.

4   Experiments

4.1   Basic Configuration

Datasets COCO-VQA-v1. The COCO-VQA-v1 dataset [3] is built over ~200k images from MSCOCO [16]. Each of these questions is answered by 10 workers, yielding a list of 10 ground-truth answers. The dataset is split into three parts: training set (~248k questions), validation set (~121k questions) and testing set (~244k questions). In the dataset, there are three types of questions: "Yes/No", "Number" and "Other" but their answers are not strictly limited by the question types (e.g., the answers for "Yes/No" questions may not have to be {yes, no}, the ground-truth may also be "I don't know" or some other words). The ground truth answers are available only for training and validation sets while the evaluation for testing set can be only done on server of the dataset.

COCO-VQA-v2. The COCO-VQA-v2 dataset is an expanded version of COCO-VQA-v1. It shares the same images with COCO-VQA-v1 but doubles the questions and makes them more balanced [7]. In COCO-VQA-v2, there are ~443k questions in train split, ~214k in val split and ~453k in test split.

Experimental Setup. In COCO-VQA-v1, we use ResNet152 [8] as the visual encoder whose output feature map is in size of $14 \times 14 \times 2048$ (196×2048). While in COCO-VQA-v2, we use fixed-bottom-up features [1] of size 36×2048 for their high performance in VQA Challenge 2017. As for question encoding, we use a word-embedding (dim = 300) module followed by a GRU whose hidden size is 512. $|\mathcal{A}|$ is fixed to 3000 most frequent answers as in [6] and we train all of our model using ADAM [13] with fixed learning rate. When training ADN, the learning rate is 7e–4 while for answer prediction network, the learning rate is 1e–4.

4.2   Results

Answer Distillation Network. For answer distillation, we only care about whether the ground-truth answers are in the answer candidates. We use Recall

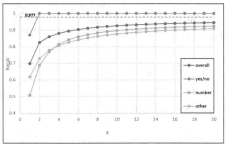 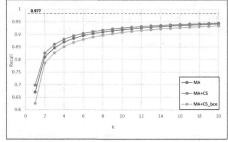

(a) Recall@K curves of different question types.   (b) Recalls of different model modifications.

Fig. 4. Recall rate curves of the results of answer distillation network.

Rate as the metric to measure the performance of our answer distillation network. It is formulated as follow:

$$r = \frac{\sum_{(a,Q,I) \in \mathcal{D}} \delta(a \in c(Q))}{|\mathcal{D}|} \qquad (22)$$

where $(a, Q, I) \in \mathcal{D}$ is a sample in dataset, $c(Q)$ is the answer candidate set for question Q and $\delta()$ is indicative function whose value is in $\{0, 1\}$. The experiments are conducted on COCO-VQA-v2 and models are trained on training set and tested on validation set.

The first factor to influence recall is the number of answer candidates (i.e. K), which is shown in Fig. 4(a). In this figure, we show the curves of the *overall* recall as well as recalls of three types of questions. As expected, the recalls of all kinds of questions increase along with K. For the *overall* recall, almost 93% questions can retrieve its ground-truth answers by the answer candidates at $K = 10$, and it rises to 94% when K is 20. As a contrast, the recall rate of the answer vocabulary (i.e. $K = 3000$) is 97.7% (see the dotted line in Fig. 4(a)), which is the upper bound performance of our model. We see that the gap between our recall and the upper bound is very small, verifying the efficiency of our distillation model. Besides, the recall of *yes/no* questions achieves a very high level of 99.9% just at $K = 2$. The reason is that the candidate answers for *yes/no* questions are very simple: either "yes" or "no". On the contrary, the recalls of *number* and *other* questions are under the level of *overall*, since their potential answers are complicated.

Figure 4(b) shows the recalls of three models with different modifications, providing an ablation analysis. "MA" stands for a model only trained with multiple-answer task, and "MA+CS" means the model is trained with both tasks of multiple-answer and common sense. In "MA+CS_bce", the loss function of common sense task is a standard binary cross-entropy instead of ours in Eq. 8. Comparing "MA" and "MA+CS", we notice that the recall is improved by adding common sense task to the network. Besides, "MA+CS_bce" gets the

Table 1. Recalls of the answer candidates. n_candidate denotes the number of candidate answers. TFA denotes that the candidates are the top frequent answers in the answer vocabulary. ADN denotes that the candidates are generated by our Answer Distillation Network.

n_candidates	10	20	500	1000	3000
TFA	57.27	63.27	92.21	95.23	97.77
ADN	**92.72**	**94.63**	**97.588**	**97.69**	97.77

Fig. 5. Examples of answer candidates. There are three boxes from left to right demonstrating results of ADN for "Yes/No", "Number" and "Other" questions, respectively. For each question, only top 9 candidates are displayed. The red ones are the ground-truth answers. The candidate answers for every question are ranked by their predicted relevance from left to right, top to bottom. (Color figure online)

worst performance, indicating that the loss function in Eq. 8 is very important for common sense learning.

We also compare the recall rates of the results of ADN and the top frequent answers in answer vocabulary $\mathcal{A}$ and the results are listed in Table 1. We find that when the number of reserved candidates decreases, the recall rate of our ADN results is still at a high level, which is similar to the conclusion in Fig. 4. On the contrary, for the top frequent answers (TFA), the recall rate is hard to be maintained if the number of reserved candidates becomes small. In fact, as we can see in Table 1, when there are only 10 answers reserved, the recall of TFA drops to 57.27% while ours is still at 97.77%. This demonstrates the superiority of our answers distillation network.

Figure 5 shows some result examples of our answer distillation. The first box shows the answer candidates for some "Yes/No" questions and we can see that the candidate answers are similar with different questions. The second box is for "Number" questions whose candidates are mostly number-related. It is noteworthy that the candidate numbers are around the ground-truth numbers. We also notice that if the ground-truth number is too big (e.g. the answer of the second question in this box is 27), the ADN may fail to predict a correct set of candidates. The last box shows examples of "Other" question. The predicted candidate set can effectively limit the answer space of such questions.

Answer Prediction Network. In this section, we compare models with different configurations to show how our answer prediction network works. All the

Table 2. Comparison between Bottom-Up model [1] and different modifications of our model.

Model	Yes/No	Number	Other	Overall
Bottom-Up	80.37	42.06	54.44	62.48
Baseline	81.72	42.56	53.20	62.31
AD	81.53	**43.59**	53.79	62.88
AD+AGVA	**81.94**	43.05	**54.46**	**63.27**

Table 3. Performances of different answer candidate sets. n_candidate denotes the number of candidate answers. ADN denotes the candidates are generated by our Answer Distillation Network. TFA denotes the candidates are the top frequent answers in the answers vocabulary.

n_candidates	10	500	1000	3000
ADN	**62.88**	**62.45**	**62.37**	62.31
TFA	40.59	60.65	62.06	62.31
ADN - TFA	+22.29	+1.80	+0.31	+0.00

models are trained on training set and the results are reported on validation set of COCO-VQA-v2. The answer candidates for each question are provided by the best model of answer distillation, in which K is set to 10 to balance recall and computation. During training, some questions are abandoned if their ground-truth answers are not in the corresponding answer candidates. We represent the answer candidates with a word-embedding module whose feature dimension is 300. Table 2 shows the comparison of the performances of different models. The models are:

Bottom-Up. This model is proposed by the state-of-the-art method in VQA Challenge 2017. We use their performance reported in [26] under the same configuration: trained on training split and tested on validation split without extended data from Visual Genome [14].

Baseline. In order to obtain the baseline, we bring two modifications to our standard answer prediction network described in Fig. 2(c): (1) setting the answer candidates as the whole answer vocabulary, i.e., let $K = |\mathcal{A}| = 3000$; (2) removing the answer guided visual attention module. The first modification removes the influence of answer distillation and the second one disables the interactions between images and answers.

AD. Based on **Baseline model**, we only let $K = 10$, to see the effectiveness of answers distillation.

AD+AGVA. This is our standard model as described in Fig. 2(c).

Firstly, when compared with **Bottom-Up**, our **Baseline** is similar to their overall performance. This is reasonable because when let $K = |\mathcal{A}|$, the answer

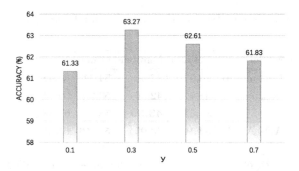

Fig. 6. The influence of γ on VQA performance.

representations a just act as the same role of the linear transformation in **Bottom-Up**'s classifier. Secondly, comparing **AD** with **Baseline**, we notice that the answer distillation provides about 0.5% improvements to the baseline. The most definite improvement happens on "other" questions which verifies the effectiveness of answer distillation. Finally, when further introducing the answer guided visual attention module, the performance keeps increasing, demonstrating that the knowledge from answer candidates is helpful for VQA.

Besides, when considering the performance improvements from baseline to final model on different question types, we can see that "other" questions (1.26%) rank the first, "number" (0.49%) behind and "yes/no" (0.22%) the last. This is because the answer candidates for "yes/no" questions are fixed and simple thus the effect of answer distillation is limited. But for the other two types of questions, distilling answers can effectively decrease the number of answers and the effect is significant. This observation is also consistent with the results shown in Fig. 4(a).

In order to further analyze the effectiveness of answer distillation, we use answer candidates from different sources (i.e. from ADN or from TFA) and change the number of candidates to train the answer prediction network. We compare their final performances in Table 3. Note that all models in Table 3 are based on **AD** model where the AGVA module is not included. The only difference among these models is the input answer candidates. The results show that the performances with answer distillation is significantly higher than those without answer distillation (i.e., TFA), especially when the number of candidates is small. Meanwhile, when the number of candidates decreases, models with answer distillation can get better performances because irrelevant answers are excluded, while the models with TFA suffer serious decrease in performance.

In Fig. 6, we evaluate the influence of γ in Eq. 8 on model performance. As we can see, the γ is supposed to be neither too big nor too small. The best performance relies on the careful tuning of gamma, or it may cause some side effects. This implies that the effectiveness of Eq. 8 is a result of the tradeoff between the two parts.

Table 4. Comparison with the state-of-the-art approaches on test-dev split of COCO-VQA-v1 and COCO-VQA-v2. "Ver." indicates the version of COCO-VQA dataset. The results are reported in terms of accuracy in %. "W.E." indicates where the approaches uses pre-trained word embedding models. "V.G." indicates training data is augmented with Visual Genome dataset. "Bottom-up" indicates the model uses bottom-up features rather than ResNet152 features.

Model	Ver.	W.E.	V.G.	Bottom-Up	Test-dev			
					Yes/No	Number	Other	Overall
NMN [2]	v1				81.2	38.0	44.0	58.6
SAN [28]	v1				79.3	36.6	46.1	58.7
HieCoAtt [17]	v1				79.7	38.7	51.7	61.8
RAU [20]	v1				81.9	39.0	53.0	63.3
MCB [6]	v1				82.2	37.7	54.8	64.2
DAN [19]	v1				83.0	39.1	53.9	64.3
MFB [29]	v1				83.2	38.8	**55.5**	65.1
Ours	v1				**83.9**	40.4	55.0	**65.3**
MCB+GloVe [6]	v1	✓			82.5	37.6	55.6	64.7
MLB+StV [12]	v1	✓			84.1	38.2	54.9	65.1
MFB+GloVe [29]	v1	✓			84.0	39.8	56.2	65.9
Ours+GloVe	v1	✓			**84.8**	41.2	**56.7**	**66.2**
MCB+GloVe+VG [6]	v1	✓	✓		82.3	37.2	57.4	65.4
MLB+StV+VG [12]	v1	✓	✓		83.9	37.9	56.8	65.8
MFB+GloVe+VG [29]	v1	✓	✓		84.1	39.1	58.4	66.9
Ours+GloVe+VG	v1	✓	✓		**85.5**	41.4	**58.7**	**67.6**
BottomUp+GloVe+VG [1]	v2	✓	✓	✓	81.8	44.2	56.1	65.3
Ours+GloVe+VG	v2	✓	✓	✓	**83.4**	45.3	**58.0**	**67.0**

Comparison with State-of-the-Art Methods. Table 4 compares our approaches with the current state-of-the-art methods under different conditions. All the models employ attention mechanisms and are evaluated on "test-dev" by official server. The table is split into four parts over the rows according to different settings.

From Table 4, we have the following observations: Firstly, our model with answer distillation outperforms all the comparative methods significantly, which can verify the effectiveness of our proposed methods. Secondly, GloVe model and Visual Genome can further improve the performance which implies that good question embeddings are important. Finally, the improvement on COCO-VQA-v2 is more significant than COCO-VQA-v1. This is also reasonable because COCO-VQA-v2 doubles the number of questions asked for each image and balances the answers for different images, which means that there is more Common Sense knowledge and Multiple Answers knowledge for Answer Distillation.

5   Conclusion

We propose a method based on answer distillation to convert an open-ended question to a multiple-choice one, which decreases the difficulty for answering a question. In order to make full use of the knowledge from answer candidates, we develop an answer guided visual attention mechanism to refine the joint representations of image. By jointly taking account of three modalities of question, image and answer candidates, our architecture gives further improvements on real-world VQA dataset compared to state-of-the-art.

Acknowledgements. This work was supported by National Natural Science Foundation of China (61872366 and 61472422).

References

1. Anderson, P., et al.: Bottom-up and top-down attention for image captioning and visual question answering (2017). http://arxiv.org/abs/1707.07998
2. Andreas, J., Rohrbach, M., Darrell, T., Klein, D.: Deep compositional question answering with neural module networks. CoRR abs/1511.02799 (2015). http://arxiv.org/abs/1511.02799
3. Antol, S., et al.: VQA: visual question answering. In: International Conference on Computer Vision (ICCV) (2015)
4. Ben-younes, H., Cadene, R., Cord, M., Thome, N.: MUTAN: multimodal tucker fusion for visual question answering, pp. 2612–2620 (2017). https://doi.org/10.1109/ICCV.2017.285, http://arxiv.org/abs/1705.06676
5. Cho, K., et al.: Learning phrase representations using RNN encoder-decoder for statistical machine translation. arXiv preprint arXiv:1406.1078 (2014)
6. Fukui, A., Park, D.H., Yang, D., Rohrbach, A., Darrell, T., Rohrbach, M.: Multimodal compact bilinear pooling for visual question answering and visual grounding. arXiv:1606.01847 (2016)
7. Goyal, Y., Khot, T., Summers-Stay, D., Batra, D., Parikh, D.: Making the V in VQA matter: elevating the role of image understanding in visual question answering. In: Conference on Computer Vision and Pattern Recognition (CVPR) (2017)
8. He, K., Zhang, X., Ren, S., Sun, J.: Deep residual learning for image recognition. CoRR abs/1512.03385 (2015). http://arxiv.org/abs/1512.03385
9. Hochreiter, S., Schmidhuber, J.: Long short-term memory. Neural Comput. $9(8)$, 1735–1780 (1997)
10. Jabri, A., Joulin, A., van der Maaten, L.: Revisiting visual question answering baselines. In: Leibe, B., Matas, J., Sebe, N., Welling, M. (eds.) ECCV 2016. LNCS, vol. 9912, pp. 727–739. Springer, Cham (2016). https://doi.org/10.1007/978-3-319-46484-8_44
11. Jordan, M.I.: Serial order: a parallel distributed processing approach. In: Advances in Psychology, vol. 121, pp. 471–495. Elsevier (1997)
12. Kim, J.H., On, K.W., Lim, W., Kim, J., Ha, J.W., Zhang, B.T.: Hadamard product for low-rank bilinear pooling. In: The 5th International Conference on Learning Representations (2017)
13. Kingma, D.P., Ba, J.: Adam: a method for stochastic optimization. CoRR abs/1412.6980 (2014). http://arxiv.org/abs/1412.6980
14. Krishna, R., et al.: Visual genome: connecting language and vision using crowdsourced dense image annotations. CoRR abs/1602.07332 (2016). http://arxiv.org/abs/1602.07332
15. Krizhevsky, A., Sutskever, I., Hinton, G.E.: Imagenet classification with deep convolutional neural networks. In: Advances in Neural Information Processing Systems, pp. 1097–1105 (2012)

16. Lin, T., et al.: Microsoft COCO: common objects in context. CoRR abs/1405.0312 (2014). http://arxiv.org/abs/1405.0312
17. Lu, J., Yang, J., Batra, D., Parikh, D.: Hierarchical question-image co-attention for visual question answering (2016)
18. Malinowski, M., Fritz, M.: A multi-world approach to question answering about real-world scenes based on uncertain input. CoRR abs/1410.0210 (2014). http://arxiv.org/abs/1410.0210
19. Nam, H., Ha, J., Kim, J.: Dual attention networks for multimodal reasoning and matching. CoRR abs/1611.00471 (2016). http://arxiv.org/abs/1611.00471
20. Noh, H., Han, B.: Training recurrent answering units with joint loss minimization for VQA. CoRR abs/1606.03647 (2016). http://arxiv.org/abs/1606.03647
21. Ren, M., Kiros, R., Zemel, R.S.: Image question answering: a visual semantic embedding model and a new dataset. CoRR abs/1505.02074 (2015). http://arxiv.org/abs/1505.02074
22. Ren, S., He, K., Girshick, R.B., Sun, J.: Faster R-CNN: towards real-time object detection with region proposal networks. CoRR abs/1506.01497 (2015). http://arxiv.org/abs/1506.01497
23. Schwartz, I., Schwing, A.G., Hazan, T.: High-order attention models for visual question answering (Nips) (2017). http://arxiv.org/abs/1711.04323
24. Simonyan, K., Zisserman, A.: Very deep convolutional networks for large-scale image recognition. arXiv preprint arXiv:1409.1556 (2014)
25. Szegedy, C., et al.: Going deeper with convolutions. In: CVPR (2015)
26. Teney, D., Anderson, P., He, X., van den Hengel, A.: Tips and tricks for visual question answering: learnings from the 2017 challenge (2017). http://arxiv.org/abs/1708.02711
27. Wu, Q., Shen, C., van den Hengel, A., Wang, P., Dick, A.R.: Image captioning and visual question answering based on attributes and their related external knowledge. CoRR abs/1603.02814 (2016). http://arxiv.org/abs/1603.02814
28. Yang, Z., He, X., Gao, J., Deng, L., Smola, A.J.: Stacked attention networks for image question answering. CoRR abs/1511.02274 (2015). http://arxiv.org/abs/1511.02274
29. Yu, Z., Yu, J., Fan, J., Tao, D.: Multi-modal factorized bilinear pooling with co-attention learning for visual question answering (2017). https://doi.org/10.1109/ICCV.2017.202, http://arxiv.org/abs/1708.01471
30. Zhou, B., Tian, Y., Sukhbaatar, S., Szlam, A., Fergus, R.: Simple baseline for visual question answering. CoRR abs/1512.02167 (2015). http://arxiv.org/abs/1512.02167
31. Zhu, Y., Groth, O., Bernstein, M.S., Fei-Fei, L.: Visual7W: grounded question answering in images. CoRR abs/1511.03416 (2015). http://arxiv.org/abs/1511.03416

Spiral-Net with F1-Based Optimization for Image-Based Crack Detection

Takumi Kobayashi[✉]

National Institute of Advanced Industrial Science and Technology, Tsukuba, Japan
takumi.kobayashi@aist.go.jp

Abstract. Detecting cracks on concrete surface images is a key inspection for maintaining infrastructures such as bridge and tunnels. From the viewpoint of computer vision, the task of automatic crack detection poses two challenges. First, since the cracks are visually depicted by subtle patterns and also exhibit similar appearance to the other structural patterns, it is difficult to discriminatively characterize such less distinctive and finer defects. Second, the cracks are scarcely found, making the number of training samples for cracks significantly smaller than that of the other normal samples to be distinguished from the cracks. This is regarded as a class imbalance problem where the classifier is highly biased toward majority classes. In this study, we propose two methods to address these issues in the framework of deep learning for crack detection: a novel network, called Spiral-Net, and an effective optimization method to train the network. The proposed network is extended from U-Net to extract more detailed visual features, and the optimization method is formulated based on F1 score (F-measure) for properly learning the network even on the highly imbalanced training samples. The experimental results on crack detection demonstrate that the two proposed methods contribute to performance improvement individually and jointly.

1 Introduction

Crack detection on concrete surfaces is a primary task for inspecting infrastructures such as bridges and tunnels [23]. Since the degradation of those concrete structures is assessed by the length, width and density of the cracks [38], it is critical for the maintenance to finely record the situation of the cracks. As the number of concrete structures has been rapidly grown, an automatic crack detection attracts keen attention to reduce and replace the manual inspection, especially based on still images captured by digital cameras.

Cracks are rather related to lower-level image characteristics, being a bit apart from semantic objects of the targets in object detectors [27]. Thus, crack

Electronic supplementary material The online version of this chapter (https://doi.org/10.1007/978-3-030-20887-5_6) contains supplementary material, which is available to authorized users.

C. V. Jawahar et al. (Eds.): ACCV 2018, LNCS 11361, pp. 88–104, 2019.
https://doi.org/10.1007/978-3-030-20887-5_6

detection has been addressed mainly in the field of image processing by utilizing some heuristics of image derivative [8], wavelets [31] and morphological operations [37,42]. The image processing technique, however, is not so enough to well distinguish the characteristics of cracks, producing lots of false positives, and therefore we demand the more discriminative approach for crack detection.

Discriminatively detecting cracks in an image is formulated into a pixel-wise binary classification task where each pixel is classified into the category either of *crack* or *non-crack* (normal). A naive approach toward the pixel-wise classification is a patch-based one as in most of neural network based crack detection methods [4,40]. It predicts a class label at each pixel by classifying the features extracted from the patch and then processes whole image by means of sliding window. The method, however, has difficulty in detecting cracks minutely at pixel level, which is thus unsuitable for the purpose of finely depicting cracks.

The deep neural networks are successfully applied even to the pixel-wise classification. It is mainly addressed as semantic segmentation [20] by means of an encoder-decoder network [26,34] which directly estimates a class label map of the same spatial dimensions as the input image. The encoder-decoder network whose shape resembles a hourglass has been further extended into U-Net [28] efficiently exploiting multi-resolution features via *skip connections* between the encoding and decoding layers. For such a semantic segmentation, fully convolutional network (FCN) [20] is also successfully applied with promising performance, leveraging discriminative object classification network [30].

In the crack detection task of our focus, there are mainly two difficulties from the perspective of computer vision. (1) In contrast to semantic segmentation as well as object classification, where the targets exhibit distinctive image patterns, the crack detection on images has difficulty in charactering/describing the targets. The cracks are of less distinctive and finer patterns on image pixels, being vulnerable to confusion with the other structural patterns and superficial scratches which are irrelevant to degeneration of the (concrete) structures. (2) The task of detecting cracks also poses another challenge regarding class *imbalance* problem [9]. In the standard classification benchmark datasets, the distribution of training samples across classes are carefully designed so as to be close to uniform for facilitating classifier learning. On the other hand, the number of pixels belonging to cracks of the detection target is inherently too small compared with the other non-crack (normal) ones, which results in imbalanced training samples across two classes. Cracks are shown as *thin* lines and *rarely* found in healthy concrete images, causing the more highly imbalanced data than those used in the other semantic segmentation tasks and even edge detection [1,29,35]. The classifier trained on so imbalanced samples is biased toward the majority class while ignoring the characteristics of samples in the minor class.

In this paper, we propose two methods to address those two challenges naturally posed in the crack detection. For pixel-wise classification, we extend the U-Net [28] to extract detailed image characteristics in the encoder-decoder framework. The proposed network, called *Spiral-Net*, can produce a label map finely at pixel level by effectively exploiting diverse-level image features with keeping finer patterns to distinguish cracks from the others. And then the network is effectively learned on the imbalanced training samples where the crack pixels are significantly fewer than the non-crack pixels. While a standard loss such as binary cross-entropy usually employed in training networks suffers from the imbalanced-class samples, we propose an optimization approach based on *F1* score (F-measure) which has been mainly employed as an evaluation metric robust against the imbalanced classes. We derive from maximizing the F1 score an effective form of gradients for properly training a neural network over the imbalanced classes through back-propagation. These two proposed methods, Spiral-Net and the optimization approach, work individually and jointly in an end-to-end learning to improve performance of image-based crack detection.

1.1   Related Works

Network Architecture. In the deep learning framework, the encoder-decoder network [26, 34] is successfully applied to such as semantic segmentation tasks. It is further extended to U-Net [28] by adding skip connections between the encoding and decoding layers to extract diverse-level image features of multi resolution. Both the encoding and decoding processes are improved in some works [5, 10, 34] to provide effective building blocks of the encoder-decoder networks. On the other hand, the overall network architecture is also improved beyond the simple encoder-decoder network. In [25], the encoder-decoder networks are sequentially stacked, being closely related to our method in terms of sequencing encoder-decoder networks (Sect. 2). It, however, differs in the two architectural points regarding skip connections and depths of the encoder-decoder networks. Those two characteristics in our network are useful for extracting finer image patterns of cracks. FCN [20] is also extended to cope with an edge detection task in [35] leveraging object classification network [30], and is empirically compared to our method on the crack detection task in the experiments (Sect. 4).

Imbalance Problem. The methods to cope with the imbalanced classes are mainly categorized into two approaches of re-sampling and weighted loss.

Re-sampling. The imbalanced sample distribution can be corrected by either down-sampling samples in the majority class or over-sampling those of the minority class [6, 16, 21, 22, 41], which is sophisticated by [2] in the deep learning literature and is related to hard-negative mining [7, 12]. However, over-sampling can easily introduce undesirable noise and also have a risk of overfitting, while in down-sampling valuable information in training samples would be lost, which is a critical issue in training deep neural networks, a data-hungry procedure.

Weighted loss. On the other hand, there are methods to (re-)weight loss functions, called cost-sensitive approaches [9]. By assigning asymmetric weights on the losses across classes, one can remedy the high bias toward the majority class; that is, the losses for the majority classes are less-weighted, while those in the minor classes get larger weights to attract higher attention. They have been formulated for shallow learning methods, such as SVM [32], boosting [33] and

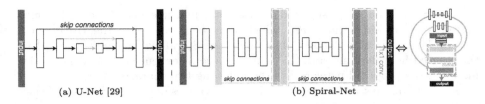

Fig. 1. Network architecture of the proposed Spiral-Net in comparison to U-Net [28]. The U-Net (a) is unfolded in terms of its multiple skip connections into the Spiral-Net (b) where the encoder-decoder modules are stacked with increasing their depths and stacking their output feature maps, which results in a *spiral* shape; the stacked feature maps are finally classified into the class labels by 1×1 convolution.

random forest [17]. Such methods are recently investigated in the deep learning literature [3,24,36], though most of them follow the approaches applied to the shallow models; they take a relatively simple cost-sensitive approach based on re-balancing scheme using an inverse class frequency [35]. In [29], the target class is divided into sub-classes and the cross-entropy loss is regularized by using those sub-classes to cope with the imbalanced samples. In the method, however, the categorization into sub-classes is carefully designed and the number of sub-classes is a hyper-parameter to be tuned by users. The recent method [19] focusing on dense object detection defines a focal loss to suppress the contributions of easily classified samples while shifting up the importance of hard samples, though introducing tunable hyper-parameters. Our approach to remedy the imbalance among classes is derived from F1 score through reformulating it toward differentiable loss from a probabilistic viewpoint [15], which thus exhibits clear difference from those previous works. It should be noted that our method is parameter-free and thereby easily embedded into the end-to-end learning.

2   Spiral-Net

We first describe our network architecture inspired from U-Net [28]. In recent years, image segmentation is often formulated in the framework of an encoder-decoder network [11] where an input image is encoded into effective features at the coarser resolutions and then decoded with increasing resolutions into the same spatial dimensions as the input but in the different domain such as labels. The simple encoder-decoder architecture is extended to U-Net [28] by introducing the skip connections between encoding and decoding layers, as shown in Fig. 1a, in order to leverage diverse image features of finer and coarser levels to predict the pixel-wise labels. In the U-Net (Fig. 1a), there are multiple paths from an input image to an output label map through skip connections with sharing convolution layers. For exploiting more detailed image characteristics, however, it may be necessary to apply respective encoding processes to extract features of various levels without sharing them. Therefore, we propose a network architecture, *Spiral-Net* (Fig. 1b), by unfolding the multiple paths in the U-Net.

The Spiral-Net is constructed by stacking multiple encoder-decoder subnetworks (modules) *sequentially*, which have been folded in the U-Net with sharing layers via skip connections. While the Spiral-Net is closely related to the stacked hourglass network [25] which is also a sequence of encoder-decoder modules, the proposed network has the following two characteristics.

First, it contains various encoder-decoder modules of diverse *depths* and they are sequentially aligned in an *increasing order* regarding their depths. In such an architecture, the first shallowest encoder-decoder is expected to work as rather simple image preprocessing, and then the more discriminative features of larger receptive fields are gradually extracted by the deeper encoder-decoder modules stacked in the latter positions. The former encoder-decoder module is of shallower depth, containing less parameters, so as to be effectively trained even though it is far from the loss layer, the source of gradients in back-propagation.

Second, as in DenseNet [13], we densely string skip connections between the feature maps produced by the encoder-decoder modules for directly exploiting features of diverse levels. It should be noted that the feature maps have the same spatial dimensions as in an input image, thus being fed into the successive encoder-decoder modules (Fig. 1b). Through concatenating the previous feature maps, the deeper encoder-decoder module receives the *wider* feature maps, which is favorable for extracting discriminative features. The input image is not propagated via the skip connection since the raw pixel values exhibit different characteristics from the other features. At the final classification layer, the 1×1 convolution is applied to predict pixel-wise class labels from the densely concatenated feature maps. Through these dense skip connections, the gradient information for updating parameters can be effectively back-propagated into the former encoder-decoder modules [13]. We can say that the proposed Spiral-Net is different from the DenseNet [13] in that the *encoder-decoder* module is embedded in each block, being also distinctive compared to FCN [20] which applies decoders (up-sampling) just as outgoing branches to output a map of class labels.

Based on these characteristics of the network, we can conceptually fold it into the *spiral* shape (Fig. 1b). In the Spiral-Net, the features of various depths can be extracted by the respective encoder-decoder modules unlike the U-Net which shares parts of the encoding processes, and our deeper encoder-decoder can effectively extract the features from the wider input feature map composed of diverse-level features. The Spiral-Net has flexibility in the encoder-decoder module so that we can choose various types of networks such as the ones based on dilated convolutions [39] and residual blocks [10,34]; in this work, for computational efficiency, we employ the simple hourglass encoder-decoder which applies 3×3 convolutions to output a one-channel feature map, as shown in Table 1.

3 F1-Based Optimization

The standard cross-entropy on imbalanced training samples biases the network toward the majority classes. To alleviate the class imbalance problem naturally found in the crack detection task, we propose an F1-based optimization method which is applicable to gradient-based optimization, i.e., back-propagation.

3.1   F1-Based Loss

The F1 score (F-measure) is a standard metric to evaluate the classification performance in a robust manner against the imbalanced classes. Suppose two-class problem comparing *crack* (positive) with *non-crack* (negative) where the negative class is a majority. The F1 score is computed by

$$\mathtt{F1} = \frac{2\,\mathtt{prec}\cdot\mathtt{rec}}{\mathtt{prec}+\mathtt{rec}}, \tag{1}$$

where $\mathtt{prec}$ and $\mathtt{rec}$ indicate precision and recall rates based on the binary classification results, respectively. The $\mathtt{F1}$ depends on empirical counts of such as false positives, being obviously not differentiable, and thus has been applied mainly to evaluate the performance of the trained classifier. Toward a loss function, we first reformulate the definition (1) in a similar manner to [15]. Note, however, that our method is clearly different from [15] via the weighting scheme (Sect. 3.2).

At the i-th sample (pixel) assigned with the ground truth label $l_i \in \{-1, 1\}$, the posterior probability is computed by

$$\mathrm{p}(\hat{l}_i = 1) = \sigma(x_i) = \frac{1}{1 + \exp(-x_i)} \triangleq \sigma_i, \tag{2}$$

where $\hat{l}_i \in \{-1, 1\}$ indicates the predicted label by applying the sigmoid function σ to the feature x_i extracted at the i-th sample. Let N_1 and N_{-1} be the numbers of samples belonging to positive and negative classes, respectively, and the prior probabilities can be empirically estimated as

$$\mathrm{p}(l = 1) = \frac{N_1}{N_{-1} + N_1}, \quad \mathrm{p}(l = -1) = \frac{N_{-1}}{N_{-1} + N_1}. \tag{3}$$

Then, the precision and recall rates in the F1-score (1) are described as

$$\mathtt{rec} = \frac{\mathrm{p}(\hat{l} = 1, l = 1)}{\mathrm{p}(l = 1)} = \mathrm{p}(\hat{l} = 1 | l = 1) = \frac{1}{N_1} \sum_{j | l_j = 1} \sigma_j, \tag{4}$$

$$\mathtt{prec} = \frac{\mathrm{p}(\hat{l} = 1, l = 1)}{\mathrm{p}(\hat{l} = 1)} = \frac{\mathrm{p}(\hat{l} = 1 | l = 1)\mathrm{p}(l = 1)}{\sum_{c \in \{-1,1\}} \mathrm{p}(\hat{l} = c | l = c)\mathrm{p}(l = c)} = \frac{\sum_{j | l_j = 1} \sigma_j}{\sum_j \sigma_j}. \tag{5}$$

Therefore, the F1 score (1) is reformulated into

$$\mathtt{F1} = \frac{2\sum_{j | l_j = 1} \sigma_j}{N_1 + \sum_j \sigma_j}, \tag{6}$$

where $0 \le \mathtt{F1} \le 1$. Beyond this naive F1 loss (6) which is also found in [15], we construct the loss by negative logarithm of the F1 score, $L = -\log(\mathtt{F1})$, of which the derivative is given by

$$\frac{\partial L}{\partial \sigma_i} = \begin{cases} -\left(\frac{1}{\sum_{j | l_j = 1} \sigma_j} - \frac{1}{N_1 + \sum_j \sigma_j} \right), & l_i = 1 \\ \frac{1}{N_1 + \sum_j \sigma_j}, & l_i = -1 \end{cases}, \tag{7}$$

where $\frac{1}{\sum_{j|l_j=1} \sigma_j} - \frac{1}{N_1+\sum_j \sigma_j} > 0$ due to that $N_1 > 0$ and $\sum_{j|l_j=1} \sigma_j \leq \sum_j \sigma_j$.

3.2 F1-Guided Gradient Weighting

It is possible to directly compute the gradient of the F1-based loss L w.r.t x_i by combining (7) and $\frac{\partial \sigma(x_i)}{\partial x_i} = \sigma(x_i)\{1 - \sigma(x_i)\}$ into the chain rule $\frac{\partial L}{\partial x_i} = \frac{\partial L}{\partial \sigma_i}\frac{\partial \sigma_i}{\partial x_i}$;

$$\frac{\partial L}{\partial x_i} = \begin{cases} -\left(\frac{1}{\sum_{j|l_j=1}\sigma_j} - \frac{1}{N_1+\sum_j \sigma_j}\right)\sigma_i(1-\sigma_i), & l_i = 1 \\ \frac{1}{N_1+\sum_j \sigma_j}\sigma_i(1-\sigma_i), & l_i = -1 \end{cases}. \quad (8)$$

This straightforward approach, however, is not favorable for gradient-based optimization (back-propagation) since the gradient (8) contains $\sigma(x_i)\{1 - \sigma(x_i)\}$ which unfavorably vanishes at the extreme predictions, $\sigma(x_i) \rightarrow \{1,0\}$; it is empirically shown in Sect. 4.4.

On the other hand, the most naive loss directly derived from the posterior probabilities is described by[1]

$$\tilde{L} = -\sum_{y \in \{-1,1\}}\sum_{i|l_i=y} \mathrm{p}(\hat{l}_i = y), \quad \frac{\partial \tilde{L}}{\partial \sigma_i} = \begin{cases} -1, & l_i = 1 \\ 1, & l_i = -1 \end{cases}. \quad (9)$$

We can regard the gradient (7) as the weighted version of (9) by introducing the weights derived from the F1 score (6), leading to the reformulation of $\frac{\partial L}{\partial x_i} = \left|\frac{\partial L}{\partial \sigma_i}\right|\frac{\partial \tilde{L}}{\partial \sigma_i}\frac{\partial \sigma_i}{\partial x_i}$. This point of view inspires us to apply the similar weighting approach to the commonly used cross-entropy loss of

$$\bar{L} = -\sum_{y \in \{-1,1\}}\sum_{i|l_i=y} \log(\mathrm{p}(\hat{l}_i = y)), \quad \frac{\partial \bar{L}}{\partial \sigma_i} = \begin{cases} -\frac{1}{\sigma_i}, & l_i = 1 \\ \frac{1}{1-\sigma_i}, & l_i = -1 \end{cases}. \quad (10)$$

Thus, we propose the following pseudo[2] gradients by weighting (10) with (7);

$$\mathrm{g}(x_i) = \left|\frac{\partial L}{\partial \sigma_i}\right|\frac{\partial \bar{L}}{\partial \sigma_i}\frac{\partial \sigma_i}{\partial x_i} = \begin{cases} -\left(\frac{1}{\sum_{j|l_j=1}\sigma_j} - \frac{1}{N_1+\sum_j \sigma_j}\right)(1-\sigma_i), & l_i = 1 \\ \frac{1}{N_1+\sum_j \sigma_j}\sigma_i, & l_i = -1 \end{cases}. \quad (11)$$

In contrast to most of cost-sensitive methods [3,24,36], we directly impose weights on the gradients, not on the losses, though from the viewpoint of gradient-based optimization the cost-sensitive methods also produce weighted gradients through the weighted loss. In our end-to-end learning, the pseudo gradient (11) is back-propagated to update the parameters of the neural network.

---

[1] Actually, in the training, we divide the losses $\tilde{L}$ and $\bar{L}$ by the number of samples $N = N_1 + N_{-1}$, which is here omitted for simplicity.

[2] Unfortunately, there is no analytic loss function that produces the derivative (11); see the supplementary material.

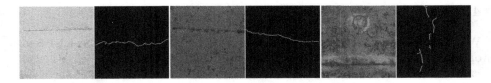

Fig. 2. Example images of cracks with label maps composed of binary values indicating crack pixels. They are of 512×512 pixels sampled from whole image of 5472×3678 pixels to focus on cracks. There are some structural patterns such as traces of concrete frame molds which are similar to but *not* cracks. Best viewed on the screen.

The adaptive weights (7) work on the imbalance issue in a manner derived from optimizing the F1 score (6). The weight of the positive class is rewritten to

$$\left|\frac{\partial L}{\partial \sigma_i}\right|_{l_i=1} = \frac{1}{\sum_{j|l_j=1}\sigma_j} - \frac{1}{N_1 + \sum_j \sigma_j} = \frac{N_1 + \sum_{j|l_j=-1}\sigma_j}{\sum_{j|l_j=1}\sigma_j}\frac{1}{N_1 + \sum_j \sigma_j}, \quad (12)$$

and its ratio to the weight $\left|\frac{\partial L}{\partial \sigma_i}\right|_{l_i=-1} = \frac{1}{N_1+\sum_j \sigma_j}$ of the negative class is given by

$$r \triangleq \frac{\left|\frac{\partial L}{\partial \sigma_i}\right|_{l_i=1}}{\left|\frac{\partial L}{\partial \sigma_i}\right|_{l_i=-1}} = \frac{N_1 + \sum_{j|l_j=-1}\sigma_j}{\sum_{j|l_j=1}\sigma_j} \geq 1, \quad (13)$$

where the inequality comes from $\sum_{j|l_j=1}\sigma_j \leq N_1$. Thus, we can see that as is the case with the cost-sensitive methods, the minority samples belonging to the positive class are highly weighted while the negative ones are assigned with less weights. It is noteworthy that such weighting is theoretically induced from the probabilistic formulation of the F1 score (6), while a weighting scheme has been heuristically designed in the previous methods.

In particular, our weighing has the following properties which facilitate learning networks. In case that the classifier is biased toward the negative class (majority), resulting in $\sum_j \sigma_j \to 0$, the ratio r in (13) becomes larger to highly encourage the learning for the positive class. On the other hand, approaching to favorable classification of $\sum_{j|l_j=-1}\sigma_j \to 0$ and $\sum_{j|l_j=1}\sigma_j \to N_1$, the ratio r is close to 1, which realizes the equal weighting across positive and negative classes as is the case with the standard cross-entropy loss. This adaptive weighting scheme in the optimization enables the end-to-end learning to enjoy whole samples for effectively training networks unlike the re-sampling based methods.

In the case of mini-batch based optimization, we can consider the statistics on the mini-batch, and thereby all the ingredients in (11), $N_1, \sum_j \sigma_j$ and $\sum_{j|l_j=1}\sigma_j$, are computed over those samples within the mini-batch[3]. Since the proposed method (11) merely produces the weighted gradients, we can apply various types of effective optimization techniques used in the end-to-end learning.

---

[3] In the preliminary experiment, we confirmed that the optimization using the globally cumulative statistics does not provide any performance improvement.

4   Experimental Results

We evaluate the proposed methods of Spiral-Net (Sect. 2) and the F1-based optimization (Sect. 3.1) on a crack detection task; we assign a label either of *crack* or *non-crack* to every pixel in an image, which naturally induces the class imbalance problem while requiring finer image feature extraction to capture the visual characteristics of cracks.

4.1   Crack Dataset

We have collected still images at various locations such as tunnels, pillars and slabs of concrete bridges which are actually subject to inspection. The RGB-color images of 5472×3678 pixels show the concrete surface containing a few cracks somewhere (Fig. 2), as well as the other objects, *e.g.*, pipes and steels, which are not eliminated for fairly evaluating the performance in the wild. It is noteworthy that the images are captured in the unconstrained situation, exhibiting high variations in terms of such as illumination and concrete colors. Then, the experts assigned the positive (*crack*) labels in a pixel-wise manner tracing the cracks by lines of roughly 3 pixel width, while the other pixels are regarded as belonging to the negative (*non-crack*) class. The crack pixels are scarce and the ratio of the numbers of crack and non-crack pixels is 1:450. In addition, as shown in Fig. 2, the crack patterns are not so distinct with exhibiting high similarity to the other structural patterns derived from such as molds and superficial scars. We used 278 images for training and 14 images for test which are picked up by crack inspectors, *not* expert of computer vision, to make fair evaluation of the performance on crack detection in real conditions; note that this dataset is as large as BSDS dataset [1] of edge detection in terms of number of pixels. The performance of crack detection is measured on each test image by average precision as well as precision, recall and F1 score which are computed based on

Table 1. Building blocks in U-Net and Spiral-Net. The encoders and decoders are implemented by convolution (`conv`) and transposed convolution (`convT`) of 3×3 filter without any padding nor cropping, respectively, which are followed by BatchNormalization and ReLU. We apply 2-pixel stride in `conv` and upsampling factor of 2 in `convT`.

Layer	Encoder (`conv`)	Decoder (`convT`)		
	Output dim.	Input size	Output size	Output dim.
1	64		255×255	1
2	128		127×127	64
3	256		63×63	128
4	512		31×31	256
5	512		15×15	512
6	512		7×7	512
7	512		3×3	512

the output (sigmoid function) of the network with the threshold of 0.5, and then we report those evaluation scores averaged across all the test images. Note that the these performance metrics are computed in a pixel-wise manner since the ground-truth label is assigned to each pixel.

4.2 Implementation Details

Inspired by the model used in pix2pix [14], we construct a vanilla U-Net which gradually downsizes an input image of 255×255 pixels by a factor of 2 as shown in Table 1; the 1st~7th encoders and the 7th~1st decoders are sequentially stacked with skip connections (Fig. 1a). The Spiral-Net is also composed of the same building blocks as in Table 1; the encoder-decoder module of d depth is built by sequentially stacking the 1st~d-th encoders and the d-th~1st decoders.

In training, we randomly pick up 32 image patches of 255×255 pixels from training images to shape the mini-batch with random flipping either horizontally or vertically. At each epoch, such sampling is repeated 256 times *per image* so as to roughly cover the whole image of 5472×3678 pixels by using the patches; thereby, we receive $2224 = 256 \cdot 278/32$ mini-batches per epoch in the training. The network is trained by applying Adam optimizer [18] with the learning rate of 0.0001 and momentum of 0.9 over 200 epochs.

Table 2. Performance results (%) by Spiral-Net of various depth orders. The numbers in the second column indicate the depths of the stacked encoder-decoder modules.

Architecture	Order of depths	mAP	F1	Precision	Recall
Increasing ╱	1-2-3-4-5-6-7	**80.91**	**71.76**	86.56	62.89
Decreasing ╲	7-6-5-4-3-2-1	0	0	0	0
Triangle ⋀	1-2-3-4-3-2-1	78.92	69.35	87.33	59.23
	1-3-5-7-6-4-2	79.03	70.82	85.39	62.38
Uniform ─	1-1-1-1-1-1-1	53.67	43.85	82.16	32.06
	2-2-2-2-2-2-2	77.56	67.51	87.29	57.05
	3-3-3-3-3-3-3	78.33	67.99	88.71	58.50
	4-4-4-4-4-4-4	74.77	66.96	81.34	58.79
	5-5-5-5-5-5-5	72.08	63.77	84.22	55.02
	6-6-6-6-6-6-6	0	0	0	0
	7-7-7-7-7-7-7	0	0	0	0

Table 3. Performance results on Spiral-Net with various types of skip connections.

Connection	mAP	F1	Precision	Recall
Concatenation	**80.91**	**71.76**	86.56	62.89
Sum	79.16	71.14	87.12	61.46
None	0	0	0	0

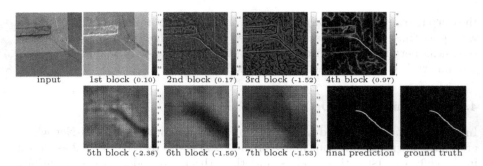

Fig. 3. Intermediate feature maps by the 1st~7th encoder-decoder modules. The number in parentheses indicates the aggregation weight at the final 1×1 convolution layer.

4.3   Performance Analysis on Spiral-Net

We evaluate the Spiral-Net (Sect. 2) in terms of the network configuration, by training all the networks based on the standard binary cross-entropy loss.

The Spiral-Net stacks the encoder-decoder modules of different depths (Fig. 1b). Thus, the network architecture is controlled by the sequential order of those modules; the depths of the stacked encoder-decoder can be designed as follows.

- As described in Sect. 2, the modules are sequentially aligned so that their depths are in an increasing order.
- In contrast, it is also possible to stack them in a decreasing order.
- An intermediate design between those two could be the one in which the depths are first increasing and then decreasing like a triangle shape.
- On the other hand, the simplest architecture is that all the modules have the uniform (identical) depth.

The performance results are shown in Table 2. We can see that the performance is significantly degraded by locating the deeper encoder-decoder module early in the network (decreasing and deeper uniform). As discussed in Sect. 2, such deeper module can not be properly trained since it is far from the loss layer and receives *narrower* feature maps. Actually, the networks of decreasing and uniform with depths of 6 and 7 are improperly learned to always output labels of negative class which is the majority in the dataset; in those cases, the performances are shown as all 0's. In the uniform architecture, the moderately deep encoder-decoders work well while the shallower and deeper ones provide poor performance. On the other hand, the networks gradually increasing depths in increasing and triangle produce the better performance than the uniform one, and especially, the best performance is achieved by the increasing order.

Next, we evaluate the following types of skip connection on the Spiral-Net of the increasing depth order.

- The feature maps propagated via skip connections are concatenated along the channel dimension to increase the number of channels in a way of DenseNet [13].
- The propagated feature maps are summed up as in ResNet [10] with keeping the number of channels. Note that any encoder-decoder modules output one-channel feature maps (see Table 1).
- We do not apply any skip connections in the network in a similar way to [25].

The performance comparison is shown in Table 3. Without any skip connections (none), in this imbalanced data, the network is not properly learned due to that the gradient information is not effectively back-propagated. The skip connections based on sum and concatenation remedy the issue, producing favorable performance. In comparison with sum-based connection, the concatenation one provides the *wider* feature maps which contribute to further performance improvement, as described in Sect. 2.

These experimental results quantitatively validate the proposed Spiral-Net that stacks encoder-decoder modules of increasing depths with providing wider feature maps due to concatenation-based skip-connections.

We then qualitatively analyze how the Spiral-Net detects cracks by showing the intermediate one-channel feature maps together with the last 1×1 convolution weights to merge them; the respective encoder-decoder modules produce the non-negative one-channel maps as shown in Fig. 3. The 1st&2nd modules work as lower-level image processing like pixel-value enhancement and derivative computation; these outputs less contribute to the final prediction due to their smaller weights. Based on those low-level features, the 3rd&4th modules detect crack-like structures, extracting candidates for cracks with the positive weight at the 4th module while suppressing the other regions by the negative weight at the 3rd one. Finally, the 5th~7th modules detect cracks rather semantically and effectively eliminate the false positives by assigning large negative weights on the non-crack regions. These sequential processes are quite reasonable by integrating lower-level image processing and higher-level classification in the Spiral-Net.

4.4   Comparison to Other Methods

We compare the proposed methods, Spiral-Net and the F1-based optimization (11), to the other methods in terms of a network and a loss function. In the training, the F1-based optimization is applied by simply replacing the gradients of the cross-entropy loss with the pseudo weighted gradients (11), and thus is applicable to any types of networks including the proposed Spiral-Net. The optimization method is formulated so as to cope with the class imbalance problem, which is particularly found in this crack detection task; note again that the number of crack pixels is far smaller ($\approx 1/450$) than that of non-crack pixels. Table 4 shows the performance results of various networks trained on various losses.

As to networks, we compare the Spiral-Net to the U-Net [28] and HED [35] all of which are trained on the crack dataset. HED is proposed based on FCN [20] for detecting (semantic) *edges* whose shapes are formed as thin lines similarly to cracks. In the work [35] which tackles edge detection tasks, HED is fine-tuned from the VGG pre-trained model [30] based on the cross-entropy loss weighted by the inverse of class frequency. However, we can see that, on any types of losses, so fine-tuned HED is inferior to the one trained from scratch on this crack dataset. While the semantic edge detection is closely related to object recognition on which the VGG pre-trained model works, the crack detection is not so dependent on the object recognition but is rather formulated as lower-level image processing taking into account the finer image structure, though both tasks aim to produce thin lines, *edge* and *crack*. Thus, although the HED fine-tuned from the image classification model (VGGnet) is suitable for edge detection tasks, it largely degrades performance on this crack detection task. On the other hand, while the U-net is comparable to the HED trained from scratch, the proposed Spiral-Net outperforms those on diverse types of losses, demonstrating that the network effectively extracts detailed image characteristics of crack patterns.

Next, the F1-based optimization method is compared to the other types of loss functions: the widely used cross-entropy loss $\bar{L}$ in (10), the one weighted by inverse of class frequency, and the focal loss [19]. The latter two losses are developed from the cross-entropy loss via weighting; the weights by inverse class frequency are introduced as $\check{L} = -\sum_{y \in \{-1,1\}} \frac{1}{N_y} \sum_{i|l_i=y} \log(\mathrm{p}(\hat{l}_i = y))$, and the simple weighting scheme is recently more sophisticated in the focal loss [19] as $\acute{L} = -\sum_{y \in \{-1,1\}} \alpha_y \sum_{i|l_i=y} \{1 - \mathrm{p}(\hat{l}_i = y)\}^\gamma \log(\mathrm{p}(\hat{l}_i = y))$ where $\alpha_1 = \alpha$, $\alpha_{-1} = 1 - \alpha$, and γ, α are the parameters to be determined by users; we set $\gamma = 2$ and $\alpha = 0.25$ as suggested in [19] and then tuned it to $\alpha = 0.5$. These losses are applied to the above-mentioned networks, and the performance results are shown in Table 4.

The cross-entropy loss makes the detector focus on the majority class of *non-crack* pixels, which results in relatively high precision and low recall as shown in Table 4a. On the contrary, through weighting by the inverse class frequency, the detector is highly biased to the minor class of *crack* pixels, producing high recall and low precision (Table 4b); it shows the difficulty in manually tuning the class weight in this highly imbalanced data. And, even the focal loss [19] degrades performance compared to the cross-entropy loss (Table 4cd). Note that the focal loss indirectly corrects the class imbalance through suppressing the effect of *easy* negative samples which would occupy most of training samples causing the class imbalance. Such assumption holds on the object detection task addressed in [19] where the target objects exhibit clear difference in their visual appearance compared with most of background samples. In the crack detection, however, the target cracks are less distinctive in comparison with the other image patterns, reducing the number of *easy* samples, which would be the main reason why the focal loss is inferior even to the cross-entropy loss. In contrast, the proposed method favorably improves the performance of all the networks, including Spiral-Net (Table 4f), while the straightforward F1-based loss deteriorates performance (Table 4e). As discussed in Sect. 3.2, the gradients (8) of the F1-based loss contains the term of $\sigma_i(1 - \sigma_i)$ which unfavorably hampers

Table 4. Performance comparison in terms of networks and loss functions.

Network	(a) Cross-entropy $\bar{L}$				(b) Cross-entropy weighted by inverse class frequency $\check{L}$			
	mAP	F1	Precision	Recall	mAP	F1	Precision	Recall
Spiral-Net (ours)	80.91	71.76	86.56	62.89	80.11	29.19	17.48	97.99
U-net [28]	73.12	64.78	78.69	55.63	73.02	21.80	13.83	90.83
HED(fine-tune) [35]	63.47	18.83	90.83	11.16	67.44	22.51	12.98	94.49
HED(scratch) [35]	72.47	61.70	85.85	51.61	72.39	28.87	18.66	92.29
Network	(c) Focal loss $\acute{L}$ [19] ($\alpha = 0.2$)				(d) Focal loss $\acute{L}$ [19] ($\alpha = 0.5$)			
	mAP	F1	Precision	Recall	mAP	F1	Precision	Recall
Spiral-Net (ours)	68.01	44.20	91.24	31.08	75.77	65.17	86.87	54.46
U-net [28]	63.40	41.45	90.46	28.98	64.99	55.54	82.10	45.84
HED(fine-tune) [35]	53.86	0.98	64.21	0.50	56.34	12.37	90.94	7.01
HED(scratch) [35]	69.99	31.09	95.25	19.77	71.22	57.18	85.68	45.98
Network	(e) F1 in (8)				(f) Pseudo-F1 (ours) in (11)			
	mAP	F1	Precision	Recall	mAP	F1	Precision	Recall
Spiral-Net (ours)	68.81	74.92	81.90	69.95	**85.61**	**79.04**	78.44	79.81
U-net [28]	63.06	68.97	81.63	64.85	77.16	69.48	77.67	68.34
HED(fine-tune) [35]	0	0	0	0	69.02	62.32	66.05	61.14
HED(scratch) [35]	0	0	0	0	76.96	69.00	70.43	71.20

Table 5. Parameter sizes of networks. The *wide* U-Net is constructed by increasing the number of channels in the U-Net.

Network	HED	U-Net	Spiral-Net	*wide* U-Net
# of parameter	14.1M	22.4M	39.7M	39.9M
mAP	76.96	77.16	85.61	77.01
F1	69.00	69.48	79.04	69.69

learning; especially, the HEDs are improperly learned since it is trapped in the state extremely biased toward negative class (Table 4e). The proposed method adaptively tune the weights for the gradients based on the optimization of F1 score while avoiding the unfavorable formulation in the gradients to effectively improve performance (Table 4f); it is also balanced in terms of precision/recall.

The Spiral-Net trained by the F1-based optimization method achieves 85.61% which significantly outperforms 73.12% of the baseline method of U-Net trained by the cross-entropy loss. It is noteworthy that the performance improvement by the Spiral-Net comes from the architecture itself, not the increased size of parameters, as shown in Table 5. The examples of the detected cracks are shown in Fig. 4, demonstrating that even less distinctive cracks can be detected while being insensitive to the other patterns similar to cracks.

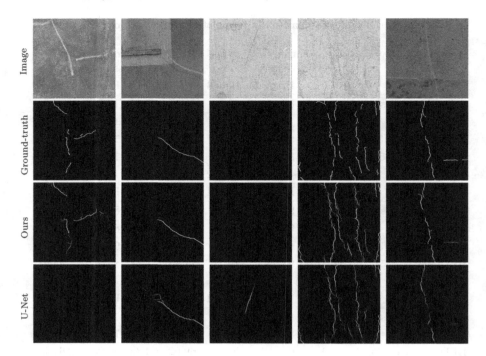

Fig. 4. Examples of detected cracks on the test images, showing the sigmoid output [0, 1] in gray scale. The baseline method of U-Net trained by cross-entropy loss failed to detect some cracks while producing the false positives. In contrast, the proposed method of Spiral-Net trained by F1-based optimization favorably detects cracks exhibiting well correspondence with the ground truth. Best viewed on the screen.

5   Conclusion

We have proposed the Spiral-Net and the optimization method based on F1 score for detecting cracks in an image. The cracks are of finer patterns and can be scarcely found on concrete surfaces, posing a class imbalance problem. The Spiral-Net is constructed by sequentially stacking encoder-decoder modules of increasing depths with skip connections for feature maps in order to extract detailed image features of cracks. In learning the network on the highly imbalanced training samples, we adaptively weight the gradients of the cross-entropy loss and the weights are theoretically derived from optimizing F1 score which is robust against the imbalanced classes. The experimental results on image-based crack detection demonstrate the effectiveness of the two proposed methods, respectively, as well as their joint contribution to performance improvement.

Acknowledgment. The author thanks Takeshi Nagami, Hisashi Sato and Yohei Hayasaka for their great effort to build the crack dataset. This work is based on a project commissioned by the New Energy and Industrial Technology Development Organization (NEDO).

References

1. Arbelaez, P., Maire, M., Fowlkes, C., Malik, J.: Contour detection and hierarchical image segmentation. PAMI **33**(5), 898–916 (2011)
2. Bulo, S.R., Neuhold, G., Kontschieder, P.: Loss max-pooling for semantic image segmentation. In: CVPR, pp. 7082–7091 (2017)
3. Caesar, H., Uijlings, J.R.R., Ferrari, V.: Joint calibration for semantic segmentation. In: BMVC (2015)
4. Cha, Y.J., Choi, W., Büyüköztürk, O.: Deep learning-based cracking damage detection using CNNs. Comput. Aided Civ. Infrastruct. Eng. **32**(5), 361–378 (2017)
5. Chatfield, K., Simonyan, K., Vedaldi, A., Zisserman, A.: Return of the devil in the details: delving deep into convolutional nets. In: BMVC (2014)
6. Chawla, N.V., Bowyer, K.W., Hall, L.O., Kegelmeyer, W.P.: SMOTE: synthetic minority over-sampling technique. J. Artif. Intell. Res. **16**, 321–357 (2002)
7. Dong, Q., Gong, S., Zhu, X.: Class rectification hard mining for imbalanced deep learning. In: ICCV, pp. 1869–1878 (2017)
8. Fujita, Y., Hamamoto, Y.: A robust automatic crack detection method from noisy concrete surfaces. Mach. Vis. Appl. **22**, 245–254 (2011)
9. He, H., Garcia, E.A.: Learning from imbalanced data. IEEE Trans. Knowl. Data Eng. **21**(9), 1263–1284 (2009)
10. He, K., Zhang, X., Ren, S., Sun, J.: Deep residual learning for image recognition. In: CVPR, pp. 770–778 (2016)
11. Hinton, G.E., Salakhutdinov, R.R.: Reducing the dimensionality of data with neural networks. Science **313**(5786), 504–507 (2006)
12. Huang, C., Li, Y., Loy, C.C., Tang, X.: Learning deep representation for imbalanced classification. In: CVPR, pp. 5375–5384 (2016)
13. Huang, G., Liu, Z., Maaten, L., Weinberger, K.Q.: Densely connected convolutional networks. In: CVPR, pp. 2261–2269 (2017)
14. Isola, P., Zhu, J.Y., Zhou, T., Efros, A.A.: Image-to-image translation with conditional adversarial networks. In: CVPR, pp. 5967–5976 (2017)
15. Jansche, M.: Maximum expected F-measure training of logistic regression models. In: HLT, pp. 692–699 (2005)
16. Jeatrakul, P., Wong, K.W., Fung, C.C.: Classification of imbalanced data by combining the complementary neural network and SMOTE algorithm. In: Wong, K.W., Mendis, B.S.U., Bouzerdoum, A. (eds.) ICONIP 2010. LNCS, vol. 6444, pp. 152–159. Springer, Heidelberg (2010). https://doi.org/10.1007/978-3-642-17534-3_19
17. Khoshgoftaar, T.M., Golawala, M., Hulse, J.V.: An empirical study of learning from imbalanced data using random forest. In: ICTAI, pp. 310–317 (2007)
18. Kingma, D., Ba, J.: Adam: a method for stochastic optimization. In: ICLR (2015)
19. Lin, T.Y., Goyal, P., Girshick, R., He, K., Dollar, P.: Focal loss for dense object detection. In: ICCV, pp. 2999–3007 (2017)
20. Long, J., Shelhamer, E., Darrell, T.: Fully convolutional networks for semantic segmentation. In: CVPR, pp. 3431–3440 (2015)
21. Maciejewski, T., Stefanowski, J.: Local neighborhood extension of smote for mining imbalanced data. In: ICDM, pp. 104–111 (2011)

22. Mani, I., Zhang, I.: KNN approach to unbalanced data distributions: a case study involving information extraction. In: Workshop on Learning from Imbalanced Datasets (2003)
23. Mohan, A., Poobal, S.: Crack detection using image processing: a critical review and analysis. Alexandria Eng. J. (2017). https://doi.org/10.1016/j.aej.2017.01.020
24. Mostajabi, M., Yadollahpour, P., Shakhnarovich, G.: Feed-forward semantic segmentation with zoom-out features. In: CVPR, pp. 3376–3385 (2015)
25. Newell, A., Yang, K., Deng, J.: Stacked hourglass networks for human pose estimation. In: Leibe, B., Matas, J., Sebe, N., Welling, M. (eds.) ECCV 2016. LNCS, vol. 9912, pp. 483–499. Springer, Cham (2016). https://doi.org/10.1007/978-3-319-46484-8_29
26. Noh, H., Hong, S., Han, B.: Learning deconvolution network for semantic segmentation. In: ICCV, pp. 1520–1528 (2015)
27. Redmon, J., Divvala, S., Girshick, R., Farhadi, A.: You only look once: unified, real-time object detection. In: CVPR, pp. 779–788 (2016)
28. Ronneberger, O., Fischer, P., Brox, T.: U-Net: convolutional networks for biomedical image segmentation. In: Navab, N., Hornegger, J., Wells, W.M., Frangi, A.F. (eds.) MICCAI 2015. LNCS, vol. 9351, pp. 234–241. Springer, Cham (2015). https://doi.org/10.1007/978-3-319-24574-4_28
29. Shen, W., Wang, X., Wang, Y., Bai, X., Zhang, Z.: DeepContour: a deep convolutional feature learned by positive-sharing loss for contour detection. In: CVPR, pp. 3982–3991 (2015)
30. Simonyan, K., Zisserman, A.: Very deep convolutional networks for large-scale image recognition. CoRR abs/1409.1556 (2014)
31. Taha, M.M.R., Noureldin, A., Lucero, J.L., Baca, T.J.: Wavelet transform for structural health monitoring: a compendium of uses and features. Struct. Health Monit. 5, 267–295 (2006)
32. Tang, Y., Zhang, Y.Q., Chawla, N.V., Krasser, S.: SVMs modeling for highly imbalanced classification. IEEE Trans. Syst. Man. Cybern. 39(1), 281–288 (2009)
33. Ting, K.M.: A comparative study of cost-sensitive boosting algorithms. In: ICML, pp. 983–990 (2000)
34. Wojna, Z., et al.: The devil is in the decoder. In: BMVC (2017)
35. Xie, S., Tu, Z.: Holistically-nested edge detection. In: ICCV, pp. 1395–1403 (2015)
36. Xu, J., Schwing, A.G., Urtasun, R.: Learning to segment under various forms of weak supervision. In: CVPR, pp. 3781–3790 (2015)
37. Yamaguchi, T., Hashimoto, S.: Fast crack detection method for large-size concrete surface images using percolation-based image processing. Mach. Vis. Appl. 21, 797–809 (2010)
38. Yang, Y.S., Yang, C.M., Huang, C.W.: Thin crack observation in a reinforced concrete bridge pier test using image processing and analysis. Adv. Eng. Softw. 83, 99–108 (2015)
39. Yu, F., Koltun, V.: Multi-scale context aggregation by dilated convolutions. In: ICLR (2016)
40. Zhang, L., Yang, F., Zhang, Y.D., Zhu, Y.J.: Road crack detection using deep convolution neural network. In: ICIP, pp. 2791–2799 (2016)
41. Zhou, Z.H., Liu, X.Y.: Training cost-sensitive neural networks with methods addressing the class imbalance problem. IEEE Trans. Knowl. Data Eng. 18(1), 63–77 (2006)
42. Zou, Q., Cao, Y., Li, Q., Mao, Q., Wang, S.: CrackTree: automatic crack detection from pavement images. Pattern Recogn. Lett. 33(3), 227–238 (2012)

Flex-Convolution
Million-Scale Point-Cloud Learning Beyond Grid-Worlds

Fabian Groh[1][(✉)] , Patrick Wieschollek[1,2] , and Hendrik P. A. Lensch[1]

[1] University of Tübingen, Tübingen, Germany
fabian.groh@uni-tuebingen.de
[2] Max Planck Institute for Intelligent Systems, Tübingen, Germany

Abstract. Traditional convolution layers are specifically designed to exploit the natural data representation of images – a fixed and regular grid. However, unstructured data like 3D point clouds containing irregular neighborhoods constantly breaks the grid-based data assumption. Therefore applying best-practices and design choices from 2D-image learning methods towards processing point clouds are not readily possible. In this work, we introduce a natural generalization *flex-convolution* of the conventional convolution layer along with an efficient GPU implementation. We demonstrate competitive performance on rather small benchmark sets using fewer parameters and lower memory consumption and obtain significant improvements on a million-scale real-world dataset. Ours is the first which allows to efficiently process 7 million points *concurrently*.

1 Introduction

Deep Convolutional Neural Networks (CNNs) shine on tasks where the underlying data representations are based on a regular grid structure, e.g., pixel representations of RGB images or transformed audio signals using Mel-spectrograms [11]. For these tasks, research has led to several improved neural network architectures ranging from VGG [24] to ResNet [9]. These architectures have established state-of-the-art results on a broad range of classical computer vision tasks [29] and effortlessly process entire HD images (∼2 million pixels) within a single pass. This success is fueled by recent improvements in hardware and software stacks (*e.g.* TensorFlow), which provide highly efficient implementations of layer primitives [15] in specialized libraries [6] exploiting the grid-structure of the data. It seems appealing to use grid-based structures (*e.g.* voxels) to process higher-dimensional data relying on these kinds of layer implementations. However, grid-based approaches are often unsuited for processing irregular point clouds and unstructured data. The grid resolution on equally spaced grids poses

Electronic supplementary material The online version of this chapter (https://doi.org/10.1007/978-3-030-20887-5_7) contains supplementary material, which is available to authorized users.

C. V. Jawahar et al. (Eds.): ACCV 2018, LNCS 11361, pp. 105–122, 2019.
https://doi.org/10.1007/978-3-030-20887-5_7

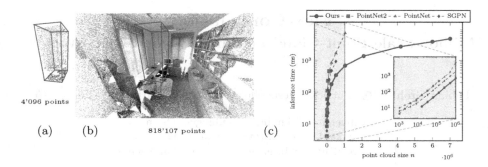

Fig. 1. Processing full-resolution point clouds is an important ingredient for successful semantic segmentation. Previous methods [17,19,28] subsample small blocks (a), while ours (b) processes the entire room and can (c) handle inputs up to 7 Million points in a single forward-pass with the *same* accuracy. Previous methods *could* handle at most 1 Million points – but training is not feasible on today's hardware.

a trade-off between discretization artifacts and memory consumption. Increasing the granularity of the cells is paid by higher memory requirements that even grows exponentially due to the curse of dimensionality.

While training neural networks on 3D voxel grids is possible [16], even with hierarchical octrees [20] the maximum resolution is limited to 256^3 voxels—large data sets are currently out-of-scope. Another issue is the discretization and resampling of continuous data into a fixed grid. For example, depth sensors produce an arbitrarily oriented depth map with different resolution in x, y and z. In Structure-from-Motion, the information of images with arbitrary perspective, orientation and distance to the scene—and therefore resolution—need to be merged into a single 3D point cloud. This potentially breaks the grid-based structure assumption completely, such that processing such data in full resolution with conventional approaches is infeasible by design. These problems become even more apparent when extending current data-driven approaches to handle higher-dimensional data. A solution is to learn from unstructured data directly. Recently, multiple attempts from the PointNet family [17,19,28] amongst others [10,14,25] proposed to handle *irregular* point clouds directly in a deep neural network. In contrast to the widely successful general purpose 2D network architectures, these methods propose very particular network architectures with an optimized design for very specific tasks. Also, these solutions only work on rather small point clouds, still lacking support for processing million-scale point cloud data. Methods from the PointNet family subsample their inputs to 4096 points per $1\,\text{m}^2$ as depicted in Fig. 1. Such a low resolution enables single object classification, where the primary information is in the global shape characteristics [30]. Dense, complex 3D scenes, however, typically consist of millions of points [2,8]. Extending previous learning-based approaches to *effectively* process larger point clouds has been infeasible (Fig. 1(c)).

Inspired by commonly used CNNs architectures, we hypothesize that a simple convolution operation with a small amount of learnable parameters is advantageous when employing them in deeper network architectures—against recent trends of proposing complex layers for 3D point cloud processing.

To summarize our main contributions: (1) We introduce a novel convolution layer for arbitrary metric spaces, which represents a natural generalization of traditional grid-based convolution layers along (2) with a highly-tuned GPU-based implementation, providing significant speed-ups. (3) Our empirical evaluation demonstrates substantial improvements on large-scale point cloud segmentation [2] *without* any post-processing steps, and competitive results on small benchmark sets using fewer parameters and less memory.

2   Related Work

Recent literature dealing with learning from 3D point cloud data can be organized into three categories based on their way of dealing with the input data.

Voxel-based methods [16,18,20,30] discretize the point cloud into a voxel-grid enabling the application of classical convolution layers afterwards. However, this either loses spatial information during the discretization process or requires substantial computational resources for the 3D convolutions to avoid discretization artifacts. These approaches are affected by the curse of dimensionality and will be infeasible for higher-dimensional spaces. Interestingly, ensemble methods [22,26] based on classical CNNs still achieve state-of-the-art results on common benchmark sets like ModelNet40 [30] by rendering the 3D data from several viewing directions as image inputs. As the rendered views omit some information (*i.e.* occlusions) Cao *et al.* [4] propose to use a spherical projection.

Graph-based methods are geared to process social networks or knowledge graphs, particular instances of unstructured data where each node locations is solely defined by its relation to neighboring nodes in the absence of absolute position information. Recent research [13] proposes to utilize a sparse convolution for graph structures based on the adjacency matrix. This effectively masks the output of intermediate values in the classical convolution layers and mimics a diffusion process of information when applying several of these layers.

Euclidean Space-based methods deal directly with point cloud data featuring absolute position information but *without* explicit pair-wise relations. PointNet [17] is one of the first approaches yielding competitive results on ModelNet40. It projects each point *independently* into some learned features space, which then is transformed by a spatial transformer module [12] – a rather costly operation for higher feature dimensions. While the final aggregation of information is done effectively using a max-pooling operation, keeping all high dimensional features in memory beforehand is indispensable and becomes infeasible for larger point clouds by hardware restrictions. The lack of granularity during features aggregation from local areas is addressed by the extension PointNet++ [19] using "mini"-PointNets for each point neighborhood across different resolutions and later by [28]. An alternative way of introducing a structure in point clouds relies on kD-trees [14], which allows to share convolution layers depending on the kD-tree splitting orientation. Such a structure is affected by the curse of dimensionality can only fuse point pairs in each hierarchy level. Further, defining splatting and slicing operations [25] has shown promising results on segmenting a facade datasets. Dynamic Edge-Condition Filters [23] learn parameters in the fashion of Dynamic Filter-Networks [7] for each single point neighborhood. Note, predicting a neighborhood-dependent filter can become quickly expensive for reasonably large input data. It is also noted by the authors, that tricks like BatchNorm are required during training.

Our approach belongs to the third category proposing a natural extension of convolution layers (see next section) for unstructured data which can be considered as a scalable special case of [7] but allows to evaluate point clouds and features more efficiently "in one go" – without the need of additional tricks.

3   Method

The basic operation in convolutional neural networks is a discrete 2D convolution, where the image signal[1] $I \in \mathbb{R}^{H \times W \times C}$ is convolved with a filter-kernel w. In deep learning a common choice of the filter size is $3 \times 3 \times C$ such that this mapping can be described as

$$(w \circledast f)[\ell] = \sum_{c \in C} \sum_{\tau \in \{-1,0,1\}^2} w_{c'}(c, \tau) f(c, \ell - \tau), \qquad (1)$$

where $\tau \in \{-1, 0, 1\}^2$ describes the 8-neighborhood of ℓ in regular 2D grids. One usually omits the location information ℓ as it is given implicitly by arranging the feature values on a grid in a canonical way. Still, each pixel information is a *pair* of a feature/pixel value $f(c, \ell)$ and its location ℓ.

In this paper, we extend the convolution operation $\circledast$ to support irregular data with real-valued locations. In this case, the kernel w needs to support arbitrary relative positions $\ell_i - \tau_i$, which can be potentially unbounded. Before discussing such potential versions of w, we shortly recap the grid-based convolution layer in more detail to derive desired properties of a more generic convolution operation.

3.1   Convolution Layer

For a discrete $3 \times 3 \times C$ convolution layer such a filter mapping[2]

$$w_{c'} \colon C \times \{-1, 0, 1\}^2 \to \mathbb{R}, \quad (c, \tau) \mapsto w_{c'}(c, \tau) = \sum_{\tau' \in \{-1,0,1\}^2} 1_{\{\tau = \tau'\}} w_{c,c',\tau'} \quad (2)$$

---

[1] $c \in C$ represents the RGB, where we abuse notation and write C for $\{0, 1, \ldots, C - 1\} \subset \mathbb{N}$ as well.

[2] 1_M is the indicator function being 1 iff $M \neq \emptyset$.

is based on a lookup table with 9 entries for each (c, c') pair. These values $w_{c,c',\tau'}$ of the box-function $w_{c'}$ can be optimized for a specific task, *e.g.* using back-propagation when training CNNs. Typically, a single convolution layer has a filter bank of multiple filters. While these box functions are spatially invariant in ℓ, they have a bounded domain and are neither differentiable nor continuous wrt. τ by definition. Specifically, the 8-neighborhood in a 2D grid always has exactly the same underlying spatial layout. Hence, an implementation can exploit the implicitly given locations. The same is also true for other filter sizes $k_h \times k_w \times C$.

Processing irregular data requires a function $w_{c'}$, which can handle an *unbounded* domain of arbitrary—potentially real-valued—relations between τ and ℓ, besides retaining the ability to share parameters across different neighborhoods. To find potential candidates and identify the required properties, we consider a point cloud as a more generic data representation

$$P = \left\{ (\ell^{(i)}, f^{(i)}) \in L \times F \mid i = 0, 1, \ldots, n - 1 \right\}. \tag{3}$$

Besides its value $f^{(i)}$, each point cloud element now carries an *explicitly* given location information $\ell^{(i)}$. In arbitrary metric spaces, *e.g.* Euclidean space $(\mathbb{R}^d, \|\cdot\|)$, $\ell^{(i)}$ can be real-valued without matching a discrete grid vertex. Indeed, one way to deal with this data structure is to *voxelize* a given location $\ell \in \mathbb{R}^d$ by mapping it to a specific grid vertex, *e.g.* $L' \subset \alpha \mathbb{N}^d, \alpha \in \mathbb{R}$. When L' resembles a grid structure, classical convolution layers can be used after such a discretization step. As already mentioned, choosing an appropriate α causes a trade-off between rather small cells for finer granularity in L' and consequently higher memory consumption.

Instead, we propose to define the notion of a convolution operation for a set of points in a local area. For any given point at location ℓ such a set is usually created by computing the k nearest neighbor points with locations $\mathcal{N}_k(\ell) = \{\ell'_0, \ell'_1, \ldots, \ell'_{k-1}\}$ for a point at ℓ, e.g. using a kD-tree. Thus, a generalization of Eq. (1) can be written as

$$f'(c', \ell^{(i)}) = \sum_{c \in C} \sum_{\ell' \in \mathcal{N}_k(\ell^{(i)})} \tilde{w}(c, \ell^{(i)}, \ell') \cdot f(c, \ell). \tag{4}$$

Note, for point clouds describing an image Eq. (4) is equivalent[3] to Eq. (1). But for the more general case we require that

$$\tilde{w}_{c'} : C \times \mathbb{R}^d \times \mathbb{R}^d \to \mathbb{R}, \quad (c, \ell, \ell') \mapsto \tilde{w}(c, \ell, \ell') \tag{5}$$

is an *everywhere* well-defined function instead of a "simple" look-up table. This ensures, we can use $\tilde{w}$ in neighborhoods of arbitrary sizes. However, a side-effect of giving up the grid-assumption is that $\tilde{w}$ needs to be differentiable in both ℓ, ℓ' to perform back-propagation during training.

---

[3] By setting $\mathcal{N}_9(\ell) = \{\ell - \tau | \tau \in \{-1, 0, 1\}^d\}$ and $\tilde{w}_{c'}(c, \ell^{(i)}, \ell') = w_{c'}(c, \ell^{(i)} - \ell')$.

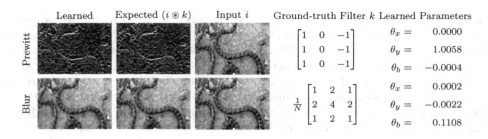

Fig. 2. Results on a toy dataset for illustration purposes. The special-case $w(x, y) = \theta_x(x - x_0) + \theta_y(y - y_0) + \theta_{b_c}$ of Eq. (6) is trained to re-produce the results of basic image operations like Prewitt or Blur.

While previous work [19,23] exert small neural networks for $\tilde{w}$ as a workaround inheriting all previously described issues, we rely on the given standard scalar product as the natural choice of $\tilde{w}$ in the Euclidean space with learnable parameters $\theta_c \in \mathbb{R}^d, \theta_{b_c} \in \mathbb{R}$:

$$\tilde{w}(c, \ell, \ell' \,|\, \theta_c, \theta_{b_c}) = \langle \theta_c, \ell - \ell' \rangle + \theta_{b_c}. \tag{6}$$

This formulation can be considered as a linear approximation of the lookup table, with the advantage of being defined everywhere. In a geometric interpretation $\tilde{w}$ is a learnable linear transformation (scaled and rotated) of a high-dimensional Prewitt operation. It can represent several image operations; Two are depicted in Fig. 2.

Hence, the mapping $\tilde{w}$ from Eq. (6) exists in all metric spaces, is *everywhere* well-defined in c, ℓ, ℓ', and continuously differentiable wrt. to *all* arguments, such that gradients can be propagated back even through the locations ℓ, ℓ'. Further, our rather simplistic formulation results in a significant reduction of the required trainable parameters and retains translation invariance. One observed consequence is a more stable training even *without* tricks like using BatchNorm as in [23]. This operation is parallel and can be implemented using CUDA to benefit from the sparse access patterns of local neighborhoods. In combination with a minimal memory footprint, this formulation is the first being able to process millions of irregular points simultaneously – a crucial requirement when applying this method in large-scale real-world settings. We experimented with slightly more complex versions of flex-conv, *e.g.* using multiple sets of parameters for one filter dependent on local structure. However, they did not lead to better results and induced unstable training.

3.2   Extending Sub-sampling to Irregular Data

While straightforward in grid-based methods, a proper and scalable sub-sampling operation in unstructured data is not canonically defined. On grids, down-sampling an input by a factor 4 is usually done by just taking every second cell in each dimension and aggregating information from a small surrounding region. There is always an implicitly well-defined connection between a point and its representative at a coarser resolution.

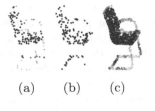

(a)   (b)   (c)

Fig. 3. IDISS (a) against random sub-sampling (b) for an object (c) with color-coded density.

For sparse structures this property no longer holds. Points being neighbors in one resolution, potentially are not in each other's neighborhood at a finer resolution. Hence, it is even possible that some points will have no representative within the next coarser level. To avoid this issue, Simonovsky et al. [23] uses the VoxelGrid algorithm which inherits all voxel-based drawbacks described in the previous sections. Qi et al. [19] utilizes Farthest point sampling (FPS). While this produces sub-samplings avoiding the missing representative issue, it pays the price of having the complexity of $\mathcal{O}(n^2)$ for *each* down-sampling layer. This represents a serious computation limitation. Instead, we propose to utilize inverse density importance sub-sampling (IDISS). In our approach, the inverse density ϕ is simply approximated by adding up all distances from one point in ℓ to its k-neighbors by $\phi(\ell) = \sum_{\ell' \in \mathcal{N}_k(\ell)} \|\ell - \ell'\|$.

Sampling the point cloud proportional to this distribution has a computational complexity of $\mathcal{O}(n)$, and thereby enables processing million of points in a very efficient way. In most cases, this method is especially cheap regarding computation time, since the distances have already been computed to find the K-nearest neighbors. Compared to pure random sampling, it produces better uniformly distributed points at a coarser resolution and more likely preserves important areas. In addition, it still includes randomness that is preferred in training of deep neural networks to better prevent against over-fitting. Figure 3 demonstrates this approach. Note, how the chair legs are rarely existing in a randomly sub-sampled version, while IDISS preserves the overall structure.

4   Implementation

To enable building complete DNNs with the presented flex-convolution model we have implemented two specific layers in TensorFlow: *flex-convolution* and *flex-max-pooling*. Profiling shows that a direct highly hand-tuned implementation in CUDA leads to a run-time which is in the range of regular convolution layers (based on cuDNN) during inference.

4.1 Neighborhood Processing

Both new layers require a known neighborhood for each incoming point. For a fixed set of points, this neighborhood is computed once upfront based on an efficient kD-tree implementation and kept fixed. For each point, the k nearest neighbors are stored as indices into the point list. The set of indices is represented as a tensor and handed over to each layer.

The *flex-convolution* layer merely implements the convolution with continuous locations as described in Eq. (6). Access to the neighbors follows the neighbor indices to lookup their specific feature vectors and location. No data duplication is necessary. As all points have the same number of neighbors, this step can be parallelized efficiently. In order to make the position of each point available in each layer of the network, we attach the point location ℓ to each feature vector.

The *flex-max-pooling* layer implements max-pooling over each point neighborhood individually, just like the grid-based version but without subsampling.

For subsampling, we exploit the IDISS approach described in Sect. 3.2. Hereby, flex-max-pooling is applied before the subsampling procedure. For the subsequent, subsampled layers the neighborhoods might have changed, as they only include the subsampled points. As the point set is static and known beforehand, all neighborhood indices at each resolution can be computed on-the-fly during parallel data pre-fetching, which is neglectable compared to the cost of a network forward+backward pass under optimal GPU utilization.

Upsampling (*flex-upsampling*) is done by copying the features of the selected points into the larger-sized layer, initializing all other points with zero, like zero-padding in images and performing the flex-max-pooling operation.

Table 1. Profiling information of diverse implementations with 8 batch of 4096 points with $C' = C = 64$ and 9 neighbors using a CUDA profiler.

Method	Timing		Memory	
	Forward	Backward	Forward	Backward
flex-convolution (pure TF)*	1829 ms	2738 ms	34015.2 MB	63270.8 MB
flex-convolution (Ours)	24 ms	265 ms	8.4 MB	8.7 MB
flex-convolution (TC [27])	42 ms	-	8.4 MB	-
grid-based conv. (cuDNN)	16 ms	1.5 ms	1574.1 MB	153.4 MB
flex-max-pooling (Ours)	1.44 ms	15 us	16.78 MB	8.4 MB

4.2 Efficient Implementation of Layer Primitives

To ensure a reasonably fast training time, highly efficient GPU-implementations of flex-convolution and flex-max-pooling as a custom operation in TensorFlow are required. We implemented a generic but hand-tuned CUDA operation, to ensure optimal GPU-throughput. Table 1 compares our optimized CUDA kernel against a version (pure TF) containing exclusively existing operations

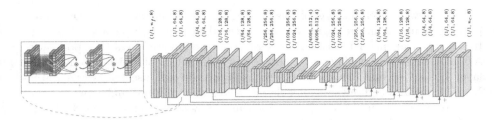

Fig. 4. Network architecture for semantic 3D point cloud segmentation. The annotations (a, d_f, k) represent the spatial resolution factor a (*i.e.* using $a \cdot n$ points) and feature length d_f with n_f input features and n_c classes. The used neighborhood size is given by k. In each step, the position information ▨ and neighborhood information ▢ is required besides the actual learned features. After flex-convolution layers ▨ , each downsampling step ▢ (flex-max-pool) has a skip-connection to the corresponding decoder block with flex-upsampling layer ▢.

provided by the TensorFlow framework itself and its grid-based counterpart in cuDNN [6] using the CUDA profiler for a *single* flex-convolution layer on a set of parameters, which fits typical consumer hardware (Nvidia GTX 1080Ti). As the grid-based convolution layer typically uses a kernel-size of $3 \times 3 \times C$ in the image domain, we set $k = 9$ as well – though we use $k = 8$ in all subsequent point cloud experiments. We did some experiments with a quite recent polyhedral compiler optimization using TensorComprehension (TC) [27] to automatically tune a flex-convolution layer implementation. While this approach seems promising, the lack of supporting flexible input sizes and slower performance currently prevents us from using these automatically generated CUDA kernels in practice.

An implementation of the flex-convolution layer by just relying on operations provided by the TensorFlow framework requires data duplication. We had to spread the pure TensorFlow version across 8 GPUs to run a *single* flex-convolution layer. Typical networks usually consist of several such operations. Hence, it is inevitable to recourse on tuning custom implementations when applying such a technique to larger datasets. Table 1 reveals that the grid-based version (cuDNN) prepares intermediate values in the forward pass resulting in larger memory consumption and faster back-propagation pass—similar to our flex-max-pooling.

4.3   Network Architecture for Large-Scale Semantic Segmentation

With the new layers at hand, we can directly transfer the structure of existing image processing networks to the task of processing large point clouds. We will elaborate on our network design and choice of parameters for the task of semantic point cloud segmentation in more detail. Here, we draw inspiration from established hyper-parameter choices in 2D image processing.

Our network architecture follows the SegNet-Basic network [3] (a 2D counterpart for semantic image segmentation) with added U-net skip-connections [21]. It has a typical encoder-decoder network structure followed by a final point-wise soft-max classification layer. To not obscure the effect of the flex-convolution layer behind several other effects, we explicitly do *not* use tricks like Batch-Normalization, weighted soft-max classification, or computational expensive preresp. post-processing approaches, which are known to enhance the prediction quality and could further be applied to the results presented in the Sect. 5.

The used architecture and output sizes are given in Fig. 4. The encoder network is divided into six stages of different spatial resolutions to process multi-scale information from the input point cloud. Each resolution stage consists of two ResNet-blocks. Such a ResNet block chains the following operations: 1×1-convolution, flex-convolution, flex-convolution (compare Fig. 4). Herewith, the output of the last flex-convolution layer is added to the incoming feature following the common practice of Residual Networks [9]. To decrease the point cloud resolution across different stages, we add a flex-max-pooling operation with subsampling as the final layer in each stage of the encoder. While a grid-based max-pooling is normally done with stride 2 in x/y dimension, we use the flex-max-pooling layer to reduce the resolution n by factor 4. When the spatial resolution decreases, we increase the feature-length by factor two.

Moreover, we experimented with different neighborhood sizes k for the flex-convolution layers. Due to speed considerations and the widespread adoption of 3×3 filter kernels in image processing we stick to a maximal nearest neighborhood size of $k = 8$ in all flex-convolution layers. We observed no decrease in accuracy against $k = 16$ but a drop in speed by factor 2.2 for 2D-3D-S [2].

The decoder network mirrors the encoder architecture. We add skip connections [21] from each stage in the encoder to its related layer in the decoder. Increasing spatial resolution at the end of each stage is done via flex-upsampling. We tested a trainable flex-transposed-convolution layer in some preliminary experiments and observed no significant improvements. Since pooling irregular data is more light-weight (see Table 1) regarding computation effort, we prefer this operation. As this is the first network being able to process point clouds in such a large-scale setting, we expect choosing more appropriate hyperparameters is possible when investing more computation time.

5   Experiments

We conducted several experiments to validate our approach. These show that our flex-convolution-based neural network yields competitive performance to previous work on synthetic data for single object classification ([30], 1024 points) using fewer resources and provide some insights about human performance on this dataset. We improve single instance part segmentation ([31], 2048 points). Furthermore, we demonstrate the effectiveness of our approach by performing semantic point cloud segmentation on a large-scale real-world 3D scan ([2], 270 Mio. points) improving previous methods in both accuracy and speed.

5.1 Synthetic Data

To evaluate the effectiveness of our approach, we participate in two benchmarks that arise from the ShapeNet [5] dataset, which consists of synthetic 3D models created by digital artists.

ModelNet40 [30] is a single object classification task of 40 categories. We applied a smaller version of the previously described encoder network-part followed by a fully-connected layer and a classification layer. Following the official test-split [17] of randomly sampled points from the object surfaces for

Table 2. Classification accuracy on ModelNet40 (1024 points) and 256 points*.

Method	Accuracy	#params.
PointNet [17]	89.2	1'622'705
PointNet2 [19]	90.7	1'658'120
KD-Net [14]	90.6	4'741'960
D-FilterNet [23]	87.4	345'288
Human	64.0	-
Ours	90.2	346'409
Ours (1/4)	89.3	171'048

object classification, we compare our results in Table 2. Our predictions are provided from by a single forward-pass in contrast to a voting procedure as in the KD-Net [14]. This demonstrates that a small flex-convolution neural network with significant fewer parameters provides competitive results on this benchmark set. Even when using just 1/4th of the point cloud and thus an even smaller network the accuracy remains competitive. To put these values in a context to human perception, we conducted a user study asking participants to classify point clouds sampled from the official test split. We allowed them to rotate the presented point cloud for the task of classification without a time limit. Averaging all 2682 gathered object classification votes from humans reveals some difficulties with this dataset. This might be related to the relatively unconventional choice of categories in the dataset, *i.e.* plants and their flower pots and bowls are sometimes impossible to separate. Please refer to the Supplementary for a screenshot of the user study, a confusion matrix, saliency maps and an illustration of label ambiguity.

ShapeNet Part Segmentation [31] is a semantic segmentation task with per-point annotations of 31963 models separated into 16 shape categories. We applied a smaller version of the previously described segmentation network that receives the (x, y, z) position of 2048 points per object. For the evaluation, we follow the procedure of [25] by training a network for per category. Table 3 contains a comparison of methods using only point cloud data as input. Our method demonstrates an improvement of the average mIoU while being able to process a magnitude more shapes per second. Examples of ShapeNet part segmentation are illustrated in Fig. 5. These experiments on rather small synthetic data confirm our hypothesis that even in three dimensions simple filters with a small amount of learnable parameters are sufficient in combination with deeper network architectures. This matches with the findings that are known from typical CNN architectures of preferring deeper networks with small 3×3 filters. The resulting smaller memory footprint and faster computation time enable processing more points in reasonable time. We agree with [25,28] on the data labeling issues.

Table 3. ShapeNet part segmentation results per category and mIoU (%) for different methods and inference speed (on a Nvidia GeForce GTX 1080 Ti).

	Airpl.	Bag	Cap	Car	Chair	Earph.	Guitar	Knife	Lamp	Laptop	Motorb.	Mug	Pistol	Rocket	Skateb.	Table	mIoU	shapes/sec
Kd-Network [14]	80.1	74.6	74.3	70.3	88.6	73.5	90.2	**87.2**	81.0	94.9	57.4	86.7	78.1	51.8	69.9	80.3	77.4	n.a.
PointNet [17]	83.4	78.7	82.5	74.9	89.6	73.0	91.5	85.9	80.8	95.3	65.2	93.0	81.2	57.9	72.8	80.6	80.4	n.a.
PointNet++ [19]	82.4	79.0	87.7	77.3	**90.8**	71.8	91.0	85.9	83.7	95.3	71.6	94.1	81.3	58.7	76.4	82.6	81.9	2.7
SPLATNet3D [25]	81.9	83.9	88.6	**79.5**	90.1	73.5	91.3	84.7	**84.5**	96.3	69.7	95.0	81.7	59.2	70.4	81.3	82.0	9.4
SGPN [28]	80.4	78.6	78.8	71.5	88.6	**78.0**	90.9	83.0	78.8	95.8	**77.8**	93.8	**87.4**	60.1	**92.3**	**89.4**	82.8	n.a.
Ours	**83.6**	**91.2**	**96.7**	**79.5**	84.7	71.7	**92.0**	86.5	83.2	**96.6**	71.7	**95.7**	86.1	**74.8**	81.4	84.5	**85.0**	**489.3**

Fig. 5. Our semantic segmentation results on ShapeNet (ground-truth (left), prediction (right)) pairs. Please refer to the supplementary for more results at higher resolution.

5.2 Real-World Semantic Point Cloud Segmentation

To challenge our methods at scale, we applied the described network from Sect. 4.3 to the 2D-3D-S dataset [2]. This real-world dataset covers 3D scanning information from six square kilometers of several building complexes collected by a Matterport Camera. Previous approaches are based on sliding windows, either utilizing hand-crafted feature, *e.g.* local curvature, occupancy and point density information per voxel [1,2] or process small sub-sampled chunks PointNet [17], SGPN [28] (4096 points, Fig. 1). We argue, that a neural network as described in Sect. 4.3 can learn all necessary features directly from the data – just like in the 2D case and at *full* resolution.

An ablation study on a typical room reveals the effect of different input features f. Besides neighborhood information, providing only constant initial features $f = 1$ yields 0.31 mAP. Hence, this is already enough information to perform successful semantic segmentation. To account for the irregularity in the data, it is however useful to use normalized position data $f = (1, x, y, z)$ besides the color information $f = (1, x, y, z, r, g, b)$ which increases the accuracy to 0.39 mAP resp. 0.50 mAP. Our raw network predictions from a single inference forward pass out-performs previous approaches given the same available information and approaches using additional input information but lacks precision in categories like beam, column, and door, see Table 4. Providing features like local curvature besides post-processing [2] greatly simplify detecting these kinds of objects. Note, our processing of point clouds at full resolution benefits the handling of smaller objects like chair, sofa and table.

Consider Fig. 6, the highlighted window region in room A is classified as wall because the blinds are closed, thus having a similar appearance. In room B, our network miss-classifies the highlighted column as "wall", which is not surprising as both share similar geometry and color. Interestingly, in room C our network classifies the beanbag as "sofa", while its ground-truth annotation is "chair". For more results please refer to the accompanying video.

Training is done on two Nvidia GTX 1080Ti with batch-size 16 for two days on point cloud chunk with 128^2 points using the Adam-Optimizer with learning-rate $3 \cdot 10^{-3}$.

To benchmark inference, we compared ours against the author's implementations of previous work [17,19,28] on different point clouds sizes n. Memory requirements limits the number of processed points to at most 131k [19], 500k [28], 1Mio [17] points (highlighted region in Fig. 1). We failed to get meaningful performance in terms of accuracy from these approaches when increasing $n > 4096$. In contrast, ours – based on a fully convolutional network – can process up to 7 Mio. points concurrently providing the *same* performance during inference within 4.7 s. Note, [17] can at most process 1 Mio. points within 7.1 s. Figure 1 further reveals an exponential increase of runtime for the Point-Net family [17,19,28], ours provides significant faster inference and shows better utilization for larger point clouds with a linear increase of runtime.

Table 4. Class specific average precision (AP) on the 2D-3D-S dataset. (‡) uses additional input features like local curvature, point densities, surface normals. (*) uses non-trivial post-processing and (**) a mean filter post-processing.

	Table	Chair	Sofa	Bookc.	Board	Ceiling	Floor	Wall	Beam	Col.	Wind.	Door	mAP
Armeni *et al.* [2]*	46.02	16.15	6.78	54.71	3.91	71.61	88.70	72.86	66.67	**91.77**	25.92	54.11	49.93
Armeni *et al.* [2]‡	39.87	11.43	4.91	**57.76**	3.73	50.74	80.48	65.59	68.53	85.08	21.17	45.39	44.19
PointNet [17]*	46.67	33.80	4.76	n.a.	11.72	n.a.	n.a.	n.a.	n.a.	n.a.	n.a.	n.a.	n.a.
SGPN [28]*	46.90	40.77	6.38	47.61	11.05	79.44	66.29	**88.77**	**77.98**	60.71	**66.62**	**56.75**	54.35
Ours	66.03	51.75	15.59	39.03	43.50	87.20	96.00	65.53	54.76	52.74	55.34	35.81	55.27
Ours**	**67.02**	**52.75**	**16.61**	39.26	**47.68**	**87.33**	**96.10**	65.52	56.83	55.10	57.66	36.76	**56.55**

Limitation. As we focus on static point cloud scans ours is subject to the same limitations as [17,19,25,28], where neighborhoods are computed during parallel data pre-fetching. Handling dynamic point clouds, *e.g.* completion or generation, requires an approximate nearest-neighborhood layer. Our prototype implementation suggests this could be done *within* the network. For 2 Million points it takes around 1 s which is still faster by a factor of 8 compared to the used kd-Tree, which however has neglectable costs being part of parallel pre-fetching.

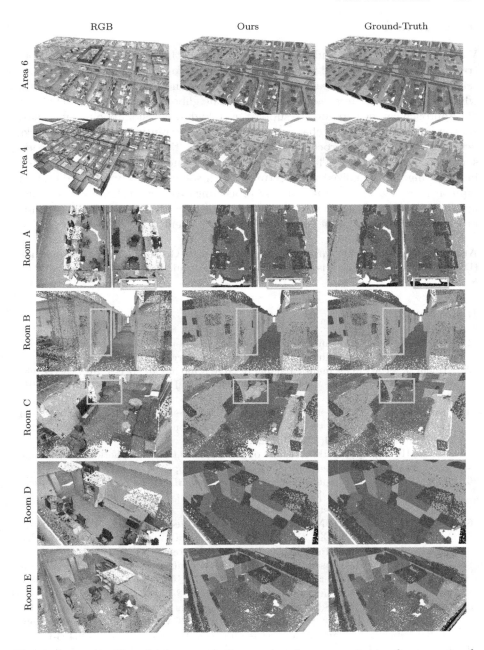

Fig. 6. Semantic point cloud segmentation produced as raw outputs of our proposed network from the held-out validation set. In this point-based rendering, surfaces might not be illustrated as opaque.

6 Conclusion

We introduced a novel and natural extension to the traditional convolution, transposed convolution and max-pooling primitives for processing irregular point sets. The novel sparse operations work on the local neighborhood of each point, which is provided by indices to the k nearest neighbors. Compared to 3D CNNs our approach can be extended to support even high-dimensional point sets easily. As the introduced layers behave very similar to convolution layers in networks designed for 2D image processing, we can leverage the full potential of already successful architectures. This is against recent trends in point cloud processing with highly specialized architectures which sometimes rely on hand-crafted input features, or heavy pre- and post-processing. We demonstrate state-of-the-art results on small synthetic data as well as large real-world datasets while processing millions of points concurrently and efficiently.

Acknowledgment. This work was supported by the German Research Foundation (DFG): SFB 1233, Robust Vision: Inference Principles and Neural Mechanisms, TP 01 & 02.

References

1. Armeni, I., Sax, A., Zamir, A.R., Savarese, S.: Joint 2D-3D-semantic data for indoor scene understanding. arXiv e-prints, February 2017
2. Armeni, I., et al.: 3D semantic parsing of large-scale indoor spaces. In: Proceedings of the IEEE Conference on Computer Vision and Pattern Recognition (CVPR) (2016)
3. Badrinarayanan, V., Kendall, A., Cipolla, R.: SegNet: a deep convolutional encoder-decoder architecture for image segmentation. IEEE Trans. Pattern Anal. Mach. Intell. (PAMI) **39**(12), 2481–2495 (2017)
4. Cao, Z., Huang, Q., Karthik, R.: 3D object classification via spherical projections. In: International Conference on 3D Vision (3DV), pp. 566–574. IEEE (2017)
5. Chang, A.X., et al.: ShapeNet: an information-rich 3D model repository. Technical report arXiv:1512.03012 [cs.GR], Stanford University — Princeton University — Toyota Technological Institute at Chicago (2015)
6. Chetlur, S., et al.: cuDNN: efficient primitives for deep learning. CoRR (2014)
7. De Brabandere, B., Jia, X., Tuytelaars, T., Van Gool, L.: Dynamic filter networks. In: Advances in Neural Information Processing Systems (NIPS) (2016)
8. Groh, F., Resch, B., Lensch, H.P.A.: Multi-view continuous structured light scanning. In: Roth, V., Vetter, T. (eds.) GCPR 2017. LNCS, vol. 10496, pp. 377–388. Springer, Cham (2017). https://doi.org/10.1007/978-3-319-66709-6_30
9. He, K., Zhang, X., Ren, S., Sun, J.: Identity mappings in deep residual networks. In: Leibe, B., Matas, J., Sebe, N., Welling, M. (eds.) ECCV 2016. LNCS, vol. 9908, pp. 630–645. Springer, Cham (2016). https://doi.org/10.1007/978-3-319-46493-0_38
10. Hermosilla, P., Ritschel, T., Vázquez, P.P., Vinacua, À., Ropinski, T.: Monte Carlo convolution for learning on non-uniformly sampled point clouds. arXiv preprint arXiv:1806.01759 (2018)
11. Hershey, S., et al.: CNN architectures for large-scale audio classification. In: IEEE International Conference on Acoustics, Speech and Signal Processing (ICASSP), pp. 131–135. IEEE (2017)

12. Jaderberg, M., Simonyan, K., Zisserman, A., Kavukcuoglu, K.: Spatial transformer networks. In: Cortes, C., Lawrence, N.D., Lee, D.D., Sugiyama, M., Garnett, R. (eds.) Advances in Neural Information Processing Systems (NIPS), pp. 2017–2025. Curran Associates, Inc., Red Hook (2015)
13. Kipf, T.N., Welling, M.: Semi-supervised classification with graph convolutional networks. In: International Conference on Learning Representations (ICLR) (2017)
14. Klokov, R., Lempitsky, V.: Escape from cells: deep Kd-networks for the recognition of 3D point cloud models. In: Proceedings of the IEEE International Conference on Computer Vision (ICCV), pp. 863–872, October 2017
15. Lavin, A., Gray, S.: Fast algorithms for convolutional neural networks. In: Proceedings of the IEEE Conference on Computer Vision and Pattern Recognition (CVPR), pp. 4013–4021 (2016)
16. Maturana, D., Scherer, S.: VoxNet: a 3D convolutional neural network for real-time object recognition. In: International Conference on Intelligent Robots and Systems (2015)
17. Qi, C.R., Su, H., Mo, K., Guibas, L.J.: PointNet: deep learning on point sets for 3D classification and segmentation. In: Proceedings of the IEEE Conference on Computer Vision and Pattern Recognition (CVPR) (2017)
18. Qi, C.R., Su, H., Niessner, M., Dai, A., Yan, M., Guibas, L.J.: Volumetric and multi-view CNNs for object classification on 3D data. In: Proceedings of the IEEE Conference on Computer Vision and Pattern Recognition (CVPR) (2016)
19. Qi, C.R., Yi, L., Su, H., Guibas, L.J.: PointNet++: deep hierarchical feature learning on point sets in a metric space. In: Guyon, I., et al. (eds.) Advances in Neural Information Processing Systems (NIPS), pp. 5099–5108. Curran Associates, Inc., Red Hook (2017)
20. Riegler, G., Ulusoy, A.O., Bischof, H., Geiger, A.: OctNetFusion: learning depth fusion from data. In: International Conference on 3D Vision (3DV), October 2017
21. Ronneberger, O., Fischer, P., Brox, T.: U-Net: convolutional networks for biomedical image segmentation. In: Navab, N., Hornegger, J., Wells, W.M., Frangi, A.F. (eds.) MICCAI 2015. LNCS, vol. 9351, pp. 234–241. Springer, Cham (2015). https://doi.org/10.1007/978-3-319-24574-4_28
22. Sfikas, K., Pratikakis, I., Theoharis, T.: Ensemble of PANORAMA-based convolutional neural networks for 3D model classification and retrieval. Comput. Graph. **71**, 208–218 (2017)
23. Simonovsky, M., Komodakis, N.: Dynamic edge-conditioned filters in convolutional neural networks on graphs. In: Proceedings of the IEEE Conference on Computer Vision and Pattern Recognition (CVPR) (2017). https://arxiv.org/abs/1704.02901
24. Simonyan, K., Zisserman, A.: Very deep convolutional networks for large-scale image recognition. CoRR (2014)
25. Su, H., et al.: SPLATNet: sparse lattice networks for point cloud processing. In: Proceedings of the IEEE Conference on Computer Vision and Pattern Recognition (CVPR), pp. 2530–2539 (2018)
26. Su, H., Maji, S., Kalogerakis, E., Learned-Miller, E.G.: Multi-view convolutional neural networks for 3D shape recognition. In: Proceedings of the IEEE International Conference on Computer Vision (ICCV) (2015)
27. Vasilache, N., et al.: Tensor comprehensions: framework-agnostic high-performance machine learning abstractions (2018)
28. Wang, W., Yu, R., Huang, Q., Neumann, U.: SGPN: similarity group proposal network for 3D point cloud instance segmentation. In: Proceedings of the IEEE Conference on Computer Vision and Pattern Recognition (CVPR), pp. 2569–2578 (2018)

29. Wieschollek, P., Schölkopf, M.H.B., Lensch, H.P.A.: Learning blind motion deblur-
    ring. In: International Conference on Computer Vision (ICCV), October 2017
30. Wu, Z., et al.: 3D ShapeNets: a deep representation for volumetric shapes. In:
    Proceedings of the IEEE Conference on Computer Vision and Pattern Recognition
    (CVPR), pp. 1912–1920 (2015)
31. Yi, L., et al.: A scalable active framework for region annotation in 3D shape col-
    lections. ACM Trans. Graph. (SIGGRAPH ASIA) 35(6), 210 (2016)

Poster Session P1

Extreme Reverse Projection Learning for Zero-Shot Recognition

Jiechao Guan, An Zhao, and Zhiwu Lu$^{(\boxtimes)}$

Beijing Key Laboratory of Big Data Management and Analysis Methods,
School of Information, Renmin University of China, Beijing 100872, China
luzhiwu@ruc.edu.cn

Abstract. Zero-shot learning (ZSL) aims to transfer knowledge from a set of seen classes to a set of unseen classes so that the latter can be recognised without any training samples. It faces two main challenges: the projection domain shift problem and the hubness problem. Existing models have been proposed to address one of the two problems, but not both. In this paper, we propose a novel ZSL model termed extreme reverse projection learning (ERPL) to solve both problems. It has a simple linear formulation that casts ZSL into a min-min optimization problem. An efficient and robust gradient-based algorithm is also introduced to find an extremely generalizable projection for ZSL. Extensive experiments on five benchmark datasets show that the proposed model outperforms the state-of-the-art alternatives on all datasets, often by significant margins. Furthermore, the obtained improvements are particularly salient under the more challenging but realistic 'pure' and generalized ZSL settings.

Keywords: Zero-shot learning · Projection domain shift · Hubness · Reverse projection · Generalized zero-shot learning

1 Introduction

Large-scale visual recognition [1,33] has recently drawn much attention from the computer vision community. This is primarily due to the emergence of large-scale image datasets such as ImageNet [30] and the advances in deep learning [8,18]. However, most existing visual recognition models are based on supervised learning which requires hundreds of samples to be collected for each object category. This may not be possible for some fine-grained and rare categories, thus limiting the scalability of these visual recognition models. To address this scalability issue, zero-shot learning (ZSL) models [3,4,6,10,12,13] have received rapidly increasing interest, which aims to recognise unseen object classes without the need of collecting any training samples.

The key idea of zero-shot recognition is to transfer knowledge from a set of seen classes to a set of unseen classes. This is made possible by introducing a semantic space, which can be a semantic attribute space [15,19] or a semantic word vector space [10,34]. In this semantic space, the names of both seen and

© Springer Nature Switzerland AG 2019
C. V. Jawahar et al. (Eds.): ACCV 2018, LNCS 11361, pp. 125–141, 2019.
https://doi.org/10.1007/978-3-030-20887-5_8

unseen classes are embedded as high dimensional vectors called class prototypes, and their semantic relationship can be measured by the distance between these vectors, e.g., 'horse' and 'donkey' would be close in an attribute space because they share many attributes; and both are far away from 'chair'. With this semantic space and a visual feature space representing the appearance of an object in an image, existing ZSL models choose or learn a joint embedding space, which in most cases is the semantic space, and learn a projection function so that both the visual feature vector and the semantic vector are embedded in the same space. This projection is learned with the seen class training samples only. But once learned, it is used to project the unseen class samples, and the class label of a test unseen class sample is assigned to the nearest unseen class.

ZSL faces a number of challenges. The first one is the *hubness problem* [25]. It arises when nearest neighbor search is performed in a high dimensional space, typical for the joint embedding space used in ZSL. In such a space, some class prototypes become hubs which have the shortest distance to most unseen class samples, resulting in mis-classification. The second one is the *projection domain shift problem* [11]. Specifically, considering the seen and unseen classes as two domains, the projection function is learned from the seen class domain but applied to the unseen class domain. Containing completely different classes, a big domain gap exists; as a result, the projection is often biased towards the seen class prototypes. This problem is particularly acute when ZSL is carried out in a more realistic setting, e.g., the recently proposed generalized ZSL setting [7], under which both seen and unseen class samples need to be recognized during the test time. The bias towards seen class domain makes most test samples being classified as seen classes even if they belong to unseen classes.

Most recent ZSL models aim to solve one of the two problems above. In particular, reverse projection learning (RPL) is proposed to solve the hubness problem in ZSL [31, 32]. It is shown that, by projecting the class prototypes to the feature space, rather than the other way around, the hubness problem can be alleviated. This is despite the fact that the feature space is often of higher dimension. As for the projection domain shift problem, most existing solutions [11,13,14,16,21,28,32,36,39,40] are based on transductive learning. ZSL is treated as an unsupervised domain adaptation problem and the unlabeled unseen class samples are used to adapt the projection function towards the unseen class domain. However, none of the existing approaches attempt to solve both problems in a unified framework, even though the two problems are clearly connected: A better projection function could potentially reduce the hubness – hubs can be formed by a biased projection function, and being less sensitive to hubness could also mean being more robust against the projection domain shift.

In this paper, for the first time, a unified framework is proposed to solve both problems. More specifically, we introduce an extreme reverse projection learning (ERPL) formulation under the transductive setting. In contrast to existing transductive ZSL models [14,16,21,32,36,40] which attempt to solve two or three subtasks simultaneously and as a results have to introduce additional intermediate variables to bridge the two domains, our ERPL model focuses on learning an *extremely generalizable* projection function and keeps the same linear regression formulation as in the original RPL models without any additional variables. Importantly, since all transductive learning methods are related in spirit to the self-training paradigm, which employs predicted labels on the unseen class samples for model adaptation, robustness to model drift is of paramount importance when the predicted labels are inaccurate. This problem is solved by an extremely generalizable solver developed for ERPL in this paper. Specifically, we develop an efficient and robust gradient-based algorithm for training our ERPL model. Since the proposed ERPL algorithm has a linear time complexity with respect to the data size, it can be applied to different ZSL tasks (and even large-scale tasks) with very low computational cost.

Our contributions are: (1) For the first time, both the projection domain shift and hubness problems in ZSL are addressed in a unified framework. (2) We formulate ZSL as a min-min optimization problem with a simple linear formulation that can be solved by an efficient and robust gradient-based algorithm. Extensive experiments are carried out on five benchmark datasets. The results show that the proposed ERPL model yields state-of-the-art results on all datasets. The improvements over alternative ZSL models are especially significant under the more challenging pure and generalised ZSL settings.

2   Related Work

Semantic Space. Various semantic spaces are used as representations of class names for ZSL. The attribute space [37,41] is the most widely used. However, for large-scale problems, annotating attributes for each class becomes very difficult. Recently, semantic word vector space has begun to be popular especially in large-scale problems [10], since no manually defined ontology is required and any class name can be represented as a word vector for free.

Projection Learning. Depending on the selection of the joint embedding space and how the projection function is established between the feature/semantic spaces and the embedding space, existing ZSL models can be divided into three groups: (1) The first group learns a projection function from a visual feature space to a semantic space by leveraging conventional regression/ranking models [3,19] or deep neural network regression/ranking models [10,20,27,34]. (2) The second group chooses the reverse projection direction [16,31,32], i.e. from the semantic space to the feature space, to alleviate the hubness problem commonly suffered by nearest neighbour search in a high dimensional space [25]. (3) The third group learns an intermediate space as the embedding space, where both the feature space and the semantic space are projected to [6,22,42]. An exception is the semantic autoencoder proposed in [17] which can be considered as a combination of the first and second groups. Our ERPL model falls into the second group in order to tackle the hubness problem, but differs from the existing RPL based models in that it is formulated for transductive learning so that the projection domain shift problem can be attacked jointly.

Transductive ZSL. Transductive ZSL is proposed to address the projection domain shift problem [11,16,28] by learning with not only the training set of labelled data from seen classes but also the test set of unlabelled data from unseen classes. Based on whether the predicted labels of the test images are iteratively used for model learning, existing transductive ZSL models fall into two categories: (1) The first category of models [11,13,28,39] first construct a graph in the semantic space and then transfer to the test set by label propagation. A variant is the structured prediction model [43] which utilize a Gaussian parameterization of the unseen class domain label predictions. (2) The second category of models [14,16,21,32,36,40] involve using the predicted labels of the unseen class data in an iterative model update/adaptation process as in self-training [38]. Although our ERPL model belongs to the second category, the need of accurately predicting the labels of the test images is significantly reduced by developing a robust gradient-based solver. Importantly, we adopt a similar linear formulation intrinsic to the RPL models, so that our model can be trained efficiently for transductive learning.

Generalized ZSL. Different from the standard ZSL setting that takes only unseen classes for test process, the generalized ZSL setting [5,7,10,26,34] assumes that the test images come from both seen and unseen classes. This setting is clearly more reflective of the real-world application scenarios. However, it also makes ZSL more challenging, precisely because of the projection domain shift problem that transductive ZSL attempts to address. In particular, as shown in [6], since the projection is learned using the seen classes only, during test, most of the test images from unseen classes would be projected to be close to the seen class prototypes, thus misclassified. This in turn aggravates the hubness problem. To address this projection bias problem, novelty detection [34] has been used as a preprocessing step to predict whether a test image is from seen/unseen classes. Alternatively, calibrated stacking [6] has been proposed to postprocess the results of ZSL. Our ERPL model is naturally suited to the generalized setting: First, its transductive learning formulation enables adaptation of the projection towards the unseen class domain. Second, the reverse projection formulation reduces the likelihood of seen class prototypes forming hubs.

Pure ZSL. Almost all ZSL models compute the visual features from images using deep convolutional neural network (CNN) models such as those proposed in [8,18], which are pre-trained on the 1K classes in ImageNet ILSVRC 2012 [30]. However, it is noted in [37,44] that many unseen classes in the existing ZSL benchmarks test data splits are actually part of the ImageNet 1K classes. The reported results are thus not obtained under a strict zero-shot setting. Therefore, they are less meaningful because the domain gap between seen and unseen classes are smaller than what it should be. To rectify this, the overlapped ImageNet classes are removed from the test set of unseen classes for the new benchmark ZSL dataset splits [37,44], resulting in a 'pure' ZSL setting. In this paper, our ERPL model is also evaluated under the same pure ZSL setting, and we can obtain state-of-the-art results (Sect. 4.3) due to the ability of ERPL to handle large projection domain shift.

3   The Proposed Model

3.1   Problem Definition

Let $\mathcal{S} = \{s_1, ..., s_p\}$ denote a set of seen classes and $\mathcal{U} = \{u_1, ..., u_q\}$ denote a set of unseen classes, where p and q are the total numbers of seen and unseen classes, respectively. These two sets of classes are disjoint, i.e. $\mathcal{S} \cap \mathcal{U} = \phi$. Similarly, $\mathbf{Y}_s = [\mathbf{y}_1^{(s)}, ..., \mathbf{y}_p^{(s)}] \in \mathbb{R}^{k \times p}$ and $\mathbf{Y}_u = [\mathbf{y}_1^{(u)}, ..., \mathbf{y}_q^{(u)}] \in \mathbb{R}^{k \times q}$ denote the corresponding seen and unseen class semantic representations (e.g. k-dimensional attribute vector). We are given a set of labelled training images $\mathcal{D}_s = \{(\mathbf{x}_i^{(s)}, l_i^{(s)}, \mathbf{y}_{l_i^{(s)}}^{(s)}) : i = 1, ..., N_s\}$, where $\mathbf{x}_i^{(s)} \in \mathbb{R}^{d \times 1}$ is the d-dimensional visual feature vector of the i-th image in the training set, $l_i^{(s)} \in \{1, ..., p\}$ is the label of $\mathbf{x}_i^{(s)}$ according to $\mathcal{S}$, $\mathbf{y}_{l_i^{(s)}}^{(s)}$ is the semantic representation of $\mathbf{x}_i^{(s)}$, and N_s denotes the total number of labeled images. Let $\mathcal{D}_u = \{(\mathbf{x}_i^{(u)}, l_i^{(u)}, \mathbf{y}_{l_i^{(u)}}^{(u)}) : i = 1, ..., N_u\}$ denote a set of unlabelled test images, where $\mathbf{x}_i^{(u)} \in \mathbb{R}^{d \times 1}$ is the d-dimensional visual feature vector of the i-th image in the test set, $l_i^{(u)} \in \{1, ..., q\}$ is the unknown label of $\mathbf{x}_i^{(u)}$ according to $\mathcal{U}$, $\mathbf{y}_{l_i^{(u)}}^{(u)}$ is the unknown semantic representation of $\mathbf{x}_i^{(u)}$, and N_u denotes the total number of unlabeled images. The goal of zero-shot learning is to predict the labels of test images by learning a classifier $f : \mathcal{X}_u \rightarrow \mathcal{U}$, where $\mathcal{X}_u = \{\mathbf{x}_i^{(u)} : i = 1, ..., N_u\}$.

3.2   Projection Learning

Many existing ZSL models learn a projection function from a visual feature space to a semantic space. Taking a linear ridge regression model, we can formulate the projection learning problem in ZSL as follows:

$$\min_{\mathbf{W}} \sum_{i=1}^{N_s} \|\mathbf{W}^T \mathbf{x}_i^{(s)} - \mathbf{y}_{l_i^{(s)}}^{(s)}\|_2^2 + \lambda \|\mathbf{W}\|_F^2, \tag{1}$$

where $\mathbf{W} \in \mathbb{R}^{d \times k}$ is a projection matrix, and λ is a regularization parameter. When the best projection matrix $\mathbf{W}^*$ is learnt, we can embed a test image $\mathbf{x}_i^{(u)}$ into the k-dimensional semantic space as $\hat{\mathbf{y}}_i^{(u)} = \mathbf{W}^{*T} \mathbf{x}_i^{(u)}$. The nearest neighbor search is then performed in the semantic space to predict the label of $\mathbf{x}_i^{(u)}$:

$$l_i^{(u)} = \arg\min_j \|\hat{\mathbf{y}}_i^{(u)} - \mathbf{y}_j^{(u)}\|_2^2. \tag{2}$$

To alleviate the hubness problem commonly suffered by the nearest neighbour search in a high dimensional space [25], a reverse projection learning (RPL) model was proposed in [31] to project the semantic prototypes (stored in $\mathbf{Y}_s$) into the feature space as follows:

$$\min_{\mathbf{W}} \sum_{i=1}^{N_s} \|\mathbf{x}_i^{(s)} - \mathbf{W} \mathbf{y}_{l_i^{(s)}}^{(s)}\|_2^2 + \lambda \|\mathbf{W}\|_F^2. \tag{3}$$

When the best projection matrix $\mathbf{W}^*$ is learnt, we can project an unseen semantic prototype $\mathbf{y}_j^{(u)}$ from the test set into the feature space as $\hat{\mathbf{x}}_j^{(u)} = \mathbf{W}^* \mathbf{y}_j^{(u)}$. The nearest neighbor search is then performed in the feature space to predict the label of a test image $\mathbf{x}_i^{(u)}$:

$$l_i^{(u)} = \arg\min_j \|\mathbf{x}_i^{(u)} - \hat{\mathbf{x}}_j^{(u)}\|_2^2. \tag{4}$$

3.3   Model Formulation

Although the RPL model is shown to yield impressive results in the ZSL tasks [16,31,32,39], it does not address the projection domain shift problem. To overcome this limitation, our extreme RPL or ERPL model is formulated as a transductive learning model which exploits unlabeled unseen class data to adapt the projection toward the unseen class domain for ZSL.

Concretely, our ERPL model solves the following optimisation problem:

$$\min_{\mathbf{W}} \sum_{i=1}^{N_s} \|\mathbf{x}_i^{(s)} - \mathbf{W}\mathbf{y}_{l_i^{(s)}}^{(s)}\|_2^2 + \lambda\|\mathbf{W}\|_F^2 + \gamma \sum_{i=1}^{N_u} \min_j \|\mathbf{x}_i^{(u)} - \mathbf{W}\mathbf{y}_j^{(u)}\|_2^2, \tag{5}$$

where γ is a weighting coefficient that controls the importance of the first and third terms, which correspond to the losses on the seen and unseen class samples respectively.

Different from existing transductive ZSL models, our model focuses on learning the projection function, i.e., a single task. In contrast, most existing transductive ZSL models [11,14,21,28,36,43] have to solve two or more subtasks at once including projection learning, label prediction, and semantic embedding. This requires one or more intermediate variables to be introduced into the models, complicating the optimisation problem. We believe that focusing on the projection learning problem only with a simple formulation enables our ERPL model to better overcome the projection domain shift problem.

3.4   Optimization

Since the third term of the objective function in Eq. (5) is denoted as a sum of minimums, it becomes difficult to solve the optimisation problem in Eq. (5). In the following, we formulate our gradient-based solver. Note that the contentional alternating optimization algorithms (like k-means) have been employed for solving this type of min-min optimization problems in many existing transductive ZSL models [14,16,21,36,40]. However, our gradient-based solver is clearly shown to perform better than these contentional algorithms (see Tables 2 and 3). This is also the *main contribution* of this paper.

Given the projection matrix $\mathbf{W}^{(t)}$ at iteration t during model learning, we define $\mathbf{f}_i^{(t)} = [f_{i1}^{(t)}, ..., f_{iq}^{(t)}]^T$ for the test image $\mathbf{x}_i^{(u)}$ $(i = 1, ..., N_u)$, where

$f_{ij}^{(t)} = \|\mathbf{x}_i^{(u)} - \mathbf{W}^{(t)}\mathbf{y}_j^{(u)}\|_2^2$ $(j = 1, ..., q)$. For the minimum function $\min \mathbf{f}_i^{(t)}$, we define its gradient $\eta_i^{(t)} = [\eta_{i1}^{(t)}, ..., \eta_{iq}^{(t)}]^T$ with respect to $\mathbf{f}_i^{(t)}$ as follows:

$$
\eta_{ij}^{(t)} = \begin{cases} 1/n_i^{(t)}, & \text{if } f_{ij}^{(t)} = \min \mathbf{f}_i^{(t)} \\ 0, & \text{otherwise} \end{cases}, \tag{6}
$$

where $n_i^{(t)}$ is the number of $f_{ij}^{(t)}$ $(j = 1, ..., q)$ being equal to $\min \mathbf{f}_i^{(t)}$. Taking the Taylor expansion, we have:

$$
\min_j \|\mathbf{x}_i^{(u)} - \mathbf{W}^{(t+1)}\mathbf{y}_j^{(u)}\|_2^2 = \min \mathbf{f}_i^{(t+1)}
$$

$$
\approx \min \mathbf{f}_i^{(t)} + \eta_i^{(t)^T}(\mathbf{f}_i^{(t+1)} - \mathbf{f}_i^{(t)})
$$

$$
= (\min \mathbf{f}_i^{(t)} - \eta_i^{(t)^T}\mathbf{f}_i^{(t)}) + \eta_i^{(t)^T}\mathbf{f}_i^{(t+1)}
$$

$$
= \eta_i^{(t)^T}\mathbf{f}_i^{(t+1)} = \sum_{j=1}^{q} \eta_{ij}^{(t)}\|\mathbf{x}_i^{(u)} - \mathbf{W}^{(t+1)}\mathbf{y}_j^{(u)}\|_2^2. \tag{7}
$$

According to the above approximation, the objective function in Eq. (5) at iteration $t + 1$ can be estimated as:

$$
\mathcal{F}(\mathbf{W}^{(t+1)}) = \sum_{i=1}^{N_s} \|\mathbf{x}_i^{(s)} - \mathbf{W}^{(t+1)}\mathbf{y}_{l_i^{(s)}}^{(s)}\|_2^2 + \lambda\|\mathbf{W}^{(t+1)}\|_F^2
$$

$$
+ \gamma \sum_{i=1}^{N_u} \sum_{j=1}^{q} \eta_{ij}^{(t)}\|\mathbf{x}_i^{(u)} - \mathbf{W}^{(t+1)}\mathbf{y}_j^{(u)}\|_2^2. \tag{8}
$$

Let $\frac{\partial \mathcal{F}(\mathbf{W}^{(t+1)})}{\partial \mathbf{W}^{(t+1)}} = 0$, we obtain the following linear equation:

$$
\mathbf{W}^{(t+1)}\mathbf{A}^{(t)} = \mathbf{B}^{(t)}, \tag{9}
$$

$$
\mathbf{A}^{(t)} = \sum_{i=1}^{N_s} \mathbf{y}_{l_i^{(s)}}^{(s)}\mathbf{y}_{l_i^{(s)}}^{(s)^T} + \lambda I + \gamma \sum_{j=1}^{q} \sum_{i=1}^{N_u} \eta_{ij}^{(t)}\mathbf{y}_j^{(u)}\mathbf{y}_j^{(u)^T}, \tag{10}
$$

$$
\mathbf{B}^{(t)} = \sum_{i=1}^{N_s} \mathbf{x}_i^{(s)}\mathbf{y}_{l_i^{(s)}}^{(s)^T} + \gamma \sum_{j=1}^{q} \sum_{i=1}^{N_u} \eta_{ij}^{(t)}\mathbf{x}_i^{(u)}\mathbf{y}_j^{(u)^T}, \tag{11}
$$

which has a closed-form solution: $\mathbf{B}^{(t)}\mathbf{A}^{(t)^{-1}}$. Let $\alpha = \gamma/(1 + \gamma) \in (0, 1)$ and $\beta = \lambda/(1 + \gamma) \in (0, +\infty)$, we have:

$$
\widehat{\mathbf{A}}^{(t)} = (1 - \alpha) \sum_{i=1}^{N_s} \mathbf{y}_{l_i^{(s)}}^{(s)}\mathbf{y}_{l_i^{(s)}}^{(s)^T} + \beta I + \alpha \sum_{j=1}^{q} \sum_{i=1}^{N_u} \eta_{ij}^{(t)}\mathbf{y}_j^{(u)}\mathbf{y}_j^{(u)^T}, \tag{12}
$$

$$
\widehat{\mathbf{B}}^{(t)} = (1 - \alpha) \sum_{i=1}^{N_s} \mathbf{x}_i^{(s)}\mathbf{y}_{l_i^{(s)}}^{(s)^T} + \alpha \sum_{j=1}^{q} \sum_{i=1}^{N_u} \eta_{ij}^{(t)}\mathbf{x}_i^{(u)}\mathbf{y}_j^{(u)^T}. \tag{13}
$$

Algorithm 1. Extreme Reverse Projection Learning

Input: Training and test sets $\mathcal{D}_s, \mathcal{X}_u$
         Sematic prototypes $\mathbf{Y}_s, \mathbf{Y}_u$
         Parameter α
Output: $\mathbf{W}^*$
1. Initialize $t = 0$;
2. Initialize $\mathbf{W}^{(0)}$ with the RPL model [31];
while *a stopping criterion is not met* **do**
    3. With the learnt matrix $\mathbf{W}^{(t)}$, compute $\eta_{ij}^{(t)}$ with Eq. (15);
    4. Compute $\widehat{\mathbf{A}}^{(t)}$ and $\widehat{\mathbf{B}}^{(t)}$ with Eq. (12) & Eq. (13);
    5. Update $\mathbf{W}^{(t+1)}$ by solving Eq. (14);
    6. Update $t = t + 1$;
end
7. $\mathbf{W}^* = \mathbf{W}^{(t)}$.

In this paper, we empirically set $\beta = 0.01$ in all experiments. The linear equation in Eq. (9) is then reformulated as:

$$\mathbf{W}^{(t+1)}\widehat{\mathbf{A}}^{(t)} = \widehat{\mathbf{B}}^{(t)}. \tag{14}$$

Considering that the predicted unseen class data labels are inevitably noisy, we choose to estimate the number $n_i^{(t)}$ used in Eq. (6) under a looser condition and redefine the gradient as follows:

$$\eta_{ij}^{(t)} = \begin{cases} 1/n_i^{(t)}, & \text{if } \frac{f_{ij}^{(t)} - \min \mathbf{f}_i^{(t)}}{\min \mathbf{f}_i^{(t)}} < \epsilon, \\ 0, & \text{otherwise} \end{cases} \tag{15}$$

where $n_i^{(t)}$ is the number of $f_{ij}^{(t)}$ ($j = 1, ..., q$) satisfying $(f_{ij}^{(t)} - \min \mathbf{f}_i^{(t)})/\min \mathbf{f}_i^{(t)} < \epsilon$. In this paper, we empirically set the small threshold $\epsilon = 0.001$ in all experiments. More importantly, the above robust gradient-based method is shown to achieve about 2–4% performance improvements over the original method based on the gradient given by Eq. (6) (see Fig. 2). This provides evidence that our robust gradient-based method is indeed effective for ZSL.

The full ERPL algorithm is outlined in Algorithm 1. Note that any non-transductive ZSL model can be used to obtain the initial projection matrix $\mathbf{W}^{(0)}$. In this paper, we choose the RPL model [31] for this initialization. Once learned, given the optimal projection matrix $\mathbf{W}^*$ found by the proposed ERPL algorithm, we first project the semantic prototypes of unseen classes into the feature space, and then predict the label of a test image with Eq. (4).

3.5   Discussion

We first provide the time complexity analysis of the proposed ERPL algorithm. Specifically, the computation of $[\eta_{ij}^{(t)}]_{N_u \times q}$, $\widehat{\mathbf{A}}^{(t)}$, and $\widehat{\mathbf{B}}^{(t)}$ has a time

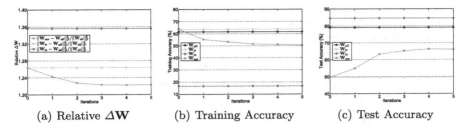

(a) Relative $\Delta\mathbf{W}$ (b) Training Accuracy   (c) Test Accuracy

Fig. 1. Convergence analysis under the standard ZSL setting. $\mathbf{W}_{all}$, $\mathbf{W}_{tr}$, and $\mathbf{W}_{te}$ are learnt by RPL [31] using the whole dataset (all are labelled), only using the training set, and only using the test set (but labelled), respectively. $\mathbf{W}_{our}$ is learnt by our ERPL using the test set (unlabeled) and the training set. Only one seen/unseen split (150/50) of the CUB-200-2011 Birds (CUB) dataset [35] is used here.

complexity of $O(qN_u)$, $O(k^2 N_s + k^2 N_u)$, and $O(dkN_s + dkN_u)$, respectively. Here, the sparsity of $[\eta_{ij}^{(t)}]$ is used to reduce the cost of computing $\widehat{\mathbf{A}}^{(t)}$ and $\widehat{\mathbf{B}}^{(t)}$. Moreover, since $\widehat{\mathbf{A}}^{(t)} \in \mathbb{R}^{k \times k}$ and $\widehat{\mathbf{B}}^{(t)} \in \mathbb{R}^{d \times k}$, solving Eq. (14) has a time complexity of $O(dk^2)$. To sum up, the time complexity of one iteration is $O(qN_u + (d+k)k(N_s + N_u) + dk^2)$ $(d, k, q \ll (N_s + N_u))$. Given that the proposed algorithm is shown to converge very quickly (<10 iterations) in our experiments, it thus has a linear time complexity with respect to the data size.

We further give the converge analysis of the proposed ERPL algorithm. To this end, we define three baseline projection matrices based on the RPL model [31]: (1) $\mathbf{W}_{all}$ – learnt by RPL using the whole dataset (all are labelled); (2) $\mathbf{W}_{tr}$ – learnt by RPL only using the training set; (3) $\mathbf{W}_{te}$ – learnt by RPL only using the test set (but labelled). Let $\mathbf{W}_{our}$ be learnt by our ERPL using the test set (unlabeled) and the training set. Firstly, we can directly compare $\mathbf{W}_{our}$, $\mathbf{W}_{tr}$, and $\mathbf{W}_{te}$ to $\mathbf{W}_{all}$ by computing the matrix distances among these matrices. Note that $\mathbf{W}_{all}$ is considered to be the best possible projection matrix (upper bound). Secondly, since label prediction can be performed on the test set (or training set) with Eq. (4) given a projection matrix, we compute the test accuracy (or training accuracy) to measure how good the projection matrix is in ZSL.

The results of convergence analysis for the standard ZSL setting are shown in Fig. 1. Here, one seen/unseen split (150/50) of the CUB-200-2011 Birds (CUB) dataset [35] is used for transductive ZSL, where the semantic space is denoted with the attributes and the visual representation is extracted by ImageNet pre-trained GoogLeNet. We have the following observations: (1) The proposed ERPL algorithm converges very quickly (≤ 5 iterations). (2) According to Fig. 1(a), $\mathbf{W}_{our}$ gets closer to $\mathbf{W}_{all}$ with more iterations, and it is the closest to $\mathbf{W}_{all}$ at convergence among $\mathbf{W}_{our}$, $\mathbf{W}_{tr}$, and $\mathbf{W}_{te}$. (3) For the training set (see Fig. 1(b)), as expected, $\mathbf{W}_{tr}$ performs the best, $\mathbf{W}_{all}$ the second best, and $\mathbf{W}_{te}$ the worst. As for $\mathbf{W}_{our}$, it gradually adapts to the test/unseen class domain, thus its performance gets worse on the training domain. (4) For the test set (see Fig. 1(c)),

Table 1. Five benchmark datasets used for performance evaluation. Notations: 'SS' – semantic space, 'SS-D' – the dimension of semantic space, 'A' – attribute, and 'W' – word vector. The two splits of SUN are separated by '|'.

Dataset	# images	SS	SS-D	# seen/unseen
AwA [19]	30,475	A	85	40/10
CUB [35]	11,788	A	312	150/50
aPY [9]	15,339	A	64	20/12
SUN [24]	14,340	A	102	707/10|645/72
ImNet [30]	218,000	W	1,000	1,000/360

again as expected, $\mathbf{W}_{te}$ performs the best, $\mathbf{W}_{all}$ the second best, and $\mathbf{W}_{tr}$ the worst. Importantly, $\mathbf{W}_{our}$ improves and approaches $\mathbf{W}_{all}$ with iterations. (5) Overall, the proposed ERPL algorithm tends to learn an *extremely generalizable* projection matrix that is close to $\mathbf{W}_{all}$ not only on the training set but also on the test set, i.e., it can alleviate the projection domain shift problem in ZSL by not overfitting to the training domain.

4   Experiments

4.1   Datasets and Settings

Datasets. Five widely-used benchmark datasets are selected. Four of them are of medium-size: Animals with Attributes (AwA) [19], CUB-200-2011 Birds (CUB) [35], aPascal&Yahoo (aPY) [9], and SUN Attribute (SUN) [24]. One large-scale dataset is ILSVRC2012/ILSVRC2010 [30] (ImNet), where the 1,000 classes of ILSVRC2012 are used as seen classes and 360 classes of ILSVRC2010 (not included in ILSVRC2012) are used as unseen classes, as in [12]. This ImNet dataset split thus gives a pure ZSL setting. The details of these benchmark datasets are given in Table 1.

Semantic Spaces. We employ attributes as the semantic space for the four medium-scale datasets, all of which provide the attribute annotations. The semantic representation based on word vectors is used for the large-scale ImNet. We train a skip-gram text model on a corpus of 4.6M Wikipedia documents to obtain the word2vec word vectors.

Visual Features. All recent ZSL models make use of the visual features extracted by deep CNN models. In this paper, we extract the GoogLeNet features which are the 1,024-dimensional activations of the final pooling layer as in [3,17]. We further reduce the dimension of CNN feature vectors to 600 by principle component analysis (PCA).

Evaluation Metrics. For the four medium-scale datasets, we compute the multi-way classification accuracy as in many previous works on ZSL. For the

Table 2. Comparative accuracies (%) on the four medium-scale datasets under the standard ZSL setting. For the SUN dataset, the results are obtained for the 707/10 and 645/72 splits respectively, separated by '|'.

Model	SS	Trans.?	AwA	CUB	aPY	SUN
DAP [19]	A	N	60.1	–	38.2	72.0|44.5
ESZSL [29]	A	N	75.3	48.7	24.3	82.1|18.7
SSE [41]	A	N	76.3	30.4	46.2	82.5|–
SJE [3]	A+W	N	73.9	51.7	–	–|56.1
JLSE [42]	A	N	80.5	42.1	50.4	83.8|–
SynC^{struct} [6]	A	N	72.9	54.7	–	–|62.7
MLZSC [4]	A	N	77.3	43.3	53.2	84.4|–
DS-SJE [27]	W	N	–	56.8	–	–|–
DeViSE [10]	A	N	56.7	33.5	–	–|–
RPL [31]	A	N	80.4	52.4	48.8	84.5|–
SAE [17]	A	N	84.7	61.4	55.4	91.5|65.2
AMP [13]	A+W	Y	66.0	–	–	–|–
DSRL [39]	A	Y	87.2	57.1	56.3	85.4|–
UDA [16]	A+W	Y	75.6	40.6	–	–|–
SMS [14]	A	Y	78.5	–	39.0	82.0|–
TSTD [40]	A	Y	90.3	58.2	–	–|–
Li [21]	A	Y	40.1	–	24.7	–|–
SSZSL [32]	A	Y	88.6	58.8	49.9	86.2|–
BiDiLEL [36]	A	Y	95.0	62.8	–	–|–
SP-ZSR [43]	A	Y	92.1	55.3	69.7	89.5|–
ERPL (ours)	A	Y	**96.2**	**63.5**	**77.4**	**93.0|67.6**

large-scale ImNet dataset, the flat hit@5 classification accuracy is computed as in [10,17], where hit@5 means that a test image is classified to a 'correct label' if it is among the top 5 labels.

Parameter Settings. Our ERPL model has only one free parameter to tune: $\alpha \in (0,1)$ (see Eqs. (12) and (13)). Note that the other parameters of our model are empirically fixed as: $\beta = 0.01$ (see Eq. (12)), $\epsilon = 0.001$ (see Eq. (15)). As in [17], the parameter α is selected by class-wise cross-validation using the training data. Moreover, only the CUB and SUN datasets have multiple seen/unseen splits. We take the same 4 splits used in [3] for CUB and the same 10 splits used in [6] for SUN, and report the average accuracies.

Compared Methods. A wide range of existing ZSL models are selected for comparison. Under each ZSL setting, we focus on the recent and representative ZSL models that have achieved the state-of-the-art results.

Table 3. Comparative accuracies (%) on the large-scale ImNet dataset under the standard ZSL setting. The notations are exactly the same as in Table 2.

Model	SS	Trans.?	hit@5
DeViSE [10]	W	N	12.8
AMP [13]	W	Y	13.1
ConSE [23]	W	N	15.5
SS-Voc [12]	W	Y	16.8
SAE [17]	W	N	27.2
ERPL (ours)	W	Y	**30.6**

Table 4. Comparative accuracies (%) on the four medium-scale datasets under the pure ZSL setting (as in [37,44]). For the SUN dataset, only the 645/72 split is used.

Model	AwA	CUB	aPY	SUN
DAP [19]	44.1	40.0	33.8	39.9
ConSE [23]	45.6	34.3	26.9	38.8
CMT [34]	39.5	34.6	28.0	39.9
SSE [41]	60.1	43.9	34.0	51.5
ALE [2]	59.9	54.9	39.7	58.1
DeViSE [10]	54.2	52.0	39.8	56.5
SJE [3]	65.6	53.9	32.9	53.7
ESZSL [29]	58.2	53.9	38.3	54.5
SynC^{struct} [6]	54.0	55.6	23.9	56.3
CLN+KRR [44]	68.2	58.1	44.8	60.0
ERPL (ours)	**91.8**	**62.1**	**51.5**	**65.1**

4.2 Results Under Standard ZSL Setting

The comparative results under the standard ZSL setting are shown in Tables 2 and 3. Here, both transductive and non-transductive ZSL models are included. We can make the following observations: (1) Our ERPL model yields the best results on all five datasets, due to its ability of leaning extremely generalizable projection for ZSL. (2) On the four medium-scale datasets (see Table 2), the gains obtained by our ERPL model over the strongest competitor range from 0.7% to 7.7%. This is really impressive considering that most of the compared models employ far complicated nonlinear models and some of them even combine two or more semantic spaces. (3) On the large-scale ImNet dataset (see Table 3), our model yields a 3.4% improvement over the state-of-the-art SAE [17]. This suggests that our model scales up well to large-scale ZSL problems.

Table 5. Comparative results (%) of generalized ZSL on the AwA and CUB datasets with the same setting used in [7].

Model	AwA			CUB		
	acc_s	acc_u	HM	acc_s	acc_u	HM
DAP [19]	77.9	2.4	4.7	55.1	4.0	7.5
IAP [19]	76.8	1.7	3.3	69.4	1.0	2.0
ConSE [23]	75.9	9.5	16.9	69.9	1.8	3.5
SynC^{struct} [6]	81.0	0.4	0.8	**72.0**	13.2	22.3
APD [26]	43.2	**61.7**	50.8	23.4	**39.9**	29.5
Bucher [5]	**81.3**	32.3	46.2	**72.0**	26.9	39.2
SAE [17]	67.6	43.3	52.8	36.1	28.0	31.5
ERPL (ours)	66.4	50.5	**57.4**	43.7	37.2	**40.2**

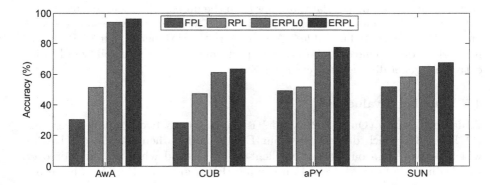

Fig. 2. Effectiveness evaluation for the main components of our ERPL model under the standard ZSL setting. Only the 645/72 split is used for SUN.

4.3 Results Under Pure ZSL Setting

Following the same 'pure' ZSL setting as in [37,44], we remove the overlapped ImageNet ILSVRC2012 1K classes from the test set of unseen classes for the four medium-scale datasets. The results in Table 4 show that, as expected, under this more strict and challenging ZSL setting, all models suffer from performance degradation. However, the performance of our ERPL model drops the least among all the ZSL models. As a result, the gap to the second-best compared method becomes bigger for each of the four datasets.

4.4 Results Under Generalized ZSL Setting

The generalized ZSL setting has gained popularity recently, under which the test set contains data samples from both the seen and unseen classes. We follow the same setting of [7]. Specifically, we hold out 20% of the data samples from the

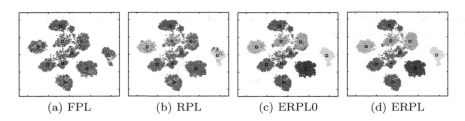

(a) FPL   (b) RPL   (c) ERPL0   (d) ERPL

Fig. 3. The tSNE visualization of the visual features of test images on the AwA dataset under the standard ZSL setting. The predicted unseen class labels of test images are obtained by FPL, RPL, ERPL0, and ERPL, respectively (see the ablation study).

seen classes and mix them with the data samples from the unseen classes. The evaluation metrics are: (1) acc_s – the accuracy of classifying the data samples from the seen classes to all the classes (both seen and unseen); (2) acc_u – the accuracy of classifying the data samples from the unseen classes to all the classes; (3) HM – the harmonic mean of acc_s and acc_u. The comparative results on AwA and CUB are presented in Table 5. According to the HM metric, our ERPL model is clearly shown to have the best overall performance due to its ability of leaning extremely generalizable projection for ZSL.

4.5 Further Evaluations

Ablation Study. Our ERPL model is compared to two simplified versions: (1) RPL [31] which derives from our ERPL model when $\alpha = 0$; (2) ERPL0 which derives from our robust gradient-based method when $\epsilon = 0$. Moreover, we also include forward projection leaning (FPL) given by Eq. (1). The results are presented in Fig. 2. We make three observations: (1) The transductive learning induced in our model achieves about 10–40% improvements (see ERPL0 vs. RPL), i.e., the domain shift problem in ZSL has been effectively tackled. (2) Our robust gradient-based method yields about 2–3% gains (see ERPL vs. ERPL0), validating its effectiveness. (3) The 3–20% improvements achieved by RPL over FPL show that the hubness problem in ZSL has been alleviated.

Qualitative Results. We provide qualitative results to show why adding more components into our model benefits ZSL. Figure 3 presents the tSNE visualization of the visual features of test images. The predicted unseen class labels (marked with different colors) of test images are obtained by FPL, RPL, ERPL0, and ERPL, respectively. It can be clearly seen that the test images are distributed more compactly and more centered around the class prototype when more components are added, resulting in better recognition results.

5   Conclusion

We have proposed a novel ERPL model for zero-shot learning which addresses both the projection domain shift and hubness. A robust and efficient gradient-based algorithm has also been developed for model optimization, followed by detailed algorithm analysis. We presented extensive experiments on five bench-mark datasets and demonstrated that the proposed ERPL model yields state-of-the-art results under all the three ZSL settings. The performance gain over the alternative ZSL models is particularly big under the more challenging yet more realistic pure and generalized ZSL settings. In the ongoing research, we will apply the proposed ERPL model to other ZSL-related problems such as ZSL with superclasses and large-scale few-shot learning.

Acknowledgements. This work was supported by National Natural Science Foundation of China (61573363), and the Fundamental Research Funds for the Central Universities and the Research Funds of Renmin University of China (15XNLQ01).

References

1. Akata, Z., Perronnin, F., Harchaoui, Z., Schmid, C.: Good practice in large-scale learning for image classification. TPAMI **36**(3), 507–520 (2014)
2. Akata, Z., Perronnin, F., Harchaoui, Z., Schmid, C.: Label-embedding for image classification. TPAMI **38**(7), 1425–1438 (2016)
3. Akata, Z., Reed, S., Walter, D., Lee, H., Schiele, B.: Evaluation of output embeddings for fine-grained image classification. In: CVPR, pp. 2927–2936 (2015)
4. Bucher, M., Herbin, S., Jurie, F.: Improving semantic embedding consistency by metric learning for zero-shot classification. In: Leibe, B., Matas, J., Sebe, N., Welling, M. (eds.) ECCV 2016. LNCS, vol. 9909, pp. 730–746. Springer, Cham (2016). https://doi.org/10.1007/978-3-319-46454-1_44
5. Bucher, M., Herbin, S., Jurie, F.: Generating visual representations for zero-shot classification. In: ICCV Workshops: Transferring and Adapting Source Knowledge in Computer Vision, pp. 2666–2673 (2017)
6. Changpinyo, S., Chao, W.L., Gong, B., Sha, F.: Synthesized classifiers for zero-shot learning. In: CVPR, pp. 5327–5336 (2016)
7. Chao, W.-L., Changpinyo, S., Gong, B., Sha, F.: An empirical study and analysis of generalized zero-shot learning for object recognition in the wild. In: Leibe, B., Matas, J., Sebe, N., Welling, M. (eds.) ECCV 2016. LNCS, vol. 9906, pp. 52–68. Springer, Cham (2016). https://doi.org/10.1007/978-3-319-46475-6_4
8. Donahue, J., et al.: DeCAF: a deep convolutional activation feature for generic visual recognition. In: ICML, pp. 647–655 (2014)
9. Farhadi, A., Endres, I., Hoiem, D., Forsyth, D.: Describing objects by their attributes. In: CVPR, pp. 1778–1785 (2009)
10. Frome, A., et al.: DeViSE: a deep visual-semantic embedding model. In: NIPS, pp. 2121–2129 (2013)
11. Fu, Y., Hospedales, T.M., Xiang, T., Gong, S.: Transductive multi-view zero-shot learning. TPAMI **37**(11), 2332–2345 (2015)
12. Fu, Y., Sigal, L.: Semi-supervised vocabulary-informed learning. In: CVPR, pp. 5337–5346 (2016)
13. Fu, Z., Xiang, T., Kodirov, E., Gong, S.: Zero-shot object recognition by semantic manifold distance. In: CVPR, pp. 2635–2644 (2015)
14. Guo, Y., Ding, G., Jin, X., Wang, J.: Transductive zero-shot recognition via shared model space learning. In: AAAI, pp. 3494–3500 (2016)
15. Kankuekul, P., Kawewong, A., Tangruamsub, S., Hasegawa, O.: Online incremental attribute-based zero-shot learning. In: CVPR, pp. 3657–3664 (2012)

16. Kodirov, E., Xiang, T., Fu, Z., Gong, S.: Unsupervised domain adaptation for zero-shot learning. In: ICCV, pp. 2452–2460 (2015)
17. Kodirov, E., Xiang, T., Gong, S.: Semantic autoencoder for zero-shot learning. In: CVPR, pp. 3174–3183 (2017)
18. Krizhevsky, A., Sutskever, I., Hinton, G.E.: ImageNet classification with deep convolutional neural networks. In: NIPS, pp. 1097–1105 (2012)
19. Lampert, C.H., Nickisch, H., Harmeling, S.: Attribute-based classification for zero-shot visual object categorization. TPAMI **36**(3), 453–465 (2014)
20. Lei Ba, J., Swersky, K., Fidler, S., et al.: Predicting deep zero-shot convolutional neural networks using textual descriptions. In: ICCV, pp. 4247–4255 (2015)
21. Li, X., Guo, Y., Schuurmans, D.: Semi-supervised zero-shot classification with label representation learning. In: ICCV, pp. 4211–4219 (2015)
22. Lu, Y.: Unsupervised learning on neural network outputs: with application in zero-shot learning. arXiv preprint arXiv:1506.00990 (2015)
23. Norouzi, M., et al.: Zero-shot learning by convex combination of semantic embeddings. In: ICLR (2014)
24. Patterson, G., Xu, C., Su, H., Hays, J.: The sun attribute database: beyond categories for deeper scene understanding. IJCV **108**(1), 59–81 (2014)
25. Radovanović, M., Nanopoulos, A., Ivanović, M.: Hubs in space: popular nearest neighbors in high-dimensional data. JMLR **11**(9), 2487–2531 (2010)
26. Rahman, S., Khan, S.H., Porikli, F.: A unified approach for conventional zero-shot, generalized zero-shot and few-shot learning. arXiv preprint arXiv:1706.08653 (2017)
27. Reed, S., Akata, Z., Lee, H., Schiele, B.: Learning deep representations of fine-grained visual descriptions. In: CVPR, pp. 49–58 (2016)
28. Rohrbach, M., Ebert, S., Schiele, B.: Transfer learning in a transductive setting. In: NIPS, pp. 46–54 (2013)
29. Romera-Paredes, B., Torr, P.H.S.: An embarrassingly simple approach to zero-shot learning. In: ICML, pp. 2152–2161 (2015)
30. Russakovsky, O., et al.: ImageNet large scale visual recognition challenge. IJCV **115**(3), 211–252 (2015)
31. Shigeto, Y., Suzuki, I., Hara, K., Shimbo, M., Matsumoto, Y.: Ridge regression, hubness, and zero-shot learning. In: Appice, A., Rodrigues, P.P., Santos Costa, V., Soares, C., Gama, J., Jorge, A. (eds.) ECML PKDD 2015. LNCS (LNAI), vol. 9284, pp. 135–151. Springer, Cham (2015). https://doi.org/10.1007/978-3-319-23528-8_9
32. Shojaee, S.M., Baghshah, M.S.: Semi-supervised zero-shot learning by a clustering-based approach. arXiv preprint arXiv:1605.09016 (2016)
33. Simonyan, K., Zisserman, A.: Very deep convolutional networks for large-scale image recognition. arXiv preprint arXiv:1409.1556 (2014)
34. Socher, R., Ganjoo, M., Manning, C.D., Ng, A.: Zero-shot learning through cross-modal transfer. In: NIPS, pp. 935–943 (2013)
35. Wah, C., Branson, S., Welinder, P., Perona, P., Belongie, S.: The caltech-UCSD birds-200-2011 dataset. Technical report CNS-TR-2011-001, California Institute of Technology (2011)
36. Wang, Q., Chen, K.: Zero-shot visual recognition via bidirectional latent embedding. IJCV **124**(3), 356–383 (2017)
37. Xian, Y., Schiele, B., Akata, Z.: Zero-shot learning - the good, the bad and the ugly. In: CVPR, pp. 4582–4591 (2017)
38. Xu, X., Hospedales, T., Gong, S.: Transductive zero-shot action recognition by word-vector embedding. IJCV **123**(3), 309–333 (2017)

39. Ye, M., Guo, Y.: Zero-shot classification with discriminative semantic representation learning. In: CVPR, pp. 7140–7148 (2017)
40. Yu, Y., et al.: Transductive zero-shot learning with a self-training dictionary approach. arXiv preprint arXiv:1703.08893 (2017)
41. Zhang, Z., Saligrama, V.: Zero-shot learning via semantic similarity embedding. In: ICCV, pp. 4166–4174 (2015)
42. Zhang, Z., Saligrama, V.: Zero-shot learning via joint latent similarity embedding. In: CVPR, pp. 6034–6042 (2016)
43. Zhang, Z., Saligrama, V.: Zero-shot recognition via structured prediction. In: Leibe, B., Matas, J., Sebe, N., Welling, M. (eds.) ECCV 2016. LNCS, vol. 9911, pp. 533–548. Springer, Cham (2016). https://doi.org/10.1007/978-3-319-46478-7_33
44. Zhao, A., Ding, M., Guan, J., Lu, Z., Xiang, T., Wen, J.: Domain-invariant projection learning for zero-shot recognition. In: NIPS (2018)

A Defect Inspection Method for Machine Vision Using Defect Probability Image with Deep Convolutional Neural Network

Chanhee Jang, Sangyun Yun, Hyejin Hwang, Hyunmin Shin,
SeongSoo Kim, and Yangsub Park[✉]

Advanced Technology Inc., 112, Gaetbeol-ro, Yeonsu-gu, Incheon, South Korea
{chan,syyun,hjhwang,hmshin,sskim,yspark}@ati2000.co.kr

Abstract. Deep learning is replacing many traditional machine vision techniques. However, defect inspection systems still rely on traditional methods due to difficulties in obtaining training data and the absence of color images. Thus, overall performance heavily depends on individual human skill in tuning hundreds of parameters. This paper presents a defect inspection technique using a defect probability image (DPI) and a deep convolutional neural network (CNN). DPIs are the estimated probability of a defect in given image and can be obtained from traditional inspection techniques. The DPI and gray image are stacked as input to the CNN. Performance was compared with a conventional CNN model using RGB or grayscale images, and ViDi, an artificial intelligence software for industry. The proposed method outperforms the other methods, works well on small dataset, and removes the requirement for human skill.

Keywords: Machine vision · Defect inspection ·
Defect probability image · Convolutional Neural Network

1 Introduction

Deep learning has gained tremendous popularity in recent years, not only in academic fields, but also in industrial applications, and there are many deep learning application products in the commercial market. Many traditional technologies in various fields have been either replaced or integrated with deep learning systems to take advantage of its powerful performance. The machine vision industry, is also rapidly adopting deep learning systems in various fields from surface detection [3,12] to traffic flow control [10].

However, defect inspection systems have seen little impact from deep learning developments. Most studies and industrial applications are limited to automatic defect classification or wafer map defect classification [11], where the training images are generated after conventional detection processes are completed. There are two main reasons for this lack of adoption.

© Springer Nature Switzerland AG 2019
C. V. Jawahar et al. (Eds.): ACCV 2018, LNCS 11361, pp. 142–154, 2019.
https://doi.org/10.1007/978-3-030-20887-5_9

1. Difficulty in gathering training data. The power of deep learning depends on the quantity of data available to train on. Defects are abnormal events from the perspective of an inspection system, with higher quality items implying less defects will be found. The limited number of expect defects are insufficient for conventional deep learning models, causing data imbalance [14], where the number of defect images is much smaller than good images.
2. Most state-of-the-art deep neural network models require 3 channel RGB images, but single channel gray images are more commonly used for defect inspection system to enable higher throughput.

Consequently, conventional defect inspection systems use hundreds of algorithms and parameters to amplify the defect signal in an image while reducing background noise. Although conventional techniques have reasonable performance, they also have several underlying limitations: they are very environment sensitive, leading to significant false positive problems; and overall performance depends strongly on the knowledge, experience, and judgment of the operator.

Therefore, this study proposes a defect inspection system that

- detects the defect without requiring tuning hundreds of parameters,
- requires only a small training dataset, and
- requires no RGB images.

The proposed method exploits conventional defect inspection techniques to estimate defect probability in each image, presented as a defect probability image (DPI). The DPI and original gray image are then input to a deep convolutional neural network (CNN). The proposed approach was validated with real-world data, and shown to be effective for conventional defect inspection systems.

The remainder of this paper is organized as follows. Section 2 discusses the situation and challenges the machine vision industry faces, particularly for defect inspection systems. Section 3 discusses related studies, and Sect. 4 describes the proposed defect detection method. Section 5 presents experimental results using real-world data, and Sect. 6 summarizes and concludes the paper.

2   Machine Vision System

Machine vision is an application of computer vision, and is widely used in many industrial fields. The most significant utilization has been for factory automation, such as assembly, identification, visual inspection, quality control, etc. Machine vision systems are particularly promoted as a human substitute due to reliable performance, avoiding human mistakes due to tiredness, distraction, unskilled, etc. However, this does not mean that humans are no longer required in these systems. Rather, they require more demanding tasks from skilled humans, since the system performance heavily relies on their experience and understanding of the product/process. For example, in wafer/PCB defect inspection, the operator must be capable of handling all parameters of the system depending on the type of product, inspection algorithms, or environmental changes.

Wafer/PCB defect inspection systems at micro level ($\sim$1 μm) can be categorized into image acquisition, image processing, and decision rules.

1. Image acquisition. Aside from image resolution, camera speed is very important because directly relates to system throughput. Therefore, many inspection systems use single channel cameras, since they are almost three times faster than 3 channel color cameras, e.g. Piranha camera from Teledyne DALSA: Piranha XL XDR 16k (grayscale), 125 kHz, Piranha XL 16k (color), 40 kHz. Where color images are required, many systems use additional color camera to image only specific regions of interest.
2. Image processing. Many calibrations and compensations, such as intrinsic camera parameters, lens distortion, rotation compensation, etc., are applied to the obtained images. The images are subsequently processed to suppress irrelevant background while highlighting defective regions using various techniques, including image subtraction, cross correlation, filtering, etc. Finally, defect features are extracted by decision rules.
3. Decision rules. It is a set of parameters that determines a defect by extracted features. It includes minimum pixel value, maximum defect size, defect length, region of interest, which can be either set by the operator or design layout file, i.e., graphic data system (GDS), etc. This process determines the inspection system's detection ability, and is highly dependent on the operator's knowledge and experience. Although same image processing techniques (step 2) can be applied regardless of the inspecting item, the decision rules differ for every different item. This is the most time consuming process, particularly when developing a new inspection system, because few samples are available prior to mass production, but the parameters must be valid for all possible defects that may occur during production.

3   Related Work

Cross correlation is a straightforward and promising method in machine vision, and widely used for traditional defect detection. However, the method is very sensitive to environmental changes, which often result in false positives, and carries a heavy calculation burden. Hence, many studies have investigated ways to make cross correlation faster and more stable. Wang et al. proposed calculating correlation coefficients only on those regions with some level of difference in the subtracted image to reduce false positives [16]. Bai et al. proposed a phase-only Fourier transform to effectively remove periodic signals [4]. Tsai et al. proposed a normalized cross correlation in a smoothed color image to alleviate false positives [15]. However, although such systems reduce false positive detections, human skill remains important to devise the decision rules even with current best practice algorithms.

Some leading systems exploit design information or design layout for wafer defect detection and classification [2,8,9]. Examples of design information include, but not limited to, length, width, structural details, material properties, acceptable dimension tolerances, etc. Utilizing design information allows defects to be classified by type as well as severity. However, such product information is rarely available beforehand.

In terms of artificial intelligence (AI) applications for inspection systems, Haddad et al. proposed a multiple feature, sparse based approach to extract various features from the gray image, and generate sparse code to develop a defect classifier [7]. The study showed deep understanding of industrial challenges to apply AI for defect inspection, and is a good example of implementing AI with gray scale images and small database. However, study data was quite simple, defects were assumed to be detected using simple image processing techniques, and the main focus was on defect classification.

Many studies have investigated defect classification, with or without design information. However, to the best of our knowledge, no previous studies integrated machine learning with conventional inspection techniques without design information or sufficient dataset.

4   Method

Defect inspection systems now commonly include AI aspects, but their main purpose is generally to either classify the defects into defined categories or reduce false positives by judging each detected result. Thus, the AI systems are only applied once the traditional inspection sequence completes. The images employed are usual taken after inspection, and do not contain any information about the defect other than RGB or grayscale pixel values. Various defect inspection system limitations make it difficult to adopt conventional CNN classifiers, such as most best practice CNN classifiers require RGB images, whereas most inspection systems use grayscale images; training datasets are normally very small; etc. To resolve these problems, we propose integrating conventional inspection techniques with a CNN model. To the best of our knowledge, this is the first work to consider this integration for defect detection.

4.1   Defect Probability Image

Figure 1 shows a typical conventional inspection process, and how DPIs can be obtained within the process. To detect a defect, the acquired target image (Fig. 1a) must be compared with a defect-free image. Recent semiconductor manufacturing is very advanced that adjacent dies have almost zero deviation, i.e., they look identical, as shown in Fig. 1b. Therefore, a golden image, i.e., a reference image without defects, can be obtained by taking mean or median pixel values from four neighboring images, as shown in Fig. 1c. Although a defect may exist in one of these neighboring images, it will have no significant effect to the golden image after mean or median filtering. The difference image can then be obtained by subtracting the target image from the golden image, as shown in Fig. 1d. Appropriate subsequent filtering can be applied to suppress the background noise and highlight any defects, as shown in Fig. 1e. Conventional techniques determine defect presence in a noise-filtered image (Fig. 1e) from decision rules. However, the DPI is not noise filtered to avoid human judgment required for this filtering process. Note that DPI is a difference image in this specific example (Fig. 1), but it may not always be the case.

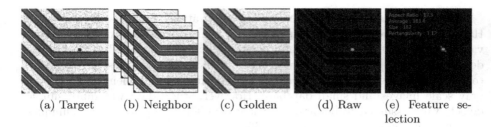

(a) Target   (b) Neighbor   (c) Golden   (d) Raw   (e) Feature se-
                                                           lection

Fig. 1. Wafer defect detection

We define the difference image as $Raw(x, y)$ where (x, y) is an image coordinate, and the defect probability can be expressed as

$$G(x, y) = \frac{1}{n} \sum_{i=1}^{n} N_i(x, y), \tag{1}$$

$$Raw(x, y) = G(x, y) - T(x, y), \tag{2}$$

and

$$p(Defect|Raw(x, y)) = \frac{1}{R(x, y)} \times \left| \frac{1}{n} \sum_{i=1}^{n} N_i(x, y) - T(x, y) \right| \tag{3}$$

$$= \frac{1}{R(x, y)} \times Raw(x, y). \tag{4}$$

The underlying concept is that larger differences between corresponding target and golden image pixels imply higher defect probability. We assume that observations from the sample follow a Gaussian distribution θ, with $N(\mu_\theta, \sigma_\theta)$, and are independent and identically distributed. $N_i(x, y)$ is the set of neighboring images to the target image, $T(x, y)$. Having no design layout as a golden image, we predict it using maximum likelihood estimation (MLE) on neighboring images. Since maximizing MLE is the same as minimizing mean squared error for a Gaussian distribution, we use the mean as the estimated golden image pixel value, $G(x, y)$. Differences in pixel values are then derived by subtracting the target image from the golden image as in Eq. (2). $R(x, y)$ is a scaling factor, e.g. $R(x, y) = 255$ for grayscale images, but may differ for each algorithm. Since larger $Raw(x, y)$ means increased difference between the golden and target images, the defect probability $p(Defect|Raw(x, y))$ also increases because $R(x, y)$ is fixed.

For IC substrate in Fig. 2a, aforementioned technique using a golden image is not adequate because of high variance of solder balls on each chip. Conventional technique assumes that defects have larger differences from adjacent pixels and often defines region of interest (ROI) for inspection as shown in Fig. 2b.

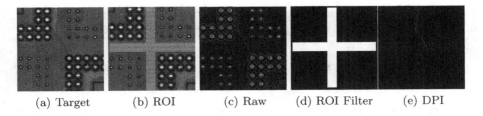

(a) Target   (b) ROI   (c) Raw   (d) ROI Filter   (e) DPI

Fig. 2. Defect probability image (DPI) generation process for a cracked IC substrate

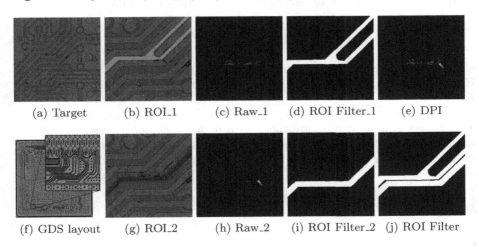

(a) Target   (b) ROI_1   (c) Raw_1   (d) ROI Filter_1   (e) DPI

(f) GDS layout   (g) ROI_2   (h) Raw_2   (i) ROI Filter_2   (j) ROI Filter

Fig. 3. Defect probability image (DPI) generation process using design layout file containing two regions of interest

To obtain raw image in Fig. 2c, high-pass filter, i.e., Laplacian of Gaussian (LoG), is applied to target image as

$$LoG(x, y) = -\frac{1}{\pi\sigma^4} \left[1 - \frac{x^2 + y^2}{2\sigma^2} \right] e^{-\frac{x^2+y^2}{2\sigma^2}}, \tag{5}$$

$$Raw_i(x, y) = LoG(x, y) * T(x, y). \tag{6}$$

ROI filter (Fig. 2d) flags if a given pixel in raw image (Fig. 2c) is in the ROI (1 or 0) and DPI is obtained as shown in Fig. 2e. This particular example shows ROI is effective to filter out irrelevant background, consequently generating better DPI.

Products with complex pattern such as Fig. 3a require more than one ROI. Such ROIs can be easily obtained by design layout file, such as graphic data system (GDS) as shown in Fig. 3f or operator can designate them. Figure 3b is an example ROI for dark region and Fig. 3g for bright region. Different inspection conditions are used to find defects for each ROI, where defects are assumed to have larger difference from average pixel value within corresponding ROI. For multiple ROIs inspection technique, raw image can be derived as

$$Raw_i(x,y) = \left| T(x,y) - \frac{1}{n} \sum_{j=1}^{n} T(x,y)_j \right|, \quad \forall(x,y)_j \in ROI_i. \tag{7}$$

After such operation, raw images are obtained as in Figs. 3c and h, and along with their corresponding ROI filters, Figs. 3d and i, DPI is obtained (Fig. 3e). Its defect probability can be calculated in a generalized form as

$$P_i(x,y) = \frac{1}{R_i(x,y)} \times Raw_i(x,y), \tag{8}$$

where $P_i(x,y)$ refers to the generalized definition of defect probability with $R_i(x,y)$ and $Raw_i(x,y)$ being different for the number of techniques and their ROI_i.

Although defect inspection techniques differ from case to case, the underlying logic is very simple: defects are those with higher difference/deviation and the greater the pixel value in DPI, the higher the possibility of defect. Regardless of algorithm type or number of ROI filter, DPI can be obtained using generalized equation as

$$DPI(x,y) = \frac{\sum_{i=1}^{n} Raw_i(x,y) \times F_i(x,y) \times w_i}{\sum_{i=1}^{N} F_i(x,y)}, \tag{9}$$

where

$$F_i(x,y) = \begin{cases} 1 & if(x,y) \in ROIFilter_i, \\ 0 & else. \end{cases} ; \tag{10}$$

DPI indicates the probability of defect at (x, y); N is total inspection number defined by number of techniques and their ROIs; $Raw_i(x,y)$ is the ith difference/deviation value calculated from $T_i(x,y)$; $T_i(x,y)$ is target image; $F_i(x,y)$ is i^{th} ROI binary function; and w_i is a weighting factor for i^{th} technique/ROI.

The numerator in Eq. (9) is a weighted sum of each raw image, flagged by the ROI binary function in Eq. (10), and the denominator is the occurrence of ROI filter (with $F_i(x,y) = 1$) on each pixel. A non-zero pixel value means the defect probability at its corresponding location.

4.2   Machine Vision and Deep Convolutional Neural Network Integration

Unlike previous studies, where machine learning techniques were regarded irrelevant to inspection system, the proposed method integrates them into a system as shown in Fig. 4. Left-side shows conventional inspection process where input image is adequately processed, according to the product types, into defect probability image (DPI). A DPI generated from each algorithm provides enough

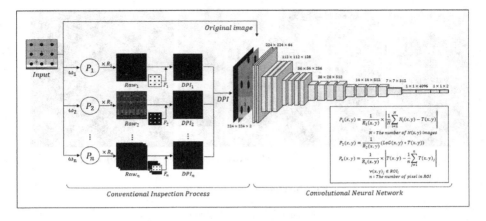

Fig. 4. Proposed inspection system

information to CNN to train, instead of color image. Obtained DPI is, then, stacked with grayscale image to form an input data for a CNN. CNN model on right-side is a VGGNet [13] and its first and the last layer were modified to fit input data and class number, respectively.

5   Experiments

This section evaluates the performance of the proposed defect inspection system incorporating DPI and CNN, including model stability depending on training dataset size. The proposed system is also compared with ViDi Suite 2.0 [5].

5.1   Database and Experimental Setup

Training and test data were obtained from a real-world chip-size package (CSP) product line, providing 1500 defect and 1796 good images after inspecting more than 3000 units. Approximately 50% were used for training and the other 50% for testing, as shown in Table 1. Smaller subsets of the training dataset (training_#.1) were randomly selected to evaluate the influence training data size. Note that good images were undersampled in data collection to avoid data imbalance.

We used Tensorflow [1] with TFlearn [6] for the deep learning framework and VGGNet-16 [13] for the CNN model. Models were fine-tuned with a pretrained VGGNet model. Other hyper parameters were fixed throughout the experiment, e.g. epoch = 30, zero-centered = True, augmentation = rotation & flip & crop, validation set = 15%, and Optimizer = stochastic gradient descent (SGD).

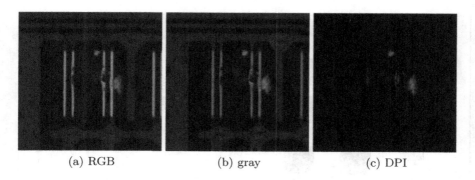

(a) RGB   (b) gray   (c) DPI

Fig. 5. Defect image example

Table 1. Image datasets

Data set	Defect	Good	Total	Percentage
training_#.1	715	896	1611	100%
training_#.2	287	360	647	40%
training_#.3	143	180	323	20%
training_#.4	72	90	162	10%
Test	785	900	1685	-

5.2   Results and Discussion

Four different models were evaluated on the same training and test dataset: VGG16 with RGB images (VGG16-RGB, Fig. 5a), VGG16 with gray images and DPI (VGG16-DPI, Figs. 5b and c), VGG16 with gray images (VGG16-Gray, Fig. 5b), and ViDi with gray image (ViDi-Gray, Fig. 5b). To obtain RGB images, 3 different grayscale images were taken under red, green, and blue lighting conditions. The gray image used in this study is taken under green light.

Table 2 shows the results for the training_#.1 dataset. VGG16-RGB and VGG16-DPI both outperformed ViDi-Gray by more than 5.1% in accuracy. Adding the DPI to the gray image, VGG16-DPI outperformed VGG16-Gray and VGG16-RGB by 1.6% and 0.1% in accuracy. The result implies that the DPI, a byproduct of conventional defect inspection technique, contains more defect information than gray or RGB images.

In practice, it is very difficult to acquire large training data sets for various reasons. Therefore, we evaluated performance impacts for reduced training datasets at 40%, 20%, and 10% of the original set (training_#.1), as shown in Table 3 and Fig. 6, and the experiment exactly reflects the reality that industry faces.

Table 2. Input data influences

Model	Recall		Precision		F-Score		Accuracy
	Defect	Good	Defect	Good	Defect	Good	
VGG16-RGB	0.980	0.993	0.992	0.982	0.986	0.988	98.694
VGG16-DPI	0.982	0.993	0.992	0.985	0.987	0.989	98.813
VGG16-Gray	0.980	0.963	0.959	0.982	0,969	0.973	97.092
ViDi-Gray	0.901	0.970	0.963	0.918	0.931	0.943	93.769

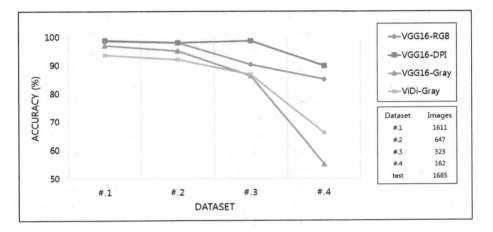

Fig. 6. Performance change on reduced dataset

1. Dataset size matters. Performance of all four models suffered as training dataset reduces. It is an obvious statement; however, it is the reality. Figures 7 and 8 are examples of real-world data used in this study. Despite high similarities among images, small training dataset (160–200 images) were insufficient to train conventional CNN model.
2. More channels, more information. Performance between VGG16-RGB and VGG16-Gray does not differ much on sufficient dataset (training #.1). But performance of VGG16-Gray decreases rapidly as dataset reduces as shown in Fig. 6. Considering conventional defect inspection systems using grayscale camera for higher throughput, conventional CNN model cannot be a solution without throughput sacrifice.
3. ViDi in industry. All conventional CNN models outperform ViDi when trained well with enough data. However, ViDi outperformed VGG16-Gray on dataset #.3 and #.4. This explains why this software is widely used in industry: it works better when there are neither RGB images nor sufficient data.

VGG16-DPI showed slightly better performance than conventional CNN model, VGG16-RGB, on both datasets, #.1 and #.2 in term of accuracy. Although detection performance is similar, VGG16-DPI will have much higher throughput than RGB system. on dataset #.4, its accuracy was about 80% higher than that of VGG16-Gray and 5% higher than that of VGG16-RGB.

The proposed method was experimentally validated that DPI contains high level defect information that it enhances CNN performance more than employing RGB images, and also works well on small training datasets.

Table 3. Training data size influence

Model	Training set	Recall		Precision		F-Score		Accuracy
		Defect	Good	Defect	Good	Defect	Good	
VGG16-RGB	#.1	0.980	0.993	0.992	0.982	0.986	0.988	98.694
	#.2	0.961	0.998	0.997	0.967	0.979	0.982	98.042
	#.3	0.883	0.922	0.908	0.900	0.895	0.911	90.386
	#.4	0.930	0.783	0.789	0.928	0.854	0.849	85.163
VGG16-DPI	#.1	0.982	0.993	0.992	0.985	0.987	0.989	98.813
	#.2	0.977	0.983	0.981	0.980	0.979	0.982	98.042
	#.3	0.980	0.993	0.992	0.982	0.986	0.988	98.694
	#.4	0.980	0.829	0.833	0.979	0.900	0.898	89.911
VGG16-Gray	#.1	0.980	0.963	0.959	0.982	0.969	0.973	97.092
	#.2	0.995	0.913	0.909	0.995	0.950	0.952	95.134
	#.3	0.966	0.774	0.789	0.963	0.868	0.858	86.350
	#.4	0.503	0.592	0.518	0.578	0.511	0.585	55.074
ViDi-Gray	#.1	0.901	0.970	0.963	0.918	0.931	0.943	93.769
	#.2	0.945	0.900	0.892	0.950	0.918	0.924	92.106
	#.3	0.959	0.789	0.799	0.957	0.872	0.865	86.824
	#.4	0.442	0.851	0.721	0.636	0.548	0.728	66.053

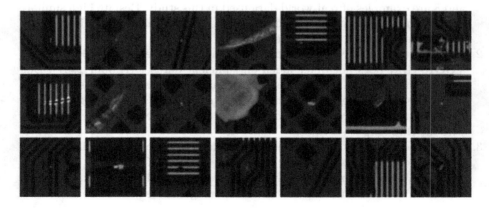

Fig. 7. Defect image dataset examples

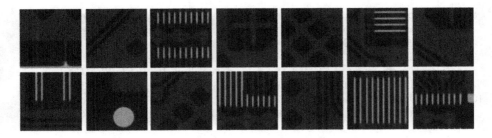

Fig. 8. Good image dataset examples

6   Conclusion

This paper proposed a defect inspection method combining defect probability images (DPIs) and a deep convolutional neural network (CNN). To the best of our knowledge, this is the first work that integrates conventional inspection system with deep CNN classifier for inspection process. In our method, DPI is extracted from conventional inspection process, and along with grayscale image it is trained on deep CNN classifier to replace conventional detection process.

Proposed method was verified experimentally with real-world data. Proposed method outperformed conventional CNN model with RGB input, implying that DPI can provide more information to CNN. Our model was validated on different dataset size and maintained its performance with little loss, while other models suffer from heavy performance drop. Lastly, proposed method eliminates parameter/design rule handling process, the most time consuming process in conventional defect inspection system. Thus, this study offers a major advance toward practical artificial intelligence application for defect inspection systems.

References

1. Abadi, M., et al.: TensorFlow: large-scale machine learning on heterogeneous systems (2015). Software https://www.tensorflow.org/
2. Abbott, G., Fouquet, C., Tadmor, O., Tada, T.: Use of design information and defect image information in defect classification. US Patent 8,175,373, 8 May 2012
3. Ak, R., Ferguson, M., Lee, Y.T.T., Law, K.H.: Automatic localization of casting defects with convolutional neural networks. In: 2017 IEEE International Conference on Big Data (BigData 2017) (2017)
4. Bai, X., Fang, Y., Lin, W., Wang, L., Ju, B.F.: Saliency-based defect detection in industrial images by using phase spectrum. IEEE Trans. Industr. Inf. **10**(4), 2135–2145 (2014)
5. COGNEX: Vidi suite (2.0) (2018). https://www.cognex.com/products/machine-vision/deep-learning-based-software
6. Damien, A., et al.: TFlearn (2016). https://github.com/tflearn/tflearn
7. Haddad, B.M., Yang, S., Karam, L.J., Ye, J., Patel, N.S., Braun, M.W.: Multifeature, sparse-based approach for defects detection and classification in semiconductor units. IEEE Trans. Autom. Sci. Eng. (2016)

8. Hayakawa, K., et al.: Semiconductor defect classifying method, semiconductor defect classifying apparatus, and semiconductor defect classifying program. US Patent 8,595,666, 26 November 2013
9. Jansen, S., Florence, G., Perry, A., Fox, S.: Utilizing design layout information to improve efficiency of SEM defect review sampling. In: IEEE/SEMI Advanced Semiconductor Manufacturing Conference, ASMC 2008, pp. 69–71. IEEE (2008)
10. Lv, Y., Duan, Y., Kang, W., Li, Z., Wang, F.Y.: Traffic flow prediction with big data: a deep learning approach. IEEE Trans. Intell. Transp. Syst. 16(2), 865–873 (2015)
11. Nakazawa, T., Kulkarni, D.V.: Wafer map defect pattern classification and image retrieval using convolutional neural network. IEEE Trans. Semicond. Manuf. 31(2), 309–314 (2018)
12. Pastor-López, I., Santos, I., Santamaría-Ibirika, A., Salazar, M., de-la Pena-Sordo, J., Bringas, P.G.: Machine-learning-based surface defect detection and categorisation in high-precision foundry. In: 2012 7th IEEE Conference on Industrial Electronics and Applications (ICIEA), pp. 1359–1364. IEEE (2012)
13. Simonyan, K., Zisserman, A.: Very deep convolutional networks for large-scale image recognition. arXiv preprint arXiv:1409.1556 (2014)
14. Tan, S.C., Watada, J., Ibrahim, Z., Khalid, M.: Evolutionary fuzzy artmap neural networks for classification of semiconductor defects. IEEE Trans. Neural Netw. Learn. Syst. 26(5), 933–950 (2015)
15. Tsai, D.M., Lin, C.T., Chen, J.F.: The evaluation of normalized cross correlations for defect detection. Pattern Recogn. Lett. 24(15), 2525–2535 (2003)
16. Wang, C.C., Jiang, B.C., Lin, J.Y., Chu, C.C.: Machine vision-based defect detection in IC images using the partial information correlation coefficient. IEEE Trans. Semicond. Manuf. 26(3), 378–384 (2013)

3D Pick & Mix: Object Part Blending in Joint Shape and Image Manifolds

Adrian Penate-Sanchez[1,2]([✉]) [iD] and Lourdes Agapito[1] [iD]

[1] Department of Computer Science, University College London, London, UK
[2] Oxford Robotics Institute, University of Oxford, Oxford, UK
adrian@robots.ox.ac.uk

Abstract. We present **3D Pick & Mix,** a new 3D shape retrieval system that provides users with a new level of freedom to explore 3D shape and Internet image collections by introducing the ability to reason about objects at the level of their constituent parts. While classic retrieval systems can only formulate simple searches such as *"find the 3D model that is most similar to the input image"* our new approach can formulate advanced and semantically meaningful search queries such as: *"find me the 3D model that best combines the design of the legs of the chair in image 1 but with no armrests, like the chair in image 2"*. Many applications could benefit from such rich queries, users could browse through catalogues of furniture and **pick** and **mix** parts, combining for example the legs of a chair from one shop and the armrests from another shop.

Keywords: Shape blending · Image embedding · Shape retrieval

1 Introduction

As databases of images and 3D shapes keep growing in size and number, organizing and exploring them has become increasingly complex. While most tools so far have dealt with shape and appearance modalities separately, some recent methods [10,15] have begun to exploit the complementary nature of these two sources of information and to reap the benefits of creating a common representation for images and 3D models. Once images and 3D shapes are linked together, many possibilities open up to transfer what is learnt from one modality to another. Creating a joint embedding allows to retrieve 3D models based on image queries (or vice-versa) or to align images of similar 3D shapes. However, recent retrieval methods still fall short of being flexible enough to allow

This work was supported by the SecondHands project, funded from the EU Horizon 2020 Research and Innovation programme under grant agreement 643950 and by the EPSRC grants RAIN and ORCA (EP/R026084/1, EP/R026173/1).

Electronic supplementary material The online version of this chapter (https://doi.org/10.1007/978-3-030-20887-5_10) contains supplementary material, which is available to authorized users.

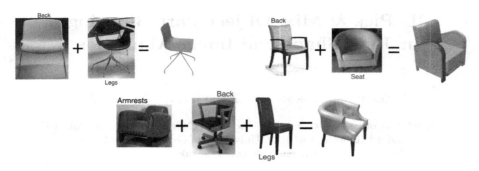

Fig. 1. Our approach takes two (or more) inputs, either RGB images or 3D models, and part labels such as *"legs"* or *"backrest"* and retrieves a shape that combines object parts or properties from all inputs, via a cross-manifold optimization technique.

advanced queries. Crucially, they are limited to reasoning about objects as a whole – taking a single query image (or shape) as input at test time prevents them from combining object properties from different inputs. **3D Pick & Mix** overcomes this limitation by reasoning about objects at the level of parts. It can formulate advanced queries such as: *"find me the 3D model that best combines the design of the backrest of the chair in image 1 with the shape of the legs of the chair in image 2"* (see Fig. 1) or *"retrieve chairs with wheels"*. The ability to reason at the level of parts provides users with a new level of freedom to explore 3D shape and image datasets. Users could browse through catalogues of furniture and **pick** and **mix** parts, combining for example the legs of a favourite chair from one catalogue and the armrests from another (see Fig. 1).

Our system first builds independent manifold shape spaces for each object part (see Fig. 4). A CNN-based manifold coordinate regressor is trained to map real images of an object to the part manifolds. Our novel deep architecture jointly performs semantic segmentation of object parts and coordinate regression for each part on the corresponding shape manifold. This network is trained using only synthetic data, Fig. 2 illustrates the architecture. At test time the user provides two (or more) images (or 3D models) as input and determines which parts they would like to pick from each (note that this only requires a label name such as *'legs'*). Our system retrieves the model that best fits the arrangement of parts by performing a cross-manifold optimization (see Fig. 1). The main **contributions** of our 3D Pick & Mix system are:

- We learn embeddings (manifolds) for object parts (for instance the legs or the armrests of chairs) which allow us to retrieve images or 3D models of objects with similarly shaped parts.
- We propose a new deep architecture that can map RGB images onto these manifolds of parts by regressing their coordinates. Crucially, the input to the network is simply an RGB image and the name (label) of the object part. The CNN learns to: *(i)* segment the pixels that correspond to the chosen part, and *(ii)* regress its coordinates on the shape manifold.

- At query time our retrieval system can combine object parts from multiple input images, enabled by a cross-manifold optimization technique.

2 Related Work

Joint 3D Model/Image Embeddings: While most shape retrieval methods had traditionally dealt with shape and appearance modalities separately, a recent trend has emerged that exploits the complementary nature of appearance and shape by creating a common representation for images and 3D models. [10] exploits the different advantages of shape and images by using the robustness of 3D models for alignment and pose estimation and the reliability of image labels to identify the objects. While they do not explicitly create a joint manifold based on shape similarity they do rely on image representations for both modalities. Another example of 3D model/image embedding is [15] who first builds a manifold representation of 3D shapes and then trains a CNN to recognize the shape of the object in an image. Unlike our approach, both [10,15], limit their representations to objects as a whole preventing the combination of properties taken from different inputs. [23] perform shape retrieval from sketches, words, depth maps and real images by creating a manifold space that combines the different inputs. Since intra-class similarity is not the main focus, most instances of the same class tend to appear clustered. [16] learn a manifold-space metric by using triplets of shapes where the first is similar to the third but dissimilar to the second. Similarly to our approach, the metric space is defined based on shape and not image similarity. [9] first generates voxel representations of the objects present in the RGB image inputs. A shared latent shape representation is then learnt for both images and the voxelized data. At test time RGB convolutions and volume generation deconvolution layers are used to produce the 3D shape.

3D Shape Blending/Mixing: Much in the line of the work presented in this paper, there has been fruitful research in shape blending in recent years. The *"3D model evolution"* approach of [27] takes a small set of 3D models as input to generate many. Parts from two models cross-over to form a new 3D model, continuing to merge original models with new ones to generate a large number of 3D models. In [1] new shapes are generated by interpolating and varying the topology between two 3D models. The photo-inspired 3D modeling method of [28] takes a single image as input, segments it into parts using an interactive model-driven approach, then retrieves a 3D model candidate that is finally deformed to match the silhouette of the input photo. The probabilistic approach of [11] learns a model that describes the relationship between the parts of 3D shapes which then allows to create an immense variety of new blended shapes by mixing attributes from different models. The sketch driven method of [26] edits a pre- segmented 3D shape using user-drawn sketches of new parts. The sketch is used to retrieve a matching 3D part from a catalogue of 3D shapes which is then snapped onto the original 3D shape to create a new blended 3D shape. Note that the above approaches use only 3D shapes as input for shape blending, with the exception of [28] who use a single photograph and [26] who use sketches.

However, unlike ours, neither of these approaches can combine different input images to retrieve a shape that blends parts from each input.

Modeling of 3D Object Parts: We will differentiate between 3D segmentation approaches that seek to ensure consistency in the resulting segmentation across different examples of the same object class (co-segmentation) and those that seek a semantically meaningful segmentation (semantic segmentation). Some recent examples of approaches that perform co-segmentation can be found in [8,24], but as we seek to describe parts that have meaning to humans we will focus on the later. We can find examples of semantic 3D parts in approaches like [29]. [29] provides accurate semantic region annotations for large geometric datasets with a fraction of the effort by alternating between using few manual annotations from an expert and a system that propagates labels to new models. We exploit the ShapeNet annotations provided by [29] as the ground truth part shape when constructing our joint manifold.

Recognition of 3D Structure from Images: The exemplar-based approach of [2] performs joint object category detection viewpoint estimation, exploiting 3D model datasets to render instances from different viewpoints and then learn the combination of viewpoint-instance using exemplar SVMs. [5] uses 3D Convolutional LSTMs to extract the 3D shape of an object from one or more viewpoints. By using LSTM blocks that contain memory, they progressively refine the shape of the object. [7] learn to generate a 3D point cloud from a single RGB image, it learns purely from synthetic data. By using a point cloud instead of a volumetric representation better definition of the details of the shape are obtained. Their novel approach learns how to generate several plausible 3D reconstructions from a single RGB image at test time if the partial observation of the image is ambiguous. [22] learn to recognize the object category and the camera viewpoint for an image using synthetically generated images for training. This work showed that datasets of real images annotated with 3D information were not required to learn shape properties from images as this could be learnt from synthetically generated renderings. [21] obtain good depth estimates for an image given a set of 3D models of the same class.

3   Overview

In this section we provide a high level overview of our **3D Pick & Mix** retrieval system. Our system requires a training stage in which: *(i)* manifolds of 3D shapes of object parts are built (see Fig. 4) and *(ii)* a CNN is trained to take as input an image and regress the coordinates of each of its constituent parts on the shape manifolds (illustrated in Fig. 2). At query time the system receives an image or set of images as input and obtains the corresponding coordinates on the part manifolds. If the user chooses object parts from different images a cross-manifold optimization is carried out to retrieve a single shape that blends together properties from different images.

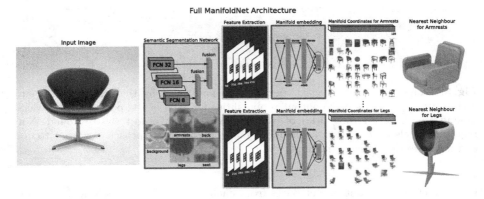

Fig. 2. Summary of the architecture of **ManifoldNet**, our new deep network that takes an image as input and learns to regress the coordinates of each object part in the different part manifolds. The architecture has 3 sections: the first set of layers performs semantic segmentation of the image pixels into different semantic parts (such as *"backrest"*, *"seat"*, *"armrests" or "legs"* in the case of chairs). The second section learns an intermediate feature representation for manifold coordinate regression. The final section learns to regress the shape coordinates in each of the part manifolds. We show the nearest neighbour shapes found on the *"armrests"* and *"legs"* manifolds for the depicted input image.

Training: At training time, our method takes as input a class-specific collection of 3D shapes (we used *ShapeNet* [3]) for which part label annotations are available. The first step at training time is to *learn a separate shape manifold for each object part* (see Fig. 4). Each shape is represented with a Light Field descriptor [4] and characterized with a pyramid of HoG features. The manifolds are then built using non-linear multi-dimensional-scaling (MDS) and the L_2 norm between feature vectors as the distance metric – in each resulting low-dimensional manifold, objects that have similarly shaped parts are close to each other. So far these manifolds of object parts (for instance *back-rests, arm-rests, legs, seats* in the case of chairs) contain 3D shapes. The second step at training time is to train a CNN to embed images onto each part manifold by *regressing their coordinates*. We create a set of synthetic training images with per pixel semantic segmentation annotations for the object parts and ground truth manifold coordinates. The architecture of this novel CNN (which we denote **ManifoldNet** and is shown in Fig. 2) has three clear parts: a set of fully convolutional layers for semantic segmentation of the object into parts; a set of convolutional feature extraction layers; and a set of fully connected layers for manifold coordinate regression. This architecture can be trained end-to-end. We give an example of the produced semantic segmentation in Fig. 3.

Retrieval: At test time, given a new query image of an unseen object, **ManifoldNet** can embed it into each of the part manifolds by regressing the coordinates. More importantly, our retrieval system can take more than one image as

Image	Background	Armrest	Back	Legs	Seat	Segmentation

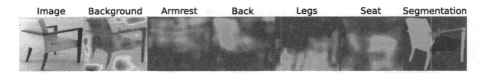

Fig. 3. Example of the semantic segmentation performed by the first stages of our architecture. We can see the output probabilities for each of the parts and the background give a very strong prior of were the parts of an object can be found. Not requiring labels for each part in the input image makes our approach very easy to use and increases dramatically its applicability.

input, picking different object parts from each image. Note that **ManifoldNet** only needs the input images and the name of the object part that will be used from each image. The network learns jointly to segment the image into parts and to regress the manifold coordinates and therefore it does not require any manual annotations as input. A **cross-manifold optimization** will then take the coordinates on each of the part manifolds as input and return the coordinates of a unique 3D shape that blends the different object parts together. This is achieved through an energy optimization approach, described in Sect. 4.4.

4   Methodology

4.1   Building Shape Manifolds for Object Parts

We choose to create an embedding space that captures the similarity between the shape of object parts based exclusively on the 3D shapes. The reason behind this choice is that 3D models capture a more complete, pure and reliable representation of geometry as opposed to images that often display occlusions, or other distracting factors such as texture or shading effects. We then rely on our new CNN architecture to map images onto the same embedding by regressing their coordinates on the corresponding manifolds.

Defining a Smooth Similarity Measure Between 3D Shapes: Shape similarity between object parts is defined as follows. Given a shape S_i, we define its Light-field Descriptor (LfD) [4] L_i as the concatenation of the HoG responses [6] $L_i = [H_1; H_2; ...; H_k]$. The value of k is fixed to $k = 20$ throughout this work. The light field descriptor H_k for each view k is defined as $H_k = [H_k^{mid}; H_k^{low}] \in \mathbb{R}^{2610}$. The L_2 distance between feature vectors is then used as the similarity measure between a pair of shapes S_i and S_j: $d_{ij} = ||L_i - L_j||_2$ where $L_i \in \mathbb{R}^{52200}$. We found that using only the mid and low frequency parts of the HoG pyramid leads to smoother transitions in shape similarity. For this reason we do not use the original 3 level HoG pyramids but just the 2 higher levels of the pyramids. This allows for smooth transitions in shape similarity between parts making the shape blending possible. Due to most 3D models available in Internet datasets not being watertight but only polygon soups we are required to use projective

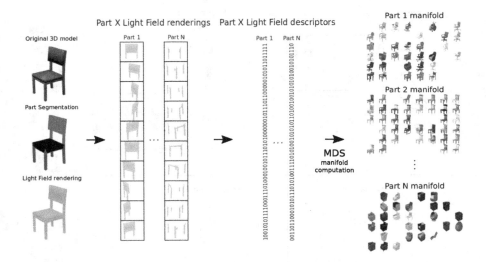

Fig. 4. Shape manifold construction. The shape of each part for each 3D model is rendered from different viewpoints and represented with a Light Field descriptor [4]. The manifolds are then built using non-linear multi-dimensional-scaling (MDS) and the L_2 norm between feature vectors as the distance metric. In each low-dimensional manifold, objects that have similarly shaped parts appear close to each other.

shape features like [4]. We now build separate manifolds for each object part. Each shape S_i is therefore split into its constituent parts $\forall S_i : \exists \{S_i^1; S_i^2; ...; S_i^P\}$, where P is the total number of parts and S_i^p is the shape of part p of object i. If a part is not present in an object (for instance, chairs without arm-rests) we set all the components of the vector L_i^p to zero, which is equivalent to computing the HoG descriptor of an empty image.

4.2   Building Shape Manifolds of Parts

Using the similarity measure between the shape of object parts we use it to construct a low dimensional representation of the shape space. In principle, the original $L_i \in \mathbb{R}^{52200}$ feature vectors could have been used to represent each shape, since distances in that space reflect well the similarity between shapes. We reduce the dimensionality from $52,200$ to 128 dimensions and we use non-linear Multi-Dimensional Scaling (MDS) [14] to build the shape manifolds. We compute the distance matrix $D^p \in \mathbb{R}^{n \times n}$ as $D^p(i, j) = d_{ij}^p$, were p is the index of the part and n is the total number of shapes. The manifold is built using MDS by minimizing a Sammon Mapping error [19] defined as

$$E^p = \frac{1}{\sum_{i<j} D^p(i,j)} \sum_{i<j} \frac{(D^p(i,j) - D'^p(i,j))^2}{D^p(i,j)},\qquad(1)$$

where D^p is the distance matrix between shapes in the original high dimensional feature space L^p; and D'^p is the distance matrix between shapes in the new

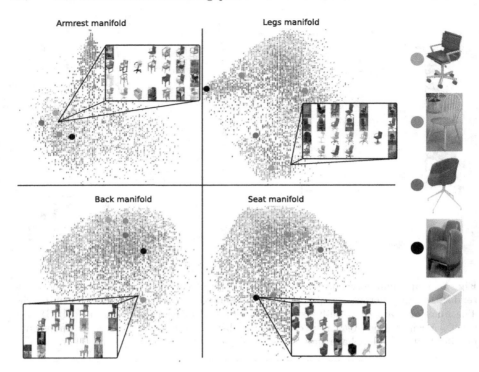

Fig. 5. Two dimensional visualizations of the four low dimensional part manifolds *"backrest"*, *"seat"*, *"armrests"* and *"legs"* for the chairs class. Probabilistic PCA has been used to provide a 2D visualization of the 128 dimensional manifolds. Both images and 3D models have been represented in the manifolds. Objects with similarly shaped parts lie close to each other on the manifold. Several shapes are tagged in all four manifolds to show how vicinity changes for each part. All shapes and images exist in all manifolds.

low dimensional manifold L'^p. With the different manifolds L'^p for each part p computed, a low dimensional representation of shape similarity exists and all 3D shapes are already included in it. Adding new 3D shapes to the manifold is done by solving an optimization that minimizes the difference between the distances between all previous shapes and the new shape in D^p and the distances in D'^p with respect to the predicted embedding point. To understand the shape of the produced manifolds we provide a 2D visualization in Fig. 5. In this figure we can better understand how the manifolds relate the different parts of an object.

4.3   Learning to Embed Images into the Shape Manifolds

Building the shape manifolds L'^p for each part p based only on 3D models, we have successfully abstracted away effects such as textures, colours or materials. The next step is to train a deep neural network that can map RGB images onto

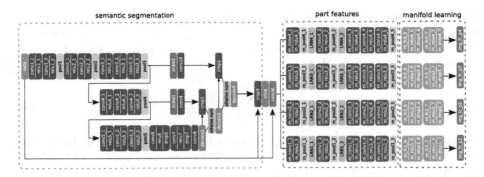

Fig. 6. Detailed **ManifoldNet** architecture. The first set of layers take care of assigning semantic part labels to each of the pixels in the image. The second stage extracts intermediate features to help the manifold learning. The final stage uses a set of fully convolutional layers to regress the manifold coordinates.

each manifold by regressing the coordinates on each part manifold directly from RGB inputs. Crucially, the input to the network must be simply the RGB image and the name (label) of the object part p selected from that image – for instance *embed this image of a chair into the manifold of "chair legs"*.

We propose a novel deep learning architecture, which we call **ManifoldNet**, it performs three tasks: first, it learns how to estimate the location of different parts in the image by performing semantic segmentation, it then uses the semantic labeling and the original input image to learn p different intermediate feature spaces for each object part and finally, p different branches of fully connected layers will learn the final image embedding into the respective part shape manifold. The network has a general core that performs semantic segmentation and specialized branches for each of the manifolds in a similar fashion as [13].

ManifoldNet: A Multi-manifold Learning Architecture: A summary of our new architecture is shown in Fig. 2 and a detailed description of all the layers in Fig. 6. The common part of the architecture, which performs the semantic segmentation, is a fully convolutional approach closely related to [17]. The fully convolutional architecture uses a VGG-16 architecture [20] for its feature layers. A combination of FCN-32, FCN-16 and FCN-8 is used to obtain more detailed segmentations but all sub-parts are trained together for simplicity.

The other two parts of the architecture shown in Fig. 2 take care of: *(1)* creating an intermediate feature space, and *(2)* learning the manifold embedding. The intermediate feature layers take as input the concatenation of the original RGB image and the heat maps given as output by the semantic segmentation layers to learn a feature representation that eases the manifold learning task. Finally, the manifold coordinate regression module is formed by 3 fully connected layers (the first two use relu non-linearities). A dropout scheme is used to avoid over-fitting to the data. Trained models and code are already public but due to anonymity we cannot disclose the URL.

Details of the Training Strategy: The training of such a deep architecture requires careful attention. First, to avoid vanishing gradients and to improve convergence, the semantic segmentation layers are trained independently by using a standard cross-entropy classification loss:

$$L_{seg} = \frac{-1}{N} \sum_{n=1}^{N} \log\left(p_{n,l_n}\right) \tag{2}$$

where p_{n,l_n} is the softmax output class probability for each pixel. A batch size of only 20 is used at this stage due to memory limitations on the GPU and the high number of weights to be trained.

When trying to train the manifold layers we found out that convergence heavily depended on big batch sizes and many iterations. At this point we used a learning scheme that allowed us to have bigger batch sizes during training and faster computation of each iteration. The trick is quite simple really, we precompute for all training images the output of the semantic layers and only train the part branches of the network. By doing this we are training a substantially shallower network allowing for significantly bigger batch sizes. The network is trained by minimizing the following euclidean loss:

$$L_{mani} = \frac{1}{2N} \sum_{i=1}^{N} \|x_{est}^i - x_{gt}^i\|_2^2 \tag{3}$$

where x_{est}^i are the manifold coordinates estimated by the network and x_{gt}^i are the ground truth manifold coordinates. The Euclidean loss is chosen since the part shape manifolds are themselves Euclidean spaces. With this good initialization of the weights we finally perform an end-to-end training of all layers using only the final euclidean loss.

Training Data and Data Augmentation: The training images are generated synthetically by rendering models from *ShapeNet* [3]. We use the 3D part annotations on the 3D models, available from [29], to provide ground truth values for the semantic segmentation. We generate 125 training images per model from different poses, and a random RGB image taken from the Pascal 3D [25] dataset is added as background. To recap, the proposed approach is invariant to pose and manages to learn solely from rendered synthetic images.

4.4   Shape Blending Through Cross-Manifold Optimization

Once the manifold coordinates for the different object parts have been estimated all the information needed to blend them into a single 3D model is available. We formulate this as a 3D shape retrieval problem: *"find the 3D model, from the existing shapes represented in the manifolds, that best fits the arrangement of parts"*. The user selects two (or more) images (or 3D models) and indicates the part they wish to select from each one (note that no annotations are needed, only the name/label of the part). The cross-manifold optimization now finds the

3D shape in the collection that minimizes the sum of the distances to each of the parts. In more detail – first, all manifolds need to be normalized to allow a meaningful comparison of distances. Then, given the set of manifold coordinates for the selected parts, a shape prediction b can be defined as the concatenation of the respective part coordinates $b = \{b^1; ...; b^m\}$. The goal is now to retrieve a 3D model from the shape collection whose coordinates $a = \{a^1; ...; a^m\}$ are closest to this part arrangement by minimizing the following distance:

$$B = \min_{a \in \mathbb{S}} \sum_{k=1}^{m} \|a^m - b^m\| \tag{4}$$

where $\mathbb{S}$ is the set of existing shapes. Note that not all parts need to be selected to obtain a blended shape, we define m as the subset of parts to be blended, where $m \subseteq p$. Also, notice that blending can be done by combining any number of parts from any number of sources (shapes/images).

5   Results

We perform a set of qualitative and quantitative experiments to evaluate the performance of our approach. Although our approach performs shape blending from several inputs this can be understood as a retrieval task.

5.1   Quantitative Results on Image-Based Shape Retrieval

The proposed approach will be at a disadvantage when trying to retrieve whole shapes and the same will happen when approaches that model the object as a whole try to retrieve parts. What has been have done is a experimental comparison of both approaches on both tasks, as a clean unbiased comparison cannot be done on a single experiment both approaches will be used to solve both tasks. This is possible as all compared approaches can be used as a similarity measure between images and 3D models. By doing this, how much is lost can be measured when either modeling the whole object or the individual parts.

Image-Based Whole Shape Retrieval: We perform whole shape image-based retrieval on the **ExactMatch** dataset [15]. The experiment compares against Li [15], a state-of-the-art deep volumetric approach in Girdhar [9], HoG, AlexNet and Siamese networks. Two versions of the proposed approach using the original three level HoG pyramid features to build the manifold and the two level HoG manifold features that have been shown to be better fitted for a smoother shape similarity measure. Our approach predicts the part coordinates separately in each of the part manifolds. The estimations of each part are then used to solve the blending optimization and obtain a single shape prediction. The fact that the neural network estimates the coordinates individually means that all the part co-occurrence information that is implicitly encoded in the approaches that model the object as a whole is lost during training, nevertheless, the proposed approach can still yield good results that are comparable to those that model the

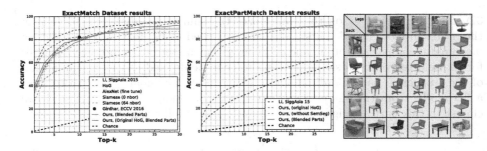

Fig. 7. Left: Image-based whole shape retrieval results obtained on the ExactMatch dataset. **Center**: Shape blending and retrieval from image-based part descriptions on the ExactPartMatch dataset. **Right**: Matrix showing combination results. Experiment performed by generating all possible combinations of legs and backs from 10 test shapes.

whole shape. It can also be seen that for the exact shape retrieval the original three level pyramid HoG features perform better which is to be expected. The results of these experiments are included in Fig. 7 (first image).

Image-Based Part Shape Blending: To test the performance of part blending and retrieval a new dataset has been created using a subset of the shapes from the Shapenet database [3] and the images from the ExactMatch dataset [15] to create the **Exact*Part*Match** dataset. The task is to find the correct 3D shape out of all the annotated 3D models using the parts from the specified inputs. As in Shapenet many of the 3D models are repeated many times (e.g. ikea chairs) we need to control that there is only one correct match with the part mix. The dataset contains **845 test cases**. Each test case if assigned to one of 187 hand annotated ground truth 3D models, the same part shape combination is tested using images from different angles, textures and lighting conditions. Each **test case** is a combination of parts from two different images for which a 3*D* model exists in Shapenet. There are examples of all possible part combinations. The whole dataset is already public online, the link is not disclosed due to anonymity. The results of these experiments are included in Fig. 7 (second image). Also in Fig. 7 (third image) we include an experiment that shows the product of crossing 5 leg samples with 5 back samples densely, this shows the kind of results to expect from our approach.

For all approaches similarity is estimated in shape manifold space and then the multi-manifold optimization is performed, which is required to obtain the most similar shape. The approaches shown are *Ours*, with and without semantic segmentation, *Ours* with semantic segmentation but using the original three level HoG features, *Li'sSiggAsia*15 [15] and random chance. [15] struggles to get results as good as the ones obtained by modeling the parts. Their holistic representation enables them to better model a whole object but loose substantial performance when trying to identify individually the parts. In contrast our approach looses the information of part co-occurrence in favour of being capable to

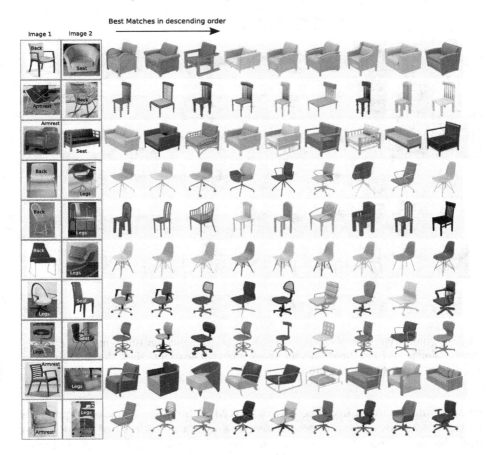

Fig. 8. Qualitative results extracted to show performance on different part arrangements using parts from two different image inputs. The results show the closest matching shapes in the Shapenet dataset. If the part is not present, like in the second row, an X in the colour of the desired part is used to label the non-presence of that part.

model the parts individually. Also, exploiting the semantic segmentation of the input is consistently better as it defines the actual interest zones in the image. If considering top-5 results without semantic segmentation 65% is obtained but when using semantic segmentation 76% is obtained, which is a substantial 11% improvement in performance. It can also be seen that trying to blend shapes when using shape similarities that do not correctly model smoothness over shape has a tremendous impact in performance (Ours_original_HoG).

5.2 Qualitative Results on Image-Based Shape Retrieval

To further asses the quality of the results examples are shown depicting the performance of the approach being applied to real images. In Figs. 8 and 9 many

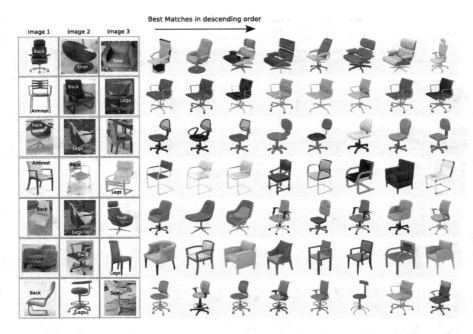

Fig. 9. Qualitative results extracted to show performance on different part arrangements using parts from three different image inputs. The results show the closest matching shapes in the Shapenet dataset.

example images taken from the **Exact*Part*Match** dataset detailed in previous sections are shown. Our approach searches over the entirety of Shapenet to find the closest matches to the input part arrangement. Many different part arrangements are accounted for in the figure to show that the proposed approach can capture not only the big differences but also more subtle differences like the number of legs in the base of a specific swivel chair, the fact that a wooden chair with a back made of bars has a round top or a flat top, capturing the detailed shape of the interconnecting supports of the four legs, etc.

6   Conclusions

An approach capable of modelling and blending object parts from images and 3D models has been presented. This approach has demonstrated that by using a common manifold representation very elaborate queries can be done in massive Internet databases. It has also shown to be capable to produce accurate shape retrieval which proves its understanding of the underlying shapes. This provides a natural link between shape and image datasets and opens numerous possibilities of application to similar tasks. Also, by understanding the parts that give semantic meaning to an object volumetric approaches like [5,7,12,18] could potentially address their difficulties to produce details in their volumetric estimations by either enforcing the shape of the part explicitly or balancing the voxel occupancy probabilities for each part independently.

References

1. Alhashim, I., Li, H., Xu, K., Cao, J., Ma, R., Zhang, H.: Topology-varying 3D shape creation via structural blending. ACM Trans. Graph. **33**(4) (2014). Article No. 158
2. Aubry, M., Maturana, D., Efros, A., Russell, B., Sivic, J.: Seeing 3D chairs: exemplar part-based 2D-3D alignment using a large dataset of cad models. In: Computer Vision and Pattern Recognition (CVPR) (2014)
3. Chang, A.X., et al.: ShapeNet: an information-rich 3D model repository. Technical report arXiv:1512.03012 [cs.GR], Stanford University – Princeton University – Toyota Technological Institute at Chicago (2015)
4. Chen, D.Y., Tian, X.P., Shen, Y.T., Ouhyoung, M.: On visual similarity based 3D model retrieval. Comput. Graph. Forum **22**(3), 223–232 (2003)
5. Choy, C.B., Xu, D., Gwak, J.Y., Chen, K., Savarese, S.: 3D-R2N2: a unified approach for single and multi-view 3D object reconstruction. In: Leibe, B., Matas, J., Sebe, N., Welling, M. (eds.) ECCV 2016. LNCS, vol. 9912, pp. 628–644. Springer, Cham (2016). https://doi.org/10.1007/978-3-319-46484-8_38
6. Dalal, N., Triggs, B.: Histograms of oriented gradients for human detection. In: Computer Vision and Pattern Recognition (CVPR) (2005)
7. Fan, H., Su, H., Guibas, L.J.: A point set generation network for 3D object reconstruction from a single image. In: Computer Vision and Pattern Recognition (CVPR) (2017)
8. Fish, N., van Kaick, O., Bermano, A., Cohen-Or, D.: Structure-oriented networks of shape collections. ACM Trans. Graph. **35**(6), 171 (2016)
9. Girdhar, R., Fouhey, D.F., Rodriguez, M., Gupta, A.: Learning a predictable and generative vector representation for objects. In: Leibe, B., Matas, J., Sebe, N., Welling, M. (eds.) ECCV 2016. LNCS, vol. 9910, pp. 484–499. Springer, Cham (2016). https://doi.org/10.1007/978-3-319-46466-4_29
10. Hueting, M., Ovsjanikov, M., Mitra, N.: Crosslink: joint understanding of image and 3D model collections through shape and camera pose variations. ACM Trans. Graph. **34**(6), 233 (2015). Proc. SIGGRAPH Asia
11. Kalogerakis, E., Chaudhuri, S., Koller, D., Koltun, V.: A probabilistic model for component-based shape synthesis. ACM Trans. Graph. **31**(4), 55 (2012)
12. Kar, A., Tulsiani, S., Carreira, J., Malik, J.: Category-specific object reconstruction from a single image. In: Computer Vision and Pattern Recognition (CVPR) (2015)
13. Kokkinos, I.: UberNet: training a 'universal' convolutional neural network for low-, mid-, and high-level vision using diverse datasets and limited memory. In: Computer Vision and Pattern Recognition (CVPR) (2017)
14. Kruskal, J.B.: Multidimensional scaling by optimizing goodness of fit to a nonmetric hypothesis. Psychometrika **29**(1), 1–27 (1964)
15. Li, Y., Su, H., Qi, C.R., Fish, N., Cohen-Or, D., Guibas, L.J.: Joint embeddings of shapes and images via CNN image purification. ACM Trans. Graph. **34**(6), 234 (2015)
16. Lim, I., Gehre, A., Kobbelt, L.: Identifying style of 3D shapes using deep metric learning. Comput. Graph. Forum **35**(5), 207–215 (2016)

17. Long, J., Shelhamer, E., Darrell, T.: Fully convolutional networks for semantic segmentation. In: Computer Vision and Pattern Recognition (CVPR) (2015)
18. Qi, C.R., Su, H., Niessner, M., Dai, A., Yan, M., Guibas, L.J.: Volumetric and multi-view CNNs for object classification on 3D data. In: Computer Vision and Pattern Recognition (CVPR) (2016)
19. Sammon, J.W.: A nonlinear mapping for data structure analysis. IEEE Trans. Comput. **100**(5), 401–409 (1969)
20. Simonyan, K., Zisserman, A.: Very deep convolutional networks for large-scale image recognition. In: International Conference on Learning Representations (2015)
21. Su, H., Huang, Q., Mitra, N.J., Li, Y., Guibas, L.: Estimating image depth using shape collections. ACM Trans. Graph. **33**(4), 37 (2014)
22. Su, H., Qi, C.R., Li, Y., Guibas, L.J.: Render for CNN: viewpoint estimation in images using CNNs trained with rendered 3D model views. In: International Conference on Computer Vision (ICCV) (2015)
23. Tasse, F.P., Dodgson, N.: Shape2vec: semantic-based descriptors for 3D shapes, sketches and images. ACM Trans. Graph. **35**(6) (2016). Article No. 208
24. Tulsiani, S., Su, H., Guibas, L.J., Efros, A.A., Malik, J.: Learning shape abstractions by assembling volumetric primitives. In: Computer Vision and Pattern Recognition (CVPR) (2017)
25. Xiang, Y., Mottaghi, R., Savarese, S.: Beyond Pascal: a benchmark for 3D object detection in the wild. In: IEEE Winter Conference on Applications of Computer Vision (WACV) (2014)
26. Xie, X., et al.: Sketch-to-design: context-based part assembly. Comput. Graph. Forum **32**(8), 233–245 (2013)
27. Xu, K., Zhang, H., Cohen-Or, D., Chen, B.: Fit and diverse: set evolution for inspiring 3D shape galleries. ACM Trans. Graph. **31**(4), 57 (2012)
28. Xu, K., Zheng, H., Zhang, H., Cohen-Or, D., Liu, L., Xiong, Y.: Photo-inspired model-driven 3d object modeling. ACM Trans. Graph. **30**(4) (2011). Article No. 80
29. Yi, L., et al.: A scalable active framework for region annotation in 3D shape collections. In: SIGGRAPH Asia (2016)

Minutiae-Based Gender Estimation for Full and Partial Fingerprints of Arbitrary Size and Shape

Philipp Terhörst[1,2]([✉]), Naser Damer[1,2], Andreas Braun[1], and Arjan Kuijper[1,2]

[1] Fraunhofer Institute for Computer Graphics Research IGD, Darmstadt, Germany
{philipp.terhoerst,naser.damer,andreas.braun,
arjan.kuijper}@igd.fraunhofer.de
[2] Mathematical and Applied Visual Computing Group, TU Darmstadt,
Darmstadt, Germany

Abstract. Since fingerprints are one of the most widely deployed biometrics, accurate fingerprint gender estimation can positively affect several applications. For example, in criminal investigations, gender classification may significantly minimize the list of potential subjects. Previous work mainly offered solutions for the task of gender classification based on complete fingerprints. However, partial fingerprint captures are frequently occurring in many applications, including forensics and the fast growing field of consumer electronics. Due to its huge variability in size and shape, gender estimation on partial fingerprints is a challenging problem. Therefore, in this work we propose a flexible gender estimation scheme by building a gender classifier based on an ensemble of minutiae. The outputs of the single minutia gender predictions are combined by a novel adjusted score fusion approach to obtain an enhanced gender decision. Unlike classical solutions this allows to deal with unconstrained fingerprint parts of arbitrary size and shape. We performed investigations on a publicly available database and our proposed solution proved to significantly outperform state-of-the-art approaches on both full and partial fingerprints. The experiments indicate a reduction in the gender estimation error by 19.34% on full fingerprints and 28.33% on partial captures in comparison to previous work.

1 Introduction

Fingerprints are one of the biggest commonly used biometric modalities [18], while gender is one of the most widely deployed soft biometrics [24]. Including gender information from fingerprints can positively affect a wide variety of applications, such as context-based indexing [6,7], human-computer interactions, or simply to establish a persons's identity with a high degree of reliability [11]. In addition, it can enhance forensic investigations and the rapidly growing field of consumer electronics, such as mobile devices. In forensic investigations [12], latent fingerprints typically contain ridge information from only partial areas of the fingerprints and gender estimation on this captures may significantly reduce

© Springer Nature Switzerland AG 2019
C. V. Jawahar et al. (Eds.): ACCV 2018, LNCS 11361, pp. 171–186, 2019.
https://doi.org/10.1007/978-3-030-20887-5_11

the list of potential subjects. Moreover, a number of consumer electronic devices, such as smartphones, are beginning to incorporate fingerprint sensors to enable more personalized and user-friendly environments. However, the size of these sensors is limited [9, 29] and thus, only a partial area of the fingerprint can be used. Previous work mainly focused on solutions based on complete fingerprint images [20]. Consequently, these methods perform poorly when it comes to partial fingerprints, as recent work has shown [28].

Unlike previous work, we propose a gender estimation scheme for full and partial fingerprint images. Decisions based on a variable ensemble size of minutiae offer the required flexibility to efficiently deal with unconstrained fingerprint captures of arbitrary size and shape. Due to minutia alignments of the single minutia gender estimators, this scheme also avoids the common rotation and translation invariance problems of fingerprints. In this work, we evaluate several minutia-based gender estimators and fusion schemes on the public available NIST Special Database 4 [10] and build a comparison with state-of-the-art approaches. A convolutional neural network (CNN) approach in combination with our novel adjusted score fusion (ASF) achieved a gender decision performance of 80.07% on full fingerprints and 73.51% on partial fingerprint images and thus, significantly outperform previous work as demonstrated in the rest of the paper.

2   Related Work

In forensics and consumer electronics, applications can benefit from an accurate gender estimation of full and partial fingerprints. Consequently, a lot of research was conducted in the area of biometrics, and related fields such as anthropology, forensics, and medicine.

Most of the biological and anthropological studies came to the conclusion that ridge configurations are controlled by genetics. The influence of genes on the dermatoglyphic development was studied in [2] and a correlation between the sex chromosomes and the total number of ridges per finger was exposed. In [13, 14], Jantz et al. found a similar correlation and formulated the assumption that the Y-chromosome may play a role in dermal ridge development. Further studies observed a difference in the number of ridges that occurs in a certain area [26]. Due to finer epidermal details, they reported a higher ridge density for females compared to males. This hypothesis was supported by Gungadin et al. [30] who reported a decision boundary of 13 ridges/25 mm^2. Ridge densities over this value are more likely to have a female origin while for lower values it is more probable to come from a male origin. Studies on different population groups [17, 22, 25] confirmed similar results.

Experiments were also conducted in the field of biometrics. In 1999, Acree investigated gender estimation by manually counting ridges in a well-defined area [1]. In [16], the mean epidermal ridge breadth (MRB) was used to identify the gender and the results point out that males have a 9% higher MRB than females. On basis of the manually extracted features proposed by Acree, the gender of fingerprints was estimated in [3, 23]. Badawi et al. [3] also used the manual extraction of five ridge features (ridge count, ridge thickness to valley thickness ratio, white lines count, pattern type concordance and ridge count asymmetry) for estimating the gender of a fingerprint. On a set of 2200 fingerprint images, they reported a gender decision accuracy of 88.8% with the use of a neural network approach.

More recent research tackled the problem of manual extracted features and proposed approaches on automated feature extraction. Most of these approaches followed a ridge analysis in the spatial domain [15]. A method based on discrete wavelet transform (DWT) and singular value decomposition was proposed by Gnanasivam et al. [8]. They reported a correct gender classification rate of 88.28%. However, the test/train split was not performed on identity level. Another approach based on DWT was proposed in [27]. They applied a contrast limited adaptive histogram equalization as a preprocessing step before training a neural network and reported an overall accuracy of 96%. Nevertheless, 7.5% of the images were kept out for the experiments and again the test/train split was not performed on identity level. An approach on the basis of local binary pattern (LBP) and local phase quantization (LPQ) operators was proposed by Marasco et al. [19]. They reported a fingerprint gender decision accuracy of 88.7% on full fingerprint images. In [28], however, Rattani et al. demonstrated that these texture descriptors (LBP & LPQ) struggle when it comes to partial fingerprint images. More precisely, they analysed quarter of the fingerprint images and observed a significant drop in the gender decision accuracy down to 54.5% (LBP) and 62.9% (LPQ). Using binary statistical image features (BSIF), they also reported a correct overall gender accuracy of 70.5% on the same partial fingerprint images. However, this experiment was conducted on a database with five times more males than female captures. In their other experiments investigating the gender decision performance for BSIF, they report a gender accuracy which is 22% to 40% higher for male fingerprint captures than for female. Therefore, this metric might be biased.

In [32] and [31], gender estimation was investigated on a small variable portion of fingerprints. They investigated gender classification on the level of a single minutia and analysed its dependency on various attributes as the minutia area, minutia type and its reliability [32]. Using more advanced deep and multi-algorithmic approaches a gender classification rate of 62.47% was reported in [31]. However, they only analysed the gender decision performance on the small-scale level of a single minutia.

So far, automated gender estimators were either optimized on full fingerprint images or on the small-scale level of a single minutia. This work proposes a gender estimator scheme that can intrinsically deal with partial fingerprint captures of arbitrary size and shape.

3   Methodology

Estimating the gender of partial fingerprints requires an algorithm that can deal with their significant variance in shape and size. Therefore, the proposed methodology builds a gender classifier based on an ensemble of minutiae to obtain the needed flexibility. In Sect. 3.1, the proposed concept is described and the interaction of the different components are illustrated. Section 3.2 describes the utilized extraction of minutia region features as well as how these are used to build a single minutia gender estimator. The fusion of the confidence scores of this estimator for each minutia is further explained in Sect. 3.3.

3.1 Proposed Concept

The novelty in the concept proposed in this work is based on classifying ensembles of minutiae and therefore, significantly differs from the concepts used in previous work. In Fig. 1, the different concepts are illustrated as workflows. Previous work followed a classical workflow in which a full fingerprint image is captured and preprocessed. The preprocessing includes an enhancement of the image quality (e.g. by applying a histogram equalisation [27]), normalisation and alignment steps. Next, features are extracted and forwarded to a binary gender classifier to produce a decision score. In the decision making step, this score is used to obtain a final gender decision. This classical workflow is well established for full fingerprint images. However, the feature extraction is highly dependent on the preprocessing step, which faces problems when it comes to partial fingerprint images.

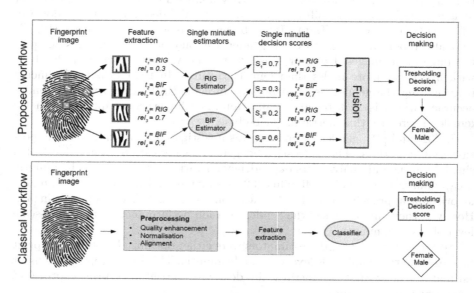

Fig. 1. Comparison of the workflows: (1) on the top the proposed ensemble-based workflow is shown including the in-depth information as minutia type t and reliability rel. (2) on the bottom the classical workflow is illustrated. It consists of a linear flow of preprocessing steps, feature exaction and classification.

In this work, a novel fingerprint gender estimation concept is proposed. This concept is built on an ensemble of single minutia gender estimators. Therefore, it is intrinsically able to deal with partial fingerprints of arbitrary size and shape. Given a full or partial fingerprint image, the feature extraction process includes creating the minutia region features for each minutia of the fingerprint as well as determining its type and reliability. Depending on the type of each minutia, each minutia feature vector is fed into its single minutia estimator and outputs a score s. As a result, each minutia can be described as a tuple (s, rel, t) consisting of a minutia score s, its reliability rel and its minutia type t. During fusion, these information is combined in order to obtain a final gender decision score and thus, a final decision.

3.2   Feature Extraction and Model Building

In order to extract discriminate features, this work utilizes features extracted from single minutia regions [32]. Given a grey-scale fingerprint image, the location and orientation of each minutia m is determined using the NIST Fingerprint Image Software (NFIS) MINDTCT [33] as well as its reliabilities $rel_m \in [0, 1]$ and minutia types $t_m \in \{BIF, RIG\}$. To extract the minutia region features f^m, an area of 20×20 pixels around each minutia m is captured such that each minutia location is in the center and each minutia orientation shows in the same direction. Extracting features with such an alignment concerning the minutia position and orientation, makes them robust to rotation and translation invariance problems, which most fingerprints algorithms suffer. In order to make the scale of the features comparable across different minutia regions, a z-score normalisation is further applied.

With the goal of obtaining a single gender decision score for each minutia, several single minutia gender estimator models are trained on the minutia region features. The models were selected from the most stable single minutia estimators from previous work [31,32]. This includes logistic regression (LogReg), support vector machines (SVM), random forest (RF) [32] and a convolutional neural network (CNN) approach [31].

3.3   Minutiae Ensemble Fusion

The main contribution of this work deals with the high flexibility required for gender estimation of partial fingerprint captures. In this work, weak single minutia gender estimators are utilized as a building block to create a strong ensemble classifier. For fusing this ensemble three approaches were considered. These are illustrated in Fig. 3. First, a simple sum fusion approach was used combing the estimator scores only. However, reliability and minutia type information is available and in Fig. 2 it can be seen that these attributes have a big impact on the performance. In order to include this additional information in the decision process, two procedures were applied. First, the additional information is used to create a weight for each minutia score (weighted-sum approach). Second, this information is used to adjust the minutia scores (adjusted score fusion).

Given a dataset $\mathcal{D}$ consisting of information tuples (s_m, rel_m, t_m) for each minutia m, where s_m is the minutia estimator score given the region features f^m, rel_m describes the minutia reliability and t_m represents its minutia type. The dataset is divided in a training set $\mathcal{D}_{train}$, a development set $\mathcal{D}_{dev}$, and a test set $\mathcal{D}_{test}$. Moreover, a z-score normalisation is trained on the scores of $\mathcal{D}_{dev}$, in order to scale the confidence scores of the estimator.

Sum fusion - One of the simplest score-level fusion approaches can be created by using the non-parametric sum fusion

$$SUM(\{s\}) = \frac{1}{n} \sum_{i=1}^{n} s_i \qquad (1)$$

which gets a set of n normalized minutia scores $\{s\}$ from the estimator and outputs the average score for the final gender decision. However, this approach ignores some of the available information and adding this information can further enhance the overall performance.

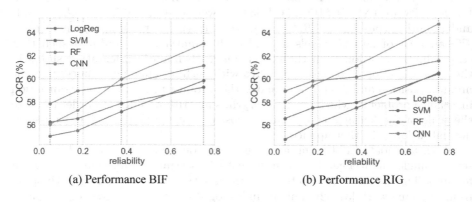

(a) Performance BIF   (b) Performance RIG

Fig. 2. Analysis of the reliability dependence on the minutia type: in (a) the COCR is shown over four different reliability values for BIF, while in (b) the same is shown for RIG. These reliability values were seleced by clustering the reliability densities and choosing the mean values of the four resulting clusters as evaluation points [31].

Weighted-sum fusion - In the weighted-sum fusion approach [5]

$$WSUM(\{s, rel, t\}) = \frac{1}{n} \sum_{i=1}^{n} \omega_{t_i}(rel_i) \cdot s_i, \qquad (2)$$

the confidence scores $\{s\}$ for each minutia is associated with a weight

$$\omega_t(rel) = \frac{f_t(rel)}{\sum_{t \in \mathcal{T}} \int_0^1 f_t(r) \, dr} \qquad (3)$$

capturing the gender decision performance dependent on the minutia type $t \in \mathcal{T} = \{BIF, RIG\}$ and its reliability rel. The function

$$f_t(rel) = \frac{\Delta COCR_t}{rel_2 - rel_1} (rel - rel_1) + COCR_t(rel_1) \tag{4}$$

with

$$\Delta COCR_t = COCR_t(rel_2) - COCR_t(rel_1) \tag{5}$$

describes a linear fit on the development set performance. The values $rel_1 = 0.05$ and $rel_2 = 0.75$ describe the mean values for the first (lowest) and the last (largest) groups of reliabilities [31] (see Fig. 2) in $\mathcal{D}_{dev}$ and calibrate the fit. Using the weights $\omega_t(rel)$ enables to dynamically weight the estimator outputs. It assumes a linear correlation between the reliability of a minutia and its expected performance in gender classification, which is a justified assumption concerning Fig. 2.

Adjusted Score Fusion - In order to include additional information in the decision process, the adjusted score fusion (ASF) approach is proposed, which adjusts the estimator scores for each minutia using a logistic regression model. For each minutia a vector $x^{(i)} = (s, rel, t)$ is defined, consisting of an estimator score s, its reliability rel and its minutia type t. The adjusted score fusion

$$ASF\left(\{x^{(i)}\}_{i=1,\ldots,N}\right) = \frac{1}{n} \sum_{i=1}^{n} \sigma(w^T \cdot x^{(i)} + b) \tag{6}$$

is defined as the sum fusion over adjusted scores $\hat{s}_i = \sigma(w^T x^{(i)} + b)$, where $\sigma(\cdot)$ describes the logistic function. The parameters w and b can be determined by maximum likelihood estimation of the used logistic regression model.

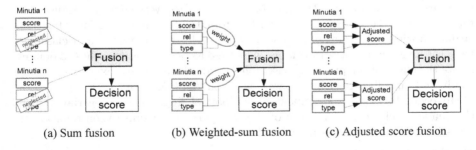

(a) Sum fusion   (b) Weighted-sum fusion   (c) Adjusted score fusion

Fig. 3. Illustration of the different methods used for fusing the minutia ensemble. In (a) the sum fusion approach is shown, which neglects additional minutia information. In (b) its weight-based extension is shown, processing the additional information in a weight for each minutia. (c) presents the proposed adjusted score fusion approach. This approach refines the score values for each minutia by processing all minutia information together.

4 Experimental Setup

Database - In order to investigate gender information of full and partial fingerprints, a database is required which contains gender information of each fingerprint
image as well as it has to resemble fingerprint captures occurring in forensic scenarios. For this reasons, the public available NIST Special Database 4 [10] was used
for the experiments. This non-sensor database consists of 8-bit grey scale rolled
fingerprint images from 2000 subjects. One finger per subject was captured in two
sessions such that each finger is approximately captured an equal number of times.
The images were split into 375 pairs of female and 1625 pairs of male captures. With
a resolution of 500 dpi (19.7 pixels per millimeter) each image consists of 512×512
pixels. On average, a single fingerprint in the database contains 63 bifurcations
(BIF) and 66 ridge endings (RIG). This allows to conduct experiments with more
than half a million minutia regions.

Evaluation Metric - In this work, all experiments were evaluated in terms of
correct overall classification rate (COCR). This metric was proposed by Rattani
et al. [28] and describes the percentage of test samples whose gender was correctly
classified. However, the utilized database only contains 18.75% of female captures
and for such unbalanced classes, the COCR will be significantly biased from the
majority class. In order to prevent this bias, all results are reported in COCR at
an operating point (decision threshold) at which the correct classification rate
of female captures equals the correct classification rate of male captures.

Investigations - The proposed ensemble approaches are evaluated on full and
partial fingerprint images. The evaluation aims at investigating the effect of the
different minutia types with different base algorithms and fusion rules.

   In order to enable a fair comparison between state-of-the-art and the proposed methodology for fingerprint gender estimation, two approaches from previous work were evaluated on our database and compared against a variant of
the proposed gender estimation scheme. For the state-of-the-art approaches, the
works from Rattani *et al.* [28] and Marasco *et al.* [19] were chosen, because they
reported the most accurate results with automated feature extraction which was
evaluated with test/train splits on identity level. Furthermore, the four classifiers
from the single minutia performance were used as baseline approaches to demonstrate the effect of the proposed ensemble-based scheme. These experiments were
conducted on full fingerprints as well as on partial fingerprint captures.

   In order to make a statement about partial fingerprints of variable sizes,
the gender decision performance of the non-parametric sum fusion approach is
investigated over a various number of minutiae.

Workflow Details - For each experiment, a 10 fold cross-validation is applied
on identity level over all finger types. In each fold and per finger type and
gender, 15 identity pairs were chosen for the training set, 8 identity pairs were
chosen for the development set, and the remaining ones were used in a test set.
By doing that over 80k minutia region samples (600 images) can be used for the
training set, 40k samples (320 images) for the development set, and 400k samples

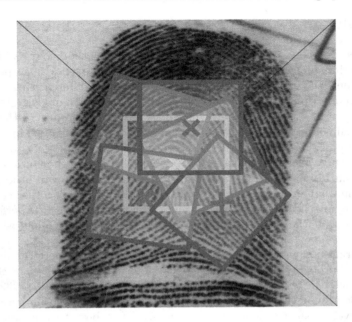

Fig. 4. Partial fingerprint construction: five points were randomly chosen from a 150×150 pixel area around the center of the image (green). For each of these points a window of size 150×150 pixels (blue) were captured with a random direction and used as a partial fingerprint. (Color figure online)

(3080 images) for the testing set. Choosing the same number of female and male identities enables estimator and fusion rule training on balanced data sets, while the chosen decision threshold for the COCR allows an unbiased performance estimate on the test set.

The experiments with the full fingerprints consider the whole fingerprint images as input. For the experiments with partial fingerprints, five points per image were randomly chosen from a 150×150 pixel area around the center of the image. Around each of these points a window of size 150×150 pixels were captured with a random direction and used as a partial fingerprint. A visualisation of this process can be seen in Fig. 4. This window size was adjusted to approximately match the size of the Apple Touch ID sensor of 8×8 mm^2 [29].

For training, all estimators are trained on the training set. The hyperparameters of the baseline estimators were optimised using 20 steps of Bayesian optimisation, while for the state-of-the-art approaches the reported hyperparameters were used. For algorithms working on minutia-level, estimators are trained on the sets of BIF and RIG separately. The normalisation and fusion schemes were trained on the development set. The final performance of the trained models were evaluated on the test set. In order to make the results comparable, every experiment was performed on the same cross-validation split.

5 Results

The gender decision performance of the proposed methodology is shown in Table 1 for full fingerprint images and in Table 2 for partial fingerprint captures. Further, the experiment were conducted on four different single minutia estimator models and two minutia types. These estimator models are logistic regression (LogReg), support vector machines (SVM), random forest (RF) and convolutional neural network (CNN). The minutia types are divided into bifurcations (BIF) and ridge endings (RIG). For fusing the decision scores of the base estimators, three fusion schemes are investigated, namely sum fusion (SUM), weighted-sum fusion (WSUM), and our proposed adjusted score fusion (ASF). These schemes are evaluated on BIF only, on RIG only and together (Both). In order to compare the general effect of using a minutia ensemble, the performance of the single minutia gender estimators is also shown (single minutia) for BIF and RIG.

In general, it can be observed that RIG achieve higher performances than BIF. This is probably because most gender information is stored in the ridge density, which can be more accurately predicted in the case of RIG [32]. Using an ensemble of minutiae significantly improves the gender decision performance over the level of such a single base estimator. The best base estimator also achieves the best results in the ensemble settings. Furthermore, including more information in the fusion process enhances the final performance as it can be seen in the improved results for WSUM and ASF. At this point, adjusting the score (ASF) leads to better results then assigning weights to the scores (WSUM), probably because it includes additional information in a more flexible way.

Comparing the full and partial fingerprint results shows that full fingerprint images achieve a better performance than partial captures due to the fact that they contain more (minutiae) information. In both cases, the best performance was achieved by using CNN with ASF. For partial fingerprints this lead to a COCR of 73.51% and for full fingerprint images a COCR of 80.07% was achieved.

In Table 3, the state-of-the-art and baseline results are shown and compared with the proposed approach for full fingerprints as well as for partial fingerprint images. The first approach is from Rattani et al. [28] and uses local binary pattern (LBP) as a texture descriptor in combination with SVM. The second approach is similar to Marasco et al. [19] and works with a discrete wavelet transform (DWT) in combination with the k-nearest neighbours algorithm (KNN). It can be seen that in all cases the gender decision performance on partial fingerprints is significantly lower than for full fingerprint captures. For the state-of-the-art and baseline approaches this performance decrease is about 16% while for the proposed approach the decrease amounts to 9%. Further, the performance of state-of-the-art and baseline approaches on partial fingerprints, containing around 16 minutiae per capture, is similar or worse to the gender decision performance achieved by a single minutia. It can be seen that approaches from previous work achieve higher performance than the baselines without preprocessing steps (None). Our proposed approach reduced the classification error by 19.93% and by 26.49% on full and partial fingerprint images in comparison to the

Table 1. Performances of the proposed methodology evaluated on full fingerprint images. The results are divided into the used minutia types (BIF/RIG/Both), the used fusion scheme (SUM/WSUM/ASF) and the estimator used for evaluating the minutia regions. For comparison reasons also the single minutia estimator performance is shown for BIF and RIG.

		COCR (%)			
		LogReg	SVM	RF	CNN
BIF	SUM	71.17 ± 1.92	70.34 ± 1.80	71.45 ± 1.39	74.41 ± 2.38
	WSUM	71.24 ± 1.80	70.35 ± 1.72	71.45 ± 1.39	74.48 ± 2.20
	ASF	71.31 ± 1.72	70.34 ± 2.02	72.21 ± 1.74	74.48 ± 2.25
RIG	SUM	71.99 ± 0.77	69.72 ± 2.12	71.93 ± 1.66	74.69 ± 2.53
	WSUM	72.07 ± 0.77	69.99 ± 2.01	71.93 ± 1.66	74.75 ± 2.73
	ASF	71.86 ± 1.23	69.93 ± 1.88	72.41 ± 1.57	74.12 ± 2.27
Both	SUM	76.62 ± 2.17	75.10 ± 1.86	76.21 ± 2.14	79.66 ± 1.70
	WSUM	76.76 ± 2.09	75.10 ± 1.99	76.21 ± 2.12	80.00 ± 1.72
	ASF	**77.03 ± 2.23**	**75.40 ± 1.88**	**77.10 ± 1.99**	**80.07 ± 1.84**
Single minutia	BIF	58.48 ± 0.30	59.39 ± 0.50	60.81 ± 0.62	60.94 ± 0.93
	RIG	59.13 ± 0.25	59.27 ± 0.61	61.42 ± 0.47	61.85 ± 0.74

Table 2. Performances of the proposed methodology evaluated on partial fingerprint images. The results are divided into the used minutia types (BIF/RIG/Both), the used fusion scheme (SUM/WSUM/ASF) and the estimator used for evaluating the minutia regions. For comparison reasons also the single minutia estimator performance is shown for BIF and RIG.

		COCR (%)			
		LogReg	SVM	RF	CNN
BIF	SUM	66.83 ± 1.00	66.69 ± 1.90	68.35 ± 1.04	70.55 ± 1.24
	WSUM	67.03 ± 2.04	67.45 ± 1.10	68.41 ± 1.41	70.83 ± 1.48
	ASF	67.38 ± 1.90	67.17 ± 2.07	69.38 ± 1.29	70.41 ± 1.65
RIG	SUM	67.86 ± 1.58	66.41 ± 1.57	69.51 ± 1.44	71.66 ± 1.29
	WSUM	68.14 ± 1.54	66.90 ± 1.45	68.62 ± 1.67	71.51 ± 1.07
	ASF	68.34 ± 1.17	66.55 ± 2.27	69.03 ± 1.29	71.58 ± 1.68
Both	SUM	68.90 ± 1.96	69.31 ± 1.28	71.62 ± 1.67	73.03 ± 2.71
	WSUM	**70.14 ± 1.38**	69.51 ± 1.77	71.10 ± 1.56	73.10 ± 1.95
	ASF	69.31 ± 2.07	**70.34 ± 2.16**	**71.72 ± 0.93**	**73.51 ± 1.35**
Single minutia	BIF	58.48 ± 0.30	59.39 ± 0.50	60.81 ± 0.62	60.94 ± 0.93
	RIG	59.13 ± 0.25	59.27 ± 0.61	61.42 ± 0.47	61.85 ± 0.74

approaches from previous work [19,28]. In a deployment scenario, aging effects will influence this performance. The aging of the skin causes it to be more dry and loose in comparison to younger skin. This effects the quality of fingerprints [21] and thus, the detection reliability of the minutiae. Our proposed methodology inherently take that into account by considering the detection reliability of each minutia. Experimentally, this can only be proven on gender and age labelled data, which is currently unavailable.

Table 3. Comparison of state-of-the-art approaches and the proposed approach based CNN and ASF. The performance on full and partial fingerprints is reported in COCR.

	COCR (%)	
	Full FP	Partial FP
Our approach	**80.07** ± 1.84	**73.51** ± 1.35
DWT + KNN [19]	75.29 ± 3.28	63.04 ± 1.22
LBP + SVM [28]	74.67 ± 0.97	61.94 ± 1.15
None + LogReg	65.82 ± 1.08	50.23 ± 0.60
None + SVM	65.64 ± 0.70	53.29 ± 3.62
None + RF	69.82 ± 1.08	59.02 ± 0.89
None + CNN	62.27 ± 4.76	53.43 ± 1.20

In order to make a performance estimate of partial fingerprints of arbitrary sizes, Fig. 5 shows the gender decision performance over the number of minutiae used. A certain number of minutiae were chosen randomly and combined using the sum fusion (SUM) with different base estimators. In order to show the efficiency of the general concept of classifying an ensemble of minutiae, the sum fusion approach was chosen, because it combines the single minutia gender scores in a non-parametric manner. The red and blue areas represent the usual number of minutiae in partial and full fingerprint captures [12]. The darker shaded areas indicate the performance of previous work [19,28] for this cases. It can be seen that for all base estimators, the performance increases sharply with an rising number of minutiae used and saturates around 40 minutiae. For full fingerprint captures, it can be seen that the basic sum fusion approach can substantially surpass state-of-the-art performance. This trend can be observed more clearly for partial fingerprint captures, since the proposed approach can intrinsically deal with missing minutiae [4]. Here, the performance from previous work can be exceeded by fusing only two random minutiae. Combining all available minutiae from a partial fingerprint capture leads to a more significant performance improvement compared to previous work.

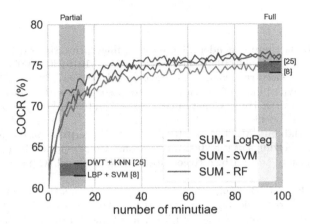

Fig. 5. Performance achieved by number of minutiae using sum fusion and different base estimators. The red and blue area represent the number of minutiae of partial and full fingerprints. The darker shaded areas shows the performance of previous work. (Color figure online)

6 Conclusion

A wide range of applications can benefit from an accurate fingerprint gender classification. Previous work mainly offered solutions for this task using complete fingerprint captures. However, applications in forensics or consumer electronics require dealing with partial fingerprints of variable sizes. In this work, we presented a flexible gender estimation scheme based on minutia ensemble gender estimator models. We jointly analysed several single minutia gender estimators and fusion methods on a publicly available database. The best performance was achieved by a convoluational neural network in combination with a novel adjusted score fusion approach that adapts the single estimator scores from each minutia by including information about the minutia type and the minutia detection reliability. This lead to a COCR of 80.07% for full fingerprint images. For partial fingerprint captures of size of a smartphone sensor, the proposed approach reached a COCR of 73.51%. Compared to previous work evaluated on this database these results represent a reduction in the gender classification error of 19.34% and 28.33% for full and partial fingerprint images. The proposed methodology is therefore characterised by its enhanced accuracy and its intrinsic ability to deal with partial fingerprints of arbitrary size and shape.

Acknowledgement. This work was supported by the German Federal Ministry of Education and Research (BMBF) as well as by the Hessen State Ministry for Higher Education, Research and the Arts (HMWK) within the Center for Research in Security and Privacy (CRISP).

References

1. Acree, M.A.: Is there a gender difference in fingerprint ridge density? Forensic Sci. Int. **102**(1), 35 – 44 (1999). https://doi.org/10.1016/S0379-0738(99)00037-7, http://www.sciencedirect.com/science/article/pii/S0379073899000377

2. Alter, M.: Is hyperploidy of sex chromosomes associated with reduced total finger ridge count? Am. J. Hum. Genet. **17**(6), 473–475 (1965). http://europepmc.org/articles/PMC1932642

3. Badawi, A., Mahfouz, M., Tadross, R., Jantz, R.: Fingerprint-based gender classification. In: Proceedings of the 2006 International Conference on Image Processing, Computer Vision, and Pattern Recognition IPCV2006, vol. 1, pp. 41–46, January 2006

4. Damer, N., Fúhrer, B., Kuijper, A.: Missing data estimation in multi-biometric identification and verification. In: 2013 IEEE Workshop on Biometric Measurements and Systems for Security and Medical Applications, pp. 41–45, September 2013. https://doi.org/10.1109/BIOMS.2013.6656147

5. Damer, N., Opel, A., Nouak, A.: CMC curve properties and biometric source weighting in multi-biometric score-level fusion. In: 17th International Conference on Information Fusion (FUSION), pp. 1–6, July 2014

6. Damer, N., Terhörst, P., Braun, A., Kuijper, A.: General borda count for multi-biometric retrieval. In: 2017 IEEE International Joint Conference on Biometrics (IJCB), pp. 420–428, October 2017. https://doi.org/10.1109/BTAS.2017.8272726

7. Damer, N., Terhörst, P., Braun, A., Kuijper, A.: Indexing of single and multi-instance iris data based on LSH-forest and rotation invariant representation. In: Felsberg, M., Heyden, A., Krüger, N. (eds.) CAIP 2017. LNCS, vol. 10425, pp. 190–201. Springer, Cham (2017). https://doi.org/10.1007/978-3-319-64698-5_17

8. Gnanasivam, P., Muttan, S.: Fingerprint gender classification using wavelet transform and singular value decomposition. CoRR abs/1205.6745 (2012). http://arxiv.org/abs/1205.6745

9. Han, B., Marciniak, C., Westerman, W.: Fingerprint sensing and enrollment. US Patent 9,715,616, 25 July 2017. http://www.google.nl/patents/US9715616,

10. Watson, C.I., Wilson, C.: NIST special database 4, November 1992

11. Jain, A.K., Dass, S.C., Nandakumar, K.: Can soft biometric traits assist user recognition? In: Proceedings of SPIE, vol. 5404, pp. 561–572 (2004). http://www.cse.msu.edu/biometrics/Publications/SoftBiometrics/JainDassNandakumar_SoftBiometrics_SPIE04.pdf

12. Jain, A.K., Nandakumar, K., Ross, A.: 50 years of biometric research: accomplishments, challenges, and opportunities. Pattern Recognit. Lett. **79**(Supplement C), 80–105 (2016). https://doi.org/10.1016/j.patrec.2015.12.013, http://www.sciencedirect.com/science/article/pii/S0167865515004365

13. Jantz, R.L., Hawkinson, C.H., Brehme, H., Hitzeroth, H.W.: Finger ridge-count variation among various subsaharan african groups. Am. J. Phys. Anthropol. **57**(3), 311–321 (1982). https://doi.org/10.1002/ajpa.1330570308

14. Jantz, R.L.: Sex and race differences in finger ridge-count correlations. Am. J. Phys. Anthropol. **46**(1), 171–176 (1977). https://doi.org/10.1002/ajpa.1330460122

15. Kaur, R., Mazumdar, S.G., Tech, M., Bhilai, R.: Fingerprint based gender identification using frequency domain analysis. Int. J. Adv. Eng. Technol. **3**, 295 (2012)

16. Kralik, M., Novotný, V.: Epidermal ridge breadth: an indicator of age and sex in paleodermatoglyphics. Variability Evol. **11**, 5–30 (2003)

17. Kunter, M., Ruehl, M.: Laterality and sex differences in quantitative fingerprint ridge analysis in a middle european sample (giessen, hessen). Anthropologischer Anzeiger; Bericht ueber die biologisch-anthropologische Literatur **53**(1), 79–90 (1995)

18. Maltoni, D., Maio, D., Jain, A.K., Prabhakar, S.: Handbook of Fingerprint Recognition, 2nd edn. Springer, London (2009). https://doi.org/10.1007/978-1-84882-254-2

19. Marasco, E., Lugini, L., Cukic, B.: Exploiting quality and texture features to estimate age and gender from fingerprints. In: Proceedings of SPIE - The International Society for Optical Engineering, vol. 9075, p. 90750F, May 2014

20. Mason, S., Gashi, I., Lugini, L., Marasco, E., Cukic, B.: Interoperability between fingerprint biometric systems: an empirical study. In: 2014 44th Annual IEEE/IFIP International Conference on Dependable Systems and Networks, pp. 586–597, June 2014. https://doi.org/10.1109/DSN.2014.60

21. Modi, S.K., Elliott, S.J., Whetsone, J., Kim, H.: Impact of age groups on fingerprint recognition performance. In: 2007 IEEE Workshop on Automatic Identification Advanced Technologies, pp. 19–23, June 2007. https://doi.org/10.1109/AUTOID.2007.380586

22. Nayak, V.C., et al.: Sex differences from fingerprint ridge density in chinese and malaysian population. Forensic Sci. Int. **197**(1), 67 – 69 (2010). https://doi.org/10.1016/j.forsciint.2009.12.055, http://www.sciencedirect.com/science/article/pii/S0379073809005568

23. Nithin, M., Manjunatha, B., Preethi, D., Balaraj, B.: Gender differentiation by finger ridge count among south indian population. J. Forensic Legal Med. **18**(2), 79–81 (2011). https://doi.org/10.1016/j.jflm.2011.01.006, http://www.sciencedirect.com/science/article/pii/S1752928X11000126

24. Nixon, M.S., Correia, P.L., Nasrollahi, K., Moeslund, T.B., Hadid, A., Tistarelli, M.: On soft biometrics. Pattern Recognit. Lett. **68**(Part 2), 218–230 (2015). https://doi.org/10.1016/j.patrec.2015.08.006, http://www.sciencedirect.com/science/article/pii/S0167865515002615., special Issue on "Soft Biometrics"

25. Oktem, H., Kurkcuoglu, A., Pelin, I.C., Yazici, A.C., Aktaş, G., Altunay, F.: Sex differences in fingerprint ridge density in a Turkish young adult population: a sample of baskent university. J. Forensic Legal Med. **32**(Supplement C), 34–38 (2015). https://doi.org/10.1016/j.jflm.2015.02.011, http://www.sciencedirect.com/science/article/pii/S1752928X15000323

26. Omidiora, E., Ojo, O., Nureni, Y., Tubi, T.O.: Analysis, design and implementation of human fingerprint patterns system towards age & gender determination, ridge thickness to valley thickness ratio (RTVTR) & ridge count on gender detection. Int. J. Adv. Res. Artif. Intell. **1**, 57–63 (2012)

27. Prabha, P., Sheetlani, J., Pardeshi, R.: Fingerprint based automatic human gender identification. Int. J. Comput. Appl. **170**(7), 1–4 (2017). https://doi.org/10.5120/ijca2017914910, http://www.ijcaonline.org/archives/volume170/number7/28079-2017914910

28. Rattani, A., Chen, C., Ross, A.: Evaluation of texture descriptors for automated gender estimation from fingerprints. In: Agapito, L., Bronstein, M.M., Rother, C. (eds.) ECCV 2014. LNCS, vol. 8926, pp. 764–777. Springer, Cham (2015). https://doi.org/10.1007/978-3-319-16181-5_58

29. Roy, A., Memon, N., Ross, A.: Masterprint: exploring the vulnerability of partial fingerprint-based authentication systems. IEEE Trans. Inf. Forensics Secur. **12**(9), 2013–2025 (2017). https://doi.org/10.1109/TIFS.2017.2691658

30. Gungadin, S.: Sex determination from fingerprint ridge density. Internet J. Med. Update **2**, 4–7 (2007)
31. Terhörst, P., Damer, N., Braun, A., Kuijper, A.: Deep and multi-algorithmic gender classification of single fingerprint minutiae. In: 21th International Conference on Information Fusion, FUSION 2018. IEEE, Cambridge, 10–13 July 2018
32. Terhörst, P., Damer, N., Braun, A., Kuijper, A.: What can a single minutia tell about gender? In: 2018 6th International Workshop on Biometrics and Forensics (IWBF), June 2018
33. Watson, C., et al.: User's guide to NIST biometric image software (NBIS) (2007)

Simultaneous Face Detection and Head Pose Estimation: A Fast and Unified Framework

Tingfeng Li and Xu Zhao$^{(\boxtimes)}$

Department of Automation, Shanghai Jiao Tong University, Shanghai, China
{litingfeng,zhaoxu}@sjtu.edu.cn

Abstract. In this paper, we present a fast and unified framework for simultaneous face detection and 3D pose (pitch, yaw, roll) estimation of unconstrained faces using deep convolutional neural networks (CNN). Face detection is implemented with region-based framework as previous work like Faster RCNN. We model the pose estimation as a classification and regression problem: first divide continuous head poses into several discrete clusters, then adjust poses within each class with a class-specific regressor to achieve more accurate results. All classification and regressions for the two tasks are trained and tested simultaneously in one unified network. Our approach runs at 10 fps, which is the fastest implementation among the recently proposed methods as far as we know. Moreover, it is able to predict pose without using any 3D information. Extensive evaluations on several challenging benchmarks such as AFLW and AFW demonstrate the effectiveness of the proposed method with competitive results.

Keywords: Face detection · Head pose estimation ·
Convolutional neural networks

1 Introduction

For the past decades, cameras and smartphones have spread widely and countless photos are captured to record people's daily life. Making full use of these photos helps improve user experience, for example, users can look for photos taken with a particular friend [5]. Other Human Computer Interaction (HCI) devices like smart home requires devices to understand the expression of a person. And all the performances of other various face based applications, from face identification and verification to face clustering, tagging and retrieval, rely on accurate and efficient **face detection**. On the other hand, as another challenging task correlated with face analysis, **head pose estimation** has been found to be useful in human-robot interaction [14], driver attention detection [3] and social behavior analysis [24].

X. Zhao—This research is supported by the funding from NSFC programs (61673269, 61273285, U1764264).

© Springer Nature Switzerland AG 2019
C. V. Jawahar et al. (Eds.): ACCV 2018, LNCS 11361, pp. 187–202, 2019.
https://doi.org/10.1007/978-3-030-20887-5_12

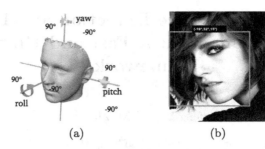

(a)   (b)

Fig. 1. Illustration of head pose estimation problem. (a) The pose of the head is described in the form of three rotation angles: pitch, yaw and roll [18]. (b) An example of predicted head pose.

These two tasks, face detection and head pose estimation, have traditionally been approached as independent problems, yet some methods solve them at the same time as separate components. As having common shared visual appearance of face regions, however, it will be more effective to solve them in a unified framework. And, pose angles could play as the latent variables and have immediate impact on visual appearance of face. It makes these two tasks closely correlated to each other and will be beneficial to model them in a unified framework. Recently, methods are proposed to solve the two tasks in a unified model [25,36,37]. However, these methods utilize facial landmarks to boost performance either explicitly or implicitly. In addition, multiple stages for training or testing makes their system not pure end-to-end, which discounts the final performance in the cost of time and accuracy.

In this paper, benefiting from deep Convolutional Neural Networks (CNN) architecture, our framework is designed to be more compact and efficient and completely end-to-end (see Fig. 2), that is, input an image, detect face and estimate head pose with only one network, without using face landmark information. In recent years, deep CNN have achieved significant performance improvement on benchmarks [5,23,26,34] for face detection. Furthermore, as a general deep CNN architecture for object detection, Ren *et al.* proposed an efficient region-based network Faster RCNN [27], showing outstanding performance in object detection. We adopt this architecture as our basic face detection network, then embed pose estimation module into this network, forming an end-to-end framework for both tasks. Specifically, we propose a novel implicit coarse-to-fine search scheme for pose estimation, which is embedded into the deep CNN architecture following multi-task learning pipeline.

For 3D head pose estimation, we expect the system to infer the orientation of person's head relative to camera coordinate, described by the rotation angles: pitch, yaw and roll, as shown in Fig. 1. Essentially, head pose estimation can be formulated as a regression problem, however, these three angles have to be recovered from monocular image to find the location of a pose in 3D pose space, precisely. Learning such a mapping function is challenging, as it is expected to be very sensitive to subtle change of face appearance, meanwhile, robust enough to noise interference. Besides, directly searching in the original space requires highly descriptive features and will be inefficient considering computational complexity. Hence we adopt a coarse-to-fine scheme, that is, we model head pose estimation as a process of discrete classification followed by fine-grained continuous regression. By firstly inferring the coarse pose through classification, a rough range of pose angles corresponding to a type of face appearance (e.g. frontal face) is determined. Then the pose is tuned again by a regressor with respect to this class to obtain a fine estimation. In so doing, we achieve state-of-the-art performance on challenging datasets with high computational efficiency.

To sum up, in this paper, we propose a fast and unified framework for accurate face detection and head pose estimation, and the main contributions can be recapped as follows.

- An effective unified framework is designed as an end-to-end network for both tasks: the input is an image containing faces, while the outputs are bounding boxes of faces and three orientation angles of each face, i.e., the head pose. The proposed method can run at 10 fps, which is fast enough for many real time applications.
- A novel pose-class-specific coarse-to-fine scheme is proposed to be embedded properly into the deep CNN network, for simultaneous pose estimation and face detection.
- New state-of-the-art performance is achieved on challenging unconstrained datasets such as AFLW and AFW.

2   Related Work

2.1   Face Detection

As a basic problem for face analysis, face detection has been studied for a long time. In numerous work, there exists two major categories, namely, rigid-templates based algorithm and Deformable Parts Model (DPM) based ones [35]. The early representative of the first category include the Viola-Jones face detection algorithm and its variations. DPM based model exploit abundant information on face, such as a potential deformation between facial parts, which enables them to combine face detection with facial part localization. Recent CNN based algorithms have shown exceptional results on general object detection, such architecture also have been investigated for face detection [5,23,26,34], in both categories.

In [26], DP2MFD based on Deformable Part Models [6] and deep pyramid features are proposed, reducing gap of DPM in training and testing on deep features. In [5], a single model is trained to fully capture faces in all orientations without pose or landmark annotation and shows good performance on popular benchmark datasets. In [34], facial parts responses by their spatial structure and arrangement are exploited. This method is very good at detecting faces under severe occlusion and variant poses.

In this work, we adopt the deep CNN framework for general object detection [27] as the basic architecture of face detection. Then we further extend it to an end-to-end system for simultaneous face detection and head pose estimation.

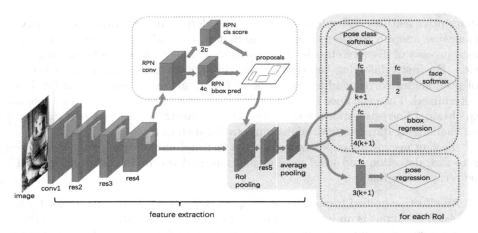

Fig. 2. The architecture of the proposed network. This network consists of four modules: feature extraction module (from conv1 layer to average pooling layer), RPN module (yellow dotted line), face detection module (purple dotted line, which is darker for white-black print), face pose estimation module (red dotted line). Best viewed in color. (Color figure online)

2.2   Head Pose Estimation

Head pose estimation aims at inferring the orientation of person's head relative to camera coordinate, which is very useful to many real life applications such as HCI and surveillance.

Several categories that describe the fundamental approaches underling its implementation have been used to estimate head pose [21]. We just briefly review some typical methods here. Appearance template methods [29] make comparison between a new head image with existing images labeled with discrete pose and find the most likely view. Detector array methods [13] train multiple detectors and assign a discrete pose to the detector with the greatest support. Nonlinear regression methods [22] use nonlinear regression tools to develop a functional mapping from the image or feature data to a head pose measurement. Manifold embedding methods [7] seek low-dimensional manifolds that model the continuous variation in head pose. New images can be embedded into these manifolds and then used for embedded template matching or regression. Recently, CNN-based model for head pose estimation had been developed, for example, in [20], deep CNN network is used to capture head pose in low-resolution RGB-D data.

Even more, it has been shown that learning correlated tasks simultaneously can boost the performance of individual tasks under the context of face detection and head pose estimation [2,25,36,37]. The first work jointly addresses these two tasks is [36]. This method models each facial mark as a part then construct a mixture of trees to capture topological changes, much like DPM based model. Later on, this work was extended to [37]. Then Ranjan *et al.* proposed a multi-task learning framework, HyperFace [25], for face detection, landmark localization, pose estimation and gender recognition. Nevertheless, this work is not exactly end-to-end, since it requires a preprocess step such as Selective Search [31] to generate region of interests (RoIs). Counting all steps together it costs $3s$ per image in total. The most recent work KEPLER [16] is much faster and trained jointly, however, it is an iterative method requiring 5-stage training procedure to achieve adequate performance, which is very time consuming.

As all tasks work under deep CNN architectures, when compares to [16, 25], our network is purely end-to-end for both training and testing. Furthermore, as we adopt a novel pose-class-specific coarse-to-fine scheme, our method is computational efficient while maintaining high accuracy for both face detection and pose estimation.

3 Proposed Framework

The architecture of the proposed network is shown in Fig. 2. We devise the framework based on Faster RCNN [27], which is a region-based framework in general object detection. To further speed up and reduce computation cost, Region Proposal Network (RPN) is proposed to compute proposals that share convolutional layers with object detection networks. Following [27], we train both classifier and bounding box regressor with multi-task loss, which is convenient and has been demonstrated to facilitate object detection performance. In our method, pose-class-specific face detection and pose estimation is integrated naturally into this framework. A two-stage classification and bounding box regression solves face detection, where the first stage is to discriminate pose class defined by clustered angles and the second stage is to classify whether it is a face or not. As for pose estimation, the classification in the first stage helps predict rotation angles more accurately. In this section, we provide brief overview of the system and then describe each component in detail.

3.1 Network Architecture

Our network is constructed based on ResNet-50 [11]. Figure 2 presents the architecture of the whole framework. This framework consists of four major modules: feature extraction module, region proposal network (RPN) module, face detection module, and face pose estimation module. The first module is the backbone from input image to average pooling layer, and the rest three modules are denoted within dotted lines.

RPN Module. In the scenario of multi-category object detection, the RPN in Faster R-CNN [27] was developed as a class-agnostic detector. For single-category (face) detection, RPN is naturally a detector for the only category concerned. We specially tailor the RPN for face detection, as introduced in the following. We adopt the ResNet-50 pre-trained on the ImageNet dataset [15] as the backbone network rather than VGG-16 net [30]. The RPN is built on top of the $res4f$ layer, followed by an intermediate 3×3 convolutional layer with 256 channels and two sibling 1×1 convolutional layers for classification and bounding box regression (more details can be found in [27]). To deal with different scales and aspect ratios of objects, anchors were introduced in the RPN. An anchor is at each sliding location of the convolutional maps and thus at the center of each spatial window. Each anchor is associated with a scale and an aspect ratio. Following the default setting of [27], we use 3 scales (128^2, 256^2, and 512^2 pixels) and 3 aspect ratios (1:1, 1:2, and 2:1), leading to $c = 9$ anchors at each location. Therefore, for a convolutional feature map of size $W \times H$, we have at most $W \times H \times c$ possible proposals.

Face Detection Module. As in [27], this module is based on the Fast R-CNN detector [8] that uses the proposed regions. After feature map generated by *res4f* layer, with the input of RoIs supplied by RPN, an RoI pooling layer is appended to extract a fixed-length feature vector in each RoI. Moreover, this layer helps to excavate local information. By taking full advantage of global and local information, the network is able to classify various scales of objects. After RoI pooling, features are extracted and down sampled by several *res5* conv layers and the following average pooling layer respectively in every region. In the original ResNet, after the average pooling, a *fc* layer of 1000 output is added to classify 1000 categories in ImageNet. In our implementation, this last layer is removed to adapt our two tasks. Specifically, we replace it with three sibling *fc* layers: the first for two-stage classification, the second for bounding box regression and the third for pose regression. The two-stage classification is inspired by SubCNN, in which an additional *subcategory fc* layer is inserted before the *fc* layer for object class classification. In our network, pose class can be interpreted as subcategory, and the corresponding *fc* layer provides pose-class-specific feature for further face classification. For k pose classes, the corresponding *fc* layer outputs $k + 1$ dimensional vector with one additional dimension for the background class. As for the second sibling, we apply class-specific regression, that is, every regressor is tuned for each pose class. Finally, the detection module terminates at three output layers. A softmax function is applied at the first output layer directly based on the output of the "pose class fc" layer for pose class classification. The second output layer operates on the vector generated by "pose class fc" layer. And the last one outputs four real-valued numbers for each of the $k + 1$ pose classes. Each set of 4 values encodes refined bounding-box positions for one of the $k + 1$ classes.

Face Pose Estimation Module. As mentioned previously, estimation can be considered to have two steps: coarse estimation applied as classification and fine estimation implemented as regression. At first thought regression should come after classification. However, inspired by the design of prevalent class-specific bounding box regressors in object detection network architecture RCNN series [8,9,27], we are able to set these two steps as siblings leveraging the natural architecture of neural network. Specifically, we append a *fc* layer for the two parallel steps, i.e., classification and regression (see regions in red dotted line in Fig. 2). Note that we re-utilize the "pose class fc" layer with $k + 1$ outputs for pose classification. While "pose regression fc" layer has $3(k + 1)$ outputs, each

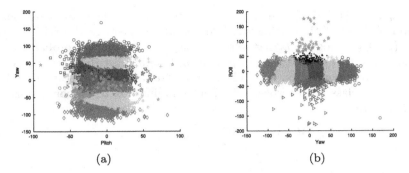

(a)   (b)

Fig. 3. Scatter plot of clusters projected. In this plot, only 9 poses have distinct location on this plane, while the remaining 3 clusters seems like can not be separated clearly. However, these pose classes can be differentiated easily when projected in roll direction.

set of 3 real values encodes the 3 refined rotation angles. For testing, we select the regressor corresponding to the highest pose class score to refine its pose in post process.

3.2 Pose Class

Subcategory has been widely utilized to facilitate object detection. Some methods discover subcategories by clustering objects according to the viewpoint [10]. In [32], Xiang *et al.* utilized subcategory to improve CNN-based detection, where clustering is preformed according to the orientation of the object for pedestrian and cyclist, and each cluster is considered to be a subcategory. Motivated by SubCNN, we introduce pose class for each face, which can be considered as subcategory in face detection. The pose class has two functions: first, it facilitates face detection. Second, it provides coarse pose information, narrowing down search space for further pose angle regression.

We apply kmeans [17] on AFLW dataset to form 12 clusters. Figure 3 shows an example scatter plot of clusters. In this plot, only 9 poses have distinct locations on Pitch-Yaw plane, while the remaining 3 clusters can not be separated clearly. However, when projected in roll direction, these 3 pose classes can be differentiated easily. Thus, on the whole, it is viable to divide continuous real valued pose angles into discrete classes.

3.3 Training

For RPN, we adopt loss function following [27]. While for the detection and estimation network, we utilize multi-task loss for joint pose class classification, face or non-face classification, bounding box regression and head pose regression (Eq. 1),

$$Loss = \lambda_1 L_{posecls}(p,k) + \lambda_2 L_{cls}(p',k')$$
$$+\lambda_3[k \geqslant 1]L_{loc}(t^k,v) + \lambda_4[k \geqslant 1]L_{pose}(a,a^*), \tag{1}$$

where $\lambda_i, i \in \{1, 2, 3, 4\}$ are loss weights to balance their contributions to the overall loss. Both classification losses are implemented with log loss: $L_{posecls}(p, k) = -\log p_k$, $L_{cls}(p', k') = -\log p'_{k'}$, where p is a probability distribution over $K + 1$ pose classes, p' is a probability distribution over $K' + 1$ classes, in our experiments, $K = 12, K' = 1$, k and k' are the truth pose label and the truth class label respectively. For bounding box regression, we use the *smooth* $L1$ loss.

$$L_{loc}(t^k, v) = \sum_{i \in \{x, y, w, h\}} smooth_{L_1}(t_i^k - v_i), \tag{2}$$

in which $t^k = (t_x^k, t_y^k, t_w^k, t_h^k)$ denotes a predicted bounding box for class k, which specifies the pixel coordinates of the center together with width and height of each proposal. And $v = (v_x, v_y, v_w, v_h)$ indicates true bounding box regression targets for pose class k. Each ground-truth bounding box $G = (G_x, G_y, G_w, G_h)$ is specified in the same way. Thus, target $v_i, i \in \{x, y, w, h\}$ is computed as follows,

$$v_x = (G_x - t_x^k)/t_w^k \tag{3}$$

$$v_y = (G_y - t_y^k)/t_h^k \tag{4}$$

$$v_w = \log(G_w/t_w^k) \tag{5}$$

$$v_h = \log(G_h/t_h^k) \tag{6}$$

The *smooth* $L1$ loss is defined as,

$$smooth_{L_1}(x) = \begin{cases} 0.5x^2 & if\ |x| < 1 \\ |x| - 0.5 & otherwise, \end{cases} \tag{7}$$

Similarly, the loss function for pose regression is defined as:

$$L_{pose}(a, a^*) = \sum_{i \in \{p, y, r\}} smooth_{L_1}(d_i - d_i^*), \tag{8}$$

where d_i^* and d_i denotes truth angles and estimated ones respectively. In back-propagation training, derivatives for the multi-task loss are back-propagated to the previous layers.

3.4   Testing

From a given test image, we first extract convolutional features. Then, RoIs are generated by the RPN module. As features in each RoI are pooled by RoI pooling layer from the last conv layer, the subsequent network is able to make predictions for all tasks. In post-process, we select top N boxes according to face softmax output. Then for each box, assign the correspongding pose class with the highest score out of k+1 pose class scores. After determining pose class, more accurate rotation angles are obtained by appling the pose-class-specific pose regression.

With the sharing of convolutional layers among RPN, detection and estimation networks, computation time is reduced substantially. In our experiments, we also found that even without sharing convolutional layers, the computation speed summed up by the two separate networks is still faster than [25] thanks to the RPN proposed by Ren *et al.* [27].

4   Experimental Validations

4.1   Datasets and Evaluation Metric

For face detection, we evaluate the proposed framework on the challenging AFW [36] dataset. For face pose estimation, we carry out experiments on both AFW and AFLW [18] datasets.

AFLW. To test the robustness of our approach for images from real senarios with challenging scale variations, cluttering background and significant shape changes, we utilize AFLW for training and testing to evaluate the estimation performance. AFLW contains 24386 faces annotated with pose (yaw, pitch and roll) and truth bounding box in 21997 real-world wild images. Head poses ranging from 0° to 120° for yaw and upto 90° for pitch and roll exhibiting a large variety in appearance (e.g., pose, expression, ethnicity, age, gender) as well as general imaging and environmental conditions. We use exactly the same 1000 images randomly selected by [25] for testing, all other images for training.

AFW is a very popular benchmark for evaluation of both face detection and head pose estimation algorithms. This dataset provides 205 images with 468 faces in the wild with yaw degree up to 90°. The images tend to contain cluttered background with large variations in face viewpoint, illumination and appearance (aging, sunglasses, make-ups, skin color, expression etc.). Each face is labeled with a bounding box, and a discretized viewpoint (−90° to 90° every 15°) along pitch and yaw directions and (left, center, right) viewpoints along the roll direction. In consistent with other methods, we select 341 faces with height greater than 150 pixels following the protocol of [36].

Evaluation Metric. Following previous works, we use Average Precision (AP) [4], and precision-recall curves to test our face detector on AFW. We demonstrate our estimation results with mean error over all samples and Cumulative Error Distribution (CED) curve. This curve provides the fraction of faces with predicted pose within ±15° error tolerance. Note that we only evaluate on discretized pose predictions rounded to the nearest 15° on AFW, e.g., if a predicted angle is 85°, then it is rounded to 90° rather than 75° for evaluation.

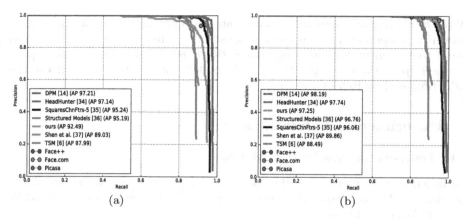

Fig. 4. Precision-recall curves for face detection on AFW dataset. (a) evaluation IoU threshold 0.5, (b) evaluation IoU threshold 0.3. The numbers in the legend are the average precision for the corresponding IoU threshold.

4.2 Setting

The proposed approach is tested on Intel(R) Xeon(R) 2.10 GHz CPU with 32 GB RAM, NVIDIA TITAN X GPU. As in Faster RCNN, we use image-centric sampling, each SGD mini-batch is constructed from a single image. In RPN subnetwork, a mini-batch is expected to have 64 positive RoIs and 64 negative RoIs selected by random sampling for simplicity. For detection and estimation network, a mini-batch is constructed from a single image with 128 RoIs with 1:3 ratio of positive and negative samples. We use a learning rate of 0.001 in the beginning and drop 0.1 every 50k mini-batch iterations for all 200k mini-batches. We use a weight decay of 0.0005 and a momentum of 0.9 [15]. All the experiments including training and testing were performed using the Caffe [12] framework.

4.3 Face Detection

Face Detection Results and Error Analysis. Notice that we evaluate on different training and testing dataset. As pointed out by Mathias *et al.* [19], one important problem in the evaluation of face detection methods is the mismatch of face annotations in the training and testing stages. Specifically, it is not so obvious about how to define the rectangle around face for profile and semi-profile views. In AFLW, the rectangle tends to be larger and more like a square. When it comes to profile face, the rectangle extends to non-face areas. While in AFW, the rectangle is likely to be tighter.

To solve this problem, we follow the remedial method proposed by Mathias *et al.* [19] to search for a global rigid transformation of the detection outputs to maximize the overlapping ratio with the ground-truth annotations with the shared evaluation tool [19]. However, our annotations cannot be simply linearly

Table 1. AP comparison of our method with different configurations.

IoU 0.3	IoU 0.5	Transformation	AP (%)
√		√	**97.25**
	√	√	92.49
	√		81.55
√			96.85

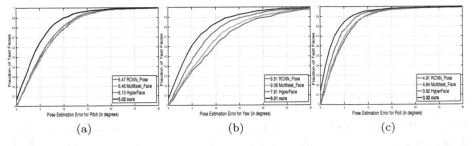

(a)   (b)   (c)

Fig. 5. Performance evaluation of pose estimation on AFLW dataset for (a) roll (b) pitch and (c) yaw angles respectively. The numbers in the legend are the mean error in degrees for the respective pose angles.

mapped to AFW annotations. After the global transformation step the mismatches still exist. According to our analysis, the inferior performance on AFW may primarily be caused by poor localization resulting from annotation mismatch. To demonstrate this, we set evaluation IoU to 0.3 and 0.5 on all methods to make a fair comparison. Table 1 presents results with different IoU thresholds with or without transformation. When IoU is decreased from 0.5 to 0.3, we can see that with transformation, our method is improved a lot from 92.49% to 97.25%, and accuracy is improved even more without transformation, from 81.55% to 96.85%. Moreover, even with consistent IoU threshold 0.3, AP of our method increases to 0.4 by applying transformation, suggesting that our method tends to predict results following the annotation style of AFLW dataset. We also provide precision-recall curves produced by different methods with transformation under IoU 0.5 and 0.3 in Fig. 4. We compare our approach with the following: (1) Deformable part model (DPM) [6], (2) HeadHunter [19], (3) SquaresChnFtrs-5 [1], (4) Structured Models [33], (5) shen *et al.* [28], (6) TSM [36], (7) Google Picasa's face detector, manually scored by inspection, (8) Face.com's face detector, (9) Face++'s face detector. As can be seen, AP of our method increases largely while others only have slightly improvement. Therefore, we can conclude that the inferiority of our method comparing to others is mainly due to annotation mismatch between training and testing datasets.

Table 2. Comparison of head pose estimation with other methods on AFLW dataset, where speed is for both tasks, i.e., face detection and pose estimation. HyperFace is based on AlexNet, while HF-ResNet is based on ResNet-101. ours_vgg is based on VGG16, while ours_res is based on ResNet-50.

Methods	Mean absolute error			Speed
	Pitch	Yaw	Roll	
HyperFace [25]	6.13	7.61	3.92	0.33 fps
HF-ResNet [25]	5.33	6.24	**3.29**	0.33 fps
KEPLER [16]	5.85	6.45	8.75	3∼4 fps
ours_vgg	5.36	6.51	3.41	**10 fps**
ours_res	**5.02**	**6.01**	3.32	3 fps

4.4 Head Pose Estimation

AFLW. We compare MAE of different pose angles and speed with other state-of-the-arts in Table 2. As we can see, except for roll, the proposed method achieves the best performance with the speed of 3 fps on MAE when using ResNet-50. In addition, when we apply VGG16 as backbone, the method is accelerated to 10 fps while maintains relatively lower MAE in both pitch and roll comparing to HyperFace, only slightly higher than KEPLER in yaw. This indicates that our method is the best in both accuracy and efficiency. Figure 5 shows the cumulative error distribution curves on AFLW dataset in the order of pitch, yaw and roll respectively. It can be seen that the proposed method outperforms others with more accurate predictions. It is worth noticing that either in Table 2 or Fig. 5, the estimated yaw angle is less good as other two angles. One possible explanation is that the distribution of yaw in AFLW dataset is more decentralized. We will discuss it in detail in Subsect. 4.5.

Table 3. Comparison on AFW dataset.

Methods	Accuracy ($\leq 15°$)
Multi.HoG	74.6%
Multi.AAMs	36.8%
Face.com	64.3%
FaceDPL [37]	89.4%
KEPLER [16]	96.67%
HyperFace [25]	97.7%
HF-ResNet [25]	98.5%
Ours	98.24%

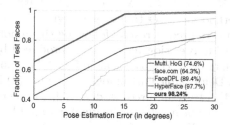

Fig. 6. Cumulative error distribution curves for pose estimation on AFW dataset.

Table 4. Mean average error comparison on AFLW, where nocls indicates no clustering applied, 12clsAll means clustering into 12 groups by three directions, and 12clsYaw denotes clustering into 12 groups by yaw.

Methods	Mean absolute rrror			AP
	Pitch	Yaw	Roll	
nocls	5.21	6.46	3.33	94.12%
12clsAll	5.13	6.16	**3.32**	94.62%
12clsYaw	**5.02**	**6.01**	**3.32**	**95.03%**

AFW. To demonstrate the capability of generalization of our method, we conduct experiments on AFW dataset. Because the ground-truth yaw angles are provided in multiples of $15°$, we round-off our predicted yaw to the nearest $15°$ for evaluation in consistent with the previous works. In Table 3, we compare our method with the following: (1) Multiview AAMs: an AAM trained for each viewpoint, (2) face.com, (3) Multiview HoG, (4) FaceDPL [37], (5) HyperFace [25], (6) KEPLER [16]. The proposed method scores 98.24% when allowing $±15°$ error tolerance, achieving comparable performance. Note that HF-ResNet [25] uses ResNet-101, a deeper and better backbone, while our implementation is based on ResNet-50. This difference might potentially affects the final performance. Figure 6 shows cumulative error distribution curves of the proposed method as well as some of other methods. It is clear that the proposed algorithm achieves the state-of-the-art, and is able to predict yaw in the range of $±15°$ for more than 98% of the faces.

4.5 Does Pose Class Help?

To test whether pose class boost performance or not in our work, we conduct experiments with different settings. First we calculate the covariance of three angles on AFLW dataset,

$$cov_{pose} = \begin{pmatrix} 0.0547 & 0.0097 & 0.0002 \\ 0.0097 & 0.5324 & -0.0168 \\ 0.0002 & -0.0168 & 0.0601 \end{pmatrix}$$

The 3×3 matrix is the covariance matrix of angles in the order of pitch, yaw and roll. From the diagonal elements, we can see that the variances of pitch and roll are 0.0547 and 0.0601 respectively. It is very small comparing to that of yaw, which is 0.5324. Other elements indicate that the three angles are almost independent to each other. Therefore, yaw contributes most to the variance of 3D pose and clustering in this direction should be more effective. We verify this guess by making comparison between *no clustering, clustering only on the basis of yaw direction* and *of all directions*. When *no clustering* is applied, the *fc* layer before pose class softmax is removed, and *fc* layers before bbox regression and pose regression are set to 8 and 6 respectively. Table 4 shows the results of MAE

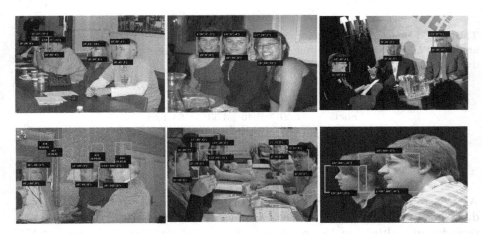

Fig. 7. Qualitative results of our method. Pose estimates for each face are shown on top of the boxes in the order (pitch, yaw, roll), groundtruth angles are shown at the bottom. Predictions within ±15° error tolerance could be considered as adequate results.

and AP on AFLW dataset. From this table, we observe that either on MAE or AP, clustering by yaw achieves the best and clustering by all directions performs better than no clustering, demonstrating that pose class indeed has impact on both pose estimation and face detection. Furthermore, generating pose classes along the principle direction significantly boosts the performance (Fig. 7).

5   Conclusions

In this paper, we propose an accurate and cost efficient framework for simultaneous face detection and 3D head pose estimation. The entire face detection module is a region-based method as Faster RCNN. While for head pose estimation, pose angles are firstly clustered into discrete groups. Then the regression problem becomes a classification problem. Finally a class-specific regressor is trained to refine the pose in each class. The estimation module is combined to detection module by sharing convolutional features in early layers. Extensive results on challenging unconstrained datasets demonstrates the effectiveness of our method.

References

1. Benenson, R., Mathias, M., Tuytelaars, T., Van Gool, L.: Seeking the strongest rigid detector. In: Proceedings of the IEEE Conference on Computer Vision and Pattern Recognition, pp. 3666–3673 (2013)
2. Chen, D., Ren, S., Wei, Y., Cao, X., Sun, J.: Joint cascade face detection and alignment. In: Fleet, D., Pajdla, T., Schiele, B., Tuytelaars, T. (eds.) ECCV 2014. LNCS, vol. 8694, pp. 109–122. Springer, Cham (2014). https://doi.org/10.1007/978-3-319-10599-4_8

3. Doshi, A., Trivedi, M.M.: Head and eye gaze dynamics during visual attention shifts in complex environments. J. Vis. **12**(2), 9–9 (2012)
4. Everingham, M., Van Gool, L., Williams, C., Winn, J., Zisserman, A.: The pascal visual object classes challenge 2012 (voc2012) results (2012) (2010). http://www. pascal-network.org/challenges/VOC/voc2011/workshop/index.html
5. Farfade, S.S., Saberian, M.J., Li, L.J.: Multi-view face detection using deep convolutional neural networks. In: Proceedings of the 5th ACM on International Conference on Multimedia Retrieval, pp. 643–650. ACM (2015)
6. Felzenszwalb, P.F., Girshick, R.B., McAllester, D., Ramanan, D.: Object detection with discriminatively trained part-based models. IEEE Trans. Pattern Anal. Mach. Intell. **32**(9), 1627–1645 (2010)
7. Fu, Y., Huang, T.S.: Graph embedded analysis for head pose estimation. In: 7th International Conference on Automatic Face and Gesture Recognition FGR 2006, p. 6. IEEE (2006)
8. Girshick, R.: Fast R-CNN. In: Proceedings of the IEEE International Conference on Computer Vision, pp. 1440–1448 (2015)
9. Girshick, R., Donahue, J., Darrell, T., Malik, J.: Rich feature hierarchies for accurate object detection and semantic segmentation. In: Proceedings of the IEEE Conference on Computer Vision and Pattern Recognition, pp. 580–587 (2014)
10. Gu, C., Ren, X.: Discriminative mixture-of-templates for viewpoint classification. In: Daniilidis, K., Maragos, P., Paragios, N. (eds.) ECCV 2010. LNCS, vol. 6315, pp. 408–421. Springer, Heidelberg (2010). https://doi.org/10.1007/978-3-642-15555-0_30
11. He, K., Zhang, X., Ren, S., Sun, J.: Deep residual learning for image recognition. In: Proceedings of the IEEE Conference on Computer Vision and Pattern Recognition, pp. 770–778 (2016)
12. Jia, Y., et al.: Caffe: convolutional architecture for fast feature embedding. In: Proceedings of the 22nd ACM International Conference on Multimedia, pp. 675–678. ACM (2014)
13. Jones, M., Viola, P.: Fast multi-view face detection. Mitsubishi Electric Res. Lab TR-20003-96 **3**(14), 2 (2003)
14. Katzenmaier, M., Stiefelhagen, R., Schultz, T.: Identifying the addressee in human-human-robot interactions based on head pose and speech. In: International Conference on Multimodal Interaction, pp. 144–151 (2004)
15. Krizhevsky, A., Sutskever, I., Hinton, G.E.: Imagenet classification with deep convolutional neural networks. In: Advances in Neural Information Processing Systems, pp. 1097–1105 (2012)
16. Kumar, A., Alavi, A., Chellappa, R.: Kepler: keypoint and pose estimation of unconstrained faces by learning efficient H-CNN regressors. In: 2017 12th IEEE International Conference on Automatic Face Gesture Recognition (FG 2017), pp. 258–265, May 2017. https://doi.org/10.1109/FG.2017.149
17. Lloyd, S.: Least squares quantization in PCM. IEEE Trans. Inf. Theor. **28**(2), 129–137 (1982)
18. Koestinger, M., Wohlhart, P., Roth, P.M., Bischof, H.: Annotated facial landmarks in the wild: a large-scale, real-world database for facial landmark localization. In: Proceedings of First IEEE International Workshop on Benchmarking Facial Image Analysis Technologies (2011)
19. Mathias, M., Benenson, R., Pedersoli, M., Van Gool, L.: Face detection without bells and whistles. In: Fleet, D., Pajdla, T., Schiele, B., Tuytelaars, T. (eds.) ECCV 2014. LNCS, vol. 8692, pp. 720–735. Springer, Cham (2014). https://doi.org/10.1007/978-3-319-10593-2_47

20. Mukherjee, S.S., Robertson, N.M.: Deep head pose: gaze-direction estimation in multimodal video. IEEE Trans. Multimedia **17**(11), 2094–2107 (2015)
21. Murphy-Chutorian, E., Trivedi, M.M.: Head pose estimation in computer vision: a survey. IEEE Trans. Pattern Anal. Mach. Intell. **31**(4), 607–626 (2009)
22. Osadchy, M., Cun, Y.L., Miller, M.L.: Synergistic face detection and pose estimation with energy-based models. J. Mach. Learn. Res. **8**(May), 1197–1215 (2007)
23. Qin, H., Yan, J., Li, X., Hu, X.: Joint training of cascaded CNN for face detection. In: Proceedings of the IEEE Conference on Computer Vision and Pattern Recognition, pp. 3456–3465 (2016)
24. Ramanathan, S., Yan, Y., Staiano, J., Lanz, O., Sebe, N.: On the relationship between head pose, social attention and personality prediction for unstructured and dynamic group interactions. In: ICMI, pp. 3–10 (2013)
25. Ranjan, R., Patel, V.M., Chellappa, R.: Hyperface: a deep multi-task learning framework for face detection, landmark localization, pose estimation, and gender recognition. In: IEEE Transactions on Pattern Analysis and Machine Intelligence, p. 1 (2017). https://doi.org/10.1109/TPAMI.2017.2781233
26. Ranjan, R., Patel, V.M., Chellappa, R.: A deep pyramid deformable part model for face detection. In: 2015 IEEE 7th International Conference on Biometrics Theory, Applications and Systems Biometrics Theory, Applications and Systems (BTAS), pp. 1–8. IEEE (2015)
27. Ren, S., He, K., Girshick, R., Sun, J.: Faster R-CNN: towards real-time object detection with region proposal networks. In: Advances in Neural Information Processing Systems, pp. 91–99 (2015)
28. Shen, X., Lin, Z., Brandt, J., Wu, Y.: Detecting and aligning faces by image retrieval. In: Proceedings of the IEEE Conference on Computer Vision and Pattern Recognition, pp. 3460–3467 (2013)
29. Sherrah, J., Gong, S., Ong, E.J.: Face distributions in similarity space under varying head pose. Image Vis. Comput. **19**(12), 807–819 (2001)
30. Simonyan, K., Zisserman, A.: Very deep convolutional networks for large-scale image recognition. arXiv preprint arXiv:1409.1556 (2014)
31. Uijlings, J.R., Van De Sande, K.E., Gevers, T., Smeulders, A.W.: Selective search for object recognition. Int. J. Comput. Vis. **104**(2), 154–171 (2013)
32. Xiang, Y., Choi, W., Lin, Y., Savarese, S.: Subcategory-aware convolutional neural networks for object proposals and detection. arXiv preprint arXiv:1604.04693 (2016)
33. Yan, J., Zhang, X., Lei, Z., Li, S.Z.: Face detection by structural models. Image Vis. Comput. **32**(10), 790–799 (2014)
34. Yang, S., Luo, P., Loy, C.C., Tang, X.: From facial parts responses to face detection: a deep learning approach. In: Proceedings of the IEEE International Conference on Computer Vision, pp. 3676–3684 (2015)
35. Zafeiriou, S., Zhang, C., Zhang, Z.: A survey on face detection in the wild: past, present and future. Comput. Vis. Image Underst. **138**, 1–24 (2015)
36. Zhu, X., Ramanan, D.: Face detection, pose estimation, and landmark localization in the wild. In: Proceedings of the IEEE Conference on Computer Vision and Pattern Recognition, pp. 2879–2886. IEEE (2012)
37. Zhu, X., Ramanan, D.: FACEDPL: detection, pose estimation, and landmark localization in the wild. Preprint **1**(2), 6 (2015)

Progressive Feature Fusion Network for Realistic Image Dehazing

Kangfu Mei[1], Aiwen Jiang[1]([⊠]), Juncheng Li[2], and Mingwen Wang[1]

[1] School of Computer and Information Engineering, Jiangxi Normal University,
Nanchang, China
{meikangfu,jiangaiwen,mwwang}@jxnu.edu.cn
[2] Department of Computer Science and Technology, East China Normal University,
Shanghai, China
51164500049@stu.ecnu.edu.cn

Abstract. Single image dehazing is a challenging ill-posed restoration problem. Various prior-based and learning-based methods have been proposed. Most of them follow a classic atmospheric scattering model which is an elegant simplified physical model based on the assumption of single-scattering and homogeneous atmospheric medium. The formulation of haze in realistic environment is more complicated. In this paper, we propose to take its essential mechanism as "black box", and focus on learning an input-adaptive trainable end-to-end dehazing model. An U-Net like encoder-decoder deep network via progressive feature fusions has been proposed to directly learn highly nonlinear transformation function from observed hazy image to haze-free ground-truth. The proposed network is evaluated on two public image dehazing benchmarks. The experiments demonstrate that it can achieve superior performance when compared with popular state-of-the-art methods. With efficient GPU memory usage, it can satisfactorily recover ultra high definition hazed image up to 4K resolution, which is unaffordable by many deep learning based dehazing algorithms.

Keywords: Single image dehazing · Image restoration ·
End-to-end dehazing · High resolution · U-like network

1 Introduction

Haze is a common atmospheric phenomena produced by small floating particles such as dust and smoke in the air. These floating particles absorb and scatter the light greatly, resulting in degradations on image quality. Under severe hazy conditions, many practical applications such as video surveillance, remote sensing, autonomous driving etc. are easily put in jeopardy, as shown in Fig. 1. High-level computer vision tasks like detection and recognition are hardly to be completed. Therefore, image dehazing (a.k.a haze removal) becomes an increasingly desirable technique.

© Springer Nature Switzerland AG 2019
C. V. Jawahar et al. (Eds.): ACCV 2018, LNCS 11361, pp. 203–215, 2019.
https://doi.org/10.1007/978-3-030-20887-5_13

Fig. 1. Examples of realistic hazy images

Being an ill-posed restoration problem, image dehazing is a very challenging task. Similar to other ill-posed problem like super-resolution, earlier image dehazing methods assumed the availability of multiple images from the same scene. However, in practical settings, dehazing from single image is more realistic and gains more dominant popularity [1]. Therefore, in this paper, we focus on the problem of single image dehazing.

Most state-of-the-art single image dehazing methods [2–8] are based on a classic atmospheric scattering model [9] which is formulated as following Eq. 1:

$$I(x) = J(x)t(x) + A \cdot (1 - t(x))$$ (1)

where, $I(x)$ is the observed hazy image, $J(x)$ is the clear image. $t(x)$ is called medium transmission function. A is the global atmospheric light. x represents pixel locations.

The physical model explained the degradations of a hazy image. The medium transmission function $t(x) = e^{-\beta \cdot d(x)}$ is a distance dependent factor that reflects the fraction of light reaching camera sensor. The atmospheric light A indicates the intensity of ambient light. It is not difficult to find that haze essentially brings in non-uniform, signal-dependent noise, as the scene attenuation caused by haze is correlated with the physical distance between object's surface and the camera.

Apart from a few works that focused on estimating the atmospheric light [10], most of popular algorithms concentrate more on accurately estimation of transmission function $t(x)$ with either prior knowledge or data-driven learning. Based on the estimated $\hat{t}(x)$ and $\hat{A}$, the clear image $\hat{J}$ is then recovered by using following Eq. 2 .

$$J(x) = \frac{I(x) - \hat{A} \cdot (1 - \hat{t}(x))}{\hat{t}(x)} = \frac{1}{\hat{t}(x)} I(x) - \frac{\hat{A}}{\hat{t}(x)} + \hat{A}$$ (2)

Though tremendous improvements have been made, as we know, the traditional separate pipeline does not directly measure the objective reconstruction errors. The inaccuracies resulted from both transmission function and atmospheric light estimation would potentially amplify each other and hinder the overall dehazing performance.

The recently proposed AOD-Net [11] was the first end-to-end trainable image dehazing model. It reformulated a new atmospheric scattering model from the classic one by leveraging a linear transformation to integrate both the transmission function and the atmospheric light into an unified map $K(x)$, as shown in Eq. 3.

$$J(x) = K(x)I(x) - K(x) + b \qquad (3)$$

where the $K(x)$ was an input-dependent transmission function. A light-weight CNN was built to estimate the $K(x)$ map, and jointly trained to further minimize the reconstruction error between the recovered output $J(x)$ and the ground-truth clear image.

Going deeper, we consider the general relationship between observed input I and recovered output J as $J(x) = \Phi(I(x); \theta)$, where $\Phi(*)$ represents some potential highly nonlinear transformation function whose parameters set is θ. Then the relationship represented by AOD-Net could be viewed as a specific case of the general function Φ.

In this paper, we argue that the formation of hazy image has complicated mechanism, and the classic atmospheric scattering model [9] is just an elegant simplified physical model based on the assumption of single-scattering and homogeneous atmospheric medium. There potentially exists some highly nonlinear transformation between the hazy image and its haze-free ground-truth. With that in mind, instead of limitedly learning the intermediate transmission function or its reformulated one from classic scattering model as AOD-Net did, we propose to build a real complete end-to-end deep network from an observed hazy image I to its recovered clear image J. To avoid making efforts on find "real" intermediate physical model, our strict end-to-end network pay much concerns on the qualities of dehazed output.

We employ an encoder-decoder architecture similar to the U-Net [12] to directly learn the input-adaptive restoration model Φ. The encoder convolves input image into several successive spatial pyramid layers. The decoder then successively recovers image details from the encoded feature mappings. In order to make full use of input information and accurately estimate structural details, progressive feature fusions are performed on different level mappings between encoder and decoder. We evaluate our proposed network on two public image dehazing benchmarks. The experimental results have shown that our method can achieve great improvements on final restoration performance, when compared with several state-of-the-art methods.

The contributions of this paper are two-fold:

- We have proposed an effective trainable U-Net like end-to-end network for image dehazing. The encoder-decoder architecture via progressive feature fusion directly learns the input-adaptive restoration model. The essential formulation mechanism of a hazy image is taken as "black box", and efforts are made on restoring the final high quality, clear output. At this viewpoint, our proposed network is in a real sense the first end-to-end deep learning based image dehazing model.

– Our proposed network can directly process ultra high-definition realistic hazed image up to 4K resolution with superior restoration performance at a reasonable speed and memory usage. Many popular deep learning based image dehazing network can not afford image of such high resolution on a single TITAN X GPU. We owe our advantage to the effective encoder-decoder architecture.

2   Related Work

Single image dehazing is a very challenging ill-posed problem. In the past, various prior-based and learning-based methods have been developed to solve the problem. On basis of the classic atmospheric scattering model proposed by Cantor [9], most of image dehazing methods followed a three-step pipeline: (a) estimating transmission map $t(x)$; (b) estimating global atmospheric light A; (c) recovering the clear image J via computing Eq. 2. In this section, we would focus on some representative methods. More related works can be referred to surveys [13–15].

A milestone work was the effective dark channel prior (DCP) proposed by He et al. [2] for outdoor images. They discovered that the local minimum of the dark channel of a haze-free image was close to zero. Base on the prior, transmission map could be reliably calculated. Zhu et al. [3] proposed a color attenuation prior by observing that the concentration of the haze was positively correlated with the difference between the brightness and the saturation. They created a linear model of scene depth for the hazy image. Based the recovered depth information, a transmission map was well estimated for haze removal. Dana et al. [6] proposed a non-local prior that colors of a haze-free image could be well approximated by a few hundred distinct color clusters in RGB space. On assumption that each of these color clusters became a line in the presence of haze, they recovered both the distance map and the haze-free image.

With the success of convolutional neural network in computer vision area, several recent dehazing algorithms directly learn transmission map $t(x)$ fully from data, in order to avoid inaccurate estimation of physical parameters from a single image. Cai et al. [7] proposed a DehazeNet, an end-to-end CNN network for estimating the transmission with a novel BReLU unit. Ren et al. [8] proposed a multi-scale deep neural network to estimate the transmission map. The recent AOD-Net [11] introduced a newly defined transmission variable to integrate both classic transmission map and atmospheric light. As AOD-Net needed learn the new intermediate transmission map, it still fell into a physical model. The latest proposed Gated Fusion Network (GFN) [16] learned confidence maps to combine several derived input images into a single one by keeping only the most significant features of them. We should note that, for GFN, handcrafted inputs were needed to be specifically derived for fusion and intermediate confidence maps were needed to be estimated. In contrast, our proposed network directly learns the transformation from input hazy image to output dehazed image, needn't learning any specific intermediate maps.

3 Progressive Feature Fusion Network (PFFNet) for Image Dehazing

In this section, we will describe our proposed end-to-end image dehazing network in details. The architecture of our progressive feature fusion network is illustrated in Fig. 2. It consists of three modules: encoder, feature transformation and decoder.

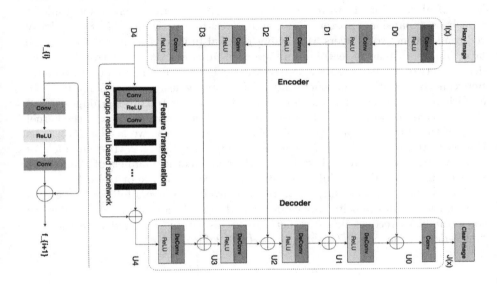

Fig. 2. The architecture of the progressive feature fusion network for image dehazing

The encoder module consists of five convolution layers, each of which is followed by a nonlinear ReLU activation. For the convenience of description, we denote the i-th "conv+relu" layer as $Conv_{en}^i$, $i = \{0, 1, 2, 3, 4\}$. The first layer $Conv_{en}^0$ is for aggregating informative features on a relatively large local receptive field from original observed hazy image I. The following four layers then sequentially perform down-sampling convolutional operations to encode image's information in pyramid scale.

$$D_i = Conv_{en}^i (D_{i-1}), i = \{0, 1, 2, 3, 4\}, \text{where}, D_{-1} = I \qquad (4)$$

We denote k_i, s_i, c_i as the receptive field size, step size, and output channels of layer $Conv_{en}^i$ respectively. In this paper, empirically, for $conv_{en}^0$, we set $k_0 = 11, s_0 = 1, c_i = 16$. Consequently, the corresponding output D_0 keeps the same spatial size as input I. For $conv_{en}^i$, $i = \{1, 2, 3, 4\}$ layers, we keep their receptive field size and step size the same. And each one learns feature mappings with double channels more than its previous layer. The super-parameters are set $k_i = 3, s_i = 2, c_i = 2c_{i-1}, i = \{1, 2, 3, 4\}$. As a result, we can easily calculate that if the size of input hazy image is $w \times h \times c$, the size of the output of encoder D_4 is consequently $\frac{1}{16}w \times \frac{1}{16}h \times 256$, where w, h, c are image width, image height, image channels in respective. That means if an input image is with 4K resolution level, the resulted feature map is with 256 spatial resolution level after encoder module, which benefits greatly the following processing stages for reducing memory usage.

The feature transformation module denoted as $\Psi(*)$ consists of residual based subnetworks. As we know, the main benefit of a very deep network is that it can represent very complex functions and also learn features at many different levels of abstraction. However, traditional deep networks often suffer gradient vanish or expansion disaster. The popular residual networks [17,18] explicitly reformulated network layers as learning residual functions with reference to the layer inputs, instead of learning unreferenced functions. It allows training much deeper networks than were previously practically feasible. Therefore, to balance between computation efficiency and GPU memory usage, in this module, we empirically employed eighteen wide residual blocks for feature learning.

Let $B(M)$ denotes the structure of a residual block, where M lists the kernel sizes of the convolutional layers in a block. In this paper, we accept $B(3,3)$ as the basic residual block, as shown in the left part of Fig. 2. The channels of convolution layer are all 256, which are the same as the channels of the feature map D_4 from encoder module. The step size is constantly kept to be 1.

The decoder module consists of four deconvolution layers followed by a convolution layer. In opposite to encoder, the deconvolution layers of decoder are sequentially to recover image structural details. Similarly, we denote the j-th "relu+deconv" layer as $DeConv_{dec}^{j}, j = \{4, 3, 2, 1\}$.

$$F_{j-1} = DeConv_{dec}^{j}(U_j), j = \{4, 3, 2, 1\} \qquad (5)$$

where, U_j is an intermediate feature map.

Through deconvolution (a.k.a transposed) layer, the $DeConv_{dec}^{j}$ performs up-sample operations to obtain intermediate feature mappings with double spatial size and half channels than its previous counterpart. Concretely, the receptive field size, step size and output channels are set $k_j = 3, s_j = 2, c_{j-1} = \frac{1}{2}c_j, j = \{4, 3, 2, 1\}$. It is not difficult to find that, in our network setting, the output map F_j from $DeConv_{dec}^{j}$ enjoys the same feature dimensions as corresponding input D_i of $Conv_{en}^{i}$ has, when $i = j \in \{3, 2, 1, 0\}$.

In order to maximize information flow along multi-level layers and guarantee better convergence, skip connections are employed between corresponding layers of different level from encoder and decoder. A global shortcut connection is applied between input and output of the feature transformation module, as shown in Fig. 2.

$$\begin{aligned} U_i &= D_i \oplus F_i, i = \{3, 2, 1, 0\} \\ U_4 &= D_4 \oplus \Psi(D_4) \end{aligned} \qquad (6)$$

where $\oplus$ is an channel-wise addition operator.

The dimension of the transposed feature map U_0 is therefore $w \times h \times 16$, as the same as D_0. A convolution operation is further applied on U_0, and generates the final recovered clear image J. Herein, for this convolution layer, the kernel size k is 3; the step size is 1; and the channels is the same as J.

The proposed image dehazing network progressively performs feature fusion on spatial pyramid mappings between encoder and decoder, which enables maximally preserved structural details from inputs for deconvolution layers, and further makes the dehazing network more input-adaptive.

4   Experiments

4.1   Dataset

We evaluate the effectiveness of our proposed method on two public dehazing benchmarks. The source code is available on GitHub[1].

NTIRE2018 Image Dehazing Dataset. The dataset was distributed by NTIRE 2018 Challenge on image dehazing [19]. Two novel subsets (I-HAZE [20] and O-HAZE [21]) with real haze and their ground-truth haze-free images were included. Hazy images were both captured in presence of real haze generated by professional haze machines. The I-HAZE dataset contains 35 scenes that correspond to indoor domestic environments, with objects of different colors and speculates. The O-HAZE contains 45 different outdoor scenes depicting the same visual content recorded in haze-free and hazy conditions, under the same illumination parameters. All images are ultra high definition images on 4K resolution level.

RESIDE [22]. The REISDE is a large scale synthetic hazy image dataset. The training set contains 13990 synthetic hazy images generated by using images from existing indoor depth datasets such as NYU2 [23] and Middlebury [24]. Specifically, given a clear image J, random atmospheric lights $A \in [0.7, 1.0]$ for each channel, and the corresponding ground-truth depth map d, function $t(x) = e^{-\beta \cdot d(x)}$ is applied to synthesize transmission map first, then a hazy image is generated by using the physical model in Eq. (1) with randomly selected scattering coefficient $\beta \in [0.6, 1.8]$. In RESIDE dataset, images are on 620×460 resolution level.

The *Synthetic Objective Testing Set* (SOTS) of RESIDE is used as our test dataset. The SOTS contains 500 indoor images from NYU2 [23] (non-overlapping with training images), and follows the same process as training data to synthesize hazy images.

4.2   Comparisons and Analysis

Several representative state-of-the-art methods are compared in our experiment: Dark-Channel Prior (DCP) [2], Color Attenuation Prior (CAP) [3], Non-Local Dehazing (NLD) [6], DehazeNet [7], Multi-scale CNN (MSCNN) [8], AOD-Net [11], and Gated Fusion Network (GFN) [16]. The popular full-reference PSNR and SSIM metrics are accepted to evaluate the dehazing performance.

---

[1] source code: https://github.com/MKFMIKU/PFFNet.

Training Details. As our PFF-Net was initially proposed to take part in the NTIRE2018 challenge on image dehazing, the GPU memory usage of our network is efficient so that we can directly recover an ultra high definition realistic hazy image on 4K resolution level on a single TITAN X GPU.

In this paper, we train our network both on I-HAZE and O-HAZE training images, which has 80 scenes in all. Based on these scenes, we further perform data augmentation for training. We first use sliding window to extract image crops of 520×520 size from the realistic hazy images. The stride is 260 pixels. For each image crop, we obtain its 12 variants at four angles $\left\{0, \frac{\pi}{2}, \pi, \frac{3}{2}\pi\right\}$ and three mirror flip cases $\{NoFlip, HorizontalFlip, VerticalFlip\}$. In consequence, about 190K patches are augmented as the training dataset.

The ADAM [25] is used as the optimizer. The initial learning rate we set is $\eta = 0.0001$, and kept a constant during training. Mean Square Errors (MSE) between recovered clear image and haze-free ground-truth is taken as our objective loss. The batch-size is 32. During training, we recorded every 2000 iterations as an epoch and the total num of training epoches is empirically 72 in practice. The testing curve on PSNR performance is shown in Fig. 3. We found that the network started to converge at the last 10 epoches.

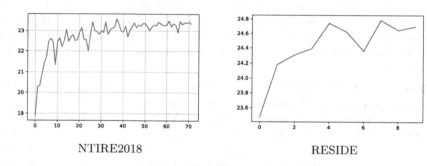

NTIRE2018   RESIDE

Fig. 3. The testing curves of our proposed PFF-Net on NTIRE2018 (*Left*) and RESIDE (*Right*). In both sub-figures, the horizontal axis shows training epochs. The vertical axis shows PSNR performance tested on training model at corresponding epoch.

Ablation Parameter Comparisons on Networks Settings. Before fixing the architecture of our PFFNet in this paper, we have done several ablation experiments on paremeters setting. We have experimented four different blocks sizes in feature transformation module: $\{6, 12, 18, 24\}$. The testing performances are shown in following Fig. 4-Left. Increasing the size of residual blocks would improve the testing performance. By considering the balance between the performance and the available computing resources, we finally adopted the feature transformation module with 18 residual blocks in this paper.

We have also compared the performance of networks with/without skip connections between encoder module and encoder module. The comparisons was experimented through training our network with 12 residual blocks in feature transformation module. In terms of the speed of convergence and the performance, network without skip connections is much worse than network with skip connections, as shown in Fig. 4-Right. The conclusion is consistent with observations in many residual based learning models and also validates the necessaries of progressive feature fusion between encoder and decoder stages.

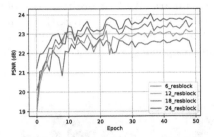 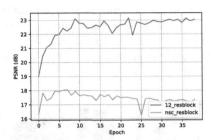

Fig. 4. *Left:* The testing performance comparisons on NTIRE2018 outdoor scenes in different block size cases. *Right:* The testing performance comparisons of network with/without skip connection between encoder and decoder module on NTIRE2018 outdoor scenes. The "12_resblock" represents network with skip connections; "nsc_resblock" represents network without skip connections. The network is trained about 40 epoches.

Experiment Results. We have taken part in the NTIRE2018 challenge on image dehazing based on the proposed network. In the final testing phase, our network has achieved top 6 ranking out of 21 teams on I-HAZE track without using any data from O-HAZE and won the NTIRE 2018 honorable mentioned award. It should be noted that our network is very straight-forward and we haven't applied any specific training trick to further boost performance during training period and we haven't re-trained our model for O-HAZE track at that submission time.

Compared with other top methods in NTIRE Dehazing Challenge, our proposed network has several distinguished differences: (1) Most of other top methods use denseblock in their networks while we just use simple residual block then. Empirically, denseblock has better learning power and will have much potentials to boost output performance. (2) Multi-scale or multi-direction ensemble inferences at testing stage are used in some top methods to achieve better performance. In contrary, we haven't applied this tricky strategy. We just use the single output for testing. Using ensemble inference strategy empirically has great potentials to achieve better performance. (3) The last is not the least. As the images used in NTIRE Dehazing Challenge are very large with 4K high-resolution, all these top methods use patch based training strategy without taking entire image as input. Their network can not afford such large image. In contrary, our proposed network can directly process the ultra high-definition realistic hazed image up to 4K resolution with superior restoration performance at a reasonable speed and memory usage.

Several dehazed examples on realistic images from NTIRE2018 are shown in Figs. 5 and 6. All these images are at 4K resolution level which most current dehazed model cannot afford. Though challenging these examples are, our network still can obtain relatively satisfactory dehazed results with natural color saturation and acceptable perceptual quality.

With the aim to compare with state-of-the-art methods, we evaluate our method on the commonly referred public benchmark. We first pre-train our network on DIV2K [26], then fine-tune the pre-trained network on RESIDE training data without data augmentation. The training curve of fine-tuning process on RESIDE is shown in Fig. 3. After about 8 epoches, the network begins to converge.

Fig. 5. Several challenging realistic dehazed examples from I-HAZE by using PFF-Net

We evaluate the performance of our network on the SOTS. The comparison results on SOTS are shown in Table 1. From the experimental comparisons, it has been demonstrated that our proposed network outperforms the current state-of-the-art methods, and achieves superior performance with great improvements.

Table 1. The dehazing performance evaluated on SOTS of RESIDE

	DCP	CAP	NLD	DehazeNet	MSCNN	AOD-Net	GFN	PFF-Net(ours)
PSNR	16.62	19.05	17.29	21.14	17.57	19.06	22.30	**24.78**
SSIM	0.8179	0.8364	0.7489	0.8472	0.8102	0.8504	0.88	**0.8923**

Fig. 6. Several challenging realistic dehazed examples from O-HAZE by using PFF-Net

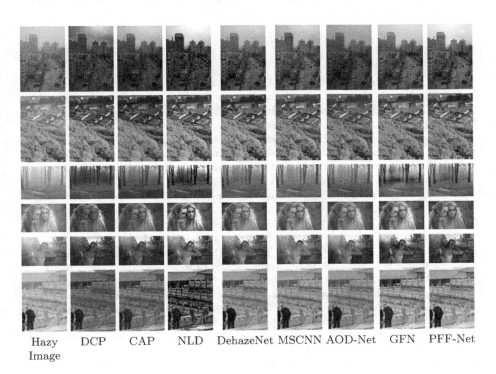

Hazy   DCP   CAP   NLD   DehazeNet MSCNN AOD-Net   GFN   PFF-Net
Image

Fig. 7. Comparisons with state-of-the-art methods on some real-world hazy images.

Some qualitative comparisons on real-world hazy image are further shown in Fig. 7. These collected hazy images are at resolution around 500 × 600 pixels and captured from natural environment, best viewed on high-resolution display. As shown, the dehazed results from our method are clear and the details of the scenes are enhanced moderately better with natural perceptual qualities.

5   Conclusion

In this paper, we have proposed an effective trainable U-Net like end-to-end network for image dehazing. Progressive feature fusions are employed to learn input adaptive restoration model. Owing to the proposed U-Net like encoder-decoder architecture, our dehazing network has efficient memory usage and can directly recover ultra high definition hazed image up to 4K resolution. We evaluate our proposed network on two public dehazing benchmarks. The experimental results demonstrate that our network can achieve superior performance with great improvements when compared with several popular state-of-the-art methods.

Acknowledgment. This work was supported by National Natural Science Foundation of China (Grant No. 61365002 and 61462045) and Provincial Natural Science Foundation of Jiangxi (Grant No. 20181BAB202013).

References

1. Fattal, R.: Single image dehazing. ACM Trans. Graph. **27**, 72 (2008)
2. He, K., Sun, J., Tang, X.: Single image haze removal using dark channel prior. IEEE Trans. Pattern Anal. Mach. Intell. **33**, 2341–2353 (2011)
3. Zhu, Q., Mai, J., Shao, L.: A fast single image haze removal algorithm using color attenuation prior. IEEE Trans. Image Process. **24**, 3522–3533 (2015)
4. Meng, G., Wang, Y., Duan, J., Xiang, S., Pan, C.: Efficient image dehazing with boundary constraint and contextual regularization. In: Proceedings of IEEE International Conference on Computer Vision. IEEE (2013)
5. Ancuti, C.O., Ancuti, C.: Single image dehazing by multi-scale fusion. IEEE Trans. Image Process. **22**, 3271–3282 (2013)
6. Dana, B., Tali, T., Shai, A.: Non-local image dehazing. In: Proceedings of IEEE Conference on Computer Vision and Pattern Recognition. IEEE (2016)
7. Cai, B., Xu, X., Jia, K., Qing, C., Tao, D.: Dehazenet: an end-to-end system for single image haze removal. IEEE Trans. Image Process. **25**, 5187–5198 (2016)
8. Ren, W., Liu, S., Zhang, H., Pan, J., Cao, X., Yang, M.-H.: Single image dehazing via multi-scale convolutional neural networks. In: Leibe, B., Matas, J., Sebe, N., Welling, M. (eds.) ECCV 2016. LNCS, vol. 9906, pp. 154–169. Springer, Cham (2016). https://doi.org/10.1007/978-3-319-46475-6_10
9. Cantor, A.: Optics of the atmosphere-scattering by molecules and particles. IEEE J. Quant. Electron. **14**, 698–699 (1978)
10. Sulami, M., Glatzer, I., Fattal, R., Werman, M.: Automatic recovery of the atmospheric light in hazy images. In: Proceedings of IEEE International Conference on Computational Photography. IEEE (2014)

11. Li, B., Peng, X., Wang, Z., Xu, J., Feng, D.: AOD-Net: all-in-one dehazing network. In: Proceedings of IEEE International Conference on Computer Vision. IEEE (2017)
12. Ronneberger, O., Fischer, P., Brox, T.: U-Net: convolutional networks for biomedical image segmentation. In: Navab, N., Hornegger, J., Wells, W.M., Frangi, A.F. (eds.) MICCAI 2015. LNCS, vol. 9351, pp. 234–241. Springer, Cham (2015). https://doi.org/10.1007/978-3-319-24574-4_28
13. Wu, D., Zhu, Q.: The latest research progress of image dehazing. Acta Automatica Sinica **41**, 221–239 (2015)
14. Xu, Y., Wen, J., Fei, L., Zhang, Z.: Review of video and image defogging algorithms and related studies on image restoration and enhancement. IEEE Access **4**, 165–188 (2016)
15. Li, Y., You, S., Brown, M.S., Tan, R.T.: Haze visibility enhancement: a survey and quantitative benchmarking. Comput. Vis. Image Underst. **165**, 1–16 (2017)
16. Ren, W., et al.: Gated fusion network for single image dehazing. In: Proceedings of IEEE Conference on Computer Vision and Pattern Recognition. IEEE (2018)
17. He, K., Zhang, X., Ren, S., Sun, J.: Deep residual learning for image recognition. In: Proceedings of IEEE Conference on Computer Vision and Pattern Recognition. IEEE (2016)
18. Zagoruyko, S., Komodakis, N.: Wide residual networks. In: Proceedings of British Machine Vision Conference (2016)
19. Ancuti, C., Cosmin, A., Radu, T., Luc, V.G., Zhang, L., Yang, M.H., et al.: NTIRE 2018 challenge on image dehazing: methods and results. In: The IEEE Conference on Computer Vision and Pattern Recognition (CVPR) Workshops (2018)
20. Ancuti, C., Cosmin, A., Radu, T., Christophe, V.: I-haze: a dehazing benchmark with real hazy and haze-free indoor images. arXiv preprint arXiv:1804.05091 (2018)
21. Ancuti, C.O., Cosmin, A., Radu, T., Christophe, D.V.: O-HAZE: a dehazing benchmark with real hazy and haze-free outdoor images. In: Proceedings of IEEE Conference on Computer Vision and Pattern Recognition Workshops. IEEE (2018)
22. Li, B., et al.: Benchmarking single image dehazing and beyond. IEEE Trans. Image Process. **28**(1), 492–505 (2018)
23. Silberman, N., Hoiem, D., Kohli, P., Fergus, R.: Indoor segmentation and support inference from RGBD images. In: Fitzgibbon, A., Lazebnik, S., Perona, P., Sato, Y., Schmid, C. (eds.) ECCV 2012. LNCS, vol. 7576, pp. 746–760. Springer, Heidelberg (2012). https://doi.org/10.1007/978-3-642-33715-4_54
24. Scharstein, D., Szeliski, R.: High-accuracy stereo depth maps using structured light. In: Proceedings of IEEE Conference on Computer Vision and Pattern Recognition. IEEE (2003)
25. Kingma, D.P., Ba, J.: Adam: a method for stochastic optimization. In: Proceedings of International Conference for Learning Representations (2015)
26. Radu, T., et al.: NTIRE 2017 challenge on single image super-resolution: methods and results. In: Proceedings of IEEE Conference on Computer Vision and Pattern Recognition Workshops. IEEE (2017)

Semi-supervised Learning for Face Sketch Synthesis in the Wild

Chaofeng Chen[1]([✉]), Wei Liu[1], Xiao Tan[2], and Kwan-Yee K. Wong[1]

[1] The University of Hong Kong, Hong Kong, China
{cfchen,wliu,kykwong}@cs.hku.hk
[2] Baidu Research, Beijing, China
tanxchong@gmail.com

Abstract. Face sketch synthesis has made great progress in the past few years. Recent methods based on deep neural networks are able to generate high quality sketches from face photos. However, due to the lack of training data (photo-sketch pairs), none of such deep learning based methods can be applied successfully to face photos in the wild. In this paper, we propose a semi-supervised deep learning architecture which extends face sketch synthesis to handle face photos in the wild by exploiting additional face photos in training. Instead of supervising the network with ground truth sketches, we first perform patch matching in feature space between the input photo and photos in a small reference set of photo-sketch pairs. We then compose a *pseudo sketch feature* representation using the corresponding sketch feature patches to supervise our network. With the proposed approach, we can train our networks using a small reference set of photo-sketch pairs together with a large face photo dataset without ground truth sketches. Experiments show that our method achieves state-of-the-art performance both on public benchmarks and face photos in the wild. Codes are available at https://github.com/chaofengc/Face-Sketch-Wild.

1 Introduction

Face sketch synthesis targets at generating a sketch from an input face photo. It has many useful applications. For instance, police officers often have to rely on face sketches to identify suspects, and face sketch synthesis makes it feasible for matching sketches against photos in a mugshot database automatically. Artists can also employ face sketch synthesis to simplify the animation production process [1]. Many people prefer using sketches as their profile pictures in social media networks [2], and face sketch synthesis allows them to produce sketches without the help of a professional artist.

Much effort has been devoted to face sketch synthesis. In particular, exemplar based methods dominated in the past two decades. These methods can achieve good performance without explicitly modeling the highly nonlinear mapping between face photos and sketches. They commonly subdivide a test photo into overlapping patches, and match these test patches with the photo patches in a reference set of photo-sketch pairs. They then compose an output sketch using

© Springer Nature Switzerland AG 2019
C. V. Jawahar et al. (Eds.): ACCV 2018, LNCS 11361, pp. 216–231, 2019.
https://doi.org/10.1007/978-3-030-20887-5_14

the corresponding sketch patches in the reference set. Although promising results have been reported [1, 3–5], these methods have several drawbacks. For example, sketches in Fig. 4(c)(d)(e)(f) are over-smoothed and fail to preserve subtle contents such as strands of hair on the forehead. Moreover, the patch matching and optimization processes are often very time-consuming. Recent methods exploited Convolutional Neural Networks (CNNs) to learn a direct mapping between photos and sketches, which is, however, a non-trivial task. The straight forward CNN based method produces blurry sketches (see Fig. 4(g)), and methods based on Generative Adversary Networks (GAN) [6] introduce undesirable artifacts (see Fig. 4(h),(i)). Besides, all these CNN based methods do not generalize well to face photos in the wild due to the lack of large training datasets of photo-sketch pairs. Although unpaired GAN based methods such as Cycle-GAN [7] can use unpaired data to transfer images between different domains, they fail to well preserve the facial content because of the weak content constraint (see Fig. 8).

In this paper, we propose a semi-supervised learning framework for face sketch synthesis that takes advantages of the exemplar based approach, the perceptual loss and GAN. We design a residual net [8] with skip connections as our generator network. Suppose we have a small reference set of photo-sketch pairs and a large face photo dataset without ground truth sketches. Similar to the exemplar based approach, we subdivide the VGG-19 [9] feature maps of the input photo into overlapping patches, and match them with the photo patches (in feature space) in the reference set. We then compose a *pseudo sketch feature* representation using the corresponding sketch patches (in feature space) in the reference set. We can then supervise our generator network using a perceptual loss based on the mean squared error (MSE) between the feature maps of the generated sketch and the corresponding pseudo sketch feature of the input photo. An adversary loss is also utilized to make the generated sketches more realistic.

In summary, our main contributions are three folds: (1) A semi-supervised learning framework for face sketch synthesis. Our framework allows us to train our networks using a small reference set of photo-sketch pairs together with a large face photo dataset without ground truth sketches. This enables our networks to generalize well to face photos in the wild. (2) A perceptual loss based on pseudo sketch feature. We show that the proposed loss is critical in preserving both facial content and texture details in the generated sketches. Extensive experiments are conducted to verify the effectiveness of our model. Both qualitative and quantitative results illustrate the superiority of our method. (3) To the best of our knowledge, our method is the first work that can generate visually pleasant sketches for face photos in the wild.

2   Related Works

2.1   Exemplar Based Methods

Tang and Wang [10] first introduced the exemplar based method based on eigentransformation. They projected an input photo onto the eignspace of the training photos, and then reconstruct a sketch from the eignspace of the training sketches using the same projection. Liu *et al.* [11] observed that the linear model holds better locally, and therefore proposed a nonlinear model based on local linear embedding (LLE). They first subdivided an input photo into over-lapping patches and reconstructed each photo patch as a linear combination of the training photo patches. They then obtained the sketch patches by applying the same linear combinations to the corresponding training sketch patches. Wang and Tang [5] employed a multi-scale markov random fields (MRF) model to improve the consistency between neighboring patches. By introducing shape priors and SIFT features, Zhang et al. [12] proposed an extended version of MRF which can handle face photos under different illuminations and poses. However, these MRF based methods are not capable of synthesizing new sketch patches since they only select the best candidate sketch patch for each photo patch. To tackle this problem, Zhou et al. [3] presented the markov weight fields (MWF) model which produces a target sketch patch as a linear combination of K best candidate sketch patches. Considering that patch matching based on traditional image features (e.g., PCA and SIFT) is not robust, a recent method [4] used CNN feature to represent the training patches and computed more accurate combination coefficients. To accelerate the synthesis procedure, Song et al. [1] formulated face sketch synthesis as a spatial sketch denoising (SSD) problem, and Wang et al. [13] presented an offline random sampling strategy for nearest neighbor selection of patches.

2.2 Learning Based Methods

Recent works applied CNN to synthesize sketches and produced promising results. Zhang et al. [14] proposed a 7-layer fully convolutional network (FCN) to directly transfer an input photo to a sketch. Although their model can roughly estimate the outline of a face, it fails to capture texture details with the use of intensity based mean squared error (MSE) loss. Zhang et al. [15] utilized a branched fully convolutional network (BFCN) consisting of a content branch and a texture branch. Because the face content and texture are predicted separately with different loss metrics, the final sketch looks disunited. Chen et al. [16] proposed the pyramid column feature and used it to compose a reference style for a test photo from the training sketches. They utilized a CNN to create a content image from the photo, and then transferred the reference style to introduce shadings and textures in the output sketch. Wang et al. [17] presented the multi-scale generative adversarial networks (GANs) to generate sketches from photos and vice versa. Multiple discriminators at different hidden layers are used to supervise the synthesis process. Gao et al. [18] took advantage of the facial pars-ing map and proposed a composition-aided stack GAN. All these deep learning based methods require ground truth photo-sketch pairs for training, and they do not generalize well to face photos in the wild due to the lack of training data.

3   Semi-supervised Face Sketch Synthesis

3.1   Overview

Our framework is composed of three main parts, namely a generator network G, a pseudo sketch feature generator and a discriminator network D (see Fig. 1). The generator network is a deep residual network with skip connections. It is used to generate a synthesized sketch $\hat{y}$ for each input photo $\mathbf{x}$. The pseudo sketch feature generator is the key to our semi-supervised learning approach. Instead of training the generator network directly with ground truth sketches, we construct a pseudo sketch feature for each input photo to supervise the synthesis of $\hat{y}$. In this way, we can train our network on any face photo datasets, and generalize our model to face photos in the wild. We further adopt a discriminator network D to minimize the gap between generated sketches and real sketches drawn by artists.

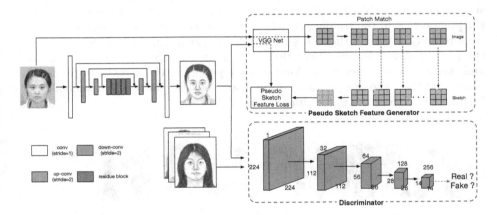

Fig. 1. Framework of the proposed method. The generator network is a deep residual network with skip connections. It generates a synthesized sketch from an input photo. The pseudo sketch feature generator utilizes patch matching in the deep feature space to generate a pseudo sketch feature for an input photo in training. The discriminator network tries to distinguish between generated sketches and sketches drawn by artists.

3.2   Pseudo Sketch Feature Generator

Given a reference set $\mathcal{R} = \{(\mathbf{x}_i^{\mathcal{R}}, \mathbf{y}_i^{\mathcal{R}})\}_{i=1}^N$, the pseudo sketch feature generator targets at constructing a pseudo sketch feature $\Phi'(\mathbf{x})$ for a test photo $\mathbf{x}$ which is used to supervise the synthesis of the sketch $\hat{y}$. We follow MRF-CNN [19] to extract a local patch representation of an image. We first feed $\mathbf{x}$ into a pretrained VGG-19 network and extract the feature map $\Phi^l(\mathbf{x})$ at the l-th layer. Similarly, we obtain $\{\Phi^l(\mathbf{x}_i^{\mathcal{R}})\}_{i=1}^N$ and $\{\Phi^l(\mathbf{y}_i^{\mathcal{R}})\}_{i=1}^N$. Let us denote a $k \times k$ patch centered at a point j of $\Phi^l(\mathbf{x})$ as $\Psi_j\left(\Phi^l(\mathbf{x})\right)$, and the same definition applies to $\Psi_j\left(\Phi^l(\mathbf{x}_i^{\mathcal{R}})\right)$

and $\Psi_j(\Phi^l(y_i^{\mathcal{R}}))$. Now for each patch $\Psi_j\left(\Phi^l(x)\right)$, where $j = 1, 2, \ldots, m$ and $m = (H^l - 2 \times \lfloor \frac{k}{2} \rfloor) \times (W^l - 2 \times \lfloor \frac{k}{2} \rfloor)$ with H^l and W^l being the height and width of $\Phi^l(x)$, we find its best match $\Psi_{j'}\left(\Phi^l(x_{i'}^{\mathcal{R}})\right)$ in the reference set based on cosine distance, i.e.,

$$(i', j') = \underset{\substack{i^*=1 \sim N \\ j^*=1 \sim m}}{\arg\max} \frac{\Psi_j\left(\Phi^l(x)\right) \cdot \Psi_{j^*}\left(\Phi^l(x_{i^*}^{\mathcal{R}})\right)}{\left\| \Psi_j\left(\Phi^l(x)\right) \right\|_2 \left\| \Psi_{j^*}\left(\Phi^l(x_{i^*}^{\mathcal{R}})\right) \right\|_2}. \tag{1}$$

Since the photos and the corresponding sketches in $\mathcal{R}$ are well aligned, we directly apply (i', j') to index the corresponding sketch feature patch $\Psi_{j'}\left(\Phi^l(y_{i'}^{\mathcal{R}})\right)$ for $\Psi_{j'}\left(\Phi^l(x_{i'}^{\mathcal{R}})\right)$, and use it as the pseudo sketch feature patch $\Psi_j'\left(\Phi^l(x)\right)$ for $\Psi_j\left(\Phi^l(x)\right)$. Finally, a pseudo sketch feature representation (at layer l) for x is given by $\{\Psi_j'\left(\Phi^l(x)\right)\}_{j=1}^m$. Figure 2 visualizes an example of the pseudo sketch feature. It can be seen that the pseudo sketch feature provides a good approximation of the real sketch feature (see Fig. 2(a)). We also show a naïve reconstruction in Fig. 2(b) obtained by directly using the matching index to index the pixel values in the training sketches. We can see such a naïve reconstruction does roughly resemble the real sketch, which also justifies the effectiveness of the pseudo sketch feature. Note that we only need alignment between photos and sketches in $\mathcal{R}$. Since we perform a dense patch matching between the input photo and the reference photos, we can also generate reasonable pseudo sketch features for input faces under different poses (see Fig. 2(c)).

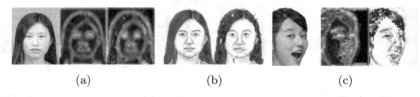

(a)   (b)   (c)

Fig. 2. (a) Ground truth sketch feature (middle) and pseudo sketch feature of the relu3_1 layer (right). (b) Ground truth sketch (left) and pixel level projection of the patch matching result (right). (c) Photos in the wild without ground truth sketches. (*Note that the pixel level results are only for visualization, and they are not used in training.*)

3.3   Loss Functions

Pseudo Sketch Feature Loss. We define our pseudo sketch feature loss as

$$L_p(x, \hat{y}) = \sum_{l=3}^{5} \sum_{j=1}^{m} \left\| \Psi_j\left(\Phi^l(\hat{y})\right) - \Psi_j'\left(\Phi^l(x)\right) \right\|_2^2, \tag{2}$$

where $l = 3, 4, 5$ refer to layers relu3_1, relu4_1, and relu5_1 respectively. High level features after relu3_1 are better representations of textures and more robust to appearance changes and geometric transforms [19]. Figure 3 shows the results of using different layers in L_p. As expected, low level features (*e.g.*, relu1_1 and relu2_1) fail to generate sketch textures. While high level features (*e.g.*, relu5_1) can better preserve textures, they produce artifacts in terms of details (see the eyes of sketches in Fig. 3). To get better performance and reduce the computation cost of patch matching, we set $l = 3, 4, 5$.

Photo & Sketch   relu1_1   relu2_1   relu3_1   relu4_1   relu5_1

Fig. 3. Results of using different layers in pseudo sketch feature loss.

GAN Loss. For easier convergence, we use the least square loss when training the GAN, known as LSGAN [20]. The objective functions of LSGAN are given by

$$L_{GAN\_D} = \frac{1}{2}\mathbb{E}_{y \sim p_{sketch}(y)}[(D(y) - 1)^2] + \frac{1}{2}\mathbb{E}_{x \sim p_{photo}(x)}[(D(G(x)))^2] \quad (3)$$

$$L_{GAN\_G} = \mathbb{E}_{x \sim p_{photo}(x)}[(D(G(x)) - 1)^2] \quad (4)$$

Total Variation Loss. Sketches generated by CNN may be unnatural and noisy. Following previous works [19,21,22], we adopt the *total variation loss* as a natural image prior to further improve the sketch quality,

$$L_{tv}(\hat{\mathbf{y}}) = \sum_{u,v} \left((\hat{\mathbf{y}}_{u+1,v} - \hat{\mathbf{y}}_{u,v})^2 + (\hat{\mathbf{y}}_{u,v+1} - \hat{\mathbf{y}}_{u,v})^2\right), \quad (5)$$

where $\hat{\mathbf{y}}_{u,v}$ denotes the intensity value at (u, v) of the synthesized sketch $\hat{\mathbf{y}}$.

Based on the above loss terms, we can train our generator network G and discriminator network D using the following two loss functions respectively:

$$L_G = \lambda_p L_p + \lambda_{adv} L_{GAN\_G} + \lambda_{tv} L_{tv}, \quad (6)$$

$$L_D = L_{GAN\_D} \quad (7)$$

where L_G and L_D are minimized alternatively until convergence. λ_p, λ_{adv} and λ_{tv} are trade-off weights for each loss term respectively.

4 Implementation Details

4.1 Datasets

Photo-Sketch Pairs. We use two public datasets: the CUFS dataset(consisting of the CUHK student dataset [10], the AR dataset [23], and the XM2VTS dataset [24]) and the CUFSF dataset [25], to evaluate our model[1]. The CUFSF dataset is more challenging than the CUFS dataset because (1) the photos were captured under different lighting conditions and (2) the sketches exhibit strong deformation in shape and cannot be aligned with the photos well. Details of these datasets are summarized in Table 1.

Table 1. Details of benchmark datasets. Align: whether the sketches are well aligned with photos. Var: whether the photos have lighting variations.

Dataset		Total pairs	Train	Test	Align	Var
CUFS	CUHK	188	88	100	✓	✗
	AR	123	80	43	✓	✗
	XM2VTS	295	100	195	✗	✗
CUFSF		1194	250	944	✗	✓

Face Photos. We use the VGG-Face dataset [26] to evaluate our model on photos in the wild. There are 2,622 persons in this dataset and each person has 1,000 photos. We randomly select 2,000 persons for training and the rest for testing. For each person in the training split, we randomly select $\mathcal{N}$ photos and named the resulting dataset VGG-Face$\mathcal{N}$[2], where $\mathcal{N} = 01, 02, \ldots, 10$. We also randomly select 2 photos for each person in the testing split to construct a VGG test set of 1,244 photos.

Preprocessing. For photos/sketches which have already been aligned and have a size of 250×200, we leave them unchanged. For the rest, we first detect 68 face landmarks on the image using dlib[3], and calculate a similarity transform to warp the image into one with the two eyes located at $(75, 125)$ and $(125, 125)$ respectively. We then crop the resulting image to a size of 250×200. We simply drop those photos/sketches from which we fail to detect face landmarks.

---

[1] Data comes from http://www.ihitworld.com/RSLCR.html.

[2] The dataset will be made available.

[3] http://dlib.net/.

4.2   Patch Matching

As in exemplar based methods, patch matching is a time-consuming process. We accelerate this process in three ways. First, we precompute and store the feature patches for the photos and sketches in the reference set (i.e., $\{\Psi_j\left(\Phi^l\left(\mathbf{x}_i^{\mathcal{R}}\right)\right)\}$ and $\{\Psi_j\left(\Phi^l\left(\mathbf{y}_i^{\mathcal{R}}\right)\right)\}$). Second, instead of searching the whole reference feature set, we first identify k best matched reference photos for each input photo based on the cosine distance of their relu5_1 feature maps. Patch matching is then restricted within these k reference photos (we set $k = 5$ in the whole training process). Third, Eq. 1 is implemented as a convolution operator which can be computed efficiently on GPU.

4.3   Training Details

We updated the generator and discriminator alternatively at every iteration. The trade-off weights λ_p, λ_{adv} were set to 1 and 10^3, and λ_{tv} was set to 10^{-5} when using CUFS as reference set and 10^{-2} when using CUFSF. We implemented our model using PyTorch[4], and trained it on a Nvidia Titan X GPU. We used Adam [27] with learning rates from 10^{-3} to 10^{-5}, decreasing with a factor of 10^{-1}. Data augmentation was done online in the color space (brightness, contrast, saturation and sharpness). Each iteration took about 2 s with a batch size of 6, and the model converged after about 5 h of training.

5   Evaluation on Public Benchmarks

In this section, we evaluate our model using two public benchmarks, namely CUFS and CUFSF, which were captured under laboratory conditions. We use the training photos from CUFS∪CUFSF to train our networks. When evaluating on CUFS, the reference photo-sketch pairs only comes from CUFS, and the same applies to CUFSF. To demonstrate the effectiveness of our model, we compare our results both qualitatively and quantitatively with seven other methods, namely MWF [3], SSD [1], RSLCR [13], DGFL [4], FCN [14], Pix2Pix-GAN [28], and Cycle-GAN [7]. We also compare our results quantitatively with the latest GAN based sketch synthesis methods, i.e., PS^2-MAN [29] and stack-CA-GAN [18]. Since the models of their work are not available, we can only compare with the results that are directly taken from their published papers.

5.1   Qualitative Comparison

As we can observe in Fig. 4, exemplar based methods (see Fig. 4(c),(d),(e) in general perform worse than learning based methods (see Fig. 4(g),(h),(i),(j)), especially in preserving contents of the input photos. Using deep features in exemplar based methods helps to alleviate the problem, but the results are over-smoothed (see Fig. 4(f)). Due to the lack of training data, FCN produces bad

---

[4] http://pytorch.org/.

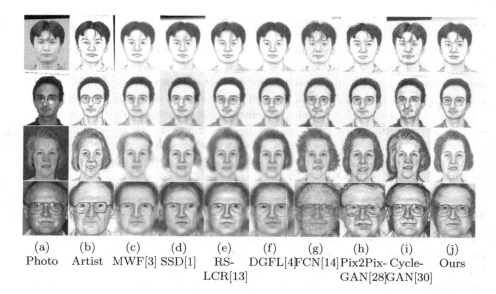

(a)	(b)	(c)	(d)	(e)	(f)	(g)	(h)	(i)	(j)
Photo	Artist	MWF[3]	SSD[1]	RS-LCR[13]	DGFL[4]	FCN[14]	Pix2Pix-GAN[28]	Cycle-GAN[30]	Ours

Fig. 4. Sketches generated using different methods. First 3 rows: test photos from CUFS. Last row: test photo from CUFSF.

results when the photos are taken under very different lighting conditions (see last two rows of Fig. 4(g)). Although the two GANs can produce much better results than FCN, they also introduce many artifacts and noise. Thanks to the pseudo sketch feature loss, our method does not suffer from the above problems. In particular, our semi-supervised strategy allows us to incorporate more training photos without ground truth in training, which helps to improve the generalization ability.

5.2   Quantitative Comparison

Image Quality Assessment. For datasets with ground truth sketches (e.g., CUFS and CUFSF), previous work [4,13,18] typically used structural similarity (SSIM) [31] as an image quality assessment metric to measure the similarity between a generated sketch and the ground truth sketch. However, many researchers (e.g., in super resolution [32] and face sketch synthesis [29,30]) pointed out that SSIM is not always consistent with the perceptual quality. One main reason is that SSIM favors slightly blurry images when the images contain rich textures. To demonstrate this, we show some sketches generated using different methods together with their SSIM scores in Fig. 5. It can be seen that sketch generated by RSLCR is smoother than those by Pix2Pix-GAN and our model, but have higher SSIM scores. We applied a bilateral filter to smooth all the sketches. It can be observed that the SSIM scores of the sketch generated by RSLCR remain roughly the same after smoothing, whereas those of the sketches generated by Pix2Pix-GAN and our model improve by more than 1.5%. In Fig. 6(a), we show the average SSIM scores of the sketches generated by different methods on CUFS, together with the average SSIM scores of their smoothed counterparts. As expected, the average SSIM scores of most of the methods improve after smoothing, same for a few exemplar based methods which produce over-smoothed sketches. The average SSIM score of our smoothed results is comparable to that of the state-of-the-art method. In Fig. 6(b), we show the corresponding results on CUFSF. Similar conclusions can be drawn.

(b) RSLCR	(c) Pix2Pix-GAN	(d) Ours.
SSIM: 0.5970/0.5903.	SSIM: 0.5648/0.5953.	SSIM: 0.5814/0.6055.
FSIM: 0.7488/0.7362.	FSIM: 0.7559/0.7506.	FSIM: 0.7692/0.7557.

Fig. 5. SSIM and FSIM scores of some generated sketches (left) and their smoothed counterparts (right).

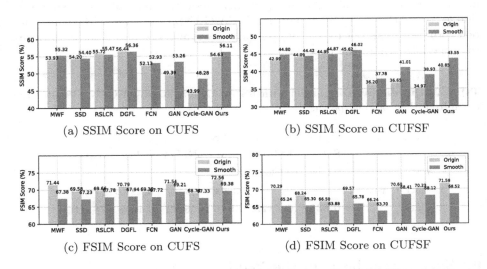

(a) SSIM Score on CUFS

(b) SSIM Score on CUFSF

(c) FSIM Score on CUFS

(d) FSIM Score on CUFSF

Fig. 6. Average SSIM and FSIM scores of the sketches generated by different methods on CUFS and CUFSF. The proposed method achieves state-of-the-art FSIM score on both datasets.

Due to the drawback of SSIM, we use feature similarity (FSIM) [33] as our image quality assessment metric. FSIM is better at evaluating detailed textures compared with SSIM. It can be observed from Fig. 5 that the FSIM scores of the sketches decrease after smoothing. The average FSIM scores of the sketches generated by different methods on CUFS and CUFSF are shown in Fig. 6(c) and (d) respectively. It can be seen that our method achieves the state-of-the-art in terms of FSIM score on both CUFS and CUFSF.

Face Sketch Recognition. Sketch recognition is an important application of face sketch synthesis. We follow the same practice of Wang *et al.* [13] and employ the null-space linear discriminant analysis (NLDA) [34] to perform the recognition experiments. Figure 7 shows the recognition accuracy of different methods on the two datasets. Our method achieves the best result when the dimension of the reduced eigenspace is less than 100, and achieves a competitive result to the state-of-the-art method [4] when the dimension is above 100.

Comparison with PS^2-MAN and stack-CA-GAN. To further demonstrate the effectiveness of the proposed method, we compare it with two latest GAN methods, namely PS^2-MAN [29] and stack-CA-GAN [18], which are specially designed for sketch synthesis. As shown in Table 2, our method achieves the best performance on almost all datasets, except for the SSIM score in CUFSF. However, we obtain a better performance on NLDA which indicates that our model can better preserve the identify information. Note that both of these GAN methods use extra information to train their network, i.e., multi-scale supervision (PS^2-MAN) and parsing map (stack-CA-GAN). Compared with them, our perceptual loss can not only avoid producing artifacts but also help to improve the generalization of the network.

Table 2. Quantitative comparison with PS^2-MAN and stack-CA-GAN. Results are taken from their original papers.

	CUHK		CUFS		CUFSF	
	SSIM	FSIM	SSIM	NLDA	SSIM	NLDA
PS^2-MAN	0.6156	0.7361	—		—	
stack-CA-GAN	—		0.5266	96.04	**0.4106**	77.31
Ours	**0.6328**	**0.7423**	**0.5463**	**98.22**	0.4085	**78.04**

6   Sketch Synthesis in the Wild

There are two challenges for sketch synthesis in the wild. The first challenge is how to deal with real photos captured under uncontrolled environments with varying pose and lighting, and cluttered backgrounds. The second is the computation time. Our method tackles the first challenge by introducing more training

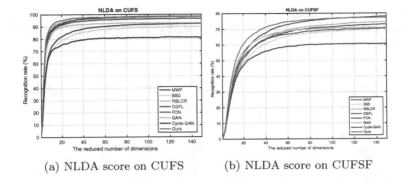

(a) NLDA score on CUFS   (b) NLDA score on CUFSF

Fig. 7. Face recognition rate against feature dimensions on CUFS and CUFSF.

photos through the construction of pseudo sketch features. Regarding computation time, our CNN based model can generate a sketch in a single feed forward pass which takes about 7 ms on a GPU for a 250 × 200 photo. We compared our method with five other methods, including SSD[5], FCN, Pix2Pix-GAN[6], Cycle-GAN[7], and Fast-RSLCR[8]. In this experiment, we trained our model using CUFS as the reference set and all the training photos from the CUFS, CUFSF and VGG-Face10 as the training set. Since there are no ground truth sketches for the test photos, we carried out a mean opinion score (MOS) test to quantitatively evaluate the results.

6.1 Qualitative Comparison

As photos in the wild are captured under uncontrolled environments, their appearance may vary largely. Figure 8 shows some photos sampled from our VGG-Face test dataset and the sketches generated by different methods. It can be observed that these photos may show very different lightings, poses, image resolutions, and hair styles. Besides, some photos may be incomplete and people may also use a cartoon as their photos for entertainment (see the last row of Fig. 8). It is therefore very difficult, if not impossible, for a method which only learns from a small set of photo-sketch pairs to generate sketches for photos in the wild. Among the results of other methods, exemplar based methods (see Fig. 8(b)(c)) fail to deal with pose changes and different hair styles. FCN produces sketches (see Fig. 8(d)) that can roughly preserve the contour of the face but lose important facial components (e.g., nose and eyes). Although GANs can generate some sketch like textures, none of them can well preserve the contents.

---

[5] http://www.cs.cityu.edu.hk/~yibisong/eccv14/index.html.

[6] https://github.com/phillipi/pix2pix.

[7] https://github.com/junyanz/pytorch-CycleGAN-and-pix2pix.

[8] http://www.ihitworld.com/RSLCR.html.

The face shapes are distorted and the key facial parts are lost. It can be seen from Fig. 8(g) that our model can handle photos in the wild well and generate pleasant results.

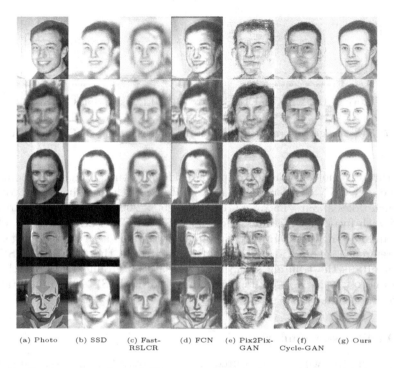

(a) Photo   (b) SSD   (c) Fast-RSLCR   (d) FCN   (e) Pix2Pix-GAN   (f) Cycle-GAN   (g) Ours

Fig. 8. Qualitative comparison of different methods for images in the wild. Benefit from the additional training photos, the proposed method can deal with various photos.

6.2 Effectiveness of Additional Training Photos

Introducing more training photos from VGG-Face dataset is the key to improve the generalization ability of our model. As demonstrated in Fig. 9, the model trained without additional photos from VGG-Face has difficulty in handling uncontrolled lightings and different hair colors (see Fig. 9(b)). As we add more photos to the training set, the results improve significantly (see the eyes and hair in Fig. 9).

6.3 Mean Opinion Score Test

Since there are no ground truth sketches for the photos in the wild, we performed a MOS test to assess the perceptual quality of the sketches generated by different methods. Specifically, we randomly selected 30 photos from the VGG test set,

(a) Photo   (b) VGGFace00   (c) VGGFace05   (d) VGGFace10

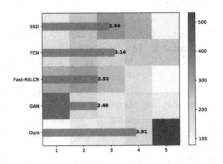

Fig. 9. Effectiveness of additional training photos. The results improve a lot when more and more photos are added to the training set.

Fig. 10. Results of MOS test on the quality of sketches generated by SSD, FCN, Fast-RSLCR, Pix2Pix-GAN and our model on photos in the wild.

and then generated the sketches for these photos using SSD, FCN, Fast-RSLCR, Pix2Pix-GAN and our method respectively. Given the example photo-sketch pairs from public benchmarks as reference, 108 raters were asked to rank 10 groups of randomly selected sketches synthesized by the five different methods. We assigned a score of 1–5 to the sketches based on their rankings (5 being the best). The results are presented in Fig. 10. It can be observed that the MOS of our results significantly outperforms that of the other methods. This demonstrates the superiority of our method on photos in the wild.

7 Conclusion

In this paper, we propose a semi-supervised learning framework for face sketch synthesis in the wild. We design a residual network with skip connections to transfer photos to sketches. Instead of supervising our network using ground truth sketches, we construct a novel pseudo sketch feature representation for each input photo based on feature space patch matching with a small reference set of photo-sketch pairs. This allows us to train our model using a large face photo dataset (without ground truth sketches) with the help of a small reference set of photo-sketch pairs. Training with a large face photo dataset enables our model to generalize better to photos in the wild. Experiments show that our method can produce sketches comparable to those produced by other state-of-the-art methods on four public benchmarks (in terms of SSIM and FSIM), and outperforms them on photos in the wild.

Acknowledgment. We thank Nannan Wang, Hao Zhou and Yibing Song for providing their codes and data. We also gratefully acknowledge the support of NVIDIA Corporation with the donation of the Titan X Pascal GPU used for this research.

References

1. Song, Y., Bao, L., Yang, Q., Yang, M.-H.: Real-time exemplar-based face sketch synthesis. In: Fleet, D., Pajdla, T., Schiele, B., Tuytelaars, T. (eds.) ECCV 2014. LNCS, vol. 8694, pp. 800–813. Springer, Cham (2014). https://doi.org/10.1007/978-3-319-10599-4_51
2. Berger, I., Shamir, A., Mahler, M., Carter, E., Hodgins, J.: Style and abstraction in portrait sketching. ACM Trans. Graph. (TOG) **32**, 55 (2013)
3. Zhou, H., Kuang, Z., Wong, K.Y.K.: Markov weight fields for face sketch synthesis. In: IEEE Conference on Computer Vision and Pattern Recognition, pp. 1091–1097 (2012)
4. Zhu, M., Wang, N., Gao, X., Li, J.: Deep graphical feature learning for face sketch synthesis. In: Proceedings of the Twenty-Sixth International Joint Conference on Artificial Intelligence, pp. 3574–3580 (2017)
5. Wang, X., Tang, X.: Face photo-sketch synthesis and recognition. IEEE Trans. Pattern Anal. Mach. Intell. **31**, 1955–1967 (2009)
6. Goodfellow, I., et al.: Generative adversarial nets. In: Advances in Neural Information Processing Systems, pp. 2672–2680 (2014)
7. Zhu, J.Y., Park, T., Isola, P., Efros, A.A.: Unpaired image-to-image translation using cycle-consistent adversarial networks. In: 2017 IEEE International Conference on Computer Vision (ICCV) (2017)
8. He, K., Zhang, X., Ren, S., Sun, J.: Deep residual learning for image recognition. arXiv:1512.03385 (2015)
9. Simonyan, K., Zisserman, A.: Very deep convolutional networks for large-scale image recognition. arXiv:1409.1556 (2014)
10. Tang, X., Wang, X.: Face sketch synthesis and recognition. In: IEEE International Conference on Computer Vision, pp. 687–694 (2003)
11. Liu, Q., Tang, X., Jin, H., Lu, H., Ma, S.: A nonlinear approach for face sketch synthesis and recognition. In: IEEE Conference on Computer Vision and Pattern recognition, vol. 1, pp. 1005–1010 (2005)
12. Zhang, W., Wang, X., Tang, X.: Lighting and pose robust face sketch synthesis. In: Daniilidis, K., Maragos, P., Paragios, N. (eds.) ECCV 2010. LNCS, vol. 6316, pp. 420–433. Springer, Heidelberg (2010). https://doi.org/10.1007/978-3-642-15567-3_31
13. Wang, N., Gao, X., Li, J.: Random sampling for fast face sketch synthesis. arXiv:1701.01911 (2017)
14. Zhang, L., Lin, L., Wu, X., Ding, S., Zhang, L.: End-to-end photo-sketch generation via fully convolutional representation learning. In: Proceedings of the 5th ACM on International Conference on Multimedia Retrieval (ICMR), pp. 627–634 (2015)
15. Zhang, D., Lin, L., Chen, T., Wu, X., Tan, W., Izquierdo, E.: Content-adaptive sketch portrait generation by decompositional representation learning. IEEE Trans. Image Process. (TIP) **26**, 328–339 (2017)
16. Chen, C., Tan, X., Wong, K.Y.K.: Face sketch synthesis with style transfer using pyramid column feature. In: IEEE Winter Conference on Applications of Computer Vision (2018)
17. Wang, N., Zhu, M., Li, J., Song, B., Li, Z.: Data-driven vs. model-driven: fast face sketch synthesis. Neurocomputing **257**, 214–221 (2017)
18. Gao, F., Shi, S., Yu, J., Huang, Q.: Composition-aided sketch-realistic portrait generation. arXiv:1712.00899 (2017)

19. Li, C., Wand, M.: Combining markov random fields and convolutional neural networks for image synthesis. In: Proceedings of the IEEE Conference on Computer Vision and Pattern Recognition, pp. 2479–2486 (2016)

20. Mao, X., Li, Q., Xie, H., Lau, R.Y., Wang, Z., Smolley, S.P.: Least squares generative adversarial networks. In: 2017 IEEE International Conference on Computer Vision (ICCV), pp. 2813–2821. IEEE (2017)

21. Johnson, J., Alahi, A., Fei-Fei, L.: Perceptual losses for real-time style transfer and super-resolution. In: Leibe, B., Matas, J., Sebe, N., Welling, M. (eds.) ECCV 2016. LNCS, vol. 9906, pp. 694–711. Springer, Cham (2016). https://doi.org/10.1007/978-3-319-46475-6_43

22. Kaur, P., Zhang, H., Dana, K.J.: Photo-realistic facial texture transfer. arXiv:1706.04306 (2017)

23. Martinez, A., Benavente, R.: The AR face database. Technical report, CVC Technical Report (1998)

24. Messer, K., Matas, J., Kittler, J., Jonsson, K.: Xm2vtsdb: the extended m2vts database. In: Second International Conference on Audio and Video-based Biometric Person Authentication, pp. 72–77 (1999)

25. Zhang, W., Wang, X., Tang, X.: Coupled information-theoretic encoding for face photo-sketch recognition. In: IEEE Conference on Computer Vision and Pattern Recognition, pp. 513–520. IEEE (2011)

26. Parkhi, O.M., Vedaldi, A., Zisserman, A.: Deep face recognition. In: British Machine Vision Conference (2015)

27. Kingma, D., Ba, J.: Adam: A method for stochastic optimization. arXiv:1412.6980 (2014)

28. Isola, P., Zhu, J.Y., Zhou, T., Efros, A.A.: Image-to-image translation with conditional adversarial networks. In: CVPR (2017)

29. Wang, L., Sindagi, V.A., Patel, V.M.: High-quality facial photo-sketch synthesis using multi-adversarial networks. arXiv:1710.10182 (2017)

30. Wang, N., Zha, W., Li, J., Gao, X.: Back projection: an effective postprocessing method for GAN-based face sketch synthesis. Pattern Recognit. Lett. **107**, 59–65 (2017)

31. Karacan, L., Erdem, E., Erdem, A.: Structure-preserving image smoothing via region covariances. ACM Trans. Graph. **32**, 176 (2013)

32. Ledig, C., et al.: Photo-realistic single image super-resolution using a generative adversarial network. arXiv:1609.04802 (2016)

33. Zhang, L., Zhang, L., Mou, X., Zhang, D.: FSIM: a feature similarity index for image quality assessment. IEEE Trans. Image Process. **20**, 2378–2386 (2011)

34. Chen, L.F., Liao, H.Y.M., Ko, M.T., Lin, J.C., Yu, G.J.: A new LDA-based face recognition system which can solve the small sample size problem. Pattern Recognit. **33**, 1713–1726 (2000)

Evolvement Constrained Adversarial Learning for Video Style Transfer

Wenbo Li[1]([⊠]), Longyin Wen[2], Xiao Bian[3], and Siwei Lyu[1]

[1] University at Albany, SUNY, Albany, USA
{wli20,slyu}@albany.edu
[2] JD Finance AI Lab, San Francisco, USA
lywen.cv.workbox@gmail.com
[3] GE Global Research, Niskayuna, USA
xiao.bian@ge.com

Abstract. Video style transfer is a useful component for applications such as augmented reality, non-photorealistic rendering, and interactive games. Many existing methods use optical flow to preserve the temporal smoothness of the synthesized video. However, the estimation of optical flow is sensitive to occlusions and rapid motions. Thus, in this work, we introduce a novel evolve-sync loss computed by evolvements to replace optical flow. Using this evolve-sync loss, we build an adversarial learning framework, termed as Video Style Transfer Generative Adversarial Network (VST-GAN), which improves upon the MGAN method for image style transfer for more efficient video style transfer. We perform extensive experimental evaluations of our method and show quantitative and qualitative improvements over the state-of-the-art methods.

1 Introduction

Great artists in history can render scenes with their distinct styles. It is the unique artistic style that differs Van Gogh from Picasso. We wonder if an algorithm can also acquire such styles? For instance, would it be able to re-render the scenes in The Avengers (2012) as if it were the oeuvre of Francis Picabia? Such an interesting question can be formulated as the video style transfer problem as shown in Fig. 1, *i.e.*, given a *style image* (Francis Picabia's Udnie) and a *source video* (a clip from The Avengers), the "synthesizer" should automatically produce a video combining both the style of Udnie and the content of The Avengers. Such an algorithm can find applications in many areas, such as augmented reality, computer games and nonphotorealistic rendering.

Many recent works in the computer vision and computer graphics community have focused on the problem of image style transfer [8,9,14,17,25]. However, these methods cannot be readily extended to videos, since independently generating each video frame leads to artifacts such as flickering and jagging in the synthesized videos. To this end, existing video style transfer algorithms [1,3,24]

W. Li and L. Wen—Equally contributed.

© Springer Nature Switzerland AG 2019
C. V. Jawahar et al. (Eds.): ACCV 2018, LNCS 11361, pp. 232–248, 2019.
https://doi.org/10.1007/978-3-030-20887-5_15

Fig. 1. Given a video and an image, our algorithm aims to synthesize a video combining the style of the image and the content of the video. To preserve the temporal smoothness of the synthesized video, we use **evolvements** derived from the source and synthesized video, and further compute the evolve-sync loss as the replacement of the optical flow constraints. This loss ensures that the textures at the same location in the image plane of the source and synthesized video evolve synchronously. For the illustration purpose, we only show the order two loss of the evolvements at one patch.

rely on signals that are estimated by a given motion model such as optical flows computed from adjacent frames to preserve temporal smoothness. We call such signals as model-driven signals. Although more visually pleasing results are achieved with these methods, optical flow estimation methods are known to be sensitive to occlusions and rapid and abrupt motions [23,27], and such limitations affects the qualities of the synthesized videos. Two recent methods [3,24] attempt to remedy these problems by introducing occlusion masks to filter out low-confidence optical flow, but the generation of occlusion masks is also error-prone and can lead to further artifacts.

In this work, we aim to exploit model-free signals (against the model-driven ones) in the source video for video style transfer, and synthesize video to match with such signals in the source video to preserve the temporal smoothness. To this end, we introduce **evolvements**, a form of inter-frame variations, as such a model-free signal, the acquisition of which is illustrated in Fig. 1. As the source and synthesized videos are synchronous in time, it is natural to require that the textures in the source and synthesized videos evolve synchronously, which we term as the *evolve-sync assumption*. The evolve-sync assumption is incorporated in our method with the *evolve-sync loss*, which encourages the evolvements from the source domain and those from the synthesized domain to be the same. As we need to preserve the temporal smoothness at both the microscopic and macroscopic levels, we extend the evolve-sync loss to be multi-level by regarding the evolvements as distributions and employing encoders (*e.g.*, a pre-trained CNN) to extract samples from these probability distributions. Thus, the evolve-sync loss encourages samples of different distributions at the corresponding level to be the same. We use the maximum mean discrepancy (MMD) [11] as the distance measure between probability distributions.

The evolve-sync loss can be combined with an image style transfer method to form the basis of video style transfer algorithms. We choose a state-of-the-art image style transfer method, *i.e.*, Markovian Generative Adversarial Network (MGAN)

Fig. 2. Saturated vs. Desaturated. The desaturated color occurs in the synthesized results for the unpaired frames.

[17], and develop the *Video Style Transfer Generative Adversarial Network* (VST-GAN). MGAN consists of two major components: (i) A Markovian Deconvolutional Adversarial Network (MDAN) denoted by D, and (ii) a generator that is a feed-forward convolutional neural network and denoted by G. G synthesizes the frames of video according to the content of the source video and the style of the style image, while D plays two roles: it creates real training samples for G with a deconvolutional process driven by the adversarial training, and acts as the adversary to G. Besidess the evolve-sync loss, our modifications to MGAN are presented as follows.

As noted in [17], generating images using the MDAN model D can be slow, and this becomes more problematic for video synthesis. Thus, we design an accelerating training strategy for VST-GAN. Specifically, we only apply D to every other frame to generate real training samples for G, leading to that the real training samples are unpaired with the synthesized frames. We expect that G can synthesize desirable textures for the unpaired frames. However, we observe the desaturated color (similar problem arose in [31]) in the synthesized results of the unpaired frames, which is illustrated in Fig. 2. We therefore modify G by adding a convolutional recurrent layer as its final output layer, which alleviates the desaturation problem, as the recurrent connection makes it possible to propagate the saturation of the paired frames to the unpaired ones.

The main contributions of our work can be summarized as follows: we introduce the evolve-sync loss, which is based on the evolvement that is more reliable than the estimated optical flow in preserving the temporal smoothness of the synthesized video. Applying the evolve-sync loss at both the microscopic and microscopic levels, we develop VST-GAN, an adversarial learning framework for video style transfer, by adapting the MGAN image style transfer method. Specifically, we add a convolutional recurrent layer as the output layer to resolve the desaturation problem in the synthesized video, which is caused by the trade-off between the training speed and the sufficiency of the real samples. Experimental results demonstrate the effectiveness of the evolve-sync loss and VST-GAN.

2   Related Works

Image and Video Style Transfer. There has been an extensive literature on image style transfer methods, which synthesize images based on sampling low-level features in the given source and style images. The extensions to video style transfer [2,12,18,32] rely on optical flow to maintain the temporal smoothness of sampling. See [16] for a comprehensive survey.

Recently, deep neural networks have been proved effective for both image [4,7–9,13,14,17,25] and video [1,3,24] style transfer. Gatys *et al.* [8,9] used the convolutional neural network (CNN) to model the patch statistics with a global Gaussian model of the higher-level feature vectors (*e.g.*, activations of CNN), and transferred the style by minimizing the feature reconstruction loss in an iterative deconvolutional process. Two follow-up works, *i.e.*, Johnson *et al.* [14] and Ulyanov *et al.* [25], proposed fast implementations of Gatys *et al.*'s method. Both methods employed precomputed decoders trained with a perceptual style loss and obtained significant runtime benefits. In contrast to these three works, Li and Wand [17] argued that real-world contextually related patches do not always comply with a Gaussian distribution, but a complex nonlinear manifold, and proposed MGAN, where a feed-forward generator is adversarially learned to project the contextually related patches to the manifold of patches.

Anderson *et al.* [1] extended Gatys *et al.*'s method to video style transfer. To preserve the temporal smoothness of the synthesized video, they used optical flow to initialize the style transfer optimization, and incorporated the flow explicitly into the loss function. To further reduce artifacts at the boundaries and occluded regions, Ruder *et al.* [24] introduced masks to filter out optical flows with low confidences in the loss function. Chen *et al.* [3,5] extended Johnson *et al.*'s method to a feed-forward network for video style transfer. To preserve the temporal smoothness, this method first obtained the current result via a learned flow, and then reduced the artifacts at the occluded regions by fusing the warped result with the independently synthesized result via a learned occlusion mask. In summary, all existing video style transfer methods rely on using optical to preserve temporal smoothness, and use the occlusion mask to stabilize the results. As such, these methods suffer from the common problems in estimating optical flow, *i.e.*, the sensitivity to occlusion and abrupt motion in video.

Generative Adversarial Network (GAN). GANs [10] have achieved impressive results for various tasks in image processing, such as style transfer [17], generation [6], editing [33], representation learning [21], and translation [34], *etc.* The key to GANs' success is the idea of an adversarial loss that forces the generated images/videos to be indistinguishable from the real ones. Only a few works develop GANs for videos, *i.e.*, generation [28–30] and prediction [20,26]. The real samples of the existing GANs for video generation and prediction are available. However, this is not the case for video style transfer. This is because the qualified real samples for this task should contain both the desirable style and the required content. In VST-GAN, we generate such samples with the deconvolutional model in MGAN constrained by the evolve-sync loss. However, the iterative deconvolutional optimization for videos is slow. Thus, we design a strategy to accelerate the training process of the GAN framework for video style transfer while maintaining the quality of the synthesized videos.

3 Overview

We first formally define the video style transfer problem as following: given a source video $\mathcal{X} = \{X_1, \cdots, X_i, \cdots\}$ and a style image S, we aim to produce a video $\mathcal{Y} = \{Y_1, \cdots, Y_i, \cdots\}$ with the style of S, and the content of $\mathcal{X}$.

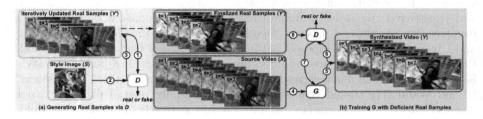

Fig. 3. Pipeline. Our method first uses a MDAN model D to generate real samples every other frame within an iterative deconvolutional process. $\mathcal{Y}'$ is initialized with the downsampled source video $\mathcal{X}'$. The generated real samples are used to train G using a GAN model. Then, G transfers the image style to the whole video. The numbers on the arrows indicate the order of operations, which are explained in detail in Sect. 3.

We design VST-GAN, an adversarial learning framework based on MGAN [17], to build a video style transfer algorithm incorporating the evolve-sync loss, with an aim to preserve temporal smoothness of frames without using optical flow. VST-GAN consists of a deconvolutional model D, and a feed-forward generator G. Both D and G are integrated with the evolve-sync loss. Figure 3 illustrates the overall pipeline of our method, which includes two steps:

Step (i) D generates real samples for G within a deconvolutional process that is constrained by the evolve-sync loss and driven by the adversarial training. Considering the efficiency issue, we accelerate the generation process by applying it to every other frame. In Fig. 3(a), steps ① and ② correspond to the convolutional forward pass, where D determines how real $\mathcal{Y}'$ is. Step ③ represents the deconvolutional backward pass, where D acts as the generator and the losses are back-propagated to pixels of $\mathcal{Y}'$.

Step (ii) Given the unpaired training samples, G is trained to transfer the style of S to the generated video, using D as the adversary in the manner of GAN. However, the original generator in MGAN suffers from the lack of real samples, which can cause the desaturation effect (or grey image tone) in the generated videos (see an example in Fig. 2). We therefore modify G by adding a convolutional recurrent layer as its final output layer, which reduces the desaturation issue effectively. In Fig. 3(b), in the runtime of updating G, $\mathcal{X}$ is fed into G (step ④) to generate $\mathcal{Y}$ (step ⑤). Then, D determines how real is the synthesized video $\mathcal{Y}$ (step ⑥), and the losses are back-propagated to update G (step ⑦). During the updating of D, $\mathcal{Y}$ and $\mathcal{Y}'$ are used as real and fake samples[1] to train D (steps ⑥ and ⑧), respectively.

---

[1] The naming fashion of real and fake samples follow the convention of GAN: the output $\mathcal{Y}$ of G is considered to be fake, while the precomputed $\mathcal{Y}'$ is real.

The evolve-sync loss is based on a more reliable signal than the optical flow estimated from the input video, thus it can better preserve the temporal smoothness. Our accelerating training strategy and the added convolutional recurrent structure effectively reduce the training complexity of VST-GAN.

4   The Evolve-Sync Loss

One basic requirement of video style transfer is to preserve the temporal smoothness between generated frames, as human visual systems are sensitive to the flickering artifacts. This means that the simple approach of generating each frame independently using existing image style transfer algorithms is not effective, as it will lead to visually displeasing results due to two factors. First, as many image style transfer methods (e.g., [8,9]) are iterative, their results are affected by different initializations and the local minima of the style loss function. Second, a small perturbation in the source images may cause large variations in the synthesized results that are not temporally smooth.

As such, in order to generate temporally smooth frames with spatially rich style patterns, existing methods [1,3,24] modify the image style transfer algorithms by incorporating optical flows estimated from the source video as supervisory signals. The reliability of the estimated optical flow is often problematic due to the problems related with the common optical flow algorithms, i.e., sensitivity to occlusions and rapid motions. This motivates us to turn to a different source of model-free signal directly acquired from the source video itself to capture inter-frame variations, which we term as *evolvement*. Given two frames X_i and $X_{i-k} \in \mathbb{R}^{h \times w \times 3}$, we define the evolvement from X_{i-k} to X_i as a distribution $\mathcal{E}(X_{i-k}, X_i)$. Figure 4 illustrates the sampling process from evolvements. We compute an evolvement sample $\mathcal{E}(X_{i-k}, X_i)_m \sim \mathcal{E}(X_{i-k}, X_i)$ as:

$$\mathcal{E}(X_{i-k}, X_i)_m = z(|g(X_i)_m - g(X_{i-k})_m|), \tag{1}$$

where $g(\cdot)$ denotes an encoder function that extracts samples from evolvements. The standardization function is represented as $z(x) = \frac{x-\mu}{\sigma}$, where the input x is a 2D matrix, and μ and σ are the mean and standard deviation of elements in x, respectively. Index m indicates the mth sample generated by $g(\cdot)$.

Our method is based on the *evolve-sync assumption*, which states that $\mathcal{X}$ and the synthesized video $\mathcal{Y}$ are synchronous in time, so their evolvements, $\mathcal{E}(X_{i-k}, X_i)$ and $\mathcal{E}(Y_{i-k}, Y_i)$, can be viewed as two synchronized signals. As seen in Fig. 4, the brighter a pixel in an evolvement sample is, the more drastic variation occurs at that pixel. The rationality behind the evolve-sync assumption can be understood by contradiction: if it does not hold for a certain pixel, it means that the drastic variation occurs in $\mathcal{E}(X_{i-k}, X_i)$ while the mild variation occurs in $\mathcal{E}(Y_{i-k}, Y_i)$, or vice versa. This suggests that the content at that location has not been properly preserved, which contradicts the problem formulation in Sect. 3.

Given $\mathcal{X}$ with a certain temporal smoothness degree, preserving the evolve-sync is equivalent to forcing the temporal smoothness of $\mathcal{Y}$ to be the same as that of $\mathcal{X}$. Consequently, we introduce the *evolve-sync loss* L_{es} to enforce the

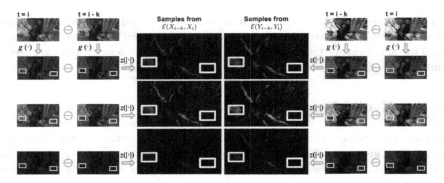

Fig. 4. **Illustration of the computation of evolvement and evolve-sync assumption.** $g(\cdot)$ represents an encoder, which splits the image in R, G and B color channels herein. $z(\cdot)$ represents a standardization function meaning subtracting mean and dividing by standard deviation. $\mathcal{E}(X_{i-k}, X_i)$ represents the evolvement from frame X_{i-k} to X_i. The yellow/white boxes highlight a spot in the image plane where drastic/mild variations occur. (Color figure online)

evolve-sync assumption in $\mathcal{Y}$ that measures the distance between $\mathcal{E}(X_{i-k}, X_i)$ and $\mathcal{E}(Y_{i-k}, Y_i)$. To this end, we employ the Maximum Mean Discrepancy [11] as the metric between two probability distributions:

$$L_{es}(\mathcal{F}, \mathcal{X}, \mathcal{Y}) = \sum_{|i-j|<\delta} \sup_{f \in \mathcal{F}} (\mathbf{E}_{x \sim \mathcal{E}(X_i, X_j)}[f(x)] - \mathbf{E}_{y \sim \mathcal{E}(Y_i, Y_j)}[f(y)]), \qquad (2)$$

where δ is a preset parameter determining the order of L_{es}, $\mathcal{F}$ is a Gaussian kernel and we set to $\delta = 3$ in our experiments.

We aim to preserve the temporal smoothness of $\mathcal{Y}$ at the microscopic level where the synthesized textures are temporally continuous, and at the macroscopic level where the synthesized textures and the video content are synchronized. To this end, we use two encoders for each level (i) the microscopic encoder $g_1(\cdot)$, which splits the image in R, G and B color channels for the microscopic level, and (ii) the macroscopic encoder $g_2(\cdot)$, which is a pretrained VGG network (sampled from $Relu3\_1$). As such, the overall evolve-sync loss is given as:

$$L_{es}(\mathcal{G}, \mathcal{F}, \mathcal{X}, \mathcal{Y}) = \sum_{r=1}^{|\mathcal{G}|} \alpha_r \cdot \sum_{|i-j|<\delta} \sup_{f \in \mathcal{F}} (\mathbf{E}_{x_r \sim \mathcal{E}(X_i, X_j)}[f(x)] - \mathbf{E}_{y_r \sim \mathcal{E}(Y_i, Y_j)}[f(y)]), \qquad (3)$$

where x_r and y_r are determined by $g_r(\cdot)$, and we set $\alpha_1 = 0.005$ and $\alpha_2 = 100$.

5   Video Style Transfer GAN (VST-GAN)

In this section, we describe the architecture of VST-GAN in Sect. 5.1 and the training of VST-GAN in Sect. 5.2.

5.1   Architecture

We build VST-GAN by adapting Markovian GAN (MGAN) [17], a state-of-the-art image style transfer framework that does not rely on the implicit assumption that the real-world textures comply with a Gaussian distribution. We show the architecture of VST-GAN in Fig. 5, where grey blocks indicate the intrinsic architecture of MGAN, and blocks with other colors indicate our modifications. MGAN consists of two major components: (i) A Markovian Deconvolutional Adversarial Network (MDAN) denoted by D, and (ii) a feed-forward generator denoted by G. D plays two roles: it creates real training samples for G with a deconvolutional process that is driven by the adversarial training, and acts as the adversary to G.

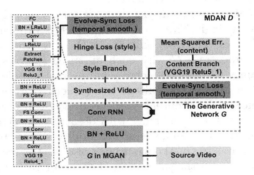

Fig. 5. Architecture of VST-GAN. Grey blocks indicate the intrinsic architecture of MGAN, and blocks with other colors indicate the input and our modifications.

MDAN D. As shown in Fig. 5, D has the *style branch* and the *content branch*. The style branch learns to distinguish the feature patches extracted from the feature maps output by VGG19 $Relu3\_1$ of the source video from those of the synthesized videos. D outputs a classification score $s = 1$ or 0 for each patch, indicating how "real" the patch is (with $s = 1$ being sampled from the style image S, or real patch). For each patch sampled from the synthesized frame, we minimize its style loss (*i.e.*, $1 - s$). Like Radford *et al.* [22], we use batch normalization (BN) and leaky ReLU (LReLU) to improve the training of D. The content branch encourages the content of the synthesized image to be similar to that of the source image, and is constructed from VGG19 features on the same image from higher and more abstract layer $Relu5\_1$. The content dissimilarity is measured by a content loss given by the mean squared error between two feature maps obtained from the source video and the synthesized one, respectively. When using D to generate real samples for G, the deconvolution process back-propagates both the style and content loss to pixels. When D acts as the adversary to G, the style and content loss are back-propagated to train G.

The Generative Network G. D requires many iterations and a separate run for each source image, so Li and Wand [17] further developed G, which consists of a pre-trained VGG encoder and a decoder. The VGG encoder of G takes the source image as input, and outputs a feature map from $Relu4\_1$. The decoder of G takes the output of the encoder, and decodes an image through a ordinary convolution followed by a cascade of fractional-strided convolutions (FS Conv in Fig. 5). Note that the content loss is used to measure the content dissimilarity between the synthesized image and its corresponding real sample. Although being trained with fixed-size input, G can be naturally extended to images of arbitrary sizes. VGG encoders in MGAN are fixed during training.

When adapting MGAN to the video style transfer, we make two major modifications to its architecture. First, in order to preserve the temporal smoothness at both the microscopic and macroscopic level, we integrate D with the proposed evolve-sync loss at two levels, $i.e.$, the synthesized video (microscopic) and the VGG encoder of the style branch (macroscopic).

Unfortunately, D has a slow running time – it takes D nearly 4 h to synthesize a 50-frame video on a single Titan X GPU – which is problematic to generate videos with more frames. Thus, we only apply D every other frame to generate real samples for G, and train G with such unpaired samples. This way, there will be a half of frames without the corresponding real samples, so these frames will not be used to compute the content loss (inherited from standard regression problems). Since such a content loss encourages conservative predictions, it makes G generate synthesized frames with desaturation artifacts (Fig. 2). To alleviate the desaturation problem, we further modify MGAN by adding a convolutional recurrent layer as the final output layer of G, as the recurrent connection makes it possible to smooth the saturation of consecutive frames.

5.2   Training

The training process of VST-GAN includes two steps: (i) generating real training samples for G using D on every other frame, and (ii) training G adversarilly against D with the unpaired training samples.

Generate Real Samples via D. In order to train G adversarially, we need qualified real samples that contain both the style of S and the content of $\mathcal{X}$. As such real samples are not accessible to us during training, we generate them using D on every other frame of the videos as described in Sect. 3. We denote the downsampled source video as $\mathcal{X}' = \{X_1, X_3, \cdots, X_{i-2}, X_i, \cdots\}$ and denote its corresponding real samples as $\mathcal{Y}' = \{Y_1', Y_3', \cdots, Y_{i-2}', Y_i', \cdots\}$. Then, we perform deconvolution with D iteratively to update $\mathcal{Y}'$ (initialized with $\mathcal{X}'$), so that the following loss is minimized:

$$
\hat{\mathcal{Y}}' = \underset{\mathcal{Y}'}{\arg\min} \sum_{Y_i' \in \mathcal{Y}'} [L_t(\Phi_t(Y_i'), \ell_{real}) + L_c(\Phi_c(X_i), \Phi_c(Y_i')) + \omega \Upsilon(Y_i')] + \\
L_{es}(\mathcal{G}, \mathcal{F}, \mathcal{X}', \mathcal{Y}'),
$$

(4)

where L_t denotes the style loss. L_c denotes the content loss, which is a mean squared error. Φ_t and Φ_c denote the VGG encoder in the style and content branch, respectively. L_{es} denotes the evolve-sync loss defined in (3). The regularizer Υ is a smoothness prior for pixels [19]. We sample patches from $\Phi_t(Y_i')$, and compute L_t as the hinge loss with their labels fixed to one, i.e., $\ell_{real} = 1$:

$$L_t(\Phi_t(Y_i'), \ell_{real}) = \frac{1}{N} \sum_{j=1}^{N} \max(0, 1 - \ell_{real} \cdot s_j), \qquad (5)$$

where s_j denotes the score (output by D) of the jth patch, and N is the total number of sampled patches in $\Phi_t(Y_i')$.

The model D is trained in tandem: its parameters are randomly initialized, and then updated after each deconvolution, so it improves as $\mathcal{Y}'$ improves. The objective of updating D is:

$$\hat{D} = \underset{D}{\arg\min}\, L_t(\Phi_t(S), \ell_{real}) + \sum_{Y_i' \in \mathcal{Y}'} L_t(\Phi_t(Y_i'), \ell_{fake}). \qquad (6)$$

$\ell_{real} = 1$ and $\ell_{fake} = 0$. Like [17], we set $\omega = 0.00001$ in (4), and minimize (4) and (6) using back-propagation with ADAM [15] (learning rate 0.02, momentum 0.5). The optimization in (6) is memory intensive. To make it feasible and efficient for a machine with a Titan X GPU with 12 GB onboard memory, we divide X' into multiple non-overlapped segments of 3 frames, and synthesize frames within one segment after another. In this way, L_{es} in (4) will only preserve the temporal smoothness within each segment. In order to preserve the inter-segment smoothness, we use the last 2 frames of the previous segment to compute L_{es}, and leave these 2 frames unchanged during the optimization for the current segment. The segment size can be adaptively enlarged with increased GPU memory capacity.

Train G Against D with Unpaired Real Samples. Given the unpaired real samples $\mathcal{Y}'$, we aim to train G against D in a GAN model. G takes $\mathcal{X}$ as input and outputs the synthesized video $\mathcal{Y} = \{Y_1, \cdots, Y_i, \cdots\}$, with $Y_i = G(X_i)$. Thus, our objective herein is as follows:

$$L(G, D, \mathcal{X}, \mathcal{Y}, \mathcal{Y}') = \sum_{Y_i \in \mathcal{Y}} [L_t(\Phi_t(Y_i), \ell_{real}) + \omega \Upsilon(Y_i)] + L_{es}(\mathcal{G}, \mathcal{F}, \mathcal{X}, \mathcal{Y}) +$$
$$\sum_{Y_i' \in \mathcal{Y}'} [L_t(\Phi_t(Y_i'), \ell_{real}) + L_c(\Phi_c(Y_i), \Phi_c(Y_i'))]. \qquad (7)$$

We therefore aim to solve:

$$\hat{G} = \arg\min_{G} \max_{D} L(G, D, \mathcal{X}, \mathcal{Y}, \mathcal{Y}'), \qquad (8)$$

where D and G are trained from scratch using back-propagation with ADAM (learning rate 0.02, momentum 0.5). Same notations as those in (7) can be found in (4), (5) and (6). Note that L_c is only valid for the paired frames.

6   Experiments

Implementation Details. We implement VST-GAN and MGAN using Tensorflow, and conduct the experiments on a computer with an Intel Xeon X5570 CPU with 16 cores of 2.93 GHz each, 94.4 GB memory, and one NVIDIA TITAN X GPU with 12 GB onboard memory. For the real sample generation process, D is trained for each segment (3 frames) for 3, 000 iterations. With a batch size of 3, G is trained for 20, 000 iterations. For a 50-frame video, it takes D approximately 2 h to generate the real samples, and approximately a further 1 h to train G.

Datasets. We use 8 classical style images, $i.e.$, *starry night, the scream, udnie; la muse, wave, composition vii, mosaic,* and *candy*, several of which are used in [3] or [24]. For the source videos, we choose 8 videos with diverse contents, including natural scenes, action scenes, close-up portraits, *etc.* Lengths of these videos vary from 40 to 300 frames, with 91 frames on average. All videos have the image resolutions of 640×360 and were captured at 23 frame per second.

Compared Methods. We compare VST-GAN with ASTV [24], a state-of-the-art neural network based video style transfer method. ASTV uses optical flow and occlusion mask to preserve temporal smoothness in the synthesized video, so it suffers from the common problems in estimating optical flow, $i.e.$, the sensitivity to occlusion and abrupt motion in video. We also create a baseline method based on MGAN [17], which uses image style transfer method in [17] to create individual frames independently. Comparison with these baseline methods demonstrates the advantage of our method in preserving temporal smoothness.

6.1   Qualitative Comparison

In Fig. 6, we show two consecutive frames from two synthesized videos produced by VST-GAN and MGAN, respectively, with two highlighted regions in each frame. The close-up regions demonstrate the effectiveness of evolve-sync loss in preserving the temporal smoothness. As mentioned in the beginning of Sect. 4, the image style transfer methods ($e.g.$, MGAN) are ineffective in preserving the temporal smoothness, which is evident from comparing the two close-up regions.

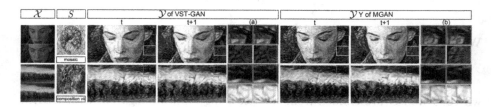

Fig. 6. Qualitative comparison with MGAN [17] The marked regions highlight that the temporal smoothness of our results is higher. The dilated marked regions are shown in (a) and (b). This figure is best viewed in color. (Color figure online)

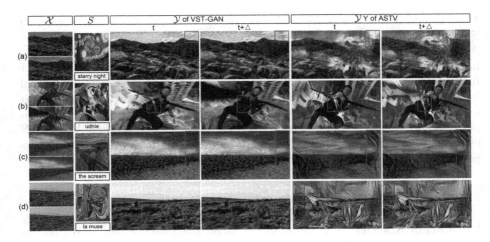

Fig. 7. Qualitative comparison of the results of VST-GAN with those of ASTV [24] Row (a) corresponds to a scene with the rapid camera motion, and we highlight the newly entering regions where the artifacts appear for ASTV's result. Row (b) displays a video associated with the rapid object motions and occlusions. The marked regions show that the artifacts appear after rapid motions and occlusions for ASTV's result. Row (c) presents an example where the ghosting salient content exists in ASTV's result but not in $\mathcal{X}$. Row (d) illustrates an example where the content of $\mathcal{X}$ is not preserved properly by ASTV. This figure is best viewed in color. (Color figure online)

We further compare the synthesis results using VST-GAN with ASTV [24] in Fig. 7, with four video clips with a variety of challenging factors including camera motions, rapid object motions, and occlusions, *etc.* Fig. 7(a) shows an epic natural scene with lateral camera motions. Note that artifacts emerge as the new content enters the scene at the top right corner. Considering the contextual information on the left side, the sky color within the marked region is supposed to be either blue, white or yellow. However, the actual color is brownish-grey, which is partially caused by optical flow's intrinsic limitation. Specifically, since the camera view moves from left to right, the estimated direction of the optical flow is the opposite. This leads to an ill-posed problem that the new content at the top right corner solely depends on the pixel values along image's right border. Thus, colors for the sky region (blue, white and yellow) on the left side have no effect on the synthesized video. Figure 7(b) displays a video in which the arrow moves rapidly and its movement causes the occlusion. Artifacts arise in the videos synthesized by ASTV due to the instability of the estimated optical flow in the presence of rapid motions and occlusions. On the other hand, because we introduce the evolve-sync loss to replace optical flow, VST-GAN is not affected by problems of the optical flow estimation.

In Fig. 7(c), we present an example where ASTV synthesized video contains some "ghost" salient content that does not exist in the source video.

The preservation of the source video content is worse in the ASTV's result in Fig. 7(d), *e.g.*, the contextually related patches that constitute the grassland in $\mathcal{X}$ become unrelated in $\mathcal{Y}$. This is because ASTV [24] models textures in the style image with a Gaussian distribution. As a result, the synthesized video does not further improve once two distribution matches and the synthesis quality of local image regions cannot be guaranteed. In contrast, VST-GAN preserves the content more properly by relaxing the above assumption to that the textures follow a complicated non-linear manifold. Furthermore, the adversarial training of VST-GAN can recognize such a manifold with its discriminative network, and strengthen its generative power with a projection on the manifold.

6.2 Quantitative Comparison

Evaluation Metrics. Existing methods [3, 24] measure the temporal smoothness of the synthesized video using the ground-truth optical flow and occlusion mask. Specifically, they warp the i-th frame in the synthesized video to be synchronized with the ground truth flow and compute the difference with the $(i - 1)$-th synthesized frame in non-occluded regions. Although this metric is straightforward to compute, it has two drawbacks. First, it restricts the choice of the evaluated videos to those with ground truth optical flow, which are very difficult to generate and scarce in number. Second, it does not allow for the evaluation of long-term temporal smoothness due to the lack of long-term ground truth optical flow. To this end, we use the *averaging evolve-sync loss* (AESL) (averaged by the video length) as a new metric that is free of the optical flow, occlusion mask and the short-term restrictions. We compute the multi-order AESL to evaluate the temporal smoothness for short (order 2 and 4)/medium (order 6 and 8)/long-term (order 10 and 12).

Comparing with the State-of-the-art Methods. The comparison results are presented in Table 1. These result show that VST-GAN outperforms MGAN significantly in terms of the temporal smoothness of the synthesized videos. The comparison between VST-GAN and ASTV based on AESL of order 2 and 4 suggests comparable performance of our method using the evolve-sync loss to those based on optical flow and occlusion mask in preserving the short-term temporal smoothness. In addition, the evolve-sync loss is more effective than optical flow in preserving medium/long-term temporal smoothness, which is demonstrated by the comparison based on AESL of order 6, 8, 10 and 12. This is due the lack of long-term optical flows, which cannot be reliably estimated using current methods. In contrast, the high-order evolve-sync loss can be more easily computed and compared.

Effects of the Evolve-Sync Loss. To investigate the impact of the evolve-sync loss on preserving the temporal smoothness, we remove it from the objective of training G (7). As a result, we observe significant increase in AESL of all orders, which indicates the retrogression on the preservation of the temporal smoothness. Nonetheless, VST-GAN still preserves the temporal smoothness

Table 1. Comparison on temporal smoothness for synthetic videos using ours and state-of-the-art video style transfer methods.

Sequence	Method	AESL					
		2-order	4-order	6-order	8-order	10-order	12-order
starry night	ASTV [24]	**45.22**	**132.74**	221.95	301.32	366.93	463.47
	MGAN [17]	97.67	245.43	344.66	443.56	523.67	598.88
	VST-GAN (ours)	60.42	143.62	**220.32**	**289.87**	**344.98**	**412.83**
	VST-GAN w/o ESL	72.13	181.80	274.85	379.43	440.05	533.49
	VST-GAN w/o RNN	65.36	175.23	271.84	347.59	375.17	422.74
the scream	ASTV [24]	32.45	105.31	178.48	237.00	308.06	368.63
	MGAN [17]	91.83	220.06	292.33	362.17	420.37	478.57
	VST-GAN (ours)	**31.33**	**96.74**	**139.05**	**179.07**	**204.43**	**240.20**
	VST-GAN w/o ESL	65.65	143.11	197.96	249.81	293.99	348.53
	VST-GAN w/o RNN	47.62	108.26	146.45	193.29	221.19	252.69
udnie	ASTV [24]	**48.38**	84.83	120.23	154.45	187.89	220.71
	MGAN [17]	81.14	121.00	149.36	173.43	226.53	252.21
	VST-GAN (ours)	48.69	**72.36**	**93.31**	**112.74**	**131.06**	**148.46**
	VST-GAN w/o ESL	81.00	116.75	148.03	177.20	204.52	230.32
	VST-GAN w/o RNN	54.60	89.00	119.84	148.73	176.09	202.08
la muse	ASTV [24]	84.18	266.25	459.09	650.63	838.49	1021.58
	MGAN [17]	193.32	493.94	737.08	940.96	1120.69	1282.29
	VST-GAN (ours)	**79.40**	**234.34**	**322.61**	**396.55**	**480.32**	**541.78**
	VST-GAN w/o ESL	131.70	309.85	460.34	595.70	721.95	841.39
	VST-GAN w/o RNN	96.43	247.61	325.74	413.65	493.38	568.11
wave	ASTV [24]	59.97	147.53	233.54	321.63	413.23	506.86
	MGAN [17]	180.76	434.75	636.53	813.08	978.61	1134.02
	VST-GAN (ours)	**58.21**	**138.57**	**193.25**	**254.14**	**306.07**	**361.79**
	VST-GAN w/o ESL	130.35	314.50	466.62	603.03	732.35	855.11
	VST-GAN w/o RNN	123.84	294.62	429.18	546.39	656.17	759.13
comp. vii	ASTV [24]	**32.92**	**99.67**	172.53	234.40	305.62	345.03
	MGAN [17]	94.61	222.30	304.92	357.22	417.04	456.96
	VST-GAN (ours)	39.23	104.42	**157.65**	**208.61**	**254.18**	**293.65**
	VST-GAN w/o ESL	92.03	245.12	373.81	470.22	578.90	642.62
	VST-GAN w/o RNN	65.00	159.93	239.70	303.14	354.22	413.45
mosaic	ASTV [24]	**30.44**	103.48	187.32	275.91	364.91	451.67
	MGAN [17]	99.35	260.26	400.07	528.03	645.29	751.34
	VST-GAN (ours)	36.84	**98.88**	**157.79**	**214.84**	**253.38**	**301.66**
	VST-GAN w/o ESL	63.13	179.14	289.34	395.64	497.12	591.47
	VST-GAN w/o RNN	53.84	144.03	225.46	301.84	374.40	442.21
candy	ASTV [24]	29.98	90.81	148.01	200.58	249.69	296.37
	MGAN [17]	33.88	84.39	126.82	164.90	200.51	233.95
	VST-GAN (ours)	**19.26**	**52.34**	**87.74**	**110.95**	**144.60**	**170.47**
	VST-GAN w/o ESL	31.07	76.31	114.84	149.95	182.66	213.52
	VST-GAN w/o RNN	23.67	67.95	107.51	143.22	175.31	203.78

better than MGAN even without the use of the evolve-sync loss. This is because we maintain the evolve-sync loss in D for generating real training samples, which further demonstrates the effectiveness of the evolve-sync loss.

Effects of the Recurrent Structure. We remove the convolutional recurrent layer from VST-GAN to study its impact on preserving the temporal smoothness. Consequently, the AESL increases slightly, but the increment is much smaller compared to that after removing the evolve-sync loss. This indicates that the recurrent structure is also useful for preserving the temporal smoothness, but its impact is less prominent than that of the evolve-sync loss.

Runtime Efficiency. The runtime speed of our VST-GAN in synthesizing videos is 18.18 fps, which is comparable to the image style transfer method MGAN (19.33 fps), and much efficient than the deconvolutional video style transfer method ASTV (0.03 fps).

7 Conclusion

In this work, we propose VST-GAN as an adversarial learning framework for video style transfer based on the evolve-sync loss. We show that the evolve-sync loss is able to preserve the temporal smoothness effectively without using optical flow. Our accelerating training strategy and the convolutional recurrent structure significantly reduce the training complexity of VST-GAN. Experimental evaluations show that VST-GAN outperforms the state-of-the-art methods based on optical flow in both running time efficiency and visual quality.

References

1. Anderson, A.G., Berg, C.P., Mossing, D.P., Olshausen, B.A.: Deepmovie: Using optical flow and deep neural networks to stylize movies. CoRR abs/1605.08153 (2016)
2. Bousseau, A., Neyret, F., Thollot, J., Salesin, D.: Video watercolorization using bidirectional texture advection. TOG **26**(3), 104 (2007)
3. Chen, D., Liao, J., Yuan, L., Yu, N., Hua, G.: Coherent online video style transfer. In: ICCV (2017)
4. Chen, D., Yuan, L., Liao, J., Yu, N., Hua, G.: Stylebank: an explicit representation for neural image style transfer. In: CVPR, pp. 1897–1906 (2017)
5. Chen, D., Yuan, L., Liao, J., Yu, N., Hua, G.: Stereoscopic neural style transfer. In: CVPR, pp. 1–9 (2018)
6. Denton, E.L., Chintala, S., Szlam, A., Fergus, R.: Deep generative image models using a laplacian pyramid of adversarial networks. In: NIPS, pp. 1486–1494 (2015)
7. Fan, Q., Chen, D., Yuan, L., Hua, G., Yu, N., Chen, B.: Decouple learning for parameterized image operators. In: Ferrari, V., Hebert, M., Sminchisescu, C., Weiss, Y. (eds.) ECCV 2018. LNCS, vol. 11217, pp. 455–471. Springer, Cham (2018). https://doi.org/10.1007/978-3-030-01261-8_27
8. Gatys, L.A., Ecker, A.S., Bethge, M.: Texture synthesis using convolutional neural networks. In: NIPS, pp. 262–270 (2015)

9. Gatys, L.A., Ecker, A.S., Bethge, M.: Image style transfer using convolutional neural networks. In: CVPR, pp. 2414–2423 (2016)
10. Goodfellow, I.J., et al.: Generative adversarial nets. In: NIPS, pp. 2672–2680 (2014)
11. Gretton, A., Borgwardt, K.M., Rasch, M.J., Schölkopf, B., Smola, A.J.: A kernel two-sample test. JMLR **13**, 723–773 (2012)
12. Hays, J., Essa, I.A.: Image and video based painterly animation. In: NPAR, pp. 113–120 (2004)
13. He, M., Chen, D., Liao, J., Sander, P.V., Yuan, L.: Deep exemplar-based colorization. TOG **37**(4), 47:1–47:16 (2018)
14. Johnson, J., Alahi, A., Fei-Fei, L.: Perceptual losses for real-time style transfer and super-resolution. In: Leibe, B., Matas, J., Sebe, N., Welling, M. (eds.) ECCV 2016. LNCS, vol. 9906, pp. 694–711. Springer, Cham (2016). https://doi.org/10.1007/978-3-319-46475-6_43
15. Kingma, D.P., Ba, J.: Adam: A method for stochastic optimization. CoRR abs/1412.6980 (2014)
16. Kyprianidis, J.E., Collomosse, J.P., Wang, T., Isenberg, T.: State of the "art": a taxonomy of artistic stylization techniques for images and video. TVCG **19**(5), 866–885 (2013)
17. Li, C., Wand, M.: Precomputed real-time texture synthesis with markovian generative adversarial networks. In: Leibe, B., Matas, J., Sebe, N., Welling, M. (eds.) ECCV 2016. LNCS, vol. 9907, pp. 702–716. Springer, Cham (2016). https://doi.org/10.1007/978-3-319-46487-9_43
18. Lu, J., Sander, P.V., Finkelstein, A.: Interactive painterly stylization of images, videos and 3D animations. In: SI3D, pp. 127–134 (2010)
19. Mahendran, A., Vedaldi, A.: Understanding deep image representations by inverting them. In: CVPR, pp. 5188–5196 (2015)
20. Mathieu, M., Couprie, C., LeCun, Y.: Deep multi-scale video prediction beyond mean square error, pp. 1–14 (2016)
21. Mathieu, M., Zhao, J.J., Sprechmann, P., Ramesh, A., LeCun, Y.: Disentangling factors of variation in deep representation using adversarial training. In: NIPS, pp. 5041–5049 (2016)
22. Radford, A., Metz, L., Chintala, S.: Unsupervised representation learning with deep convolutional generative adversarial networks. CoRR abs/1511.06434 (2015)
23. Revaud, J., Weinzaepfel, P., Harchaoui, Z., Schmid, C.: EpicFlow: edge-preserving interpolation of correspondences for optical flow. In: CVPR, pp. 1164–1172 (2015)
24. Ruder, M., Dosovitskiy, A., Brox, T.: Artistic style transfer for videos. In: GCPR, pp. 26–36 (2016)
25. Ulyanov, D., Lebedev, V., Vedaldi, A., Lempitsky, V.S.: Texture networks: feedforward synthesis of textures and stylized images. In: ICML, pp. 1349–1357 (2016)
26. Vondrick, C., Pirsiavash, H., Torralba, A.: Generating videos with scene dynamics. In: NIPS, pp. 613–621 (2016)
27. Weinzaepfel, P., Revaud, J., Harchaoui, Z., Schmid, C.: Deepflow: large displacement optical flow with deep matching. In: ICCV, pp. 1385–1392 (2013)
28. Xu, T., et al.: AttnGAN: fine-grained text to image generation with attentional generative adversarial networks, pp. 1–9 (2018)
29. Zhang, H., Xu, T., Li, H.: StackGAN: text to photo-realistic image synthesis with stacked generative adversarial networks. In: ICCV, pp. 5908–5916 (2017)
30. Zhang, H., et al.: StackGAN++: Realistic image synthesis with stacked generative adversarial networks. CoRR abs/1710.10916 (2017)

31. Zhang, R., Isola, P., Efros, A.A.: Colorful image colorization. In: Leibe, B., Matas, J., Sebe, N., Welling, M. (eds.) ECCV 2016. LNCS, vol. 9907, pp. 649–666. Springer, Cham (2016). https://doi.org/10.1007/978-3-319-46487-9_40
32. Zhang, S., Li, X., Hu, S., Martin, R.R.: Online video stream abstraction and stylization. TMM **13**(6), 1286–1294 (2011)
33. Zhu, J.-Y., Krähenbühl, P., Shechtman, E., Efros, A.A.: Generative visual manipulation on the natural image manifold. In: Leibe, B., Matas, J., Sebe, N., Welling, M. (eds.) ECCV 2016. LNCS, vol. 9909, pp. 597–613. Springer, Cham (2016). https://doi.org/10.1007/978-3-319-46454-1_36
34. Zhu, J., Park, T., Isola, P., Efros, A.A.: Unpaired image-to-image translation using cycle-consistent adversarial networks. In: ICCV

An Unsupervised Deep Learning Framework via Integrated Optimization of Representation Learning and GMM-Based Modeling

Jinghua Wang[ID] and Jianmin Jiang[(✉)][ID]

Research Institute for Future Media Computing, College of Computer Science
and Software Engineering, Shenzhen University, Shenzhen, China
{wang.jh,jianmin.jiang}@szu.edu.cn

Abstract. While supervised deep learning has achieved great success in
a range of applications, relatively little work has studied the discovery
of knowledge from unlabeled data. In this paper, we propose an unsupervised deep learning framework to provide a potential solution for the
problem that existing deep learning techniques require large labeled data
sets for completing the training process. Our proposed introduces a new
principle of joint learning on both deep representations and GMM (Gaussian Mixture Model)-based deep modeling, and thus an integrated objective function is proposed to facilitate the principle. In comparison with
the existing work in similar areas, our objective function has two learning targets, which are created to be jointly optimized to achieve the best
possible unsupervised learning and knowledge discovery from unlabeled
data sets. While maximizing the first target enables the GMM to achieve
the best possible modeling of the data representations and each Gaussian component corresponds to a compact cluster, maximizing the second term will enhance the separability of the Gaussian components and
hence the inter-cluster distances. As a result, the compactness of clusters
is significantly enhanced by reducing the intra-cluster distances, and the
separability is improved by increasing the inter-cluster distances. Extensive experimental results show that the propose method can improve the
clustering performance compared with benchmark methods.

Keywords: Unsupervised clustering · Representation learning ·
Gaussian Mixture Model · Deep learning

1 Introduction

With the advanced machine learning technologies, we can process the explosion
data effectively. Deep learning is one of the most popular techniques, and has
been successfully applied in many computer vision tasks, such as image classification [12,36], semantic segmentation [24,37], and object detection [26]. However,

© Springer Nature Switzerland AG 2019
C. V. Jawahar et al. (Eds.): ACCV 2018, LNCS 11361, pp. 249–265, 2019.
https://doi.org/10.1007/978-3-030-20887-5_16

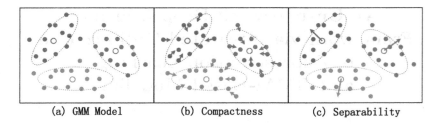

(a) GMM Model   (b) Compactness   (c) Separability

Fig. 1. GMM modeling (best viewed in color). (a) Traditional GMM method fits the data points by a set of Gaussian components. (b) Adjust the sample representations towards the Gaussian center to improve the compactness. (c) Adjust the Gaussian centers to enhance the separability between different components. (Color figure online)

these techniques [12,24,26,37] heavily rely on a huge number of high quality labeled training data to learn a good model. Yet, manually labeling the training data is extremely time-consuming. Thus, it is necessary to develop unsupervised techniques that can discover knowledge from the easily available unlabeled data.

Clustering is one of the most popular unsupervised machine learning techniques. Traditional clustering methods, such as k-means [19] and Gaussian Mixture Model (GMM) [2], categorize samples by investigating their similarities directly in the original data space. Thus, their performances heavily depend on the distribution of the data samples [1].

In order to achieve robustness against the data sample distributions, researchers propose to extract features or learn representations before conducting the clustering procedure [15,44]. While the supervised methods learn representations which are closely correlated with the class labels, unsupervised representation learning is more difficult due to the unavailability of label information. Through representation learning, the methods [15,44] can explicitly or implicitly discover the hidden variables which are more discriminative than the original data sample. While the methods [15,44] successfully learn discriminative representations for various tasks, the resulting representations are not necessarily the optimal choice for the clustering task. To learn representations of data samples that are catered for the clustering task, researchers [40,42,43] propose to integrate the representation learning with clustering.

In this paper, we propose a new joint optimization approach for unsupervised representation learning and clustering. We aim at formulating a GMM out of the whole data representations and the center of each Gaussian component represents a cluster center. Modeling the representations by a GMM significantly alleviates the constraints in the work [31] and [38], which model the representations by a single Gaussian model. Our approach aims to learn data representations which are intra-cluster compact and inter-cluster separable.

As in all of the GMM-based methods, we maximize the GMM likelihood to discover a feasible GMM for the whole data representations, as shown in Fig. 1(a). An important but rarely mentioned point in a GMM model is: a larger GMM likelihood also means a smaller distance between a sample representation and its associating cluster center (in addition to a set of well positioned Gaussian centers). In our approach, both the cluster center and the representation are iteratively updated in the training procedure. Thus, by maximizing the GMM likelihood, we can not only well position the Gaussian centers, but also adjust the representations towards their associating cluster centers, as shown in Fig. 1(b). In this way, we can enhance the compactness of the clusters. We also explicitly maximize the distance between the Gaussian centers, which is achieved by iteratively update the Gaussian centers to make them far away from each other, as shown in Fig. 1(c). By doing this, we can enhance the separability of the clusters. This can also implicitly enlarge the inter-cluster distance between data sample representations.

Fundamentally, our contributions can be highlighted as: (i) we propose a new network structure for joint optimization of both deep representation learning and GMM-based modeling; (ii) the proposed framework can learn representations which are inter-cluster separable and intra-cluster compact; (iii) we model a deep representation of the whole data set as a GMM, and expect the data from the same cluster share a Gaussian component.

2   Related Work

Clustering has been widely applied in many computer vision tasks. Popular clustering methods include k-means [19], GMM [2], non-negative matrix factorization [9], and spectral clustering [45]. Based on the low-level features, the clustering methods can perform well on a limited number of tasks [1].

Since the popularity of deep learning, researchers tend to conduct clustering based on the deep features. Hinton and Salakhutdinov [15] propose to train deep autoencoder networks and take the outputs of a bottleneck central layer as the representations. Schroff et al. [28] train FaceNet to extract features that can reveal the similarity between face images. Bruna and Mallat [3] propose a wavelet scattering network to learn image representations that are stable to deformation and feasible for both classification and clustering.

The generic deep features can be applied in many different tasks and achieve better performance than the hand-crafted features [3,15,28]. However, they are not necessarily optimal for the task of clustering. To further improve the clustering performance, researchers [40,42,43] propose to integrate the representation learning with clustering. For joint optimization, Yang et al. [42] introduce an objective function consisting of three parts, i.e. dimension reduction, data reconstruction, and cluster structure regularization. Based on the idea of agglomerative clustering, Yang et al. [43] propose a recurrent framework for unsupervised clustering. By introducing an auxiliary distribution, Xie et al. [40] propose another method for joint optimization.

While all the existing approaches for joint optimization have achieved certain level of success as reported in the literature, none of them directed the joint optimization towards improving the compactness and separability in clustering, which remains crucial for unsupervised deep learning among unlabeled data sets.

To this end, we propose a new joint optimization approach for both representation learning and clustering, which simultaneously increases the separability and compactness for all the evolved clusters.

Significant research efforts have been reported to model the distribution of image representations, and achieve good performance in a variety of computer vision tasks, such as scene categorization [21] and image classification [29]. In our newly proposed approach, we model the distribution of data representations (from many different clusters) with a GMM, and expect the data from the same cluster share a Gaussian component.

Modeling the features by a GMM has been studied in the research area of automatic speech recognition [8,27]. The work [25] proposes a framework for bottleneck feature extraction. However, this work does not update the GMM parameter in the back-propagation procedure. Based on the observation that log-linear mixture model (LMM) is equivalent to GMM [14], the work [33] transforms GMM to LMM and implements it using popularly used neural network elements. As reported [13], the soft-max layer in CNN is equivalent to a single Gaussian model with a globally pooled covariance matrix. The work [34] applied a joint optimization strategy of feature extraction and classification in the task of automatic speech recognition. However, to the best of our knowledge, the joint optimization of CNN and GMM has not been studied in the area of unsupervised clustering.

3   The Proposed Approach

3.1   GMM

A Gaussian Mixture Model (GMM) expresses the probability as a weighted sum of a finite number of Gaussian component densities, as follows

$$p(x|\lambda) = \sum_{k=1}^{m} \omega_k g(x|\mu_k, \Sigma_k) \tag{1}$$

where $x \in R^d$ is a continuous-valued feature vector, m is the number of Gaussian components, $\omega_k (k = 1, \cdots, m)$ are the mixture weights, and $g(x|\mu_k, \Sigma_k)(k = 1, \cdots, m)$ are the Gaussian densities. The mixture weights satisfy the constraint that $\Sigma_{k=1}^{m} \omega_k = 1$. Each Gaussian component density is a Gaussian function, i.e.

$$g(x|\mu_k, \Sigma_k) = \frac{1}{\sqrt{(2\pi)^d |\Sigma_k|}} exp\{-\frac{1}{2}(x - \mu_k)^T \Sigma_k^{-1}(x - \mu_k)\} \tag{2}$$

with mean vector μ_k and covariance matrix Σ_k. For simplicity, let λ denote a combinational group of the mean, covariance, and the mixture weight of the Gaussian components, i.e. $\lambda = \{\omega_k, \mu_k, \Sigma_k\}, k = 1, \cdots, m$.

In order to estimate the parameter λ, we normally maximize the GMM likelihood formulated as

$$p(X|\lambda) = \prod_{n=1}^{N} p(x_n|\lambda) \tag{3}$$

where $X = \{x_1, x_2, \cdots, x_N\}$ is a set of independent observations. To solve this maximization problem, the expectation-maximization (EM) algorithm is widely applied, which improves the parameters iteratively with the following two steps.
Expectation. Fix the parameter λ, and calculate the posteriori probability of every sample belonging to each component.
Maximization. With the above probability, update the parameter of each Gaussian component to maximize the GMM likelihood (expressed in Eq. 3).

3.2 Representation Learning and GMM-Based Modeling

Inspired by the fact that a proper representation learning procedure can significantly improve the clustering results [16,42], we propose in this paper a new approach for integrated optimization of representation learning and clustering.

Regarding the representation learning, the distribution of the representations itself is another important factor to consider, in addition to the correlation between the representations and class labels. The work [31] shows that we can learn a neural network that transforms arbitrary data distribution into a Gaussian distribution. The Gaussian distributed data representations are successfully applied in different computer vision tasks [21,29,38,39]. Inspired by this, we propose to model the representations of the unlabeled data samples by a GMM and expect that the representations from the same cluster share a Gaussian component.

Let $f_\theta(x)$ denote the representation of data sample x extracted by a convolutional neural network (with θ as the parameter). We model the distribution of $f_\theta(x)$ as follows

$$f_\theta(x) \sim p(f_\theta(x)|\lambda) \tag{4}$$

where the probability function p is a GMM formulated in Eq. 1 and λ denotes a combinational group of the GMM parameters, i.e. $\lambda = \{\omega_k, \mu_k, \Sigma_k\}$.

For joint optimization of both deep representation learning and GMM-based modeling, we maximize the following objective function

$$O = log(P(f_\theta(X)|\lambda)) + \eta S(\mu) = log(\prod_{n=1}^{N} p(f_\theta(x_n)|\lambda)) + \eta \sum_{k=1}^{m} \sum_{j \in n(k)} d(\mu_k, \mu_j) \tag{5}$$

Here, $X = \{x_1, x_2, \cdots, x_N\}$ represents the whole data sample set and N is the number of data samples. The parameter η is nonnegative to balance the two terms. Let $n(k)$ denote the neighboring Gaussian components of the kth component (measured by the distance $d(x,y)$ between the centers of different components). While the first term in Eq. 5 calculates the log GMM likelihood of the representations, the second term assesses the separability between different Gaussian components.

The first term in Eq. 5 has two sets of parameters, i.e. the CNN parameter θ and the GMM parameter λ. With a fixed parameter θ (and thus the representations of the data samples), a larger likelihood means the GMM can better model the distributions of the representations. With a fixed parameter λ, a larger likelihood means the sample representations are closer to their associated Gaussian centers. At the learning stage, we update the GMM parameter λ to better model the data representations, and update the CNN parameter θ to adjust the data representations towards their associating centers (which can enhance the compactness of each Gaussian component). Correspondingly, maximizing the first term in Eq. 5 guarantees: (1) that the GMM can well model the data representations; and (2) that each Gaussian component corresponds to a compact cluster in the data representation space.

By maximizing the second term in Eq. 5, we can enhance the separability of the Gaussian components and thus improve the clustering performance. In addition to compactness, separability is another important criteria in clustering tasks. When we enlarge the distance between the Gaussian centers, we implicitly increase the distances between the data representations belonging to different components. In addition, the introduction of the separability term also guarantees that the GMM model and the data representations are not trivial (i.e. all of data samples sharing the same representation).

To optimize our proposed framework for CNN-based representation learning, we update the parameters θ and λ iteratively based on the the evaluations of data samples. With a data sample x_n as the input, specifically, we introduce and maximize the following objective function

$$O(x_n|\lambda, \theta) = log(p(f_\theta(x_n)|\lambda)) + \eta \sum_{k=1}^{m} \sum_{j \in n(k)} \|\mu_k - \mu_j\|^2 \qquad (6)$$

To speed up the parameter learning process, we restrict the covariance matrix to be diagonal, i.e. $\Sigma = diag(\sigma_1^2, \sigma_2^2, \cdots, \sigma_D^2)$, where D is the dimensionality of the representations. Let y denote the representation of x, i.e. $y = f(x)$. Out of the basic mathematical derivations, we achieve the following deviations as given in Eqs. 7–10 for the objective function regarding the parameters, where the index $1 \leq k \leq m$ corresponds to the Gaussian components, and $1 \leq d \leq D$ denotes the dimension of the representations (or the parameters), i.e. μ_{kd} and y_d respectively denotes the dth dimension of the mean vector and the representation.

$$\frac{\partial O(x|\lambda, \theta)}{\partial y} = \sum_{k=1}^{m} p(c_k|y)\Sigma_k^{-1}(\mu_k - y) \tag{7}$$

$$\frac{\partial O(x|\lambda, \theta)}{\partial \mu_{kd}} = p(c_k|y)\frac{y_d - \mu_{kd}}{\sigma_{kd}^2} + 2\eta \sum_{j \in n(k)} (\mu_{kd} - \mu_{jd}) \tag{8}$$

$$\frac{\partial O(x|\lambda, \theta)}{\partial \sigma_{kd}} = p(c_k|y)[\frac{(y_d - \mu_{kd})^2}{\sigma_{kd}^2} - 1] \tag{9}$$

$$\frac{\partial O(x|\lambda, \theta)}{\partial \omega_k} = p(c_k|y) - \omega_k \tag{10}$$

where $p(c_k|y) = \omega_k g(y|\mu_k, \Sigma_k)/p(y|\lambda)$ denotes the probability that sample x_i belonging to the kth Gaussian component. In a backpropagation stage, we can use Eq. 7 to update the parameter θ and thus the representation y. The parameters of the GMM are updated based on Eqs. 8, 9, and 10.

3.3 Network Structure

To complete our proposed framework for integrated representation learning and clustering, we propose a network structure (as shown in Fig. 2) consisting of three components, i.e. the encoder, the decoder, and the representation modeling network (RMN). Three steps are designed for its training. Firstly, we train the encoder and the decoder by the data samples. Secondly, we initialize the RMN by a GMM that best captures the distribution of data representations produced by the encoder. Finally we jointly optimize the encoder and the RMN.

It is widely recognized that autoencoders can learn representations that are semantically meaningful [15]. This work trains an autoencoder and uses the encoder subnetwork to initialize the representation extraction network. We train the denoising autoencoder layer-by-layer. At the training stage, we first randomly corrupt the input data sample, and then use the denoising autoencoder to reconstruct the clean sample. The mathematical expression of a one layer denosing autoencoder is given as follows

$$\tilde{x} \sim q_D(\tilde{x}|x) \tag{11}$$

$$y = f_{\theta_1}(\tilde{x}) = s(W_1\tilde{x} + b_1) \tag{12}$$

$$z = g_{\theta_1'}(y) = s(W_1'y + b_1') \tag{13}$$

and the objective function is the squared distance between the input data sample and the reconstruction result, i.e. $||x - z||^2$. The function q_D in Eq. 11 denotes a stochastic mapping, i.e. randomly chooses a portion of data sample dimensions and set them to be 0. Let y be the representation extracted by the parameters W_1 and b_1, and the reconstruction result is denoted by z and the decoder parameters by W_1' and b_1'. The autoencoder can be easily extended to multiple layers.

The RMN consists of two layers, one Λ layer corresponding to the parameters of the m Gaussian components and one G layer corresponding to the mixture

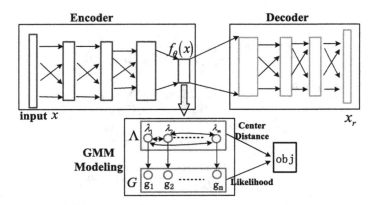

Fig. 2. The proposed network structure. The network consists of three components, i.e. the encoder, the decoder, and the representation modeling network. The representation modeling network models the data representations by a GMM. Each node λ represents the weight, the mean and covariance matrix of a Gaussian component. The Λ layer is initialized based on the deep features $f_\theta(x)$ produced by the encoder. Note that, the Gaussian components are mutually influenced by each other.

of Gaussian components. In the network, a node $\lambda_k = \{\mu_k, \Sigma_k, \omega_k\}$ denotes the parameters of the kth Gaussian component. A node $g_k(1 \leq k \leq m)$ denotes the value of the kth Gaussian component evaluated by Eq. 2. Both the two layers take the data representations produced by the encoder as input. For the Λ layer, the data representations are used to initialized the parameters. For the G layer, the data representations are used to evaluated the likelihood. Note that, due to the second term in Eq. 5, the centers of the Gaussian components are mutually influenced by each other. Each of these two layers contributes to a term of Eq. 6, i.e. the Λ layer corresponds to the separability between clusters and the G layer corresponds to the compactness of clusters in addition to the feasibility of the GMM. Thus, both of these two layers contribute to the objective function.

4   Experiments

4.1   Dataset

To evaluate the proposed method, we conduct experiments on six datasets, i.e. MNIST, USPS, COIL20, COIL100, STL-10, and Reuters.

The MNIST dataset [17] is one of the most popular image datasets. It consists of 60,000 training samples and 10,000 testing samples from 10 classes (from 0 to 9). Each image in the dataset represents a handwritten digit. The images are centered with size of 28×28.

USPS[1] is a another handwritten digits dataset produced by the USPS postal service. In total, this data set contains 11,000 samples belonging to the 10 different classes, where the image size is 16×16.

---

[1] https://cs.nyu.edu/~roweis/data.html.

COIL20 [23] and COIL100 [22] are two datasets built by Columbia University, which respectively contain 1,440 gray images of 20 objects and 7,200 color images of 100 objects. The images are captured under different views.

The STL-10 dataset [7] consists of images from 10 different classes: *airplane, bird, car, cat, deer, dog, horse, monkey, ship, truck*. Each class has 1300 labeled images. In addition, there are also 100,000 unlabeled images. Note that, the unlabeled image set contains images not belonging to the above 10 classes. The size of the images is 96×96. While we use the labeled data to test our method, the unlabeled data are also used to train the autoencoder network. Following [11], we calculate the 8×8 color map and the HOG features of each image, and take the concatenation of them as the input.

The REUTERS dataset [18] contains 804,414 documents from 103 different topics. We use a subset of this dataset, which contains 365,968 documents from 20 different topics. As in [30], we use the tf-idf features of those most frequently used words to represent the documents.

4.2 Implementation Details

In the autodecoder, we adopt the widely used rectified linear units (ReLUs), except the layer where the data samples are reconstructed and the layer where the representations are produced [35]. Inspired by the work [20], the encoder consists of 3 fully connected layers (excluding the input layer), and their numbers of nodes are respectively 500, 500, and 2000. The representations of the input is extracted by a fully connected layer with 10 nodes. The decoder is a mirrored version of the encoder. After the encoder and decoder are trained layer-by-layer in a greedy manner, we concatenate them together and fine-tune the whole network. We then use the encoder subnetwork to produce the representations of data samples. In other words, the encoder can be considered as the initialization of the representation learning network.

The RMN is initialized based on the distribution of data sample representations. Specifically, we first extract the representations of all the data samples using the encoder, and then learn the initialized GMM based on these representations. To maintain a fair comparison with the benchmark methods, the number of the Gaussian components in the GMM is equal to the number of clusters in the dataset. The parameters of this GMM are adopted to initialize the RMN, and each node λ_i corresponds to the three parameters corresponding to a Gaussian component, i.e. the coefficient, the mean and the covariance. After that, we adopt the SGD (stochastic gradient descent) method to jointly optimize the encoder and RMN. We set the base learning rate to be 0.01 and take the step policy to update the learning rate. We set the cardinality $|n(k)|$ (in Eq. 5) to be half of the number of clusters. In this way, the center of one cluster is influenced by half of the remaining clusters that are nearby. For the parameter η in Eq. 5, we choose the best one from $\{0.1, 0.01, 0.001, 0.0001\}$.

4.3   Benchmarks

We compare the proposed method with a number of unsupervised clustering methods. Firstly, we take K-means [19] and GMM [2] as the baseline benchmarks. They can either take the low-level feature, or the deep autoencoder feature (AEF) as input.

Secondly, we compare our method with two agglomerative methods, i.e. agglomerative clustering (AC-GDL) [46] and agglomerative clustering via path integral (AC-PIC) [47]. We also take two subspace-based clustering methods as benchmarks, including large-scale spectral clustering (SC-LS) [6] and NMF with deep model (NMF-D) [32].

Thirdly, we compare our method with four other benchmarks, which also jointly optimize the representation learning and clustering, i.e. DEC [40], Joint unsupervised learning (JULE) [43], DCN [42], and DEPICT [10].

Finally, we set the parameter η in Eq. 5 to be zero and produce another benchmark (denoted as DeepGMM in this paper) to assess the effectiveness of the second term in Eq. 5, in terms of improving the separability between different Gaussian components. The only difference from the proposed method is that DeepGMM does not explicitly enlarge the distances between the Gaussian centers.

Table 1. The accuracy of the proposed method and the benchmarks on six datasets

Dataset	MNIST	USPS	COIL 20	COIL 100	STL-10	REUTERS
K-Means [19]	53.5%	46.0%	48.3%	51.4%	28.4%	32.3%
GMM [2]	47.6%	64.2%	54.3%	67.5%	20.3%	26.6%
AEF+KM	80.0%	64.3%	54.1%	67.5%	29.7%	35.8%
AEF+GMM	64.1%	71.3%	69.8%	73.8%	22.2%	31.6%
AC-GDL [46]	11.3%	86.7%	76.5%	80.5%	26.8%	36.1%
AC-PIC [47]	11.5%	85.5%	70.3%	84.6%	24.1%	29.4%
SC-LS [6]	71.4%	65.9%	76.4%	82.6%	20.4%	37.2%
NMF-D [32]	17.5%	38.2%	64.3%	70.2%	30.6%	39.5%
DEC [40]	84.4%	61.9%	83.6%	75.5%	35.9%	14.0%
JULE [43]	90.6%	91.4%	80.0%	77.4%	17.7%	38.1%
DCN [42]	83.0%	77.8%	76.5%	69.7%	34.1%	47.0%
DEPICT [10]	91.2%	91.4%	81.3%	76.4%	32.8%	29.9%
DeepGMM	72.5%	65.4%	72.3%	52.9%	27.6%	51.3%
The proposed	**93.9%**	**94.7%**	**88.5%**	**85.1%**	**36.3%**	**56.9%**

Table 2. The NMI of the proposed method and the benchmarks on six datasets

Dataset	MNIST	USPS	COIL 20	COIL 100	STL-10	REUTERS
K-Means [19]	0.50	0.45	0.74	0.78	0.25	0.17
GMM [2]	0.46	0.63	0.51	0.75	0.16	0.38
AEF+KM	0.73	0.59	0.77	0.82	0.26	0.41
AEF+GMM	0.59	0.68	0.60	0.81	0.20	0.48
AC-GDL [46]	0.12	0.82	0.80	0.78	0.21	0.38
AC-PIC [47]	0.12	0.84	0.79	0.81	0.18	0.27
SC-LS [6]	0.71	0.68	0.77	0.83	0.16	0.34
NMF-D [32]	0.15	0.29	0.69	0.72	0.24	0.31
DEC [40]	0.80	0.58	0.84	0.79	0.31	0.28
JULE [43]	0.87	0.88	0.85	0.83	0.14	0.36
DCN [42]	0.81	0.85	0.79	0.74	0.30	0.51
DEPICT [10]	0.87	0.88	0.84	0.84	0.36	0.48
DeepGMM	0.64	0.51	0.74	0.51	0.21	0.49
The proposed	**0.87**	**0.92**	**0.89**	**0.90**	**0.34**	**0.56**

4.4 Performance

Three popular standard metrics in evaluating the clustering algorithms are adopted, which include clustering accuracy (ACC) [41], normalized mutual information (NMI) [4], Calinski-Harabaz score (CH) [5].

The ACC is defined as $ACC = \frac{1}{N}\sum_{i=1}^{N}\delta(l_i, map(r_i))$, where N is the total number of data samples, l_i denotes the ground truth cluster label, and $\delta(x,y)$ is the delta function which equals 1 iff its two parameters are the same.

Let C and R respectively denote the clustering results and the groundtruth clusters, the NMI is defined as $NMI(C,R) = MI(C,R)/max(H(R),H(C))$ where $MI(C,R)$ is the mutual information between C and R, and $H(R)$ and $H(C)$ are the entropies.

Let k denote the number of clusters, the CH score is defined based on the between-clusters dispersion mean matrix B_k and within-cluster dispersion matrix $s(k) = \frac{tr(B_k)}{tr(W_k)} \times \frac{N-k}{k-1}$. The CH score is higher when the resulting clusters are compact and well separated.

Tables 1 and 2 respectively list the accuracy and NMI of the proposed methods in comparison with the benchmarks. As seen, the proposed method achieves the highest ACC and NMI on all of five image datasets and one text dataset, which indicates that the proposed method can learn feasible deep representations for the clustering task in different applications. Generally speaking, the methods involving deep representations perform better than the ones with low-level features. Taking the popular k-means and GMM as the examples, we can always improve the accuracy and NMI by learning deep representations in our experiments. Thus, it is necessary to proposed unsupervised deep representation learning methods for clustering.

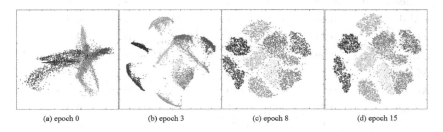

(a) epoch 0   (b) epoch 3   (c) epoch 8   (d) epoch 15

Fig. 3. The clusters of MNIST in different epochs. While the initial representations are mixture together, they gradually evolve into separable clusters. In addition, the compactness of the clusters is also gradually improved.

The results in these two tables also illustrate that the proposed method always performs better than DeepGMM, which validates that the introduction of the second term in Eq. 5 can indeed improve the separability of the data representations. In other words, by maximizing the distance between the Gaussian centers, we enlarge the distances of representations belonging to different classes. In Fig. 3, we visualize the data representations of a MNIST subset (with $10,000$ images) in different epochs using t-distributed stochastic neighbor embedding (t-SNE) [20]. We can clearly see that the clusters are gradually become more compact and more separable. To explicitly assess the compactness and separability of the resulting clusters, we list the CH-score of different deep learning methods in Table 3. As seen, the proposed method achieves the highest CH-score, indicating the resulting clusters are more compact and more separable.

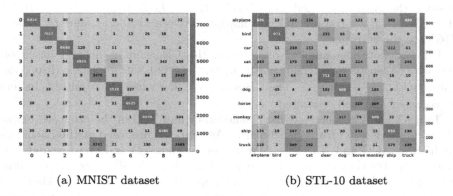

(a) MNIST dataset   (b) STL-10 dataset

Fig. 4. The confusion matrices of MNIST dataset and STL-10 dataset. Each row corresponds to a resulting cluster and each column corresponds to a groundtruth cluster.

The proposed method significantly outperforms the benchmark AE+GMM, which indicates that our proposed joint optimization can produce more clustering-friendly representations. While autoencoder can extract semantic meaningful representations, the proposed method can significantly enhance their discriminant ability.

Table 3. The CH-score of different deep learning methods on six datasets

Dataset	MNIST	USPS	COIL 20	COIL 100	STL-10	REUTERS
DEC [40]	2172	274	72	64	41	88
JULE [43]	1977	228	68	57	82	112
DCN [42]	2270	304	57	48	48	66
DeepGMM	1684	199	64	42	64	55
The proposed	2441	327	86	79	119	150

4.5 Discussion

This subsection discusses the experimental results on MNIST and STL-10. Figure 4 shows the confusion matrices of the proposed method on MNIST and STL-10. As seen in Fig. 4(a), we can know that the difficulty of the MNIST dataset mainly lies in the separability of 4 and 9 from each other. For the STL-10 dataset, on the other hand, we can achieve relatively better performance on the clusters whose background and pose do not change significantly, such as *airplane* and *bird*.

To identify the difficulty examples from the easier ones, we visualize the data samples which are far away from the centers (in Fig. 5 for MNIST and STL-10). For MNIST, the center samples are the ones which are similar to the standard written characters. In other words, if the digit is well written, we can easily categorize it into the right cluster. On the contrary, the images which are far from their associating centers are not well written (if judged by common sense), as shown in Fig. 5(a). Most of them are visually distorted. For some images, it is even difficult for human beings to recognize it correctly.

The complex background heavily affects the clustering performances for the STL-10 dataset. In addition, the pose of the foreground is another important factor for correct clustering. If the target foreground is well posed, its whole body is visible and can be easily clustered. Take the cluster of *car* as an example, the difficult images only contain a small portion of a car due to heavily side view capture or occlusion, as shown in Fig. 5(b). In these cases, it is difficult for our method to identify their similarities with the cars which are well posed. In our examples, if an image is captured from a side view, its representation is far from the corresponding cluster center and thus are more likely to be mis-clustered.

We vary the parameter η to show the robustness of our method on MNIST. With $\eta = 10^{-1}, 10^{-2}, 10^{-3}$, and 10^{-4}, the average accuracies of 10 times running

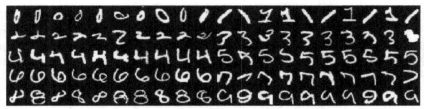

(a) Images in MNIST that are far from their associating centers

(b) Images in STL-10 that are far away from their associating centers

Fig. 5. The MNIST and STL-10 images which are far away from their associating centers

are $90.3\% \pm 3.9\%$, $93.6\% \pm 2.3\%$, $81.6\% \pm 5.6\%$, and $72.5\% \pm 4.8\%$, respectively. The average NMIs are 0.84 ± 0.02, 0.86 ± 0.03, 0.80 ± 0.04, and 0.68 ± 0.05, respectively.

5   Conclusion

In this paper, we have proposed and described an unsupervised deep learning framework by integrating deep representation with GMM-based modeling and joint optimization of representation learning and clustering. The deep representation learning procedure not only optimizes the compactness of each cluster corresponding to an individual Gaussian component inside the GMM, but also optimizes the separability across different clusters. As a result, the proposed network structure as shown in Fig. 2 can jointly optimize these two learning targets, and especially learn representations which are catered for the task of clustering. In addition, the optimization process simultaneously minimizes the distance between the representations and their associating centers, and maximizes the distances across different Gaussian centers. In this way, our proposed achieves the advantage that not only the compactness within individual clusters is improved, but also the separability across different clusters is enhanced, leading to significant improvements over the compared existing benchmarks.

Acknowledgment. The authors wish to acknowledge the financial support from: (i) Natural Science Foundation China (NSFC) under the Grant No. 61620106008;

(ii) Natural Science Foundation China (NSFC) under the Grant No. 61802266; and (iii) Shenzhen Commission for Scientific Research & Innovations under the Grant No. JCYJ20160226191842793.

References

1. Aggarwal, C.C., Reddy, C.K.: Data Clustering: Algorithms and Applications, 1st edn. Chapman & Hall/CRC, Boca Raton (2013)
2. Bishop, C.M.: Pattern Recognition and Machine Learning. Information Science and Statistics. Springer, New York (2006)
3. Bruna, J., Mallat, S.: Invariant scattering convolution networks. TPAMI **35**(8), 1872–1886 (2013)
4. Cai, D., He, X., Han, J.: Document clustering using locality preserving indexing. TKDE **17**(12), 1624–1637 (2005)
5. CaliåSki, T., Harabasz, J.: A dendrite method for cluster analysis. Commun. Stat. **3**(1), 1–27 (1974)
6. Chen, X., Cai, D.: Large scale spectral clustering with landmark-based representation. In: AAAI, pp. 313–318 (2011)
7. Coates, A., Ng, A., Lee, H.: An analysis of single-layer networks in unsupervised feature learning **15**, 215–223 (2011)
8. Deng, L., Chen, J.: Sequence classification using the high-level features extracted from deep neural networks. In: ICASSP, pp. 6844–6848 (2014)
9. Ding, C., Li, T., Jordan, M.I.: Convex and semi-nonnegative matrix factorizations. TPAMI **32**(1), 45–55 (2010)
10. Dizaji, K.G., Herandi, A., Huang, H.: Deep clustering via joint convolutional autoencoder embedding and relative entropy minimization. In: ICCV, pp. 5747–5756 (2017)
11. Doersch, C., Singh, S., Gupta, A., Sivic, J., Efros, A.A.: What makes paris look like paris? ACM Trans. Graph. **31**(4), 101:1–101:9 (2012)
12. He, K., Zhang, X., Ren, S., Sun, J.: Deep residual learning for image recognition. In: CVPR, pp. 770–778 (2016)
13. Heigold, G., Ney, H., Lehnen, P., Gass, T., Schluter, R.: Equivalence of generative and log-linear models. IEEE Trans. Audio Speech Lang. Process. **19**(5), 1138–1148 (2011)
14. Heigold, G.: A log-linear discriminative modeling framework for speech recognition. Ph.D. dissertation, Rwth Aachen (2010)
15. Hinton, G., Salakhutdinov, R.: Reducing the dimensionality of data with neural networks. Science **313**(5786), 504–507 (2006)
16. Law, M.T., Urtasun, R., Zemel, R.S.: Deep spectral clustering learning. In: ICML, vol. 70, pp. 1985–1994 (2017)
17. Lecun, Y., Bottou, L., Bengio, Y., Haffner, P.: Gradient-based learning applied to document recognition. Proc. IEEE **86**(11), 2278–2324 (1998)
18. Lewis, D.D., Yang, Y., Rose, T.G., Li, F.: RCV1: a new benchmark collection for text categorization research. J. Mach. Learn. Res. **5**, 361–397 (2004)
19. Lloyd, S.: Least squares quantization in PCM. IEEE Trans. Inf. Theory **28**(2), 129–137 (1982)
20. Maaten, L.: Learning a parametric embedding by preserving local structure. In: Proceedings of the Twelfth International Conference on Artificial Intelligence and Statistics, pp. 384–391 (2009)

21. Nakayama, H., Harada, T., Kuniyoshi, Y.: Global Gaussian approach for scene categorization using information geometry, pp. 2336–2343 (2010)
22. Nene, S.A., Nayar, S.K., Murase, H.: Columbia university image library (coil-100) (1996)
23. Nene, S.A., Nayar, S.K., Murase, H.: Columbia university image library (coil-20) (1996)
24. Noh, H., Hong, S., Han, B.: Learning deconvolution network for semantic segmentation. In: CVPR, pp. 1520–1528 (2015)
25. Paulik, M.: Lattice-based training of bottleneck feature extraction neural networks. In: INTERSPEECH (2013)
26. Ren, S., He, K., Girshick, R., Sun, J.: Faster R-CNN: towards real-time object detection with region proposal networks. In: NIPS, pp. 91–99 (2015)
27. Sainath, T.N., Kingsbury, B., Ramabhadran, B.: Auto-encoder bottleneck features using deep belief networks. In: ICASSP, pp. 4153–4156 (2012)
28. Schroff, F., Kalenichenko, D., Philbin, J.: FaceNet: a unified embedding for face recognition and clustering. In: CVPR, pp. 815–823 (2015)
29. Serra, G., Grana, C., Manfredi, M., Cucchiara, R.: Gold: Gaussians of local descriptors for image representation. Comput. Vis. Image Underst. **134**, 22–32 (2015)
30. Srivastava, N., Hinton, G., Krizhevsky, A., Sutskever, I., Salakhutdinov, R.: Dropout: a simple way to prevent neural networks from overfitting. JMLR **15**, 1929–1958 (2014)
31. Stuhlsatz, A., Lippel, J., Zielke, T.: Feature extraction with deep neural networks by a generalized discriminant analysis. IEEE Trans. Neural Netw. Learn. Syst. **23**, 596–608 (2012)
32. Trigeorgis, G., Bousmalis, K., Zafeiriou, S., Schuller, B.W.: A deep semi-NMF model for learning hidden representations. In: ICML, pp. II-1692–II-1700 (2014)
33. Tüske, Z., Tahir, M.A., Schlüter, R., Ney, H.: Integrating Gaussian mixtures into deep neural networks: softmax layer with hidden variables. In: ICASSP, pp. 4285–4289 (2015)
34. Variani, E., Mcdermott, E., Heigold, G.: A Gaussian mixture model layer jointly optimized with discriminative features within a deep neural network architecture. In: ICASSP, pp. 4270–4274 (2015)
35. Vincent, P., Larochelle, H., Lajoie, I., Bengio, Y., Manzagol, P.A.: Stacked denoising autoencoders: learning useful representations in a deep network with a local denoising criterion. JMLR **11**, 3371–3408 (2010)
36. Wang, J., Wang, G.: Hierarchical spatial sum-product networks for action recognition in still images. IEEE Trans. Circuits Syst. Video Technol. **28**(1), 90–100 (2018)
37. Wang, J., Wang, Z., Tao, D., See, S., Wang, G.: Learning common and specific features for RGB-D semantic segmentation with deconvolutional networks. In: Leibe, B., Matas, J., Sebe, N., Welling, M. (eds.) ECCV 2016. LNCS, vol. 9909, pp. 664–679. Springer, Cham (2016). https://doi.org/10.1007/978-3-319-46454-1_40
38. Wang, Q., Li, P., Zhang, L.: G^2DeNet: global gaussian distribution embedding network and its application to visual recognition. In: CVPR (2017)
39. Wang, Q., Li, P., Zuo, W., Zhang, L.: RAID-G: robust estimation of approximate infinite dimensional Gaussian with application to material recognition. In: CVPR, pp. 4433–4441 (2016)
40. Xie, J., Girshick, R., Farhadi, A.: Unsupervised deep embedding for clustering analysis. In: ICML, pp. 478–487
41. Xu, W., Liu, X., Gong, Y.: Document clustering based on non-negative matrix factorization. In: Proceedings of the ACM SIGIR 2003, pp. 267–273 (2003)

42. Yang, B., Fu, X., Sidiropoulos, N.D., Hong, M.: Towards k-means-friendly spaces: simultaneous deep learning and clustering. ICML **70**, 3861–3870 (2017)

43. Yang, J., Parikh, D., Batra, D.: Joint unsupervised learning of deep representations and image clusters. In: CVPR, pp. 5147–5156 (2016)

44. You, C., Robinson, D.P., Vidal, R.: Scalable sparse subspace clustering by orthogonal matching pursuit. In: CVPR, pp. 3918–3927, June 2016

45. Zelnik-Manor, L.: Self-tuning spectral clustering. NIPS **17**, 1601–1608 (2004)

46. Zhang, W., Wang, X., Zhao, D., Tang, X.: Graph degree linkage: agglomerative clustering on a directed graph. In: Fitzgibbon, A., Lazebnik, S., Perona, P., Sato, Y., Schmid, C. (eds.) ECCV 2012. LNCS, vol. 7572, pp. 428–441. Springer, Heidelberg (2012). https://doi.org/10.1007/978-3-642-33718-5_31

47. Zhang, W., Zhao, D., Wang, X.: Agglomerative clustering via maximum incremental path integral. Pattern Recognit. **46**(11), 3056–3065 (2013)

Recovering Affine Features from Orientation- and Scale-Invariant Ones

Daniel Barath[1,2]([✉])

[1] Centre for Machine Perception, Czech Technical University,
Prague, Czech Republic
[2] Machine Perception Research Laboratory, MTA SZTAKI, Budapest, Hungary
barath.daniel@sztaki.mta.hu

Abstract. An approach is proposed for recovering affine correspondences (ACs) from orientation- and scale-invariant, e.g. SIFT, features. The method calculates the affine parameters consistent with a pre-estimated epipolar geometry from the point coordinates and the scales and rotations which the feature detector obtains. The closed-form solution is given as the roots of a quadratic polynomial equation, thus having two possible real candidates and fast procedure, i.e. <1 ms. It is shown, as a possible application, that using the proposed algorithm allows us to estimate a homography for every single correspondence independently. It is validated both in our synthetic environment and on publicly available real world datasets, that the proposed technique leads to accurate ACs. Also, the estimated homographies have similar accuracy to what the state-of-the-art methods obtain, but due to requiring only a single correspondence, the robust estimation, e.g. by Graph-Cut RANSAC, is an order of magnitude faster.

1   Introduction

This paper addresses the problem of recovering fully affine-covariant features [13] from orientation- and scale-invariant ones obtained by, for instance, SIFT [11] or SURF [6] detectors. This objective is achieved by considering the epipolar geometry to be known between two images and exploiting the geometric constraints which it implies.[1] The proposed algorithm requires the epipolar geometry, i.e. characterized by either a fundamental $\mathbf{F}$ or an essential $\mathbf{E}$ matrix, and an orientation and scale-invariant feature as input and returns the affine correspondence consistent with the epipolar geometry.

Nowadays, a number of solutions is available for estimating geometric models from affine-covariant features. For instance, Perdoch et al. [17] proposed techniques for approximating the epipolar geometry between two images by generating point correspondences from the affine features. Bentolila and Francos [7] showed a method to estimate the exact, i.e. with no approximation, $\mathbf{F}$ from three correspondences. Raposo et al. [19] proposed a solution for essential matrix estimation

---

[1] Note that the pre-estimation of the epipolar geometry, either that of a fundamental $\mathbf{F}$ or essential matrix $\mathbf{E}$, is usual in computer vision applications.

© Springer Nature Switzerland AG 2019
C. V. Jawahar et al. (Eds.): ACCV 2018, LNCS 11361, pp. 266–281, 2019.
https://doi.org/10.1007/978-3-030-20887-5_17

using two feature pairs. Baráth et al. [5] proved that even the semi-calibrated case, i.e. when the objective is to find the essential matrix and a common focal length, is solvable from two correspondences. Homographies can also be estimated from two features [10] without any a priori knowledge about the camera movement. In case of known epipolar geometry, a single affine correspondence is enough for estimating a homography [3]. Also, local affine transformations encode the surface normals [15]. Therefore, if the cameras are calibrated, the normal can be unambiguously estimated from a single affine correspondence [10]. Pritts et al. [18] recently showed that the lens distortion parameters can also be retrieved.

Affine correspondences encode higher-order information about the underlying scene geometry and this is what makes the listed algorithms able to estimate geometric models, e.g. homographies and fundamental matrices, exploiting only a few correspondences – significantly less than what point-based methods require. This however implies the major drawback of the previously mentioned techniques: obtaining affine correspondences accurately (for example, by applying Affine-SIFT [16], MODS [14], Hessian-Affine, or Harris-Affine [13] detectors) in real image pairs is time consuming and thus, these methods are not applicable when real time performance is necessary. In this paper, the objective of the proposed method is to bridge this problem by recovering the full affine correspondence from only a part of it, i.e. the feature rotation and scale. This assumption is realistic since some of the widely-used feature detectors, e.g. SIFT or SURF, return these parameters besides the point coordinates.

Interestingly, the exploitation of these additional affine components is not often done in geometric model estimation applications, even though, it is available without demanding additional computation. Using only a part of an affine correspondence, e.g. solely the rotation component, is a well-known technique, for instance, in wide-baseline feature matching [12,14]. However, to the best of our knowledge, there are only two papers [1,2] involving them to geometric model estimation. In [1], F is assumed to be known a priori and a technique is proposed for estimating a homography using two SIFT correspondences exploiting their scale and rotation components. Even so, an assumption is made, considering that the scales along the horizontal and vertical axes are equal to that of the SIFT features – which generally does not hold. Thus, the method yields only an approximation. The method of [2] obtains the fundamental matrix by first estimating a homography using three point correspondences and the feature rotations. Then F is retrieved from the homography and two additional correspondences.

The contributions of the paper are: (i) we propose a technique for estimating affine correspondences from orientation- and scale-invariant features in case of known epipolar geometry.[2] The method is fast, i.e. <1 ms, due to being solved in closed-form as the roots of a quadratic polynomial equation. (ii) It is validated

---

[2] Note that the proposed method can straightforwardly be generalized for multiple fundamental matrices, i.e. multiple rigid motions.

both in our synthetic environment and on more than 9000 publicly available real image pairs that homography estimation is possible from a single SIFT correspondence accurately. Benefiting from the number of correspondence required, robust estimation, by e.g. GC-RANSAC [4], is an order of magnitude faster than by combining it with the standard techniques, e.g. four- and three-point algorithms [9].

2 Theoretical Background

Affine Correspondences. In this paper, we consider an affine correspondence (AC) as a triplet: $(\mathbf{p}_1, \mathbf{p}_2, \mathbf{A})$, where $\mathbf{p}_1 = [u_1 \quad v_1 \quad 1]^T$ and $\mathbf{p}_2 = [u_2 \quad v_2 \quad 1]^T$ are a corresponding homogeneous point pair in the two images (the projections of point $\mathbf{P}$ in Fig. 1a), and

$$\mathbf{A} = \begin{bmatrix} a_1 & a_2 \\ a_3 & a_4 \end{bmatrix}$$

is a 2×2 linear transformation which we call *local affine transformation*. To define $\mathbf{A}$, we use the definition provided in [15] as it is given as the first-order Taylor-approximation of the 3D $\rightarrow$ 2D projection functions. Note that, for perspective cameras, the formula for $\mathbf{A}$ simplifies to the first-order approximation of the related *homography* matrix

$$\mathbf{H} = \begin{bmatrix} h_1 & h_2 & h_3 \\ h_4 & h_5 & h_6 \\ h_7 & h_8 & h_9 \end{bmatrix}$$

as follows:

$$a_1 = \frac{\partial u_2}{\partial u_1} = \frac{h_1 - h_7 u_2}{s}, \qquad a_2 = \frac{\partial u_2}{\partial v_1} = \frac{h_2 - h_8 u_2}{s},$$
$$a_3 = \frac{\partial v_2}{\partial u_1} = \frac{h_4 - h_7 v_2}{s}, \qquad a_4 = \frac{\partial v_2}{\partial v_1} = \frac{h_5 - h_8 v_2}{s}, \tag{1}$$

where u_i and v_i are the directions in the ith image ($i \in \{1, 2\}$) and $s = u_1 h_7 + v_1 h_8 + h_9$ is the so-called projective depth.

Fundamental Matrix

$$\mathbf{F} = \begin{bmatrix} f_1 & f_2 & f_3 \\ f_4 & f_5 & f_6 \\ f_7 & f_8 & f_9 \end{bmatrix}$$

is a 3×3 transformation matrix ensuring the so-called epipolar constraint $\mathbf{p}_2^T \mathbf{F} \mathbf{p}_1 = 0$ for rigid scenes. Since its scale is arbitrary and $\det(\mathbf{F}) = 0$, matrix $\mathbf{F}$ has seven degrees-of-freedom (DoF). The geometric relationship of $\mathbf{F}$ and $\mathbf{A}$ was first shown in [5] and interpreted by formula $\mathbf{A}^{-T}(\mathbf{F}^T \mathbf{p}_2)_{(1:2)} + (\mathbf{F} \mathbf{p}_1)_{(1:2)} = 0$, where lower index $(i : j)$ selects the sub-vector consisting of the elements from the ith to the jth. These properties will help us to recover the full affine transformation from a SIFT correspondence.

3   Recovering Affine Correspondences

In this section, we show how can affine correspondences be recovered from rotation- and scale-invariant features in case of known epipolar geometry. Even though we will use SIFT as an alias for this kind of features, the derived formulas hold for the output of every scale- and orientation-invariant detector. First, the affine transformation model is described in order to interpret the SIFT angles and scales. Then this model is substituted into the relationship of affine transformations and fundamental matrices. Finally, the obtained system is solved in closed-form to recover the unknown affine parameters.

3.1   Affine Transformation Model

The objective of this section is to define an affine transformation model which interprets the scale and rotation components of the features. Suppose that we are given a rotation α_i and scale q_i in each image ($i \in \{1,2\}$) besides the point coordinates. For the geometric interpretation of the features, see Fig. 1a. Reflecting the fact that the two rotations act on different images, we interpret $\mathbf{A}$ as follows:

$$\mathbf{A} = \mathbf{R}_{\alpha_2}\mathbf{U}\mathbf{R}_{-\alpha_1} =$$

$$\begin{bmatrix} a_1 & a_2 \\ a_3 & a_4 \end{bmatrix} = \begin{bmatrix} c_2 & -s_2 \\ s_2 & c_2 \end{bmatrix} \begin{bmatrix} q_u & w \\ 0 & q_v \end{bmatrix} \begin{bmatrix} c_1 & -s_1 \\ s_1 & c_1 \end{bmatrix},$$

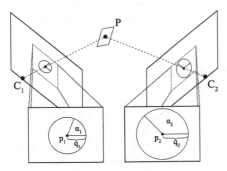
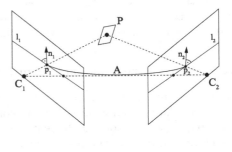

(a) Point $\mathbf{P}$ and the surrounding patch projected into cameras $\mathbf{C}_1$ and $\mathbf{C}_2$. A window showing the projected points $\mathbf{p}_1 = [u_1 \; v_1 \; 1]^{\mathrm{T}}$ and $\mathbf{p}_2 = [u_2 \; v_2 \; 1]^{\mathrm{T}}$ are cut out and enlarged. The rotation of the feature in the ith image is α_i and the size is q_i ($i \in \{1,2\}$). The scaling from the 1st to the 2nd image is calculated as $q = q_2/q_1$.

(b) The geometric interpretation of the relationship of a local affine transformations and the epipolar geometry (Eq. 4; proposed in [5]). Given the projection $\mathbf{p}_i$ of $\mathbf{P}$ in the ith camera $\mathbf{C}_i$, $i \in \{1,2\}$. The normal $\mathbf{n}_1$ of epipolar line $\mathbf{l}_1$ is mapped by affinity $\mathbf{A} \in \mathbb{R}^{2\times2}$ into the normal $\mathbf{n}_2$ of epipolar line $\mathbf{l}_2$.

Fig. 1. (a) The geometric interpretation of orientation- and scale-invariant features. (b) The relationship of local affine transformation and epipolar geometry.

where α_1 and α_2 are the two rotations in the images obtained by the feature detector; q_u and q_v are the scales along the horizontal and vertical axes; w is the shear parameter; and $c_1 = \cos(-\alpha_1)$, $s_1 = \sin(-\alpha_1)$, $c_2 = \cos(\alpha_2)$, $s_2 = \sin(\alpha_2)$. Thus $\mathbf{A}$ is written as a multiplication of two rotations ($\mathbf{R}_{-\alpha_1}$ and $\mathbf{R}_{\alpha_2}$) and an upper triangle matrix $\mathbf{U}$. Since a uniform scale q_i is given for each image, we calculate the scale between the images as $q = q_2/q_1$. Even though q is thus considered as a known parameter, q_u and q_v are still unknowns. However, the scales obtained by scale-invariant feature detectors are calculated from the sizes of the corresponding regions. Therefore, it can be easily seen that,

$$q = \det \mathbf{A} = q_u q_v. \tag{2}$$

After the multiplication of the matrices, the affine elements are as follows:

$$
\begin{aligned}
a_1 &= c_1 c_2 q_u - s_1 s_2 q_v + c_1 s_2 w, \\
a_2 &= -c_1 s_2 q_u - s_1 c_2 q_v + c_1 c_2 w, \\
a_3 &= s_1 c_2 q_u + c_1 s_2 q_v + s_1 s_2 w, \\
a_4 &= -s_1 s_2 q_u + c_1 c_2 q_v + s_1 c_2 w.
\end{aligned}
\tag{3}
$$

Using these equations, the rotations and scales which the rotation- and scale-invariant feature detectors obtain are interpretable in terms of perspective geometry. Note that this decomposition is not unique, as it is discussed in the appendix. However, in this way, the rotations are interpreted affecting independently in the two images.

Also note that [1] also proposes a decomposition for $\mathbf{A}$, however, their method assumes that the rotation equals to $\alpha_2 - \alpha_1$ and therefore does not reflect the fact that feature detectors obtain these affine components separately in the images. Also, due to assuming that $q_u = q_v$, the algorithm of [1] obtains only an approximation – the error is not zero even in the noise-free case.

3.2 Affine Correspondence from Epipolar Geometry

Estimating the epipolar geometry as a preliminary step, either a fundamental or an essential matrix, is often done in computer vision applications. In the rest of the paper, we consider fundamental matrix $\mathbf{F}$ to be known in order to exploit the relationship of epipolar geometry and affine correspondences proposed in [5]. Note that the formulas proposed in this paper hold for essential matrix $\mathbf{E}$ as well.

For a local affine transformation $\mathbf{A}$ consistent with $\mathbf{F}$, formula

$$\mathbf{A}^{-T}\mathbf{n}_1 = -\mathbf{n}_2, \tag{4}$$

holds, where $\mathbf{n}_1$ and $\mathbf{n}_2$ are the normals of the epipolar lines in the first and second images regarding to the observed point locations (see Fig. 1b). These normals are calculated as follows: $\mathbf{n}_1 = (\mathbf{F}^T\mathbf{p}_2)_{(1:2)}$ and $\mathbf{n}_2 = (\mathbf{F}\mathbf{p}_1)_{(1:2)}$, where lower index $(1:2)$ selects the first two elements of the input vector. This relationship can be written by a linear equation system consisting of two equations, one for each coordinate of the normals, as

$$(u_2 + a_1u_1)f_1 + a_1v_1f_2 + a_1f_3 + (v_2 + a_3u_1)f_4 + a_3v_1f_5 + a_3f_6 + f_7 = 0,$$
$$a_2u_1f_1 + (u_2 + a_2v_1)f_2 + a_2f_3 + a_4u_1f_4 + (v_2 + a_4v_1)f_5 + a_4f_6 + f_8 = 0,$$
(5)

where f_i is the ith ($i \in [1,9]$) element of the fundamental matrix in row-major order, a_j is the jth element of the affine transformation and each u_k and v_k are the point coordinates in the kth image ($k \in \{1,2\}$). Assuming that $\mathbf{F}$ and point coordinates (u_1, v_1), (u_2, v_2) are known and the only unknowns are the affine parameters, Eq. 5 are reformulated as follows:

$$(u_1f_1 + v_1f_2 + f_3)a_1 + (u_1f_4 + v_1f_5 + f_6)a_3 = -u_2f_1 - v_2f_4 - f_7,$$
$$(u_1f_1 + v_1f_2 + f_3)a_2 + (u_1f_4 + v_1f_5 + f_6)a_4 = -u_2f_2 - v_2f_5 - f_8.$$
(6)

These equations are linear in the affine components. Let us replace the constant parameters by variables and thus introduce the following notation:

$$B = u_1f_1 + v_1f_2 + f_3, \qquad C = u_1f_4 + v_1f_5 + f_6,$$
$$D = -u_2f_1 - v_2f_4 - f_7, \qquad E = -u_2f_2 - v_2f_5 - f_8.$$

Therefore, Eq. 6 become

$$Ba_1 + Ca_3 = D, \qquad Ba_2 + Ca_4 = E.$$
(7)

By substituting Eq. 3 into Eq. 7 the following formula is obtained:

$$B(c_1c_2q_u - s_1s_2q_v + c_1s_2w) + C(s_1c_2q_u + c_1s_2q_v + s_1s_2w) = D,$$
$$B(-c_1s_2q_u - s_1c_2q_v + c_1c_2w) + C(-s_1s_2q_u + c_1c_2q_v + s_1c_2w) = E.$$
(8)

Since the rotations, and therefore their sinuses (s_1, s_2) and cosines (c_1, c_2), are considered to be known, Eq. 8 are re-arranged as follows:

$$(Bc_1c_2 + Cs_1c_2)q_u + (Cc_1s_2 - Bs_1s_2)q_v + (Bc_1s_2 + Cs_1s_2)w = D,$$
$$(-Bc_1s_2 - Cs_1s_2)q_u + (Cc_1c_2 - Bs_1c_2)q_v + (Bc_1c_2 + Cs_1c_2)w = E.$$
(9)

Let us introduce new variables encapsulating the constants as

$$G = Bc_1c_2 + Cs_1c_2, \qquad H = Cc_1s_2 - Bs_1s_2,$$
$$I = Bc_1s_2 + Cs_1s_2, \qquad J = -Bc_1s_2 - Cs_1s_2.$$
$$K = Cc_1c_2 - Bs_1c_2,$$

Eq. 9 are then become
$$Gq_u + Hq_v + Iw = D,$$
$$Jq_u + Kq_v + Gw = E.$$
(10)

From the first equation, we express w as follows:

$$w = \frac{D}{I} - \frac{G}{I}q_u - \frac{H}{I}q_v.$$
(11)

Let us notice that q_u and q_v are dependent due to Eq. 2 as $q_u = q/q_v$. By substituting this formula and Eq. 11 into the second equation of Eq. 10, the following quadratic polynomial equation is given:

$$\left(K + \frac{GD - GH}{I}\right) q_v^2 - \left(\frac{G^2 q}{I} + E\right) q_v + Jq = 0.$$

After solving the polynomial equation which has two solutions $q_{v,1}$ and $q_{v,2}$, all the other parameters of each solution can be straightforwardly calculated as

$$q_{u,i} = \frac{q}{q_{v,i}}, \qquad w_i = \frac{D}{I} - \frac{G}{I}q_{u,i} - \frac{H}{I}q_{v,i}, \qquad i \in \{1, 2\}.$$

Consequently, each SIFT correspondence lead to two possible affine correspondences. Therefore, *we recovered the local affine transformation from an orientation- and scale-invariant correspondence in case of known epipolar geometry.* Note that a good heuristics for rejecting invalid affinities is to discard those having extreme scaling or shearing.

Fig. 2. Inlier (circles) and outlier (black crosses) correspondences found by the proposed method on image pairs from the `AdelaideRMF` (1st and 2nd columns) and `Multi-H` datasets (3rd). Every 5th correspondence is drawn.

4   Experimental Results

In this section, we compare the affine correspondences obtained by the proposed method with techniques approximating them. Then it is demonstrated that by using the affinities recovered from a SIFT correspondence, a homography can be estimated from a single correspondence. Due to requiring only a single correspondence, robust homography estimation becomes significantly faster, i.e. an order of magnitude, than by using the traditional techniques, for instance, the four- or three-point algorithms [9].

4.1   Comparing Techniques to Estimate Affine Correspondences

For testing the accuracy of the affine correspondences obtained by the proposed method, first, we created a synthetic scene consisting of two cameras represented by their 3×4 projection matrices $\mathbf{P}_1$ and $\mathbf{P}_2$. They were located in random surface points of a 10-radius center-aligned sphere. A plane with random normal was generated in the origin and ten random points, lying on the plane, were projected into both cameras. To get the ground truth affine transformations, we first calculated homography $\mathbf{H}$ by projecting four random points from the plane to the cameras and applying the normalized direct linear transformation [9] algorithm to them. The local affine transformation regarding to each correspondence were computed from the ground truth homography as its first order Taylor-approximation by Eq. 1. Note that $\mathbf{H}$ could have been calculated directly from the plane parameters as well. However, using four points promised an indirect but geometrically interpretable way of noising the affine parameters: by adding zero-mean Gaussian-noise to the coordinates of the four projected points which implied $\mathbf{H}$. Finally, after having the full affine correspondence, $\mathbf{A}$ was decomposed to $\mathbf{R}_{-\alpha}$, $\mathbf{R}_\beta$ and $\mathbf{U}$ in order to simulate the SIFT output. The decomposition is discussed in the appendix in depth. Since the decomposition is ambiguous, due to the two angles, β was set to a random value. Zero-mean Gaussian noise was added to the point coordinates and the affine transformations were noised in the previously described way. The error of an estimated affinity is calculated as $|\mathbf{A}_{est} - \mathbf{A}_{gt}|_F$, where $\mathbf{A}_{est}$ is the estimated affine matrix, $\mathbf{A}_{gt}$ is the ground truth one and norm $|.|_F$ is the Frobenious-norm. Out of the at most two real solutions of the proposed algorithm, we selected the one which is the closest to the ground truth.

Figure 3a reports the error of the estimated affinities plotted as the function of the noise σ. The affine transformations were estimated by the proposed method (red curve), approximated as $\mathbf{A} \approx \mathbf{R}_{\beta-\alpha}\mathbf{D}$ (green; proposed in [1]) and as $\mathbf{A} \approx \mathbf{R}_\beta \mathbf{D} \mathbf{R}_{-\alpha}$ (blue), where $\mathbf{R}_\theta$ is 2D rotation matrix rotating by θ degrees and $\mathbf{D} = \mathrm{diag}(q, q)$. Note that $\mathbf{D}\mathbf{R}_{\beta-\alpha}$ does not have to be tested since $\mathbf{R}_{\beta-\alpha}\mathbf{D} = \mathbf{D}\mathbf{R}_{\beta-\alpha}$. To Fig. 3a, approximating the affine transformation by the tested ways lead to inaccurate affine estimates – the error is not zero even in the noise-free case. The proposed method behaves reasonably as the noise σ increases and the error is zero if there is no noise.

4.2   Application: Homography Estimation

In [3], a method, called HAF, was published for estimating the homography from a single affine correspondence. The method requires the fundamental matrix and an affine correspondence to be known between two images. Assuming that $\mathbf{P}$ is on a continuous surface, HAF estimates homography $\mathbf{H}$ which the tangent plane of the surface at point $\mathbf{P}$ implies. The solution is obtained by first exploiting the fundamental matrix and reducing the number of unknowns in $\mathbf{H}$ to four. Then the relationship written in Eq. 1 is used to express the remaining homography parameters by the affine correspondences. The obtained inhomogeneous linear system consists of six equations for four unknowns. The problem to solve is $\mathbf{Cx} = \mathbf{b}$, where $\mathbf{x} = [h_7,\ h_8,\ h_9]^T$ is the vector of unknowns, i.e. the last row of $\mathbf{H}$, vector $\mathbf{b} = [f_4,\ f_5,\ -f_1,\ -f_2,\ -u_1 f_4 - v_1 f_5 - f_6,\ u_1 f_1 + v_1 f_2 - f_3]$ is the inhomogeneous part and $\mathbf{C}$ is the coefficient matrix as follows:

$$\mathbf{C} = \begin{bmatrix} a_1 u_1 + u_2 - e_u & a_1 v_1 & a_1 \\ a_2 u_1 & a_2 v_1 + u_2 - e_u & a_2 \\ a_3 u_1 + v_2 - e_v & a_3 v_1 & a_3 \\ a_4 u_1 & a_4 v_1 + v_2 - e_v & a_4 \\ u_1 e_u - u_1 u_2 & v_1 e_u - v_1 u_2 & e_u - u_2 \\ u_1 e_v - u_1 v_2 & v_1 e_v - v_1 v_2 & e_v - v_2 \end{bmatrix},$$

where e_u and e_v are the coordinates of the epipole in the second image. The optimal solution in the least squares sense is $\mathbf{x} = \mathbf{C}^\dagger \mathbf{b}$, where $\mathbf{C}^\dagger = (\mathbf{C}^T \mathbf{C})^{-1} \mathbf{C}^T$ is the Moore-Penrose pseudo-inverse of $\mathbf{C}$.

According to the experiments in [3], the method is often superior to the widely used solvers and makes robust estimation significantly faster due to the low number of points needed. However, its drawback is the necessity of the affine features which are time consuming to obtain in real world. By applying the algorithm proposed in this paper, it is possible to use the HAF method with SIFT features as input. Due to having real time SIFT implementations, e.g. [20], the method is easy to be applied online. In this section, we test HAF getting its input by the proposed algorithm both in our synthesized environment and on publicly available real world datasets.

Synthesized Tests. For testing the accuracy of homography estimation, we used the same synthetic scene as for the previous experiments. For Fig. 3b, the homographies were estimated from affine correspondences recovered or approximated from the two rotations and the scale in different ways (similarly as in the previous section). Homography $\mathbf{H}$ was calculated from the recovered and also from the approximated affine features by the HAF method [3]. In order to measure the accuracy of $\mathbf{H}$, ten random points were projected to the cameras from the 3D plane inducing $\mathbf{H}$ and the average re-projection error was calculated (vertical axis; average of 1000 runs) and plotted as the function of the noise σ (horizontal axis). To Fig. 3b, these approximations lead to fairly rough results. The error is significant even in the noise-free case. Also, it can be seen that the proposed method leads to perfect results in the noise-free case and the error behaves reasonably as the noise increases.

In Fig. 3c and d, the HAF method with its input got from the proposed algorithm is compared with the normalized four-point [9] (4PT) and three-point [3] (3PT) methods. The re-projection error (vertical axis; average of 1000 runs) is plotted as the function of the noise σ (horizontal axis). For Fig. 3c, the ground truth fundamental matrix was used. For Fig. 3d, $\mathbf{F}$ was estimated from the noisy correspondences by the normalized eight-point algorithm [9]. It can be seen that the HAF algorithm using the calculated affinities as input leads to the most accurate homographies in both cases.

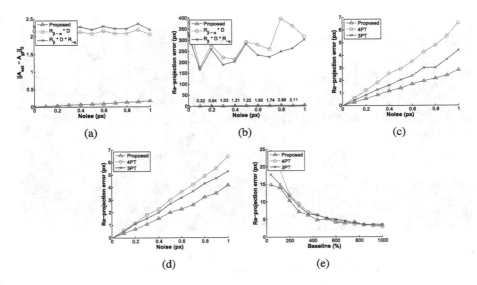

Fig. 3. (a) The error, i.e. $\|\mathbf{A}_{\text{est}} - \mathbf{A}_{\text{gt}}\|_{\text{F}}$, of the estimated affinities plotted as the function of the noise σ added to the point coordinates. The affinities were recovered by the proposed method (red curve), approximated as $\mathbf{A} \approx \mathbf{R}_{\beta-\alpha}\mathbf{D}$ (green) and as $\mathbf{A} \approx \mathbf{R}_{\beta}\mathbf{D}\mathbf{R}_{-\alpha}$ (blue), where $\mathbf{R}_{\theta}$ is 2D rotation by θ degrees and $\mathbf{D} = \text{diag}(q,q)$. (b–e) Homography estimation from synthetic data. The horizontal axes show the noise σ (px) added to the point coordinates. The vertical axes report the re-projection error (px; avg. of 1000 runs) computed from correspondences not used for the estimation. (b) Estimation by the HAF method [3] from affine correspondences recovered in different ways. (c–e) Comparison of estimators: the proposed, the normalized four- (4PT) and three-point (3PT) [3] algorithms. For (c), the ground truth fundamental matrix was used. For (d), it was estimated from the noisy correspondences by the normalized eight-point algorithm [9]. For (d), the fundamental matrix was estimated, the noise σ was 0.5 pixels, and the errors are plotted as the function of the baseline between the cameras. The baseline is the ratio (%) of the distances of the 1st camera from the 2nd one and from the origin. (Color figure online)

Figure 3e shows the sensitivity to the baseline (horizontal axis; in percentage). The baseline is considered as the ratio of the distance of the two cameras and the distance of the first one from the observed point. It can be seen that all methods are fairly sensitive to the baseline and obtain instable results for image pairs with short one. However, HAF method is slightly more accurate than the other competitors.

Real World Tests. In order to test the proposed method on real world data, we used the AdelaideRMF[3], Multi-H[4] and Malaga[5] datasets (see Figs. 2 and 4

---

[3] cs.adelaide.edu.au/~hwong/doku.php?id=data.

[4] web.eee.sztaki.hu/~dbarath.

[5] www.mrpt.org/MalagaUrbanDataset.

Fig. 4. Inlier (circles) and outlier (black crosses) correspondences found by the proposed method on image pairs from the `Malaga` dataset. Every 5th point is drawn.

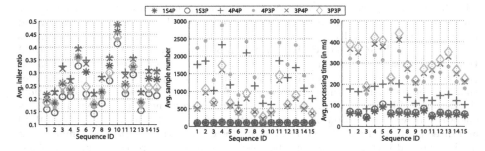

Fig. 5. Homography estimation on the `Malaga` dataset. The horizontal axis shows the identifier of the image sequence. Each column reports the mean result on the image pairs. In total, 9064 pairs were used. See Table 1 for the description of the methods. The shown properties are: the inlier ratio (left), the number of samples drawn by GC-RANSAC (middle) and the processing time of homography estimation plus that of fundamental matrix estimation except for method 4P4P (in milliseconds; right).

for example image pairs). `AdelaideRMF` and `Multi-H` consist of image pairs of resolution from 455×341 to 2592×1944 and manually annotated (assigned to a homography, i.e. a plane, or to the outlier class) correspondences. Since the reference point sets do not contain rotations and scales, we detected and matched points applying the SIFT detector. Ground truth homographies were estimated from the manually annotated correspondences. For each of them, we selected the points out of the detected SIFT correspondences which are closer than a manually set inlier-outlier threshold, i.e. 2 pixels. The `Malaga` dataset was gathered entirely in urban scenarios with a car equipped with several sensors, including a high-resolution camera and five laser scanners. We used the 15 video sequences taken by the camera and every 10th image from each sequence. As a robust estimator, we chose GC-RANSAC [4][6] with PROSAC [8] sampling and inlier-outlier threshold set to 2 pixels. For the other parameters, we used the setting proposed by the authors. All the real solutions of the proposed method were considered and validated as different homographies against the input correspondences.

Given an image pair, the procedure to evaluate the estimators, i.e. the proposed, the four- and three-point algorithms, is as follows:

1. A fundamental matrix was estimated by GC-RANSAC using the seven-point algorithm as a minimal method, the normalized eight-point algorithm for least-squares fitting, the Sampson-distance as residual function, and a threshold set to 2 pixels.
2. The ground truth homographies, estimated from the manually annotated correspondence sets, were selected one by one. For each homography:
   (a) The correspondences which did not belong to the selected homography were replaced by completely random correspondences to reduce to probability of finding a different plane than what was currently tested.
   (b) GC-RANSAC combined with the compared estimators was applied to the point set consisting of the inliers of the current homography and outliers.
   (c) The estimated homography is compared against the ground truth one estimated from the manually selected inliers.

In GC-RANSAC (step 2b), there are two cases when a model is estimated: (i) when a minimal sample is selected and (ii) during the local optimization by fitting to a set of inliers. To select the estimators used for these steps, there is a number of possibilities exists. For instance, it is possible to use the proposed algorithm for fitting to a minimal sample and the normalized four-point algorithm for the least-squares fitting. The tested combinations are reported in Table 1.

Table 1. The compared settings for Graph-Cut RANSAC [4]. The first column shows the abbreviations of the combinations, the second one contains the methods applied for fitting a homography to a minimal sample and the last one consists of the algorithms used for the least-squares fitting. The two light gray rows are the ones which use the proposed method.

Abbreviations		
Name	**Minimal method**	**Least-squares fitting**
1S4P	Proposed algorithm	Four-point method
1S3P	Proposed algorithm	Three-point method
4P4P	Four-point method	Four-point method
4P3P	Four-point method	Three-point method
3P4P	Three-point method	Four-point method
3P3P	Three-point method	Three-point method

---

[6] https://github.com/danini/graph-cut-ransac.

Table 2 reports the results on the `AdelaideRMF` and `Multi-H` datasets (first column). The number of planes in each dataset is written into brackets. The tested methods are shown in the second column (see Table 1). The blocks, each consisting of four columns, show the percentage of the homographies not found (FN, in %); the mean re-projection error of the found ones computed from the manually annotated correspondences and the estimated homographies (ϵ, in pixels); the number of samples drawn by GC-RANSAC (s); and the processing time (t, in milliseconds). We considered a homography as a not found one if the re-projection error was higher than 10 pixels. For the first block, the required confidence of GC-RANSAC in the results was set to 0.95, for the second one, it was 0.99. The values were computed as the means of 100 runs on each homography selected. The methods which applied the proposed technique are 1S4P and 1S3P. It can be seen that the proposed algorithm, in terms of accuracy and number of planes found, leads to similar results to that of the competitor algorithms – sometimes worse sometimes more accurate. However, due to requiring only a sole correspondence, both its processing time and number of samples required are almost an order of magnitude lower than that of the other methods.

The results on the `Malaga` dataset are shown in Fig. 5. In total, 9064 image pairs were tested. The confidence of GC-RANSAC was set to 0.99 and the inlier-outlier threshold to 2.0 pixels. The reported properties are the ratio of the inliers found (left plot; vertical axis), the number of samples drawn by GC-RANSAC

Table 2. Homography estimation on the `AdelaideRMF` (18 pairs; 43 planes; from the 3rd to the 8th rows) and `Multi-H` (4 pairs; 33 planes; from 9th to 14th rows) datasets by GC-RANSAC [4] combined with minimal methods. Each row reports the results of a method (gray color indicates the proposed one). See Table 1 for the abbreviations. The required confidence of GC-RANSAC was set to 0.95 for the 3–6th and to 0.99 for the 7–10th columns. The reported properties are: the ratio of homographies not found (FN, in percentage); the mean re-projection error (ϵ, in pixels); the number of samples drawn by GC-RANSAC (s); and the processing time (t, in milliseconds). Average of 100 runs.

		Confidence 95%				Confidence 99%			
		FN	ϵ	s	t	FN	ϵ	s	t
Adelaide (43#)	1S4P	2.28	1.62	125	168	2.15	1.60	147	174
	1S3P	1.09	1.60	**118**	**67**	0.86	1.60	**129**	**68**
	4P4P	1.39	1.56	2172	321	1.17	**1.57**	2351	338
	4P3P	0.62	**1.57**	1928	173	0.72	1.59	2179	188
	3P4P	**0.08**	1.60	977	250	**0.12**	1.60	1224	260
	3P3P	0.39	1.63	1070	107	0.51	1.62	1278	113
Multi-H (33#)	1S4P	11.31	1.91	1872	222	10.13	**1.55**	1993	250
	1S3P	9.14	**1.89**	**1816**	**198**	8.26	1.57	**1936**	**218**
	4P4P	12.10	2.73	4528	538	12.40	2.68	4581	528
	4P3P	10.30	2.46	4069	483	10.13	2.29	4103	480
	3P4P	10.05	2.33	4312	447	9.09	2.09	4244	450
	3P3P	**8.74**	1.94	4105	385	**8.11**	1.80	4110	377

(middle; vertical axis) and the processing time in milliseconds (right; vertical axis). We added the processing time the fundamental matrix estimation to the times of methods which use $\mathbf{F}$ for the estimation (all but 4P4P). The horizontal axes show the identifiers of the image sequences in the dataset. The same trend can be seen as for the other datasets. The ratio of inliers found is similar to that of the competitor algorithms. The number of samples used and the processing time is significantly lower than that of the other techniques.

5   Conclusion

An approach is proposed for recovering affine correspondences from orientation- and scale-invariant features obtained by, for instance, SIFT or SURF detectors. The method estimates the affine correspondence by enforcing the geometric constraints which the pre-estimated epipolar geometry implies. The solution is obtained in closed-form as the roots of a quadratic polynomial equation. Thus the estimation is extremely fast, i.e. <1 ms, and leads to at most two real solutions. It is demonstrated on synthetic and publicly available real world datasets – containing more than 9000 image pairs – that by using the proposed technique correspondence-wise homography estimation is possible. The geometric accuracy of the obtained homographies is similar to that of the state-of-the-art algorithms. However, due to requiring only a single correspondence, the robust estimation, e.g. by GC-RANSAC, is an order of magnitude faster than by using the four- or three-point algorithms.

Acknowledgement. D. Barath acknowledges the support of the OP VVV funded project CZ.02.1.01/0.0/0.0/16_019/0000765 and that of the Hungarian Scientific Research Fund (No. OTKA/ NKFIH 120499).

Affine Decomposition

The decomposition of local affine transformation $\mathbf{A}$ to two rotations ($\mathbf{R}_\gamma \in \mathbb{R}^{2\times2}$ and $\mathbf{R}_\delta \in \mathbb{R}^{2\times2}$) and an upper triangle matrix ($\mathbf{U} \in \mathbb{R}^{2\times2}$) is discussed in this section. The problem is as follows:

$$\mathbf{A} = \mathbf{R}_\gamma \mathbf{U} \mathbf{R}_\delta.$$

Note that we write the decomposition generally, thus the angles, denoted by γ and δ, does not correspond to the orientation of any features. This is the reason why we do not write $\mathbf{R}_{-\alpha_1}$ instead of $\mathbf{R}_\delta$. After multiplying the three matrices, the following system is given for the affine components:

$$
\begin{aligned}
a_1 &= c_\gamma c_\delta q_u + c_\gamma s_\delta w - s_\gamma q_v s_\delta, \\
a_2 &= c_\gamma c_\delta w - c_\gamma s_\delta q_u - s_\gamma c_\delta q_v, \\
a_3 &= s_\gamma c_\delta q_u + s_\gamma s_\delta w + c_\gamma s_\delta q_v, \\
a_4 &= s_\gamma c_\delta w - s_\gamma s_\delta q_u + c_\gamma c_\delta q_v,
\end{aligned}
\tag{12}
$$

where $c_\gamma = \cos(\gamma)$, $s_\gamma = \sin(\gamma)$, $c_\delta = \cos(\delta)$, $s_\delta = \sin(\delta)$; scalars q_u and q_v are the scales along axes x and y; and w is the shear. Since there are four equations for five unknowns, the decomposition is not unique. Considering that in real world setting, the used features are usually orientation-invariant ones, we chose angle $\gamma \in [0, 2\pi)$ to parameterize the possible decompositions. Thus, for each γ, there will be a unique decomposition of $\mathbf{A}$.

Solving Eq. 12 lead to two possible δs, each providing a valid solution, as follows:

$$\delta_{12} = \cos^{-1}(\pm(c_\gamma a_4 - s_\gamma a_2)$$

$$\frac{\sqrt{c_\gamma^2(a_4^2 + a_3^2) - 2c_\gamma s_\gamma(a_2 a_4 + a_1 a_3) + s_\gamma^2(a_2^2 + a_1^2)}}{c_\gamma^2(a_4^2 + a_3^2) - 2c_\gamma s_\gamma(a_2 a_4 + a_1 a_3) + s_\gamma^2(a_2^2 + a_1^2)})$$

Since both δs are valid, the computation of the remaining affine components falls apart to two cases – they have to be calculated for each δ independently. The formulas for the scales (q_u and q_v) and the shear (w) are as follows:

$$q_u = -\frac{a_\delta s_\delta - a_\gamma c_\delta}{c_\gamma s_\delta^2 + c_\gamma c_\delta^2}$$

$$q_v = -\frac{s_\gamma a_\gamma - c_\gamma a_3}{(s_\gamma^2 + c_\gamma^2)s_\delta}$$

$$w = \frac{(c_\gamma s_\gamma a_3 + c_\gamma^2 a_\gamma)s_\delta^2 + (s_\gamma^2 a_\delta + c_\gamma^2 a_\delta)c_\delta s_\delta + (c_\gamma s_\gamma a_3 - s_\gamma^2 a_\gamma)c_\delta^2}{(c_\gamma s_\gamma^2 + c_\gamma^3)s_\delta^3 + (c_\gamma s_\gamma^2 + c_\gamma^3)c_\delta^2 s_\delta},$$

where $\delta \in \{\delta_1, \delta_2\}$. Finally, the decomposition is selected for which $\|\mathbf{A} - \mathbf{R}_\gamma^i \mathbf{U}^i \mathbf{R}_\delta^i\|_2$ is minimal ($i \in \{1, 2\}$).

References

1. Barath, D.: P-HAF: homography estimation using partial local affine frames. In: International Conference on Computer Vision Theory and Applications (2017)
2. Barath, D.: Five-point fundamental matrix estimation for uncalibrated cameras. In: Conference on Computer Vision and Pattern Recognition (2018)
3. Barath, D., Hajder, L.: A theory of point-wise homography estimation. Pattern Recognit. Lett. **94**, 7–14 (2017)
4. Barath, D., Matas, J.: Graph-Cut RANSAC. In: Conference on Computer Vision and Pattern Recognition (2018)
5. Barath, D., Toth, T., Hajder, L.: A minimal solution for two-view focal-length estimation using two affine correspondences. In: Conference on Computer Vision and Pattern Recognition (2017)
6. Bay, H., Tuytelaars, T., Van Gool, L.: SURF: speeded up robust features. In: Leonardis, A., Bischof, H., Pinz, A. (eds.) ECCV 2006. LNCS, vol. 3951, pp. 404–417. Springer, Heidelberg (2006). https://doi.org/10.1007/11744023_32
7. Bentolila, J., Francos, J.M.: Conic epipolar constraints from affine correspondences. Comput. Vis. Image Underst. **122**, 105–114 (2014)
8. Chum, O., Matas, J.: Matching with PROSAC-progressive sample consensus. In: Computer Vision and Pattern Recognition (2005)

9. Hartley, R., Zisserman, A.: Multiple view Geometry in Computer Vision. Cambridge University Press, Cambridge (2003)

10. Köser, K.: Geometric estimation with local affine frames and free-form surfaces. Shaker (2009)

11. Lowe, D.G.: Object recognition from local scale-invariant features. In: International Conference on Computer Vision (1999)

12. Matas, J., Chum, O., Urban, M., Pajdla, T.: Robust wide-baseline stereo from maximally stable extremal regions. Image Vis. Comput. **22**, 761–767 (2004)

13. Mikolajczyk, K., et al.: A comparison of affine region detectors. Int. J. Comput. Vis. **65**(1–2), 43–72 (2005)

14. Mishkin, D., Matas, J., Perdoch, M.: MODS: fast and robust method for two-view matching. Comput. Vis. Image Underst. **141**, 81–93 (2015)

15. Molnár, J., Chetverikov, D.: Quadratic transformation for planar mapping of implicit surfaces. J. Math. Imaging Vis. **48**, 176–184 (2014)

16. Morel, J.M., Yu, G.: ASIFT: a new framework for fully affine invariant image comparison. SIAM J. Imaging Sci. **2**(2), 438–469 (2009)

17. Perdoch, M., Matas, J., Chum, O.: Epipolar geometry from two correspondences. In: International Conference on Pattern Recognition (2006)

18. Pritts, J., Kukelova, Z., Larsson, V., Chum, O.: Radially-distorted conjugate translations. In: Conference on Computer Vision and Pattern Recognition (2018)

19. Raposo, C., Barreto, J.P.: Theory and practice of structure-from-motion using affine correspondences. In: Computer Vision and Pattern Recognition (2016)

20. Sinha, S.N., Frahm, J.M., Pollefeys, M., Genc, Y.: GPU-based video feature tracking and matching. In: Workshop on Edge Computing Using New Commodity Architectures, vol. 278, p. 4321 (2006)

Totally Looks Like - How Humans Compare, Compared to Machines

Amir Rosenfeld[(✉)], Markus D. Solbach, and John K. Tsotsos

York University, Toronto, ON M3J 1P3, Canada
{amir,solbach,tsotsos}@cse.yorku.ca

Abstract. Perceptual judgment of image similarity by humans relies on rich internal representations ranging from low-level features to high-level concepts, scene properties and even cultural associations. However, existing methods and datasets attempting to explain perceived similarity use stimuli which arguably do not cover the full breadth of factors that affect human similarity judgments, even those geared toward this goal. We introduce a new dataset dubbed **Totally-Looks-Like** (TLL) after a popular entertainment website, which contains images paired by humans as being visually similar. The dataset contains 6016 image-pairs from the wild, shedding light upon a rich and diverse set of criteria employed by human beings. We conduct experiments to try to reproduce the pairings via features extracted from state-of-the-art deep convolutional neural networks, as well as additional human experiments to verify the consistency of the collected data. Though we create conditions to artificially make the matching task increasingly easier, we show that machine-extracted representations perform very poorly in terms of reproducing the matching selected by humans. We discuss and analyze these results, suggesting future directions for improvement of learned image representations.

1   Introduction

Human perception of images goes far beyond objects, shapes, textures and contours. Viewing a scene often elicits recollection of other scenes whose global properties or relations resemble the currently observed one. This relies on a rich representation of image space in the brain, entailing scene structure and semantics, as well as a mechanism to use the representation of an observed scene to recollect similar ones from the profusion of those stored in memory. Though not fully understood, the capacity of the human brain to memorize images is surprisingly large [3,12]. The recent explosion in the performance and applicability

This research was supported through grants to the senior author, for which all authors are grateful: Air Force Office of Scientific Research (FA9550-18-1-0054), the Canada Research Chairs Program (950-219525), the Natural Sciences and Engineering Research Council of Canada (RGPIN-2016-05352) and the NSERC Canadian Network on Field Robotics (NETGP417354-11).

© Springer Nature Switzerland AG 2019
C. V. Jawahar et al. (Eds.): ACCV 2018, LNCS 11361, pp. 282–297, 2019.
https://doi.org/10.1007/978-3-030-20887-5_18

of deep-learning models in all fields of computer vision [14, 19, 25] (and others), including image retrieval and comparison [26], can tempt one to conclude that the representational power of such methods approaches that of humans, or perhaps even exceeds them. We aim to explore this by testing how deep neural networks fare on the challenge of similarity judgment between pairs of images from a new dataset, dubbed "**Totally-Looks-Like**" (TLL); See Fig. 1. It is based on a website for entertainment purposes, which hosts pairs of images deemed by users to appear similar to each other, though they often share little common appearance, if judging by low-level visual features. These include pairs of images out of (but not limited to) objects, scenes, patterns, animals, and faces across various modalities (sketch, cartoon, natural images). The website also includes user ratings, showing the level of agreement with the proposed resemblances. Though it is not very large, the diversity and complexity of the images in the dataset implicitly captures many aspects of human perception of image similarity, beyond current datasets which are larger but at the same time narrower in scope. We evaluate the performance of several state-of-the-art models on this dataset, cast as a task of image retrieval. We compare this with human similarity judgments, forming not only a baseline for future evaluations, but also revealing specific weaknesses in the strongest of the current learned representations that point the way for future research and improvements. We conduct human experiments to validate the consistency of the collected data. Even though in some experiments we allow very favorable conditions for the machine-learned representations, they still often fall short of correctly predicting the human matches.

The next section overviews related work. This is followed by a description of our method, experiments and analysis. We close the paper with discussion about the large gaps between what is expected of state-of-the art learned representations and suggestions for future work. The dataset is available at the following address: https://sites.google.com/view/totally-looks-like-dataset

2   Related Work

This paper belongs to a line of work that compares machine and human vision (in the context of perception) or attempts to perform some vision related task that is associated with high-level image attributes. As ourselves, others also tapped the resources of social media/online entertainment websites to advance research in high-level image understanding. For example, Deza and Parikh [6] collected datasets from the web in order to predict the virality of images, reporting superhuman capabilities when five high-level features were used to train an SVM classifier to predict virality.

Several lines of work measure and analyze differences between human and machine perception. The work of [17] collected 26k perceived dissimilarity measurements from 2,801 visual objects across 269 human subjects. They found several discrepancies between computational models and human similarity measurements. The work of [10] suggests that much of human-perceived similarity can readily be accounted for by representations emerging in deep-learned models.

Fig. 1. The *Totally-Looks-Like* dataset: pairs of perceptually similar images selected by human users. The pairings shed light on the rich set of features humans use to judge similarity. Examples include (but are not limited to): attribution of facial features to objects and animals *(a, b)*, global shape similarity *(c, d)*, near-duplicates *(e)*, similar faces *(f)*, textural similarity *(g)*, color similarity *(h)*

Others modify learned representations to better match this similarity, reporting a high-level of success in some cases [16], and near-perfect in others [2]. The work of [2] is done in a context which reduces similarity to categorization. Very recently, Zhang et al. [24] have shown that estimation of human perceptual similarity is dramatically better using deep-learned features, whether they are learned in a supervised or unsupervised manner, than more traditional methods. Their evaluation involved comparing images to their distorted versions. The distortions tested were quite complex and diverse. Akin to ours, there are works who question the behavioral level of humans vs. machines. For instance, Das et al. [5] compare the attended image regions in Visual Question Answering (VQA, [1]) to that of humans and report a rather low correlation. Other works tackle high level tasks such as understanding image aesthetics [23] or even humor [4]. The authors of [7] compare the robustness of humans vs. machines to image degradations, showing that DNN's that are not trained on noisy data are more error-prone than humans, as well as having a very different distribution of non-class predictions when confronted with noisy images. Matching images and recalling them are two very related subjects, as it seems unlikely for a human (or any other system storing a non-trivial amount of images) to perform exhaustive search over the entire collection of images stored in memory. Studies of image memorability [11] have successfully produced computational models to predict which images are more memorable than others.

The works of [10,16,17,24] show systematic results on large amounts of data. However, most of the images within them either involve objects with a blank background [10,17] or of a narrow type (e.g., animals [16]). Our dataset is smaller in scale than most of them, but it features images from the "wild", requiring similarities to be explained by features ranging from low-level to abstract scene properties. In [24], a diverse set of distortions is applied to images, however, the source image always remains the same, whereas the proposed dataset shows pairs of images of different scenes and objects, still deemed similar by human observers. In this context, the proposed dataset does not contradict the systematic evaluations performed by prior art, but rather complements them and broadens the scope to see where modern image representations still fall short. Finally, many insights are provided by psychological studies of factors of perceived similarity, such as the Gestalt principles [22].

3   Method

The main source of data for the reported experiments is a popular website called *TotallyLooksLike*[1]. The website describes itself simply as "Stuff That Looks Like Other Stuff". For the purpose of amusement, users can upload pairs of images which, in their judgment, resemble each other. Such images may be have any content, such as company logos, household objects, art-drawing, faces of celebrities and others. Figure 1 shows a few examples of such image pairings. Each submission is shown on the website, and viewers can express their agreement (or disagreement) about the pairing by choosing to up-vote or down-vote. The total number of up-votes and down-votes for each pair of images is displayed.

Little do most of the casual visitors of this humorous website realize that it is in fact a hidden treasure: humans encounter an image in the wild and recall another image which not only do they deem similar, but so do hundreds of other site users (according to the votes). This provides a dataset of thousands of such image pairings, by definition collected from the wild, that may aid to explore the cognitive drive behind judgment of image similarity. Beyond this, it contains samples of images that one recollects when encountering others, allowing exploration in the context of long-term visual memory and retrieval.

While other works have explored image memorability [11], in this work we focus on the aspects of similarity judgment. We next describe the dataset we created from this website.

3.1   Dataset

We introduce the *Totally-Looks-Like* (**TLL**) dataset. The dataset contains a snapshot of 6016 image-pairs along with their votes downloaded from the website in Jan. 2018 (a few images are added each day). The data has been downloaded with permission from the web-site's administrators to make it publicly available for research purposes. For each image pair, we simply refer to the two images as the "left image" and the "right image", or more concisely as $<L_i, R_i>, i \in 1 \ldots N$ where N is the total number of images in the dataset. We plan to make the data available on the project website, along with pre-computed features which will be listed below.

---

[1] http://memebase.cheezburger.com/totallylookslike.

3.2 Image Retrieval

The TLL dataset is the basis for our experiments. We wish to test to what degree similarity metrics based on generic machine-learned representations are able to reproduce the human-generated pairings.

We formulate this as a task of image retrieval: Let $\mathcal{L} = (L_i)_i$ be the set of all left images and similarly let $\mathcal{R}$ be the set of all right images. For a given image L_i we measure the distance $\phi(L_i, R_j)$ between L_i and each $R_j \in \mathcal{R}$. This induces a ranking $r_1, \ldots r_n$ over $\mathcal{R}$ by sorting according to the distance $\phi(\cdot, \cdot)$. A perfect ranking returns $r_1 = i$. Calculating distances using ϕ over all pairs of the dataset allows us to measure its overall performance as a distance metric for retrieval. For imperfect rankings, we can measure the recall up to some ranking k, which is the average number of times the correct match was in the top-k ranked images. In practice, we measure distances between feature representations extracted via state-of-the-art DCNN's, either specialized for generic image categorization or face identification, as detailed in the experiments section.

Direct Comparison vs. Recollection: We note that framing the task as image retrieval may be unfair to both sides: when humans encounter an image and recollect a perceptually similar one to post on the website, they are not faced with a forced choice task of selecting the best match out of a predetermined set. Instead, the image triggers a recollection of another image in their memory, which leads to uploading the image pair. On one hand, this means that the set of images from which a human selects a match is dramatically larger than the limited-size dataset we propose, so the human can potentially find a better match. On the other hand, the human does not get to scrutinize each image in memory, as the process of recollection likely happens in an associative manner, rather than by performing an exhaustive search on all images in memory. In this regard, the machine is more free to spend as many computational resources as needed to determine the similarity between a putative match. Another advantage for the machine is that the "correct" match already exists in the predetermined dataset; possibly finding it will be easier than in an open-ended manner as a human does. Nevertheless, we view the task of retrieval from this closed set as a first approximation. In addition, we suggest below some ways to make the comparison more fair.

4 Experiments

We now describe in detail our experiments, starting from data collection and pre-processing, through various attempts to reproduce the human data and accompanying analysis.

Data Preprocessing. All images if the TLL (Totally-Looks-Like) dataset were automatically downloaded along with their up-votes and down-votes from the website. Each image pair $<L_i, R_i>$ appears on the website as a single image showing L_i and R_i horizontally concatenated, of constant width of 401 pixels and height of 271 pixels. We discard for each image the last column and split it equally to left and right images. In addition, the bottom 26 pixels of each image contains for each side a description of the content. While none of the methods we apply explicitly use any kind of text detection/recognition, we discard these rows as well to avoid the possibility of "cheating" in the matching process.

4.1   Feature Extraction

We extract two kinds of features from each image: generic and facial.

Generic Features: we extract "generic" image features by recording the output of the penultimate layer of various state-of-the-art network architectures for image categorization, trained on the ImageNet benchmark [18], which contains more than a million training images spread over a thousand object categories. Training on such a rich supervised task been shown many times to produce features which are transferable across many tasks involving natural images [20]. Specifically, we use various forms of Residual Networks [8], Dense Residual Networks [9], AlexNet [13] and VGG-16 from [21], giving rise to feature-vector dimensionalities ranging from a few hundred to a few thousands, dependent on the network architecture. We extract the activations of the penultimate layer of each of these networks for each of the images and store them for distance computations.

Facial Features: many of the images contain faces, or objects that resemble faces. Faces play an important role in human perception and give rise to many of the perceived similarities. We run a face detector on all images, recording the location of the face. For each detected face in each image, we extract features using a deep neural network which was specifically designed for face recognition. The detector and features both use an off-the-shelf implementation[2]. The dimensionality the extracted face descriptor is 128. Figure 5(c) shows the distribution of the number of detected faces in images, as well as the agreement between the number of detected faces in human-matched pairs. The majority of images have a face detected in them, which very few containing more than one face. When a face is detected in a left image of a given pair, it is likely that a face will be detected in the right one as well.

Generic-Facial Features: very often in the TLL dataset, we can find objects that resemble faces and play an important role in these images, being the main object which led to the selection of an image pair. To allow comparing such objects to one another, we extract generic image features from them, as described above, to complement the description by specifically tailored facial features. We do this under the likely assumption that while a facial feature extractor might not produce reliable features for comparison from a face-like object (because the network was not trained on such images), a generic feature extractor might.

We denote by G_i, F_i, and GF_i the set of generic features, facial features and generic-facial features extracted from each image. Note that for some images

---

[2] https://github.com/ageitgey/face_recognition.

faces are not at all detected, and so F_i and GF_i are empty sets. For others, possibly more than one face is detected, in which F_i and GF_i can be sets of features.

We next describe how we take all of these features into account.

4.2   Matching Images

The distance between a pair of images L_i, R_j is calculated using the corresponding features A, B as follows. We either use the ℓ_2 (Euclidean) distance between a pair of features, i.e., $\phi_l^f(A, B) = \|A - B\|_2$ or the cosine distance, i.e., $\phi_c^f(A, B) = 1 - \frac{A \cdot B}{\|A\|\|B\|}$. The subscripts l, c specify ℓ_2 norm or cosine distance. The superscript f specifies the kind of representation, i.e, $f \in \{G, F, GF\}$. For facial features (F) we use only the euclidean distance, as is designated by the applied facial recognition method. Each distance function ϕ_l^f induces a distance matrix $\Phi_l^f \in \mathcal{R}^N$ with the i, j location representing the distance between L_i, R_j using this function. For image pairs with more than one face in either image we assign minimal distance between all pairs of features extracted from the corresponding faces. For image pairs where at least one image has no detected face we assign the corresponding distance to $+\infty$. Armed with Φ_l^f, we may now test how the distance-induced ranking aligns with the human-selected matches.

Evaluating Generic Features: first, we evaluate which metric (Euclidean vs. cosine) better matches the pairings in TLL. We noted that the recall for a given number of candidates using the cosine distance is always higher compared to that of the Euclidean distance. We calculated recall for each of the nets as a function of the number of retrieved candidates. Figure 2(a) shows the difference for each k between the recall using the cosine vs. Euclidean distances. Due to the clear advantage of the cosine distance we choose to use it for all subsequent experiments (except when using facial features).

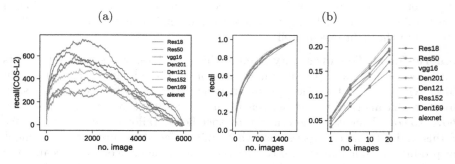

Fig. 2. (a) Difference between recall per number of images retrieved for cosine and ℓ_2-distance based retrieval. Recall is always improved if we use the cosine distance over the ℓ_2 distance between representations. (b) Retrieval performance by various learned representations in the TLL dataset. Left: all images. Right: showing recall only for the top 1 (first place), 5, 10, 20 images.

Near duplicates: visualizing some of the returned nearest neighbors revealed that there are duplicate (or near duplicate images) within the $\mathcal{L}$ and $\mathcal{R}$ image sets. As this could cause an ambiguity and hinder retrieval scores, we removed all pairs where either the left or the right image was part of the duplicate. We did this for both generic features and face-based features. For generic ones, this corresponds to a cosine distance of ≥ 0.15 (using Densenet121); virtually all images below a distance of 0.1 were near-duplicate, so we set the threshold conservatively to avoid accidental duplicates. For faces we set the threshold to 0.5. We also removed duplicates across pairs, meaning that if L_i and R_j were found to be near-duplicates then we removed them, as an identical copy R_j of L_i may be a better match for it than R_i. Removing all such duplicates leaves us with a subset we name TLL_d, containing 1828 valid image pairs. The results of Tables 1 and 2(a) are calculated based on this dataset. This does not, however, reduce the importance of the full dataset of 6016 images as it still contains many interesting and useful image pairs to learn from. The reduction of the dataset size is only done for evaluation purposes.

Faces: many images in the dataset contain faces, as indicated by Fig. 5(c). In fact, the figure represents an underestimation of the number of faces as some faces we not detected. Such images seem qualitatively different from the ones containing faces, in that the similarities are more about global shape, texture, or face-like properties, though there are no actual faces in them in the strict sense. Hence, we create another partition of the data without any detected faces, and without the duplicate images according to the generic feature criteria. This subset, TLL_{obj}, contains 1622 images. Both TLL_d and TLL_{obj} are used in Sect. 4.3 where we report additional results of human experiments.

Table 1. Retrieval performance (percentage retrieved after varying number of candidates) by various learned representations in the TLL dataset.

	R@1	R@5	R@10	R@2	R@50	R@100
AlexNet	3.67	9.19	12.09	15.37	22.59	30.63
vgg16	3.77	8.97	12.58	16.90	24.02	32.39
Res50	4.38	11.43	15.04	19.91	28.77	36.71
Res152	4.98	11.16	14.61	18.82	26.20	35.61
Den201	5.47	12.91	16.63	21.44	30.47	38.18
Res18	5.53	12.14	15.10	19.47	28.06	35.61
Den169	5.69	13.07	16.19	19.31	28.67	37.53
Den121	5.80	13.84	16.90	21.94	29.92	38.89

Next, we evaluate the retrieval performance as a function of the number of returned image candidates. This can be seen graphically in Fig. 2(b). The left sub-figure shows the recall for the entire dataset and the right sub-figure shows it for the first, 5th, 10th and 20th returned candidates. Table 1 shows these values numerically. For face features the retrieval accuracy using one retrieve item was slightly better than the generic features, reaching 6.1%. Using generic features extracted on faces performed quite poorly, at 2.6%. Evidently, none of the networks we tested performed well on this benchmark. Such a direct comparison is problematic for several reasons. Next, we attempt to ease the retrieval task for the machine-based features.

Simulating Associative Recall: As mentioned in Sect. 3.2, directly comparing to all images in the dataset is perhaps unfair to the machine-learning test. Arguably, a human recalling an image first narrows down the search given the query image, so only images with relevant features are retrieved from memory. Though we do not speculate about how this may be done, we can test how retrieval improves if such a process were available. To do so, we sample for each left image L_i a random set $R(L_i)$ of size m which includes the correct right image R_i and an additional $m - 1$ images. This simulates a state where viewing the image L_i elicited a recollection of m candidates (including the correct one) from which the final selection can be made. We do this for varying sizes of a recollection set $m \in \{1 - 5, 10, 20, 50, 100\}$, with 10 repetitions each. Table 2(a) summarizes the mean performance obtained here. Although these are almost "perfect" conditions, the retrieval accuracy falls to less than 50% if we use as little as ten examples as the test set. The variance (not shown) was close to 0 in all conditions.

Comparing Distances to Votes: we test whether there is any consistency between the feature-based distances and the number of votes assigned by human users. Assuming that a similar number of users viewed each uploaded image pair, a higher number of votes suggests higher agreement that the pairing is indeed a valid one. Possibly, this could also suggest that the images should be easier to match by automatically extracted features. We calculate the correlation between number of up-votes and down-votes vs the cosine-distance resulting from the Densenet121 network. Unfortunately, there seems very little correlation, with a Pearson coefficient of $0.023/-0.068$ for up/down-votes respectively.

4.3   Human Experiments

We conducted experiments both in-lab and using Amazon Mechanical Turk (AMT). We chose 120 random pairs of images from the dataset, as follows: 40 pairs were selected TLL_{obj} and 80 from TLL_d. From each pair, we displayed the left image to the user, along with 4 additional selected images and the correct right image. The images were shuffled in random order. Human subjects were requested to select the most similar image to the query (left) image. We allocated 20 images to each sub-experiment. The names of the experiments are **random, generic, face** and **face-generic**, indicating the type of features used to select the subset, if any. For **random** we simply chose a subset of 5 images randomly, similarly to what is described in Sect. 4.2. For each of the others, we ordered the images from the corresponding subset using each feature type

Fig. 3. Automatic retrieval errors: distances between state-of-the-art deep-learned representations often do not do well in reproducing human similarity judgments. Each row shows a query image on the left, five retrieved images and the ground-truth on the right. Perceptual similarity can be attributed to similarity between cartoonish and real faces (first three rows), flexible transfer of facial expression (4th row), visually similar subregions (last two rows, hair of person on row 5 resembles spider legs, hair of person on last row resembles waves). Though the queries and the retrieved images may be more similar to each other in a strict sense, humans still consistently agree on the matched ones (first, last columns).

and retained the top-5. If the top-5 images retrieved did not contain the correct answer, we randomly replaced one of them with it. A correct answer in this sense is selecting the correct right image, for the human, and ranking it highest for the machine. In each experiment, the four images except the correct match are regarded as *distractors*. Distractors generated using feature similarity (as opposed to random selection) pose a greater challenge for human participants, as they tend to resemble, in some sense, the "correct" answer. Table 2(b) summarizes the overall accuracy rates. In lab settings (12 participants, ages 28–39) answered all 120 questions each (labeled human1, human2 in the table). For AMT, we repeated each experiment 20 times, where an experiment is answering a single query, making an overall of 2400 experiments. A payment of 5 cents was rewarded for the completion of each experiment. Only "master" workers were used in the experiment, for increased reliability. We next highlight several immediate conclusions from this data.

Table 2. (a) Modeling Associative Recall: percentage of correct matches using conv-net derived features for the TLL dataset when a random sample of m images including the correct one is used. For 10 images, the performance is less that 50%. (b) man-versus-machine image matching accuracy for the perceptual similarity task. †The relatively high accuracy for "random" is because a small subset is selected which contains the correct answer, highly increasing the chance for correct guessing.

m	% correct
1	100.00
2	73.35
3	61.54
4	54.30
5	50.49
10	37.99
20	27.23
50	13.37

(a)

	TLL_{obj}		TLL_d			
	random†	generic	random	face	face-generic	generic
human(lab)	83.3	70	82.5	63.3	64.5	83.3
human(AMT)	84	68.25	90.25	59	60.5	74.5
machine	20	20	25	0	0	5

(b)

Data Verification: the first utility of the collected human data is to validate the consistency of that collected from the website. Though not quite perfect, there is large consistency between the human workers on AMT and the users that uploaded the original TLL images. The performance of the lab-tested humans seems to be higher on average than the AMT workers, hinting that either the variability in human answers is rather large or that the AMT results contain some noise. Indeed, when we count the number of votes given to each of the five options, we note a trend to select the first option the most, persisting through options 2–4. The number of times each option was selected was 627, 522, 465, 395, 391; option 1 selected 30% more times than the expected probability. Nevertheless, we see quite a high agreement rate throughout the table.

Human vs Machine Performance: the average human performance is generally lower when distractors are selected non-randomly, as expected. This is especially true for face images, where deep-learned features are used to select the distractor set; here AMT humans achieve around 60% agreement with the TLL dataset. This is not very surprising, as deep-learned face representations have already been reported to surpass human performance several years ago [15]. This may suggest that for faces, distractor images brought by the automatic retrieval seemed like better candidates to the humans than the original matches. The very low consistency of the machine retrieval with humans is consistent with what is reported in Table 1; the less than 6% performance rates translated to 0, in this specific sample of twenty examples for each test case. The relatively high performance in the "random" cases is due to selection of random distractors which were likely no closer in feature-space than the nearest neighbors of the query, hence resulting in seemingly high performance. We further show the consistency among human users by counting the number of agreements on answers. We count for each query the frequency of each answer and test how many times humans agreed between themselves. In 87% of the cases, the majority of users (at least 11 out of 20) agreed on the answer. In fact, the most frequent event, occurring 30% of the time, was a total agreement -20 out of 20 identical answers. Moreover, the Pearson correlation coefficient between user agreement and a correct matching to TLL was 0.94. The plot of agreement frequencies is shown in Fig. 5(a). This large agreement is not in contradiction to the lower rates of success in reproducing the TLL results, because the TLL dataset was generated by a different process of unconstrained recollection, rather than forced choice as in our experiments. Figure 5(b) shows the relation between user agreement ratios and the distribution of correctly answered images.

Correlation with Voting: We further test a connection between votes and human/machine success rates as reported in Table 2. We calculate for each experiment of Sect. 4.3 the correlation between either the number of upvotes U, downvotes D, or the difference $U - D$ and between the average human/machine accuracy. None resulted in a correlation of significance (p-values $\gg 0.1$). We conclude that the votes involve too many nuisance factors, such as the will of an individual to vote at all and the content of one of the images eliciting a vote regardless of the similarity to the other (e.g., presence of a favorite character).

Finally, Fig. 4 shows four queries from the dataset, with a query image (left column) and five candidates (remainder of columns). Two of the rows show cases where there was a perfect human agreement and two show cases where the answers were almost uniformly spread over the candidates. It is not difficult to guess which rows represent each case.

5   Discussion

We have looked into a high-level task of human vision: perceptual judgment of image similarity. The new dataset offers a glimpse into images which are matched by human beings in the "wild", in a less controlled fashion, but one that sheds a different light on various factors compared to previous work. Prior art in image retrieval deals with near-duplicate images, or images which mostly depict the same type of concept. We explored the ability of existing state-of-the-art deep-learned features to reproduce the matchings in the dataset. Though one would expect this to produce a reasonable baseline, neither features resulting from object classification networks and ones tailored for face verification seem to be able to remotely reproduce the matchings between the image pairs. We verified this using additional human experiments, both in-lab and using Amazon Mechanical Turk. Tough the collected data from AMT was not cleaned and clearly showed signs of existence of biases, the statistics still clearly show that humans are quite consistent in choosing image pairs, even when faced with a fair amount of distractors. Emulating easier scenarios for machines (for example, Table 2(a)) yielded improved results, but ones which are still very far from reproducing the consistency observed among humans.

One could argue that fine-tuning the machine learned representation with a subset of images in this dataset will reduce the observed gap. However, we believe

Fig. 4. Sample queries with varying user agreement. Each row shows on the left column a query image and 5 images from which to select a match. Some queries are very much agreed upon and on some the answers are evenly distributed. We show two rows of the first case, and two of the second. We encourage the reader to guess which images were of each kind.

that sufficiently generic visual features should be able to reproduce the same similarity measurements without being explicitly trained to do so, just as humans do. Moreover, the set of various features employed by humans is likely rather large; previous attempts to reproduce human similarity measurements resulted in datasets much larger than the proposed one, though they were narrower in scope in terms of image variability (for example [17]). This raises the question, how many images will an automatic method require to reproduce this rich set of similarities demonstrated by humans?

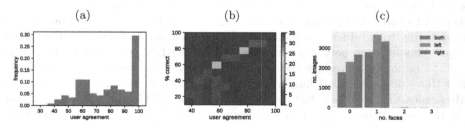

Fig. 5. (a) Probability of agreement between human users on the AMT experiment. Humans tend to be highly consistent in their answers. (b) user agreement ratio vs. correct matching with TLL. (c) Distribution of number of detected faces and agreement on detected faces between left-right image pairs.

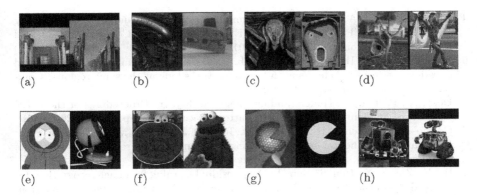

Fig. 6. Additional examples. Perceived image similarities can be abstract/symbolic: cats ↔ guards, doorway ↔ mountain passageway *(a)*, low-level (colors, *(d, e, f)*, 2D shape *(b, c, e, g)*, 3D-shape *(e)*, related to well-known iconic images from pop-culture *(b, e, f, h)* art *(c)* or pose-transfer across very different objects/domains *(b, c, d)*

We do not expect strong retrieval systems to reproduce the matchings in TLL. On the contrary, a cartoon figure should not be automatically associated with the face of Nicolas Cage Fig. 3 (2nd row), this would likely constitute a retrieval error in normal conditions and lead to additional unexpected ones. However, we do expect a high-level representation to report that of all the images in that row, the most similar one is indeed that of the said actor. Humans can easily point to the facial features in which the cartoon and the natural face image bear resemblance. Similarity factors include (1) facial features (2) facial expressions (Fig. 3, 3rd row), a robust comparison between facial expressions in different modalities (3) texture/structure of part of the image (last row, person's hair) (4) comparison between different objects (5) familiarity with iconic images or characters as (Fig. 6).

We believe that for similarity judgments to be consistent with those of humans (note there is no "correct" or "incorrect"), they should be multi-modal and conditioned on *both* images. Information extracted from one image - such as presence of a face, waves, facial expression, or spider-legs (Fig. 3) - is necessary to produce a basis for comparison and feature extraction from the other. We leave further development of this direction to future work.

References

1. Antol, S., et al.: VQA: visual question answering. In: Proceedings of the IEEE International Conference on Computer Vision, pp. 2425–2433 (2015)
2. Battleday, R.M., Peterson, J.C., Griffiths, T.L.: Modeling human categorization of natural images using deep feature representations. arXiv preprint arXiv:1711.04855 (2017)
3. Brady, T.F., Konkle, T., Alvarez, G.A., Oliva, A.: Visual long-term memory has a massive storage capacity for object details. Proc. Nat. Acad. Sci. **105**(38), 14325–14329 (2008)

4. Chandrasekaran, A., et al.: We are humor beings: understanding and predicting visual humor. In: Proceedings of the IEEE Conference on Computer Vision and Pattern Recognition, pp. 4603–4612 (2016)

5. Das, A., Agrawal, H., Zitnick, L., Parikh, D., Batra, D.: Human attention in visual question answering: do humans and deep networks look at the same regions? Comput. Vis. Image Underst. **163**, 90–100 (2017)

6. Deza, A., Parikh, D.: Understanding image virality. In: Proceedings of the IEEE Conference on Computer Vision and Pattern Recognition, pp. 1818–1826 (2015)

7. Geirhos, R., Janssen, D.H., Schütt, H.H., Rauber, J., Bethge, M., Wichmann, F.A.: Comparing deep neural networks against humans: object recognition when the signal gets weaker. arXiv preprint arXiv:1706.06969 (2017)

8. He, K., Zhang, X., Ren, S., Sun, J.: Deep residual learning for image recognition. In: Proceedings of the IEEE Conference on Computer Vision and Pattern Recognition, pp. 770–778 (2016)

9. Huang, G., Liu, Z., Weinberger, K.Q., van der Maaten, L.: Densely connected convolutional networks. arXiv preprint arXiv:1608.06993 (2016)

10. Jozwik, K.M., Kriegeskorte, N., Storrs, K.R., Mur, M.: Deep convolutional neural networks outperform feature-based but not categorical models in explaining object similarity judgments. Front. Psychol. **8**, 1726 (2017). https://doi.org/10.3389/fpsyg.2017.01726. https://www.frontiersin.org/article/10.3389/fpsyg.2017.01726

11. Khosla, A., Raju, A.S., Torralba, A., Oliva, A.: Understanding and predicting image memorability at a large scale. In: International Conference on Computer Vision (ICCV) (2015)

12. Konkle, T., Brady, T.F., Alvarez, G.A., Oliva, A.: Scene memory is more detailed than you think: the role of categories in visual long-term memory. Psychol. Sci. **21**(11), 1551–1556 (2010)

13. Krizhevsky, A., Sutskever, I., Hinton, G.E.: ImageNet classification with deep convolutional neural networks. In: Advances in Neural Information Processing Systems, pp. 1097–1105 (2012)

14. Liu, W., Wang, Z., Liu, X., Zeng, N., Liu, Y., Alsaadi, F.E.: A survey of deep neural network architectures and their applications. Neurocomputing **234**, 11–26 (2017)

15. Lu, C., Tang, X.: Surpassing human-level face verification performance on LFW with GaussianFace. In: AAAI, pp. 3811–3819 (2015)

16. Peterson, J.C., Abbott, J.T., Griffiths, T.L.: Adapting deep network features to capture psychological representations. arXiv preprint arXiv:1608.02164 (2016)

17. Pramod, R., Arun, S.: Do computational models differ systematically from human object perception? In: Proceedings of the IEEE Conference on Computer Vision and Pattern Recognition, pp. 1601–1609 (2016)

18. Russakovsky, O., et al.: Imagenet large scale visual recognition challenge. Int. J. Comput. Vision **115**(3), 211–252 (2015)

19. Schmidhuber, J.: Deep learning in neural networks: an overview. Neural Netw. **61**, 85–117 (2015)

20. Sharif Razavian, A., Azizpour, H., Sullivan, J., Carlsson, S.: CNN features off-the-shelf: an astounding baseline for recognition. In: Proceedings of the IEEE Conference on Computer Vision and Pattern Recognition Workshops, pp. 806–813 (2014)

21. Simonyan, K., Zisserman, A.: Very deep convolutional networks for large-scale image recognition. arXiv preprint arXiv:1409.1556 (2014)

22. Wertheimer, M.: Laws of organization in perceptual forms. Psychologische Forschung **4**, 301–350 (1923)

23. Workman, S., Souvenir, R., Jacobs, N.: Quantifying and predicting image scenic-ness. arXiv preprint arXiv:1612.03142 (2016)
24. Zhang, R., Isola, P., Efros, A.A., Shechtman, E., Wang, O.: The unreasonable effec-tiveness of deep features as a perceptual metric. arXiv preprint arXiv:1801.03924 (2018)
25. Zhou, P., Feng, J.: The landscape of deep learning algorithms. arXiv preprint arXiv:1705.07038 (2017)
26. Zhou, W., Li, H., Tian, Q.: Recent advance in content-based image retrieval: a literature survey. arXiv preprint arXiv:1706.06064 (2017)

Generation of Virtual Dual Energy Images from Standard Single-Shot Radiographs Using Multi-scale and Conditional Adversarial Network

Bo Zhou[1]([✉]), Xunyu Lin[1], Brendan Eck[2], Jun Hou[2], and David Wilson[2]

[1] Robotics Institute, School of Computer Science, Carnegie Mellon University,
Pittsburgh, USA
bzhou2@cs.cmu.com
[2] Department of Biomedical Engineering, Case Western Reserve University,
Cleveland, USA

Abstract. Dual-energy (DE) chest radiographs provide greater diagnostic information than standard radiographs by separating the image into bone and soft tissue, revealing suspicious lesions which may otherwise be obstructed from view. However, acquisition of DE images requires two physical scans, necessitating specialized hardware and processing, and images are prone to motion artifact. Generation of virtual DE images from standard, single-shot chest radiographs would expand the diagnostic value of standard radiographs without changing the acquisition procedure. We present a Multi-scale Conditional Adversarial Network (MCA-Net) which produces high-resolution virtual DE bone images from standard, single-shot chest radiographs. Our proposed MCA-Net is trained using the adversarial network so that it learns sharp details for the production of high-quality bone images. Then, the virtual DE soft tissue image is generated by processing the standard radiograph with the virtual bone image using a cross projection transformation. Experimental results from 210 patient DE chest radiographs demonstrated that the algorithm can produce high-quality virtual DE chest radiographs. Important structures were preserved, such as coronary calcium in bone images and lung lesions in soft tissue images. The average structure similarity index and the peak signal to noise ratio of the produced bone images in testing data were 96.4 and 41.5, which are significantly better than results from previous methods. Furthermore, our clinical evaluation results performed on the publicly available dataset indicates the clinical values of our algorithms. Thus, our algorithm can produce high-quality DE images that are potentially useful for radiologists, computer-aided diagnostics, and other diagnostic tasks.

B. Zhou and X. Lin—Equally contribute to this work.

Electronic supplementary material The online version of this chapter (https://doi.org/10.1007/978-3-030-20887-5_19) contains supplementary material, which is available to authorized users.

C. V. Jawahar et al. (Eds.): ACCV 2018, LNCS 11361, pp. 298–313, 2019.
https://doi.org/10.1007/978-3-030-20887-5_19

Keywords: Dual energy radiography · GAN · Bone suppression

1 Introduction

Chest radiography (CR) is the most ordered clinical imaging procedure for the initial detection of abnormality in the chest. As a noninvasive, low radiation dose, and low cost imaging modality, CR is commonly used for detecting lung disease, such as lung cancer, and diagnosing conditions such as tuberculosis and pneumonia. However, some lung lesions are extremely difficult to detect due to overlapping bone structures such as ribs and clavicles. Dual energy subtraction (DES) can separate high-density material, such as bone, from soft tissue and thus improve detection [1,2]. Recent studies have also shown that DES can detect and assess coronary disease by visualizing coronary calcifications in the DE bone image [3–5] and computer-aided disease detection tasks [6]. DE chest radiography requires two x-ray exposures at two different tube voltages to capture two radiographs which are then linearly combined to generate bone images and soft-tissue images [7]. However, specialized equipment is required for DES and radiation dose is nearly doubled as compared to traditional CR. In addition, the organ motion artifacts caused by heartbeat and lung breathing motion during the time-lapse between the two kVp x-ray exposures can contaminate the image quality of DE radiographs. The ability to generate bone and soft tissue images from standard, single-shot CR would benefit this diagnostic imaging procedure.

In this work, our goal is to develop a deep learning algorithm for generating virtual DE images from a standard, single-shot chest radiograph. We collected a large number of real two-exposure DE chest radiographs as training data for this task. The Convolutional Neural Network (CNN) has achieved significant success in medical image analysis field for tasks, such as chest disease detection/classification [8]; CT pulmonary nodule detection [9]; automatic organ segmentations [10], etc. However, predicting DE images using deep models remains a challenge. The structure and contextual information of a large receptive field in standard radiograph should be extracted by CNN model in a maximal manner to determine whether bony component are present and to predict the corresponding distribution. If the CNN model for fine-scale prediction is in a fully convolutional form, the size of CNN model would become very large with excessive number of parameters to learn, making it difficult to train.

In order to avoid training a very large CNN model and efficiently train a CNN model to predict fine-scale DE information, we propose a deep model to generate high-quality virtual DE images from a standard, single-shot chest radiograph. Our model is based on a multi-scale and conditional adversarial network. The general pipeline of the algorithm is shown in Fig. 1. The algorithm is comprised of two parts: (1) the bone image generator using the multi-scale fully convolutional network, and (2) the soft tissue image generator which applies bone suppression on standard images. We introduce the concept of conditional adversarial loss in training the bone image generator [11] so that high frequency information and details are learned and preserved in the virtual bone image. To produce the virtual soft tissue image, an adapted edge and shadow suppression algorithm using cross projection tensor [12] is applied to suppress bone in the standard radiograph. Outputs from the algorithm are compared to a set of test data comprised of DE patient images. The algorithm performance is also evaluated by comparing to other algorithms that are applicable for virtual DE image generation.

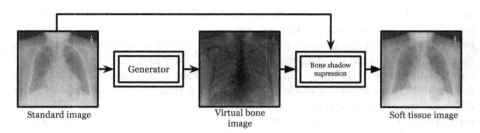

Fig. 1. Overall processing pipeline. A virtual DE bone image is generated from the MCA-Net generator. Given the bone image, bone in the standard/high kVp image is suppressed to produce a virtual soft tissue image.

1.1 Related Work

Bone suppression techniques have been developed to improve the diagnostic quality of CR exams for interpretation of disease. Current methods can be summarized into two types: learning-based methods and statistical analysis based methods. For learning-based methods, one of the early works was MTANN, proposed in [13,14], which uses traditional fully-connected neural networks with one hidden layer to predict the bone signal from a standard image. Then, the predicted bone signal can be subtract from standard image to generate an image similar to DE soft tissue image. Similar to one of the ideas in our network structure, MTANN was trained under a multi-resolution style. Each MTANN was trained separately for a certain resolution with corresponding re-sampled image using multi-resolution decomposition, and the prediction targets were the intensity values of single pixels from the DE bone images. At the testing stage, the trained MTANNs generate multi-resolution bone images and these images are combined together to produce a high-resolution bone image. Later, [15] further improved the MTANN to separate bone from soft tissue by training MTANN in different anatomical regions and producing DE images using total variation optimization. Such methods were extended to be used in portable CR systems [16]. Another one of the early works was the k-nearest neighbors (kNN) regression with optimized local feature [17] which used a linear dimensionality reduction method for local image features to optimize the performance of kNN regression for the prediction of bone images. However, this method cannot completely suppress the rib signal and requires a relatively long computation time for kNN regression. Our proposed deep model aims to address the limitations of previous work by generating high-resolution virtual DE images with relatively short computation times.

Besides from the learning-based methods, several statistical analysis based methods for suppressing bone structures without the supervision from training data have been studied. [18] proposed a clavicle suppression algorithm which works by first generating a bone image from a gradient map modified along the bone border direction and then creating a soft-tissue image by subtraction of the bone image from the standard image. Later, [19,20] proposed blind-source signal separation algorithms for suppression of bone structures in standard chest radiography. [4] presented a ribcage segmentation algorithm based on Active Appearance Model that can accurately estimate the rib border and suppress the bone signal from standard radiography based on this prediction. In general, these statistical based methods require accurate segmentation and border annotations for the target structures, which is challenging to acquire. Although all have shown improvement in thoracic disease detection, many image details are suppressed and these methods require substantial hyper-parameter tuning.

In this case, deep learning has been successfully applied in image classification, image segmentation, and image-to-image translation [21]. Deep learning also yields similar improvements in performance in the medical vision field for anatomical and pathological structures detection and segmentation tasks [8–10,22]. The boost of hardware and algorithms for deep learning have given the possibility of training CNN with many layers on large-scale datasets. Our work is closely related to the current state-of-the-art CNN models for image synthesis [11] and image transformation [23]. [11] proposed a conditional GAN (cGAN) that can generate realistic images given the outlines of the target image as the prior condition. [23] employed a GAN structure for generating high-resolution images from nominal resolution images. Our proposed method for generation of virtual DE images benefits from these excellent works on the application of GAN. The standard radiographs were used as a strong prior condition for generating virtual DE images using structure similar to cGAN [11]. The adversarial network structure also helps generation of super-resolution DE images (normally around 2022 pixels × 2022 pixels). In this paper, a customized GAN structure was designed for the generation of virtual DE bone images.

1.2 Contributions

In summary, our contributions are listed as follows:

- We propose a novel algorithm for generation of virtual dual energy images from single-shot standard radiographs based on a customized adversarial network.
- We collected a relatively large number of clinical images and demonstrated our deep model's superior performance compared to current approaches.
- We evaluated our proposed algorithm on a public chest radiograph dataset and obtained clinical values on diagnostic task of our algorithm.

2   Methods

We propose a novel algorithm to generate virtual bone and soft tissue images from a standard, single-shot CR. It consists of two parts: (1) generation of a virtual bone image from a standard CR using Multi-scale and Conditional Adversarial network (MCA-Net); (2) generation of a virtual soft tissue image using bone suppression of the standard CR with the virtual bone image. In order to generate bone images with both well-shaped general appearance and fine-grained details, we introduced MCA-Net with multi-scale generations and patch-wise adversarial learning [11]. The architecture of MCA-Net is shown in Fig. 2, with details described in the following sections.

2.1   Multi-scale and Conditional Adversarial Network (MCA-Net)

Multi-scale Generator. The multi-scale architecture is motivated by the observation that a coarse-to-fine generation can be beneficial for generating high resolution images [24]. Therefore, the multi-scale generator is designed to first generate a low resolution image capturing coarse appearances, then to add finer details generated from a higher resolution image of higher network level outputs. The final generation utilizes all of the images at different resolutions.

As shown in 2(a), the multi-scale generator follows the encoder-decoder bottleneck architecture. The encoder is formed by a set of convolution operations to map standard images to deep features. The decoder incorporates both convolution and deconvolution operations to generate bone images at different scales from these features. Images at different scales are added together element-wise and followed by a tanh activation function to form the final output. Skip connections [25] between encoder-decoder and multi-scale supervision in the network provide more feedback to aid back-propagation and reduce the effect of gradient vanishing in the shallow layers.

Adversarial Learning and Conditional Patch-Discriminator. Adversarial learning [26] is a game-theory-based learning scheme consisting of two models, a discriminator and a generator, which are trained to combat each other. As shown in Fig. 2(b), the discriminator tries to distinguish ground truth bone images from virtual bone images (output by the generator) by conditioning on the standard images. Adopted from [11], the discriminator only takes image patches as input which gives two benefits: (1) less memory usage, and (2) improving network capability to identify finer details in one image patch.

Instead of feeding raw image patches, we feed the discriminator with image gradients computed with a Sobel filter in both x and y directions. Using an image gradient can enhance high frequency signals and result in sharper generated images. It has been shown that adversarial learning can help to generate high frequency signals that are usually not captured by L_1 loss, such as sharp edges and fine textures [11].

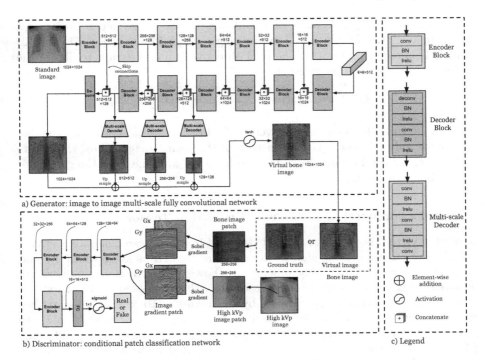

Fig. 2. Illustration of the MCA-Net architecture: (a) the multi-scale generator computes a virtual bone image from a standard/high kVp image; (b) the conditional patch discriminator distinguishes real and generated images by looking at the bone image gradient patch and conditioning on the standard image gradient patch; (c) legend of operations.

Hybrid Loss Function. With only adversarial learning, artifacts may be present in the generated DE bone image [11]. To alleviate this, we combine L_1 loss together with adversarial loss to form a hybrid loss function, where L_1 and adversarial loss contribute to the low and high frequency information to the image generation, respectively. Specifically, L_1 loss helps in learning the general appearance of the skeleton while adversarial loss emphasizes sharp edges and aims to preserve subtle calcium signals. We denote L_1 loss as L_{l_1} and adversarial loss as L_{adv}, calculated according to:

$$L_{adv}(G, D) = -\mathbb{E}_{I_H, I_B} \left[\log(D(I_H, I_B))\right] - \mathbb{E}_{I_H} \left[\log(1 - D(I_H, G(I_H)))\right] \quad (1)$$

$$L_{l_1}(G) = \mathbb{E}_{I_H, I_B} \left[\|I_B - G(I_H)\|_1\right] \quad (2)$$

where D and G denote the discriminator and generator, respectively. G is a function of the standard, high x-ray tube voltage chest radiograph, I_H, and generates a virtual bone image for comparison by the discriminator. D conditions on I_H and takes either the ground truth bone image, I_B, or virtual bone image generated by G as input and determines whether the input is a real bone image or virtual bone image. A value of 1 indicates a real image and 0 indicates a virtual image. Overall, we seek G minimizing the hybrid loss of Eqs. (1) and (2) while D is maximizing Eq. (1). The final objective of our system is to obtain the optimal generator G^* that satisfies:

$$G^* = arg \min_{G}(\max_{D} L_{adv}(G, D) + \lambda L_{l_1}(G)) \qquad (3)$$

where λ controls the importance given to L_1 loss. As discussed above, the discriminator only processes patches instead of the entire image. Thus, the inputs to D are three patches: one from I_H, one from I_B, and one from the generated virtual bone image, $G(I_H)$.

2.2 Bone suppression with Cross Projection Tensor

Given the virtual bone image $\hat{I}_B$ and standard image I_H, we generate a virtual soft tissue image $\hat{I}_S$. Bone signals from $\hat{I}_B$ are treated as shadows in I_H and are removed by our processing pipeline as shown in Fig. 3. Bone shadows are removed using the cross projection tensor algorithm proposed in [12]. First, we estimate the general soft tissue profile by convolving a 201×201 Gaussian filter with $\sigma = 50$, to I_H. Then, the Gaussian filtered I_H is subtracted from I_H to yield ΔI_H which only contains high frequency signal from soft tissue and bone. After that, the cross projection tensors are computed with the gradient images of $\hat{I}_B$ and ΔI_H. The gradient field of ΔI_H is transformed by the cross projection tensor, removing edges from ΔI_H that are also present in $\hat{I}_B$. Finally, we add the general soft tissue profile with the high frequency soft tissue to produce the virtual soft tissue image $\hat{I}_S$.

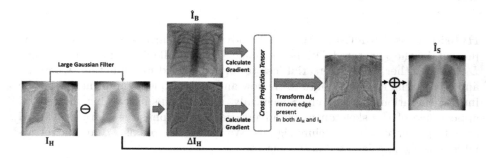

Fig. 3. Image processing pipeline of bone suppression with cross projection tensor.

3   Data and Training Details

We collected 210 posterior-anterior DE chest radiographs with a two-shot DE digital radiography system (Discovery XR656, GE Healthcare). The data was acquired using a 60 kVp exposure followed by 120 kVp exposure procedure with

100 ms between exposures. The sizes of the images ranged from 1300×1400 to 2022×2022 pixels (approximately 39.4 cm by 39.4 cm). We split the dataset into a training set of 170 cases and a testing set of 40 cases. More information about our DE dataset is summarized in Table 1. Data augmentation was performed on the training set by randomly (1) translating the images in x,y directions from $[-80, 80]$ pixels; and (2) rotating the images from $[-15, 15]$ degrees about the image center. A total of 3906 cases were augmented to train the network, and evaluations were made on the testing set.

Table 1. Statistical information of our DE datasets

Dataset	Total	Age				Gender	
		$\leqslant 20$	20–40	40–60	$\geqslant 60$	Male	Female
Training	170	5	36	78	51	114	56
Test	40	1	9	17	13	32	8

The training scheme of our adversarial network is depicted in Algorithm 1. Before training, we normalized all images to the range of $[-1, 1]$ per image by Eq. (4), where I_{raw} and I_{norm} are unnormalized and normalized images, respectively. Our network directly takes normalized standard images to generate normalized virtual bone images. Ground truth intensity maximum and minimum is then applied to the normalized virtual bone image to recover the final result.

$$I_{norm}^{x,y} = 2 \times \frac{I_{raw}^{x,y} - \min_{\forall i,j} I_{raw}^{i,j}}{\max_{\forall i,j} I_{raw}^{i,j}} - 1 \tag{4}$$

We observed synthesis artifacts in virtual images when λ was too small. We found that $\lambda = 1000$ provided a good balance between low and high frequency information in generations. N_G and N_D were chosen to balance the ability of two networks. Training the generator with more iterations than the discriminator, $N_G > N_D$, gave us better results (including sharper images and fewer artifacts). Although optimal values for N_D and N_G could be further optimized, we observed that a ratio $\frac{N_G}{N_D}$ around 3 generally gave good quality generations. The network was trained with batch size of 3 and learning rate 10^{-4} for 100 epochs.

4 Experiments and Results

4.1 Algorithm Evaluation

Figure 4 shows a testing data example of virtual DE bone and soft tissue images from MCA-net compared to ground truth DE images. Cardiac motion artifacts are significantly reduced in the virtual DE images. High-quality and high-resolution virtual bone and soft tissue images with subtle details are produced.

Algorithm 1. Training MCA-Net for one epoch

Input: Weights of G: W_G; weights of D: W_D; i-th mini-batch of standard
images and bone images: I_H^i, I_B^i $(i = 1, \ldots, n)$
Initialize: $i = 1$; # of iterations to train G: $N_G = 3$; # of iterations to train D:
$N_D = 1$; hybrid loss weight: $\lambda = 1000$;
while $i \leq n$ **do**

    for $k = 1$ *to* N_D **do**

        $L_D \leftarrow -\mathbb{E}_{I_H^i, I_B^i}\left[\log(D(I_H^i, I_B^i))\right] - \mathbb{E}_{I_H^i}\left[\log(1 - D(I_H^i, G(I_H^i)))\right]$;

        $W_D \leftarrow Optimizer(L_D, W_D)$;

        $i \leftarrow i + 1$;

    for $l = 1$ *to* N_G **do**

        $L_G \leftarrow -\mathbb{E}_{I_H^i}\left[\log(D(I_H^i, G(I_H^i)))\right] + \lambda\mathbb{E}_{I_H^i, I_B^i}\left[\left\|I_B^i - G(I_H^i)\right\|_1\right]$;

        $W_G \leftarrow Optimizer(L_G, W_G)$;

        $i \leftarrow i + 1$;

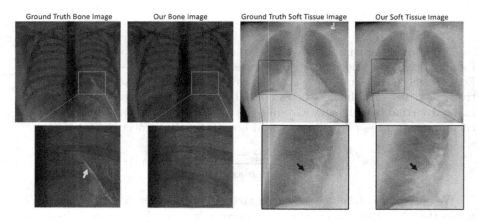

Fig. 4. Dual energy bone and soft tissue images of a 64-year-old male patient with cardiogenic pulmonary edema (CPE: green arrows) generated from DES and the virtual DES algorithm. Motion artifact (red arrow) in DES, due to cardiac motion between exposures, is significantly suppressed using from the virtual DES algorithm. CPE in the virtual soft tissue image has sharper visualization of pulmonary veins compared to ground truth. (Color figure online)

Two evaluation metrics were used to assess the performance of our algorithm. Given a virtual image $\hat{I}$ and ground truth image I, we can calculate the Peak Signal-to-Noise Ratio (PSNR) and Structure Similarity index (SSIM). PSNR is commonly used for evaluating quality of image compression/reconstruction, where our algorithm can be viewed as a compression/reconstruction process. SSIM is used for measuring the structure similarity between two images. We evaluated the improvement in virtual image generation from the conditional adversarial loss and multi-scale generator architectures. We compared the PSNR

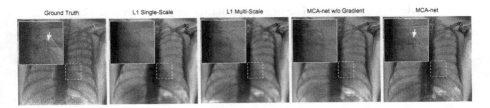

Fig. 5. Comparison of DE bone images generated with or without multi-scale or adversarial training. MCA-Net can preserve subtle calcium signal such as coronary calcium (red arrow). (Color figure online)

Table 2. Comparison of algorithm performance with different training settings

Network structure	PSNR (dB)		SSIM	
	I_{Bone}	I_{Soft}	I_{Bone}	I_{Soft}
l_1 loss + single scale	28.2 ± 4.5	22.2 ± 3.4	86.4 ± 3.1	80.3 ± 3.4
l_1 loss + multi-scale	34.9 ± 3.6	29.2 ± 4.1	88.3 ± 2.8	81.4 ± 3.7
MCA-Net w/o gradients	35.2 ± 5.1	29.8 ± 3.8	88.5 ± 2.5	81.2 ± 4.3
MCA-Net w/ gradients	$\mathbf{41.5 \pm 2.1}$	$\mathbf{39.7 \pm 1.8}$	$\mathbf{93.4 \pm 1.4}$	$\mathbf{88.4 \pm 3.4}$

and SSIM values of images generated with and without those network components. The quantitative results are listed in Table 2 using the testing dataset. The average PSNR values of virtual bone and soft tissue images produced by MCA-net are 41.5 and 39.7. The average SSIM values of bone and soft tissue images generated by MCA-net are 93.4 and 88.4. Adding the conditional adversarial loss and the multi-scale output in generator, the PSNR and SSIM were significantly improved. A visual comparison of images generated with and without those network components is shown in Fig. 5. Important calcium signals, such as coronary artery calcification, are preserved in the virtual bone image which are potentially useful for evaluating cardiovascular diseases.

We compared the prediction performance of the models by comparing to other methods (Table 3) using the same data. In addition to SSIM and PSNR, the Relative Mean-Absolute-Error (RMAE) is used as an additional metric. Please note MTANN [14] and Vis-CAC [4] don't generate virtual DE bone image, so there is no comparison available for virtual DE bone images. The results show that our algorithm has better performance in the generation of virtual DE images and is able to produce high-quality DE bone images.

We included additional examples of comparison results between ground truth DE images and virtual DE images as shown in Fig. 6. Overall image characteristics are captured by the algorithm, such as delineation of the spine, ribs, and clavicles in the virtual bone images, and visualization of pulmonary vessels in the soft tissue images. Cardiac motion artifact was present in all cases and is significantly suppressed in the generated virtual dual energy images.

Table 3. Comparison of algorithm performance with previous works

Methods	RMAE		SSIM		PSNR (dB)	
	I_{Bone}	I_{Soft}	I_{Bone}	I_{Soft}	I_{Bone}	I_{Soft}
MTANN [6]	N/A	28.3 ± 8.4	N/A	84.3 ± 4.7	N/A	39.2 ± 3.1
Vis-CAC [4]	N/A	16.8 ± 7.1	N/A	80.3 ± 3.4	N/A	38.8 ± 4.2
Ours	8.3 ± 2.3	12.1 ± 4.3	93.4 ± 1.4	88.4 ± 3.4	41.5 ± 2.1	39.7 ± 1.8

4.2   Clinical Evaluation

To evaluate the clinical application of our algorithm, we used the publicly available Japanese Society of Radiological Technology (JSRT) database [27]. The Posterior-Anterior standard chest radiographs in the database were collected from 14 medical institutions by use of screen-film systems over a period of three years. All nodules present in the radiographs were confirmed by CT, and the locations of the nodules were confirmed by chest radiologists. The images were digitized to yield 12-bit images with a resolution of 2048×2048 pixels. The size of a pixel was 0.175×0.175 mm. This database contains 93 normal cases and 154 cases with confirmed lung nodules.

We recruited one chest radiologist and one experienced radiology resident to evaluate our algorithm's clinical application of lung nodule detection. The performance was evaluated by use of the Free-response Receiver Operating Characteristic (FROC) analysis [28]. The radiologists were asked to classify and score the data into lesion and non-lesion localizations. The scoring was done by choosing an acceptance-radius and classifying marks within the acceptance-radius of lesion centers as lesion localizations, and all other marks are classified as non-lesion localizations. The scored data is plotted as a FROC curve as shown in Fig. 7, essentially a plot of appropriately normalized numbers of lesion localizations vs. non-lesion localizations. For both the radiologist and the radiology resident, the virtual DE images generated from our MCA-net shows significant improvement in localization of lung nodules. For the radiologist, we achieved sensitivity $= 0.91$ using our virtual DE images that is significantly higher than the sensitivity $= 0.81$ using standard images when FP $= 1$. The detailed FROC analysis is shown in Table 4.

5   Discussion

In this paper, we presented a deep network to generate virtual dual energy images from standard chest radiographs based on a multi-scale conditional adversarial network architecture (MCA-Net). According to the quantitative and qualitative results, we demonstrated our model can generate high-quality virtual DE images,

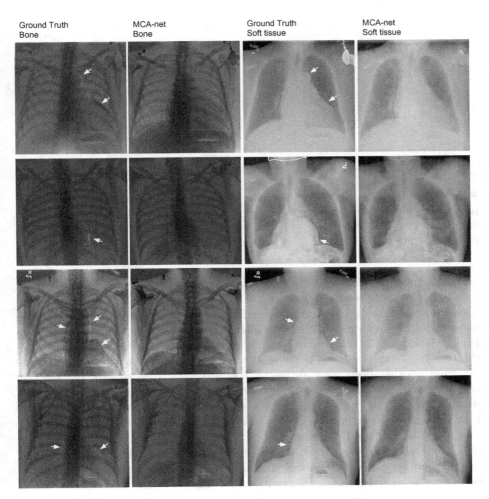

Fig. 6. Additional examples of comparison results between ground truth and MCA-net's results. The cardiac motion artifact (red arrows) along the heart boundary on both DE bone image and DE soft tissue images are significantly reduced in our virtual images. Mitigation of motion artifacts led to improvements in overall image quality. (Color figure online)

Table 4. Mean bootstrap values are given for sensitivity in our FROC analysis. 95th-percentile confidence intervals obtained through bootstrapping are shown between brackets. (FP = False positive detections per image)

FROC analysis	Radiologist		Radiology resident	
	1 FP	2 FP	1 FP	2 FP
Standard	0.81(0.69−0.85)	0.88(0.82−0.91)	0.59(0.50−0.79)	0.80(0.72−0.83)
Virtual-DE	0.91(0.81−0.96)	0.95(0.82−0.98)	0.81(0.76−0.88)	0.89(0.78−0.92)

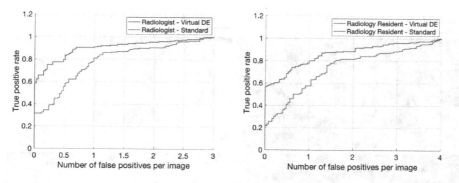

Fig. 7. Comparison of FROC curves on lung nodule localization using standard radiography and virtual DE radiography. Left: the chest radiologist results; Right: the radiology resident results. Use of the virtual DE images improved detection and localization of lung nodules.

and obtained significant clinical values. Several customized strategies may contribute to the effectiveness of our method. Firstly, the multi-scale fully connected network is used as the core unit for predicting the virtual bone image. The end-to-end training for large-scale samples allows it to extract effective image features from different resolutions and to have good prediction generalizability. Secondly, the training of the multi-scale generator was reinforced with adversarial training from another deep network, so that it was able to learn how to preserve subtle signals, such as coronary calcifications, and other high frequency signals in the virtual bone image. Thirdly, adversarial training in the gradient domain may further improve the core CNN to learn the mapping between the standard image and the virtual DE images.

Compared with the two-shot dual energy technique, the proposed MCA-net method generates DE images with fewer motion artifacts in the bone and soft tissue images. Given that majority of our dataset doesn't suffer from severe motion artifacts and was used as the training data, we assumed only the mapping between the regions of standard image and the corresponding bone information without motion artifacts was learned by the model. However, even the current DE techniques cannot perfectly separate the bone and soft tissue information from x-ray. Therefore, our method trained on current DE data were not perfect. In some case, some residual bone edge signal remained and can be observed in our virtual DE soft tissue images. Compared with similar previous works on generation of virtual DE soft tissue image through machine learning and image processing techniques, our method was trained with a relatively large scale of DE training data. For instance, there were much fewer DE training cases in [14, 15]. Our deep model trained on relatively large-scale dataset can produce more reliable and robust virtual DE image, and generate both DE bone and soft tissue images. In addition, our method doesn't require segmentation of the lung field and the contrast normalization process for the input of standard image.

Besides from evaluating the prediction performance of our model, the clinical usefulness of our method was also assessed. In the task of localizing the lung nodule on a public dataset (JSRT), our results demonstrated that our virtual DE images can significantly improve the radiologist's performance on finding the lung nodule. Given our model's ability of preserving useful signal, such as tissue contrast and object structure, we expect similar improvement on other disease detection using our technique. However, the clinical examination of a comparison of the virtual DE images with standard image for these disease should be conducted with larger population of radiologists. Moreover, we believe a more reliable evaluation of the performance could be obtained through a cross-validation or increase the test dataset to reduce the bias. Other hyper-parameters, such as filters size, number of filters, generator-discriminator training scheme, and learning rate, can be further optimized through the evaluation strategies mentioned above.

6   Conclusion

In this work, we developed and evaluated a deep network to generate virtual dual energy images from standard chest radiographs based on a multi-scale conditional adversarial network architecture (MCA-Net). With adversarial training, training of the multi-scale generator was reinforced so that it was able to learn how to preserve subtle signals in the virtual DE bone image. After obtaining the virtual bone image, high-quality virtual soft tissue images were produced by using a modified cross projection tensor algorithm to suppress the bone signal in the standard radiograph. Images from the testing data had high PSNR and SSIM values and were found to preserve clinically relevant features, indicating that the algorithm can produce virtual dual energy images comparable to ground truth dual energy images. The clinical evaluation demonstrated the clinical usefulness of this technique. Given the high cost of DES equipment and increased radiation dose from DE acquisitions, the use of virtual DE bone and soft tissue images provides a potential alternative solution for improved detection and diagnosis of disease using chest radiography.

References

1. Kelcz, F., Zink, F., Peppler, W., Kruger, D., Ergun, D., Mistretta, C.: Conventional chest radiography vs dual-energy computed radiography in the detection and characterization of pulmonary nodules. AJR Am. J. Roentgenol. **162**, 271–278 (1994)
2. Li, F., Hara, T., Shiraishi, J., Engelmann, R., MacMahon, H., Doi, K.: Improved detection of subtle lung nodules by use of chest radiographs with bone suppression imaging: receiver operating characteristic analysis with and without localization. Am. J. Roentgenol. **196**, W535–W541 (2011)
3. Zhou, B., et al.: Detection and quantification of coronary calcium from dual energy chest x-rays: phantom feasibility study. Med. Phys. **44**, 5106–5119 (2016)

4. Zhou, B., Jiang, Y., Wen, D., Gilkeson, R.C., Hou, J., Wilson, D.L.: Visualization of coronary artery calcium in dual energy chest radiography using automatic rib suppression. In: Medical Imaging 2018: Image Processing, vol. 10574, p. 105740E. International Society for Optics and Photonics (2018)

5. Wen, D., et al.: Enhanced coronary calcium visualization and detection from dual energy chest x-rays with sliding organ registration. Comput. Med. Imaging Graph. **64**, 12–21 (2018)

6. Chen, S., Suzuki, K.: Computerized detection of lung nodules by means of "virtual dual-energy" radiography. IEEE Trans. Biomed. Eng. **60**, 369–378 (2013)

7. Vock, P., Szucs-Farkas, Z.: Dual energy subtraction: principles and clinical applications. Eur. J. Radiol. **72**, 231–237 (2009)

8. Wang, X., Peng, Y., Lu, L., Lu, Z., Bagheri, M., Summers, R.M.: ChestX-ray8: hospital-scale chest x-ray database and benchmarks on weakly-supervised classification and localization of common thorax diseases. In: 2017 IEEE Conference on Computer Vision and Pattern Recognition (CVPR), pp. 3462–3471. IEEE (2017)

9. Shin, H.C., et al.: Deep convolutional neural networks for computer-aided detection: CNN architectures, dataset characteristics and transfer learning. IEEE Trans. Med. Imaging **35**, 1285–1298 (2016)

10. Roth, H.R., et al.: DeepOrgan: multi-level deep convolutional networks for automated pancreas segmentation. In: Navab, N., Hornegger, J., Wells, W.M., Frangi, A.F. (eds.) MICCAI 2015. LNCS, vol. 9349, pp. 556–564. Springer, Cham (2015). https://doi.org/10.1007/978-3-319-24553-9_68

11. Isola, P., Zhu, J.Y., Zhou, T., Efros, A.A.: Image-to-image translation with conditional adversarial networks. arXiv preprint (2017)

12. Agrawal, A., Raskar, R., Chellappa, R.: Edge suppression by gradient field transformation using cross-projection tensors. In: 2006 IEEE Computer Society Conference on Computer Vision and Pattern Recognition, vol. 2, pp. 2301–2308. IEEE (2006)

13. Suzuki, K., Abe, H., Li, F., Doi, K.: Suppression of the contrast of ribs in chest radiographs by means of massive training artificial neural network. In: Medical Imaging 2004: Image Processing, vol. 5370, pp. 1109–1120. International Society for Optics and Photonics (2004)

14. Suzuki, K., Abe, H., MacMahon, H., Doi, K.: Image-processing technique for suppressing ribs in chest radiographs by means of massive training artificial neural network (MTANN). IEEE Trans. Med. Imaging **25**, 406–416 (2006)

15. Chen, S., Suzuki, K.: Separation of bones from chest radiographs by means of anatomically specific multiple massive-training anns combined with total variation minimization smoothing. IEEE Trans. Med. Imaging **33**, 246–257 (2014)

16. Chen, S., Zhong, S., Yao, L., Shang, Y., Suzuki, K.: Enhancement of chest radiographs obtained in the intensive care unit through bone suppression and consistent processing. Phys. Med. Biol. **61**, 2283 (2016)

17. Loog, M., van Ginneken, B., Schilham, A.M.: Filter learning: application to suppression of bony structures from chest radiographs. Med. Image Anal. **10**, 826–840 (2006)

18. Simkó, G., Orbán, G., Máday, P., Horváth, G.: Elimination of clavicle shadows to help automatic lung nodule detection on chest radiographs. In: Vander Sloten, J., Verdonck, P., Nyssen, M., Haueisen, J. (eds.) 4th European Conference of the International Federation for Medical and Biological Engineering. IFMBE, vol. 22, pp. 488–491. Springer, Heidelberg (2009). https://doi.org/10.1007/978-3-540-89208-3_116

19. Hogeweg, L., Sanchez, C.I., van Ginneken, B.: Suppression of translucent elongated structures: applications in chest radiography. IEEE Trans. Med. Imaging **32**, 2099–2113 (2013)
20. Rasheed, T., Ahmed, B., Khan, M.A., Bettayeb, M., Lee, S., Kim, T.S.: Rib suppression in frontal chest radiographs: a blind source separation approach. In: 9th International Symposium on Signal Processing and Its Applications, ISSPA 2007, pp. 1–4. IEEE (2007)
21. LeCun, Y., Bengio, Y., Hinton, G.: Deep learning. Nature **521**, 436 (2015)
22. Mirza, M., Osindero, S.: Conditional generative adversarial nets. arXiv preprint arXiv:1411.1784 (2014)
23. Ledig, C., et al.: Photo-realistic single image super-resolution using a generative adversarial network. arXiv preprint (2016)
24. Karras, T., Aila, T., Laine, S., Lehtinen, J.: Progressive growing of GANs for improved quality. Stability, and Variation. arXiv preprint (2017)
25. Ronneberger, O., Fischer, P., Brox, T.: U-Net: convolutional networks for biomedical image segmentation. In: Navab, N., Hornegger, J., Wells, W.M., Frangi, A.F. (eds.) MICCAI 2015. LNCS, vol. 9351, pp. 234–241. Springer, Cham (2015). https://doi.org/10.1007/978-3-319-24574-4_28
26. Goodfellow, I., et al.: Generative adversarial nets. In: Advances in Neural Information Processing Systems, pp. 2672–2680 (2014)
27. Schilham, A.M., Van Ginneken, B., Loog, M.: A computer-aided diagnosis system for detection of lung nodules in chest radiographs with an evaluation on a public database. Med. Image Anal. **10**, 247–258 (2006)
28. Chakraborty, D., Yoon, H.J., Mello-Thoms, C.: Spatial localization accuracy of radiologists in free-response studies: inferring perceptual froc curves from mark-rating data. Acad. Radiol. **14**, 4–18 (2007)

GD-GAN: Generative Adversarial Networks for Trajectory Prediction and Group Detection in Crowds

Tharindu Fernando$^{(\boxtimes)}$, Simon Denman, Sridha Sridharan, and Clinton Fookes

Image and Video Research Laboratory, SAIVT,
Queensland University of Technology (QUT), Brisbane, Australia
{t.warnakulasuriya,s.denman,s.sridharan,c.fookes}@qut.edu.au

Abstract. This paper presents a novel deep learning framework for human trajectory prediction and detecting social group membership in crowds. We introduce a generative adversarial pipeline which preserves the spatio-temporal structure of the pedestrian's neighbourhood, enabling us to extract relevant attributes describing their social identity. We formulate the group detection task as an unsupervised learning problem, obviating the need for supervised learning of group memberships via hand labeled databases, allowing us to directly employ the proposed framework in different surveillance settings. We evaluate the proposed trajectory prediction and group detection frameworks on multiple public benchmarks, and for both tasks the proposed method demonstrates its capability to better anticipate human sociological behaviour compared to the existing state-of-the-art methods (This research was supported by the Australian Research Council's Linkage Project LP140100282 "Improving Productivity and Efficiency of Australian Airports").

Keywords: Group detection · Generative Adversarial Networks · Trajectory prediction

1 Introduction

Understanding and predicting crowd behaviour plays a pivotal role in video based surveillance; and as such is becoming essential for discovering public safety risks, and predicting crimes or patterns of interest. Recently, focus has been given to understanding human behaviour at a group level, leveraging observed social interactions. Researchers have shown this to be important as interactions occur at a group level, rather than at an individual or whole of crowd level.

As such we believe group detection has become a mandatory part of an intelligent surveillance system; however this group detection task presents several new

Electronic supplementary material The online version of this chapter (https://doi.org/10.1007/978-3-030-20887-5_20) contains supplementary material, which is available to authorized users.

C. V. Jawahar et al. (Eds.): ACCV 2018, LNCS 11361, pp. 314–330, 2019.
https://doi.org/10.1007/978-3-030-20887-5_20

challenges [31,32]. Other than identifying and tracking pedestrians from video, modelling the semantics of human social interaction and cultural gestures over a short sequence of clips is extremely challenging. Several attempts [27,31,32,34] have been made to incorporate handcrafted physics based features such as relative distance between pedestrians, trajectory shape and motion based features to model their social affinity. Hall et al. [16] proposed a proxemic theory for such physical interactions based on different distance boundaries; however recent works [31,32] have shown these quantisations fail in cluttered environments.

Furthermore, proximity doesn't always describe the group membership. For instance two pedestrians sharing a common goal may start their trajectories in two distinct source positions, however, meet in the middle. Hence we believe being reliant on a handful of handcrafted features to be sub-optimal [1,10,19].

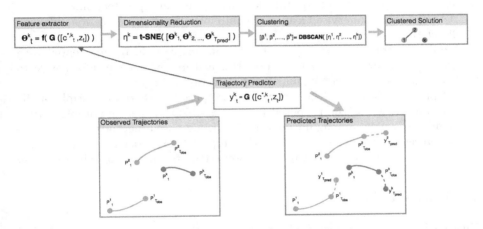

Fig. 1. Proposed group detection framework: After observing short segments of trajectories for each pedestrian in the scene, we apply the proposed trajectory prediction algorithm to forecast their future trajectories. The context representation generated at this step is extracted and compressed using t-SNE dimensionality reduction. Finally, the DBSCAN clustering algorithm is applied to detect the pedestrian groups.

To this end we propose a deep learning algorithm which automatically learns these group attributes. We take inspiration from the trajectory modelling approaches of [8] and [11], where the approaches capture contextual information from the local neighbourhood. We further augment this approach with a Generative Adversarial Network (GAN) [10,15,28] learning pipeline where we learn a custom, task specific loss function which is specifically tailored for future trajectory prediction, learning to imitate complex human behaviours.

Figure 1 illustrates the proposed approach. First, we observe short segments of trajectories from 1 to T_{obs} for each pedestrian, p^k, in the scene. Then, we apply the proposed trajectory prediction algorithm to forecast their future trajectories from $T_{obs+1} - T_{pred}$. This step generates hidden context representations for each pedestrian describing the current environmental context in the local neighbourhood of the pedestrian. We then apply t-SNE dimensionality reduction to extract the most discriminative features, and we detect the pedestrian groups by clustering these reduced features.

The simplistic nature of the proposed framework offers direct transferability among different environments when compared to the supervised learning approaches of [27,31,32,34], which require re-training of the group detection process whenever the surveillance scene changes. This ability is a result of the proposed deep feature learning framework which learns the required group attributes automatically and attains commendable results among the state-of-the-art.

Novel contributions of this paper can be summarised as follows:

- We propose a novel GAN pipeline which jointly learns informative latent features for pedestrian trajectory forecasting and group detection.
- We remove the supervised learning requirement for group detection, allowing direct transferability among different surveillance scenes.
- We demonstrate how the original GAN objective could be augmented with sparsity regularisation to learn powerful features which are informative to both trajectory forecasting and group detection tasks.
- We provide extensive evaluations of the proposed method on multiple public benchmarks where the proposed method is able to generate notable performance, especially among unsupervised learning based methods.
- We present visual evidence on how the proposed trajectory modelling scheme has been able to embed social interaction attributes into its encoding scheme.

2   Related Work

Related literature is categorised into human behaviour prediction approaches (see Sect. 2.1); and group detection architectures (see Sect. 2.2).

2.1   Human Behaviour Prediction

Social Force models [17,34], which rely on the attractive and repulsive forces between pedestrians to model their future behaviour, have been extensively applied for modelling human navigational behaviour. However with the dawn of deep learning, these methods have been replaced as they have been shown to ill represent the structure of human decision making [7,8,15].

One of the most popular deep learning methods is the social LSTM [1] model which represents the pedestrians in the local neighbourhood using LSTMs and then generates their future trajectory by systematically pooling the relevant information. This removes the need for handcrafted features and learns the required feature vectors automatically through the encoded trajectory representation. This architecture is further augmented in [8] where the authors propose a more efficient method to embed the local neighbourhood information via a soft and hardwired attention framework. They demonstrate the importance of fully capturing the context information, which includes the short-term history of the pedestrian of interest as well as their neighbours.

Generative Adversarial Networks (GANs) [10,15,28] propose a task specific loss function learning process where the training objective is a minmax game between the generative and discriminative models. These methods have shown promising results, overcoming the intractable computation of a loss function, in tasks such as autonomous driving [9,23], saliency prediction [10,25], image to image translation [19] and human navigation modelling [15,28].

Even though the proposed GAN based trajectory modelling approach exhibits several similarities to recent works in [15,28], the proposed work differs in multiple aspects. Firstly, instead of using CNN features to extract the local structure of the neighbourhood as in [28], pooling out only the current state of the neighbourhood as in [15], or discarding the available historical behaviour which is shown to be ineffective [7,8,28]; we propose an efficient method to embed the local neighbourhood context based on the soft and hardwired attention framework proposed in [8]. Secondly, as we have an additional objective of localising the groups in the given crowd, we propose an augmentation to the original GAN objective which regularises the sparsity of the generator embeddings, generating more discriminative features and aiding the clustering processes.

2.2   Group Detection

Some earlier works in group detection [5,29] employ the concept of F-formations [20], which can be seen as specific orientation patterns that individuals engage in when in a group. However such methods are only suited to stationary groups.

In a separate line of work researchers have analysed pedestrian trajectories to detect groups. Pellegrinin et al. [27] applied Conditional Random Fields to jointly predict the future trajectory of the pedestrian of interest as well as their group membership. [34] utilises distance, speed and overlap time to train a linear SVM to classify whether two pedestrians are in the same group or not. In contrast to these supervised methods, Ge et al. [13] proposed using agglomerative clustering of speed and proximity features to extract pedestrian groups.

Most recently Solera et al. [31] proposed proximity and occupancy based social features to detect groups using a trained structural SVM. In [32] the authors extend this preliminary work with the introduction of sociologically inspired features such as path convergence and trajectory shape. However these supervised learning mechanisms rely on hand labeled datasets to learn group segmentation, limiting the methods applicability. Furthermore, the above methods all utilise a predefined set of handcrafted features to describe the sociological identity of each pedestrian, which may be suboptimal. Motivated by the impressive results obtained in [8] with the augmented context embedding, we make the first effort to learn group attributes automatically and jointly through trajectory prediction.

3   Architecture

3.1   Neighbourhood Modelling

We use the trajectory modelling framework of [8] (shown in Fig. 2) for modelling the local neighbourhood of the pedestrian of interest.

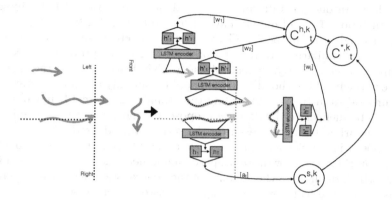

Fig. 2. Proposed neighbourhood modelling scheme [8]: A sample surveillance scene is shown on the left. The trajectory of the pedestrian of interest, k, is shown in green, and has two neighbours (in purple) to the left, one in front and none on right. The neighbourhood encoding scheme shown on the right: Trajectory information is encoded with LSTM encoders. A soft attention context vector $C_t^{s,k}$ is used to embed trajectory information from the pedestrian of interest, and a hardwired attention context vector $C_t^{h,k}$ is used for neighbouring trajectories. In order to generate $C_t^{s,k}$ we use a soft attention function denoted a_t in the above figure, and the hardwired weights are denoted by w. The merged context vector $C_t^{*,k}$ is then generated by merging $C_t^{s,k}$ and $C_t^{h,k}$. (Color figure online)

Let the trajectory of the pedestrian k, from frame 1 to T_{obs} be given by,

$$p^k = [p_1, \ldots, p_{T_{obs}}], \tag{1}$$

where the trajectory is composed of points in a Cartesian grid. Then we pass each trajectory through an LSTM [18] encoder to generate its hidden embeddings,

$$h_t^k = LSTM(p_t^k, h_{t-1}^k), \tag{2}$$

generating a sequence of embeddings,

$$h^k = [h_1^k, \ldots, h_{T_{obs}}^k]. \tag{3}$$

Following [8], the trajectory of the pedestrian of interest is embedded with soft attention such that,

$$C_t^{s,k} = \sum_{j=1}^{T_{obs}} \alpha_{tj} h_j^k, \tag{4}$$

which is the weighted sum of hidden states. The weight α_{tj} is computed by,

$$\alpha_{tj} = \frac{exp(e_{tj})}{\sum_{l=1}^{T} exp(e_{tl})}, \tag{5}$$

$$e_{tj} = a(h_{t-1}^{k}, h_{j}^{k}). \tag{6}$$

The function a is a feed forward neural network jointly trained with the other components.

To embed the effect of the neighbouring trajectories we use the hardwired attention context vector $C_t^{h,k}$ from [8]. The hardwired weight w is computed by,

$$w_j^n = \frac{1}{\text{dist}(n,j)}, \tag{7}$$

where $\text{dist}(n,j)$ is the distance between the nth neighbour and the pedestrian of interest at the jth time instant. Then we compute $C_t^{h,k}$ as the aggregation for all the neighbours such that,

$$C_t^{h,k} = \sum_{n=1}^{N} \sum_{j=1}^{T_{obs}} w_j^n h_j^n, \tag{8}$$

where there are N neighbouring trajectories in the local neighbourhood, and h_j^n is the encoded hidden state of the nth neighbour at the jth time instant. Finally we merge the soft attention and hardwired attention context vectors to represent the current neighbourhood context such that,

$$C_t^{*,k} = \tanh([C_t^{s,k}, C^{h,k}]). \tag{9}$$

3.2 Trajectory Prediction

Unlike [8], we use a GAN to predict the future trajectory. There exists a minmax game between the generator (G) and the discriminator (D) guiding the model G to be closer to the ground truth distribution. The process is guided by learning a custom loss function which generates an additional advantage when modelling complex behaviours such as human navigation, where multiple factors such as human preferences and sociological factors influence behaviour.

Trajectory prediction can be formulated as observing the trajectory from time 1 to T_{obs}, denoted as $[p_1, \ldots, p_{T_{obs}}]$, and forecasting the future trajectory for time T_{obs+1} to T_{pred}, denoted as $[y_{T_{obs+1}}, \ldots, y_{T_{pred}}]$. The GAN learns a mapping from a noise vector z to an output vector y, $G : z \rightarrow y$ [10]. Adding the notion of time, the output of the model y_t can be written as $G : z_t \rightarrow y_t$.

We augment the generic GAN mapping to be conditional on the current neighbourhood context C_t^*, $G : (C_t^*, z_t) \rightarrow y_t$, such that the synthesised trajectories follow the social navigational rules that are dictated by the environment.

This objective can be written as,

$$V = \mathbb{E}_{y_t, C_t^* \sim p_{data}}([logD(C_t^*, y_t)]) + \mathbb{E}_{C_t^* \sim p_{data}, z_t \sim noise}([1 - logD(C_t^*, G(C_t^*, z_t))]). \tag{10}$$

Our final aim is to utilise the hidden state embeddings from the trajectory generator to discover the pedestrian groups via clustering those embeddings. Hence having a sparse feature vector for clustering is beneficial as they are more discriminative compared to their dense counterparts [12]. Hence we augment the objective in Eq. 10 with a sparsity regulariser such that,

$$L_1 = ||f(G(C_t^*, z_t))||_1, \tag{11}$$

and

$$V^* = V + \lambda L_1, \tag{12}$$

where f is a feature extraction function which extracts the hidden embeddings from the trajectory generator G, and λ is a weight vector which controls the tradeoff between the GAN objective and the sparsity constraint.

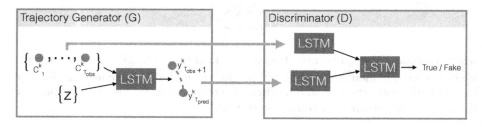

Fig. 3. Proposed trajectory prediction framework: The generator model G samples from the noise distribution z and synthesises a trajectory y_t, which is conditioned upon the local neighbourhood context C_t^*. The discriminator D considers both y_t and C_t^* when classifying the authenticity of the trajectory.

The architecture of the proposed trajectory prediction framework is presented in Fig. 3. We utilise LSTMs as the Generator (G) and the Discriminator (D) models. G samples from the noise distribution, z, and synthesises a trajectory for the pedestrian motion which is conditioned upon the local neighbourhood context, C_t^*, of that particular pedestrian. Utilising these predicted trajectories, y_t, and the context embeddings, C_t^*, D tries to discriminate between the synthesised and ground truth human trajectories.

3.3 Group Detection

Figure 1 illustrates the proposed group detection framework. We pass each trajectory in the given scene through Eqs. 2 to 9 and generate the neighbourhood embeddings, $C_t^{*,k}$. Then using the feature extraction function f we extract the hidden layer activations for each pedestrian k such that,

$$\theta_t^k = f(G(C_t^{*,k}, z_t)). \tag{13}$$

Then we pass the extracted feature vectors through a t-SNE [24] dimensionality reduction step. The authors in [12] have shown that it is inefficient to cluster dense deep features. However they have shown the t-SNE algorithm to generate discriminative features capturing the salient aspects in each feature dimension. Hence we apply t-SNE for the kth pedestrian in the scene such that,

$$\eta^k = \text{t-SNE}([\theta_1^k, \ldots, \theta_{T_{obs}}^k]). \tag{14}$$

As the final step we apply DBSCAN [6] to discover similar activation patterns, hence segmenting the pedestrian groups. DBSCAN enables us to cluster the data on the fly without specifying the number of clusters. The process can be written as,

$$[\beta^1, \ldots, \beta^N] = \text{DBSCAN}([\eta^1, \ldots, \eta^N]), \tag{15}$$

where there are N pedestrians in the given scene and $\beta^n \in [\beta^1, \ldots, \beta^N]$ are the generated cluster identities.

4   Evaluation and Discussion

4.1   Implementation Details

When encoding the neighbourhood information, similar to [8], we consider the closest 10 neighbours from each of the left, right, and front directions of the pedestrian of interest. If there are more than 10 neighbours in any direction, we take the closest 9 trajectories and the mean trajectory of the remaining neighbours. If a trajectory has less than 10 neighbours, we created dummy trajectories with hardwired weights (i.e. Eq. 7) of 0, such that we always have 10 neighbours.

For all LSTMs, including LSTMs for neighbourhood modelling (i.e. Sect. 3.1), the trajectory generator and the discriminator (i.e. Sect. 3.2), we use a hidden state embedding size of 300 units. We trained the trajectory prediction framework iteratively, alternating between a generator epoch and a discriminator epoch with the Adam [21] optimiser, using a mini-batch size of 32 and a learning rate of 0.001 for 500 epochs. The hyper parameter $\lambda = 0.2$, and the hyper parameters of DBSCAN, epsilon = 0.50, minPts = 1, are chosen experimentally.

4.2   Evaluation of the Trajectory Prediction

Datasets. We evaluate the proposed trajectory predictor framework on the publicly available walking pedestrian dataset (BIWI) [26], Crowds By Examples (CBE) [22] dataset and Vittorio Emanuele II Gallery (VEIIG) dataset [3]. The BIWI dataset records two scenes, one outside a university (ETH) and one at a bus stop (Hotel). CBE records a single video stream with a medium density

322   T. Fernando et al.

crowd outside a university (Student 003). The VEIIG dataset provides one video
sequence from an overhead camera in the Vittorio Emanuele II Gallery (gall).
The training, testing and validation splits for BIWI, CBE and VEIIG are taken
from [26], [31] and [32] respectively.

These datasets include a variety of pedestrian social navigation scenarios
including collisions, collision avoidance and group movements, hence presenting
challenging settings for evaluation. Compared to BIWI which has low crowd
densities, CBE and VEIIG contain higher crowd densities and as a result more
challenging crowd behaviour arrangements, continuously varying from medium
to high densities.

Evaluation Metrics. Similar to [15, 28] we evaluated the trajectory prediction
performance with the following 2 error metrics: Average Displacement Error
(ADE) and Final Displacement Error (FDE). Please refer to [15, 28] for details.

Baselines and Evaluation. We compared our trajectory prediction model to
5 state-of-the-art baselines. As the first baseline we use the Social Force (SF)
model introduced in [34], where the destination direction is taken as an input
to the model and we train a linear SVM model similar to [8] to generate this
input. We use the Social-LSTM (So-LSTM) model of [1] as the next baseline
and the neighbourhood size hyper-parameter is set to 32 px. We also compare to
the Soft + Hardwired Attention (SHA) model of [8] and similar to the proposed
model we set the embedding dimension to be 300 units and consider a 30 total
neighbouring trajectories. We also considered the Social GAN (So-GAN) [15]
and attentive GAN (SoPhie) [28] models. To provide fair comparisons we set
the hidden state dimensions for the encoder and decoder models of So-GAN and
SoPhie to be 300 units. For all models we observe the first 15 frames (i.e. $1-T_{obs}$)
and predicted the future trajectory for the next 15 frames (i.e. $T_{obs+1} - T_{pred}$).

When observing the results tabulated in Table 1 we observe poor performance
for the SF model due to it's lack of capacity to model history. Models So-LSTM
and SHA utilise short term history from the pedestrian of interest and the local
neighbourhood and generate improved predictions. However we observe a significant increase in performance from methods that optimise generic loss functions
such as So-LSTM and SHA to GAN based methods such as So-GAN and SoPhie.
This emphasises the need for task specific loss function learning in order to imitate complex human social navigation strategies. In the proposed method we
further augment this performance by conditioning the trajectory generator on
the proposed neighbourhood encoding mechanism.

We present a qualitative evaluation of the proposed trajectory generation
framework with the SHA and So-GAN baselines in Fig. 4 (selected based on the
availability of their implementations). The observed portion of the trajectory is
denoted in green, the ground truth observations in blue and predicted trajectories
are shown in red (proposed), yellow (SHA) and brown (So-GAN). Observing the
qualitative results it can be clearly seen that the proposed model generates better
predictions compared to the state-of-the-art considering the varying nature of the

Table 1. Quantitative results for the BIWI [26], CBE [22] and VEIIG [3] datasets. In all methods the forecast trajectories are of length 15 frames. Error metrics are as in Sect. 4.2. '-' refers to unavailability of that specific evaluation. The best values are denoted in bold.

Metric	Dataset	SF [34]	So-LSTM [1]	SHA [8]	So-GAN [15]	SoPhie [28]	Proposed
ADE	ETH (BIWI)	1.42	1.05	0.90	0.92	0.81	**0.63**
	Hotel (BIWI)	1.03	0.98	0.71	0.65	0.76	**0.55**
	Student 003 (CBE)	1.83	1.22	0.96	-	-	**0.72**
	gall (VEIIG)	1.72	1.14	0.91	-	-	**0.68**
FDE	ETH (BIWI)	2.20	1.84	1.43	1.52	1.45	**1.22**
	Hotel (BIWI)	2.45	1.95	1.65	1.62	1.77	**1.43**
	Student 003 (CBE)	2.63	1.97	1.80	-	-	**1.65**
	gall (VEIIG)	2.55	1.83	1.65	-	-	**1.45**

(a)   (b)   (c)   (d)

Fig. 4. Qualitative results for the proposed trajectory prediction framework for sequences from the CBE dataset. Given (in green), Ground Truth (in blue) and Predicted trajectories from proposed (in red), SHA model (in yellow) crom So-GAN (in brown). For visual clarity, we show only the trajectories for some of the pedestrians in the scene. (Color figure online)

neighbourhood clutter. For instance in Fig. 4(c) and (d) we observe significant deviations between the predictions for SHA and So-GAN and the ground truth. However the proposed model better anticipates the pedestrian motion with the improved context modelling and learning process. It should be noted that the proposed method has a better ability to anticipate stationary groups compared to the baselines, which is visible in Fig. 4(c).

4.3   Evaluation of the Group Detection

Datasets. Similar to Sect. 4.2 we use the BIWI, CBE and VEIIG datasets in our evaluation. Dataset characteristics are reported in Table 2.

Table 2. Dataset characteristics for different sequences in BIWI [26], CBE [22] and VEIIG [3] datasets

Dataset	ETH (BIWI)	Hotel (BIWI)	Student-003 (CBE)	gall (VEIIG)
Frames	1448	1168	541	7500
Pedestrian	360	390	434	630
Groups	243	326	288	207

Evaluation Metrics. One popular measure of clustering accuracy is the pairwise loss Δ_{pw} [35], which is defined as the ratio between the number of pairs on which β and $\hat{\beta}$ disagree on their cluster membership and the number of all possible pairs of elements in the set.

However as described in [31, 32] Δ_{pw} accounts only for positive intra-group relations and neglects singletons. Hence we also measure the Group-MITRE loss, Δ_{GM}, introduced in [31], which has overcome this deficiency. Δ_{GM} adds a fake counterpart for singletons and each singleton is connected with it's counterpart. Therefore δ_{GM} also takes singletons into consideration.

Baselines and Evaluation. We compare the proposed Group Detection GAN (GD-GAN) framework against 5 recent state-of-the-art baselines, namely [13, 30, 32, 34, 35], selected based on their reported performance in public benchmarks.

In Table 3 we report the Precision (P) and Recall (R) values for Δ_{pw} and Δ_{GM} for the proposed method along with the state-of-the-art baselines. The proposed GD-GAN method has been able to achieve superior results, especially among unsupervised grouping methods. It should be noted that methods [30, 32, 34, 35] utilise handcrafted features and use supervised learning to separate the groups. As noted in Sect. 1 these methods cannot adapt to scene variations and require hand labeled datasets for training. Furthermore we would like to point out that the supervised grouping mechanism in [32] directly optimises Δ_{GM}. However, without such tedious annotation requirements and learning strategies, the proposed method has been able to generate commendable and consistent results in all considered datasets, especially in cluttered environments[1].

In Fig. 5 we show groups detected by the proposed GD-GAN method for sequences from the CBE and VEIIG datasets. Regardless of the scene context, occlusions and the varying crowd densities, the proposed GD-GAN method generates acceptable results. We believe this is due to the augmented features that

---

[1] See the supplementary material for the results for using supervised learning to separate the groups on proposed context features.

we derive through the automated deep feature learning process. These features account for both historical and future behaviour of the individual pedestrians, hence possessing an ability to detect groups even in the presence of occlusions such as in Fig. 5(c).

Table 3. Comparative results on the BIWI [26], CBE [22] and VEIIG [3] datasets using the Δ_{GM} [31] and Δ_{PW} [35] metrics. '-' refers to unavailability of that specific evaluation. The best results are shown in bold and the second best results are underlined.

	Shao et al. [30]		Zanotto et al. [35]		Yamaguchi et al. [34]		Ge et al. [13]		Solera et al. [32]		GD-GAN	
	P	R	P	R	P	R	P	R	P	R	P	R
BIWI Δ_{GM}	67.3	64.1	-	-	84.0	51.2	89.2	90.9	_97.3_	**97.7**	**97.5**	**97.7**
Hotel Δ_{PW}	51.5	90.4	81.0	91.0	83.7	**93.9**	88.9	89.3	_89.1_	91.9	**90.2**	_93.1_
BIWI Δ_{GM}	69.3	68.2	-	-	60.6	76.4	87.0	84.2	_91.8_	**94.2**	**92.5**	**94.2**
ETH Δ_{PW}	44.5	87.0	79.0	82.0	72.9	78.0	80.7	80.7	_91.1_	_83.4_	**91.3**	**83.5**
CEB Δ_{GM}	40.4	48.6	-	-	56.7	76.0	77.2	73.6	**81.7**	**82.5**	_81.0_	_81.8_
Student-003 Δ_{PW}	10.6	**76.0**	70.0	74.0	63.9	72.6	72.2	65.1	**82.3**	_74.1_	_82.1_	63.4
VEIIG Δ_{GM}	-	-	-	-	-	-	-	-	**84.1**	**84.1**	_83.1_	_79.5_
gall Δ_{PW}	-	-	-	-	-	-	-	-	**79.7**	**77.5**	_77.6_	_73.1_

We selected the first 30 pedestrian trajectories from the VEIIG test set and in Fig. 6 we visualise the embedding space positions before (in blue) and after (in red) training of the proposed trajectory generator (G). Similar to [2] we extracted the activations using the feature extractor function f and applied PCA [33] to plot them in 2D. The respective ground truth group IDs are indicated in brackets. This helps us to gain an insight into the encoding process that G utilises, which allows us to discover groups of pedestrians. Considering the examples given, it can be seen that trajectories from the same cluster become more tightly grouped. This is due to the model incorporating source positions, heading direction, trajectory similarity, when embedding trajectories, allowing us to extract pedestrian groups in an unsupervised manner.

4.4 Ablation Experiment

To further demonstrate the proposed group detection approach, we conducted a series of ablation experiments identifying the crucial components of the proposed methodology[2]. In the same setting as the previous experiment we compare the proposed GD-GAN model against a series of counter parts as follows:

- GD-GAN/GAN: removes D and the model G is learnt through supervised learning as in [8].
- GD-GAN/cGAN: optimises the generic GAN objective defined in [14].

---

[2] See the supplementary material for an ablation study for the trajectory prediction.

(a) GVEII - Frame 2127   (b) GVEII- Frame 2320

(c) CBE - Frame 2603   (d) CBE - Frame 2910

Fig. 5. Qualitative results from the proposed GD-GAN methods for sequences from the CBE and GVEII datasets. Connected coloured blobs indicate groups of pedestrians.

- GD-GAN/L_1: removes sparsity regularisation and optimises Eq. 10.
- GD-GAN + hf: utilises features from G as well as the handcrafted features defined in [32] for clustering.

Table 4. Ablation experiment evaluations

	GD-GAN/GAN		GD-GAN/cGAN		GD-GAN/L_1		GD-GAN + hf		GD-GAN	
	P	R	P	R	P	R	P	R	P	R
CEB Δ_{GM}	73.6	75.1	76.7	76.2	77.3	78.0	**79.0**	**79.2**	78.7	**79.2**
Student-003 Δ_{PW}	74.1	52.8	75.5	60.2	78.1	65.1	**80.4**	68.0	**80.4**	**68.4**

The results of our ablation experiment are presented in Table 4. Model GD-GAN/GAN performs poorly due to the deficiencies in the supervised learning process. It optimises a generic mean square error loss, which is not ideal to guide the model through the learning process when modelling a complex behaviour such as human navigation. Therefore the resultant feature vectors do not capture the full context which contributes to the poor group detection accuracies. We observe an improvement in performance with GD-GAN/cGAN due to the GAN learning process which is further augmented and improved through GD-GAN/L_1 where the model learns a conditional behaviour depending on the neighbourhood context. L_1 regularisation further assists the group detection process via making the learnt feature distribution more discriminative.

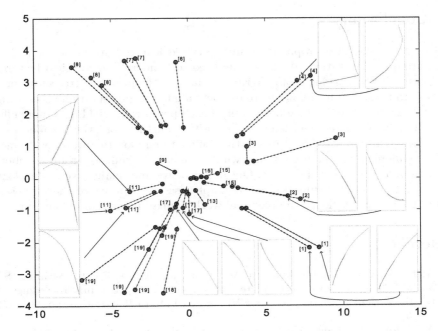

Fig. 6. Projections of the trajectory generator (G) hidden states before (in blue) and after (in red) training. Ground truth group IDs are in brackets. Each insert indicates the trajectory associated with the embedding. The given portion of the trajectory is in green, and the ground truth and prediction are in blue and red respectively (Color figure online)

In order to demonstrate the credibility of the learnt group attributes from the proposed GD-GAN model, we augment the feature vector extracted in Eq. 13 together with the features proposed in [32] and apply subsequent process (i.e. Eqs. 14 and 15) to discover the groups. We utilise the public implementation[3] released by the authors for the feature extraction.

We do not observe a substantial improvement with the group detection performance being very similar, indicating that the proposed GD-GAN model is sufficient for modelling the social navigation structure of the crowd.

4.5   Time Efficiency

We use the Keras [4] deep learning library for our implementation. The GD-GAN module does not require any special hardware such as GPUs to run and has 41.8K trainable parameters. We ran the test set in Sect. 4.3 on a single core of an Intel Xeon E5-2680 2.50 GHz CPU and the GD-GAN algorithm was able to generate 100 predicted trajectories with 30, 2 dimensional data points in each trajectory (i.e. using 15 observations to predict the next 15 data points) and complete the group detection process in 0.712 s.

---

[3] https://github.com/francescosolera/group-detection.

5 Conclusions

In this paper we have proposed an unsupervised learning approach for pedestrian group segmentation. We avoid the need to handcraft sociological features by automatically learning group attributes through the proposed trajectory prediction framework. This allows us to discover a latent representation accounting for both historical and future behaviour of each pedestrian, yielding a more efficient platform for detecting their social identities. Furthermore, the unsupervised learning setting grants the approach the ability to employ the proposed framework in different surveillance settings without tedious learning of group memberships from a hand labeled dataset. Our quantitative and qualitative evaluations on multiple public benchmarks clearly emphasise the capacity of the proposed GD-GAN method to learn complex real world human navigation behaviour.

References

1. Alahi, A., Goel, K., Ramanathan, V., Robicquet, A., Fei-Fei, L., Savarese, S.: Social LSTM: human trajectory prediction in crowded spaces. In: Proceedings of the IEEE Conference on Computer Vision and Pattern Recognition, pp. 961–971 (2016)
2. Aubakirova, M., Bansal, M.: Interpreting neural networks to improve politeness comprehension. arXiv preprint arXiv:1610.02683 (2016)
3. Bandini, S., Gorrini, A., Vizzari, G.: Towards an integrated approach to crowd analysis and crowd synthesis: a case study and first results. Pattern Recognit. Lett. **44**, 16–29 (2014)
4. Chollet, F., et al.: Keras (2015) (2017)
5. Cristani, M., et al.: Social interaction discovery by statistical analysis of f-formations. In: BMVC, vol. 2, p. 4 (2011)
6. Ester, M., Kriegel, H.P., Sander, J., Xu, X., et al.: A density-based algorithm for discovering clusters in large spatial databases with noise. In: KDD, vol. 96, pp. 226–231 (1996)
7. Fernando, T., Denman, S., McFadyen, A., Sridharan, S., Fookes, C.: Tree memory networks for modelling long-term temporal dependencies. Neurocomputing **304**, 64–81 (2018)
8. Fernando, T., Denman, S., Sridharan, S., Fookes, C.: Soft+ hardwired attention: an LSTM framework for human trajectory prediction and abnormal event detection. arXiv preprint arXiv:1702.05552 (2017)
9. Fernando, T., Denman, S., Sridharan, S., Fookes, C.: Learning temporal strategic relationships using generative adversarial imitation learning. arXiv preprint arXiv:1805.04969 (2018)
10. Fernando, T., Denman, S., Sridharan, S., Fookes, C.: Task specific visual saliency prediction with memory augmented conditional generative adversarial networks. In: 2018 IEEE Winter Conference on Applications of Computer Vision (WACV), pp. 1539–1548. IEEE (2018)
11. Fernando, T., Denman, S., Sridharan, S., Fookes, C.: Tracking by prediction: a deep generative model for mutli-person localisation and tracking. In: 2018 IEEE Winter Conference on Applications of Computer Vision (WACV), pp. 1122–1132. IEEE (2018)

12. Figueroa, J.A., Rivera, A.R.: Learning to cluster with auxiliary tasks: a semi-supervised approach. In: 2017 30th SIBGRAPI Conference on Graphics, Patterns and Images (SIBGRAPI), pp. 141–148. IEEE (2017)

13. Ge, W., Collins, R.T., Ruback, R.B.: Vision-based analysis of small groups in pedestrian crowds. IEEE Trans. Pattern Anal. Mach. Intell. **34**(5), 1003–1016 (2012). https://doi.org/10.1109/TPAMI.2011.176

14. Goodfellow, I., et al.: Generative adversarial nets. In: Advances in Neural Information Processing Systems, pp. 2672–2680 (2014)

15. Gupta, A., Johnson, J., Fei-Fei, L., Savarese, S., Alahi, A.: Social GAN: socially acceptable trajectories with generative adversarial networks. In: IEEE Conference on Computer Vision and Pattern Recognition (CVPR). No. CONF (2018)

16. Hall, E.T.: The hidden dimension (1966)

17. Helbing, D., Molnar, P.: Social force model for pedestrian dynamics. Phys. Rev. E **51**(5), 4282 (1995)

18. Hochreiter, S., Schmidhuber, J.: Long short-term memory. Neural Comput. **9**(8), 1735–1780 (1997)

19. Isola, P., Zhu, J.Y., Zhou, T., Efros, A.A.: Image-to-image translation with conditional adversarial networks. arXiv preprint (2017)

20. Kendon, A.: Conducting interaction: patterns of behavior in focused encounters, vol. 7. CUP Archive (1990)

21. Kingma, D.P., Ba, J.: Adam: a method for stochastic optimization. arXiv preprint arXiv:1412.6980 (2014)

22. Lerner, A., Chrysanthou, Y., Lischinski, D.: Crowds by example. In: Computer Graphics Forum, vol. 26, pp. 655–664. Wiley Online Library (2007)

23. Li, Y., Song, J., Ermon, S.: InfoGAIL: interpretable imitation learning from visual demonstrations. In: Advances in Neural Information Processing Systems, pp. 3815–3825 (2017)

24. Maaten, L.v.d., Hinton, G.: Visualizing data using t-SNE. J. Mach. Learn. Res. **9**, 2579–2605 (2008)

25. Pan, J., et al.: SalGAN: visual saliency prediction with generative adversarial networks. arXiv preprint arXiv:1701.01081 (2017)

26. Pellegrini, S., Ess, A., Schindler, K., Van Gool, L.: You'll never walk alone: modeling social behavior for multi-target tracking. In: 2009 IEEE 12th International Conference on Computer Vision, pp. 261–268. IEEE (2009)

27. Pellegrini, S., Ess, A., Van Gool, L.: Improving data association by joint modeling of pedestrian trajectories and groupings. In: Daniilidis, K., Maragos, P., Paragios, N. (eds.) ECCV 2010. LNCS, vol. 6311, pp. 452–465. Springer, Heidelberg (2010). https://doi.org/10.1007/978-3-642-15549-9_33

28. Sadeghian, A., Kosaraju, V., Sadeghian, A., Hirose, N., Savarese, S.: SoPhie: an attentive GAN for predicting paths compliant to social and physical constraints. arXiv preprint arXiv:1806.01482 (2018)

29. Setti, F., Lanz, O., Ferrario, R., Murino, V., Cristani, M.: Multi-scale f-formation discovery for group detection. In: 2013 20th IEEE International Conference on Image Processing (ICIP), pp. 3547–3551. IEEE (2013)

30. Shao, J., Loy, C.C., Wang, X.: Scene-independent group profiling in crowd. In: 2014 IEEE Conference on Computer Vision and Pattern Recognition, pp. 2227–2234, June 2014. https://doi.org/10.1109/CVPR.2014.285

31. Solera, F., Calderara, S., Cucchiara, R.: Structured learning for detection of social groups in crowd. In: 2013 10th IEEE International Conference on Advanced Video and Signal Based Surveillance (AVSS), pp. 7–12. IEEE (2013)

32. Solera, F., Calderara, S., Cucchiara, R.: Socially constrained structural learning for groups detection in crowd. IEEE Trans. Pattern Anal. Mach. Intell. **38**(5), 995–1008 (2016)
33. Wold, S., Esbensen, K., Geladi, P.: Principal component analysis. Chemom. Intell. Lab. Syst. **2**(1–3), 37–52 (1987)
34. Yamaguchi, K., Berg, A.C., Ortiz, L.E., Berg, T.L.: Who are you with and where are you going? In: 2011 IEEE Conference on Computer Vision and Pattern Recognition (CVPR), pp. 1345–1352. IEEE (2011)
35. Zanotto, M., Bazzani, L., Cristani, M., Murino, V.: Online Bayesian nonparametrics for group detection. In: Proceedings of BMVC (2012)

Multi-level Sequence GAN for Group Activity Recognition

Harshala Gammulle[✉], Simon Denman, Sridha Sridharan, and Clinton Fookes

Image and Video Research Laboratory, SAIVT,
Queensland University of Technology (QUT), Brisbane, Australia
pranaliharshala.gammulle@hdr.qut.edu.au,
{s.denman,s.sridharan,c.fookes}@qut.edu.au

Abstract. We propose a novel semi supervised, Multi Level Sequential Generative Adversarial Network (*MLS-GAN*) architecture for group activity recognition. In contrast to previous works which utilise manually annotated individual human action predictions, we allow the models to learn it's own internal representations to discover pertinent sub-activities that aid the final group activity recognition task. The generator is fed with person-level and scene-level features that are mapped temporally through LSTM networks. Action-based feature fusion is performed through novel gated fusion units that are able to consider long-term dependancies, exploring the relationships among all individual actions, to learn an intermediate representation or 'action code' for the current group activity. The network achieves it's semi-supervised behaviour by allowing it to perform group action classification together with the adversarial real/fake validation. We perform extensive evaluations on different architectural variants to demonstrate the importance of the proposed architecture. Furthermore, we show that utilising both person-level and scene-level features facilitates the group activity prediction better than using only person-level features. Our proposed architecture outperforms current state-of-the-art results for sports and pedestrian based classification tasks on Volleyball and Collective Activity datasets, showing it's flexible nature for effective learning of group activities (This research was supported by the Australian Research Council's Linkage Project LP140100282 "Improving Productivity and Efficiency of Australian Airports").

Keywords: Group activity recognition ·
Generative adversarial networks · Long short term memory networks

1 Introduction

The area of human activity analysis has been an active field within the research community as it can aid in numerous important real world tasks such as video surveillance, video search and retrieval, sports video analytics, etc. In such scenarios, methods with the capability to handle multi-person actions and determine the collective action being performed play a major role. Among the main

© Springer Nature Switzerland AG 2019
C. V. Jawahar et al. (Eds.): ACCV 2018, LNCS 11361, pp. 331–346, 2019.
https://doi.org/10.1007/978-3-030-20887-5_21

challenges, handling different personnel appearing at different times and capturing their contribution towards the overall group activity is crucial. Learning the interactions between these individuals further aids the recognition of the collaborative action. Methods should retain the ability to capture information from the overall frame together with information from individual agents. We argue that the overall frame is important as it provides information regrading the varying background and context, the positions of agents within the frame and objects related to the action (e.g. the ball and the net in volleyball) together with the individual agent information.

Recent works on group activity analysis have utilised recurrent neural network architectures to capture temporal dynamics in video sequences. Even though deep networks are capable of performing automatic feature learning, they require manual human effort to design effective losses. Therefore, GAN based networks have become beneficial in overcoming the limitation of deep networks as they are capable of learning both features and the loss function automatically. Furthermore extending the GAN based architecture to a semi-supervised architecture, which is obtained by combining the unsupervised GAN objective with supervised classification objective, leverages the capacity for the network to learn from both labelled and unlabelled data.

In this paper we present a semi-supervised Generative Adversarial Network (GAN) architecture based on video sequence modelling with LSTM networks to perform group activity recognition. Figure 1, shows the overall framework of our proposed GAN architecture for group activity recognition. The generator is fed with sequences of person-level and scene-level RGB features which are extracted through the visual feature extractor for each person and the scene. Then the extracted features are sent through separate LSTMs to map the temporal correspondence of the sequences at each level. We utilise a gated fusion unit inspired by [1] to map the relevance of these LSTM outputs to an intermediate action representation, an 'action code', to represent the current group action. These action codes are then employed by the discriminator model which determines the current group action class and whether the given action code is real (ground truth) or fake (generated). Overall, the generator focuses on generating action codes that are indistinguishable from the ground truth action codes while the discriminator tries to achieve the real/fake and group activity classifications. With the use of a gated fusion unit, the model gains the ability to consider all the inputs when deciding on the output. Therefore, it is able to map the relevance of each performed individual action and their interactions with the attention weights automatically, to perform the final classification task. The contributions of our proposed model are as follows: (i) we introduce a novel recurrent semi-supervised GAN framework for group activity recognition, (ii) we formulate the framework such that it incorporates both person-level and scene-level features to determine the group activity, (iii) we demonstrate a feature fusion mechanism with gated fusion units, which automatically learns an attention mechanism focusing on the relevance of each individual agent and their interactions, (iv) we evaluate the

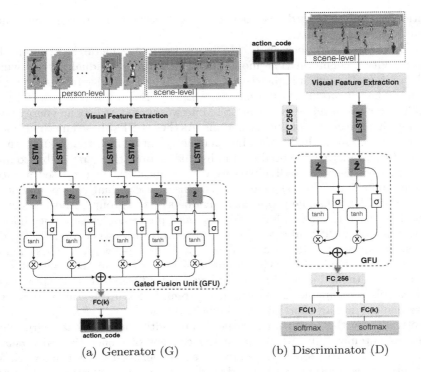

(a) Generator (G)   (b) Discriminator (D)

Fig. 1. The proposed Multi-Level Sequence GAN (*MLS-GAN*) architecture: (a) G is trained with sequences of person-level and scene-level features to learn an intermediate action representation, an 'action code'. (b) The model D performs group activity classification while discriminating real/fake data from scene level sequences and ground truth/generated action codes.

proposed model on pedestrian and sports datasets and achieve state-of-the-art results, outperforming the current baseline methods.

2 Related Work

Human action recognition is an area that has been of pivotal interest to researchers in the computer vision domain. However, a high proportion of proposed models are based on single-human actions which do not align with the nature of the real world scenarios where actions are continuous. Furthermore, many existing approaches only consider actions performed by a single agent, limiting the utility of these approaches.

Some early works [2–5] on group activity recognition have addressed the group activity recognition task on surveillance and sports video datasets with probabilistic and discriminative models that utilise hand-crafted features. As these hand-crafted feature based methods always require feature engineering, attention has shifted towards deep network based methods due to their automatic feature learning capability.

In [6] authors introduce an LSTM based two stage hierarchical model for group activity recognition. The model first learns the individual actions which are then integrated into a higher level model for group activity recognition. Shu et al. in [7] have introduced the Confidence Energy Recurrent Network (CERN) which is also a two-level hierarchy of LSTM networks that utilises an energy layer for estimating the energy of the predictions. As these LSTM based methods focus on learning each individual action independently, afterwhich the group activity is learnt by considering the predicted individual action, they are unable to map the interactions between individuals well [8]. Kim et al. [9] proposed a gated recurrent unit (GRU) based model that utilises discriminative group context features (DGCF) to handle people as individuals or sub groups. Another similar approach is suggested in [10] for classifying puck possession events in ice hockey by extracting convolutional layer features to train recurrent networks. In [8], the authors introduced the Structural Recurrent Neural Network (SRNN) model which is able to handle a varying number of individuals in the scene at each time step with the aid of a grid pooling layer. Even though these deep network based models are capable of performing automatic feature learning, they still require manual human effort to design effective losses.

Motivated by the recent advancements and with the ability to learn effective losses automatically, we build on the concept of Generative Adversarial Networks (GANs) to propose a recurrent semi-supervised GAN framework for group activity recognition. GAN based models are capable of learning an output that is difficult to discriminate from real examples, and also learn a mapping from input to output while learning a loss function to train the mapping. As a result of this ability, GANs have been used in solving different computer vision problems such as inpainting [11], product photo generation [12] etc. We utilise an extended variant of GANs, the conditional GAN [13–15] architecture where both the generator and the discriminator models are conditioned with additional data such as class labels or data from other modalities. A further enhancement of the architecture can be achieved by following the semi-supervised GAN architecture introduced in [16]. There are only a handful of GAN based methods [17,18] that have been introduced for human activity analysis. In [17] the authors train the generative model to synthesise frames in an action video sequence, and in [18] the generative model synthesises masks for humans. While these methods try to learn a distribution on video frame level attributes, no effort has been made to learn an intermediate representation at the human behaviour level. Motivated by [19,20], which have demonstrated the viability of learning intermediate representations with GANs, we believe that learning such an intermediate representation ('action code') would help the action classification process, as the classification model has to classify this discriminative action code.

To this end we make the first attempt to apply GAN based methods to the group activity recognition task, where the network jointly learns a loss function as well as providing auxiliary classification of the class.

(a)   (b)

Fig. 2. Sample ground truth action codes, with $k = 7$ (i.e. we have 7 actions). For the code in (a), $y = 5$ and for the code shown in (b) $y = 3$. Note that a green border is shown around the codes for clarify, this is not part of the code and is only included to aid display. Codes are of size $1 \times k$ pixels. (Color figure online)

3 Methodology

GANs are generative models that are capable of learning a mapping from a random noise vector z to an output vector, $y : G : z \rightarrow y$ [21]. In our work, we utilise the conditional GAN [13], an extension of the GAN that is capable of learning a mapping from the observed image x_t at time t and a random noise vector z_t to $y_t : G : \{x_t, z_t\} \rightarrow y_t$ [13,22].

GANs are composed of two main components: the Generator (G) and the Discriminator (D), which compete in a two player game. G tries to generate data that is indistinguishable from real data, while D tries to distinguish between real and generated (fake) data.

We introduce a conditional GAN based model, Multi-Level Sequence GAN (*MLS-GAN*), for group activity recognition, which utilises sequences of person-level and scene-level data for classification. In Sect. 3.1, we describe the action code format that the GAN is trained to generate; Sect. 3.2 describes the semi-supervised GAN architecture; and in Sect. 3.3 we explain the objectives that the models seek to optimise.

3.1 Action Codes

The generator network is trained to synthesise an 'action code' to represent the current group action. The generator maps dense pixel information to this action code. Hence having a one hot vector is not optimal. Therefore we scale it to a range from 0 to 255 giving more freedom for the action generator and discriminator to represent each action code as a dense vector representation,

$$y \in \mathbb{R}^{1 \times k}, \tag{1}$$

where k is the number of group action classes in the dataset. This action code representation can also be seen as an intermediate representation for the action classification. Several works have previously demonstrated the viability of these representations with GANs [19,20]. Overall, this action code generation is effected by adversarial loss as well as the classification loss, where the learnt action codes need to be informative for the classification task. In Fig. 2 we have sample action codes for a scenario where there are 7 action classes.

3.2 Semi-supervised GAN Architecture

The semi-supervised GAN architecture is achieved by combining the unsupervised GAN objective with the supervised classification objective. Unlike the standard GAN architecture, the discriminator of the semi-supervised GAN network is able to perform group action classification together with the real/fake classification task.

Generator. The generator takes person-level inputs and scene-level inputs for each video sequence for T time steps. Let $\{\hat{X}_1, \hat{X}_2, \ldots, \hat{X}_T\}$ be the full frame image sequences, while $\{x_1^n, x_2^n, \ldots, x_T^n\}$ is the person-level cropped bounding box image sequence for the nth person, where $n \in [1, \ldots, N]$. The generator input, I_G, can be defined as follows,

$$I_G = (\{x_1^1, x_2^1, \ldots, x_T^1\}, \ldots, \{x_1^N, x_2^N, \ldots, x_T^N\}, \{\hat{X}_1, \hat{X}_2, \ldots, \hat{X}_T\}). \quad (2)$$

The generator learns a mapping from the observed input I_G and a noise vector z to $y : G : \{I_G, z\} \rightarrow y$, where y is the action code. As shown in Fig. 1, to obtain this mapping the generator extracts visual features from I_G using a pre-trained deep network. Let the extracted scene-level visual feature be $\hat{\theta}_t$ and the extracted person-level features be θ_t^n for the nth person such that,

$$\hat{\theta}_t = f(\hat{X}_t), \quad (3)$$

and

$$\theta_t^n = f(x_t^n). \quad (4)$$

These extracted visual features are then passed through the respective LSTM networks,

$$\hat{Z} = \text{LSTM}([\hat{\theta}_1, \hat{\theta}_2, \ldots, \hat{\theta}_T]), \quad (5)$$
$$Z^n = \text{LSTM}([\theta_1^n, \theta_2^n, \ldots, \theta_T^n]). \quad (6)$$

Outputs for the nth person LSTM model are subsequently sent through a gated fusion unit to perform feature fusion as follows,

$$h^n = \tanh(\dot{W}^n Z^n), \quad (7)$$

where $\dot{W}^n$ is a weight vector for encoding. Next the sigmoid function, σ, is used to determine the information flow from each input stream,

$$q^n = \sigma(\bar{W}^n [Z^1, Z^2, \ldots, Z^N, \hat{Z}]), \quad (8)$$

afterwhich we multiply the embedding h^n with gate output q^n such that,

$$r^n = h^n \times q^n. \quad (9)$$

Therefore, when determining information flow from the nth person stream we attend over all the other input streams, rather than having one constant weight value for the entire stream. Using these functions we generate gated outputs, r^n, for each person level input as well as the other $\hat{r}$ for the scene level input. Given these person and scene level outputs, the fused output of the gated unit can be defined as,

$$C = \sum_{j=1}^{N} r^n + \hat{r}. \tag{10}$$

This output, C, is finally sent through a fully connected layer to obtain the action code to represent the current action,

$$y = FC(C, z), \tag{11}$$

which also utilises a latent vector z in the process.

Discriminator. The discriminator takes the scene-level inputs (I_{D1}) from each video sequence for T time steps together with real (ground truth)/fake (generated) action codes (I_{D2}). The aim of the semi-supervised GAN architecture is to perform real/fake validation together with the group action classification. The inputs to the discriminator models are as follows,

$$I_{D1} = (\{\hat{X}_1, \hat{X}_2, \ldots, \hat{X}_T\}); I_{D2} = y. \tag{12}$$

Unlike the generator, the discriminator is not fed with person-level features. The action codes provide intermediate representations of the group activities that have been generated by considering person-level features. Therefore, the activities of the individuals are already encoded in the action codes and the scene-level features are used to support the decision. Considering these scene level inputs also contain the individual people, providing the crops of every individual is redundant and greatly increases the architecture complexity. We believe that by providing the scene level features the model should be able to capture the spatial relationships and the spatial arrangements of the individuals, which is essential when deciding upon the authenticity of the generated action code.

The scene-level feature input (I_{D1}) is then sent through the visual feature extractor defined in Eq. 3 and we obtain $\hat{\theta}_t$. The scene-level features capture spatial relationships and the spatial arrangements of the people, which helps to decide whether the action is realistic given the arrangements. The action code input (I_{D2}) is sent through a fully connected layer and we obtain $\acute{\theta}_t$. These extracted features are then sent through gated fusion unit to perform feature fusion and the output of the gated unit $\acute{C}$ can be defined as,

$$\acute{C} = \acute{r}^n + \hat{r}^n \tag{13}$$

Finally $\acute{C}$ is passed through fully connected layers to perform group action classification together with the real/fake validation of the current action code.

3.3 GAN Objectives

The objective of the proposed *MLS-GAN* model can be defined as,

$$
\begin{aligned}
L_{GAN}(G,D) = \min_{G} \max_{D} \; &\mathbb{E}[\log D(I_{D1}, I_{D2})] \\
+ \; &\mathbb{E}[\log(1 - D(I_{D1}, G(I_G, z)))] + \lambda_c \mathbb{E}[\log D_c(k|I_{D1}, I_{D2})],
\end{aligned}
\tag{14}
$$

where $D_c(x)$ is the output classifier head of the discriminator and λ_c is a hyper parameter which balances the contributions of classification loss and the adversarial loss.

4 Experiments

4.1 Datasets

To demonstrate the flexibility of the proposed method we evaluate our proposed model on sports and pedestrian group activity datasets: the volleyball dataset [6] and the collective activity dataset [3]. We don't use the annotation for individual person activities in this research. Rather, we allow the model to learn it's own internal representation of the individual activities. We argue this is more appropriate for group activity recognition as the model is able to discover pertinent sub-activities rather than being forced to learn a (possibly) less informative representation that is provided by the individual activity ground truth.

Volleyball Dataset. The Volleyball dataset is composed of 55 videos containing 4,830 annotated frames. The dataset represents 8 group activities that can be found in Volleyball: right set, right spike, right pass, right win-point, left win-point, left pass, left spike and left set. The train/test splits of [6,7] are used.

Collective Activity Dataset. The collective activity dataset is composed of 44 video sequences representing five group-level activities. The group activity label is assigned by considering the most common action that is performed by the people in the scene. The train/test splits for evaluations are as in [6]. The available group actions are crossing, walking, waiting, talking and queueing.

4.2 Metrics

We perform comparisons to the state-of-the-art by utilising the same metrics used by the baseline approaches [6,23]. We use the multi-class accuracy (MCA) and the mean per class accuracy (MPCA) to overcome the imbalance in the test set (e.g. the total number of crossing examples is more than twice that of queueing and talking examples [23]) when evaluating the performance. As MPCA calculates the accuracy for each class, before taking the average accuracy values, this overcomes the accuracy bias on the imbalanced test set.

4.3   Network Architecture and Training

We extract visual features through a ResNet-50 [24] network pre-trained on Ima-geNet [25] for each set of person-level and scene-level inputs. Each input frame is resized to 224 × 224 as a preprocessing step prior to feature extraction. The features are extracted from the 40th layer of ResNet-50 and these features are then sent through the first layer of the LSTMs which have 10 time steps. The number of LSTMs for the first layer is determined by considering the maximum number of persons (with bounding boxes) in each dataset. If the maximum number of available bounding boxes for a dataset is N, then the first layer of LSTMs is composed of (N+1) LSTMs i.e. one LSTM for each person plus one for the scene level features. In cases where there are fewer than N person we create a dummy sequence with default values. We select $N = 12$ for the volleyball dataset and $N = 10$ for the collective activity dataset. The gated fusion mechanism automatically learns to discard dummy sequences when there are less than N people in the scene.

The outputs of these LSTMs are passed through the gated fusion unit (GFU) to map the correspondences among person-level and scene-level streams. For all the LSTMs we set the hidden state embedding dimension to be 300 units. For the volleyball dataset the dimensionality of the FC(k) layer is set to 8 as there are 8 group activities in the dataset, and for the collective activity dataset we set this to 5. The hyper parameter, $\lambda_c = 2.5$, is chosen experimentally.

In both datasets, the annotations are given in a consistent order. In the volleyball dataset the annotations are ordered based on player role (i.e. spiker, blocker); and in the collective dataset, persons in the frame are annotated from left to right in the scene. We maintain this order of the inputs allowing the GFU to understand the contribution of each person in the scene and learn how the individual actions affect the group action.

The training procedure is similar to [22] and alternates between one gradient decent pass for the discriminators and one for the action generators using mini-batch standard gradient decent (32 examples per mini-batch), and uses the Adam optimiser [26] with an initial learning rate of 0.1 for 250 epochs and 0.01 for the next 750 epochs.

For discriminator training, we take (batch_size)/2 generated (fake) action codes and (batch_size)/2 ground truth (real) action codes where the ground truth action codes are manually created. We use Keras [27] and Theano [28] to implement our model.

4.4   Results

Tables 1 and 2 present the evaluations for the proposed *MLS-GAN* along with the state-of-the-art baseline methods for the Collective Activity [3] and Volleyball [6] benchmark datasets respectively.

When observing the results in Table 1, we observe poor performance from the hand-crafted feature based models [23,29] as they are capable of capturing only abstract level concepts [30]. The deep structured model [31] utilising a

CNN based feature extraction scheme improves upon the handcrafted features. However it does not utilise temporal modelling to map the evolution of the actions, which we believe causes the deficiencies in it's performance.

The authors in [6,7] utilise enhanced temporal modelling through LSTMs and achieved improved performance. However we believe the two step training process leads to an information loss. First, they train a person-level LSTM model which generates a probability distribution over the individual action class for each person in the scene. In the next level only these distributions are used for deciding upon the group activities. Neither person-level features, nor the scene structure information such as the locations of the individual persons is utilised.

In contrast, by utilising features from both the person level and scene level, and further improving the learning process through the proposed GAN based learning framework, the proposed MLS-GAN model has been able to outperform the state-of-the-art models in both considered metrics.

Table 1. Comparison of the results on Collective Activity dataset [3] using MCA and MPCA. NA refers to unavailability of that evaluation.

Approach	MCA	MPCA
Latent SVM [23]	79.7	78.4
Deep structured [31]	80.6	NA
Cardinality Kernel [29]	83.4	81.9
2-layer LSTMs [6]	81.5	80.9
CERN [7]	87.2	88.3
MLS-GAN	91.7	91.2

In Fig. 3 we visualise sample frames for 4 sequences from the collective activity dataset which contain the 'Crossing' scene level activity. We highlight each pedestrian within a bounding box which is colour coded based on the individual activity performed where yellow denotes 'Crossing', green denotes 'Waiting' and blue denotes 'Walking' activity classes. Note that the group activity label is assigned by considering the action that is performed by the majority of people in the sequence. These sequences clearly illustrate the challenges with the dataset. For the same scene level activity we observe significant view point changes. Furthermore there exists a high degree of visual similarity between the action transition frames and the action frames themselves. For example in 3rd column we observe such instances where the pedestrians transition from the 'Crossing' to 'Walking' classes. However, the proposed architecture has been able to overcome these challenges and generate accurate predictions.

Comparing Tables 1 with 2, we observe a similar performance for [6,7] with the volleyball dataset due to the deficiencies in the two level modelling structure. In [8] and [32] the methods achieved improvements over [6,7] by pooling the hidden feature representation when predicting the group activities. However, these methods still utilise hand engineered loss functions for training the

Fig. 3. Sample frames from 4 example sequences (in columns) from the collective activity dataset with the 'Crossing scene level activity'. The colour of the bounding box indicates the activity class of each individual where yellow denotes 'Crossing', green denotes 'Waiting' and blue denotes 'Walking'. The sequences illustrate the challenges due to view point changes and visual similarity between the transition frames and the action frames (i.e. 3rd column. transitions from 'Crossing' to 'Walking'). (Color figure online)

model. Our proposed GAN based model is capable of learning a mapping to an intermediate representation (i.e. action codes) which is easily distinguishable for the activity classifier. The automatic loss function learning process embedded within the GAN objective synthesises this artificial mapping. Hence we are able to outperform the state-of-the-art methods in all considered metrics.

With the results presented in Table 2 we observe a clear improvement in performance over the baseline methods when considering players as 2 groups rather than 1 group. The 2 group representation first segments the players into the respective 2 teams using the ground truth annotations and then pools out the features from the 2 groups separately. Then these team level features are merged together for the group activity recognition. In contrast, the 1 group representation considers all players at once for feature extraction, rather than considering the two state approach. However this segmentation process is an additional overhead when these annotations are not readily available. In contrast the proposed MLS-GAN method receives all the player features together and automatically learns the contribution of each player for the group activity, outperforming both the 1 group and 2 group methods. We argue this is a result of the enhanced structure with the gated fusion units for the feature fusion process. Instead of learning a single static kernel for pooling out features from each player in the team, we attend over all the feature streams from both the player and scene levels, at that particular time step. This generates a system which efficiently varies the level of attention to each feature stream depending on the scene context.

342   H. Gammulle et al.

Figure 4 visualises qualitative results from the proposed MLS-GAN model for the Volleyball dataset. Irrespective of the level of clutter and camera motion, the proposed model correctly recognises the group activity.

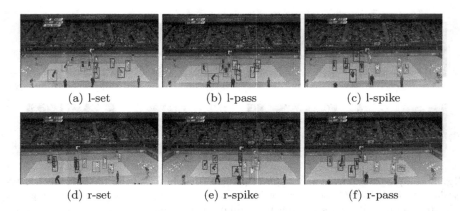

(a) l-set   (b) l-pass   (c) l-spike

(d) r-set   (e) r-spike   (f) r-pass

Fig. 4. Visualisations of the predicted group activities for the Volleyball dataset using the proposed MLS-GAN model.

Table 2. Comparisons with the state-of-the-art for Volleyball Dataset [6]. The first block of results (1 group) are for the methods considering all the players as a one group and the second block is for dividing players into two groups (i.e. each team) first and extracting features from them separately. NA refers to unavailability of results.

Approach	MCA	MPCA
2-layer LSTMs [6] (1 group)	70.3	65.9
CERN [7] (1 group)	73.5	72.2
SRNN [8] (1 group)	73.39	NA
2-layer LSTMs [6] (2 group)	81.9	82.9
CERN [7] (2 group)	83.3	83.6
SRNN [8] (2 group)	83.47	NA
Social Scene [32] (2 group)	89.90	NA
MLS-GAN	**93.0**	**92.4**

4.5   Ablation Experiments

We further experiment with the collective activity dataset by conducting an ablation experiment using a series of a models constructed by removing certain components from the proposed *MLS-GAN* model. Details of the ablation models are as follows:

a *G-GFU*: We use only the generator from *MLS-GAN* and trained it to predict group activity classes by adding a final softmax layer. This model learns through supervised learning using categorical cross-entropy loss. Further, we removed the Gated Fusion Unit (GFU) defined in Eqs. 7 to 10. Therefore this model simply concatenates the outputs from each stream.

b *G*: The generator model plus the GFU trained in a fully supervised model as per the *G-GFU* model above.

c *cGAN-(GFU and $\hat{z}$)*: a conditional GAN architecture where the generator model utilises only the person-level features (no scene-level features), and does not utilise the GFU mechanism for feature fusion. However the discriminator model D still receives the scene level image and the generated action code as the inputs.

d *cGAN-GFU*: a conditional GAN architecture which is similar to the proposed *MLS-GAN model*, however does not utilise the GFU mechanism for feature fusion.

e *MLS-GAN-$\hat{z}$*: *MLS-GAN* architecture where the generator utilises only the person-level features for action code generation. The discriminator model is as in *cGAN-(GFU and $\hat{z}$)*. As per *cGAN-(GFU and $\hat{z}$)*, the discriminator still receives the scene level image.

Table 3. Ablation experiment results on collective activity dataset [3].

Approach	MCA	MPCA
G-GFU	58.9	58.7
G	61.3	60.5
cGAN-(GFU and $\hat{z}$)	88.4	87.7
cGAN-GFU	89.5	88.3
MLS-GAN-$\hat{z}$	91.2	90.8
MLS-GAN	**91.7**	**91.2**

When analysing the results presented in Table 3 we observe significantly lower accuracies for methods *G-GFU* and *G*. Even though a slight improvement of performance is observed with the introduction of the GFU fusion strategy, still we observe a significant reduction in performance. We believe this is due to the deficiencies with the supervised learning process where we directly map the dense visual features to a sparse categorical vector. However, with the other variants and the proposed approach we learn an objective which maps the input to an intermediate representation (i.e. action codes) which is easily distinguishable by the classifier. The merit of the intermediate representation is shown by the performance gap between *G* and the *cGAN-(GFU and $\hat{z}$)*, which we further enhance in *cGAN-GFU* by including scene information alongside the features extracted for the individual agents. This allows the GAN to understand the spatial arrangements of the actors when determining the group activity. Comparing *cGAN-GFU* and *MLS-GAN- ẑ*, we can also see the value of the GFU which is able to better combine data from the individual agents. Finally by utilising both person-level and scene-level features and combining those through proposed GFUs the proposed *MLS-GAN* model attains better recognition results.

We would like to further compare the performance of non-GAN based models *G-GFU* and *G* with the results for the deep architectures in Table 1. Methods such as the 2-layer LSTMs [6] and CERN [7] have been able to attain improved performance compared to *G-GFU* and *G*, however with the added expense of the need for hand annotated individual actions in the database. In contrast, with the improved GAN learning procedure the same architectures (i.e. *cGAN-(GFU and ẑ)*, *cGAN-GFU* and *MLS-GAN- ẑ*) have been able to achieve much better performance without using those individual level annotations.

In order to further demonstrate the discriminative power of the generated action codes we directly classified the action codes generated by *cGAN-(GFU and ẑ)* model. We added a softmax layer to the generated model and tried directly classifying the action codes. We trained only this added layer by freezing the rest of the network weights. We obtained 90.7 MPCA for the collective activity dataset. Comparing this with the ablation model *G* in Table 3 (the generator without the GAN objective, trained using only the classification objective), the reported MPCA value is 60.5. Hence it is clear that the additional GAN objective makes a substantial contribution.

4.6   Time Efficiency

We tested the computational requirements of the MLS-GAN method using the test set of the Volleyball dataset [6] where the total number of persons, N, is set to 12 and each sequence contains 10 time steps. Model generates 100 predictions in 20.4 s using a single core of an Intel E5-2680 2.50 GHz CPU.

5   Conclusions

In this paper we propose a Multi-Level Sequential Generative Adversarial Network (*MLS-GAN*) which is composed of LSTM networks for capturing separate individual actions followed by a gated fusion unit to perform feature integration, considering long-term feature dependancies. We allow the network to learn both person-level and scene-level features to avoid information loss on related objects, backgrounds, and the locations of the individuals within the scene. With the inherited ability to learn both features and the loss function automatically, we employ a semi supervised GAN architecture to learn an intermediate representation of the scene and person-level features of the given scene, rendering an easily distinguishable vector representation, an action code, to represent the group activity. Our evaluations on two diverse datasets, Volleyball and Collective Activity datasets, demonstrates the augmented learning capacity and the flexibility of the proposed *MLS-GAN* approach. Furthermore, with the extensive evaluations it is evident that the combination of scene-level features with person-level features is able to enhance performance by a considerable margin.

References

1. Arevalo, J., Solorio, T., Montes-y Gómez, M., González, F.A.: Gated multimodal units for information fusion. In: 5th International Conference on Learning Representations 2017 Workshop (2017)
2. Amer, M.R., Lei, P., Todorovic, S.: HiRF: hierarchical random field for collective activity recognition in videos. In: Fleet, D., Pajdla, T., Schiele, B., Tuytelaars, T. (eds.) ECCV 2014. LNCS, vol. 8694, pp. 572–585. Springer, Cham (2014). https://doi.org/10.1007/978-3-319-10599-4_37
3. Choi, W., Shahid, K., Savarese, S.: What are they doing? Collective activity classification using spatio-temporal relationship among people. In: 2009 IEEE 12th International Conference on Computer Vision Workshops (ICCV Workshops), pp. 1282–1289. IEEE (2009)
4. Lan, T., Sigal, L., Mori, G.: Social roles in hierarchical models for human activity recognition. In: 2012 IEEE Conference on Computer Vision and Pattern Recognition (CVPR), pp. 1354–1361. IEEE (2012)
5. Ramanathan, V., Yao, B., Fei-Fei, L.: Social role discovery in human events. In: 2013 IEEE Conference on Computer Vision and Pattern Recognition (CVPR), pp. 2475–2482. IEEE (2013)
6. Ibrahim, M.S., Muralidharan, S., Deng, Z., Vahdat, A., Mori, G.: A hierarchical deep temporal model for group activity recognition. In: 2016 IEEE Conference on Computer Vision and Pattern Recognition (CVPR), pp. 1971–1980. IEEE (2016)
7. Shu, T., Todorovic, S., Zhu, S.C.: CERN: confidence-energy recurrent network for group activity recognition. In: Proceedings of CVPR, Honolulu, Hawaii (2017)
8. Biswas, S., Gall, J.: Structural recurrent neural network (SRNN) for group activity analysis. In: IEEE Winter Conference on Applications of Computer Vision (WACV) (2018)
9. Kim, P.S., Lee, D.G., Lee, S.W.: Discriminative context learning with gated recurrent unit for group activity recognition. Pattern Recogn. **76**, 149–161 (2018)
10. Tora, M.R., Chen, J., Little, J.J.: Classification of puck possession events in ice hockey. In: 2017 IEEE Conference on Computer Vision and Pattern Recognition Workshops (CVPRW), pp. 147–154. IEEE (2017)
11. Pathak, D., Krahenbuhl, P., Donahue, J., Darrell, T., Efros, A.A.: Context encoders: feature learning by inpainting. In: Proceedings of the IEEE Conference on Computer Vision and Pattern Recognition, pp. 2536–2544 (2016)
12. Yoo, D., Kim, N., Park, S., Paek, A.S., Kweon, I.S.: Pixel-level domain transfer. In: Leibe, B., Matas, J., Sebe, N., Welling, M. (eds.) ECCV 2016. LNCS, vol. 9912, pp. 517–532. Springer, Cham (2016). https://doi.org/10.1007/978-3-319-46484-8_31
13. Mirza, M., Osindero, S.: Conditional generative adversarial nets. arXiv preprint arXiv:1411.1784 (2014)
14. Fernando, T., Denman, S., Sridharan, S., Fookes, C.: Tracking by prediction: a deep generative model for mutli-person localisation and tracking. In: 2018 IEEE Winter Conference on Applications of Computer Vision (WACV), pp. 1122–1132. IEEE (2018)
15. Fernando, T., Denman, S., Sridharan, S., Fookes, C.: Task specific visual saliency prediction with memory augmented conditional generative adversarial networks. In: 2018 IEEE Winter Conference on Applications of Computer Vision (WACV), pp. 1539–1548. IEEE (2018)
16. Denton, E., Gross, S., Fergus, R.: Semi-supervised learning with context-conditional generative adversarial networks. arXiv preprint arXiv:1611.06430 (2016)

17. Ahsan, U., Sun, C., Essa, I.: Discrimnet: Semi-supervised action recognition from videos using generative adversarial networks. arXiv preprint arXiv:1801.07230 (2018)
18. Li, X., et al.: Region-based activity recognition using conditional GAN. In: Proceedings of the 2017 ACM on Multimedia Conference, pp. 1059–1067. ACM (2017)
19. Li, Y., Song, J., Ermon, S.: InfoGAIL: interpretable imitation learning from visual demonstrations. In: Advances in Neural Information Processing Systems, pp. 3815–3825 (2017)
20. Bora, A., Jalal, A., Price, E., Dimakis, A.G.: Compressed sensing using generative models. In: International Conference on Machine Learning (ICML) (2018)
21. Goodfellow, I., et al.: Generative adversarial nets. In: Advances in Neural Information Processing Systems, pp. 2672–2680 (2014)
22. Isola, P., Zhu, J.Y., Zhou, T., Efros, A.A.: Image-to-image translation with conditional adversarial networks. In: The IEEE Conference on Computer Vision and Pattern Recognition (CVPR) (2017)
23. Lan, T., Wang, Y., Yang, W., Robinovitch, S.N., Mori, G.: Discriminative latent models for recognizing contextual group activities. IEEE Trans. Pattern Anal. Mach. Intell. **34**, 1549–1562 (2012)
24. He, K., Zhang, X., Ren, S., Sun, J.: Deep residual learning for image recognition. In: Proceedings of the IEEE Conference on Computer Vision and Pattern Recognition, pp. 770–778 (2016)
25. Russakovsky, O., et al.: Imagenet large scale visual recognition challenge. Int. J. Comput. Vis. **115**, 211–252 (2015)
26. Kingma, D., Ba, J.: Adam: a method for stochastic optimization. In: International Conference on Learning Representations (ICLR) (2015)
27. Chollet, F., et al.: Keras (2015). https://keras.io
28. Al-Rfou, R., et al.: Theano: a python framework for fast computation of mathematical expressions. arXiv preprint arXiv:1605.02688, vol. 472, p. 473 (2016)
29. Hajimirsadeghi, H., Yan, W., Vahdat, A., Mori, G.: Visual recognition by counting instances: a multi-instance cardinality potential kernel. In: IEEE Computer Vision and Pattern Recognition (CVPR) (2015)
30. Gammulle, H., Denman, S., Sridharan, S., Fookes, C.: Two stream LSTM: a deep fusion framework for human action recognition. In: 2017 IEEE Winter Conference on Applications of Computer Vision (WACV), pp. 177–186. IEEE (2017)
31. Deng, Z., et al.: Deep structured models for group activity recognition. In: British Machine Vision Conference (BMVC) (2015)
32. Bagautdinov, T., Alahi, A., Fleuret, F., Fua, P., Savarese, S.: Social scene understanding: end-to-end multi-person action localization and collective activity recognition. In: Conference on Computer Vision and Pattern Recognition, vol. 2 (2017)

Spatio-Temporal Fusion Networks
for Action Recognition

Sangwoo Cho[✉] and Hassan Foroosh[✉]

University of Central Florida, Orlando, FL 32816, USA
swcho@knights.ucf.edu, foroosh@cs.ucf.edu

Abstract. The video based CNN works have focused on effective ways to fuse appearance and motion networks, but they typically lack utilizing temporal information over video frames. In this work, we present a novel spatio-temporal fusion network (STFN) that integrates temporal dynamics of appearance and motion information from entire videos. The captured temporal dynamic information is then aggregated for a better video level representation and learned via end-to-end training. The spatio-temporal fusion network consists of two set of Residual Inception blocks that extract temporal dynamics and a fusion connection for appearance and motion features. The benefits of STFN are: (a) it captures local and global temporal dynamics of complementary data to learn video-wide information; and (b) it is applicable to any network for video classification to boost performance. We explore a variety of design choices for STFN and verify how the network performance is varied with the ablation studies. We perform experiments on two challenging human activity datasets, UCF101 and HMDB51, and achieve the state-of-the-art results with the best network.

Keywords: Action recognition · Spatio-temporal fusion ·
Temporal dynamics

1 Introduction

Video-based action recognition is an active research topic due to its important practical applications in many areas, such as video surveillance, behavior analysis, and human-computer interaction. Unlike a single image that contains only spatial information, a video provides additional motion information as an important cue for recognition. Although a video provides more information, it is non-trivial to extract the information due to a number of difficulties such as viewpoint changes, camera motions, and scale variations, to name a few. It is thus crucial to design an effective and generalized representation of a video.

This work was supported in part by the National Science Foundation under grant IIS-1212948.

© Springer Nature Switzerland AG 2019
C. V. Jawahar et al. (Eds.): ACCV 2018, LNCS 11361, pp. 347–364, 2019.
https://doi.org/10.1007/978-3-030-20887-5_22

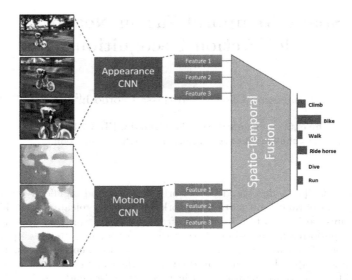

Fig. 1. An illustration of spatio-temporal fusion network for action recognition. Given multiple segments of a video, the network extracts temporal dynamics of appearance and motion cues and fuses them to build a spatio-temporal video representation via end-to-end learning. The appearance and motion ConvNets share the same weights and are employed to extract appearance and motion features, respectively.

Convolutional Neural Networks (ConvNets) [19] have been playing a key role in solving hard problems in various areas of computer vision, e.g. image classification [14,19,50] and human face recognition [30]. ConvNets also have been employed to solve the problem of action recognition [20,31,35,37,47] in recent literature. The data-driven supervised learning enables to achieve discriminating power and proper representation of a video from raw data. However, ConvNets for action recognition have not shown a significant performance gain over the methods utilizing hand-crafted features [29,42]. We speculate that the main reason for the lack of big impact is that ConvNets employed in action recognition do not take full advantage of temporal dynamics among frames.

A two-stream [31] ConvNet is one of the popular approaches used in action recognition utilizing the appearance and motion data. However, the two data streams are typically trained with separate ConvNets and only combined by averaging the prediction scores. This approach is not helpful when the two information are needed simultaneously, e.g. motions of brushing teeth and brushing hair are similar so appearance information is needed to discriminate them. Due to the lack of spatio-temporal feature for action recognition, several methods [3,8,41] attempted to incorporate both sources of information. They typically take frame-level features and integrate them using an RNN [14] network and temporal feature pooling [9,27,45] in order to incorporate temporal information. However, they still lack of extracting a representation that captures video-wide temporal information.

In this work, we aim to investigate a proper model to fuse the appearance and motion dynamics to learn video level spatio-temporal representation (Fig. 1). STFN aggregates different size of local temporal dynamics in multiple video segments and combines them to obtain video level spatio-temporal representation. STFN is mainly motivated by two components: a residual-inception module [35], and 1D convolution layers [24]. The former is suitable for extracting latent features and the latter works well in extracting temporal dynamics. We modify the original residual-inception module [35] and design a new block for spatio-temporal fusion. The new residual-inception block processes local and global temporal dynamics for each data. Given the extracted dynamic information, appearance and motion dynamics are merged with fusion operations for spatio-temporal features. This method overcomes the previous drawback, i.e. the lack of utilizing video-wide temporal information, and learning spatio-temporal features. We investigate a variety of different fusion methods and execute ablation studies to find the best networks.

Our key contributions can thus be summarized as follows: (i) A convolution block, effective to extract temporal representations, is proposed. (ii) A novel ConvNet is introduced to learn spatio-temporal features effectively by fusing two different features properly. (iii) STFN achieves state-of-the-art performance on the two challenging datasets, UCF101 (95.4%) and HMDB51 (72.1%). (iv) The entire system is easy to implement and is trained by an end-to-end learning of deep networks.

The rest of this paper is organized as follows. In Sect. 2, we discuss related works. We describe our proposed method in Sect. 3. Experimental results and analysis are presented in Sect. 4. Finally, we conclude our work in Sect. 5.

2   Related Work

Several works using ConvNets to acquire temporal information for action recognition have been studied. In [44], hand crafted features are used in the pooling layer of ConvNet to take advantage of both merits of hand-designed and deep learned features. Temporal information from optical flow is explicitly learned with ConvNets in [31] and the result is fused with the effect of the trained spatial (appearance) ConvNet. [9] connects several convolution layers of two stream ConvNets to capture spatio-temporal information. Although the aforementioned approaches capture temporal information in small time windows, they fail to capture long-range temporal sequencing information that contains long-range ordered information.

Several works modeling a video-level representation or modeling long temporal information with ConvNets have also been investigated. [10] proposes a method that employs a ranking function to generate a video-wide representation that captures global temporal information. In [38], a HMM model is used to capture the appearance transitions and a max-margin method is employed for temporal information modeling in a video. [12,26,38] utilize LSTM [14] unit in their ConvNets and attempt to capture long-range temporal information.

However, the most natural way of representing a video as long-range ordered temporal information is not fully exploited.

Recently several researches [18,24] have used frame level representations for predicting actions with temporal ConvNets. The rational behind these methods is to extract the temporal dynamics more directly by utilizing 1D convolution over time. This approach is widely used in a sentence classification [16,17,52] problem in Natural Language Processing literature. Each word is encoded to vectors and 1D convolution over a sequence of words extracts semantic information between words. For videos, two stream [31] ConvNets are typically employed to train appearance and motion features separately. Once the two streams are trained, sampled RGB or optical flow video frames are fed to each network to extract appearance and motion features respectively. This is the standard feature extraction method and each frame can be represented in a vector form. The biggest advantage of the feature representation is that the temporal information distributed over entire videos can be effectively extracted by using 1D convolutions. Our work is based on the 1D convolution layers to obtain temporal dynamics of appearance and motion cues.

Many ConvNets [13,20,32,36] for image recognition are utilized for action recognition as well. Among them, a concept of the inception is useful to our encoded data to extract more informative features. The encoded features are convoluted over time with different kernel sizes and concatenated. This process extracts local and global temporal information similar to extracting N-gram semantic information in NLP. [35] introduces an effective residual inception module, which basically has another shortcut connection to the inception module. We employ the residual inception module with 1D convolution layers as it is suitable for extracting temporal dynamics.

The critical drawback of the two-stream [31] ConvNets is the two features cannot be integrated in feature level. In order to solve this problem, different fusion methods are introduced. In [41] they try to extract spatio-temporal features directly by applying 3D convolution to a stack of input frames. [7,8] connect learned two stream ConvNets to integrate the two stream signals generating the spatio-temporal features. [3] encodes local deep features as a super vector efficiently so that spatio-temporal information can be handled with spatio-temporal ConvNets. We utilize different basic fusion operations, average, maximum, and multiply, as investigated in [7,8,45]. Since we combine the appearance and motion features, we naturally take advantage of two stream ConvNet architecture and connect them with different fusion methods. This work provides a systematic investigation of fusion methods and ablation studies to choose the best fusion methods for better performance.

3   Approach

A video contains many redundant temporal information between consecutive frames. Instead of densely sampled feature points [43,46] samples frames in different video segments, while [31] deals with multiple consecutive frames. These techniques train ConvNets for different modalities, appearance and motion, and use late fusion to combine them. However, two issues are raised from these methods: (1) multiple consecutive frames only cover local temporal dynamics not global temporal dynamics over videos, and (2) the prediction score fusion only captures dynamic of each appearance and motion cue separately not the spatio-temporal dynamics. In this section, we propose a spatio-temporal fusion network (STFN) to extract temporal dynamic information over an entire video and combine appearance and motion dynamics, using end-to-end ConvNets training, as shown in Fig. 2. The network has the following properties: (1) convolutions are computed over time so that the temporal dynamic information is extracted; (2) each convolution block extracts local and global temporal information with different feature map sizes; and (3) the extracted appearance and motion dynamic features are integrated through an injection from one to the other or with bi-direction way. More details about STFN are described in Sect. 3.1.

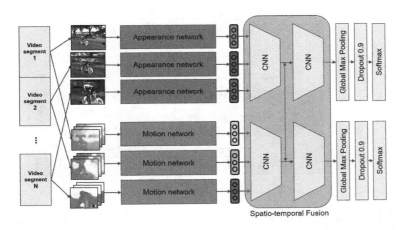

Fig. 2. The proposed spatio-temporal fusion network. The number of segments is an arbitrary number. We use three segments in the figure for illustration purpose.

3.1 Spatio-Temporal Fusion Networks

We consider the output feature maps of CNNs for N segments from a video V. Each feature map $\{F_1, F_2, \cdots, F_N\}$ is a vector of size $F \in \mathbb{R}^d$, where d is the output feature map dimension. The feature maps can be retrieved from different networks trained with different modalities such as appearance and motion. F^a, F^m, where $F^x \in \mathbb{R}^{N \times d}$, are the feature maps from appearance and motion networks, respectively. STFN is applied to the sequence of feature maps, F^a and F^m, to extract temporal dynamics of each feature map and fuse them as follows:

$$
\begin{aligned}
\mathrm{STFN}(F^a, F^m) = &\ \mathcal{H}(\mathcal{F}(\mathcal{G}(\mathcal{F}(F^a; \mathrm{W_a}), \mathcal{F}(F^m; \mathrm{W_m})); \mathrm{W_{fa}})) \\
&+ \mathcal{H}(\mathcal{F}(\mathcal{G}(\mathcal{F}(F^a; \mathrm{W_a}), \mathcal{F}(F^m; \mathrm{W_m})); \mathrm{W_{fm}}))
\end{aligned}
\tag{1}
$$

$\mathcal{F}(F^x; W_x)$, where x $\in \{a, m, fa, fm\}$ meaning appearance, motion, fused appearance, fused motion sequences, is a ConvNet function with parameters W_x which produces sequences of same input sizes for the given sequences. More details about the ConvNet are given in Sect. 3.2. The fusion aggregation function $\mathcal{G}$ combines the output sequences of appearance and motion dynamic information. $\mathcal{G}$ and the follow-up ConvNets, $\mathcal{F}(F^x; W_{fa})$, can be omitted depending on the design choice of STFN. More details are provided in the next subsection. From the learned sequences, the prediction function $\mathcal{H}$ predicts the probability of each activity class. Softmax function, which is widely used for multi-class classification, is chosen for $\mathcal{H}$.

The overall network is learned in an end-to-end scheme like TSN [46]. The sequences of feature maps are $X = F^a, F^m$ and the outputs of the $\mathcal{F}$ function are denoted by y. Also, let $\mathcal{L}$ be the loss function. The gradient of the loss function with respect to X, $\frac{d\mathcal{L}}{dX}$, during the training process is defined as:

$$\frac{d\mathcal{L}}{dF_k^x} = \mathcal{F}(F_{k'}^x; W_x) \frac{d\mathcal{L}}{dX} \qquad (2)$$

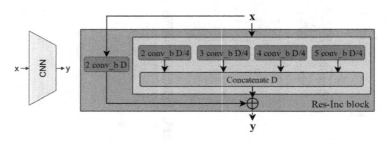

Fig. 3. A Residual Inception block. The res-inc block in the right figure shows the components of the CNN in the left figure. The number in each module inside of the Res-Inc block depicts convolution kernel size. conv_b consists of the 1D convolution, batch normalization, and relu activation layers. D represents the input vector dimension, d.

where $k \in N$ and $k' = \{1, 2, \cdots, k-1, k+1, \cdots, N\}$. In the end-to-end training, the parameters for the N segments are learned using stochastic gradient descent (SGD). The parameters are learned from the entire video with segmented temporal inputs.

3.2 STFN Components

In this subsection, we describe the ConvNets, $\mathcal{F}$, and the fusion aggregation function, $\mathcal{G}$, in detail. We also discuss different STFN architectures to find the most suitable model.

Residual Inception Block. A sequence of frame representations, F^a, F^m, inherently contains temporal dynamics between features. The consecutive features are convoluted over time with different kernel sizes to extract local temporal information. This operation is conceptually similar to an n-gram of a sentence that contains local semantic information among n words. The convoluted features are then concatenated to formulate a hierarchical feature from each input. Motivated by an inception module [20,35] that convolves an input signal with different filters, we design an inception block with different kernel sizes as shown in Fig. 3. The input signal F^x is convoluted across time using 1D convolution with four different sizes of kernels, 2,3,4,5, whose filter size is a quarter of the input dimension, d. The 1D convolution retains the same temporal length as the input. We did preliminary experiments to find out the best combination of the kernel sizes and 2,3,4,5 shows the best performance. We designed the filter size of each convolution to be a quarter of input dimension, making the concatenated feature have the same dimension as the input with same weight. We also used convolution layers with kernel size of 1 [20,35] before the conv_b block to reduce the input dimension. However, they decrease the performance since it perturbs the input signal that contains temporal dynamics, so we decided not to include them.

The concatenated multi features and the input signal F are added for residual learning [12]. We chose a convolution kernel size of 2 for the skip connection to capture the smallest local temporal information. Formally, the Residual Inception (Res-Inc) block in this paper is defined as:

$$y = \mathcal{C}(\mathcal{R}(F^x, \{W_i\})) + \mathcal{R}(F^x, \{W_j\}) \tag{3}$$

where $\mathcal{R}$ is the convolution function with weights $W_i, i \in \{2,3,4,5\}$ for the residual connection or $W_j, j \in \{2\}$ for the skip connection, and the function $\mathcal{C}(\cdot)$ represents a concatenation operation. In Fig. 3, x is identical to F^x in Eq. 3. The convolution block, conv_b, is composed of Batch normalization [15] and ReLU [20], while the convolution block in skip connection lacks the ReLU activation layer. The output signal is further activated with ReLU before it is aggregated with the other signal. The output sequence of the Res-Inc block contains more discriminative temporal dynamic information than the input sequence. Since the Res-Inc block outputs signals of same dimension of input signals, a series of Res-Inc block can be easily setup.

Spatio-Temporal Fusion. Despite the successful performance with the two-stream approach, a clear drawback is that a spatio-temporal information is not achievable with separate training of the appearance and motion data. The appearance and motion information are complementary to each other in order to discern an action of similar motion or appearance patterns e.g. brushing teeth and hammering. In order to overcome this deficiency, a number of researches have been looking into fusing two-stream networks [7–9] directly and learning spatio-temporal features [39,41]. Although, their results show improved performance, their spatio-temporal features are limited to local snippets of an entire video sequence. In contrast, STFN takes advantage of extracted temporal dynamic features that capture long term temporal information over entire video to fuse them.

We investigate three different fusion operations $\mathcal{G}$ with the output sequences of two Res-Inc blocks $\{P_1^x, P_2^x, \cdots, P_N^x\}$, where $P_n^x \in \mathbb{R}^d$, and $x \in \{a, m\}$ represent either appearance or motion features.

Element-wise Average

$$P_n' = \frac{(P_n^a + P_n^m)}{2} \tag{4}$$

where P' is the aggregated sequence and $n \in \{1, 2, \cdots, N\}$. This operation leverages all information and uses the mean activation for the fused signal. This operation may get affected by noisy input signals but since we deal with highly informative features, it is a good choice for our architecture.

Element-wise Multiplication

$$P_n' = P_n^a \times P_n^m \tag{5}$$

The intuition behind this operation is to amplify a signal when both signals are strong, i.e. similar to attention mechanism. However, the noisy strong signal may affect heavily the fused signal leading to performance decrease.

Element-wise Maximum

$$P_n' = max(P_n^a, P_n^m) \tag{6}$$

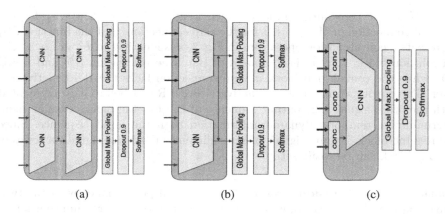

(a)   (b)   (c)

Fig. 4. Different designs of spatio-temporal fusion architecture. (a) shows our proposed architecture; (b) lacks the follow-up Res-Inc blocks after fusion; and (c) concatenation of the appearance and motion sequences in feature level before extracting temporal dynamics. The blue and red arrows represent the appearance and motion sequence inputs, respectively. (Color figure online)

The idea of max pooling is to seek the most discriminative signal among inputs. It selects either appearance or motion cue for each element of input signals. This operation may confuse the following Res-Inc block since the aggregated vectors are mixed with the appearance and motion signals.

We compare the performance of each operation in the ablation studies.

Architecture Variations of STFN. We propose different design architectures of STFN and investigate them in detail. Figure 4b is a variation of Fig. 4a where we want to learn how the additional Res-Inc blocks affect to the results. The Res-Inc blocks after fusion extract temporal dynamics of spatio-temporal features leading to better performance. In Fig. 4c, fusion is executed in feature-level by simply concatenating appearance and motion signals. This fused signal is fed to the Res-Inc block to extract temporal dynamic information.

Fusion Direction. As shown in Fig. 5, aggregating two signals can be three possible ways: appearance to motion, motion to appearance, and bi-directional fusion. The fused signals are fed to the next Res-Inc blocks and affect to the residual and skip connection along the forward an backward propagations when training. Considering the three fusion operations, only multiplication operation results in byproduct signal from partial derivatives of the fused signals when signals are back-propagated. This means the fusion with multiplication operation makes the input signal change rapidly than other operations. Thus, it is not easy to learn proper spatio-temporal features especially when there is significant gap between the discriminative abilities of appearance and motion features.

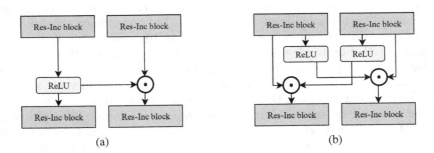

(a)   (b)

Fig. 5. Two types of fusion methods: asymmetric and symmetric fusion. (a) shows asymmetric fusion method and two fusions are possible with this method: appearance to motion features and motion to appearance features. (b) shows symmetric fusion where each fused signal is further used in following layers. Two signals are merged with the previous described fusion operations. Note that this figure only illustrates the fusion connections between two Res-Inc blocks and the rest layers are omitted.

4   Experiments

In this section, we first discuss the datasets and implementation details. Then we evaluate each design choice for STFN. Finally, we compare our best performance with the state-of-the-art methods.

4.1   Datasets

We tested our method on two large action datasets, HMDB51 [21], UCF101 [33]. The HMDB51 dataset consists of 51 action classes with 6,766 videos and more than 100 videos in each class. All videos are acquired from movies or youtube and contain various human activities and interactions with human or object. Each action class has 70 videos for training and 30 videos for testing. The UCF101 dataset consists of 101 action categories with 13,320 videos and at least 100 videos are involved in each classes. UCF101 provides large diverse videos with a fixed resolution of 320 × 240 with 5 different types of actions. All videos are gathered from youtube. Both datasets provide evaluation scheme for three training and testing splits and we follow the original evaluation method.

4.2   Implementation Details

Two-stream ConvNets: ResNet-101 [12] and Inception-V3 [37] are employed for the base networks to train appearance and motion networks. Both networks are initialized with the pre-trained weights trained on the ImageNet [4] dataset. To fine-tune the networks, we replace the classification layer with C softmax layer, where C is the number of action classes. The appearance network takes RGB images, while the motion network a stack of 10 dense optical flow frames. The input RGB or optical flow images are resized to make the smaller side as 256. We augment the input image by cropping, resizing, and mirroring in horizontal direction. The width and height of the cropped image are randomly sampled from $\{256, 224, 192, 168\}$, and the input images are cropped from the four corners and the center of the original images. The cropped images are then resized to 224×224 for the network input. This augmentation considers both scale and aspect ratio. We pre-compute the optical flows using the TVL1 method [51] before training to improve the training speed. The optical flow input is stacked with 10 frames making a $224 \times 224 \times 20$ sub-volume for x and y directions. Same data augmentation techniques are employed for the optical flow sub-volume. We use mini-batch stochastic gradient descent (SGD) to learn models with a batch size of 32 and momentum of 0.9. The learning rate is set to 10^{-3} initially and decreased by a factor of 10 when the validation error saturates, for both networks.

STFN: In order to train STFN, we retain only convolutional layers and global pooling layer of each network, similar to [5]. The feature maps for STFN are extracted from the output of the global pooling layer. The output dimension is 2048 for both ResNet-101 and Inception-V3. We apply two step training process.

Table 1. Prediction accuracy(%) on the first split of HMDB51 and UCF101 using different architectures of STFN as shown in Fig. 4.

Design	HMDB51	UCF101
Fig. 4a	**70.4**	**93.5**
Fig. 4b	69.6	93.2
Fig. 4c	69.2	92.0

We first fix the weights of trained appearance and motion networks and train only STFN. Then we train the entire networks with same methods described in Two-stream ConvNets training. For the first training, we initialize the learning rate with 10^{-4} and decrease it until 10^{-7} by a factor of 10 when the validation error saturates. RMSProp [40] optimizer is used for the STFN training. The second training is executed with same setting of the two-stream ConvNets training without fixing all weights. For training and testing, we divide the videos into $N = 5$ segments with same lengths. Note that we use $N = 5$ for all evaluation except for the experiment in Sect. 4.6. A random frame is selected from each N segment and optical flow stacks centered on the selected frames are associated for two input sequences. We apply same augmentations for selected frames and optical flow stacks in an input sequence. When testing, 5 frames are uniformly sampled from each segment making 5 sequences and the final prediction scores are averaged over each output. The experiments are performed with 5 segments, average fusion operation, and bi-directional fusion as default except for each ablation study.

4.3   Evaluation of Different Designs

As we discussed in Sect. 3.2, the performances of three proposed STFN architectures are presented in Table 1. Comparing Figs. 4a and b networks, we verify that the Res-Inc blocks make important role extracting temporal dynamics. We conjecture that the consecutive Res-Inc blocks extract temporal dynamics of fused features and they contain better video-wide discriminative features. Another architecture design, Fig. 4c, is introduced to see how the feature level fusion affects to the performance as opposed to the baseline two-stream networks. We observe the significant performance drop in both datasets and it proves the importance of the fusion scheme. Since the architecture of Fig. 4c shows the best performance, we choose it as our default STFN network.

The result with a single Res-Inc module Fig. (4b) outperforms the baseline late fusion results shown in Table 5 by 8.1% on HMDB51 and 0.2% on UCF101. This shows the effectiveness of the Res-Inc module. With another Res-Inc module and feature fusion, 0.8% and 0.3% additional gains are obtained on HMDB51 and UCF101, respectively. Note that from a preliminary experiment by increasing the number of consecutive Res-Inc blocks from two to four, we observe performance drops: 3.8%, 6.5% on HMDB51, 4.1%, 6.9% on UCF101. The signals undergo the

Table 2. Prediction accuracy(%) on the first split of HMDB51 and UCF101 using different fusion operations.

Fusion operation	HMDB51	UCF101
Average	**70.4**	**93.5**
Maximum	69.5	92.9
Multiplication	68.3	92.6

Res-Inc block contain temporally convoluted information with different kernel sizes (from residual connection). More Res-Inc blocks extract higher level temporal information, but we conjecture that signals experienced more than two levels confuse the original temporal orders, introducing noise.

4.4   Evaluation of Fusion Operations

This section presents the performances based on different fusion operations: Element-wise average, maximum, and multiplication. As shown in Table 2, the average operation outperforms other methods. It is interesting to see the performance gap between the average and the multiplication operations, 0.9% and 2.1% for HMDB51 and UCF101 respectively. We speculate the reason is due to the performance discrepancy of two networks as shown in Table 5. With multiplication, the inferior feature (appearance cue on HMDB51) could harm the fused signal. Also, it is better to take into account all data by averaging than picking the strongest signals since STFN deals with highly pre-processed signals. From the results, we take the average operation as our default choice. Note that we tried weighted average based on the normalized performances of baseline networks and automatic scaling by applying 1 x 1 2D conv to each signal before fusing. However simple average results in the best performance.

4.5   Evaluation of Fusion Directions

In Table 3, we compare the performance variation with different fusion directions. Note that A←M is a simply reflected network of A→M and we use 5 segments for all experiments. For the asymmetric fusion methods, A←M connection outperforms the other way consistently on both datasets. This effect is due to the fact that the motion stream overfits quickly with the A→M fusion and no further spatio-temporal learning occurs. This comes from the base performance different between appearance and motion features so that fusion injection to the higher discriminative feature leads to worse performance. The bi-direction fusion outperforms A←M with small margin, 0.1% on HMDB51 and 0.2% on UCF101. This makes sense since two spatio-temporal features are learned simultaneously in two streams, whereas asymmetric fusion learns spatio-temporal in the injected stream and the learned weights are propagated to the other stream only when back propagating from the fused connection. However, we argue that our proposed STFN is robust to the fusion connection based on the small performance differences on both datasets. We choose the bi-direction fusion as our base fusion method.

Table 3. Prediction accuracy(%) on the first split of HMDB51 and UCF101 using different fusion directions. A and M represent the appearance and motion features, respectively. The bottom two methods are asymmetric fusion methods whereas the top one is bi-direction fusion method.

Fusion direction	HMDB51	UCF101
A↔M	**70.4**	**93.5**
A←M	70.3	93.4
A→M	70.1	93.2

Table 4. Prediction accuracy(%) on the first split of HMDB51 and UCF101 using different numbers of segments in videos.

Number of segments	HMDB51	UCF101
3	70.3	93.2
5	70.4	93.5
7	**70.8**	**93.9**
9	70.5	93.6

4.6   Evaluation of a Number of Segments

We evaluate the number of segments according to the default fusion method and architecture. One may assume that more segments result in better performance. However, as we discussed, more redundant temporal dynamics are introduced when increasing the number of segments. The performances based on different number of segments are shown in Table 4. It turns out that 7 segments performs best and 0.4% performance increases are observed on both datasets. The STFN with 9 segments underperforms compared with the one with 7 segments. We verify our hypothesis with this experiments that sparse sampling is necessary to avoid redundant temporal dynamics over entire videos. For the best network, we determine the number of segments as 7.

4.7   Base Performance of Two-Stream Network

We compare the different ConvNet architectures for STFN. ResNet-101 [12] and Inception-V3 [37] networks are employed to train the two-stream networks. As shown in Table 5, the performance with Inception-V3 is better than ResNet-101 on both datasets. The performance gaps of the appearance and motion networks are 3.0%/1.1% on HMDB51 and 1.3%/2.1% on UCF101, respectively.

Table 5. Performance comparison(%) of two-stream networks with ResNet-101 and Inception-V3 on HMDB51 and UCF101 (split1). Inception-V3 shows consistently better prediction accuracies over ResNet-101 on both appearance and motion networks.

Dataset	Network	Appear	Motion	Late Fusion
HMDB51	ResNet-101	48.2	58.1	61.1
	Inception-V3	51.2	59.2	62.7
UCF101	ResNet-101	83.5	86.0	91.8
	Inception-V3	84.8	88.1	92.3

4.8   Comparison with the State-of-the-art

We compare STFN with the current state-ot-the-art methods in Table 6. We report the mean accuracy over three splits of the HMDB51 and UCF101. The first section of Table 6 consists of the hand-crafted features with different encoding methods. The second and third sections describe approaches using ConvNets but the methods in third section utilize additional modalities for the final prediction. STFN with the Inception-V3 achieves the best results: 72.1% on HMDB51 and 95.4% on UCF101. There is 0.9%/1.1% performance increase from STFN with ResNet-101 architecture. STFN with both networks shows the state-of-the-art performance. Comparing with baseline late fusion performance of two-stream networks, performance increases are observed as follows: 9.4%, 10.1% on HMDB51 and 3.1%, 2.5% on UCF101 with Inception-V3 and ResNet-101, respectively.

Our best results outperform TSN [46] by 1.0% on HMDB51 and 0.5% on UCF101 with same number of segments, 7. While TSN predicts scores with consensus operations and averages each score, STFN extracts temporal dynamic information and aggregates signals in feature level leading to better results. The results prove our method produces effective spatio-temporal features. DOVF [23] and TLE [5] show better results than STFN with ResNet-101 but are outperformed by STFN with Inception-V3. TLE [5] only outperforms our method with small margin, 0.2%, on UCF101 but the gap is reversed with additional hand-crafted feature score.

We combine our results with the hand-crafted MIFS[1] [22] features by averaging prediction scores. The performance gain on HMDB51, 3,0%, is larger than on UCF101, 1.6%. The combined performances, 75.1% on HMDB51 and 96.0% on UCF101, outperform all state-of-the-arts and even on par with [2,34] which employ more prediction scores from additional modalities. Note that we observe similar performance boost with iDT [42] but choose MIFS since the prediction scores are available in public.

---

[1] The prediction scores of MIFS are downloaded from HERE.

Table 6. Comparison with state-of-the-art methods on HMDB51 and UCF101. Mean accuracy over three splits. Numbers inside of parenthesis are classification accuracies with hand-crafted features. (i: iDT [42], H: HMG [6], M: MIFS [22])

Methods	HMDB51	UCF101
iDT+FV [43]	57.2	85.9
iDT+HSV [28]	61.1	87.9
Two-stream [31]	59.4	88.0
Transformation [48]	62.0	92.4
KVM [13]	63.3	93.1
Two-Stream Fusion [9]	65.4 (69.2 i)	92.5 (93.5 i)
ST-ResNet [7]	66.4 (70.3 i)	93.4 (94.6 i)
ST-Multiplier [8]	68.9 (72.2 i)	94.2 (94.9 i)
ActionVLAD [11]	66.9 (69.8 i)	92.7 (93.6 i)
ST-Vector [3]	69.5 (73.1 i+H)	93.6 (94.3 i+H)
DOVF [23]	71.7 (75.0 M)	94.9 (95.3 M)
ST-Pyramid [49]	68.9	94.6
I3D [1]	66.4	93.4
CO2FI [25]	69.0 (72.6 i)	94.3 (95.2 i)
TLE [5]	71.1	**95.6**
TSN [46]	71.0	94.9
Four-Stream [2]	72.5 (74.9 i)	95.5 (96.0 i)
OFF [34]	74.2	96.0
STFN (ResNet-101)	71.2 (73.3 M)	94.3 (95.1 M)
STFN (Inception-V3)	**72.1** (**75.1** M)	95.4 (**96.0** M)

5 Conclusion

In this paper, we introduced the spatio-temporal fusion network (STFN), a network suitable for extracting temporal dynamics of features and learning spatio-temporal features by combining them. The spatio-temporal features are learned effectively with STFN via an end-to-end learning method. In the ablation studies, we show the best fusion methods and architecture and investigate the intuition behind each method. STFN enables appearance and motion dynamic features integrate inside of the networks in a highly abstract manner and overcomes the naive fusion strategy of late fusion. STFN is applicable to any sequencial data with two different modalities and effectively fuses them into highly discriminative feature that captures dynamic information over the entire sequence. The best result of STFN achieves the state-of-the-art performance, 75.1% on HMDB51 and 96.0% on UCF101. As future work, we consider scalability of our work with larger dataset and applying more than two modalities.

References

1. Quo Vadis, Action Recognition? A New Model and the Kinetics Dataset (2017)
2. Bilen, H., Fernando, B., Gavves, E., Vedaldi, A.: Action recognition with dynamic image networks. IEEE Trans. Pattern Anal. Mach. Intell. **40**, 2799–2813 (2018)
3. Cosmin Duta, I., Ionescu, B., Aizawa, K., Sebe, N.: Spatio-temporal vector of locally max pooled features for action recognition in videos. In: The IEEE Conference on Computer Vision and Pattern Recognition (CVPR), July 2017
4. Deng, J., Dong, W., Socher, R., Li, L.J., Li, K., Fei-Fei, L.: ImageNet: a large-scale hierarchical image database. In: CVPR (2009)
5. Diba, A., Sharma, V., Gool, L.V.: Deep temporal linear encoding networks. In: 2017 IEEE Conference on Computer Vision and Pattern Recognition, CVPR (2017)
6. Duta, I.C., et al.: Histograms of motion gradients for real-time video classification. In: 2016 14th International Workshop on Content-Based Multimedia Indexing (CBMI) (2016)
7. Feichtenhofer, C., Pinz, A., Wildes, R.: Spatiotemporal residual networks for video action recognition. In: Advances in Neural Information Processing Systems (NIPS) (2016)
8. Feichtenhofer, C., Pinz, A., Wildes, R.P.: Spatiotemporal Multiplier Networks for Video Action Recognition (2017)
9. Feichtenhofer, C., Pinz, A., Zisserman, A.: Convolutional Two-stream Network Fusion for Video Action Recognition (2016)
10. Fernando, B., Gavves, E., José Oramas, M., Ghodrati, A., Tuytelaars, T.: Modeling video evolution for action recognition. In: IEEE Conference on Computer Vision and Pattern Recognition (2015)
11. Girdhar, R., Ramanan, D., Gupta, A., Sivic, J., Russell, B.: ActionVLAD: learning spatio-temporal aggregation for action classification. In: CVPR (2017)
12. He, K., Zhang, X., Ren, S., Sun, J.: Deep residual learning for image recognition. In: IEEE Conference on Computer Vision and Pattern Recognition (2016)
13. He, K., Zhang, X., Ren, S., Sun, J.: Identity mappings in deep residual networks. In: 14th European Conference ECCV (2016)
14. Hochreiter, S., Schmidhuber, J.: Long short-term memory. Neural Comput. **9**, 1735–1780 (1997)
15. Ioffe, S., Szegedy, C.: Batch normalization: accelerating deep network training by reducing internal covariate shift. In: International Conference on Machine Learning, ICML (2015)
16. Johnson, R., Zhang, T.: Effective use of word order for text categorization with convolutional neural networks. In: NAACL HLT 2015 (2015)
17. Kalchbrenner, N., Grefenstette, E., Blunsom, P.: A convolutional neural network for modelling sentences. In: Proceedings of the 52nd Annual Meeting of the Association for Computational Linguistics, ACL (2014)
18. Kim, T.S., Reiter, A.: Interpretable 3D human action analysis with temporal convolutional networks. In: 2017 IEEE Conference on Computer Vision and Pattern Recognition Workshops (CVPRW) (2017)
19. Krizhevsky, A., Sutskever, I., Hinton, G.E.: ImageNet classification with deep convolutional neural networks. In: Advances in Neural Information Processing (2012)
20. Krizhevsky, A., Sutskever, I., Hinton, G.E.: ImageNet classification with deep convolutional neural networks. Commun. ACM **60**, 84–90 (2017)
21. Kuehne, H., Jhuang, H., Garrote, E., Poggio, T., Serre, T.: HMDB: a large video database for human motion recognition. In: ICCV (2011)

22. Lan, Z., Lin, M., Li, X., Hauptmann, A.G., Raj, B.: Beyond Gaussian pyramid: multi-skip feature stacking for action recognition. In: IEEE Conference on Computer Vision and Pattern Recognition, CVPR (2015)
23. Lan, Z., Zhu, Y., Hauptmann, A.G., Newsam, S.D.: Deep local video feature for action recognition. In: 2017 IEEE Conference on Computer Vision and Pattern Recognition Workshops, CVPR Workshops (2017)
24. Lea, C., Flynn, M.D., Vidal, R., Reiter, A., Hager, G.D.: Temporal convolutional networks for action segmentation and detection. In: 2017 IEEE Conference on Computer Vision and Pattern Recognition (CVPR) (2017)
25. Lin, W., et al.: Action recognition with coarse-to-fine deep feature integration and asynchronous fusion. In: Proceedings of the Thirty-Second AAAI Conference on Artificial Intelligence (2018)
26. Ng, J.Y.H., Hausknecht, M., Vijayanarasimhan, S., Vinyals, O., Monga, R., Toderici, G.: Beyond short snippets: deep networks for video classification. In: Computer Vision and Pattern Recognition (2015)
27. Ng, J.Y., Hausknecht, M.J., Vijayanarasimhan, S., Vinyals, O., Monga, R., Toderici, G.: Beyond short snippets: deep networks for video classification. In: IEEE Conference on Computer Vision and Pattern Recognition, CVPR (2015)
28. Peng, X., Wang, L., Wang, X., Qiao, Y.: Bag of visual words and fusion methods for action recognition: comprehensive study and good practice. Comput. Vis. Image Underst. **150**, 109–125 (2016)
29. Peng, X., Zou, C., Qiao, Y., Peng, Q.: Action recognition with stacked fisher vectors. In: Fleet, D., Pajdla, T., Schiele, B., Tuytelaars, T. (eds.) ECCV 2014. LNCS, vol. 8693, pp. 581–595. Springer, Cham (2014). https://doi.org/10.1007/978-3-319-10602-1_38
30. Schroff, F., Kalenichenko, D., Philbin, J.: Facenet: a unified embedding for face recognition and clustering. In: IEEE Conference on Computer Vision and Pattern Recognition, CVPR 2015 (2015)
31. Simonyan, K., Zisserman, A.: Two-stream convolutional networks for action recognition in videos. In: NIPS (2014)
32. Simonyan, K., Zisserman, A.: Very deep convolutional networks for large-scale image recognition. CoRR (2014)
33. Soomro, K., Zamir, A.R., Shah, M.: UCF101: a dataset of 101 human actions classes from videos in the wild. CoRR (2012)
34. Sun, S., Kuang, Z., Ouyang, W., Sheng, L., Zhang, W.: Optical flow guided feature: a fast and robust motion representation for video action recognition. CoRR (2017)
35. Szegedy, C., Ioffe, S., Vanhoucke, V., Alemi, A.A.: Inception-v4, inception-resnet and the impact of residual connections on learning. In: Proceedings of the Thirty-First AAAI Conference on Artificial Intelligence (2017)
36. Szegedy, C., et al.: Going deeper with convolutions. In: IEEE Conference on Computer Vision and Pattern Recognition (2015)
37. Szegedy, C., Vanhoucke, V., Ioffe, S., Shlens, J., Wojna, Z.: Rethinking the inception architecture for computer vision. In: IEEE Conference on Computer Vision and Pattern Recognition, CVPR (2016)
38. Tang, K.D., Li, F., Koller, D.: Learning latent temporal structure for complex event detection. In: 2012 IEEE Conference on Computer Vision and Pattern Recognition (2012)
39. Taylor, G.W., Fergus, R., LeCun, Y., Bregler, C.: Convolutional learning of spatio-temporal features. In: Daniilidis, K., Maragos, P., Paragios, N. (eds.) ECCV 2010. LNCS, vol. 6316, pp. 140–153. Springer, Heidelberg (2010). https://doi.org/10.1007/978-3-642-15567-3_11

40. Tieleman, T., Hinton, G.: Lecture RmsProp: divide the gradient by a running average of its recent magnitude. COURSERA Neural Networks Mach. Learn. **4**, 26–30 (2012)
41. Tran, D., Bourdev, L.D., Fergus, R., Torresani, L., Paluri, M.: Learning spatiotemporal features with 3D convolutional networks. In: 2015 IEEE International Conference on Computer Vision, ICCV (2015)
42. Wang, H., Schmid, C.: Action recognition with improved trajectories. In: ICCV 2013 - IEEE International Conference on Computer Vision (2013)
43. Wang, H., Schmid, C.: Action recognition with improved trajectories. In: IEEE International Conference on Computer Vision, ICCV (2013)
44. Wang, L., Qiao, Y., Tang, X.: Action recognition with trajectory-pooled deep-convolutional descriptors. In: IEEE Conference on Computer Vision and Pattern Recognition (2015)
45. Wang, L., et al.: Temporal segment networks: towards good practices for deep action recognition. In: Leibe, B., Matas, J., Sebe, N., Welling, M. (eds.) ECCV 2016. LNCS, vol. 9912, pp. 20–36. Springer, Cham (2016). https://doi.org/10.1007/978-3-319-46484-8_2
46. Wang, L., et al.: Temporal segment networks: towards good practices for deep action recognition. In: ECCV (2016)
47. Wang, P., Li, W., Ogunbona, P., Wan, J., Escalera, S.: RGB-D-based human motion recognition with deep learning: a survey. CoRR (2017)
48. Wang, X., Farhadi, A., Gupta, A.: Actions ∼ transformations. In: 2016 IEEE Conference on Computer Vision and Pattern Recognition, CVPR 2016 (2016)
49. Wang, Y., Long, M., Wang, J., Yu, P.S.: Spatiotemporal pyramid network for video action recognition. In: 2017 IEEE Conference on Computer Vision and Pattern Recognition (CVPR) (2017)
50. Xiong, Y., Zhu, K., Lin, D., Tang, X.: Recognize complex events from static images by fusing deep channels. In: IEEE Conference on Computer Vision and Pattern Recognition, CVPR 2015 (2015)
51. Zach, C., Pock, T., Bischof, H.: A duality based approach for realtime TV-L^1 optical flow. In: Hamprecht, F.A., Schnörr, C., Jähne, B. (eds.) DAGM 2007. LNCS, vol. 4713, pp. 214–223. Springer, Heidelberg (2007). https://doi.org/10.1007/978-3-540-74936-3_22
52. Zhou, C., Sun, C., Liu, Z., Lau, F.C.M.: A C-LSTM neural network for text classification. CoRR (2015)

Image2Mesh: A Learning Framework for Single Image 3D Reconstruction

Jhony K. Pontes[1(✉)], Chen Kong[2], Sridha Sridharan[1], Simon Lucey[2], Anders Eriksson[1], and Clinton Fookes[1]

[1] Queensland University of Technology, Brisbane, Australia
jhonykaesemodel@gmail.com
[2] Carnegie Mellon University, Pittsburgh, USA

Abstract. A challenge that remains open in 3D deep learning is how to efficiently represent 3D data to feed deep neural networks. Recent works have been relying on volumetric or point cloud representations, but such approaches suffer from a number of issues such as computational complexity, unordered data, and lack of finer geometry. An efficient way to represent a 3D shape is through a polygon mesh as it encodes both shape's geometric and topological information. However, the mesh's data structure is an irregular graph (*i.e.* collection of vertices connected by edges to form polygonal faces) and it is not straightforward to integrate it into learning frameworks since every mesh is likely to have a different structure. Here we address this drawback by efficiently converting an unstructured 3D mesh into a regular and compact shape parametrization that is ready for machine learning applications. We developed a simple and lightweight learning framework able to reconstruct high-quality 3D meshes from a single image by using a compact representation that encodes a mesh using free-form deformation and sparse linear combination in a small dictionary of 3D models. In contrast to prior work, we do not rely on classical silhouette and landmark registration techniques to perform the 3D reconstruction. We extensively evaluated our method on synthetic and real-world datasets and found that it can efficiently and compactly reconstruct 3D objects while preserving its important geometrical aspects.

1 Introduction

Most of us take for granted the ability to effortlessly perceive our surrounding world and its objects in three dimensions with a rich geometry. In general, we have good understanding of the 3D structure only by looking at a single 2D image of an object even when there are many possible shapes that could have produced the same image. We simply rely on assumptions and prior knowledge

Electronic supplementary material The online version of this chapter (https://doi.org/10.1007/978-3-030-20887-5_23) contains supplementary material, which is available to authorized users.

© Springer Nature Switzerland AG 2019
C. V. Jawahar et al. (Eds.): ACCV 2018, LNCS 11361, pp. 365–381, 2019.
https://doi.org/10.1007/978-3-030-20887-5_23

acquired throughout our lives for the inference. It is one of the fundamental goals of computer vision to give machines the ability to perceive its surroundings as we do, for the purpose of providing solutions to tasks such as self-driving cars, virtual and augmented reality, robotic surgery, to name a few.

A specific problem that is of particular interest towards achieving this ambition of human-like machine perception is that of recovering 3D information from a single image. This exceedingly difficult and highly ambiguous problem is typically addressed by incorporating prior knowledge about the scene such as shape or scene priors [1–10]. This body of work has provided a valuable foundation for this task and it has in particular indicated that the use of shape priors is highly beneficial. With the online availability of millions of 3D CAD models across different categories, the use of 3D shape prior becomes even more attractive and motivating. This is a realisation we propose to exploit in this work.

With the recent arrival of deep learning many interesting work have been done to tackle 3D inference from 2D imagery by exploring the abundance of 3D models available online [11–14]. Most of them rely on volumetric shape representation, an approach arguably motivated by the ease of which convolutions can be generalized from 2D to 3D. A significant drawback these methods have is that the computational and memory costs scale cubically with the resolution. Octrees [13] and point cloud [15] representations have been proposed to make the learning more efficient. However, despite of its improved performance, such representations still fail to capture fine-grained geometry as a dense 3D mesh representation would might capture.

The aim of this work is to exploit a compact mesh representation to better unlock fine-grained 3D geometry reconstruction from a single image. We propose a novel learning framework based on a graph that embeds 3D meshes in a low-dimensional space and still allow us to infer compelling reconstructions with high-level details. We draw inspiration by the works in [10,16], where a graph embedding to compactly model the intrinsic variation across classes of 3D models has been proposed. Essentially, any 3D mesh can be parametrized in terms of free-form deformation (FFD) [17] and sparse linear combination in a dictionary. FFD allow us to embed a 3D mesh in a grid space where deformations can be performed by repositioning a smaller number of control points. Although the FFD conserves the objects' topology, the sparse linear combination step allows it to be modified to better generalise the 3D reconstruction to unseen data.

Our method first classifies the latent space of an image to retrieve a coarse 3D model from a graph of 3D meshes as initialization. Then, the compact shape parameters are estimated from a feedforward neural network which maps the image features to the shape parameters space - FFD and sparse linear combination parameters. The dense 3D mesh model is then recovered by applying the estimated deformations to the 3D model selected. An overview of the proposed framework is illustrated in Fig. 1. In contrast to [10,16], our proposed method neither rely on landmark and silhouette registration techniques nor manually annotated 2D semantic landmarks which would limit its applicability.

The main contributions of this paper are:

- We propose a simple and lightweight learning framework to infer a high-quality 3D mesh model from a single image through a low-dimensional shape embedding space;
- To the best of our knowledge, the proposed method is the first to estimate visual compelling 3D mesh models with fine-grained geometry from a single image neither relying on classical landmark and silhouette registration techniques nor class-specific 2D landmarks;
- We demonstrate the performance of our method and its generalization capacity through extensive experiments on both synthetic and real data.

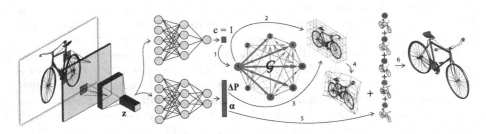

Fig. 1. Given a single image, our framework employs a convolutional autoencoder to extract the image's latent space z to be classified into an index c and regressed to a shape parametrization (ΔP, α). We use a graph embedding $\mathcal{G}$ to compactly represent 3D meshes. The estimated index c selects from $\mathcal{G}$ the closest 3D model to the image. The selected model is then deformed with the estimated parameters - FFD displacements ΔP and sparse linear combination weights α. For instance, model 1 is selected (arrows 1 and 2), FFD is then applied (arrows 3 and 4), and finally the linear combination with the nodes 3, 4, 5, 6, and 7 (blue arrows on the graph that indicates the models in dense correspondence with node 1) are performed (arrow 5) to reconstruct the final 3D mesh (arrow 6). (Color figure online)

2   Related Work

Recent advances in neural networks and the online availability of 3D models (such as ShapeNet [18]) has sparked a considerable interest in methods using deep learning to solve tasks related to geometry and 3D reconstruction. Several papers have proposed a 3D volumetric representation [11,12,19–27] so they can feed deep neural networks applying 3D convolutions, pooling, and other techniques that have been successfully applied to 2D images for the learning process. Volumetric autoencoders [12,28,29] and generative adversarial networks (GANs) have been proposed [30–32] to learn of the probabilistic latent space of object shapes for object completion, classification and reconstruction. Despite of all the great work, volumetric representation has a great drawback. The memory and the computational costs grow cubically as the voxel resolution increases which limit such works to low-resolution 3D reconstructions.

Octree-based convolutional neural networks have been presented to manage these limitations [13,14,33,34]. It splits the voxel grid by recursively subdividing it into octants thus reducing the computational complexity of the 3D convolution. The computations are then focused on regions where most of the information about the object's geometry is contained - generally on its surface. Although it allows for higher resolution outputs, around 64^3 voxels, and a more efficient training, the volumetric models still lacks fine-scaled geometry. In trying to provide some answers to these shortcomings, a more efficient input representation for 3D geometry using point clouds have been recently entertained [15,35–39]. In [36] it was proposed a generative neural network to output a set of unordered 3D points used for the 3D reconstruction from single image and shape completion tasks. These architectures have been demonstrated for the generation of low-resolution 3D models and to scale them to higher resolution is yet to be investigated. Moreover, generating 3D surfaces for point clouds is a challenging problem, specially in the case of incomplete, noisy and sparse data [40].

3D shapes can be efficiently represented by polygon meshes to encode both geometrical and topological information [41]. However, the data structure of a mesh is an irregular graph (*i.e.* set of vertices connected by edges to form polygonal faces) and it is not straightforward to integrate it into learning frameworks since every mesh is likely to have a different structure. A deep residual network to generate 3D meshes has been proposed in [42]. The authors used a regular data structure to encode an irregular mesh by employing the geometry image representation. Geometry images however can only manage simple surfaces (*i.e.* genus-0 surface). FFD has also been explored for 3D mesh representation where one can represent an object by a set of polynomial basis and a fixed number of control points used to deform the mesh. A 3D shape editing tool has been presented in [43] and it uses a volumetric network to infer per-voxel deformation flows using FFD. Their method takes a volumetric representation of a 3D mesh as input and a deformation intention label (*e.g.* sporty car) to learn the FFD displacements to be applied to the original mesh. A novel graph embedding based on the local dense correspondences between 3D meshes has been proposed in [10,16]. The method is able to reconstruct the finer geometry of a single image based on a low-dimensional 3D mesh parametrization. Although they showed impressive results for high-quality 3D mesh reconstruction, it relies on classical landmark and silhouette registration techniques that depend on manually annotated and class-specific 2D/3D landmarks which considerably limit its applicability.

To step further, our work efficiently converts a given unstructured 3D mesh into a regular and compact parametrization that is ready for machine learning applications. Nevertheless, our proposed learning framework is able to reconstruct high-quality 3D meshes from a single image.

3   Proposed Learning Framework

Given a single image from a specific category (*e.g.* bicycle, chair, etc.), our aim is to learn a model to infer a high-quality 3D mesh. Our proposed framework does not build on classical landmark/silhouette registration techniques which might constrain its applicability due to the need for image annotations. To efficiently represent the 3D mesh data to feed neural networks we employed a compact mesh representation [10,16] that embeds a 3D mesh in a low-dimensional space composed of a 3D model index c, FFD displacements $\Delta \mathbf{P}$ ($e.g.$ 32 control points), and sparse linear combination weights $\boldsymbol{\alpha}$ depending on the size of the graph $\mathcal{G}$.

To estimate the 3D shape parameters, we train a multi-label classifier to infer the index and a feedforward neural network to regress the FFD displacements and the sparse linear combination weights from the image latent space learnt from a convolutional autoencoder (CAE). A model is then selected from the graph using the estimated index and the FFD parameters are applied to initially deform the model. Once a satisfactory model candidate is selected and deformed, we have, from the graph, the information about what models are possible to establish dense correspondences. Finally, we apply the sparse linear combination parameters to refine the 3D reconstruction. The framework is shown in Fig. 1.

3.1 3D Mesh Embedding

We parametrize a 3D mesh model using the compact shape representation method presented in [10,16]. It uses a graph $\mathcal{G}$ with 3D mesh models from the same class as nodes and its edges indicate whether we can establish dense correspondences or not (see an example in Fig. 1). Note that the graph is not fully connected but sparse. So in every subgraph we can perform sparse linear combinations to deform a 3D model ($i.e.$ a union of subspaces). Mathematically, consider Ω as an index set of the nodes in a certain subgraph with $\mathcal{S}_c(\mathbf{V}_c, \mathbf{F}_c)$ as the central shape node. Here $\mathbf{V}$ and $\mathbf{F}$ stand for the shape vertices and faces. Dense correspondences will always exist for all $i \in \Omega$ allowing us to deform the model $\mathcal{S}(\mathbf{V}, \mathbf{F})$ by linear combination,

$$\mathbf{V} = \alpha_c \mathbf{V}_c + \sum_{i \in \Omega} \alpha_i \mathbf{V}_c^i, \quad \mathbf{F} = \mathbf{F}_c, \tag{1}$$

where α's are the weights. A two-step process is then performed, first we need to find a candidate model ($i.e.$ a node) from $\mathcal{G}$. Second, knowing the index that indicates the central node we can deform the model by linearly interpolating it with its dense correspondences from the subgraph. To select a good candidate in the first step, FFD is used to deform a model and pick the one which best fit the image. A 3D mesh can be represented in terms of FFD as

$$\mathbf{S}_{ffd} = \mathbf{B}\mathbf{\Phi}(\mathbf{P} + \Delta\mathbf{P}), \tag{2}$$

where $\mathbf{S}_{ffd} \in \mathbb{R}^{N \times 3}$ are the vertices of the 3D mesh, $\mathbf{B} \in \mathbb{R}^{N \times M}$ is the deformation matrix, $\mathbf{P} \in \mathbb{R}^{M \times 3}$ are the control point coordinates, N and M are the number of vertices and control points respectively, $\Delta\mathbf{P}$ are the control point displacements, and $\mathbf{\Phi} \in \mathbb{R}^{3M \times 3M}$ is a matrix to impose symmetry in the FFD grid as in [16]. The deformation matrix $\mathbf{B}$ is a set of Bernstein polynomial basis[1].

---

[1] Refer to [10,16] for more details about the graph creation and deformation process.

Once we have the graph, we can embed a 3D mesh in a low-dimensional space. We only need an index that indicates what model we should pick up from the graph, the symmetric FFD displacements $\Delta\mathbf{P}$ to apply the initial deformation to the selected model, and the sparse linear combination weights $\boldsymbol{\alpha}$ to refine the final 3D model. In this work however we tackle the following question: Could such a low-dimensional parametrization be learned to infer high-quality 3D meshes from a single image?

3.2   Collecting Synthetic Data

Since our goal is to infer a 3D mesh parametrization from a single image, we need to synthetically generate data containing 3D meshes, its compact parametrization (c, $\Delta\mathbf{P}$ and $\boldsymbol{\alpha}$ parameters) and rendered images. One may question why we do not simply parametrize the whole ShapeNet, for example, using the shape embedding proposed in [16]. One of the drawbacks of such an approach would be that it relies on 3D semantic landmarks that were manually annotated in some CAD models from ShapeNet to obtain a compact shape embedding. For this reason we decided to use the graph to generate data for each object class instead of manually annotating 3D anchors on several models which is laborious.

To generate the data, we randomly choose an index and then we deform the selected model by applying $\Delta\mathbf{P}$ and $\boldsymbol{\alpha}$ from a learned probability density function (PDF). To learn a PDF for the displacements $\Delta\mathbf{P}$ we use a Gaussian Mixture Model (GMM) to capture information about the nature of the desired deformations. Since we have from the graph creation process the FFD parameters for every pair of 3D models in the graph (obtained during the deformation process to find dense correspondences in [16]), we can learn a GMM from such prior information. For the sparse linear coefficients' PDF we simply fit a normal distribution to a set of α's from 3D reconstructions on the PASCAL3D+ dataset [44] from [16]. Armed with this, we can synthetically generate deformed 3D meshes and images with their respective low-dimensional shape parametrizations.

3.3   Learning the Image Latent Space

To obtain a lower-dimensional representation for the images, a convolutional autoencoder is proposed to extract useful features in an unsupervised manner. The encoder network consists of three fully convolution layers with numbers of kernels $\{8, 16, 32\}$, kernel sizes $\{5, 3, 3\}$, and strides $\{3, 3, 3\}$. The decoder network has three fully transposed convolutional layers with numbers of kernels $\{16, 8, 1\}$, kernel sizes $\{3, 3, 5\}$, and strides $\{3, 3, 3\}$. The input takes grayscale images of size 220×220. All layers use ReLU as activation functions except the last one that uses tanh. This gives us a network flow of size $220^2 \rightarrow 72^2 \rightarrow 24^2 \rightarrow 8^2 \rightarrow 24^2 \rightarrow 72^2 \rightarrow 220^2$, respectively. We take the image latent space $\mathbf{z} \in \mathbb{R}^{2,048}$ from the last layer of the encoder as feature representation.

3.4   Learning the Index to Select a Model

Firstly we need to retrieve a 3D model from the graph (a graph is explained in Subsect. 3.1). We treat it as a multi-label classification problem. For example, if a graph of the class 'car' would have 30 different cars/nodes it would have 30 indices/labels (1 to 30) that an image of a car could be classified as. In this way, the "closest" car to the image can be selected from the graph according to the estimated index. For this purpose, we propose a simple yet effective multi-label classifier to estimate graph indices. The input is the image latent space from the convolutional autoencoder of size 2,048 ($8 \times 3 \times 3$). The output is a one-hot encoded vector that represents the ground truth indices. The network has one hidden layer of size 1,050 and ReLU is used as activation function.

3.5   Learning the Shape Parameters

We wish to learn a mapping from the image feature representation $\mathbf{z}$ of size 2,048 to its corresponding 3D shape parameters $\mathbf{\Delta P}$ and $\boldsymbol{\alpha}$, *i.e.* $f : \mathbf{z} \rightarrow \{\mathbf{\Delta P}, \boldsymbol{\alpha}\}$. Given the training set of image features and the ground truth shape parameters, we learn this mapping using a feedforward neural network. The input is the image latent space of size 2,048. The output is a vector containing the shape parameters $\boldsymbol{\kappa} \in \mathbb{R}^{MN}$ where M and N are the number of FFD and sparse linear combination parameters $\boldsymbol{\alpha}$ respectively. For instance, $\boldsymbol{\kappa} \in \mathbb{R}^{126}$, where 96 values would be the FFD parameters (32×3) and the remaining 30 values would be the $\boldsymbol{\alpha}$ parameters for the sparse linear combination of models in the graph. To handle different number of α coefficients that might differ according to the subgraph selected, we consider a fixed-size vector according to the size of $\mathcal{G}$ and then we pick only the estimated α's corresponding to the subgraph (*i.e.* we ignore the other α's). The network has one hidden layer of size 1,500 and it uses ReLU as activation function.

4   Experiments

Dataset. To train our framework we take the approach of synthesizing 3D mesh models using the strategy discussed in the Subsect. 3.2. We generated 5,000 deformed 3D models of eight object categories (car, bicycle, motorbike, aeroplane, bus, chair, dining table, and sofa) using the graphs from [16]. Every graph has 30 CAD models sampled from ShapeNet [18] except for the bicycle and motorbike graphs that have 21 and 27 CAD models, respectively. We rendered for every 3D synthesised model a 2D view of size 256×192 using different viewpoints with a white background. We also produced uniform lighting across the surfaces of the object. With the images and the ground truth 3D meshes, indices and shape parameters, we can train our framework and evaluate its performance. The data was split in 70% for training and 30% for testing.

Evaluation Metrics. To quantify the quality of the classification step we employed the accuracy, precision and recall metrics. To evaluate the estimated shape parameters we use the mean squared error (MSE). For the 3D shape reconstruction measure we use the symmetric surface distance s_{dist} to the ground truth. s_{dist} is computed by densely sampling points on the faces and using normalized points distance to estimate the model similarity and it is defined as

$$dist_{3D} = \frac{1}{|\hat{\mathbf{V}}|} \sum_{\mathbf{v_i} \in \hat{\mathbf{V}}} dist(\mathbf{v_i}, \mathcal{S}) + \frac{1}{|\mathbf{V}|} \sum_{\mathbf{v_i} \in \mathbf{V}} dist(\mathbf{v_i}, \hat{\mathcal{S}}), \quad (3)$$

where $\hat{\mathbf{V}}$, $\hat{\mathcal{S}}$, $\mathbf{V}$, $\mathcal{S}$ are the estimated vertices and surfaces, and the ground truth vertices and surfaces, respectively.

Moreover, we use the intersection over union (IoU) as a metric to compare different voxel models as in [11] defined as $(\hat{\mathcal{V}} \cap \mathcal{V})/(\hat{\mathcal{V}} \cup \mathcal{V})$, where $\hat{\mathcal{V}}$ and $\mathcal{V}$ are the voxel models of the estimated and ground truth models respectively.

Table 1. Evaluation of our method on synthetic data. We show the performance of the convolutional autoencoder (CAE), the multi-label classification for the 3D model selection, the feedforward network for the parameters estimation, and the 3D reconstruction from single image. *Acc, Prec, Rec,* and *t* stand for the accuracy, precision, recall, and the training time, respectively.

	CAE		3D model selection				Params estimation		3D reconstruction	
	MSE	$\sim t$ (min)	Acc(%)	Prec(%)	Rec(%)	$\sim t$ (min)	MSE	$\sim t$ (min)	$dist_{3D}$	IoU
Car	**0.0012**	30	92.13	92.11	93.33	76	0.0176	197	0.006	0.664
Bicycle	0.0068	25	92.07	92.00	92.18	73	0.0102	208	0.025	0.795
motorbike	0.0045	21	**97.80**	**97.87**	**97.72**	74	0.0150	322	0.007	0.679
Aeroplane	0.0017	24	77.33	76.90	78.09	96	0.0108	209	0.023	0.551
Bus	0.0013	35	90.20	90.05	92.14	74	0.0329	156	**0.003**	**0.776**
Chair	0.0022	21	87.33	86.97	88.74	75	**0.0090**	158	0.034	0.403
Dining table	0.0020	22	73.60	73.60	74.98	75	0.0119	158	0.165	0.332
Sofa	0.0013	21	76.20	76.47	77.69	74	0.0119	156	0.063	0.402
Mean	0.0024	25	85.86	85.75	86.86	77	0.0144	195	0.041	0.575

4.1   Estimating the Image Latent Space

The first set of experiments were performed on the CAE to learn a latent representation from an image. We found the architecture described in the Subsect. 3.3 to have the better performance. The image feature representation is discriminative and performed well on the classifier and on the feedforward network to estimate the shape parameters. Table 1 shows the MSE on the test set and the time spent training the network for every class used. We used the MSE evaluated on the test set as a measure of how close the input image (ground truth) is from the image generated by the decoder. It does not mean the latent space is representative as image descriptors but we further validated it in the classification and regression steps by achieving high accuracies. This means that the descriptors learnt are indeed discriminative.

4.2   Selecting a 3D Mesh

The first stage of our learning framework after having the image latent space is the selection of a 3D mesh from the given graph. The performance of the proposed multi-label classifier is shown in Table 1. One can note that the overall performance on the test set was satisfactory in terms of accuracy (85.68%), precision (85.75%) and recall (86.86%). The best performance was achieved on the motorbike category (27 labels) with an accuracy of 97.80%. The category has very different motorbikes from each other which explains the great performance. Besides, the dining table category (30 labels) had the lowest performance, with an accuracy of 73.60%. This is presumably due to the high degree of similarity between the synthesised images as there are not many unique tables in this object class. Moreover, we fixed the network architecture for all classes. Fine tuning the classifier for specific classes would more than likely improve performance.

4.3   Estimating the Shape Parameters

The last stage of our framework before the final 3D reconstruction is the estimation of the shape parameters. The results are also shown in Table 1. The overall MSE (0.0144) on the testing set shows that the network is indeed learning a mapping function to estimate the FFD and the α parameters from the image latent space. The graphs used have about 30 mesh models each which means that once we have selected a model we can establish dense correspondences with up to 29 models (29 α values). The resolution of the FFD grid is of 4^3 that gives us 64 control points to free-deform the model. Since the majority of man-made object are symmetric, we impose a symmetry on the FFD grid so that the deformations are forced to be symmetric and more realistic. Therefore, we have to estimate only half of the FFD parameters. The feedforward network then maps the image latent space to 32 displacements of the control points in the 3D space, $\Delta\mathbf{P} \in \mathbb{R}^{32 \times 3}$, and to 30 sparse linear combinations parameters, $\alpha \in \mathbb{R}^{30}$ (29 + the model selected). The estimated parameters, in this case, is of size $\kappa \in \mathbb{R}^{126}$. One can note that it is a very low-dimensional parametrization that is efficiently learned through a simple and lightweight network architecture.

4.4   3D Reconstruction from a Single Image

Given a single image we can forward pass it to our learned framework to estimate an index c and the shape parameters κ. A 3D mesh is initially selected from the class-specific graph by the estimated index c. Afterwards, the FFD displacements $\Delta\mathbf{P}$ is applied to free-deform the model using Eq. 2, $\mathbf{S}_{ffd} = \mathbf{B}\Phi(\mathbf{P} + \Delta\mathbf{P})$. Note that we only need to add the estimated displacements to the initial grid of control points $\mathbf{P}$. Finally, we can apply the linear combination parameters α to deform the model through Eq. 1, $\mathbf{V} = \alpha_c \mathbf{V}_c + \sum_{i \in \Omega} \alpha_i \mathbf{V}_c^i$.

3D Reconstruction from Synthetic Images. The initial experiments were performed on synthetic images from the 8 classes where we have the ground

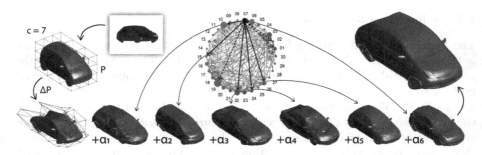

Fig. 2. Given an input image of a car and a graph $\mathcal{G}$ with 30 models as nodes, our method first selected the model 7 from $\mathcal{G}$. Then it is deformed by the FFD displacements $\mathbf{\Delta P}$ on the initial grid $\mathbf{P}$. Afterwards, the linear combination is performed with the estimated α to reconstruct the final model. The black arrows on $\mathcal{G}$ show what models are possible to perform linear combination with the selected model (*i.e.* models in dense correspondences). Note that not all the models were selected, but 6 out of 9 models in this example. For illustration, the node size in $\mathcal{G}$ is proportional to the number of edges starting from the node.

truth 3D meshes and also the shape embedding parameters. Figure 2 shows a real example of our framework flow. Table 1 summarizes the quantitative results of our 3D reconstruction on the synthetic test set. We measured the quality of the 3D reconstruction through the surface distance metric $dist_{3D}$ and the IoU between the reconstructed and the ground truth model. Our framework clearly performed well on the synthetic dataset according to the surface distance metric. The IoU for the classes aeroplane, chair, dining table, and sofa had the lowest values which means that a good voxel intersection between the reconstructed model and the ground truth was not possible. In fact, IoU between thin structures (*e.g.* chair's legs, aeroplane's wings, etc.) are low if they are not well aligned.

Qualitative results are shown in Fig. 3. Our proposed learning framework performed well at selecting a proper model to start the deformation process, and also at estimating the shape parameters to obtain the final deformed mesh. In the successful cases, one can see the final models are similar to the ground truth with slight differences that can be hard to point them out. An interesting example to show the expressiveness of our proposed method is the chair instance. One can note that the selected chair has long legs, but the estimated FFD parameters managed to deform the chair to get shorter legs before applying the linear combination parameters to get the final model. A failure case is shown on the last row in red where an "incorrect" model was selected from the graph, in this case a fighter jet instead of a commercial airplane. This can in part be explained by the challenging image perspective of this instance. Even for a human it is difficult to correctly classify such an image, in this case a fighter jet is in fact a highly plausible choice of model[2].

---

[2] More results, failure cases, and videos can be found in the supplementary material.

(a) Input   (b) Selected model   (c) FFD   (d) Final model   (e) Voxel model   (f) GT

Fig. 3. Visual results on synthetic data. (a) shows the input image; (b) the selected model from the graph; (c) the selected model deformed by FFD. The final 3D model deformed by linear combination is shown in (d). The voxelized final model is shown in (e) and the ground truth in (f). In the success cases (blues), one can note the final models are similar to the ground truth with slight differences that can be hard to point it out. A failure case is shown on the last row in red where a "wrong" model was selected from the graph. (Color figure online)

3D Reconstruction from Real World Images. To verify our framework's generalization capacity we test it on a dataset with real world images. We evaluated the performance of the proposed method on the PASCAL3D+ dataset and we compared to the results presented in [16]. We found that it is a fair comparison since we are not playing with volumetric or point cloud representations but with dense polygonal meshes. In order to forward pass the real world images to our learning framework, we used the image silhouettes provided by the PASCAL3D+ dataset since our framework was trained on images with uniform

Table 2. Evaluation of our method on the PASCAL3D+ dataset and comparison with the method presented in [16].

	[16]		Ours	
	$dist_{3D}$	IoU	$dist_{3D}$	IoU
Car	**0.174**	**0.382**	0.179	0.371
Bicycle	0.290	**0.419**	**0.282**	0.402
Motorbike	**0.084**	**0.384**	0.186	0.309
Aeroplane	0.262	**0.442**	**0.153**	0.366
Bus	0.091	**0.376**	**0.058**	0.280
Chair	**0.309**	**0.261**	0.461	0.236
Dining table	**0.353**	**0.256**	0.695	0.223
Sofa	**0.346**	**0.241**	0.573	0.207
Mean	**0.239**	**0.345**	0.323	0.299

background. Table 2 summarizes the results of our method and the results presented in [16] in terms of the surface distance $dist_{3D}$ and the IoU. Our method did not outperform the method proposed in [16], except for some classes. However, we achieved a similar performance on the real world dataset neither relying on silhouette and landmark registration algorithms nor using class-specific landmarks. Moreover, since the ground truth models in the PASCAL3D+ dataset were aligned to images by humans, the comparison metrics are not robust. As stated in [16], most of the 3D reconstructions look closer to the images than the

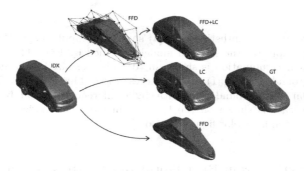

Fig. 4. Qualitative example showing the importance of every step in our method. The upper part shows the FFD and the linear combination (LC) being applied to the selected model (IDX). It shows that even with a strong FFD deformation the LC step managed to deform the mesh to look similar to the GT model. The bottom part shows the 3D reconstruction when omitting the FFD step. The LC step managed to deform the van into a compact car (perhaps SUV?) but it is still different from the GT. If we omit the LC step one can notice that the FFD model looks very different from the GT.

ground truth models themselves. This explains the high values for the surface distance and the low values for the IoU.

Qualitative results are shown in Fig. 5. One can see that our proposed method performed well on a real-world dataset. In the motorbike example it is clear when looking at the image that the motorbike does not have a backrest. The selected model was a good choice, since it shares topological similarities and although it has a backrest device, the linear combination step managed to diminish it. Another interesting example is the airplane where the selected model has a different type of wings, but the deforming process made it appear much more similar to the input image.

Ablation Study. We performed an extensive ablation experiment where the 3D reconstruction error is evaluated at each step, model selection, FFD and linear combination, to shows how sensitive the model is to the noise of each step. A qualitative example showing the importance of every step is shown in Fig. 4. Please refer to the supplementary material for the quantitative analysis.

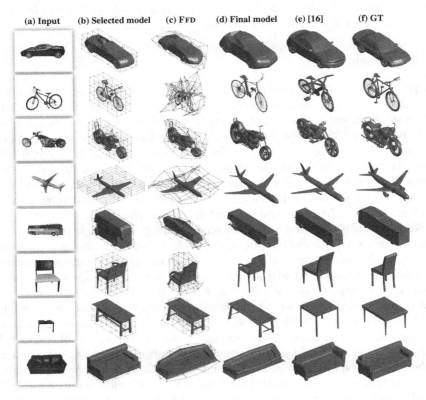

Fig. 5. Visual results on real-world data. (a) shows the input image; (b) the selected model; (c) the selected model deformed by FFD. The final 3D model reconstructed by linear combination is shown in (d). We compare with [16] in (e) and the ground truth is shown in (f).

4.5   Implementation Details and Limitations

We designed our networks using PyTorch [45] on a GPU Nvidia Tesla M40. We trained the convolutional autoencoder using a MSE loss function and the Adam optimizer [46] with a learning rate of $1e^{-3}$ and a weight decay of $1e^{-5}$ during 100 epochs. To train the multi-label classifier we used a multi-label soft margin loss function and for the feedforward neural network we trained it using a MSE loss function. For both models we used the Adam optimizer with a learning rate of $1e^{-3}$ during 1,000 epochs.

One of the main limitations, which our method inherits from [16], is the need for a good embedding graph. One can see in Fig. 2 that some models for the linear combination step have some crinkles on its surfaces. This happens during the graph construction when searching for dense correspondences between the models. This is especially important for achieving high-quality 3D reconstructions. However, finding dense correspondences between two different models that do not share the same number of vertices is still an open problem [16]. Another limitation is that we synthesise the images with white background so real images must be segmented beforehand for the 3D reconstruction. GANs can fit in this context to generate more realistic images.

5   Discussion

The proposed method is category-specific, meaning that we need one model for every class. Although [36] is able to better generalize to multiple object categories with a single model trained on 2k classes, we believe our paper makes a step in 3D reconstruction in the mesh domain which has largely been unexplored. Moreover, we would like to reinforce the generalization capacity of our method to specific classes, where a single graph embedding model is able to generalize to unseen 3D meshes by using FFD and linear combination in a small dictionary. We successfully validated this using the PASCAL3D+ dataset since its shape distribution differs from the trained models on synthetic data.

We would like to reiterate our argument that even though [16] performed better in some categories, our method is able to achieve similar fine-grained mesh reconstruction without the need to rely on any class-specific landmarks or any silhouette registration whatsoever. This is a significant improvement in the utility of these approaches with the removal of these limiting constraints.

A comparison with [36] would not be entirely straightforward or terribly informative as we propose a mesh representation with fine-grained geometry whereas [36] proposes a coarse point cloud representation. Moreover, IoU of coarse volumetric models is not a robust metric to capture fine-grained geometry contained on the surface of mesh models. From visual inspection, one can observe that our deformation of mesh models and their detailed geometry clearly outperforms [36] and [34]. We considered a comparison with [34], however the authors have not yet shared their code and the information found in the paper is not enough to reproduce their results. For this reason we decided to restrict our comparisons to competing methods as [16] (and implicitly with [10]).

6   Conclusion

We have proposed a simple yet effective learning framework to infer 3D meshes from a single image using a compact mesh representation that does not rely on class-specific object landmarks which would limit its applicability. A 3D mesh is embedded in a low-dimensional space that allows one to perform deformations by FFD and sparse linear combination. Experiments on synthetic and real-world datasets show that our method convincingly reconstructs 3D meshes from a single image with fine-scaled geometry not yet achieved in previous works that rely on volumetric and point cloud representations. Although our method relies on background segmentation, we do believe the field is mature to provide off-the-shelf segmentation techniques for a practical application. Such high quality 3D representation and reconstruction as proposed in our work is extremely important, especially to unlock virtual and augmented reality applications. Finally, we believe that this work is a great first step towards more effective mesh representations for 3D geometric learning purposes.

Acknowledgements. This research was supported by the grants ARC DP170100632, ARC FT170100072 and NSF 1526033.

References

1. Wang, C., Wang, Y., Lin, Z., Yuille, A.L., Gao, W.: Robust estimation of 3D human poses from a single image. In: CVPR (2014)
2. Zhou, X., Leonardos, S., Hu, X., Daniilidis, K.: 3D shape estimation from 2D landmarks: a convex relaxation approach. In: CVPR (2015)
3. Kar, A., Tulsiani, S., Carreira, J., Malik, J.: Category-specific object reconstruction from a single image. In: CVPR (2015)
4. Rock, J., Gupta, T., Thorsen, J., Gwak, J., Shin, D., Hoiem, D.: Completing 3D object shape from one depth image. In: CVPR (2015)
5. Zhou, X., Zhu, M., Leonardos, S., Derpanis, K.G., Daniilidis, K.: Sparseness meets deepness: 3D human pose estimation from monocular video. In: CVPR (2016)
6. Wu, J., et al.: Single image 3D interpreter network. In: Leibe, B., Matas, J., Sebe, N., Welling, M. (eds.) ECCV 2016. LNCS, vol. 9910, pp. 365–382. Springer, Cham (2016). https://doi.org/10.1007/978-3-319-46466-4_22
7. Kong, C., Zhu, R., Kiani, H., Lucey, S.: Structure from category: a generic and prior-less approach. In: 3DV (2016)
8. Bansal, A., Russell, B., Gupta, A.: Marr revisited: 2D–3D model alignment via surface normal prediction. In: CVPR (2016)
9. Han, K., Wong, K.Y.K., Tan, X.: Single view 3D reconstruction under an uncalibrated camera and an unknown mirror sphere. In: 3DV (2016)
10. Kong, C., Lin, C.H., Lucey, S.: Using locally corresponding CAD models for dense 3D reconstructions from a single image. In: CVPR (2017)
11. Choy, C.B., Xu, D., Gwak, J.Y., Chen, K., Savarese, S.: 3D-R2N2: a unified approach for single and multi-view 3D object reconstruction. In: Leibe, B., Matas, J., Sebe, N., Welling, M. (eds.) ECCV 2016. LNCS, vol. 9912, pp. 628–644. Springer, Cham (2016). https://doi.org/10.1007/978-3-319-46484-8_38

12. Sharma, A., Grau, O., Fritz, M.: VConv-DAE: deep volumetric shape learning without object labels. In: Hua, G., Jégou, H. (eds.) ECCV 2016. LNCS, vol. 9915, pp. 236–250. Springer, Cham (2016). https://doi.org/10.1007/978-3-319-49409-8_20

13. Tatarchenko, M., Dosovitskiy, A., Brox, T.: Octree generating networks: efficient convolutional architectures for high-resolution 3D outputs. In: ICCV (2017)

14. Riegler, G., Ulusoy, A.O., Geiger, A.: OctNet: learning deep 3D representations at high resolutions. In: CVPR (2017)

15. Qi, C.R., Yi, L., Su, H., Guibas, L.J.: PointNet++: deep hierarchical feature learning on point sets in a metric space. In: NIPS (2017)

16. Pontes, J.K., Kong, C., Eriksson, A., Fookes, C., Lucey, S.: Compact model representation for 3D reconstruction. In: 3DV (2017)

17. Sederberg, T., Parry, S.: Free-form deformation of solid geometric models. In: SIGGRAPH (1986)

18. Chang, A.X., et al.: ShapeNet: an information-rich 3D model repository. Technical report arXiv:1512.03012 [cs.GR] (2015)

19. Wu, Z., Song, S., Khosla, A., Tang, X., Xiao, J.: 3D ShapeNets: a deep representation for volumetric shapes. In: CVPR (2015)

20. Ulusoy, A.O., Geiger, A., Black, M.J.: Towards probabilistic volumetric reconstruction using ray potential. In: 3DV (2015)

21. Cherabier, I., Häne, C., Oswald, M.R., Pollefeys, M.: Multi-label semantic 3D reconstruction using voxel blocks. In: 3DV (2016)

22. Rezende, D.J., Eslami, S.M.A., Mohamed, S., Battaglia, P., Jaderberg, M., Heess, N.: Unsupervised learning of 3D structure from images. In: NIPS (2016)

23. Yan, X., Yang, J., Yumer, E., Guo, Y., Lee, H.: Perspective transformer nets: learning single-view 3D object reconstruction without 3D supervision. In: NIPS (2016)

24. Qi, C.R., Su, H., Nießner, M., Dai, A., Yan, M., Guibas, L.J.: Volumetric and multi-view CNNs for object classification on 3D data. In: CVPR (2016)

25. Kar, A., Häne, C., Malik, J.: Learning a multi-view stereo machine. In: NIPS (2017)

26. Zhu, R., Galoogahi, H.K., Wang, C., Lucey, S.: Rethinking reprojection: closing the loop for pose-aware shape reconstruction from a single image. In: NIPS (2017)

27. Wu, J., Wang, Y., Xue, T., Sun, X., Freeman, W.T., Tenenbaum, J.B.: MarrNet: 3D shape reconstruction via 2.5D sketches. In: NIPS (2017)

28. Liao, Y., Donné, S., Geiger, A.: Deep marching cubes: learning explicit surface representations. In: CVPR (2018)

29. Girdhar, R., Fouhey, D.F., Rodriguez, M., Gupta, A.: Learning a predictable and generative vector representation for objects. In: Leibe, B., Matas, J., Sebe, N., Welling, M. (eds.) ECCV 2016. LNCS, vol. 9910, pp. 484–499. Springer, Cham (2016). https://doi.org/10.1007/978-3-319-46466-4_29

30. Wu, J., Zhang, C., Xue, T., Freeman, W.T., Tenenbaum, J.B.: Learning a probabilistic latent space of object shapes via 3D generative-adversarial modeling. In: NIPS (2016)

31. Liu, J., Yu, F., Funkhouser, T.A.: Interactive 3D modeling with a generative adversarial network. In: 3DV (2017)

32. Gwak, J., Choy, C.B., Garg, A., Chandraker, M., Savarese, S.: Weakly supervised generative adversarial networks for 3D reconstruction. In: 3DV (2017)

33. Wang, P.S., Liu, Y., Guo, Y.X., Sun, C.Y., Tong, X.: O-CNN: octree-based convolutional neural networks for 3D shape analysis. In: SIGGRAPH (2017)

34. Häne, C., Tulsiani, S., Malik, J.: Hierarchical surface prediction for 3D object reconstruction. In: 3DV (2017)

35. Li, J., Chen, B.M., Lee, G.H.: SO-Net: self-organizing network for point cloud analysis. In: CVPR (2018)
36. Fan, H., Su, H., Guibas, L.J.: A point set generation network for 3D object reconstruction from a single image. In: CVPR (2017)
37. Qi, C.R., Su, H., Mo, K., Guibas, L.J.: PointNet: deep learning on point sets for 3D classification and segmentation. In: CVPR (2017)
38. Lin, C.H., Kong, C., Lucey, S.: Learning efficient point cloud generation for dense 3D object reconstruction. In: AAAI (2018)
39. Kurenkov, A., et al.: DeformNet: free-form deformation network for 3D shape reconstruction from a single image. In: WACV (2018)
40. Nan, L., Wonka, P.: PolyFit: polygonal surface reconstruction from point clouds. In: ICCV (2017)
41. Shin, D., Fowlkes, C.C., Hoiem, D.: Pixels, voxels, and views: a study of shape representations for single view 3d object shape prediction. In: CVPR (2018)
42. Sinha, A., Unmesh, A., Huang, Q., Ramani, K.: SurfNet: generating 3D shape surfaces using deep residual network. In: CVPR (2017)
43. Yumer, M.E., Mitra, N.J.: Learning semantic deformation flows with 3D convolutional networks. In: Leibe, B., Matas, J., Sebe, N., Welling, M. (eds.) ECCV 2016. LNCS, vol. 9910, pp. 294–311. Springer, Cham (2016). https://doi.org/10.1007/978-3-319-46466-4_18
44. Xiang, Y., Mottaghi, R., Savarese, S.: Beyond PASCAL: a benchmark for 3D object detection in the wild. In: WACV (2014)
45. Paszke, A., et al.: Automatic differentiation in PyTorch. In: NIPS-W (2017)
46. Diederik, K., Jimmy, B.: Adam: a method for stochastic optimization. In: ICLR (2014)

Face Completion with Semantic Knowledge and Collaborative Adversarial Learning

Haofu Liao[1]([✉]), Gareth Funka-Lea[2], Yefeng Zheng[3], Jiebo Luo[1], and S. Kevin Zhou[4]

[1] Department of Computer Science, University of Rochester, Rochester, USA
`hliao6@cs.rochester.edu`
[2] Digital Technology and Innovation, Siemens Healthineers, Princeton, USA
[3] Tencent X-Lab, Shenzhen, China
[4] Institute of Computing Technology, Chinese Academy of Sciences, Beijing, China

Abstract. Unlike a conventional background inpainting approach that infers a missing area from image patches similar to the background, face completion requires *semantic knowledge* about the target object for realistic outputs. Current image inpainting approaches utilize generative adversarial networks (GANs) to achieve such semantic understanding. However, in adversarial learning, the semantic knowledge is learned implicitly and hence good semantic understanding is not always guaranteed. In this work, we propose a *collaborative adversarial learning* approach to face completion to explicitly induce the training process. Our method is formulated under a novel generative framework called collaborative GAN (collaGAN), which allows better semantic understanding of a target object through collaborative learning of multiple tasks including face completion, landmark detection and semantic segmentation. Together with the collaGAN, we also introduce an inpainting concentrated scheme such that the model emphasizes more on inpainting instead of autoencoding. Extensive experiments show that the proposed designs are indeed effective and collaborative adversarial learning provides better feature representations of the faces. In comparison with other generative image inpainting models and single task learning methods, our solution produces superior performances on all tasks.

Keywords: Face completion · Image inpainting · Generative adversarial networks · Multitask learning

1 Introduction

Image inpainting is the process of reconstructing a missing region in an image such that the inpainted area is visually consistent with its neighboring pixels and the inpainted image overall looks realistic. Traditional approaches to this problem either require that the filling information is available in the image [1,6,9]

H. Liao, Y. Zheng, S. Kevin Zhou—The work was done when the authors were with Siemens Healthineers.

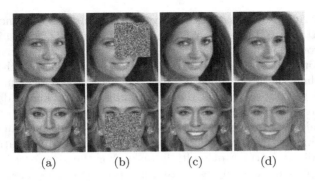

(a)   (b)   (c)   (d)

Fig. 1. Inpainting results using our method. (a) original image. (b) masked image. (c) inpainted image with collaborative learning. (d) inpainted image without collaborative learning (top) and segmentation mask from our method superimposed on the inpainted image (bottom).

or rely on the availability of a large photo database to retrieve the missing region [5, 32]. These approaches work well for the *background inpainting* problem, i.e., filling the missing part with image patches similar to its background, as inpainting can be performed through pattern matching. However, when it comes to complete a missing part of an object where no existing patches can be matched or retrieved, the traditional approaches may fail. For example, if a mouth is missing, it is not possible to synthesize the mouth using image patches from other face parts. Instead, image inpainting in this case requires *semantic knowledge* about faces, e.g., location, shape, color and texture of face parts.

To address this *object completion* problem in image inpainting, recent models [20, 25, 34] propose to use generative adversarial networks (GANs) for more semantically consistent results. However, for generative models, the semantic understanding is implicitly learned through adversarial training. There are no direct constraints on the structure of the target object and hence the inherent semantic understanding is not always guaranteed. Fortunately, in recent years, the success of deep learning has made the semantic labels of objects accessible. In this work, we investigate the possibility of introducing the semantic knowledge of face labels to the adversarial training of face completion for better induction of semantic understanding.

We focus on helping the inpainting model better understand the underlying structure of faces through the collaborative learning of other face related tasks. We argue that current approaches using generative inpainting models alone may not be able to produce structurally realistic results in some cases. For example, when an eye is missing from the image, the inpainting model should be able to predict the missing eye's location and shape based on the facial symmetry. However, as shown in the first row of Fig. 1, the generative image inpainting model trained without using our proposed collaborative method produces a structurally unrealistic face (Fig. 1(d) top) with the inpainted eye smaller and darker than the eye outside the corrupted region. In contrast, a collaboratively trained model can keep the structural consistency between the inpainted region and the nearby context (Fig. 1(c) top). In addition, we also find that models trained in this manner tend to produce visually consistent results among tasks. As demonstrated in Fig. 1(d) bottom, the segmentation result is closely aligned with the inpainting result other than the ground truth. This provides a clear evidence that they are inherently helping each other during training and the knowledge is shared instead of individually learned.

To this end, we propose an innovative image-to-image generative network for face completion. The proposed method formulates a collaborative GAN to facilitate the direct learning of multiple tasks. For the generator, the network outputs multiple channels for each task and has them share most of the network parameters for better collaborative learning. We also stand apart from the existing inpainting models by introducing skip connections between the encoder and decoder [12,29]. For the discriminators, we apply conditional GAN (cGAN) [24] for better transformation quality and have dedicated discriminators for each task. For the loss function, we introduce an inpainting concentrated scheme to allow the model focusing on the inpainting itself instead of autoencoding the context. Our experimental results demonstrate the effectiveness of the proposed design and better feature representations can be obtained with the proposed collaborative GAN. Comparing with other generative models without using collaborative adversarial learning, our approach consistently produces remarkably more realistic inpainting results. Comparing with single task adversarial learning, our joint approach produces better performances on all tasks.

2   Related Work

Generative Image to Image Transformation. Image inpainting can be seen as a special case of the image-to-image transformation problem in that image inpainting tries to transform a cropped image to a reconstructed one. One of the typical image-to-image transformation problems is autoencoding [8]. In relation to inpainting, a seminal work in this area is the denoising autoencoder [31]. It introduces an inpainting-like scheme to autoencoding, hoping the autoencoder can learn a better feature representation by recovering the damage to the input image. Another related work is [16]. It incorporates GANs into a variational autoencoder (VAE) [14] and argues that the network trained using the adversarial loss and VAE loss can give a more representative feature vector of the input image.

For other image-to-image transformation problems, [18] proposes to use GANs for image super-resolution. It has the standard generative image-to-image model setting: a generator that maps the low resolution input to high resolution and a discriminator that tells if the input image is a high resolution one or not. It also includes perceptual loss [18] to further regularize the realism. [12] proposes a general framework for image-to-image translation. It improves the generative image-to-image networks by introducing image-conditional GANs and PatchGAN.

Semantic Inpainting. The term *semantic inpainting* is first introduced by context encoder (CE) [25] to address the challenging inpainting case where a large region of the image is missing and the inpainting generally requires semantic understanding of the context. The paper proposes to learn a better autoencoder by having it recover the missing part of an input image. To our best knowledge, it is the first work that introduces adversarial loss into the inpainting problem. Another approach to semantic inpainting is [34]. The method is based on a pretrained GAN model that maps a noise vector to the generator manifold. The algorithm finds the noise vector such that the generated image through the pretrained GAN model minimizes both a contextual loss and a perceptual loss. This work is an indirect approach to image inpainting. Due to the limitation of the pretrained GAN model, it may fail to produce good results when the image resolution is high or the scene is complex. [33] proposes a multi-scale neural patch synthesis network for high-resolution image inpainting. Since our work focus on the benefit of collaborative adversarial training, high-resolution image inpainting is not within the cope of this study. In fact, our work can be easily extended to high-resolution settings for better performance.

The closest work to ours is [20]. It advances the state of the art [25] by introducing a parsing network to further regularize the inpainting through semantic parsing/segmentation and a local discriminator to ensure the generated contents are semantically coherent. However, the parsing network is pretrained and independent of the generator. Thus, the semantic parsing information is not shared with the generator during training. The semantic parsing loss is also not applied directly to the generator. As a result, the accuracy of computing such a loss is limited by the parsing network which further limits the final performance of the generator. Meanwhile, for the local discriminator, it requires the mask to be rectangular and hence cannot be generalized to other inpainting cases such as noise inpainting. [10] also proposes a similar global and local discriminator design and it suffers from the same disadvantage as [20].

Multi-Task Learning. Multi-task learning (MTL) [2] refers to the process of learning multiple tasks jointly in order to improve the generalization performance of the model. It has been widely used in deep neural networks (DNNs) for various tasks, such as object detection [4], action recognition [30], landmark detection [35], etc. In terms of generative models, many works have introduced adversarial learning in a multi-task fashion to improve the learning of the main tasks. [3] uses adversarial learning for better domain adaption of the main task. [19] proposes a perceptual GAN that learns super-resolution together with object detection for better performance of small object detection. [21] leverages GANs to generate shared features that are independent of different text classification tasks. In this work, we contribute to the literature with a novel image-to-image generative framework that collaboratively learns multiple tasks for better semantic understanding and, ultimately, yields better image inpainting.

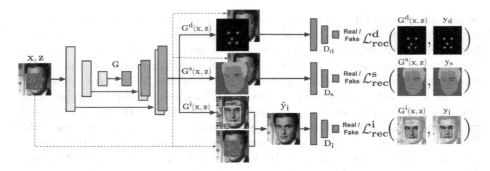

Fig. 2. The architecture of the proposed method. The network is trained collaboratively with three tasks: inpainting i, segmentation s and (landmark) detection d. The generator G takes a masked image x as input and outputs the inpainted image $G^i(x, z)$, segmentation mask $G^s(x, z)$ and detection heatmap $G^d(x, z)$ simultaneously. The discriminators D_i, D_s and D_d are used for adversarial learning. In addition, reconstruction losses $\mathcal{L}_{rec}^i$, $\mathcal{L}_{rec}^s$ and $\mathcal{L}_{rec}^d$ are also applied to the three tasks, respectively.

3   Collaborative Face Completion

The proposed collaborative face completion method is formulated under a novel GAN framework which we call *collaborative GAN* (collaGAN). The proposed framework aims to inductively improve the main generation task (face completion in our case) by incorporating the additional knowledge embedded in other tasks. In this section, we give the formal definition of collaGAN and its connection to face completion.

3.1   Collaborative GAN

Let Ω be a finite set of tasks and y_t be a data sample of task $t \in \Omega$. A collaGAN learns a mapping from an input image x and a random noise z to a set of outputs $\{y_t | t \in \Omega\}$, i.e., $G : \{x, z\} \rightarrow \{y_t | t \in \Omega\}$. The network is trained in an adversarial fashion. The generator G tries to generate data samples as "real" as possible such that, $\forall t \in \Omega$, the adversarially trained discriminator D_t cannot tell if a sample is generated by G or from the data domain of task t.

Similar to the classic GAN, the objective function of a collaGAN can be given as follows:

$$\mathcal{L}_{adv}^{\Omega} = \min_G \max_{D_\Omega} \sum_{t \in \Omega} \mathbb{E}_{x, y_t \sim p_{data}(x, y_t)}[\log D_t(x, y_t)]$$
$$+ \lambda_{adv}^t \mathbb{E}_{x \sim p_{data}(x), z \sim p_z(z)}[1 - \log D_t(x, G^t(x, z))], \tag{1}$$

Here, $D_\Omega = \{D_t | t \in \Omega\}$. p_{data} and p_z denote the data and noise distribution, respectively. $G(x, z)$ is the channel-wisely stacked generator outputs of the tasks with $G^t(x, z)$ denoting the output generated for task t. λ_{adv}^t balances the importance of the generator loss for task t. D_Ω and G play a minmax game where D_Ω tries to maximize the objective and G tries to minimize it.

In this work, as shown in Fig. 2, $\Omega = \{i, s, d\}$, denoting three tasks of *inpainting, segmentation, and (landmark) detection*. x is the occluded image, y_i is the original face, y_s is the segmentation mask and y_d is the detection heatmap (See Sect. 4.1). $D_\Omega = \{D_i, D_s, D_d\}$ where D_i, D_s and D_d are the discriminators for the inpainting, segmentation and detection tasks, respectively. $G(x, z) = [G^i(x, z), G^s(x, z), G^d(x, z)]$ are the generator outputs for the three tasks.

Note that instead of only applying adversarial loss to the inpainted image, we have dedicated discriminators for each of the tasks. Our design follows the observation that GANs can also be helpful in some non-generative tasks [23]. In our case, GANs are used to keep the long-range spatial label contiguity of the detection heatmap and the segmentation mask. We have also experimented with an optional setting that feeds the outputs of all the tasks into a single discriminator and discover that the single discriminator selectively ignores the inpainting result as the outputs from other tasks are much easier for judging the realness. Besides, the collaGAN is conditioned on multiple discriminators, one per task. Such a design, on one hand, ensures the perceptual quality of the generated image and, on the other hand, keeps the spatial consistency between the input image and generated image. This choice was also shown to be effective in [12].

3.2 Reconstruction Loss

It has been found by previous approaches [12,25] that mixing the adversarial loss with a reconstruction loss is beneficial to generative image-to-image models. A reconstruction loss gives pixel-level measurement of the errors which is a direct regularization between the output and the ground truth. It can capture the overall structure of the target object but is usually unable to give a sharp output. An adversarial loss, on the other hand, can produce perceptually better result but the output may not be structurally consistent with input. Therefore, we combine these two to achieve both realistic and coherent output.

In our case, the reconstruction loss is computed for the three tasks as denoted in Fig. 2. For the inpainting task, the reconstruction loss $\mathcal{L}^i_{rec}$ measures the L1 distance between the inpainted image and the unoccluded image. Here, we use the L1 loss instead of the L2 loss due to the observation that the L2 loss tends to give slightly blurry outputs. For the segmentation mask, we first convert the label mask to a multi-channel map with each channel denoting the binary mask of a label. Then, we apply the L2 loss $\mathcal{L}^s_{rec}$ to the multi-channel map to measure the difference between the network output and the ground truth. We use the L2 loss against the typical cross-entropy loss simply for the ease of implementation as the generator (See sect. 3.4) produces outputs with values between $(-1, 1)$ which favors a regression loss. In our experiments, we find the L2 loss gives good enough segmentation map for this study. For the detection heatmap, we also use the L2 loss $\mathcal{L}^d_{rec}$ as the regularizer adapting the choices from [26,27]. Let $\Omega = \{i, s, d\}$, the total reconstruction loss can be written as

$$\mathcal{L}^\Omega_{rec} = \sum_{t \in \Omega} \lambda^t_{rec} \mathcal{L}^t_{rec}, \tag{2}$$

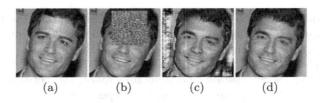

(a)   (b)   (c)   (d)

Fig. 3. Example output of the network trained with the inpainting concentrated scheme. From (a) to (d): (a) original image, (b) masked image, (c) output from the generator, and (d) the final inpainting result by combining (b) and (c).

where $\mathcal{L}_{rec}^{t}$ denotes the reconstruction loss of task t and λ_{rec}^{t} denotes the importance of each loss. The final objective function is then given by

$$\mathcal{L}^{\Omega} = \mathcal{L}_{adv}^{\Omega} + \mathcal{L}_{res}^{\Omega}. \tag{3}$$

3.3 Inpainting Concentrated Generation

In previous work [20], image inpainting is performed through an autoencoder and the unoccluded region is reconstructed along with the occluded part. Thus, the network spends significant portion of its computing power on autoencoding the already available information while the inpainting itself is not fully addressed. A direct approach [25] to this problem is having the generator only outputs the content within the mask. However, this only works for the case where the masks are rectangular. When random shaped regions are occluded, this approach fails as convolutional neural networks cannot generate non-rectangular images. To address this problem for arbitrary shaped masks, we propose an inpainting concentrated scheme. The scheme consists of two parts: an adversarial part and a reconstruction part. For the adversarial part, we modify the adversarial loss for the inpainting such that the discriminator, instead of judging the realness of the generated image, concentrates on finding the incoherence between the inpainted region and the context. Formally, let M be a binary mask where pixels in the occluded region are 0 and anywhere else are 1. The new adversarial loss for inpainting can be written as

$$\begin{aligned}
\mathcal{L}_{adv}^{i} = {}&\mathbb{E}_{x,y_i \sim p_{data}(x,y_i)}[\log D_i(x, y_i)] \\
&+ \lambda_i \mathbb{E}_{x \sim p_{data}(x), z \sim p_z(z)}[(1 - \log D_i(x, \hat{y}_i))],
\end{aligned} \tag{4}$$

where

$$\hat{y}_i = G^i(x, z) \odot (1 - M) + x \odot M. \tag{5}$$

As demonstrated in Fig. 2, $\hat{y}_i$ is nothing but the inpainting output from the generator with the unoccluded region replaced by the ground truth. Such a replacement guarantees that the discriminator does not need to worry about the unrealness of the context and thus the inpainted region is concentrated.

For the reconstruction part, we introduce an inpainting concentrated reconstruction loss that only computes the L1 distances within the occluded region. That is,

$$\mathcal{L}^i_{rec} = \|y_i \odot (1 - M) - G^i(x, z) \odot (1 - M)\|_1. \tag{6}$$

With this scheme, we make sure that no errors outside the occluded region will be backpropagated to the generator, and therefore the unnecessary autoencoding is not learned. Figure 3(c) shows an example output of the network trained with the proposed scheme. The network produces sharp and realistic results inside the occluded region and produces inferior results for the context region that contributes little to the inpainting. Figure 3(d) is obtained using Eq. (5). The inpainted region coherently fits with the context and the image overall looks realistic.

3.4 Network Architecture

For the generator and discriminator, we follow the architecture choices in [28] for stable deep convolutional GANs. Both the generator and discriminator take an input image of size 128×128. For the generator, its encoder has 7 convolution layers with each layer followed by a batch normalization layer [11] and a LeakyReLU [7] layer. The decoder has a symmetric structure with the encoder, except that it uses transposed convolution and ReLU [15]. All the convolutional layers have a 4×4 kernel size with a stride of 2. The output layer of the generator is a tanh function. We also adapt the design suggestion from [12] by adding skip connections between encoder and decoder. Skip connections shuttle low level features directly to decoder without passing through the "bottleneck layers". Such a circumvention is critical to some tasks such as semantic segmentation [29] (which is also included in the collaborative training) and we also find this helpful in improving the coherence between the context and the inpainted region. The discriminator has a similar structure to the encoder, except that it only has 5 convolutional layers.

4 Experimental Results

4.1 Datasets

The dataset used in our experiment is the CelebA [22] dataset. It has $202,599$ face images and we use the official split for training, validating and testing. Unlike the state-of-the-art works [20,34], we do not align the faces according to the eyes. In fact, we find such alignment makes the inpainting much easier for the models as they do not need to semantically learn too much about the locations of the eyes and other face parts. Hence, to avoid overfitting and achieve better generalization, when cropping the faces, we only guarantee that the eyes, nose and mouth are included and *no alignment is performed*. We also augment the dataset by random shift, scaling, rotation and flipping to further ensure the diversity of the faces during training.

Along with each face image, the CelebA dataset readily provides the locations of 5 face landmarks (the two eyes, nose and the two corners of mouth). We create heatmaps from the landmarks according to the method denoted in [26,27]. The generated heatmaps will be used during collaborative training. For a fair comparison, we obtain segmentation masks for each of the faces using the parsing network provided by [20]. The network is trained based on the Helen [17] dataset and achieves a close to state-of-the-art performance. Hence, it is considered sufficient for this study, which is supported by our experimental results. Also, using computer-generated annotations alleviates the burden of laborious manual annotation effort.

All images in the experiments are resized to 128×128. We use this size choice for a fair comparison with other approaches. It is straighforward for the proposed method to use a large image size. Following the mask generation strategies in previous works, we apply three different masks to the resized images: (1) random block mask with a 64×64 block [20]; (2) random pattern mask [25] with roughly 25% of the pixels missing; (3) random noise mask with *80% of the pixels missing* [34]. For (1) and (3) the masked region is filled with random noise. For (2), the masked region is filled with zeros as the mask itself is already noisy.

4.2 Models

To demonstrate the effectiveness of the proposed method, the performances under different model settings are investigated. We denote M_Ω as the model trained using L^Ω and M_Ω^* as the model trained in addition with the inpainting concentrated scheme. Model settings are changed by varying Ω and switching between M_Ω and M_Ω^*. All the investigated models are trained for 20 epochs. For the optimization, we use the Adam [13] optimizer with a learning rate of 0.01. For the weights of different losses, we emperically find the following settings work well:

- Models with one task: $\lambda_{adv}^i = 1.0$, $\lambda_{adv}^s = 1.0$, $\lambda_{adv}^d = 1.0$, $\lambda_{res}^i = 100$, $\lambda_{res}^s = 1000$, $\lambda_{res}^d = 1000$;
- Models with two tasks: $\lambda_{adv}^i = 0.8$, $\lambda_{adv}^s = 0.2$, $\lambda_{adv}^d = 0.2$, $\lambda_{res}^i = 100$, $\lambda_{res}^s = 200$, $\lambda_{res}^d = 200$;
- Models with three tasks: $\lambda_{adv}^i = 0.8$, $\lambda_{adv}^s = 0.1$, $\lambda_{adv}^d = 0.1$, $\lambda_{res}^i = 100$, $\lambda_{res}^s = 200$, $\lambda_{res}^d = 200$.

For models with one or two tasks, the parameters will be used only when they are applicable. For example, model M_i will only have $\lambda_{adv}^i = 1.0$ and $\lambda_{res}^i = 100$. Other parameters are not used as they are for M_s and M_d. Note for this study we are not interested in the best parameter settings for each model. Hence, the parameters are chosen when they work reasonably well. For the λ_{adv}^i, we generally find 0.8 works better in a multi-task scenario and other adversarial loss parameters are chosen such that they sum up to 1. For the reconstruction loss parameters, we find minor performance differences when they are in a reasonable range. In general, setting λ_{res}^i around 100 and λ_{res}^s and λ_{res}^d around 200 gives

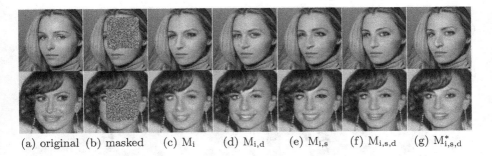

(a) original (b) masked   (c) M_i (d) $M_{i,d}$ (e) $M_{i,s}$ (f) $M_{i,s,d}$ (g) $M^*_{i,s,d}$

Fig. 4. Qualitative face completion comparison of our models with different settings and varying numbers of tasks.

good performance when several tasks are presented. Setting values close to these numbers give similar performances and the performances are degraded when the values are too large or too small.

4.3   Face Completion

Qualitative Comparison. During the experiments, we find in general the models trained with more collaborative tasks produce more realistic results. Figure 4 shows some example outputs of our models. The first row demonstrates that, for an easy inpainting task where context coherence is not critical, the models can all produce relatively good looking faces and the collaboratively trained models from (d) to (g) tend to emphasis more on landmarks (eyes in this case) to give even better inpainting outputs. In the second row, we show a challenging case that only part of the woman's left eye is present and the entire right eye is missing. This inpainting task requires the model to both complete the left eye and reconstruct the entire right eye such that the two eyes together looking realistic. Without collaborative inpainting, the M_i model does not consider the coherence between the two eyes and inpaints the two eyes independently. The other models try to more or less balance the two eyes so that they have the same shape, size and color. The two models trained with three tasks in the last two columns give relatively better inpainting of the eyes. Overall, the $M^*_{i,s,d}$ model trained using the inpainting concentrated scheme gives the most realistic output.

We then compare the proposed method with other state-of-the-art models: CE [25], SII [34] and GFC [20]. Since we use a slightly different face cropping and data augmentation strategy, we retrain those models for a fair comparison. All retrainings are based on their officially released code with the training parameters unchanged. For SII [34], all the experiments are performed using 64×64 images as DCGAN [28] only works well on images with lower resolution. For all the models, poisson blending is performed. The comparison results are given in Fig. 5. To demonstrate the generalizability of the models on various shaped masks, the random pattern masked images in columns 4–6 are inpainted using the models trained with random block masks. For the random noise mask case, since GFC

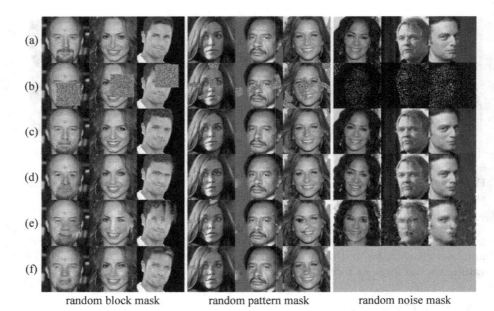

random block mask   random pattern mask   random noise mask

Fig. 5. Qualitative face completion comparison between the proposed method and the state-of-the-art methods. Different masks are applied. From (a) to (f): (a) original image, (b) masked image, and results of (c) the proposed $M^*_{i,s,d}$ model, (d) CE [25], (e) SII [34] and (f) GFC [20].

[20] requires square masks for the local loss, it cannot be used to complete random noise masked images. Hence, the inpainting results for GFC [20] in columns 7–9 are omitted. Also, for CE [25] and our method, we train a new model for the random noise mask case due to the uniqueness of the mask.

From Fig. 5, we observe that our model consistently gives better inpainting results than the state-of-the-art methods. In general, SII [34] gives the worst results in all the cases. Due to the unaligned faces and the data augmentation we performed during training, the face scene becomes more complex. Thus, the DCGAN model used in SII [34] can not learn the face data distribution very well, which as a result yields inferior inpainting performance during the inference step. The CE [25] and GFC [20] models in general produce reasonably good results, especially when the faces are aligned. However, as they either train the model without using additional structural constrains or has an indirect measurement of structural inconsistency, their synthesized images are less structurally realistic than ours.

Figure 6 shows the feature maps extracted from the last common layer in the generator G of the $M_{i,s,d}$ and M_i models. The $M_{i,s,d}$ model is more predictive in the masked region and outputs better face related features. It is interesting to notice that the $M_{i,s,d}$ model tends to treat face parts independently. Some maps contain features only about nose, mouse or eyes. While the M_i model mostly outputs features about the whole face. This indicates that the $M_{i,s,d}$ model is more discriminative about the face than the M_i model and it learns to distinguish (and generate) each of the face part due to the training with other tasks.

Quantitative Comparison. In addition to the visual comparison, we also quantitatively compare our models with the state-of-the-art models to statistically understand their inpainting performances. All the evaluated models are trained with random block masks. We use two classic metrics, PSNR and SSIM, to evaluate the similarity between the inpainted image and the ground truth. PSNR gives the similarity score at pixel-level while SSIM evaluates the similarity at perceptual level. The ground truth image used in this evaluation is the original image before the occlusion. However, since given an occluded image, there could be multiple inpainting results that are perceptually correct, we realize that neither PSNR or SSIM offers a perfect evaluation. But due to the symmetric nature and the relatively simple structure of human faces, as long as the occlusion does not hide significant portion of the face, all the possible inpainting results should be similar which gives us an opportunity to roughly evaluate the models' performance. Table 1 shows the evaluation results at 6 different mask locations. We adapt the mask location choices from [20] by masking out left eye (O1), right eye (O2), upper face (O3), left face (O4), right face (O5), and lower face (O6), respectively. We intentionally select relatively smaller mask sizes (on average 40 × 48) to limit the variation of all possible inpaintings. We observe from Table 1 that our base model M_i has already obtained comparable or slightly better performance than the state-of-the-art models which demonstrates the effectiveness of introducing skip connections into collaGAN. In general, models trained with more tasks give better numbers in both PSNR and SSIM. This shows that the knowledge among tasks is indeed collaborative during training. We also find that $M_{i,s,d}^*$ performs slightly better than $M_{i,s,d}$ which means the inpainting concentrated scheme is really helpful in getting better inpainting results. Overall, we spot a significant performance jump with the proposed method when compared with the state-of-the-art models.

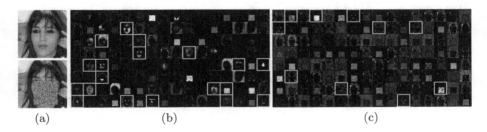

(a)   (b)   (c)

Fig. 6. Feature maps from the last common layer in the generator G. Face related features are marked with yellow squares. Features from early layers through the skip connection are omitted. (a) The original and masked images, (b) Feature maps of the $M_{i,s,d}$ model, and (c) Feature maps of the M_i model. (Color figure online)

To further understand the collaborative nature of the proposed method, we also investigate if the inpainting can help other tasks as well. Specifically, we want to know that, through the learning of inpainting, if the model can predict better of the semantical labels and landmarks in the occluded region. We use the Dice coefficient and localization error to evaluate the performance of semantic segmentation and landmark detection, respectively. The Dice coefficient measures the similarity of two segmentation masks and the localization error computes the Euclidean distances between two landmarks. Table 2 gives the Dice coefficient of different models trained with the segmentation task. The performances of the facial semantic labels are shown. It is clear that *the collaboratively trained models perform better than single task models and the models trained with three tasks better than with two tasks.* The $M_{i,s,d}$ and $M^*_{i,s,d}$ models have very close performance. This is reasonable as the inpainting concentrated scheme is designed solely for inpainting and does not introduce additional information to other tasks during training. Note that, for the segmentation task, the performance difference of the models is not significant. We speculate that this is because the "ground truth" is computer-generated from the parsing network which sometimes may not give perfect prediction. For the landmark detection task with accurately labeled landmark locations provided by the CelebA dataset, the performance boost of introducing other tasks is more significant as shown in Table 3. For example, the average landmark error is reduced from 2.38 (the M_d model) to 1.72 (the $M^*_{i,s,d}$ model), a 27.7% decrease.

Table 1. Quantitative face completion comparison of different models evaluated at 6 different mask locations: left eye (O1), right eye (O2), upper face (O3), left face (O4), right face (O5), and lower face (O6). The numbers in each cell are SSIM (%)/PSNR (dB), the higher the better.

	O1	O2	O3	O4	O5	O6
CE [25]	90.5/26.74	90.6/27.01	93.8/27.90	95.8/30.37	96.0/30.65	90.0/27.11
SII [34]	87.5/23.93	87.7/24.12	93.0/26.61	95.6/28.94	95.9/29.38	87.8/24.84
GFC [20]	90.6/27.10	90.9/27.34	94.0/28.68	96.3/31.18	96.3/31.11	90.0/27.13
M_i	90.8/27.23	91.1/27.42	94.0/28.53	96.0/30.94	96.1/31.06	90.1/27.15
$M_{i,d}$	91.7/27.57	91.6/27.81	94.5/28.76	96.4/31.33	96.5/31.36	90.7/27.37
$M_{i,s}$	91.4/27.59	91.5/27.65	94.3/28.66	96.4/31.22	96.4/31.28	90.7/27.55
$M_{i,s,d}$	91.7/27.66	91.8/27.87	94.5/28.77	96.5/31.31	96.6/31.39	90.8/27.57
$M^*_{i,s,d}$	**92.4/27.76**	**92.6/27.96**	**95.2/28.79**	**97.2/31.44**	**97.2/31.50**	**91.7/27.81**

Table 2. Semantic segmentation performance of different models. The numbers are given as Dice coefficient (%). Higher numbers are better.

	M_s	$M_{i,s}$	$M_{i,s,d}$	$M_{i,s,d*}$
Face	93.7	94.1	**94.2**	94.1
Left eyebrow	74.2	74.6	75.0	**75.2**
Right eyebrow	72.3	72.8	**73.8**	73.5
Left eye	70.7	71.9	72.2	**72.7**
Right eye	70.0	70.3	**71.2**	**71.2**
Nose	90.5	90.9	90.9	**91.0**
Upper lip	68.1	67.2	**68.3**	67.9
Teeth	64.2	66.5	**66.8**	66.7
Lower lip	81.4	82.1	**82.8**	82.4
Average	76.1	76.7	**77.2**	**77.2**

Table 3. Landmark detection performance of different models. The numbers are given as localization errors in pixels. Lower numbers are better.

	M_d	$M_{i,d}$	$M_{i,s,d}$	$M_{i,s,d*}$
Left eye	2.12	1.73	1.61	**1.60**
Right eye	2.29	1.71	**1.59**	1.62
Nose	2.53	2.15	**1.92**	1.93
Left mouth	2.39	1.80	1.74	**1.73**
Right mouth	2.57	1.86	1.77	**1.72**
Average	2.38	1.85	1.73	**1.72**

5 Conclusion

We present a novel collaborative GAN framework for face completion. The experimental results suggest that training multiple related tasks together within the proposed framework is beneficial. By infusing more knowledge, the generative model learns better about the inpainting through the knowledge sharing of the segmentation and detection tasks, whose performances are boosted vice versa. We have also found that optimizing directly toward inpainting other than autoencoding produces better inpainting results in an image-to-image network. Finally, we have demonstrated that the proposed method can give superior inpainting performance than the state-of-the-art methods.

Disclaimer: This feature is based on research, and is not commercially available. Due to regulatory reasons its future availability cannot be guaranteed.

References

1. Barnes, C., Shechtman, E., Finkelstein, A., Goldman, D.B.: PatchMatch: a randomized correspondence algorithm for structural image editing. ACM Trans. Graph. **28**(3), 24 (2009)
2. Caruana, R.: Multitask learning. In: Thrun, S., Pratt, L. (eds.) Learning to learn, pp. 95–133. Springer, Boston (1998). https://doi.org/10.1007/978-1-4615-5529-2_5
3. Ganin, Y., et al.: Domain-adversarial training of neural networks. J. Mach. Learn. Res. **17**(1), 2030–2096 (2016)
4. Girshick, R.: Fast R-CNN. arXiv preprint arXiv:1504.08083 (2015)
5. Hays, J., Efros, A.A.: Scene completion using millions of photographs. ACM Trans. Graph. (TOG) **26**, 4 (2007)

6. He, K., Sun, J.: Statistics of patch offsets for image completion. In: Fitzgibbon, A., Lazebnik, S., Perona, P., Sato, Y., Schmid, C. (eds.) ECCV 2012. LNCS, pp. 16–29. Springer, Heidelberg (2012). https://doi.org/10.1007/978-3-642-33709-3_2

7. He, K., Zhang, X., Ren, S., Sun, J.: Delving deep into rectifiers: surpassing human-level performance on ImageNet classification. In: Proceedings of the IEEE International Conference on Computer Vision, pp. 1026–1034 (2015)

8. Hinton, G.E., Salakhutdinov, R.R.: Reducing the dimensionality of data with neural networks. Science 313(5786), 504–507 (2006)

9. Huang, J.B., Kang, S.B., Ahuja, N., Kopf, J.: Image completion using planar structure guidance. ACM Trans. Graph. (TOG) 33(4), 129 (2014)

10. Iizuka, S., Simo-Serra, E., Ishikawa, H.: Globally and locally consistent image completion. ACM Trans. Graph. (TOG) 36(4), 107 (2017)

11. Ioffe, S., Szegedy, C.: Batch normalization: accelerating deep network training by reducing internal covariate shift. In: International Conference on Machine Learning, pp. 448–456 (2015)

12. Isola, P., Zhu, J.Y., Zhou, T., Efros, A.A.: Image-to-image translation with conditional adversarial networks. arXiv preprint arXiv:1611.07004 (2016)

13. Kingma, D., Ba, J.: Adam: a method for stochastic optimization. arXiv preprint arXiv:1412.6980 (2014)

14. Kingma, D.P., Welling, M.: Auto-encoding variational Bayes. arXiv preprint arXiv:1312.6114 (2013)

15. Krizhevsky, A., Sutskever, I., Hinton, G.E.: ImageNet classification with deep convolutional neural networks. In: Advances in Neural Information Processing Systems, pp. 1097–1105 (2012)

16. Larsen, A.B.L., Sønderby, S.K., Larochelle, H., Winther, O.: Autoencoding beyond pixels using a learned similarity metric. arXiv preprint arXiv:1512.09300 (2015)

17. Le, V., Brandt, J., Lin, Z., Bourdev, L., Huang, T.S.: Interactive facial feature localization. In: Fitzgibbon, A., Lazebnik, S., Perona, P., Sato, Y., Schmid, C. (eds.) ECCV 2012. LNCS, vol. 7574, pp. 679–692. Springer, Heidelberg (2012). https://doi.org/10.1007/978-3-642-33712-3_49

18. Ledig, C., et al.: Photo-realistic single image super-resolution using a generative adversarial network. arXiv preprint arXiv:1609.04802 (2016)

19. Li, J., Liang, X., Wei, Y., Xu, T., Feng, J., Yan, S.: Perceptual generative adversarial networks for small object detection. In: IEEE CVPR (2017)

20. Li, Y., Liu, S., Yang, J., Yang, M.H.: Generative face completion. arXiv preprint arXiv:1704.05838 (2017)

21. Liu, P., Qiu, X., Huang, X.: Adversarial multi-task learning for text classification. arXiv preprint arXiv:1704.05742 (2017)

22. Liu, Z., Luo, P., Wang, X., Tang, X.: Deep learning face attributes in the wild. In: Proceedings of International Conference on Computer Vision (ICCV), December 2015

23. Luc, P., Couprie, C., Chintala, S., Verbeek, J.: Semantic segmentation using adversarial networks. arXiv preprint arXiv:1611.08408 (2016)

24. Mirza, M., Osindero, S.: Conditional generative adversarial nets. arXiv preprint arXiv:1411.1784 (2014)

25. Pathak, D., Krahenbuhl, P., Donahue, J., Darrell, T., Efros, A.A.: Context encoders: feature learning by inpainting. In: Proceedings of the IEEE Conference on Computer Vision and Pattern Recognition, pp. 2536–2544 (2016)

26. Payer, C., Štern, D., Bischof, H., Urschler, M.: Regressing heatmaps for multiple landmark localization using CNNs. In: Ourselin, S., Joskowicz, L., Sabuncu, M.R., Unal, G., Wells, W. (eds.) MICCAI 2016. LNCS, vol. 9901, pp. 230–238. Springer, Cham (2016). https://doi.org/10.1007/978-3-319-46723-8_27

27. Pfister, T., Charles, J., Zisserman, A.: Flowing ConvNets for human pose estimation in videos. In: Proceedings of the IEEE International Conference on Computer Vision, pp. 1913–1921 (2015)

28. Radford, A., Metz, L., Chintala, S.: Unsupervised representation learning with deep convolutional generative adversarial networks. arXiv preprint arXiv:1511.06434 (2015)

29. Ronneberger, O., Fischer, P., Brox, T.: U-Net: convolutional networks for biomedical image segmentation. In: Navab, N., Hornegger, J., Wells, W.M., Frangi, A.F. (eds.) MICCAI 2015. LNCS, vol. 9351, pp. 234–241. Springer, Cham (2015). https://doi.org/10.1007/978-3-319-24574-4_28

30. Simonyan, K., Zisserman, A.: Two-stream convolutional networks for action recognition in videos. In: Ghahramani, Z., Welling, M., Cortes, C., Lawrence, N.D., Weinberger, K.Q. (eds.) Advances in Neural Information Processing Systems, vol. 27, pp. 568–576. Curran Associates, Inc. (2014). http://papers.nips.cc/paper/5353-two-stream-convolutional-networks-for-action-recognition-in-videos.pdf

31. Vincent, P., Larochelle, H., Bengio, Y., Manzagol, P.A.: Extracting and composing robust features with denoising autoencoders. In: Proceedings of the 25th International Conference on Machine learning, pp. 1096–1103. ACM (2008)

32. Whyte, O., Sivic, J., Zisserman, A.: Get out of my picture! internet-based inpainting. In: BMVC, vol. 2, p. 5 (2009)

33. Yang, C., Lu, X., Lin, Z., Shechtman, E., Wang, O., Li, H.: High-resolution image inpainting using multi-scale neural patch synthesis. In: The IEEE Conference on Computer Vision and Pattern Recognition (CVPR), vol. 1, p. 3 (2017)

34. Yeh, R., Chen, C., Lim, T.Y., Hasegawa-Johnson, M., Do, M.N.: Semantic image inpainting with perceptual and contextual losses. arXiv preprint arXiv:1607.07539 (2016)

35. Zhang, Z., Luo, P., Loy, C.C., Tang, X.: Facial landmark detection by deep multitask learning. In: Fleet, D., Pajdla, T., Schiele, B., Tuytelaars, T. (eds.) ECCV 2014. LNCS, vol. 8694, pp. 94–108. Springer, Cham (2014). https://doi.org/10.1007/978-3-319-10599-4_7

Water-Filling: An Efficient Algorithm for Digitized Document Shadow Removal

Seungjun Jung[1] , Muhammad Abul Hasan[2] , and Changick Kim[1]([⊠])

[1] KAIST, Daejeon, Republic of Korea
{seungjun45,changick}@kaist.ac.kr
[2] University of South Australia, Adelaide, SA, Australia
muhammad.hasan@unisa.edu.au

Abstract. In this paper, we propose a novel algorithm to rectify illumination of the digitized documents by eliminating shading artifacts. Firstly, a topographic surface of an input digitized document is created using luminance value of each pixel. Then the shading artifact on the document is estimated by simulating an immersion process. The simulation of the immersion process is modeled using a novel diffusion equation with an iterative update rule. After estimating the shading artifacts, the digitized document is reconstructed using the Lambertian surface model. In order to evaluate the performance of the proposed algorithm, we conduct rigorous experiments on a set of digitized documents which is generated using smartphones under challenging lighting conditions. According to the experimental results, it is found that the proposed method produces promising illumination correction results and outperforms the results of the state-of-the-art methods.

Keywords: Shadow removal · Document image processing ·
Diffusion equation

1 Introduction

With the progressive development of built-in cameras in smart handheld devices, people are generating more amount of digital content than ever before. Aside from capturing moments, such powerful cameras are used for capturing digitized copies of printed documents (*e.g.,* important notes, certificates, and visiting cards) for keeping personal records or for sharing the documents with others. This is a popular global trend for its ready availability and relatively easier operational methods than any other digitizing device, *e.g.,* a scanner. Although generating a digitized document using a smart handheld devices is as easy as

This research was supported by Hancom Inc.

Electronic supplementary material The online version of this chapter (https://doi.org/10.1007/978-3-030-20887-5_25) contains supplementary material, which is available to authorized users.

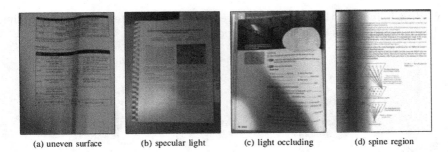

(a) uneven surface   (b) specular light   (c) light occluding   (d) spine region

Fig. 1. Examples of digitized documents having different types of illumination distortion.

pointing and shooting the document, the quality of the captured digitized image is often a concern. The paper documents having folds cause to create uneven surfaces and result in illumination distorted documents. Additionally, the specular reflections and hard shadows contribute in rendering poor quality digitized documents. Figures 1(a)-(c) show three examples of illumination distorted digitized documents captured in different lighting conditions using smartphone's cameras. Even though the scanners are specialized device for producing digitized documents, it also suffer from its own limitations. For example, scanning a page from a thick and bound book suffer from illumination distortions because of the curled area in the spine region [17] (see Fig. 1(d)). Although one can attempt to remove these distortions by pressing the spine of the book harder, such distortions can hardly be eliminated [28, 32].

Correcting illumination distortion of digitized documents is an important task for not only enhancing readability for the readers but also it is crucial for applying optical character recognition (OCR) software on the digitized documents. Various studies in the literature show the importance of rectifying illumination distortion for a significantly better OCR performance [2, 28, 33, 35]. The traditional methods for correcting the illumination distortions of the digitized documents start with estimating the background shades followed by removing the shades using a document surface reconstruction model [4, 10, 13, 18–20, 30, 34, 36]. These strategies can be divided into two prominent categories: mask-and-interpolation approaches [10, 13, 34, 36], and without-mask-interpolation approaches [4, 18–20, 30]. The former approaches figure out the location of text, and photo regions on the documents using either edge detection or binarization methods, followed by applying a mask that covers the text regions along with its adjacent area on the surface of the documents. Finally, the masked regions are interpolated to correct the illumination of the background of the digitized documents. The latter approaches use color histograms of the document image patches to aggregate the local regions belonging to the background and interpolate the rest of the regions to identify the shaded regions. As removing the shading artifacts using the well-known surface models is a straightforward task, most of the works in the literature concentrated only on extracting the background layers as accurately as possible.

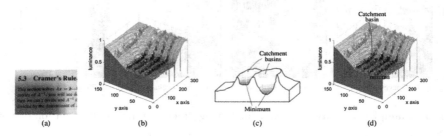

Fig. 2. A sample digitized image and its topographic surface visualization. (a) Example digitized document, (b) topographic surface representation of (a). (c) An illustration of catchment basins and corresponding minimum. (d) An example of catchment basin and the corresponding minima on a digitized document representing (c).

In this paper, we propose an efficient illumination correction algorithm for removing shading artifacts from the digitized images. The algorithm starts with constructing a topographic surface using the luminance values of the digitized image pixels irrespective of the contents of the documents (*i.e.,* text and image). Then, a model representing the dynamics of fluids simulated by a diffusion equation is applied on the topographic surface to estimate the shading artifacts on the documents. After estimating the shading artifacts, the digitized documents are reconstructed using the Lambertian surface model to correct the illumination by removing the shades. An example of a topographic surface representation of a digitized image segment is visualized in Fig. 2. In designing the proposed algorithm, we are influenced by the techniques applied in watershed transform [9,23,31]. Accordingly, we named the proposed method called *Water-filling* algorithm for digitized document illumination correction.

The rest of the paper is organized as follows. In Sect. 2, we present a review of related works in the literature. In Sect. 3, the proposed Water-filling algorithm for digitized documents illumination correction is described. The experimental results and comparison studies are presented in Sect. 4. Finally, the conclusion is presented in Sect. 5.

2   Related Work

We review the related works in the literature and categorize them into three groups. Each of the groups is discussed in the following subsections.

2.1   Classical Image Binarization Based Approaches

One of the traditional approaches for removing shading artifacts from gray-scale digitized documents is binarization technique based methods [1,5,8,21,22,24,25].

In [22, 24], the authors proposed a locally adaptive binarization technique based method which computes a threshold using local information. Bukhari *et al.* [5] improved the method proposed in [24] by introducing an automatic parameter tuning algorithm using local information. However, it is inevitable that these methods suffer from significant degradation in photographic regions of the digitized documents as such methods convert a gray level pixel to either a black or a white pixel.

2.2 Parametric Reconstruction Based Approaches

Illumination distortion on camera generate digitized documents of uneven surface is addressed in the literature using the 3D reconstruction models. Lu *et al.* [14, 15] proposed a method for removing shading artifacts on digitized documents captured by cameras using the 3D shape reconstruction of the documents. In their method, the 3D shape of the document is reconstructed by fitting the illumination values to a polynomial surface. Meng *et al.* [16] proposed another 3D shape reconstruction based method by isometric mesh construction assuming that the page shape is a general cylindrical surface (GCS). Assuming GCS form for camera captured digitized document images are proven to work well for modeling 3D shape of real documents [11]. Tian *et al.* [29] proposed another 3D reconstruction based method using text region on the digitized documents. In their method, the perspective distortion of the text region is estimated using text orientation and horizontal text line. Although their proposed methods are effective for correctly identify illumination distortions due to uneven surface of the target document, the method has a limitation in rectifying the shading artifacts on the digitized documents in general (e.g., shadow from light occluding object).

Digitized documents generated by scanners experience relatively less shading artifacts as the scanners use a single light source with an uniform light direction. For such digitized documents, the surface of the document is rendered smooth and constant. Such type of digitized document can be considered to have a parametric surface with various assumptions [17, 28, 37]. Such assumptions lead to a straightforward reconstruction of the 3D shapes for digitized documents captured by scanners. The assumption based digitized document illumination correction techniques are not applicable for correcting generalized digitized documents.

2.3 Background Shading Estimation Based Approaches

In background shading estimation based methods, a digitized document is assumed to have two separate layers. The background layer–the shading layer which contains illumination distortions, and the foreground layer–the layer which contains the text and images.

Mask and Interpolation. In mask and interpolation approach, a mask is created to cover the text region by detecting the text on a digitized document. Zhang *et al.* [34,36] proposed a method to create a mask using the Canny edge detector algorithm followed by using morphological closing operation. This method is effective only for documents which contain the text of a specific size. Documents containing varying font sizes or documents containing large photos result in poor illumination corrected documents. Lee *et al.* [13] proposed another mask and interpolation based method which detects the text and the photo regions with large rectangle masks using the connected edges. This approach is effective for closely captured digitized documents without having any surrounding information. Otherwise, the whole document would be identified as a separate rectangle, hence, that rectangle area would be processed separately. In [10], Gato *et al.* introduced a method to use the binarized image as a mask. However, the generated masks often fail to cover the photo regions on the digitized documents accurately.

Direct Interpolation. In direct interpolation based approaches, image local regions belonging to the background are detected for illumination correction. Oliveira *et al.* [18–20] assumed that a local region belonging to a digitized document background has a narrow Gaussian shape in its color histogram. With that assumption, initially, the local blocks belonging to the background were identified. Then those blocks were used to estimate the background layer using an interpolation method. Since an image with large size photo has piecewise constant color in the photo region, this method does not work correctly in such cases. In [3], a shadow map is directly generated using local background and text color estimation. Then the shadow map is applied on the document to rectify the shading artifacts, while document images having irregular shading patterns leads the algorithm to a wrong estimation of the global shadow map. Tsoi *et al.* [4,30] assumed that boundary region of a digitized document has uniform color. The shading of the interior document was estimated by a linear interpolation of shading using document boundary regions. The method also suffers when shading of the boundary regions has irregular shading patterns. Fan *et al.* [7] used a watershed-transform segmentation method on the noise-filtered luminance image to segment the background regions. Then, interpolation is applied on the textual regions to estimate illuminant values throughout the image. Since the segmented region with the largest catchment basins is automatically classified as a background region, the performance of the method from this paper is limited when two background regions are separated.

3   Water-Filling Algorithm

The proposed digitized document illumination correction algorithm, called Water-Filling algorithm, is described in this section. We start the description with a number of terminologies and relevant observation. The topography of a local area depicts the detail physical construction of land which includes elevation and depression. A *catchment basin* on a topography surface is an extent of a land where the accumulated natural water converges to a meeting point, technically called a *minima*. An illustration of catchment basins and the corresponding minimum is displayed in Fig. 2(c). A digitized document, having shading artifacts, can be represented as a topographic surface, where a depression on the surface is considered as a catchment basin. Figure 2(d) shows an example of a catchment basin appeared on a digitized document due to shading artifact.

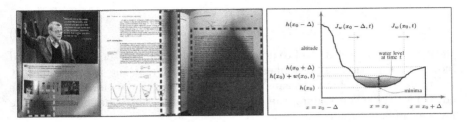

Fig. 3. (left) Examples of shaded region, marked in orange dashed rectangles, touching the boundary of the images. (right) An one dimensional cartoon topographic model of a catchment basin for illustrating the diffusion equation for document illumination correction. (Color figure online)

An illumination distorted region on a digitized document captured by a camera often occurred due to occluding light sources by the camera itself along with the camera holding hand(s) of the user. In case of a scanner, the illumination distorted region often spread from top to bottom or left to right of the document surface. In both device cases, the shaded regions touch the boundary of the digitized documents. Figure 3 shows a few examples where it can be seen that the shaded region meets the boundary. In the proposed algorithm, we exploit this general observation and summarize that the catchment basins which touch the image boundary correspond to the shading artifacts of the background regions. The other existing catchment basins which do not touch the boundary of the image are in general belonging to the text or photo regions of the documents. We solidify this idea empirically in supplementary material. Based on these observations, let us assume that $D_I \subset \mathbb{Z}^2$ be a domain of two-dimensional gray-scale image I.

In the following sub-sections, the digitized document illumination correction problem is formalized followed by the proposed solution mechanism discussion in detail.

3.1 Modeling by Diffusion Equation

In this sub-section, we model a filling mechanism of catchment basins using water. We use two different methods to simulate this water-filling task: *incrementally filling of catchment basins* method, and *flood-and-effuse* method. Ultimately, we link these two methods to provide the final version of the water-filling method.

Before modeling each method using diffusion equation, we need to set a few constraints on a few variables. Let h be the altitude of the topographic surface, $w(x, t)$ be the water level on the topographic surface x at a point of time t. Figure 3 illustrates an one-dimensional cartoon model of a catchment basin. Our objective is to restore the background of the digitized document affecting the structure of the document as low as possible. In order to meet the objective, we set two constraints on $w()$ given as follows.

$$w(x, t) \geq 0, \quad \forall t, \ \forall x \in D_I, \tag{1}$$

$$w(x, t) = 0, \quad \forall t, \ \forall x \in \partial D_I. \tag{2}$$

Equation (1) limits water level to non-negative value since water either be stored or flow out. Equation (2) describes the drop of water at the image boundary so that only the catchment basins at the interior region of the image can be filled. Under these constraints, the dynamics of water can be designed with the following diffusion equation.

$$\left. \frac{\partial w(x, t)}{\partial t} \right|_{x=x_0} = -\nabla \cdot \mathbf{J_w}(x, t) \big|_{x=x_0} = -\left. \frac{\partial J_w(x, t)}{\partial x} \right|_{x=x_0}, \tag{3}$$

where $\mathbf{J_w}$ is the flux of the diffusing water. Equation (3) states that a change in water level in any part of the structure is due to inflow and outflow of water into and out of that part of the structure.

Incremental Filling of Catchment Basins. Let $G(x_0, t)$ be the overall altitude of water at a location x_0 at time t, which can be written as $G(x_0, t) = h(x_0) + w(x_0, t)$. Assuming that there is a continuous water supply source from the peak $\hat{h}$, the total flow of the water is directly proportional to its relative height at x. Formally, we write it as follows.

$$J_w(x_0, t) \propto G(x_0, t) - G(x_0 + \Delta, t). \tag{4}$$

Similarly, for the 2D case of the flow dynamics, the partial derivative of $w()$ with respect to time t at (x_0, y_0) is proportional to the relative difference of $G()$ at (x_0, y_0) and its neighborhood, where (x, y) represents a point on the 2D topographic surface. Following Eq. (4), the 2D iterative update formula for $w()$ is given in Eq. (5).

$$
\begin{aligned}
w(x, y, t + \Delta) = &\eta \cdot \{ G(x + \Delta, y, t) + G(x - \Delta, y, t) + G(x, y + \Delta, t) \\
&+ G(x, y - \Delta, t) - 4 \cdot G(x, y, t) \} + w(x, y, t),
\end{aligned} \tag{5}
$$

where η is a hyperparameter which decides the speed of the process. The η has to be chosen carefully as an inappropriate value of η might lead to slow convergence or divergence of the water level. Since we use $G()$ values from four-neighborhood at point (x, y), the proper η value is less than or equal to 0.25. The value of η greater than 0.25 cause flooding in the whole structure. The value of $G()$ after convergence is used to estimate the background layer of an image. The $G()$ is converged at time t, if $G(x, y, t) - G(x, y, t - \Delta) < \delta, \ \forall x, \forall y$, where δ should be a small value. In this paper, we have set $\delta = 0.01$.

(a)   (b)

Fig. 4. Estimation of background layer $G(x, y, t)$ using our iterative water-filling approach (left to right: $t = 0, 50, 200$): (a) incremental filling of catchment basins and (b) flood-and-effuse method. The red color circles show 10 times magnified content at the center of the documents for better visual understanding. (Color figure online)

Flood and Effuse. In this method, the diffusion equation can be decomposed into two independent processes: water effusion, and flood. For the effusion process, we consider the dynamics of water flow on the topographic surface without any external water supply. We only consider the effusion of water through the image border. Also, the amount of water at each location never be increased in order to further boost the speed of effusion process. The formulation of this idea for the 1D case is straightforward from Eqs. (3) and (4) is given as follows.

$$w_\phi(x_0, t) \propto \min\{-G(x_0, t) + G(x_0 + \Delta, t), 0\}$$
$$+ \min\{-G(x_0, t) + G(x_0 - \Delta, t), 0\}. \tag{6}$$

The effusive term $w_\phi(x_0, t)$ has non-positive value and represents the amount of water to be effused at each location. Conversely, in the flooding process $w_\psi(x_0, t)$, the model be immersed to the same altitude for few initial moments.

$$w_\psi(x_0, t) = (\hat{h} - G(x_0, t)) \cdot e^{-t}, \tag{7}$$

where $\hat{h} = \max h(x), \forall x$. The partial derivative of the overall water level $w()$ with respect to time t is the sum of w_ϕ and w_ψ. The iterative update formula for $w()$ in the 2D case is as follows.

$$w(x, y, t + \Delta) = (\hat{h} - G(x, y, t)) \cdot e^{-t} + \eta \cdot \{\min\{G(x + \Delta, y, t) - G(x, y, t), 0\}$$
$$+ \min\{G(x - \Delta, y, t) - G(x, y, t), 0\} + \min\{G(x, y + \Delta, t) - G(x, y, t), 0\}$$
$$+ \min\{G(x, y - \Delta, t) - G(x, y, t), 0\}\} + w(x, y, t). \tag{8}$$

Once $G()$ converges, it can be used to reconstruct photometrically correct version of the digitized document. In Lambertian surface model, the illumination value of an image is equal to the product of foreground text layer and background shading layer [34]. Given the background layer I_b and foreground layer I_f, the final photometrically correct image, I_r, can be computed as follows.

$$I_r(x, y) = \ell \cdot I_f = \ell \cdot \frac{I(x, y)}{I_b(x, y)} \approx \lim_{t \to \infty} \frac{I(x, y)}{G(x, y, t)} \cdot \ell, \tag{9}$$

where $\ell \in [0,1]$ for tuning the brightness of the output image. A few example of simulation results of the above two methods are shown in Fig. 4. It is notable that the results from the flood method show the desired dynamics precisely – once the overall image is filled with water, it gradually effuses water at the image boundary.

Although the incremental filling method removes shading artifacts quite well, the photo regions are not clearly restored because the filling process at the large catchment basins are too slow. Meanwhile, it is observed that the overall image is reconstructed correctly with the flood-and-effuse method.

Boosting with Downsampling. The time complexity of the proposed algorithm is $\mathcal{O}(I_{width} \times I_{height} \times t)$. Since both of the Eqs. (5) and (8) are pixel-wise operations, the processing time increases as the image size gets larger, which hinders these methods from being used in real-time applications. Although applying either equation after downsampling the input document size at a rate of k_s can boost the processing time significantly, we observe that background shadings on small catchment basins may not be estimated correctly due to aliasing effect. However, shading artifacts in a small region do not have a drastic effect on the overall digitized document illumination correction. Thus, we apply the flood-and-effuse method on the downsampled input document only to attain the rough approximation quickly. After estimating the background artifacts, using bicubic interpolation the background layer is approximated to the original size. Then, the incremental filling method is used to reconstruct the details of the rectified digitized document. In the following section, we show that this combined scheme produces reliable results in a shorter time with acceptable peak signal-to-noise ratio (PSNR). The pseudocode of the proposed algorithm is given in Algorithm 1.

Algorithm 1. Water-Filling Algorithm

1: **procedure** WATER-FILLING(Img, k_s)
2:      $[Y, Cb, Cr] \leftarrow$ RGB_to_YCbCr (Img)
3:      $\overline{Y} \leftarrow$ Downsample(Y, k_s)   ▷ Sampling rate of k_s
4:      $\overline{w} \leftarrow$ Flood($\overline{Y}$)   ▷ Rough sketching
5:      $\overline{G} \leftarrow \overline{w} + \overline{Y}$
6:      $G \leftarrow$ Upscale($\overline{G}, k_s$)   ▷ Bicubic interpolation
7:      $w \leftarrow$ Incremental(G)   ▷ Detail reconstruction
8:      $G \leftarrow w + G$ ▷ Final background
9:      $Y \leftarrow \ell \times Y/G$ ▷ ℓ is a brightness factor
10:     $Img_c \leftarrow$ YCbCr_to_RGB($[Y, Cb, Cr]$)
11:     **return** Img_c
12: **end procedure**

---

4 Experiments and Performance Evaluation

4.1 Datasets

Since there is no publicly available digitized document illumination distortion dataset, we have created our own to compare the performance of our proposed method with the state-of-the-art methods. Primarily our dataset consists of 159 illumination distorted digitized documents, among them 109 images are captured using two smartphone's cameras and the rest 50 images are captured using a scanner. The size of camera captured images are 3264×2448, and the scanned images are 3455×2464 (72 dpi). The digitized document captured by cameras are generated under different lighting conditions. For PSNR comparison, we captured 87 ground-truth images along with its illumination distorted digitized images. In order to capture ground-truth images, a camera stand is used in a well-lit room to capture a ground-truth photo followed by capturing another image by adding shades on the document surface by occluding light sources intentionally. We present the benchmark dataset in our supplementary material. We further collected 22 business card images from a publicly available dataset [6] for OCR edit-distance comparison. The proposed algorithm is implemented using Microsoft C++ development environment. For experiments, we use a workstation having an Intel Corei7-4790k, 64 bits CPU with 16 GB memory and 128 SSD storage device.

4.2 Methods for Comparison

The performances of the proposed algorithm have been compared with several state-of-the-art methods in the literature. For comparing the performance on the digitized documents captured by smartphones, we select a binarization method (Sauvola *et al.* [24]), a mask and interpolation method (Zhang *et al.* [34]), two direct interpolation methods (Oliveira *et al.* [18], Bako *et al.* [3] and Kligler *et al.* [12]), and a rolling-ball based method (Sternberg *et al.* [27]). The rolling ball based method estimates a local background value for every pixel by taking an average over a large ball around the pixel. For comparing the performance on digitized documents captured by scanner devices, we select the methods proposed in [17,18,24,27,34].

4.3 Parameters Tuning

There are two hyperparameters to be set in our algorithm for generating optimum results, namely sampling rate k_s and the brightness factor ℓ. In order to do that, first of all, we conduct an experiment using the first 20 distorted images along with their ground-truth images. We need to set an acceptable k_s to correct shading artifacts fast by meeting our objective, that is we want estimate shadow as clearly as possible. Too large k_s makes the algorithm preform faster at a cost of poor quality output. In order to produce acceptable results using the proposed algorithm, we need to look after two crucial components, the PSNR

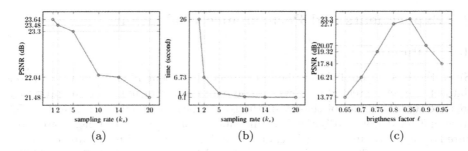

Fig. 5. (a) The trade-off between the sampling rate and PSNR, and (b) the trade-off between the sampling rate and average time taken to correct illumination distortion of a digitized document. In this experiment, we set brightness factor $\ell = 0.85$. (c) The trade-off between the brightness ℓ and PSNR. In this experiment we set the sampling rate $k_s = 5$.

in the corrected images in comparison to the ground-truths, and the average elapsed time to accomplish illumination rectification. Accordingly, we conduct two experiments for selecting an acceptable k_s which are reported in Fig. 5. It is obvious that, by selecting a larger k_s makes the algorithm perform faster by a small margin. However, that would cost the algorithm performing poorly by decreasing significantly large amount of PSNR. Accordingly, we select $k_s = 5$ for the rest of the experiments as the default sampling rate.

The brightness of the output image is also an important factor. The output document should be comfortable to the human eyes–too bright and too dark both are undesirable. To determine the correct brightness factor, we conduct another experiment to measure PSNR for changing brightness factor ℓ. Figure 5 (c) shows the experiment results. As it can be seen, for brightness factor $\ell = 0.85$, we achieve optimum results. Based on this experiment, we set $\ell = 0.85$ for the rest of the experiments to be done.

4.4   Results Analysis

After tuning the hyperparameters of the proposed method, we conduct a visual comparison of the illumination corrected results of ours method and that of a number of existing methods. Figure 6 shows an example digitized documents along with the ground-truth and the corresponding illumination corrected results performed by several methods. For [3,12,18], false detection of background region results in unclear rectification of illumination distortions along the shadow borders. Moreover, their method often adds additional noises in the text and image regions which cause in destroying the structures of the documents. Another important concern of method [18] is that the photo regions on the documents render into dark patches. For the image region on the document, the method [34] fails to preserve the structure and color, in fact, it ruins the visual quality in those regions.

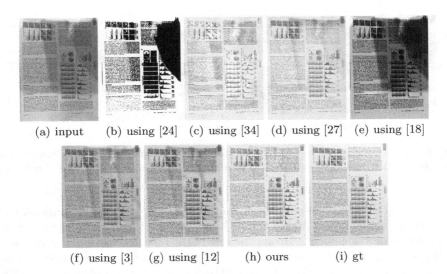

(a) input   (b) using [24]   (c) using [34]   (d) using [27]   (e) using [18]

(f) using [3]   (g) using [12]   (h) ours   (i) gt

Fig. 6. An example of illumination distorted digitized documents (generated using a smartphone camera mounted on a camera stand) and its illumination corrected results using several methods.

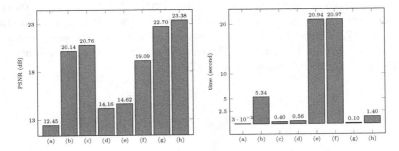

Fig. 7. (left) Ground-truth comparison in terms of average PSNR and (right) average processing time comparison (a) using [24], (b) using [34], (c) using [27], (d) using [18], (e) using [3], (f) using [12], (g) ours (using $k_s = 14$) and (h) ours (using $k_s = 5$).

In order to have a fairer comparison, the methods are applied to the ground-truth data to compare the PSNR. Figure 7 summarizes the performances of the methods on ground-truth data where it can be seen that the proposed method outperforms the other compared methods in terms of PSNR. The closest PSNR to our method is achieved by method [27] with the parameter ball size = 50. However, the PSNR difference between these two method is quite significant and our method is proved to be superior in producing better illumination corrected digitized documents.

The average time requires to produce a rectified document is another important criteria to be measured. From the experimental results shown in Fig. 7, the

(a) original images

(b) Illumination corrected images using proposed water-filling algorithm.

Fig. 8. Examples of business cards chosen from [6] and the corresponding illumination corrected, and geometrically de-warped results after applying the proposed algorithm.

method [24] take least amount of average time (3×10^{-2} sec.) to accomplish the task, while the method [3] takes the highest amount of average time to correct illumination of a document. Among the rest of the reported methods, as outputs produced by method [24] are severely distorted, we can disregard their performance. Among the rest, our proposed method, using sampling rate $k_s = 5$, takes 1.4 s to accomplish the task. Although the elapsed time (1.4 s) for this setting is higher than the elapsed time of the other remaining methods, by setting the sampling rate $k_s = 14$, our method can accomplish the task in 0.1 s. In that case, the documents have to suffer from more noise (PSNR 22.04, see Fig. 5) than that of sampling rate $k_s = 5$. However, the PSNR level is still better than that of methods [18,27,34] (see Fig. 7). That means, even if a coarse sampling rate is chosen for the proposed algorithm, the output quality of the documents is still better than the other compared methods (Fig. 8).

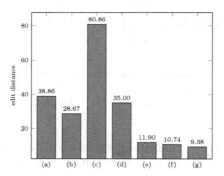

Fig. 9. Average edit-distance measure (after applying OCR) comparison (a) using original images, (b) using [27], (c) using [34], (d) using [18], (e) using [3], (g) using [12], and (f) ours.

While our method shows superior performance over other compared methods presented in the previous sub-section, it is also important to examine whether the structure of characters is well preserved after illumination correction. To this end, we measure the OCR performance for each illumination corrected output, as an indication of readability of the methods. Before applying OCR, the corrected digitized documents are geometrically de-warped using inverse perspective projective mapping by manually selecting four corner points of each document. Then, we apply an online OCR implementation of [26] to recognize texts on the documents. In order to compare the performance of our algorithm with the state-of-the-art methods, we measure the average edit-distance between the OCR results and the original text written on the cards. The edit-distance measure is reported in Fig. 9. As it can be seen, the average edit-distance measure of the text on the illumination corrected digitized business cards by our algorithm is 9.38 while the nearest edit-distance measure is performed by method [12] having the performance score 10.74. Besides, the proposed method outperforms the other reported method by a significantly large margin. Based on the experiments using OCR results, we strongly claim that the proposed method is not only capable of correcting illumination of the digitized documents but also keep the structure of the documents well preserved, which helps the user for experiencing better readability using the digitized document corrected by our proposed algorithm. We provide further experimental results (*e.g.*, scanner) in supplementary material.

5   Conclusions

In this paper, we have proposed a novel algorithm, called water-filling algorithm, for illumination correction of shaded digitized documents. The proposed water-filling algorithm is based on simulating the immersion process of a topographic surface using water. We have implemented this idea using a diffusion equation. To the best of our knowledge, no previous method used diffusion equation based approach for digitized document illumination correction. Based on the experimental results, it is found that the proposed method can outperform the performance of the previous methods under various challenging lighting conditions. The limitation of the proposed algorithm is that any text or photo connected to image border is regarded as a shadow, and eliminated after the processing. However, we have observed that this is not a critical problem since the salient text or photo is usually not connected to image borders in a document image. We also need to mention that our method often produces unsatisfactory results for eliminating specular lights in an image, because once the original luminance value of a point in the foreground layer is significantly damaged due to overexposure, it is hard to reconstruct the point using the Lambertian surface model even if we precisely estimated the background layer. These limitations will be addressed in our future works.

References

1. Athimethphat, M.: A review on global binarization algorithms for degraded document images. AU JT **14**(3), 188–195 (2011)
2. Azmi, M.H., Iqbal Saripan, M., Azmir, R.S., Abdullah, R.: Illumination compensation for document images using local-global block analysis. In: Badioze Zaman, H., Robinson, P., Petrou, M., Olivier, P., Schröder, H., Shih, T.K. (eds.) IVIC 2009. LNCS, vol. 5857, pp. 636–644. Springer, Heidelberg (2009). https://doi.org/10.1007/978-3-642-05036-7_60
3. Bako, S., Darabi, S., Shechtman, E., Wang, J., Sunkavalli, K., Sen, P.: Removing shadows from images of documents. In: Lai, S.-H., Lepetit, V., Nishino, K., Sato, Y. (eds.) ACCV 2016. LNCS, vol. 10113, pp. 173–183. Springer, Cham (2017). https://doi.org/10.1007/978-3-319-54187-7_12
4. Brown, M.S., Tsoi, Y.C.: Geometric and shading correction for images of printed materials using boundary. IEEE Trans. Image Process. **15**(6), 1544–1554 (2006)
5. Bukhari, S.S., Shafait, F., Breuel, T.M.: Foreground-background regions guided binarization of camera-captured document images. In: Proceedings of the International Workshop on Camera Based Document Analysis and Recognition, vol. 7. Citeseer (2009)
6. Chandrasekhar, V.R., et al.: The stanford mobile visual search data set. In: Proceedings of the Second Annual ACM Conference on Multimedia Systems, pp. 117–122. ACM (2011)
7. Fan, J.: Enhancement of camera-captured document images with watershed segmentation. In: CBDAR07 pp. 87–93 (2007)
8. Fan, K.C., Wang, Y.K., Lay, T.R.: Marginal noise removal of document images. Pattern Recogn. **35**(11), 2593–2611 (2002)
9. Forsyth, D.A., Ponce, J.: Computer Vision: A Modern Approach (2003)
10. Gatos, B., Pratikakis, I., Perantonis, S.J.: Adaptive degraded document image binarization. Pattern Recogn. **39**(3), 317–327 (2006)
11. Kim, B.S., Koo, H.I., Cho, N.I.: Document dewarping via text-line based optimization. Pattern Recogn. **48**(11), 3600–3614 (2015)
12. Kligler, N., Katz, S., Tal, A.: Document enhancement using visibility detection. In: Proceedings of the IEEE Conference on Computer Vision and Pattern Recognition, pp. 2374–2382 (2018)
13. Lee, J.S., Chen, C.H., Chang, C.C.: A novel illumination-balance technique for improving the quality of degraded text-photo images. IEEE Trans. Circuits Syst. Video Technol. **19**(6), 900–905 (2009)
14. Lu, S., Su, B., Tan, C.L.: Document image binarization using background estimation and stroke edges. IJDAR **13**(4), 303–314 (2010)
15. Lu, S., Tan, C.L.: Binarization of badly illuminated document images through shading estimation and compensation. In: Ninth International Conference on Document Analysis and Recognition, 2007. ICDAR 2007, vol. 1, pp. 312–316. IEEE (2007)
16. Meng, G., Pan, C., Xiang, S., Duan, J.: Metric rectification of curved document images. IEEE Trans. Pattern Anal. Mach. Intell. **34**(4), 707–722 (2012)
17. Meng, G., Xiang, S., Zheng, N., Pan, C.: Nonparametric illumination correction for scanned document images via convex hulls. IEEE Trans. Pattern Anal. Mach. Intell. **35**(7), 1730–1743 (2013)
18. Oliveira, D.M., Lins, R.D.: Generalizing tableau to any color of teaching boards. In: 2010 International Conference on Pattern Recognition, pp. 2411–2414. IEEE (2010)

19. Oliveira, D.M., Lins, R.D.: A new method for shading removal and binarization of documents acquired with portable digital cameras. In: Proceedings of Third International Workshop on Camera-Based Document Analysis and Recognition, Barcelona, Spain, pp. 3–10 (2009)

20. Oliveira, D.M., Lins, R.D., de França Pereira e Silva, G.: Shading removal of illustrated documents. In: Kamel, M., Campilho, A. (eds.) ICIAR 2013. LNCS, vol. 7950, pp. 308–317. Springer, Heidelberg (2013). https://doi.org/10.1007/978-3-642-39094-4_35

21. Otsu, N.: A threshold selection method from gray-level histograms. Automatica **11**(285–296), 23–27 (1975)

22. Rais, N.B., Hanif, M.S., Taj, R., et al.: Adaptive thresholding technique for document image analysis. In: Multitopic Conference, 2004. Proceedings of INMIC 2004. 8th International, pp. 61–66. IEEE (2004)

23. Roerdink, J.B., Meijster, A.: The watershed transform: definitions, algorithms and parallelization strategies. Fundam. Inform. **41**(1–2), 187–228 (2000)

24. Sauvola, J., Pietikäinen, M.: Adaptive document image binarization. Pattern Recogn. **33**(2), 225–236 (2000)

25. Shafait, F., van Beusekom, J., Keysers, D., Breuel, T.M.: Page frame detection for marginal noise removal from scanned documents. In: Ersbøll, B.K., Pedersen, K.S. (eds.) SCIA 2007. LNCS, vol. 4522, pp. 651–660. Springer, Heidelberg (2007). https://doi.org/10.1007/978-3-540-73040-8_66

26. Smith, R.W.: Hybrid page layout analysis via tab-stop detection. In: 10th International Conference on Document Analysis and Recognition, 2009. ICDAR 2009, pp. 241–245. IEEE (2009)

27. Sternberg, S.R.: Biomedical image processing. Computer **16**(1), 22–34 (1983)

28. Tan, C.L., Zhang, L., Zhang, Z., Xia, T.: Restoring warped document images through 3D shape modeling. IEEE Trans. Pattern Anal. Mach. Intell. **28**(2), 195–208 (2006)

29. Tian, Y., Narasimhan, S.G.: Rectification and 3D reconstruction of curved document images. In: 2011 IEEE Conference on Computer Vision and Pattern Recognition (CVPR), pp. 377–384. IEEE (2011)

30. Tsoi, Y.C., Brown, M.S.: Geometric and shading correction for images of printed materials: a unified approach using boundary. In: Proceedings of the 2004 IEEE Computer Society Conference on Computer Vision and Pattern Recognition, CVPR 2004, pp. 240–246. IEEE (2004)

31. Vincent, L., Soille, P.: Watersheds in digital spaces: an efficient algorithm based on immersion simulations. IEEE Trans. Pattern Anal. Mach. Intell. **6**, 583–598 (1991)

32. Wada, T., Ukida, H., Matsuyama, T.: Shape from shading with interreflections under proximal light source-3D shape reconstruction of unfolded book surface from a scanner image. In: Fifth International Conference on Computer Vision, 1995. Proceedings, pp. 66–71. IEEE (1995)

33. Zhang, L.: Restoring warped document images using shape-from-shading and surface interpolation. In: 18th International Conference on Pattern Recognition, 2006. ICPR 2006, vol. 1, pp. 642–645. IEEE (2006)

34. Zhang, L., Yip, A.M., Brown, M.S., Tan, C.L.: A unified framework for document restoration using inpainting and shape-from-shading. Pattern Recogn. **42**(11), 2961–2978 (2009)

35. Zhang, L., Yip, A.M., Tan, C.L.: Removing shading distortions in camera-based document images using inpainting and surface fitting with radial basis functions. In: Ninth International Conference on Document Analysis and Recognition, 2007. ICDAR 2007, vol. 2, pp. 984–988. IEEE (2007)

36. Zhang, L., Yip, A.M., Tan, C.L.: A restoration framework for correcting photometric and geometric distortions in camera-based document images. In: IEEE 11th International Conference on Computer Vision, 2007. ICCV 2007, pp. 1–8. IEEE (2007)
37. Zhang, L., Zhang, Z., Tan, C.L., Xia, T.: 3D geometric and optical modeling of warped document images from scanners. In: IEEE Computer Society Conference on Computer Vision and Pattern Recognition, 2005. CVPR 2005, vol. 1, pp. 337–342. IEEE (2005)

Matchable Image Retrieval by Learning from Surface Reconstruction

Tianwei Shen[1], Zixin Luo[1], Lei Zhou[1], Runze Zhang[1], Siyu Zhu[1],
Tian Fang[2]([✉]), and Long Quan[1]

[1] Hong Kong University of Science and Technology, Hong Kong, China
{tshenaa,zluoag,lzhouai,rzhangaj,szhu,quan}@cse.ust.hk
[2] Shenzhen Zhuke Innovation Technology (Altizure), Shenzhen, China
fangtian@altizure.com

Abstract. Convolutional Neural Networks (CNNs) have achieved superior performance on object image retrieval, while Bag-of-Words (BoW) models with handcrafted local features still dominate the retrieval of overlapping images in 3D reconstruction. In this paper, we narrow down this gap by presenting an efficient CNN-based method to retrieve images with overlaps, which we refer to as the *matchable image retrieval* problem. Different from previous methods that generates training data based on sparse reconstruction, we create a large-scale image database with rich 3D geometrics and exploit information from surface reconstruction to obtain fine-grained training data. We propose a batched triplet-based loss function combined with mesh re-projection to effectively learn the CNN representation. The proposed method significantly accelerates the image retrieval process in 3D reconstruction and outperforms the state-of-the-art CNN-based and BoW methods for matchable image retrieval. The code and data are available at https://github.com/hlzz/mirror.

Keywords: Matchable image retrieval · Image-based reconstruction

1 Introduction

Generic image retrieval is widely employed in practical Structure-from-Motion (SfM) [1–4] and visual simultaneous localization and mapping (SLAM) [5] systems to accelerate the image matching process or identify possible closed loops. Until recently, the preferred image retrieval techniques used in SfM are largely variants of the Bag-of-Words (BoW) models [6,7], despite the fact that CNN-based approaches [8–11] have shown superior efficiency and scalability for particular object retrieval.

T. Shen and Z. Luo—equal contributions.

Electronic supplementary material The online version of this chapter (https://doi.org/10.1007/978-3-030-20887-5_26) contains supplementary material, which is available to authorized users.

This discrepancy can be explained by the difference between *semantic similarity* and *geometric similarity*. For SfM tasks, geometric overlaps among images (geometric similarity), rather than information about object categories (semantic similarity), are required for later reliable image matching. We refer to this specific type of image retrieval task as *matchable image retrieval*, the goal of which is to find images with large overlaps. Two images are overlapped if they include the same area of the viewed objects or scenes. In this scenario, BoW models based on local descriptors are more robust since they serve as predictors [12] for how well the local descriptors can be matched. However, neither BoW models nor CNN-based methods perfectly solve the matchable image retrieval problem. On the one hand, BoW models generally have limited scalability as the efficiency and accuracy drop quickly with the increase of data. CNN-based methods, on the other hand, offer efficient and scalable solutions by compact global image representations distilled from intermediate feature maps, yet they lack the ability to identify regional discriminations and local information. This problem has long been overlooked because nearly all of these CNN-based methods are evaluated on object retrieval datasets such as Oxford5k [13] and Paris6k [14], in which images are organized by semantic similarity rather than geometric overlaps.

However, in a typical SfM scene (Fig. 1) consisting of overlapping images with weak semantics, current CNN-based methods are worse than BoW models because they fail to render a fine-grained ranking with respect to scene overlaps. That is probably the reason why stable SfM [1–4] and SLAM [5] solutions still adopt BoW models for matchable image retrieval. CNN-based methods should be employed because of its superior efficiency and scalability, and its previous success in object retrieval tasks. However, several problems should be addressed to get rid of the above flaws. First, we are in need of a large-scale SfM database to avoid the data bias in previous evaluations. Second, information about geometric relationships between images should be further exploited to better encode local information. Several methods such as [10] have attempted to do so but stayed in the SfM level instead of using dense correspondences. Third, the training process should be made more efficient to cope with big data.

In this paper, we present an efficient CNN-based method for matchable image retrieval that utilizes rich geometric context mined from densely reconstructed structures, namely mesh re-projection and overlap masks. Moreover, local information is taken good care of with a post-processing step that exploits regional matching. In summary, our contributions are threefold:

- We present *Geometric Learning with 3D Reconstruction (GL3D)*, a large-scale database for 3D reconstruction and geometry-related learning problems, which contains 378 different datasets with full coverage of the scenes.
- We make use of the dense correspondences mined from 3D reconstruction to develop an automatic pipeline for ground-truth data generation, which results in fine-grained training data with respect to scene overlaps.
- We propose *mask triplet loss (MTL)* with in-batch mining which utilizes the well-annotated training data combined with regional information to accelerate the training of matchable image retrieval.

2   Related Works

Local Descriptor Based Methods. In the 3D modeling of city-scale imagery, either pairwise image matching or point-cloud matching [15,16] often take a majority of computation. Since the seminal work of reconstructing Internet imagery [1], object retrieval techniques have been widely adopted in a series of SfM systems [2,3,17–19]. As a successful BoW model, vocabulary tree [7] has become indispensable in large-scale SfM, which can be regarded as a preemptive filtering step in which local descriptors vote for images that share scene overlaps. Later works focus on decreasing quantization errors [14,20,21], applying post-processing steps [22] and scaling up object retrieval by aggregating local features into compact representations. To address the very large retrieval problem, VLAD [23] was designed to be a low dimensional compact code while still preserving good performance.

CNN Methods. Different from BoW models, CNN-based image retrieval approaches mostly rely on global information. Generic deep descriptors extracted from deep convolutional neural network models are proved to be good image representations on a series of vision tasks including object retrieval. Babenko et al. [24] firstly propose a sum-pooling aggregation method utilizing a centering prior, with the knowledge that objects of interest tend to be located close to the center of images. This is not satisfied when finding similarity pairs in terms of region overlaps. Kalantidis et al. [8] later propose a feature aggregation method based on cross-dimensional weighting. It analyses the spatial weighting and the channel weighting strategies that can boost saliency and distinctiveness of the visual patterns respectively. In parallel, Tolias et al. [9] propose R-MAC (regional maximum activations of convolutions), which utilizes regional information to boost the performance. Gordo et al. [25] replace the rigid grid with a learned region proposal network (RPN). All of the above methods, evaluations and assumptions are based on images with salient semantic regions like houses or landscapes. In 3D reconstruction, however, many images in urban datasets or aerial imageries merely serve as bridges to connect partial scenes, with fragmental, discontinuous or even no semantically meaningful regions.

In terms of the network architectures, our method is most similar to Wang et al. [26] and Schroff et al. [27]. Both methods employ triplet loss to learn a similarity-based embedding. The first work uses a triplet-based hinge loss to characterize fine-grained image similarity while the second is proposed to solve face recognition problem at scale. Melekhov et al. [28] also tackle the similar whole-image matching problem while they use the 2-channel network [29], but do not go deeper into 3D reconstruction.

3   The GL3D Benchmark Dataset

We create a database, *Geometric Learning with 3D Reconstruction (GL3D)*, containing 90,590 high-resolution images in 378 different scenes. Each scene contains 50 to 1,000 images with large geometric overlaps, covering urban, rural area, or

	GL3D	Oxford 5K	Paris 6K	Holiday
Strong Semantics	N	Y	Y	Y
Sparse Correspondences	Y	Y	Y	Y
Point Clouds	Y	N	N	N
Mesh Models	Y	N	N	N
Completeness	Y	N	N	N
High Resolution	Y	N	N	N
#Image	90.5k	5k	6k	1.5k
#Image for retrieval test	9368	55	55	500

(a) Urban area   (b) Rural area   (c) Scenic spot   (d) Small object   (e) Comparisons of GL3D and object retrieval datasets

Fig. 1. (a) (b) (c) (d) show different types of scenes in the GL3D dataset, with the mesh models on the right for generating training samples. (e) Compared with existing object retrieval datasets, GL3D offers high-resolution and complete views for 379 different scenes from which 2D feature matches (sparse correspondence), 2D-3D correspondences, point clouds, and mesh models can be established

scenic spots captured by drones from multiple scales and perspectives. It also contains small objects to enrich the data diversity. Figure 1 gives an overview of various scenes in GL3D and their corresponding 3D models. We randomly select 338 datasets (81,222 images) and run 3D reconstruction pipeline (SfM → dense reconstruction → mesh reconstruction) for training sample generation as described in Sect. 4.3. To generate the 3D models, we use the incremental SfM method from [19] and the multiview stereo with surface reconstruction method from [30]. The testing is carried out on the other 40 datasets with 9,368 images as queries, which allows a thorough evaluation for the matchable image retrieval compared with only 55 queries in Oxford5k [13].

GL3D is tailored for geometry-related problems and offers rich 3D context information such as feature-track correspondences, camera poses, point cloud data and mesh models. Therefore, it has intrinsic difference with other existing object retrieval datasets such as Oxford5k [13], Paris6k [14] and Holiday [20]. The comparisons can be characterized in the following perspectives:

Full Coverage. Each dataset has full coverage of the scene, which is the major difference of GL3D from previous crowd-sourced datasets [10]. Existing object retrieval datasets usually have uneven samples of the same landmark, while GL3D are organized by densely connected images from different views.

Weak Semantics. Existing object retrieval datasets mainly contain semantically meaningful landmark buildings with intact objects. The superior CNN performance trained on object classification task is therefore suitable to be transferred to object retrieval. In contrast, GL3D has weak semantics because images only capture part of the objects or scenes without definite semantic meanings. Some query images even have texture-less patterns like lawns and rivers, which is not common in the datasets for particular object retrieval [13,14,20].

Rich Geometric Context. Since images are densely connected and have full coverage of the scenes, not only two-view feature matches, but also accurate geometric computations such as camera poses, point clouds and mesh models can be derived. Therefore, we can measure the degree of scene overlaps between images pairs from accurate mesh re-projection. This results in the proposed fine-grained ground-truth generation scheme.

GL3D is not only limited to the matchable image retrieval problem. With various geometric computations such as feature matching, camera poses, and mesh models, it is also beneficial for other geometric learning problems. For the task of matchable image retrieval, we will design and present an automatic pipeline to generate well-annotated data as described in Sect. 4.3.

4 Method

4.1 Problem Formulation

Given a set of N images $\{\mathcal{I}_i\}$ with geometric overlaps, we aim to find a rank set $\mathcal{S}_i$ for each image in $\{\mathcal{I}_i\}$. In the rank set $\mathcal{S}_i = \{\mathcal{I}_{i_1}, \mathcal{I}_{i_2}, \ldots, \mathcal{I}_{i_N}\}$, a natural ordering exists $\mathcal{O}_i = (i_1, i_2, \ldots, i_N)$ representing the similarity in terms of geometric overlaps between $\mathcal{I}_i$ and database images. To find these rank sets, one typical approach is to first map image features onto a space with lower dimension via an embedding function $f(\mathcal{I}_i)$ [9,24,29,31]. Then a similarity measurement $D(f(\mathcal{I}_i), f(\mathcal{I}_j))$ is computed and similar items are ranked by this similarity score from low to high. The similarity measurement is typically defined as the $L2$ distance between two normalized feature vectors:

$$D(f(\mathcal{I}_i), f(\mathcal{I}_j)) = \left\| \frac{f(\mathcal{I}_i)}{\|f(\mathcal{I}_i)\|} - \frac{f(\mathcal{I}_j)}{\|f(\mathcal{I}_j)\|} \right\|_2 \tag{1}$$

The most crucial part in this learning framework is to find the embedding function $f(\cdot)$. In this work, we resort to deep CNNs for embedding learning. Our objective is to train a neural network that can differentiate the degree of scene overlaps between pairs of images.

4.2 Network Architecture

We adopt three-branch networks as shown in Fig. 2, with *(anchor, positive, negative)* image triplets (denoted by $(\mathcal{I}_a, \mathcal{I}_p, \mathcal{I}_n)$) as inputs. The core of this learning method is to minimize the distance of similar image pairs and maximize the distance of dissimilar pairs to some margin. The embedding function $f(\cdot)$ is learned in three feature towers with shared parameters, which can be implemented with any commonly-used CNNs [32,33]. Different components in the networks are described in detail as follows:

Feature Tower. The three feature towers share the same parameters during training, following the essence of triplet loss. Feature tower can be fine-tuned from the widely adopted networks such as VGG [32] or GoogLeNet [33]. Though the classical networks often come with a fully-connected (FC) layer for classification, FC layers often do not work well for image retrieval tasks [9]. In addition, FC layers are often removed for testing because we would like the input image to be arbitrary size. Therefore, we make the feature tower to be fully convolutional. The feature vectors are composed by first applying pooling on each feature map and then $L2$ normalization across channels.

(a)   (b)

Fig. 2. (a) The proposed network architecture: Three-branch feature embedding towers conjoined by the ranking layer. The anchor-positive image pair is associated with a pair of down-sampled corresponding masks (the red regions), used in the loss computation (see Eq. 7) (b) The limitation of *common track ratio*. The image pair 2 survives in *mesh overlap ratio* but fails in *common track ratio* due to large perspective and scale changes. The red masks generated by mesh re-projection indicate the overlapping regions. The 3D model and overlap statistics are presented on the right (Color figure online)

Pooling Layer. We use max pooling to aggregate feature maps into a feature vector. Max pooling has the nice property of translation invariance and widely adopted by previous CNN image retrieval works [9–11, 25].

Loss Function. We use the widely adopted triplet-based loss layer for this learning-to-rank problem. Although pairwise losses such as the contrastive loss [29] based on Siamese architecture [10, 28] are also feasible, triplet-based losses are typically favored to avoid overfitting as they care about the relative ordering rather than the absolute distance [31]. We conjoin each feature tower to the ranking layer and evaluate the hinge loss of a triplet.

4.3 Fine-Grained Training Data Generation

Triplet Sampling Using SfM. As we have shown in the network architecture, the training data is composed of image triplets $(\mathcal{I}_a, \mathcal{I}_p, \mathcal{I}_n)$. Manual annotation

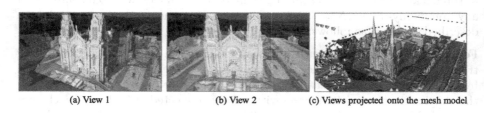

(a) View 1   (b) View 2   (c) Views projected onto the mesh model

Fig. 3. The automatic training data generation pipeline. The yellow region covers all the triangles that are seen by both (a) and (b), while the red region and the green region cover triangles that are seen by them exclusively (best view in color) (Color figure online)

for such a large quantity of training triplets is unrealistic. As is observed in [10], these triplets can be generated from SfM by computing the ratio of shared 3D tracks (which we refer to as *common track ratio*) between views in a fully automatic manner. Specifically, suppose $\mathcal{P}(i)$ is the set that contains all the 3D tracks that are observed by image i, then the *common track ratio* between the image pair $(\mathcal{I}_i, \mathcal{I}_j)$ is defined as the average of two ratio numbers:

$$\mathcal{CT}_{ij} = \mathcal{CT}_{ji} = \mathbf{Ave}(\frac{|\mathcal{P}(i) \cap \mathcal{P}(j)|}{|\mathcal{P}(i)|}, \frac{|\mathcal{P}(i) \cap \mathcal{P}(j)|}{|\mathcal{P}(j)|}) \qquad (2)$$

where the average function $\mathbf{Ave}(a, b) = \sqrt{a \cdot b}$ is the geometric mean. Though other mean functions can be used, we did not observe substantial difference.

Triplet Sampling Using Surface Reconstruction. However, the above sampling method has several drawbacks. First, the generalization power would be limited by the ability of local feature matching. As Fig. 2(b) shows, if a pair of matched images possess a large view angle change that exceeds the matching ability of SIFT ($> 30°$), this pair of images would be regarded as unmatched since few common tracks would exist. Ideally, a good retrieval algorithm should consider all geometrically overlapping pairs and get rid of this limitation. Second, hard samples as shown in Fig. 4 are helpful in matchable image retrieval, in which the triplet images are from the same scene with similar context information. But hard samples cannot be obtained from sampling using SfM, in which negative samples are constrained to be selected from two non-overlapping scenes [10], since a small ratio of shared tracks does not represent a small overlapping area.

Thus, we combine mesh model re-projection with SfM track overlaps to obtain training triplets. As shown in Fig. 3, we use triangulated mesh models to pinpoint accurate overlap regions between image pairs, which is similar to [34]. The essence is to project triangular meshes with high level-of-details (LoD) through camera projection matrices registered in SfM. Similar to *common track ratio*, we define *mesh overlap ratio* between the $(\mathcal{I}_i, \mathcal{I}_j)$ image pair as

$$\mathcal{MO}_{ij} = \mathcal{MO}_{ji} = \mathbf{Ave}(\frac{|\mathcal{T}(i) \cap \mathcal{T}(j)|}{|\mathcal{T}(i)|}, \frac{|\mathcal{T}(i) \cap \mathcal{T}(j)|}{|\mathcal{T}(j)|}) \qquad (3)$$

where $\mathcal{T}(i)$ is the set containing all the triangles that are seen by the corresponding camera of image i, and $\mathbf{Ave}$ is the same as in Eq. 2 which considers relative scale of image pairs. $\mathcal{CT}_{ij}$ and $\mathcal{MO}_{ij}$ are both in the range of $[0, 1]$.

To get a consistent overlap measurement, $\mathcal{CT}_{ij}$ and $\mathcal{MO}_{ij}$ should be carefully merged. The magnitude of $\mathcal{MO}_{ij}$ is usually larger than that of $\mathcal{CT}_{ij}$ in practice. We take a SfM-overlap-first scheme to ensure the completeness of positive samples. Namely, the *combining overlap ratio* $\mathcal{CO}$ is defined as

$$\mathcal{CO}_{ij} = \begin{cases} 1 & \text{if } \mathcal{CT}_{ij} \geq t_{sfm} \\ \mathcal{MO}_{ij} & \text{otherwise} \end{cases} \qquad (4)$$

In this work, we fix t_{sfm} to be 0.2 as is used in [10]. An image $\mathcal{I}_j$ is a **strong positive** to the anchor image $\mathcal{I}_i$ if $\mathcal{CO}_{ij} \in [t_{s1}, t_{s2}](= [0.5, 1.0])$, and a **weak** *positive* if $\mathcal{CO}_{ij} \in [t_{w1}, t_{w2}](= [0.05, 0.2])$, leaving a safe margin between strong and weak positives. Moreover, the corresponding masks generated by mesh re-projection enable a more accurate computation of the loss term, which will be detailed in the next section.

4.4 Learning with Batched Hard Mining

Triplet Loss. The original idea of triplet loss [26,27] is to push the positive distance $D_+ = D(f(\mathcal{I}_a), f(\mathcal{I}_p))$ far apart from the negative distance $D_- = D(f(\mathcal{I}_a), f(\mathcal{I}_n))$ to a certain margin α, formally known as (where $[x]_+ = \max(x, 0)$)

$$\mathcal{L}_{tl}(\mathcal{I}_a, \mathcal{I}_p, \mathcal{I}_n) = [D_+ + \alpha - D_-]_+ \tag{5}$$

Anchor Swap. For symmetric distance measurements like the one in matchable retrieval, the sample space can be halved by introducing in-triplet hard negative mining [35], which also considers the distance between the positive and the negative $D'_- = D(f(\mathcal{I}_p), f(\mathcal{I}_n))$

$$\mathcal{L}_{as}(\mathcal{I}_a, \mathcal{I}_p, \mathcal{I}_n) = [D_+ + \alpha - min(D_-, D'_-)]_+ \tag{6}$$

Mask Triplet Loss. Beyond the similar/dissimilar relations in particular object retrieval, more accurate overlap correspondences can be pin-pointed from the training data generation pipeline described in Sect. 4.3. Using the groundtruth masks associated with matched image pairs, we propose a new loss termed as *Mask Triplet Loss*

$$\mathcal{L}_{mtl}(\mathcal{I}_a, \mathcal{I}_p, \mathcal{I}_n) = [D_+^\star - \beta]_+ + \lambda[D_+ + \alpha - min(D_-, D'_-)]_+ \tag{7}$$

where $D_+^\star = D(f(\mathcal{I}_a) \odot M(\mathcal{I}_a, \mathcal{I}_p), f(\mathcal{I}_p) \odot M(\mathcal{I}_p, \mathcal{I}_a))$. $\{M(\mathcal{I}_a, \mathcal{I}_p), M(\mathcal{I}_p, \mathcal{I}_a)\}$ represents a pair of corresponding masks generated by mesh re-projection, and $\odot$ is the masking operation applied on feature maps from CNNs. In practice, we use the down-sampled corresponding region maps between the positive image pair $(\mathcal{I}_a, \mathcal{I}_p)$ as a binary filter for pooling operation. The first term in Eq. 7 penalizes the difference between the masked regions of positive pairs, with a soft margin β to prevent overfitting [36], while the second term is the triplet loss with anchor swap. β and λ are set to 0.1 and 0.5 respectively. We have found that the proposed mask triplet loss greatly accelerates the training process since it finds the accurate regions for loss computation.

Batched Hard Mining. Since the sample complexity is cubic in the number of images, which is infeasible to iterate over, triplet sampling is vital to ensure the fast convergence of the model. Therefore, a mining strategy [27] should be carefully designed to select the proper triplets. Too hard triplets would result in the collapse of the model and too easy triplets would produce no loss and slow down the training process. Inspired by the previous works used in local descriptor learning, such as structured loss [37,38], we propose a batched triplet

$N_b(=8)$ triplets from different datasets

Fig. 4. The batched triplets loss formulation

mining strategy suitable for this task which utilizes the fine-grained overlap measurement as defined in Sect. 4.3.

As shown in Fig. 4, each batch forms a matrix T of size $(3, N_b)$ where N_b is the batch size. Each triplet in the batch comes from a different dataset thus row-wise every pair of images is a ***negative*** pair. Each column itself forms a ***hard triplet sample*** meaning that the second row is more similar to anchors (the first row) than the third row, measured by the overlap ratio defined in Sect. 4.3. We call the second row ***strong positive*** and the third row ***weak positive***. The total loss is of three parts: (1) easy loss composed by (anchor, strong positive, negative), (2) weak loss composed by (anchor, weak positive, negative), (3) hard loss composed by (anchor, strong positive, weak positive), written as follows:

$$
\begin{aligned}
\mathcal{L}_T &= \mathcal{L}_{easy} + \lambda_1 \mathcal{L}_{weak} + \lambda_2 \mathcal{L}_{hard} \\
&= \frac{1}{3N_b(N_b-1)} \sum_{i=0}^{N_b-1} \sum_{j=0, j \neq i}^{N_b-1} \sum_{k=0}^{2} [\mathcal{L}_{mtl}(T_{0i}, T_{1i}, T_{kj}) + \lambda_1 \mathcal{L}_{mtl}(T_{0i}, T_{2i}, T_{kj})] \\
&\quad + \frac{\lambda_2}{N_b} \sum_{i=0}^{N_b-1} \mathcal{L}_{mtl}(T_{0i}, T_{1i}, T_{2i})
\end{aligned}
\tag{8}
$$

With this batched loss formulation, the equivalent batch size can be enlarged by an order of magnitude from $O(N_b)$ to $O(N_b^2)$, which makes the training process much more effective. In practice, we set the loss weights λ_1, λ_2 to 1.

Offline Mining with Adaptive Margins. Hard negatives are generated offline by mesh re-projection (discussed in Sect. 4.3). As mentioned in [27], we also observe that using hard negatives in the early training process can harm the performance and collapse the model. Therefore, we use adaptive margins, where we set a smaller margin for hard samples to stabilize the training process. We set $\alpha = 1$ for easy triplets, $\alpha = 0.5$ for weak triplets, and $\alpha = 0.5$ for hard triplets.

4.5   Pre-matching Regional Code (PRC)

Since matchable image retrieval needs fine-grained discrimination of overlap, it is crucial to exploit regional information. R-MAC [9] provides good insights to tackle this issue. R-MAC samples square regions on activations at different scales, then applies MAC [39] on those square regions to get regional vectors which are then combined into a single image vector by summing and L2-normalization. However, the mixed regional information may weaken its expressive power. In this work, we propose *pre-matching regional code (PRC)*, an feature aggregation method towards regional information coding based on [9, 40].

Generally, PRC can be combined with any pooling operations, such as L2 pooling, average pooling or max pooling. We use PRC with max pooling due to its translation invariance [9, 25], which is termed PR-MAC. We first sample square regions and generate regional vectors as in R-MAC. Instead of simply summing up all the regional vectors, PR-MAC does pre-matching on regional vectors and aggregates the sub-matching result. Formally, for an image pair $(\mathcal{I}_Q, \mathcal{I}_T)$ associated with regional vector sets $\{\mathcal{R}_Q\}$ and $\{\mathcal{R}_T\}$, we first obtain

$$D_T(\mathcal{R}_{Q,i}) = \min_j \{\|\mathcal{R}_{Q,i}, \mathcal{R}_{T,j}\|_2\} \tag{9}$$

as the minimum distance between a regional vector $\mathcal{R}_{Q,i}$ for the query image $\mathcal{I}_Q$, and the regional vector set $\{\mathcal{R}_T\}$ for the target image $\mathcal{I}_T$. Then we calculate

$$D(\mathcal{I}_Q, \mathcal{I}_T) = \| \sum_i (D_T(\mathcal{R}_{Q,i})) \|_2 \tag{10}$$

to represent the final distance between a pair of images. As an interpretation, PRC conducts pre-matching to find the best match for each region of the query, and computes the similarity considering the matchability of each region. As is demonstrated in extensive experiments, PRC outperforms R-MAC in both object image retrieval and matchable image retrieval.

Discussions on Efficiency and Comparison with R-MAC. PRC has the computational complexity of $O(k^2)$ where k is the number of regional vectors, which is higher than that of R-MAC. We have improved the efficiency in two ways. First, the PRC is applied on the feature map level as in R-MAC instead of on the costly image patch level [40]. Second, PRC can be applied on a shortlist (Top-200) as a re-ranking method [11]. We also compare PRC with approximate max-pooling localization (AML) [9], which replaces the sum operation in Eq. 10 with arg max.

5 Experiments

Implementation Details. We use TensorFlow [41] to train CNNs on resized 224×224 images with random contrast and color perturbation. Various methods of vocabulary tree with advanced techniques [7, 20, 21] are implemented in C++ with multi-threading and SIFT features from VLFeat [42]. Each image has 10k SIFT features on average. We use stochastic gradient descent (SGD) solver with a momentum of 0.9 and a weight decay of 0.0001. The base learning rate is 0.002 and exponentially decayed to 0.9 of the previous one for every 10k steps. All benchmarks are conducted on single Nvidia GeForce GTX 1080.

Table 1. We use 896×896 images for CNN-based methods. Dimensionality are reduced to 512 using PCA (or learned whitening L_W computed on the dataset in [11]), computed with an independent dataset with 50k images. QE means weighted query expansion (with top-10 results weighted by the distance), Holiday is not applicable for QE because queries have less than 10 ground-truth images. R-MAC and PR-MAC are used with two scales of 5 ($=1+4$) regional vectors.

(Dim $=512$ for all CNN methods)	GL3D (mAP@100)	Oxford5k	Paris6k	Holiday (top-10)	INSTRE
VocabTree [7] (depth $=6$, branch $=8$)	0.599	0.448	0.531	0.549	-
VocabTree + HE + WGC [20]	0.689	0.547	-	0.746	-
siaMAC (VGG) + MAC [10]	0.518	0.731	0.785	0.723	0.296
siaMAC (VGG) + R-MAC	0.542	0.770	0.821	0.762	0.313
siaMAC (VGG) + R-MAC (L_W)	0.553	0.779	0.810	0.767	-
siaMAC (VGG) + PR-MAC	0.617	0.786	0.832	0.782	0.389
siaMAC (VGG) + PR-MAC + QE	0.654	0.830	0.874	-	0.588
GoogLeNet + R-MAC + TL	0.636	0.711	0.794	0.821	0.243
GoogLeNet + PR-MAC + TL	0.708	0.737	0.813	0.825	0.306
GoogLeNet + PR-MAC + TL + QE	0.721	0.781	0.855	-	0.504
GoogLeNet + R-MAC + MTL	0.638	0.721	0.799	0.824	-
GoogLeNet + PR-MAC + MTL	0.711	0.740	0.816	0.841	-
GoogLeNet + PR-MAC + MTL + QE	0.722	0.789	0.862	-	-

Evaluation Protocol. We use mean Average Precision (mAP) to measure the performance. We only keep a smaller rank list of size k for each query and measure mAP@k, as only the first fewer candidate matches matter in SfM. Instead of searching the same scene dataset, each image is queried against all 9,368 test images to increase the retrieval difficulty. The ground truth overlap rank list is generated as in Sect. 4.3. We evaluate the case when $t_{pos} = 0.5$, which results in 317,090 ground-truth match pairs. It provides a challenging benchmark whose images have large scale and perspective changes unlimited by SfM results.

5.1   Distinctiveness of Matchable Image Retrieval

We first demonstrate the intrinsic difference of object retrieval and geometric overlap retrieval, by comparing vocabulary tree, which is extensively used in practical SfM systems [2–4], and various deep models on GL3D, Oxford5k, Paris6k and INSTRE [43]. Table 1 shows that siaMAC [10] achieves superior performance on object retrieval tasks but fails to beat even the naive vocabulary tree and our method on the GL3D dataset. This partially explains the prevalence of vocabulary tree in SfM, and shows that without proper care CNNs do not generalize well on the fine-grained matchable image retrieval problem.

5.2   Experiments for Matchable Image Retrieval

Below we give thorough evaluations on GL3D in the context of matchable image retrieval. If not explicitly specified, the CNN methods are tested on 896×896 images with PCA whitening and reduced feature dimensionality of 256. Different from Table 1, we use three scales of 35 ($=1+9+25$) region vectors for

Table 2. Comparison of different approaches on GL3D. For deep methods, the images are down-sampled to 896 × 896. For vocabulary tree, local descriptors are extracted from full-size 4000×3000 images. The time measurement does not count index building for BoW models. The approaches marked by ◇ are baseline models without being fine-tuned on retrieval data. The running time marked with * is evaluated on authors' public code with Matlab or Caffe, and thus may not be comparable.

Approach	GL3D		Time	Net. type	Dimension
	mAP@100	mAP@200	(min)		
CNN-based methods					
Raw + MAC ◇	0.478	0.487	11.5	VGG-16	512
SiaMAC [10] + MAC	0.519	0.527	22.6*		
SiaMAC [10] + R-MAC [9]	0.629	0.654			
SiaMAC [10] + R-MAC + diffusion [11]	0.569	0.598	60.5*		
SiaMAC [10] + PR-MAC	0.662	0.686	60.9*		
NetVLAD [44]	0.641	0.649	28.0*		256
Ours + TL + MAC	0.627	0.631	9.5		
Ours + TL + R-MAC	0.681	0.698	11.5		
Ours + MTL + R-MAC	0.691	0.707			
Ours + MTL + PR-MAC	0.724	0.731	12.3		
Fine-tuned + ROI + R-MAC [25]	0.616	0.629	12.6*	ResNet101	2048
Raw + MAC ◇	0.598	0.603	3.2	GoogLeNet	256
Ours + TL + MAC	0.625	0.638			
Ours + MTL + MAC	0.652	0.663			
Ours + MTL + SPoC [24]	0.689	0.705			
Ours + MTL + CRoW [8]	0.673	0.698			
Ours + MTL + R-MAC [9]	0.702	0.715	5.4		
Ours + MTL + AML [9]	0.630	0.637	7.2		
Ours + MTL + PR-MAC (Top-200)	0.722	0.743	5.5		
Ours + MTL + PR-MAC	**0.734**	**0.758**	8.5		
VocTree					
VocabTree [7]	0.599	0.614	44	-	2395371
VocabTree + HE [20]	0.601	0.615	726		
VocabTree + WGC [20]	0.676	0.688	144		
VocabTree + PGM [21]	0.641	0.643	173		
VocabTree + HE + WGC [20]	0.689	0.703	820		

R-MAC and PR-MAC to demonstrate the best performance. As Table 2 shows, the proposed method outperforms all the others.

Effect of Using Hard Samples. Using hard samples is the main benefit brought by our ground-truth generation method (mesh re-projection). Without hard samples, the mAP@200 of our best model drops from 0.758 to 0.717.

Effect of Triplet Loss. We compare the performance training with triplet loss (+TL) and the proposed mask triplet loss (+MTL). MTL and TL deliver similar performance after convergence, as shown in Table 2, yet it is observed that MTL converges much faster than TL.

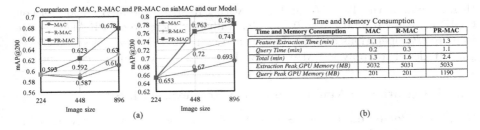

Time and Memory Consumption			
Time and Memory Consumption	MAC	R-MAC	PR-MAC
Feature Extraction Time (min)	1.1	1.3	1.3
Query Time (min)	0.2	0.3	1.1
Total (min)	1.3	1.6	2.4
Extraction Peak GPU Memory (MB)	5032	5031	5033
Query Peak GPU Memory (MB)	201	201	1190

(a)   (b)

Fig. 5. (a) Comparisons of different aggregation methods. Left: siaMAC model [10]; Right: our fine-tuned model. (b) The time and GPU peak memory consumptions for MAC, R-MAC and PR-MAC during feature extraction and query, carried out on a smaller test dataset from GL3D.

(a)   (b)   (c)

Fig. 6. (a) Comparisons of different methods on the five largest scenes in the GL3D dataset. (b) Our method works better on the datasets in the green frame while vocabulary tree does better on the datasets in the red frame. (c) Average precision (AP) for example queries, AP from left to right: siaMAC→VocabTree→Ours. (Color figure online)

Effect of PRC Feature Aggregation. Naturally, images of higher resolution provide richer information and are more likely to deliver better performance. To demonstrate that PRC can exploit information not merely from higher resolutions, we compare PR-MAC with MAC and R-MAC for different image sizes. As image size increases, PR-MAC consistently outperforms MAC and R-MAC with both siaMAC model [10] (Fig. 5(a), left) and our fine-tuned model (Fig. 5(a), right), indicating the versatility of PRC. Moreover, unlike the results in [10] where MAC and R-MAC deliver comparable improvements on object image retrieval, it shows that R-MAC is notably better than MAC in matchable image retrieval, which again demonstrates the difference between two tasks and the necessity of exploiting regional information. We have also found that manifold diffusion method [11] and approximate max-pooling localization (AML) in R-MAC [9] do not work very well on matchable image retrieval, as shown in Table 2.

Efficiency. As shown in Table 2, our best model is able to surpass above BoW models regarding both accuracy and efficiency. Furthermore, Fig. 5(b) compares the computation time and peak memory for MAC, R-MAC and PR-MAC. The higher complexity of PRC can be alleviated to some extent by using PRC as a re-ranking method. For example, by applying R-MAC on our best model and then re-ranking the Top-200 candidates with PR-MAC, the mAP@200 score on GL3D increases from 0.715 to 0.743. Generally, the increase for PR-MAC is due to more I/O operations and the fine-grained matching. However, it still achieves good trade-off to apply PR-MAC for SfM where accuracy is more concerned.

5.3 Integration of Matchable Image Retrieval with SfM

Retrieval Performance per Scene. Since SfM relies on retrieving matchable images on each independent scene, we extensively evaluate our approach on each of the 40 test sets. Both our method and vocabulary tree outperform siaMAC and again reflects the gap between object and matchable image retrieval. Figure 6(a) shows the comparisons of the five largest scenes (each more than 400 images) in GL3D. One observation (Fig. 6(b)) is that our CNN-based method is suitable for datasets with rich textures (the green frame), while vocabulary tree does better on texture-less scenes (the red frame). It indicates that vocabulary tree better encodes very local and detailed information. Performance boost for specific query images is shown in Fig. 6(c).

Table 3. Evaluation results of different retrieval methods for SfM.

		# Images	# Registered	#Pairs-to-Match	# Sparse Points	# Observations	Track Length	Reproj. Error
Madrid Metropolis	BoW	1,344	506	107,320	78,189	561K	7.18	0.59px
	siaMAC		433	103,355	69,192	510K	7.38	0.59px
	NetVLAD		467	100,876	73,724	528K	7.17	0.58px
	Ours		494	93,238	75,339	544K	7.22	0.58px
Gendarmenmarkt	BoW	1,463	1,067	110,476	222,557	1,441K	6.47	0.67px
	siaMAC		977	116,379	183,475	1,189K	6.48	0.68px
	NetVLAD		1,002	105,275	201,279	1,286K	6.39	0.67px
	Ours		1,049	103,091	212,745	1,349K	6.34	0.66px
Tower of London	BoW	1,576	780	122,534	175,452	1,441K	8.28	0.60px
	siaMAC		727	120,631	160,333	1,333K	8.31	0.59px
	NetVLAD		730	119,719	163,301	1,334K	8.35	0.59px
	Ours		740	107,044	167,426	1,386K	8.28	0.59px
Alamo	BoW	2,915	972	233,040	172,553	2,084K	12.08	0.63px
	siaMAC		904	228,021	153,483	1,948K	12.69	0.64px
	NetVLAD		912	218,617	158,686	1,994K	12.28	0.63px
	Ours		930	206,266	164,227	2,003K	12.20	0.63px
Roman Forum	BoW	2,364	1,665	179,812	357,447	2,964K	8.29	0.70px
	siaMAC		1,614	185,489	320,618	2,661K	8.30	0.69px
	NetVLAD		1,635	172,870	327,778	2,702K	8.27	0.70px
	Ours		1,653	166,474	340,396	2,796K	8.21	0.69px
ArtsQuad	BoW	6,514	6,037	505,593	1,354,474	9,227K	6.81	0.67px
	siaMAC		5,811	496,283	1,250,394	8,478K	6.78	0.65px
	Ours		5,887	448,500	1,290,811	8,757K	6.78	0.66px
	Ours@top-115		6,030	505,190	1,348,521	9,122K	6.82	0.66px

SfM Results. We conduct SfM experiments on 1DSfM [45] datasets to demonstrate the integration of the proposed method with SfM. The datasets are reconstructed using COLMAP [4] with different retrieval methods (BoW, siaMAC, NetVLAD, and ours), as shown in Table 3. We select top-100 candidates for matching, the default parameter in COLMAP. For CNN methods, the long side of image is resized to 896. siaMAC and NetVLAD are tuned to its best performance (learned whitening, query expansion etc.) as described in their papers. As shown, our method is better than siaMAC and NetVLAD, and comparable with COLMAP-BoW. However, our method generates fewer (∼10%) match pairs than COLMAP-BoW from the top-100 candidates, indicating more symmetric query results. When fixing the number of pairs to match, e.g., in the last row of table where retrieval is performed at top-115, a similar result as COLMAP-BoW can be obtained. Those experiments again validate our observation that there does exist a gap between the matchable and object image retrieval.

6 Conclusions

In this paper, we first differentiate particular object retrieval and matchable image retrieval, and present a large-scale dataset GL3D and a CNN-based method with auto-annotated training data. Based on the high-quality fine-grained training data, we utilize the overlap masks obtained from surface reconstruction and develop a batched mask triplet loss to effectively train the network. Combined with a post-processing method that exploits regional information, this method delivers state-of-the-art performance for matchable image retrieval.

Acknowledgment. This work is supported by T22-603/15N, Hong Kong ITC PSKL12EG02 and the Special Project of International Scientific and Technological Cooperation in Guangzhou Development District (No. 2017GH24).

References

1. Agarwal, S., Furukawa, Y., Snavely, N., Simon, I., Curless, B., Seitz, S.M., Szeliski, R.: Building Rome in a day. Commun. ACM **54**(10), 105–112 (2011)
2. Moulon, P., Monasse, P., Marlet, R.: Global fusion of relative motions for robust, accurate and scalable structure from motion. In: ICCV. (2013)
3. Sweeney, C., Sattler, T., Hollerer, T., Turk, M., Pollefeys, M.: Optimizing the viewing graph for structure-from-motion. In: ICCV. (2015)
4. Schonberger, J.L., Frahm, J.M.: Structure-from-motion revisited. In: CVPR. (2016)
5. Mur-Artal, R., Montiel, J.M.M., Tardos, J.D.: Orb-slam: a versatile and accurate monocular slam system. IEEE Transactions on Robotics (2015)
6. Sivic, J., Zisserman, A.: Video google: A text retrieval approach to object matching in videos. In: ICCV. (2003)
7. Nister, D., Stewenius, H.: Scalable recognition with a vocabulary tree. In: CVPR. (2006)
8. Kalantidis, Y., Mellina, C., Osindero, S.: Cross-dimensional weighting for aggregated deep convolutional features. In: ECCV Workshop. (2016)
9. Tolias, G., Sicre, R., Jégou, H.: Particular object retrieval with integral max-pooling of cnn activations. In: ICLR. (2016)
10. Radenović, F., Tolias, G., Chum, O.: Cnn image retrieval learns from bow: Unsupervised fine-tuning with hard examples. In: ECCV. (2016)
11. Iscen, A., Tolias, G., Avrithis, Y., Furon, T., Chum, O.: Efficient diffusion on region manifolds: Recovering small objects with compact cnn representations. In: CVPR. (2017)
12. Havlena, M., Schindler, K.: Vocmatch: Efficient multiview correspondence for structure from motion. In: ECCV. (2014)
13. Philbin, J., Chum, O., Isard, M., Sivic, J., Zisserman, A.: Object retrieval with large vocabularies and fast spatial matching. In: CVPR. (2007)

14. Philbin, J., Chum, O., Isard, M., Sivic, J., Zisserman, A.: Lost in quantization: Improving particular object retrieval in large scale image databases. In: CVPR. (2008)
15. Zhou, L., Zhu, S., Shen, T., Wang, J., Fang, T., Quan, L.: Progressive large scale-invariant image matching in scale space. (2017)
16. Zhou, L., Zhu, S., Luo, Z., Shen, T., Zhang, R., Zhen, M., Fang, T., Quan, L.: Learning and matching multi-view descriptors for registration of point clouds. (2018)
17. Shen, T., Zhu, S., Fang, T., Zhang, R., Quan, L.: Graph-based consistent matching for structure-from-motion. In: ECCV. (2016)
18. Zhu, S., Shen, T., Zhou, L., Zhang, R., Wang, J., Fang, T., Quan, L.: Parallel structure from motion from local increment to global averaging. arXiv preprint arXiv:1702.08601 (2017)
19. Zhu, S., Zhang, R., Zhou, L., Shen, T., Fang, T., Tan, P., Quan, L.: Very large-scale global sfm by distributed motion averaging. In: CVPR. (2018)
20. Jegou, H., Douze, M., Schmid, C.: Hamming embedding and weak geometric consistency for large scale image search. In: ECCV. (2008)
21. Li, X., Larson, M., Hanjalic, A.: Pairwise geometric matching for large-scale object retrieval. In: CVPR. (2015)
22. Chum, O., Mikulik, A., Perdoch, M., Matas, J.: Total recall ii: Query expansion revisited. In: CVPR. (2011)
23. Jégou, H., Douze, M., Schmid, C., Pérez, P.: Aggregating local descriptors into a compact image representation. In: CVPR. (2010)
24. Babenko, A., Lempitsky, V.: Aggregating local deep features for image retrieval. In: ICCV. (2015)
25. Gordo, A., Almazan, J., Revaud, J., Larlus, D.: End-to-end learning of deep visual representations for image retrieval. IJCV (2017)
26. Wang, J., Song, Y., Leung, T., Rosenberg, C., Wang, J., Philbin, J., Chen, B., Wu, Y.: Learning fine-grained image similarity with deep ranking. In: CVPR. (2014)
27. Schroff, F., Kalenichenko, D., Philbin, J.: Facenet: A unified embedding for face recognition and clustering. In: CVPR. (2015)
28. Melekhov, I., Kannala, J., Rahtu, E.: Siamese network features for image matching. In: ICPR. (2016)
29. Chopra, S., Hadsell, R., LeCun, Y.: Learning a similarity metric discriminatively, with application to face verification. In: CVPR. (2005)
30. Li, S., Siu, S.Y., Fang, T., Quan, L.: Efficient multi-view surface refinement with adaptive resolution control. In: ECCV. (2016)
31. Ustinova, E., Lempitsky, V.: Learning deep embeddings with histogram loss. In: NIPS. (2016)
32. Simonyan, K., Zisserman, A.: Very deep convolutional networks for large-scale image recognition. In: ICLR. (2015)
33. Szegedy, C., Liu, W., Jia, Y., Sermanet, P., Reed, S., Anguelov, D., Erhan, D., Vanhoucke, V., Rabinovich, A.: Going deeper with convolutions. In: CVPR. (2015)
34. Shen, T., Wang, J., Fang, T., Zhu, S., Quan, L.: Color correction for image-based modeling in the large. In: ACCV. (2016)
35. Balntas, V., Riba, E., Ponsa, D., Mikolajczyk, K.: Learning local feature descriptors with triplets and shallow convolutional neural networks. In: BMVC. (2016)
36. Lin, J., Morère, O., Veillard, A., Duan, L.Y., Goh, H., Chandrasekhar, V.: Deephash for image instance retrieval: Getting regularization, depth and fine-tuning right. In: ICMR. (2017)
37. Song, H.O., Xiang, Y., Jegelka, S., Savarese, S.: Deep metric learning via lifted structured feature embedding. In: CVPR. (2016)

38. Luo, Z., Shen, T., Zhou, L., Zhu, S., Zhang, R., Yao, Y., Fang, T., Quan, L.: Geodesc: Learning local descriptors by integrating geometry constraints. (2018)
39. Azizpour, H., Sharif Razavian, A., Sullivan, J., Maki, A., Carlsson, S.: From generic to specific deep representations for visual recognition. In: CVPR Workshops. (2015)
40. Razavian, A.S., Sullivan, J., Carlsson, S., Maki, A.: Visual instance retrieval with deep convolutional networks. ITE Transactions on Media Technology and Applications (2016)
41. Abadi, M., Barham, P., Chen, J., Chen, Z., Davis, A., Dean, J., Devin, M., Ghemawat, S., Irving, G., Isard, M., et al.: Tensorflow: A system for large-scale machine learning. In: OSDI. (2016)
42. Vedaldi, A., Fulkerson, B.: Vlfeat: An open and portable library of computer vision algorithms. In: ACM Multimedia. (2010)
43. Wang, S., Jiang, S.: Instre: a new benchmark for instance-level object retrieval and recognition. ACM Transactions on Multimedia Computing, Communications, and Applications (TOMM) (2015)
44. Arandjelovic, R., Gronat, P., Torii, A., Pajdla, T., Sivic, J.: Netvlad: Cnn architecture for weakly supervised place recognition. In: CVPR. (2016)
45. Wilson, K., Snavely, N.: Robust global translations with 1dsfm. In: ECCV. (2014)

Thinking Outside the Box: Generation of Unconstrained 3D Room Layouts

Henry Howard-Jenkins[✉] , Shuda Li , and Victor Prisacariu

Active Vision Lab, University of Oxford, Oxford, UK
{henryhj,shuda,victor}@robots.ox.ac.uk

Abstract. We propose a method for room layout estimation that does not rely on the typical box approximation or Manhattan world assumption. Instead, we reformulate the geometry inference problem as an instance detection task, which we solve by directly regressing 3D planes using an R-CNN. We then use a variant of probabilistic clustering to combine the 3D planes regressed at each frame in a video sequence, with their respective camera poses, into a single global 3D room layout estimate. Finally, we showcase results which make no assumptions about perpendicular alignment, so can deal effectively with walls in any alignment.

Keywords: Room layout estimation · Scene understanding · 3D vision

1 Introduction

3D room layout estimation aims to produce a representation of the enclosing structure of an indoor scene, one consisting of the walls, floor and ceiling which bound the environment, while removing the clutter. These supporting surfaces provide essential information for a variety of computer vision tasks, not limited to augmented reality, indoor navigation and scene reconstruction.

Almost all prior works to produce 2D and 3D room layout estimations introduce heavy constraints on scene geometry, such as the box approximation [15,34] or the Manhattan World assumption [9,32]. Where the box approximation is used, rooms are constrained to be rectangular in shape. Manhattan World geometry, on the other hand, slightly eases the constraints of a boxy world; any number of walls are instead allowed to lie along either of two perpendicular directions. Although these assumptions can still lead to highly representative layouts for a great number of indoor environments, it is trivial to produce examples of room geometries which cannot be accurately captured due to the constraints.

The aim of this paper is to relax the constraints on room shape further. In doing so, we are able to represent a wider variety of scene layouts. To accomplish this, we introduce a neural network which is able to detect the presence and the extent, as well as directly regress a representative 3D plane, of room bounding surfaces in single monocular images. By taking the per-image detections accumulated over the frames in a video sequence, along with their respective camera

© Springer Nature Switzerland AG 2019
C. V. Jawahar et al. (Eds.): ACCV 2018, LNCS 11361, pp. 432–448, 2019.
https://doi.org/10.1007/978-3-030-20887-5_27

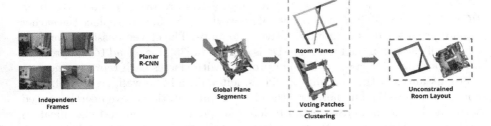

Fig. 1. An overview of our method: instantaneous room plane detections are provided by our Planar R-CNN and then combined with their respective camera pose to form a collection of global measurements. Probabilistic clustering is used to infer the number of room planes. These planes are intersected and each segment is considered a candidate wall. Candidates are then accepted into the global room layout based on the support of the measurements, thus capturing the room geometry.

poses, we are able to produce a single global map of plane measurements. From these measurements, we infer the overall geometry of the room without placing constraints on the shape. The scheme is outlined in Fig. 1.

The main contributions of this paper are: (i) a bounding surface instance detector giving the type and extent of the enclosing walls, floor and ceiling, (ii) direct plane regression for each room plane instance in an RGB image, and (iii) combining the instantaneous measurements to obtain a 3D layout from an RGB image sequence which is not limited to the boxy or Manhattan World constraints.

2   Related Work

Many researchers have sought to produce clutter-free representations of the room layout from monocular images. Early efforts largely had taken the bottom-up approach of extracting image features including colour, texture and edge information to inform vanishing point detection. In these works, a post-processing stage is used to generate, and in turn rank, a large number of room layout hypotheses with either conditional random fields (CRFs) or structured SVMs [11,14]. It is possible to recover 3D reconstruction from this approach, but in practice the final layout proves to be heavily dependant on the low level features extracted from the image, which are highly susceptible to noise and clutter. These low level features have also been used to infer 3D spatial layout through depth-ordered planes, by first grouping into lines, quadrilaterals and finally planes.

Mura *et al.* [22] recover accurate floor plans from planes detected in point clouds produced by cluttered 3D input range scans. Even when occluded, planes can be extracted and used to form a set of wall candidates which are projected into 2D. Rooms present in the floor plan are found by modelling a diffusion process over the spacial partitions caused by candidate walls. However, this method can only produce 2.5D extruded floor plans and requires a point cloud.

More recently, neural networks have been used to recover spatial layout from RGB images. In [20] fully convolutional networks are trained to replace the earlier hand-crafted features with informative edge maps. The eleven possible representations of the box model room in 2D, as defined in [35], have lead to the problem being formulated as a segmentation problem with classes each representing separate instances of the bounding surfaces: left wall, middle wall, right wall, floor and ceiling as in [7]. Furthermore, the eleven representations have been leveraged to tackle layout estimation as an ordered keypoint detection and classification problem as in [18], where the room type and the type's respective labelled corners are inferred directly from the image. These 2D representations have been used to produce 3D geometry by informing and ranking room hypotheses [15].

Panorama images have been leveraged for full scene recovery with box models in [34], and for a Manhattan World approximation in [32]. For whole room layout from more traditional camera images, incremental approaches have been used, such as that in [2]. Flint *et al.* used a combination of monocular features, with multiple-view and 3D features to infer a Manhattan World representation of the environment [9]. In [30], a RGB-D panorama is split and half of the scene is taken as input with the aim of producing reasonable, cluttered room layout estimation for the unseen portion of the panorama.

Alternatively, maps of the environment can be produced iteratively using a top down approach, such as the mapping and planning network architecture detailed in [12], where an egocentric map forms a top down 2D representation of the environment. Once again the representation of the scenes is constrained to 2D and the mapping component aims to recover the navigable space in a scene, rather than the clutter free representation which we seek.

The closest work to this paper is the direct plane regression from a single RGB image performed by Liu *et al.* in [19]. They are able to produce clutter-free room representations from single images. However, their method relies on visible planar regions and therefore room-layouts cannot be produced where a wall, floor or ceiling is completely occluded.

In this paper, rather than focusing on producing a self-complete 2D layout segmentation, we aim to recover the underlying 3D room planes in each image. Further, wall, floor and ceiling would generally be considered uncountable *stuff*, rather than countable *things* for which object detectors are most commonly used. However, there is work showing that *stuff* classes provide useful context for *thing* detection [1,21,24,27]. We take this further by treating these architectural planes as *things*, which we detect using an instance detector. These planes are then used to compile a singular representation of the scene which does not rely on any assumption about the geometric structure of the world other than that rooms are planar.

3   Planar R-CNN

To address the challenges presented by a non-boxy room layout, we approach the task of finding a room's supporting surfaces as an instance detection problem. In doing so, our network detects an arbitrary number of surfaces and outputs their extents in the form of a bounding-box. Furthermore, by removing the Manhattan World assumption, the walls, floor and ceiling can have any orientation and are no longer just aligned with one of only two dominant directions or the ground plane. Therefore, our network also outputs arbitrary alignment for each of the detection by directly regressing the 3D plane equation of each surface.

We choose to approach the task by directly targeting abstract room-planes rather than using a method which relies on visible patches of wall, floor and ceiling. The advantage gained is two-fold: (i) we are able to infer room planes that are completely or mostly occluded by furniture; (ii) we are able to disregard large planar surfaces that are irrelevant to room structure. On the first point, furniture can often block the wall behind, thus reducing the ability to capture the plane from visible pixels. Considering the second point, the occluding object itself may be the dominant plane within a region, such as the surface of a table.

In all, we are able to predict an arbitrary number of room-bounding planes, as well as their directions and extents, providing all the information required to construct an unconstrained layout.

3.1 Network Architecture

We design a network architecture capable of detecting an arbitrary number of walls, floors and ceilings, as well as their extents, positions and alignments. Inspired by Mask R-CNN [13], where Faster R-CNN [25] is augmented by an additional network head which outputs a mask for each detection, we take a similar approach by adding an additional head to regress a 3D plane equation for each region of interest (RoI). From an input of a single RGB image, the network outputs a number of detections. Each detection has an associated class (wall or floor/ceiling), bounding-box and plane equation.

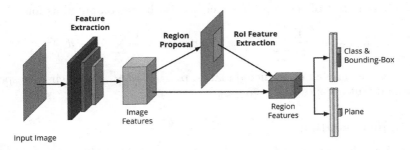

Fig. 2. An overview of the complete network from an RGB input image to the outputted classes, bounding-boxes and plane equations. The single dimensional columns in the network heads represent fully connected layers. Yellow columns have an output size of 4096, the plane has output size 4, and the class and bounding-box columns have output size equal to and four times the number of classes, respectively (Color figure online)

For clarity, we use the same nomenclature as in [13] when referring to: (i) the *backbone* which is a convolutional architecture to extract features over the entire input image, and (ii) a network *head* as an architecture applied separately to each set of RoI features, for example the Fast R-CNN head which predicts class and regresses a bounding-box.

Instance Detection. For detection, assigning of classes and determining bounding-boxes for instances of visible room planes, we use an implementation of the Faster R-CNN detector. Faster R-CNN is composed of two stages: the Region Proposal Network (RPN) and a second stage which is essentially Fast R-CNN [10].

Direct Plane Regression. In a similar fashion to Mask R-CNN, we introduce an additional network head to the second stage of Faster R-CNN. This network head directly regresses the 3D plane equation of the candidate in parallel to the classification and bounding-box regression. The plane regression head takes the same RoI features as the classification and bounding-box regression head. It outputs a 4-vector which represents the plane coefficients where a 3D point, $\mathbf{x}$, lies on plane $\mathbf{p} = [\mathbf{n} \; d]^T$ if $\mathbf{n} \cdot \mathbf{x} + d = 0$. The first 3 components of the 4-vector, which represent the normal of the plane, are normalised to unit length. The 4th plane coefficient, d, takes on a more direct physical meaning as the shortest distance between the plane and the camera origin.

The network head itself is comprised of two fully connected layers with 4096 output channels each, in addition to another fully connected layer to take the dimensions down from 4096 to the required 4 representing the plane equation coefficients, as detailed in Fig. 2. The plane head is trained using two losses, L_{norm} and L_d, which represent the loss contributed by the normal of the plane and the depth coefficient, respectively. L_{norm} is defined as the negative of the cosine similarity between predicted and ground truth normals, and L_d is the mean-squared-error of the depth coefficients. Both are computed as follows:

$$L_{norm} = -\frac{1}{m} \sum_m \mathbf{n}_p \cdot \mathbf{n}_{gt} \quad and \quad L_d = \frac{1}{m} \sum_m (d_p - d_{gt})^2 \tag{1}$$

where m is the number of predicted planes, $\mathbf{p}_p = [\mathbf{n}_p \; d_p]^T$, and their corresponding ground truth planes, $\mathbf{p}_{gt} = [\mathbf{n}_{gt} \; d_{gt}]^T$.

3.2 Implementation

We use an implementation of Faster R-CNN with a VGG-16 [28] as its backbone. The final pooling and all the fully connected layers from the VGG-16 network are removed. In addition to RoIPool from the default Faster R-CNN implementation, we also implement RoIAlign [13]. While both methods achieve the same general goal of gathering the features relevant to a proposed region, RoIAlign better preserves the spatial information of the region's features, which may lead to improved performance for the spatially sensitive task of plane regression.

The loss for each sampled RoI is defined as $L = L_{cls} + L_{box} + L_{norm} + kL_d$ where L_{norm} and L_d are as above. The classification loss, L_{cls}, and the bounding-box loss are as in [10]. We found it was necessary to introduce a weight constant, k, for L_d to keep the network stable while training. Setting $k = 0.05$ provided good stability, while ensuring that the depth coefficient still trains.

4   3D Room Layout Generation

The network described in the previous section outputs a number of plane segments representing the room geometry. However, since we do not assume that all rooms are convex, the geometric information predicted from a single image is not complete in itself. For example, there may be entire walls occluded by others. Therefore, we treat each set of plane segment predictions as local measurements from a global room layout.

We define the global room layout as a set of 3D plane segments, each produced by the intersection of underlying room planes. The first task in determining the room layout is to obtain these room planes. We transform the regressed planes and corresponding bounding box measurements from each frame into the global coordinate system. This represents an amalgamation of noisy plane segment measurements each belonging to one of an unknown number of bounding planes.

We adopt a Bayesian approach for a Gaussian mixture model, and infer the posterior distributions of the room's characteristic planes through EM training. The number of room planes is determined by placing a threshold on the weights in the mixture model.

4.1   Spatial Voting

Once we have obtained the room planes, we intersect them all to form a collection of candidate plane segments to add to the room layout. In the convex case, these intersections would provide an immediate layout, however, since we allow non-convexity we must evaluate whether a candidate segment should be added to the room set. To do this, we make use of the regressed bounding-boxes as extents for the predicted planes. Since each measured plane is spatially constrained, it is used to vote for a candidate segment's existence in the room geometry.

We assign every evidence segment to a mixture component distribution. For each of these distributions, we then define a subset of n voter segments by ranking the likelihood of the segment having come from the cluster distribution. This ensures that the measurements lie on the most similar planes to the candidate segments for which they will vote. Support for a wall segment is determined through a robust inlier weighting scheme, such as in [4]. A voting plane patch is considered an inlier for a candidate wall by thresholding the proportion of the voter that overlaps the candidate, i_{vc}, and vice versa, i_{cv}.

Every candidate is assigned an energy, $\eta = 0$, which is updated on each of the voting segments belonging to the same plane cluster. If a voting patch is considered an inlier, the candidate's energy is increased by $(1 - i_{vc})$, and the number of inliers is incremented. Once all voting patches have been tested, the ratio of inliers to the total voting on the candidate, r_c is computed and the candidate's energy is divided by r_c^a, where the exponent, a, is a tuneable parameter. To prevent cases where candidate patches are accepted with only a small total number of voter patches, we threshold total energy. The final result of the voting stage is to reject plane segments without sufficient evidence. In all, leaving only the walls that belong to the global room layout.

5   Experimental Evaluation

We train and evaluate our network's performance, both quantitative and qualitatively, capturing room geometry from an image in the SUN RGB-D dataset. We compare the regressed planes to their corresponding ground truth instances. In addition we plot the captured planes and their extents with an overlaid point cloud to demonstrate the network's ability to recover bounding planes.

Further, we evaluate our room layout estimation on single frames by testing on the NYUDv2 303 dataset [33]. We demonstrate the ability of our method to capture useful room layout information by using the output of our network to produce a prediction at the 2D room segmentation for an indoor image.

Finally, we prove the informative nature by amalgamating the independent frame-by-frame predictions on ScanNet image sequences into the global coordinate system and using these measurements to infer a whole room layout. This demonstrates the robustness of the enclosing surface detection and plane regression, since the network was never trained on any ScanNet sequence.

5.1   Training Using SUN RGB-D

To train our network, we use the SUN RGB-D dataset [29]. The dataset provides room corner coordinates on the $x - z$ plane, as well as the upper and lower y height of the room, all defined in the world coordinates. This information allows us to build a 2.5D extruded floor plan for the room for each pair of RGB and Depth images in the dataset. We transform the room model from the world coordinate system to one aligned with the camera. From this aligned floor plan, we compute the plane equation of each section of wall, floor and ceiling. The floor plan is then projected into the camera, providing us instance-level segmentation, as well as corresponding plane equations, for every image.

Since we do not test on the benchmarks provided as part of the SUN RGB-D dataset, we use our own image split comprising of 9335 training, 500 validation and 500 testing images. It was found that some of the images in the dataset do not appear to have corresponding room layout information. In addition, for some images, it was found that the picture was taken outside of the annotated room. This tended to occur when pictures were captured in a department store and the room corners were defined as the limits of the display cubicle. We removed these images from the dataset for training, validation and testing. This resulting in the final split containing 7677 training, 441 validation and 439 test images. It is worth noting that the depth images are not used in either training or prediction: instead they are only used to help visualise results.

Bounding surface instances are divided into two classes, one representing walls and the other representing floors or ceilings. This is a result of trying to keep the plane segments as generic as possible, but still maintaining the differences in their representation in the dataset. Because the rooms in the dataset are 2.5D rather than complete 3D floorplans, walls always have normals in the world $x - z$ plane, whereas floors and ceilings must have normals in the world y direction.

Training. The models were trained on the dataset for 25 epochs in total. To initialise the feature extractor and classifier in the class and location head, we use a VGG-16 model pretrained on ImageNet [16]. Other fully connected layers in both network heads are initialised from a zero mean normal distribution with variance of 0.1, 0.01, 0.001 for the final plane regressor layer, the final classifier layer and the two hidden layers in the plane head, and the bounding-box regressor respectively. We use a batch size of 1 with stochastic gradient descent, 0.5 dropout rate, 0.0005 weight decay, 0.9 momentum. The learning rate is initially set to 0.001 and decreased to 0.0001 after 20 epochs. This training schedule takes around 12 h on a single Nvidia GTX 1080Ti.

5.2   2D Room Layout Estimation with NYUDv2 303

We explore the generation of 2D layouts using our network, and thus demonstrate its ability to capture room information. As discussed in Sect. 4, our method does not require the convex box approximation. Since we do not impose any such constraints, there is no set number of possible room layouts, or wall types. We do not seek to classify left or right wall, instead we just detect the presence of a wall, floor or ceiling.

To produce predictions, we first back-project each pixel within a detection's bounding box onto its predicted plane, forming a plane patch in 3D. We then form a clutter-free 2D room segmentation by project all of an image's plane patches back into the image. Our reprojection into 2D provides insight into the usefulness of the plane detections for capturing room layout without having to place constraints on the layout, as well as providing comparison against other state-of-the-art solutions to the layout estimation problem.

Since this dataset takes a subsample of the NYUDv2 dataset [23], which is included in the SUN RGB-D dataset [29], we retrained our network ensuring that none of the 101 test images appeared in our training or validation dataset.

5.3   Whole Room Layout Estimation Using ScanNet

In order to demonstrate not only the informative nature of the measurements produced by our network, but also its versatility, we produce 3D room layouts on the ScanNet dataset [5] having never trained on it and without the network using intrinsic data for each sequence. For this task we assume that camera poses are known, as in this case they are implemented using BundleFusion [6].

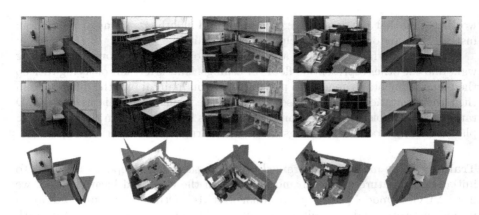

Fig. 3. The first and second rows show the image overlaid with the ground truth and network detections, respectively. The last row shows the predicted bounding-boxes projected onto their matching regressed planes to produce the informative plane segments. For additional 3D context, a point cloud from a depth image matching the input RGB image is displayed with the plane segments (Color figure online)

Implementation. We start by collecting all the plane detections in a ScanNet image sequence. Since bounding-boxes do not provide tight constraint on the extent of a plane, especially if the plane is nearly perpendicular to the image plane. Therefore, when collecting instances from images, we discard those surfaces which are less than 30° from perpendicular to the image plane.

We found that our spatial voting method for determining which candidate wall planes was sensitive to changes in set parameters. We believe that this is due, at least in part, to the lack of any training on the ScanNet dataset. Planar R-CNN does not explicitly infer the camera's intrinsics. Therefore, for spatial information recovery at full accuracy, the network should be fine-tuned on new datasets to allow the relation between image coordinates and features with the surrounding 3D environment to be re-assessed. To account for this, we adjusted parameters on a per-scene basis when constructing room layouts. Generally, the parameters were set in the vicinity of: the minimum weight for an accepted plane posterior as 0.05, the 100 most likely measurements from each wall's posterior distribution allowed to vote for existence of the plane's candidate segments, the energy update exponent set to 2, and finally the mutual overlap threshold as more than 70% of the voter's area and more than 20% of the candidate's area for the voter to be labelled an inlier for the candidate.

Almost all of the ScanNet sequences have only brief appearances of the ceiling. To provide an upper bound on the wall segments, we instead tested the normal of the dominant posterior from planes labelled as the floor or ceiling class. If the direction of the normal was in the positive z direction, we labelled the plane floor and created a ceiling plane 2 m above and with a normal in the other direction. If a downward facing dominant plane was detected, it was labelled ceiling and a floor plane was created 2 m below it.

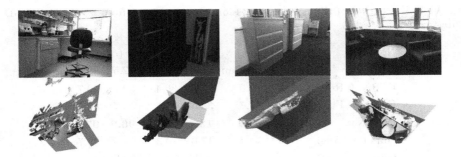

Fig. 4. Examples demonstrating the discussed advantages: The first column shows a plane being accurately regressed in a cluttered environment; The second and third columns show examples walls are regressed behind shelves and cabinets; In the forth column a wall is correctly estimated behind a chair and a non-planar curtain.

5.4 Results

Since, to our own knowledge, this network is the first to directly regress abstract room-planes, we consider its main tasks when evaluating its performance. We define these tasks as accurate detection and plane regression. The detection task is evaluated using the average precision for each of the two classes, and the mean average precision. For plane regression, the network must be able to discern the alignment of the plane, *i.e.* the normal of the surface, and the depth coefficient in order to fix the plane in 3D space. We propose a two stage evaluation of the plane regression; first, the quality of the normals, secondly, the spatial accuracy of the plane.

Instantaneous Detections on SUN RGB-D. We assess the quality of the plane normal predictions from the network in the same manner as surface normal estimation from monocular images, as in [3,8,31]. However, we compare the normal of a detected instance to its corresponding ground truth plane, instead of at each pixel. The angular disparity each pair, α, is computed. We then compute the mean, median and root-mean-square of α, as well as listing the percentage where the error falls below the thresholds 11.25°, 22.5° and 30°. These results are shown in Table 1.

To determine the accuracy of the regressed plane in terms of spacial location, we back-project all the pixels within the predicted bounding-box and on to the predicted plane. For each of the back-projected points, we compute a point-to-plane distance between the point and the corresponding ground truth plane. The mean of these distances are averaged for each instance detection in the image, providing a single distance score, δ, for every predicted bounding plane. In Table 2 we show the mean and median of these instance distances, as well as the percentage which fall within the thresholds 0.2, 0.5 and 1 m.

We compare our normal and plane location results to a baseline visible pixel-based method. This method produces a pixel-wise depth estimation using the

Table 1. Evaluation of the normal regression on our test set from the SUN RGB-D dataset.

Normals	Error			Accuracy ($\alpha < \theta$)		
	Mean	Median	RMS	11.25°	22.5°	30°
FCRN + Room	33.5	20.0	44.8	36.4	52.7	59.7
FCRN + Fine	28.6	17.7	38.5	38.7	56.7	64.7
PlaneNet	**8.35**	**6.31**	**13.4**	**82.3**	**96.9**	**97.8**
Ours (RoIAlign)	11.3	7.44	17.0	70.3	89.5	93.7
Ours (RoIPool)	9.74	6.67	15.3	76.4	92.9	94.9

Table 2. Evaluation of the location of the regressed plane compared to its corresponding ground truth plane on our test set from the SUN RGB-D dataset.

Plane location	Error		Accuracy		
	Mean	Median	$\delta < 0.2\,\mathrm{m}$	$\delta < 0.5\,\mathrm{m}$	$\delta < 1\,\mathrm{m}$
FCRN + Room	1.83	0.357	33.3	60.7	85.7
FCRN + Fine	1.01	0.242	43.7	74.0	89.2
PlaneNet	1.79	0.275	40.0	67.8	84.3
Ours (RoIAlign)	1.18	0.269	42.0	67.4	84.1
Ours (RoIPool)	**0.503**	**0.217**	**47.1**	**75.5**	**90.7**

FCRN method of [17]. The depth map is then masked by a ground truth segmentation of the image and back-projected into 3D space. We use RANSAC to fit a plane to the 3D points corresponding to each ground truth room surface. *FCRN + Room* in Tables 1 and 2 uses the ground truth room-layout segmentation to mask the estimated depth map, meaning that all depth points may influence the plane fitting. However, *FCRN + Fine* leverages the full segmentation and only points labelled wall, floor, ceiling or window are included in the plane fitting.

We also compare against PlaneNet, using the method described in [19] for inferring room layout. This takes advantage of the consistent ordering of plane inference of the PlaneNet network, where floors are predicted at a specific index etc. We fine-tuned PlaneNet on SUN RGB-D using the depth-only method mentioned in their paper, this did not outperform their pretrained model.

The results against the baseline in Tables 1 and 2 demonstrate the quality of the planes regressed using our method and further illustrate the advantages of targeting the abstract room surfaces, rather than relying on visible pixel patches. In addition, it is worth noting that the baseline makes use of ground truth segmentation. For a true comparison, segmentation predictions should be used, further reducing the quality of the fitted planes. Even with the advantage of ground truth segmentation, the baseline is out performed by our method.

Table 3. The performance of our Planar R-CNN for predicting 2D room layouts. Evaluated on the NYUDv2-303 dataset [33]. Results for other methods are as stated in [19]

2D layouts	[33] RGB + D	[33] RGB	[26]	RoomNet	PlaneNet	**Ours**
Pixel Error	*8.04%*	13.94%	13.66%	12.96%	12.64%	**12.19%**

Fig. 5. Example 2D room layouts produced by our method on the NYUDv2 303 dataset. The first row shows the input image. The second and third rows show the ground truth and predicted room layout segmentations, respectively. The fourth row shows the prediction overlaid onto the input image. It is worth noting that in the last two columns the rooms in the input are not box-shaped. We believe that our prediction better reflects the room shape in these cases than the ground truth.

While PlaneNet slightly outperforms our method for the accuracy of the regressed normals, our method predicts room planes that are closer to the groundtruth. We would suggest that the high normal accuracy shown by PlaneNet could be explained by planar objects, such as cupboards and closets, which are generally arranged to be parallel to the walls that they occlude. Further, PlaneNet tends to miss planes which are not near the camera. It is worth noting that the architectural planes which are missed by PlaneNet do not negatively impact the results shown in Tables 1 and 2. Whereas, our method performed well as a room-plane detector, with the RoIAlign and RoIPool methods each achieving a mAP of 67.9 and 69.6, respectively.

Example detections from our network, as well as their respective planes from single images, are pictured in Fig. 3. Predicted planes are limited in extent in this figure by the predicted bounding-box, in order to produce a number of plane segments. In each case, a colourised point cloud produced from the paired depth image is shown to provide context for the predictions. Also included are examples which break the box and Manhattan world assumptions. Further, in Fig. 4 we show example detections demonstrating the advantages of targeting abstract room-planes over plane-fitting to visible wall, floor or ceiling patches. These examples include walls being correctly detected and their plane being accurately regressed when it is entirely or mostly occluded.

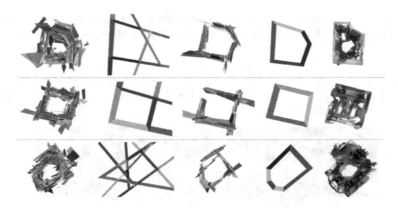

Fig. 6. Each row visualises the steps of our spatial voting method for determining the room layout from Planar R-CNN measurements on a ScanNet sequence. The first column shows a subset of the raw measurements. After inferring the underlying room planes, we intersect to find the set of wall segment candidates, shown in the second column, as well as the top n measurements from each cluster with the highest likelihood to form the group of voter patches, shown in the third column. The fourth and fifth column show the final layout estimation and the ground truth mesh, respectively.

2D Room Layout Predictions. We demonstrate the ability for our method to capture room geometry further by producing room layout predictions for individual frames on the NYUDv2 303 dataset. Since we do not assume a box layout, we do not classify the type of wall as left, right, or front. Although floor and ceilings are obvious, these wall classes present a level of ambiguity. In an image of a room corner it is often an arbitrary choice about whether the two wall should be labelled left and front, or front and right. Therefore, to evaluate our predicted layouts we must map our wall detections to the five possible plane types in the ground truth, which we do using the Hungarian algorithm.

As can be seen with the results presented in Table 3, we outperform all RGB methods tested, including PlaneNet [19] and RoomNet [18], designed for this task. We also believe our results could be improved by merging architectural plane detections if the plane equations are similar. This merging of detections would better reflect the ground truth box approximations used in the dataset.

Further, we present qualitative results of the 2D layouts that are produced by our method in Fig. 5. These examples help to illustrate that while the box representation of rooms is often very well suited to the room pictured, as shown in the first five columns of Fig. 5, there are cases where this approximation does not reflect the geometry of the scene particularly well. For example, in the last two columns of Fig. 5, we believe that our predictions offer a better representation of the room geometry than the ground truth.

Fig. 7. Manhattan boxes (magenta) fitted to the ground truth point clouds using the method provided in SUN RGB-D shown with our own unconstrained layouts (green). (Color figure online)

3D Room Layout Estimation on ScanNet. Unfortunately, ScanNet does not provide the room layout annotations required to train our network. This meant we were unable to fine-tune or provide quantitative results on this dataset. Therefore, we provide qualitative room-layouts to demonstrate that our method intended to visualise the geometric information captured by our Planar R-CNN.

The qualitative results in Fig. 6 show example room layouts obtained on the ScanNet dataset. From the noisy raw plane measurements, the clustered planes encapsulate a broad representation of the room. The fourth row provides an interesting case as the upper segment of room is missed altogether. When inspecting the ground truth mesh, it was found that there are limited vertices present in the mesh. This leads us to believe that this area was not visible for much of the video. It is worth noting that, most likely because of the level of noise in the initial measurements, fine room geometry is missed in the clustering.

Although the produced room-layout estimates lack fine detail, we believe that they provide a better representation of room geometry than a Manhattan box approximation would. To demonstrate this, we produce a qualitative baseline using the Manhattan model fitting provided in SUN RGB-D. The boxes produced in Fig. 7 were fitted to the ground truth point clouds, and thus would likely deteriorate if point clouds obtained from a monocular method were used.

We believe that the results shown in Figs. 6 and 7 demonstrate our methods ability to capture useful structure from ScanNet sequences without using depth, enforcing a Manhattan assumption, or imposing temporal consistency.

6   Conclusions

We have proposed a method for room layout estimation without a box or Manhattan world assumption. It combines instance segmentation implemented via an R-CNN with Gaussian Mixture Model-based probabilistic clustering, and allows for the effective reconstruction of walls which do not lie on either of the two principle aligned axes, or the vertical direction.

Possible extensions would include temporal coherency and consistency both in the 2D inference stage and in the clustering. We believe that this would improve the fidelity of the measurements in the global room, further increasing the fidelity of the layout estimation.

Acknowledgements. We gratefully acknowledge the European Commission Project Multiple-actOrs Virtual Empathic CARegiver for the Elder (MoveCare) for financially supporting the authors for this work.

References

1. Brahmbhatt, S., Christensen, H.I., Hays, J.: StuffNet: using 'stuff' to improve object detection. In: 2017 IEEE Winter Conference on Applications of Computer Vision (WACV), pp. 934–943. IEEE (2017)
2. Cabral, R., Furukawa, Y.: Piecewise planar and compact floorplan reconstruction from images. In: 2014 IEEE Conference on Computer Vision and Pattern Recognition (CVPR), pp. 628–635. IEEE (2014)
3. Chen, W., Xiang, D., Deng, J.: Surface normals in the wild. arXiv preprint arXiv:1704.02956 (2017)
4. Chetverikov, D., Stepanov, D., Krsek, P.: Robust Euclidean alignment of 3D point sets: the trimmed iterative closest point algorithm. Image Vis. Comput. **23**(3), 299–309 (2005)
5. Dai, A., Chang, A.X., Savva, M., Halber, M., Funkhouser, T., Nießner, M.: ScanNet: richly-annotated 3D reconstructions of indoor scenes. In: Proceedings of the IEEE Conference on Computer Vision and Pattern Recognition (CVPR), vol. 1, p. 1 (2017)
6. Dai, A., Nießner, M., Zollhöfer, M., Izadi, S., Theobalt, C.: BundleFusion: real-time globally consistent 3D reconstruction using on-the-fly surface reintegration. ACM Trans. Graph. (TOG) **36**(3), 24 (2017)
7. Dasgupta, S., Fang, K., Chen, K., Savarese, S.: Delay: robust spatial layout estimation for cluttered indoor scenes. In: Proceedings of the IEEE Conference on Computer Vision and Pattern Recognition, pp. 616–624 (2016)
8. Eigen, D., Fergus, R.: Predicting depth, surface normals and semantic labels with a common multi-scale convolutional architecture. In: Proceedings of the IEEE International Conference on Computer Vision, pp. 2650–2658 (2015)
9. Flint, A., Murray, D., Reid, I.: Manhattan scene understanding using monocular, stereo, and 3D features. In: 2011 IEEE International Conference on Computer Vision (ICCV), pp. 2228–2235. IEEE (2011)
10. Girshick, R.: Fast R-CNN. arXiv preprint arXiv:1504.08083 (2015)
11. Gupta, A., Hebert, M., Kanade, T., Blei, D.M.: Estimating spatial layout of rooms using volumetric reasoning about objects and surfaces. In: Advances in Neural Information Processing Systems, pp. 1288–1296 (2010)
12. Gupta, S., Davidson, J., Levine, S., Sukthankar, R., Malik, J.: Cognitive mapping and planning for visual navigation. arXiv preprint arXiv:1702.03920 **3** (2017)
13. He, K., Gkioxari, G., Dollár, P., Girshick, R.: Mask R-CNN. In: 2017 IEEE International Conference on Computer Vision (ICCV), pp. 2980–2988. IEEE (2017)
14. Hedau, V., Hoiem, D., Forsyth, D.: Recovering the spatial layout of cluttered rooms. In: 2009 IEEE 12th International Conference on Computer Vision, pp. 1849–1856. IEEE (2009)
15. Izadinia, H., Shan, Q., Seitz, S.M.: IM2CAD. In: 2017 IEEE Conference on Computer Vision and Pattern Recognition (CVPR), pp. 2422–2431. IEEE (2017)
16. Krizhevsky, A., Sutskever, I., Hinton, G.E.: ImageNet classification with deep convolutional neural networks. In: Advances in Neural Information Processing Systems, pp. 1097–1105 (2012)

17. Laina, I., Rupprecht, C., Belagiannis, V., Tombari, F., Navab, N.: Deeper depth prediction with fully convolutional residual networks. In: 2016 Fourth International Conference on 3D Vision (3DV), pp. 239–248. IEEE (2016)
18. Lee, C.Y., Badrinarayanan, V., Malisiewicz, T., Rabinovich, A.: RoomNet: end-to-end room layout estimation. arXiv preprint arXiv:1703.06241 (2017)
19. Liu, C., Yang, J., Ceylan, D., Yumer, E., Furukawa, Y.: PlaneNet: piece-wise planar reconstruction from a single RGB image. In: Proceedings of the IEEE Conference on Computer Vision and Pattern Recognition, pp. 2579–2588 (2018)
20. Mallya, A., Lazebnik, S.: Learning informative edge maps for indoor scene layout prediction. In: Proceedings of the IEEE International Conference on Computer Vision, pp. 936–944 (2015)
21. Mottaghi, R., et al.: The role of context for object detection and semantic segmentation in the wild. In: Proceedings of the IEEE Conference on Computer Vision and Pattern Recognition, pp. 891–898 (2014)
22. Mura, C., Mattausch, O., Villanueva, A.J., Gobbetti, E., Pajarola, R.: Automatic room detection and reconstruction in cluttered indoor environments with complex room layouts. Comput. Graph. **44**, 20–32 (2014)
23. Silberman, N., Hoiem, D., Kohli, P., Fergus, R.: Indoor segmentation and support inference from RGBD images. In: Fitzgibbon, A., Lazebnik, S., Perona, P., Sato, Y., Schmid, C. (eds.) ECCV 2012. LNCS, vol. 7576, pp. 746–760. Springer, Heidelberg (2012). https://doi.org/10.1007/978-3-642-33715-4_54
24. Rabinovich, A., Vedaldi, A., Galleguillos, C., Wiewiora, E., Belongie, S.: Objects in context. In: 2007 IEEE 11th International Conference on Computer Vision, ICCV 2007, pp. 1–8. IEEE (2007)
25. Ren, S., He, K., Girshick, R., Sun, J.: Faster R-CNN: towards real-time object detection with region proposal networks. In: Advances in Neural Information Processing Systems, pp. 91–99 (2015)
26. Schwing, A.G., Urtasun, R.: Efficient exact inference for 3d indoor scene understanding. In: Fitzgibbon, A., Lazebnik, S., Perona, P., Sato, Y., Schmid, C. (eds.) ECCV 2012. LNCS, vol. 7577, pp. 299–313. Springer, Heidelberg (2012). https://doi.org/10.1007/978-3-642-33783-3_22
27. Shi, M., Caesar, H., Ferrari, V.: Weakly supervised object localization using things and stuff transfer. In: Proceedings of the IEEE International Conference on Computer Vision (ICCV) (2017)
28. Simonyan, K., Zisserman, A.: Very deep convolutional networks for large-scale image recognition. arXiv preprint arXiv:1409.1556 (2014)
29. Song, S., Lichtenberg, S.P., Xiao, J.: SUN RGB-D: a RGB-D scene understanding benchmark suite. In: Proceedings of the IEEE Conference on Computer Vision and Pattern Recognition, pp. 567–576 (2015)
30. Song, S., Zeng, A., Chang, A.X., Savva, M., Savarese, S., Funkhouser, T.: Im2Pano3D: extrapolating 360 structure and semantics beyond the field of view. arXiv preprint arXiv:1712.04569 (2017)
31. Wang, X., Fouhey, D., Gupta, A.: Designing deep networks for surface normal estimation. In: Proceedings of the IEEE Conference on Computer Vision and Pattern Recognition, pp. 539–547 (2015)
32. Xu, J., Stenger, B., Kerola, T., Tung, T.: Pano2CAD: room layout from a single panorama image. In: 2017 IEEE Winter Conference on Applications of Computer Vision (WACV), pp. 354–362. IEEE (2017)
33. Zhang, J., Kan, C., Schwing, A.G., Urtasun, R.: Estimating the 3D layout of indoor scenes and its clutter from depth sensors. In: Proceedings of the IEEE International Conference on Computer Vision, pp. 1273–1280 (2013)

34. Zhang, Y., Song, S., Tan, P., Xiao, J.: PanoContext: a whole-room 3D context model for panoramic scene understanding. In: Fleet, D., Pajdla, T., Schiele, B., Tuytelaars, T. (eds.) ECCV 2014. LNCS, vol. 8694, pp. 668–686. Springer, Cham (2014). https://doi.org/10.1007/978-3-319-10599-4_43
35. Zhang, Y., Yu, F., Song, S., Xu, P., Seff, A., Xiao, J.: Large-scale scene understanding challenge: room layout estimation. Accessed 15 Sept 2015

VIENA²: A Driving Anticipation Dataset

Mohammad Sadegh Aliakbarian[1,2,4(✉)], Fatemeh Sadat Saleh[1,4],
Mathieu Salzmann[3], Basura Fernando[2], Lars Petersson[1,4],
and Lars Andersson[4]

[1] ANU, Canberra, Australia
[2] ACRV, Canberra, Australia
basura.fernando@anu.edu.au
[3] CVLab, EPFL, Lausanne, Switzerland
mathieu.salzmann@epfl.ch
[4] Data61-CSIRO, Canberra, Australia
{mohammadsadegh.aliakbarian,fatemehsadat.saleh,
lars.petersson,lars.andersson}@data61.csiro.au

Abstract. Action anticipation is critical in scenarios where one needs
to react before the action is finalized. This is, for instance, the case in
automated driving, where a car needs to, e.g., avoid hitting pedestrians
and respect traffic lights. While solutions have been proposed to tackle
subsets of the driving anticipation tasks, by making use of diverse, task-
specific sensors, there is no single dataset or framework that addresses
them all in a consistent manner. In this paper, we therefore introduce
a new, large-scale dataset, called VIENA², covering 5 generic driving
scenarios, with a total of 25 distinct action classes. It contains more
than 15K full HD, 5 s long videos acquired in various driving conditions,
weathers, daytimes and environments, complemented with a common
and realistic set of sensor measurements. This amounts to more than
2.25M frames, each annotated with an action label, corresponding to
600 samples per action class. We discuss our data acquisition strategy
and the statistics of our dataset, and benchmark state-of-the-art action
anticipation techniques, including a new multi-modal LSTM architecture
with an effective loss function for action anticipation in driving scenarios.

1 Introduction

Understanding actions/events from videos is key to the success of many real-
world applications, such as autonomous navigation, surveillance and sports analysis. While great progress has been made to recognize actions from complete
sequences [2,4,7,43], action anticipation, which aims to predict the observed action
as early as possible, has only reached a much lesser degree of maturity [1,39,42].

Electronic supplementary material The online version of this chapter (https://
doi.org/10.1007/978-3-030-20887-5_28) contains supplementary material, which is
available to authorized users.

C. V. Jawahar et al. (Eds.): ACCV 2018, LNCS 11361, pp. 449–466, 2019.
https://doi.org/10.1007/978-3-030-20887-5_28

Nevertheless, anticipation is a crucial component in scenarios where a system needs to react quickly, such as in robotics [17], and automated driving [12,18,19]. Its benefits have also been demonstrated in surveillance settings [26,44].

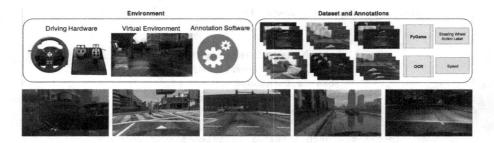

Fig. 1. Overview of our data collection. Using the GTA V environment and driving equipment depicted in the top left box, we captured a new dataset covering 5 generic scenarios, illustrated in the right box, each containing multiple action classes (samples in bottom row). For more examples and examples of the vehicles our data was gathered with, please check our supplementary material.

In this paper, we focus on the driving scenario. In this context, when consulting the main actors in the field, may they be from the computer vision community, the intelligent vehicle one or the automotive industry, the consensus is that predicting the intentions of a car's own driver, for Advanced Driver Assistance Systems (ADAS), remains a challenging task for a computer, despite being relatively easy for a human [5,12,13,23,27]. Anticipation then becomes even more complex when one considers the maneuvers of other vehicles and pedestrians [5,15,45]. However, it is key to avoiding dangerous situations, and thus to the success of autonomous driving.

Over the years, the researchers in the field of anticipation for driving scenarios have focused on specific subproblems of this challenging task, such as lane change detection [21,41], a car's own driver's intention [22] or maneuver recognition [11–13,23] and pedestrian intention prediction [18,25,27,36]. Furthermore, these different subproblems are typically addressed by making use of different kinds of sensors, without considering the fact that, in practice, the automotive industry might not be able/willing to incorporate all these different sensors to address all these different tasks.

In this paper, we study the general problem of anticipation in driving scenarios, encompassing all the subproblems discussed above, and others, such as other drivers' intention prediction, with a fixed, sensible set of sensors. To this end, we introduce the **VI**rtual **EN**vironment for **A**ction **A**nalysis (VIENA^2) dataset, covering the five different subproblems of predicting driver maneuvers, pedestrian intentions, front car intentions, traffic rule violations, and accidents.

Altogether, these subproblems encompass a total of 25 distinct action classes. VIENA^2 was acquired using the GTA V video game [30]. It contains more than

15K full HD, 5 s long videos, corresponding to more than 600 samples per action class, acquired in various driving conditions, weathers, daytimes, and environments. This amounts to more than 2.25M frames, each annotated with an action label. These videos are complemented by basic vehicle dynamics measurements, reflecting well the type of information that one could have access to in practice.

Below, we describe how VIENA^2 was collected and compare its statistics and properties to existing datasets. We then benchmark state-of-the-art action anticipation algorithms on VIENA^2, and introduce a new multi-modal, LSTM-based architecture, together with a new anticipation loss, which outperforms existing approaches in our driving anticipation scenarios. Finally, we investigate the benefits of our synthetic data to address anticipation from real images. In short, our contributions are: **(i)** a large-scale action anticipation dataset for general driving scenarios; **(ii)** a multi-modal action anticipation architecture.

VIENA^2 is meant as an extensible dataset that will grow over time to include not only more data but also additional scenarios. Note that, for benchmarking purposes, however, we will clearly define training/test partitions. A similar strategy was followed by other datasets such as CityScapes, which contains a standard benchmark set but also a large amount of additional data. VIENA^2 is publicly available, together with our benchmark evaluation, our new architecture and our multi-domain training strategy.

2   VIENA^2

VIENA^2 is a large-scale dataset for action anticipation, and more generally action analysis, in driving scenarios. While it is generally acknowledged that anticipation is key to the success of automated driving, to the best of our knowledge, there is currently no dataset that covers a wide range of scenarios with a common, yet sensible set of sensors. Existing datasets focus on specific subproblems, such as driver maneuvers and pedestrian intentions [16,25,27], and make use of different kinds of sensors. Furthermore, with the exception of [12], none of these datasets provide videos whose first few frames do not already show the action itself or the preparation of the action. To create VIENA^2, we made use of the GTA V video game, whose publisher allows, under some conditions, for the non-commercial use of the footage [31]. Beyond the fact that, as shown in [28] via psychophysics experiments, GTA V provides realistic images that can be captured in varying weather and daytime conditions, it has the additional benefit of allowing us to cover crucial anticipation scenarios, such as accidents, for which real-world data would be virtually impossible to collect. In this section, we first introduce the different scenarios covered by VIENA^2 and discuss the data collection process. We then study the statistics of VIENA^2 and compare it against existing datasets.

2.1   Scenarios and Data Collection

As illustrated in Fig. 2, VIENA^2 covers five generic driving scenarios. These scenarios are all human-centric, i.e., consider the intentions of humans, but three of

M. S. Aliakbarian et al.

them focus on the car's own driver, while the other two relate to the environment (i.e., pedestrians and other cars). These scenarios are:

1. **Driver Maneuvers (DM).** This scenario covers the 6 most common maneuvers a driver performs while driving: Moving forward (FF), stopping (SS), turning (left (LL) and right (RR)) and changing lane (left (CL) and right (CR)). Anticipation of such maneuvers as early as possible is critical in an ADAS context to avoid dangerous situations.
2. **Traffic Rules (TR).** This scenario contains sequences depicting the car's own driver either violating or respecting traffic rules, e.g., stopping at (SR) and passing (PR) a red light, driving in the (in)correct direction (WD, CD), and driving off-road (DO). Forecasting these actions is also crucial for ADAS.
3. **Accidents (AC).** In this scenario, we capture the most common real-world accident cases: Accidents with other cars (AC), with pedestrians (AP), and with assets (AA), such as buildings, traffic signs, light poles and benches, as well as no accident (NA). Acquiring such data in the real world is virtually infeasible. Nevertheless, these actions are crucial to anticipate for ADAS and autonomous driving.
4. **Pedestrian Intentions (PI).** This scenario addresses the question of whether a pedestrian is going to cross the road (CR), or has stopped (SS) but does not want to cross, or is walking along the road (AS) (on the sidewalk). We also consider the case where no pedestrian is in the scene (NP). As acknowledged in the literature [25,27,36], early understanding of pedestrians' intentions is critical for automated driving.
5. **Front Car Intentions (FCI).** The last generic scenario of VIENA^2 aims at anticipating the maneuvers of the front car. This knowledge has a strong influence on the behavior to adopt to guarantee safety. The classes are same as the ones in Driver Maneuver, but for the driver of the front car.

We also consider an additional scenario consisting of the same driver maneuvers as above but for heavy vehicles, i.e., trucks and buses. In all these scenarios, for the data to resemble a real driving experience, we made use of the equipment depicted in Fig. 1, consisting of a steering wheel with a set of buttons and a gear stick, as well as of a set of pedals. We then captured images at 30 fps with a single virtual camera mounted on the vehicle and facing the road forward. Since the speed of the vehicle is displayed at a specific location in these images, we extracted it using an OCR module [37] (see supplementary material for more detail on data collection). Furthermore, we developed an application that records measurements from the steering wheel. In particular, it gives us access to the steering angle every $1\,\mu s$, which allowed us to obtain a value of the angle synchronized with each image. Our application also lets us obtain the ground-truth label of each video sequence by recording the driver input from the steering wheel buttons. This greatly facilitated our labeling task, compared to [28,29], which had to use a middleware to access the rendering commands from which the ground-truth labels could be extracted. Ultimately, VIENA^2 consists of video sequences with synchronized measurements of steering angles and speed, and corresponding action labels.

Altogether, VIENA^2 contains more than 15K full HD videos (with frame size of 1920 × 1280), corresponding to a total of more than 2.25M annotated frames. The detailed number of videos for each class and the proportions of different weather and daytime conditions of VIENA^2 are provided in Fig. 2. Each video contains 150 frames captured at 30 frames-per-second depicting a single action from one scenario. The action occurs in the second half of the video (mostly around the 4 s mark), which makes VIENA^2 well-suited to research on action anticipation, where one typically needs to see what happens before the action starts.

Our goal is for VIENA^2 to be an extensible dataset. Therefore, by making our source code and toolbox for data collection and annotation publicly available, we aim to encourage the community to participate and grow VIENA^2. Furthermore, while VIENA^2 was mainly collected for the task of action anticipation in driving scenarios, as it contains full length videos, i.e., videos of a single drive of 30 min on average depicting multiple actions, it can also be used for the tasks of action recognition and temporal action localization.

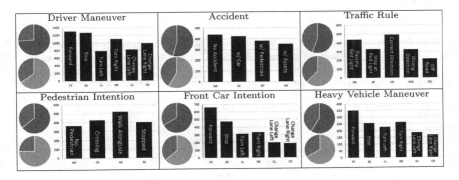

Fig. 2. Statistics for each scenario of VIENA^2. We plot the number of videos per class, and proportions of different weather conditions (clear in yellow vs rainy/snowy in gray) and different daytime (day in orange vs night in blue). Best seen in color. (Color figure online)

2.2   Comparison to Other Datasets

The different scenarios and action classes of VIENA^2 make it compatible with existing datasets, thus potentially allowing one to use our synthetic data in conjunction with real images. For instance, the action labels in the Driver Maneuver scenario correspond to the ones in Brain4Cars [12] and in the Toyota Action Dataset [23]. Similarly, our last two scenarios dealing with heavy vehicles contain the same labels as in Brain4Cars [12]. Moreover, the actions in the Pedestrian Intention scenario corresponds to those in [16]. Note, however, that, to the best of our knowledge, there is no other dataset covering our Traffic Rules and Front Car Intention scenarios, or containing data involving heavy vehicles. Similarly, there is no dataset that covers accidents involving a driver's own car. In this respect, the most closely related dataset is DashCam [3], which depicts accidents of other cars. Furthermore, VIENA^2 covers a much larger diversity of environmental conditions, such as daytime variations (morning, noon, afternoon, night, midnight), weather variations (clear, sunny, cloudy, foggy, hazy, rainy, snowy), and location variations (city, suburbs, highways, industrial, woods), than existing public datasets. In the supplementary material, we provide examples of each of these different environmental conditions. In addition to covering more scenarios and conditions than other driving anticipation datasets, VIENA^2 also contains more samples per class than existing action analysis datasets, both for recognition and anticipation. As shown in Table 1, with 600 samples per class, VIENA^2 outsizes (at least class-wise) the datasets that are considered *large* by the community. This is also the case for other synthetic datasets, such as VIPER [28], GTA5 [29], VEIS [35], and SYNTHIA [32], which, by targeting different problems, such as semantic segmentation for which annotations are more costly to obtain, remain limited in size. We acknowledge, however, that, since we target driving scenarios, our dataset cannot match in absolute size more general recognition datasets, such as Kinetics.

Table 1. Statistics comparison with action recognition and anticipation datasets. A * indicates a dataset specialized to one scenario, e.g., driving, as opposed to generic.

Recognition	Samples/Class	Classes	Videos	Anticipation	Samples/Class	Classes	Videos
UCF-101 (Soomro et al. 2012)	150	101	13.3K	UT-Interaction* (Ryoo et al. 2009)	20	6	60
HMDB/JHMDB (Kuehne et al. 2011)	120	51/21	5.1K/928	Brain4Cars* (Jain et al. 2016)	140	6	700
UCF-Sport* (Rodriguez et al. 2008)	30	10	150	JAAD* (Rasouli et al. 2017)	86	4	346
Charades (Sigurdsson et al. 2016)	100	157	9.8K				
ActivityNet (Caba et al. 2015)	144	200	15K				
Kinetics (Kay et al. 2017)	400	400	306K				
VIENA^2*	600	25	15K	VIENA^2*	600	25	15K

3   Benchmark Algorithms

In this section, we first discuss the state-of-the-art action analysis and anticipation methods that we used to benchmark our dataset. We then introduce a new multi-modal LSTM-based approach to action anticipation, and finally discuss how we model actions from our images and additional sensors.

3.1 Baseline Methods

The idea of anticipation was introduced in the computer vision community almost a decade ago by [34]. While the early methods [33,38,39] relied on handcrafted-features, they have now been superseded by end-to-end learning methods [1,12,20], focusing on designing new losses better-suited to anticipation. In particular, the loss of [1] has proven highly effective, achieving state-of-the-art results on several standard benchmarks.

Despite the growing interest of the community in anticipation, action recognition still remains more thoroughly investigated. Since recognition algorithms can be converted to performing anticipation by making them predict a class label at every frame, we include the state-of-the-art recognition methods in our benchmark. Specifically, we evaluate the following baselines:

Baseline 1: CNN+LSTMs. The high performance of CNNs in image classification makes them a natural choice for video analysis, via some modifications. This was achieved in [4] by feeding the frame-wise features of a CNN to an LSTM model, and taking the output of the last time-step LSTM cell as prediction. For anticipation, we can then simply consider the prediction at each frame. We then use the temporal average pooling strategy of [1], which has proven effective to increase the robustness of the predictor for action anticipation.

Baseline 2: Two-Stream Networks. Baseline 1 only relies on appearance, ignoring motion inherent to video (by motion, we mean explicit motion information as input, such as optical flow). Two-stream architectures, such as the one of [7], have achieved state-of-the-art performance by explicitly accounting for motion. In particular, this is achieved by taking a stack of 10 externally computed optical flow frames as input to the second stream. A prediction for each frame can be obtained by considering the 10 previous frames in the sequence for optical flow. We also make use of temporal average pooling of the predictions.

Baseline 3: Multi-Stage LSTMs. The Multi-Stage LSTM (MS-LSTM) of [1] constitutes the state of the art in action anticipation. This model jointly exploits context- and action-aware features that are used in two successive LSTM stages. As mentioned above, the key to the success of MS-LSTM is its training loss function. This loss function can be expressed as

$$\mathcal{L}(y,\hat{y}) = -\frac{1}{N}\sum_{k=1}^{N}\sum_{t=1}^{T}\left[y^t(k)\log(\hat{y}^t(k)) + w(t)(1 - y^t(k))\log(1 - \hat{y}^t(k))\right], \quad (1)$$

where $y^t(k)$ is the ground-truth label of sample k at frame t, $\hat{y}^t(k)$ the corresponding prediction, and $w(t) = \frac{t}{T}$. The first term encourages the model to predict the correct action at any time, while the second term accounts for ambiguities between different classes in the earlier part of the video.

3.2 A New Multi-modal LSTM

While effective, MS-LSTM suffers from the fact that it was specifically designed to take two modalities as input, the order of which needs to be manually defined.

As such, it does not naturally apply to our more general scenario, and must be actively modified, in what might be a sub-optimal manner, to evaluate it with our action descriptors. To overcome this, we therefore introduce a new multi-modal LSTM (MM-LSTM) architecture that generalizes the multi-stage architecture of [1] to an arbitrary number of modalities. Furthermore, our MM-LSTM also aims to learn the importance of each modality for the prediction.

Specifically, as illustrated in Fig. 3 for $M = 4$ modalities, at each time t, the representations of the M input modalities are first passed individually into an LSTM with a single hidden layer. The activations of these M hidden layers are then concatenated into an $M \times 1024$ matrix D^t, which acts as input to a time-distributed fully-connected layer (FC-Pool). This layer then combines the M modalities to form a single vector $O^t \in R^{1024}$. This representation is then passed through another LSTM whose output is concatenated with the original D^t via a skip connection. The resulting $(M+1) \times 1024$ matrix is then compacted into a 1024D vector via another FC-Pool layer. The output of this FC-Pool layer constitutes the final representation and acts as input to the classification layer.

The reasoning behind this architecture is the following. The first FC-Pool layer can learn the importance of each modality. While its parameters are shared across time, the individual, modality-specific LSTMs can produce time-varying outputs, thus, together with the FC-Pool layer, providing the model with the flexibility to change the importance of each modality over time. In essence, this allows the model to learn the importance of the modalities dynamically. The second LSTM layer then models the temporal variations of the combined modalities. The skip connection and the second FC-Pool layer produce a final representation that can leverage both the individual, modality-specific representations and the learned combination of these features.

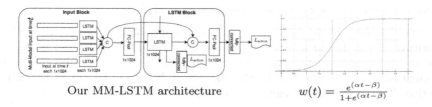

Our MM-LSTM architecture $w(t) = \frac{e^{(\alpha t - \beta)}}{1 + e^{(\alpha t - \beta)}}$

Fig. 3. (Left) Our Multi-Stage LSTM architecture. (Right) Visualization of our weighting function for the anticipation loss of Eq. 1.

Learning. To train our model, we make use of the loss of Eq. 1. However, we modify the weights as $w(t) = \frac{e^{(\alpha t - \beta)}}{1 + e^{(\alpha t - \beta)}}$, allowing the influence of the second term to vary nonlinearly. In practice, we set $\alpha = 3$ and $\beta = 6$, yielding the weight function of Fig. 3. These values were motivated by the study of [24], which shows that driving actions typically undergo the following progression: In a first stage, the driver is not aware of an action or decides to take an action. In the next stage, the driver becomes aware of an action or decides to take one. This portion of the video contains crucial information for anticipating the upcoming action. In the last portion of the video, the action has started. In this portion of the video, we do not want to make a wrong prediction, thus penalizing false positives strongly. Generally speaking, our sigmoid-based strategy to define the weight reflects the fact that, in practice and in contrast with many academic datasets, such as UCF-101 [40] and JHMDB-21 [14], actions do not start right at the beginning of a video sequence, but at any point in time, the goal being to detect them as early as possible.

During training, we rely on stage-wise supervision, by introducing an additional classification layer after the second LSTM block, as illustrated in Fig. 3. At test time, however, we remove this intermediate classifier to only keep the final one. We then make use of the temporal average pooling strategy of [1] to accumulate the predictions over time.

3.3   Action Modeling

Our MM-LSTM can take as input multiple modalities that provide diverse and complementary information about the observed data. Here, we briefly describe the different descriptors that we use in practice.

- **Appearance-based Descriptors.** Given a frame at time t, the most natural source of information to predict the action is the appearance depicted in the image. To encode this information, we make use of a slightly modified DenseNet [10], pre-trained on ImageNet. See Sect. 3.4 for more detail. Note that we also use this DenseNet as appearance-based CNN for Baselines 1 and 2.
- **Motion-based Descriptors.** Motion has proven a useful cue for action recognition [6,7]. To encode this, we make use of a similar architecture as for our appearance-based descriptors, but modify it to take as input a stack of optical flows. Specifically, we extract optical flow between L consecutive pairs of frames, in the range $[t - L, t]$, and form a $2L$ flow stack encoding horizontal and vertical flows. We fine-tune the model pre-trained on ImageNet for the task of action recognition, and take the output of the additional fully-connected layer as our motion-aware descriptor. Note that we also use this DenseNet for the motion-based stream of Baseline 2.
- **Vehicle Dynamics.** In our driving context, we have access to additional vehicle dynamics measurements. For each such measurement, at each time t, we compute a vector from its value s_t, its velocity $(s_t - s_{t-\delta})$ and its acceleration $(s_t - 2s_{t-\delta} + s_{t-2\delta})$. To map these vectors to a descriptor of size comparable to the appearance- and motion-based ones, inspired by [8], we train an LSTM with a single hidden layer modeling the correspondence between vehicle dynamics and action label. In our dataset, we have two types of dynamics measurements, steering angle and speed, which results in two additional descriptors.

When evaluating the baselines, we report results of both their standard version, relying on the descriptors used in the respective papers, and of modified versions that incorporate the four descriptor types discussed above. Specifically, for CNN-LSTM, we simply concatenate the vehicle dynamics descriptors and the motion-based descriptors to the appearance-based ones. For the Two-Stream baseline, we add a second two-stream sub-network for the vehicle dynamics and merge it with the appearance and motion streams by adding a fully-connected layer that takes as input the concatenation of the representation from the original two-stream sub-network and from the vehicle dynamics two-stream sub-network. Finally, for MS-LSTM, we add a third stage that takes as input the concatenation of the second-stage representation with the vehicle dynamics descriptors.

3.4   Implementation Details

We make use of the DenseNet-121 [10], pre-trained on ImageNet, to extract our appearance- and motion-based descriptors. Specifically, we replace the classifier with a fully-connected layer with 1024 neurons followed by a classifier with N outputs, where N is the number of classes. We fine-tune the resulting model using stochastic gradient descent for 10 epochs with a fixed learning rate of 0.001 and mini-batches of size 16. Recall that, for the motion-based descriptors, the corresponding DenseNet relies on $2L$ flow stacks as input, which requires us to also replace the first layer of the network. To initialize the parameters of this layer, we average the weights over the three channels corresponding to the original RGB channels, and replicate these average weights $2L$ times [43]. We found this scheme to perform better than random initialization.

4   Benchmark Evaluation and Analysis

We now report and analyze the results of our benchmarking experiments. For these experiments to be as extensive as possible given the available time, we performed them on a representative subset of VIENA^2 containing about 6.5K videos acquired in a large variety of environmental conditions and covering all 25 classes. This subset contains 277 samples per class, and thus still outsizes most action analysis datasets, as can be verified from Table 1. The detailed statistics of this subset are provided in the supplementary material.

To evaluate the behavior of the algorithms in different conditions, we defined three different partitions of the data. The first one, which we refer to as Random in our experiments, consists of randomly assigning 70% of the samples to the training set and the remaining 30% to the test set. The second partition considers the daytime of the sequences, and is therefore referred to as Daytime. In this case, the training set is formed by the day images and the test set by the night ones. The last partition, Weather, follows the same strategy but based on the information about weather conditions, i.e., a training set of clear weather and a test set of rainy/snowy/... weathers.

Below, we first present the results of our benchmarking on the `Random` partition, and then analyze the challenges related to our new dataset. We finally evaluate the benefits of our synthetic data for anticipation from real images, and analyze the bias of VIENA^2. Note that additional results including benchmarking on the other partitions and ablation studies of our MM-LSTM model are provided in the supplementary material. Note also that the scenarios and classes acronyms are defined in Sect. 2.1.

4.1 Action Anticipation on VIENA^2

We report the results of our benchmark evaluation on the different scenarios of VIENA^2 in Table 2 for the original versions of the baselines, relying on the descriptors used in their respective paper, and in Table 3 for their modified versions that incorporate all descriptor types. Specifically, we report the recognition accuracies for all scenarios after every second of the sequences. Note that, in general, incorporating all descriptor types improves the results. Furthermore, while the action recognition baselines perform quite well in some scenarios, such as Accidents and Traffic Rules for the two-stream model, they are clearly outperformed by the anticipation methods in the other cases. Altogether, our new MM-LSTM consistently outperforms the baselines, thus showing the benefits of learning the dynamic importance of the modalities.

Table 2. Results on the `Random` split of VIENA^2 for the original versions our three baselines: CNN+LSTM [4] with only appearance, Two-Stream [7] with appearance and motion, and MS-LSTM [1] with action-aware and context-aware features.

	CNN+LSTM [4]					Two-Stream [7]					MS-LSTM [1]				
	1''	2''	3''	4''	5''	1''	2''	3''	4''	5''	1''	2''	3''	4''	5''
DM	22.8	24.2	26.5	27.9	28.0	23.3	24.8	30.6	37.5	41.5	22.4	28.1	37.5	42.6	44.0
AC	53.6	53.6	55.0	56.3	57.0	68.5	70.0	74.5	76.3	78.0	50.3	55.6	60.4	68.3	72.5
TR	26.6	28.3	29.5	30.1	32.1	28.3	35.6	44.5	51.5	53.1	30.7	33.4	41.0	49.8	52.3
PI	38.4	40.4	41.8	41.8	42.1	36.8	37.5	40.0	40.0	41.2	50.6	52.4	55.6	56.8	58.3
FCI	33.0	36.3	39.5	39.5	39.6	37.1	38.0	35.5	39.3	39.3	44.0	45.3	51.3	60.2	63.1

Table 3. Results on the `Random` split of VIENA^2 for our three baselines with our action descriptors and for our approach.

	CNN+LSTM [4]					Two-Stream [7]					MS-LSTM [1]					Ours MM-LSTM				
	1''	2''	3''	4''	5''	1''	2''	3''	4''	5''	1''	2''	3''	4''	5''	1''	2''	3''	4''	5''
DM	24.6	25.6	28.0	30.0	30.3	26.8	30.5	40.4	53.4	62.6	28.5	35.8	57.8	68.1	78.7	32.0	38.5	60.5	71.5	83.6
AC	56.7	58.3	59.0	61.6	61.7	70.0	72.0	74.0	77.1	79.7	69.6	75.3	80.6	83.3	83.6	76.3	79.0	81.7	86.3	86.7
TR	28.0	28.7	30.6	32.2	32.8	30.6	38.7	48.0	49.6	54.1	33.3	39.4	48.3	57.1	61.0	39.8	49.8	58.8	63.7	68.8
PI	39.6	39.6	40.4	42.0	42.4	42.0	42.8	44.4	46.0	48.0	55.8	57.6	62.6	69.0	70.8	57.3	59.7	68.9	72.5	73.3
FCI	37.2	38.8	39.3	40.6	40.6	37.7	39.1	39.3	40.7	43.0	41.7	49.1	58.3	70.0	75.5	49.9	51.7	60.4	71.5	77.8

A comparison of the baselines with our approach on the `Daytime` and `Weather` partitions of VIENA^2 is provided in the supplementary material. In essence, the conclusions of these experiments are the same as those drawn above.

4.2   Challenges of VIENA^2

Based on the results above, we now study what challenges our dataset brings, such as which classes are the most difficult to predict and which classes cause the most confusion. We base this analysis on the per-class accuracies of our MM-LSTM model, which achieved the best performance in our benchmark. This, we believe, can suggest new directions to investigate in the future.

Our MM-LSTM per-class accuracies are provided in Table 4, and the corresponding confusion matrices at the earliest (after seeing 1 s) and latest (after seeing 5 s) predictions in Fig. 4. Below, we discuss the challenges of the various scenarios.

Table 4. Per-class accuracy of our approach on all scenarios of VIENA^2 (Random).

	DM						AC				TR					PI				FCI					
	FF	SS	LL	RR	CL	CR	NA	AP	AC	AA	CD	WD	PR	SR	DO	NP	CR	SS	AS	FF	SS	LL	RR	CL	CR
1″	50.7	43.8	17.8	35.0	18.7	26.1	94.9	65.7	73.2	71.3	75.5	35.0	23.7	32.8	32.2	59.3	59.1	68.8	42.2	74.5	46.3	35.6	44.6	47.8	50.9
2″	60.1	46.8	26.3	38.7	27.1	32.1	98.7	70.7	71.6	75.0	79.6	49.3	29.7	52.3	37.9	63.0	51.2	71.4	53.4	76.9	48.6	37.1	45.9	49.6	52.0
3″	81.3	75.6	54.4	63.4	42.9	45.4	100	75.4	76.1	75.2	83.7	60.0	35.1	69.5	45.8	70.4	67.6	79.2	58.6	85.7	63.9	54.1	50.7	52.7	57.3
4″	81.2	87.3	72.9	77.3	55.4	55.0	100	81.6	79.4	84.3	86.7	65.3	37.9	78.7	50.0	72.2	75.9	80.1	61.35	89.1	77.8	74.4	69.5	56.1	62.2
5″	88.0	97.2	95.8	90.4	64.9	65.4	100	80.5	86.1	80.2	85.7	75.0	40.0	95.1	48.6	74.1	78.2	76.6	63.6	91.2	83.5	84.6	81.4	59.4	66.8

1. **Driver maneuver:** After 1 s, most actions are mistaken for *Moving Forward*, which is not surprising since the action has not started yet. After 5 s, most of the confusion has disappeared, except for *Changing Lane* (left and right), for which the appearance, motion and vehicle dynamics are subject to small changes only, thus making this action look similar to *Moving Forward*.
2. **Accident:** Our model is able to distinguish *No Accident* from the different accident types early in the sequence. Some confusion between the different types of accident remains until after 5 s, but this would have less impact in practice, as long as an accident is predicted.

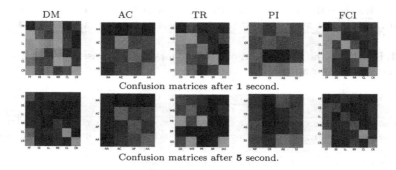

Confusion matrices after **1** second.

Confusion matrices after **5** second.

Fig. 4. Confusion Matrices. Confusion matrices of all five scenarios after observing 1 s (top) and 5 s (bottom) of each video sample.

3. **Traffic rule:** As in the maneuver case, there is initially a high confusion with *Correct Direction*, due to the fact that the action has not started yet. The confusion is then much reduced as we see more information, but *Passing a Red Light* remains relatively poorly predicted.

4. **Pedestrian intention:** The most challenging class for early prediction in this scenario is *Pedestrian Walking along the Road*. The prediction is nevertheless much improved after 5 s.

5. **Front car intention:** Once again, at the beginning of the sequence, there is much confusion with the *Forward* class. After 5 s, the confusion is significantly reduced, with, as in the maneuver case, some confusion remaining between the *Change lane* classes and the *Forward* class, illustrating the subtle differences between these actions.

4.3 Benefits of VIENA² for Anticipation from Real Images

To evaluate the benefits of our synthetic dataset for anticipation on real videos, we make use of the JAAD dataset [27] for pedestrian intention recognition, which is better suited to deep networks than other datasets, such as [16], because of its larger size (58 videos vs. 346). This dataset is, however, not annotated with the same classes as we have in VIENA², as its purpose is to study pedestrian and driver behaviors at pedestrian crossings. To make JAAD suitable for our task, we re-annotated its videos according to the four classes of our *Pedestrian Intention* scenario, and prepared a corresponding train/test split. JAAD is also heavily dominated by the *Crossing* label, requiring augmentation of both training and test sets to have a more balanced number of samples per class.

To demonstrate the benefits of VIENA² in real-world applications, we conduct two sets of experiments: (1) Training on JAAD from scratch, and (2) Pre-training on VIENA² followed by fine-tuning on JAAD. For all experiments, we use appearance-based and motion-based features, which can easily be obtained for JAAD. The results are shown in Table 5. This experiment clearly demonstrates the effectiveness of using our synthetic dataset that contains photorealistic samples simulating real-world scenarios.

Another potential benefit of using synthetic data is that it can reduce the amount of real data required to train a model. To evaluate this, we fine-tuned an MM-LSTM trained on VIENA² using a random subset of JAAD ranging from 20% to 100% of the entire dataset. The accuracies at every second of the sequence and for different percentages of JAAD data are shown in Fig. 5. Note that with 60% of real data, our MM-LSTM pre-trained on VIENA² already outperforms a model trained from scratch on 100% of the JAAD data. This shows that our synthetic data can save a considerable amount of labeling effort on real images.

Table 5. Anticipating actions on real data. Pre-training our MM-LSTM with our VIENA^2 dataset yields higher accuracy than training from scratch on real data.

Setup	After 1″	After 2″	After 3″	After 4″	After 5″
From scratch	41.01%	45.84%	51.38%	54.94%	56.12%
Fine-Tuned	45.06%	54.15%	58.10%	65.61%	66.0%

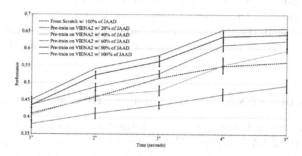

Fig. 5. Effect of the amount of real training data for fine-tuning MM-LSTM. MM-LSTM was pre-trained on VIENA^2 in all cases, except for *From Scratch w/100% of JAAD* (dashed line). Each experiment was conducted with 10 random subsets of JAAD. We report the mean accuracy and standard deviation (error bars) over 10 runs.

4.4   Bias Analysis

For a dataset to be unbiased, it needs to be representative of the entire application domain it covers, thus being helpful in the presence of other data from the same application domain. This is what we aimed to achieve when capturing data in a large diversity of environmental conditions. Nevertheless, every dataset is subject to some bias. For example, since our data is synthetic, its appearance differs to some degree from real images, and the environments we cover are limited by those of the GTA V video game. However, below, we show empirically that the bias in VIENA^2 remains manageable, making it useful beyond evaluation on VIENA^2 itself. In fact, the experiments of Sect. 4.3 on real data already showed that performance on other datasets, such as JAAD, can be improved by making use of VIENA^2. To further evaluate the bias of the visual appearance of our dataset, we relied on the idea of domain adversarial training introduced in [9]. In short, given data from two different domains, synthetic and real in our case, domain adversarial training aims to learn a feature extractor, such as a DenseNet, so as to fool a classifier whose goal is to determine from which domain a sample comes. If the visual appearance of both domains is similar, such a classifier should perform poorly. We therefore trained a DenseNet to perform action classification from a single image using both VIENA^2 and JAAD data, while learning a domain classifier to discriminate real samples from synthetic ones. The performance of the domain classifier quickly dropped down to chance, i.e., 50%. To make sure that this was not simply due to failure to effectively train the domain classifier, we then froze the parameters of the DenseNet while continuing to train the domain classifier. Its accuracy remained close to chance, thus showing that the features extracted from both domains were virtually indistinguishable. Note that the accuracy of action classification improved from 18% to 43% during the training, thus showing that, while the features are indistinguishable to the discriminator, they are useful for action classification.

Table 6. Effect of data collector on MM-LSTM performance (DM scenario).

Train, captured by	Test, captured by	After 1$''$	After 2$''$	After 3$''$	After 4$''$	After 5$''$
User 1	User 1	32.0%	38.5%	60.5%	71.5%	83.6%
User 1	User 2	32.8%	37.3%	60.7%	70.9%	82.8%

In our context of synthetic data, another source of bias could arise from the specific users who captured the data. To analyze this, we trained an MM-LSTM model from the data acquired by a single user, covering all classes and all environmental conditions, and tested it on the data acquired by another user. In Table 6, we compare the average accuracies of this experiment to those obtained when training and testing on data from the same user. Note that there is no significant differences, showing that our data generalizes well to other users.

5   Conclusion

We have introduced a new large-scale dataset for general action anticipation in driving scenarios, which covers a broad range of situations with a common set of sensors. Furthermore, we have proposed a new MM-LSTM architecture allowing us to learn the importance of multiple input modalities for action anticipation. Our experimental evaluation has shown the benefits of our new dataset and of our new model. Nevertheless, much progress remains to be done to make anticipation reliable enough for automated driving. In the future, we will therefore investigate the use of additional descriptors and of dense connections within our MM-LSTM architecture. We will also extend our dataset with more scenarios and other types of vehicles, such as motorbikes and bicycles, whose riders are more vulnerable road users than drivers. Moreover, we will extend our annotations so that every frame is annotated with bounding boxes around critical objects, such as pedestrians, cars, and traffic lights.

References

1. Aliakbarian, M.S., Saleh, F.S., Salzmann, M., Fernando, B., Petersson, L., Andersson, L.: Encouraging LSTMs to anticipate actions very early. In: ICCV (2017)
2. Bilen, H., Fernando, B., Gavves, E., Vedaldi, A., Gould, S.: Dynamic image networks for action recognition. In: CVPR (2016)
3. Chan, F.H., Chen, Y.T., Xiang, Y., Sun, M.: Anticipating accidents in dashcam videos. In: ACCV (2016)
4. Donahue, J., et al.: Long-term recurrent convolutional networks for visual recognition and description. In: CVPR (2015)
5. Dong, C., Dolan, J.M., Litkouhi, B.: Intention estimation for ramp merging control in autonomous driving. In: IV (2017)
6. Feichtenhofer, C., Pinz, A., Wildes, R.P.: Spatiotemporal multiplier networks for video action recognition. In: CVPR (2017)
7. Feichtenhofer, C., Pinz, A., Zisserman, A.: Convolutional two-stream network fusion for video action recognition. In: CVPR (2016)
8. Fernando, T., Denman, S., Sridharan, S., Fookes, C.: Going deeper: autonomous steering with neural memory networks. In: CVPR (2017)
9. Ganin, Y., Lempitsky, V.: Unsupervised domain adaptation by backpropagation. In: ICML (2015)
10. Huang, G., Liu, Z., Weinberger, K.Q., van der Maaten, L.: Densely connected convolutional networks. arXiv preprint arXiv:1608.06993 (2016)
11. Jain, A., Koppula, H.S., Raghavan, B., Soh, S., Saxena, A.: Car that knows before you do: anticipating maneuvers via learning temporal driving models. In: IV (2015)
12. Jain, A., Koppula, H.S., Soh, S., Raghavan, B., Singh, A., Saxena, A.: Brain4cars: car that knows before you do via sensory-fusion deep learning architecture. arXiv preprint arXiv:1601.00740 (2016)
13. Jain, A., Singh, A., Koppula, H.S., Soh, S., Saxena, A.: Recurrent neural networks for driver activity anticipation via sensory-fusion architecture. In: ICRA (2016)
14. Jhuang, H., Gall, J., Zuffi, S., Schmid, C., Black, M.J.: Towards understanding action recognition. In: ICCV (2013)
15. Klingelschmitt, S., Damerow, F., Willert, V., Eggert, J.: Probabilistic situation assessment framework for multiple, interacting traffic participants in generic traffic scenes. In: IV (2016)
16. Kooij, J.F.P., Schneider, N., Flohr, F., Gavrila, D.M.: Context-based pedestrian path prediction. In: Fleet, D., Pajdla, T., Schiele, B., Tuytelaars, T. (eds.) ECCV 2014. LNCS, vol. 8694, pp. 618–633. Springer, Cham (2014). https://doi.org/10.1007/978-3-319-10599-4_40
17. Koppula, H.S., Saxena, A.: Anticipating human activities using object affordances for reactive robotic response. TPAMI 38, 14–29 (2016)
18. Li, X., et al.: A unified framework for concurrent pedestrian and cyclist detection. T-ITS 18, 269–281 (2017)
19. Liebner, M., Ruhhammer, C., Klanner, F., Stiller, C.: Generic driver intent inference based on parametric models. In: ITSC (2013)
20. Ma, S., Sigal, L., Sclaroff, S.: Learning activity progression in LSTMs for activity detection and early detection. In: CVPR (2016)
21. Morris, B., Doshi, A., Trivedi, M.: Lane change intent prediction for driver assistance: on-road design and evaluation. In: IV (2011)

22. Ohn-Bar, E., Martin, S., Tawari, A., Trivedi, M.M.: Head, eye, and hand patterns for driver activity recognition. In: ICPR (2014)
23. Olabiyi, O., Martinson, E., Chintalapudi, V., Guo, R.: Driver action prediction using deep (bidirectional) recurrent neural network. arXiv preprint arXiv:1706.02257 (2017)
24. Pentland, A., Liu, A.: Modeling and prediction of human behavior. Neural Comput. **11**, 229–242 (1999)
25. Pool, E.A., Kooij, J.F., Gavrila, D.M.: Using road topology to improve cyclist path prediction. In: IV (2017)
26. Ramanathan, V., Huang, J., Abu-El-Haija, S., Gorban, A., Murphy, K., Fei-Fei, L.: Detecting events and key actors in multi-person videos. In: CVPR (2016)
27. Rasouli, A., Kotseruba, I., Tsotsos, J.K.: Agreeing to cross: how drivers and pedestrians communicate. arXiv preprint arXiv:1702.03555 (2017)
28. Richter, S.R., Hayder, Z., Koltun, V.: Playing for benchmarks. In: ICCV (2017)
29. Richter, S.R., Vineet, V., Roth, S., Koltun, V.: Playing for data: ground truth from computer games. In: Leibe, B., Matas, J., Sebe, N., Welling, M. (eds.) ECCV 2016. LNCS, vol. 9906, pp. 102–118. Springer, Cham (2016). https://doi.org/10.1007/978-3-319-46475-6_7
30. Rockstar-Games: Grand Theft Auto V: PC single-player mods (2018). http://tinyurl.com/yc8kq7vn
31. Rockstar-Games: Policy on posting copyrighted Rockstar Games material (2018). http://tinyurl.com/yc8kq7vn
32. Ros, G., et al.: Semantic segmentation of urban scenes via domain adaptation of SYNTHIA. In: Csurka, G. (ed.) Domain Adaptation in Computer Vision Applications. ACVPR, pp. 227–241. Springer, Cham (2017). https://doi.org/10.1007/978-3-319-58347-1_12
33. Ryoo, M.S.: Human activity prediction: early recognition of ongoing activities from streaming videos. In: ICCV (2011)
34. Ryoo, M.S., Aggarwal, J.K.: Spatio-temporal relationship match: video structure comparison for recognition of complex human activities. In: ICCV (2009)
35. Saleh, F.S., Aliakbarian, M.S., Salzmann, M., Petersson, L., Alvarez, J.M.: Effective use of synthetic data for urban scene semantic segmentation. In: Ferrari, V., Hebert, M., Sminchisescu, C., Weiss, Y. (eds.) ECCV 2018. LNCS, vol. 11206, pp. 86–103. Springer, Cham (2018). https://doi.org/10.1007/978-3-030-01216-8_6
36. Schulz, A.T., Stiefelhagen, R.: A controlled interactive multiple model filter for combined pedestrian intention recognition and path prediction. In: ITSC (2015)
37. Smith, R.: An overview of the tesseract OCR engine. In: ICDAR. IEEE (2007)
38. Soomro, K., Idrees, H., Shah, M.: Online localization and prediction of actions and interactions. arXiv preprint arXiv:1612.01194 (2016)
39. Soomro, K., Idrees, H., Shah, M.: Predicting the where and what of actors and actions through online action localization. In: CVPR (2016)
40. Soomro, K., Zamir, A.R., Shah, M.: UCF101: a dataset of 101 human actions classes from videos in the wild. arXiv preprint arXiv:1212.0402 (2012)
41. Tawari, A., Sivaraman, S., Trivedi, M.M., Shannon, T., Tippelhofer, M.: Looking-in and looking-out vision for urban intelligent assistance: estimation of driver attentive state and dynamic surround for safe merging and braking. In: IV (2014)

42. Vondrick, C., Pirsiavash, H., Torralba, A.: Anticipating visual representations from unlabeled video. In: CVPR (2016)
43. Wang, L., et al.: Temporal segment networks: towards good practices for deep action recognition. In: Leibe, B., Matas, J., Sebe, N., Welling, M. (eds.) ECCV 2016. LNCS, vol. 9912, pp. 20–36. Springer, Cham (2016). https://doi.org/10.1007/978-3-319-46484-8_2
44. Wang, X., Ji, Q.: Hierarchical context modeling for video event recognition. TPAMI **39**, 1770–1782 (2017)
45. Zyner, A., Worrall, S., Ward, J., Nebot, E.: Long short term memory for driver intent prediction. In: IV (2017)

Multilevel Collaborative Attention Network for Person Search

Wenbo Li[1], Ze Chen[1], Zhenyong Fu[2], and Hongtao Lu[1(✉)]

[1] Department of Computer Science and Engineering, Shanghai Jiao Tong University,
Shanghai, China
{fenglinglwb,zed,htlu}@sjtu.edu.cn
[2] Department of Computer Science, Nanjing University of Science and Technology,
Nanjing, China
z.fu@njust.edu.cn

Abstract. Person search aims to apply pedestrian detection and person re-identification simultaneously to search persons in images, which inevitably introduces pedestrian box misalignment during the procedure. And the detected boxes usually have a large variety of scales on a single image. Together with cluttered background and occlusion, all these distracting factors make it difficult to extract discriminative pedestrian representations. However, these problems are usually ignored by current person search systems. In this work, we propose a novel Multilevel Collaborative Attention Network (MCAN) to fulfill person search task efficiently. A multilevel selective learning is introduced to extract scale-aware features in different levels, and a collaborative attention module consisting of hard regional attention and soft pixel-wise attention is designed to deal with misalignment, background noise and occlusion. MCAN achieves **60.1%** top-1 accuracy and **29.1%** mAP on PRW benchmark, demonstrating its superiority over current state-of-the-art methods.

Keywords: Multilevel learning · Collaborative attention learning · Person search

1 Introduction

Person re-identification [6,39] is to conduct person matching among images captured by different cameras between query and a set of galleries. It has a broad prospect of applications in many fields including video surveillance, image retrieval and security. However, this task is implemented on an assumption that all pedestrian bounding boxes have already been hand-drawn, which is unrealistic under real-world settings. In this case, some attempts have been made to detect

Electronic supplementary material The online version of this chapter (https://doi.org/10.1007/978-3-030-20887-5_29) contains supplementary material, which is available to authorized users.

C. V. Jawahar et al. (Eds.): ACCV 2018, LNCS 11361, pp. 467–482, 2019.
https://doi.org/10.1007/978-3-030-20887-5_29

Fig. 1. Person Search is to search the target person from the whole scene image. The solid arrow points to a successful search case where it locates a pedestrian with the same identity as target person. The dotted arrow means that there is no true matching on the image.

pedestrian bounding boxes using off-the-shelf detectors firstly, and then pass these proposals to a re-identification network. However, this would inevitably bring about misalignments, misdetections and false alarms during detection which severely harms the performance of following person re-identification. Additionally, two-stage systems are usually too complicated and time-consuming to be applicable in real time. In order to deal with these problems, person search [37] as shown in Fig. 1 was proposed to accomplish pedestrian detection and re-identification simultaneously in a more simplified and faster way.

To date, there are quite limited frameworks designed to fulfill the person search task. Xu et al. [37] took advantage of a sliding window strategy to accomplish the person search task based on handcrafted features and Zheng et al. [43] explored how pedestrian detection aids re-identification performance in an end-to-end way. In [35] and [36], the authors proposed faster R-CNN [26] based person search frameworks and focused on designing elaborate losses to extract more effective person features. Additionally, Neural Person Search Machine (NPSM) [21] was proposed to shrink search area from whole image to precise localization of target person recursively.

However, various problems including large scale variation of detected pedestrian boxes, detection box misalignment, overwhelming background noise, and occlusion inevitably occur during person search process, and are usually ignored in previous works. The variety of pedestrian scales brings great difficulties during detection procedure. And misalignment will introduce much cluttered background noise and degrade the re-identification effectiveness unavoidably. Although some pedestrian bounding boxes are located accurately, there still exist various distracting factors such as other persons appearing in the background. Additionally, some instances are with large occlusion regions. In a word, all situations above contain severe distracting factors for extracting discriminative representations for person search, and a valid settlement to solve these problems is very much required.

Inspired by the observations above, we propose an end-to-end person search framework named Multilevel Collaborative Attention Network (MCAN) to deal with aforementioned problems. A multilevel feature learning strategy is introduced in our framework to locate more accurate pedestrian proposals and learn the most appropriate level features selectively. Additionally, a well-designed attention learning module including hard regional attention [13] and soft pixel-wise attention [11,32] is introduced. The hard attention aims to alleviate the negative influence of misalignment and background noise by locating several important human body regions precisely. And the soft attention learning is implemented by aggregating spatial and channel components together. It is designed to highlight the most discriminative human regions in pixel-wise level, which can effectively handle occlusion situations and ignore irrelevant background information. Besides, we define an effective Online Hard Mined Random Sampling Softmax (OHMRSS) loss to accelerate the optimization procedure and further improve the performance of the network.

In a nutshell, our contributions in this work include:

1. We propose a novel end-to-end framework named MCAN to fulfill person search task efficiently.
2. A multilevel learning strategy is introduced to deal with large variation of detected pedestrian box scales.
3. We design a collaborative hard regional and soft pixel-wise attention learning strategy to learn more refined discriminative representations for re-identification. It works well on eliminating negative influence of misalignment, cluttered background and occlusion.
4. We define an Online Hard Mined Random Sampling Softmax (OHMRSS) loss. It not only improves the re-identification performance but also speeds up the convergence of network.

2   Related Work

Person Re-identification. Apart from manually designing hand-crafted discriminative features [4,7,8,33,41] in early days, recent works concentrate more on learning distance metrics [14,18,19,24,30] and especially applying deep learning based methods to learn high-level features [2,16,17,22,25,28,29,34,44]. For the former, KISSME [14] and XQDA [18] are two of well-designed distance metrics. In terms of deep learning methods, there develop a variety of learning strategies. On the one hand, several innovative loss functions are applied to optimize the network, such as contrastive [16] and triplet [2] losses. On the other hand, numerous frameworks are designed to learn global and well-aligned part features [29,34], extract joint spatial and temporal information [28,44] as well as produce abundant samples utilizing generative models to improve re-identification performance [22,25].

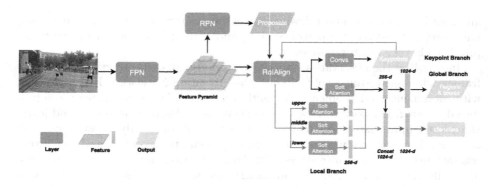

Fig. 2. Multilevel Collaborative Attention Network (MCAN) Architecture.

Pedestrian Detection. The most commonly used off-the-shelf detectors consist of Deformable Part Model (DPM) [5], Aggregated Channel Features (ACF) [3], Locally Decorrelated Channel Features (LDCF) [23], and Convolutional Channel Features (CCF) [38]. They accomplish the detection task by virtue of carefully hand-crafted features and linear classifiers. Recently, CNN based deep learning methods have obtained an immense development in pedestrian detection. [40] exploited a universal detection model, Faster R-CNN [26], in pedestrian detection tasks and obtained a large improvement in both accuracy and speed. To handle the occlusion problem, [31] proposed DeepParts taking advantage of extensive part detectors. Additionally, [1] developed an algorithm to learn complexity-aware detector cascades.

Person Search. To our knowledge, several works [15,21,35,36,43] have tried to take pedestrian detection and person re-identification together into consideration. [43] developed ID-discriminative Embedding (IDE) and Confidence Weighted Similarity (CWS) algorithms, and pointed out that joint detection and re-identification learning leads to better results than optimizing them individually. [35] and [36] designed faster R-CNN based end-to-end systems supervised by Random Sampling Softmax (RSS) and Online Instance Matching (OIM) losses. [21] proposed Neural Person Search Machine (NPSM) to shrink search area until locating the target person recursively.

All works above have developed successful networks to handle detection and re-identification tasks jointly, but they mainly focused on investigating training strategy and loss function designing rather than learning refined discriminative representations. There is quite a few efforts made to eliminate the negative influence of large scale variation of detected pedestrian boxes, detection misalignment, cluttered background noise and occlusion.

3   Multilevel Collaborative Attention Network

In this section, we present the details of Multilevel Collaborative Attention Network (MCAN), as shown in Fig. 2, which jointly realizes pedestrian detection and person re-identification. Based on the fact that both image resolutions and person scales in the wild have a large variation, we utilize the Feature Pyramid Network (FPN) [20] to learn multilevel representations. Then a region proposal network (RPN) [26] takes advantage of feature pyramid to propose pedestrian regions, and representations acquired from RoIAlign [9] are separately passed into the global branch and keypoint branch. From the former, global features are extracted to produce pedestrian bounding boxes. The latter generates 14 keypoints to locate upper, middle and lower person regions and extract relevant features. Both global and local regional features are processed by soft pixel-wise attention modules, then merged, and finally transformed into representations to infer identities discriminatively. More details of architecture are available in the supplementary material. During the training procedure, the network is trained in a multi-task manner taking advantage of a well-designed Online Hard Mined Random Sampling Softmax (OHMRSS) loss. In inference stage, person matchings are implemented by simply calculating feature distances, and then we choose the nearest query and gallery pairs.

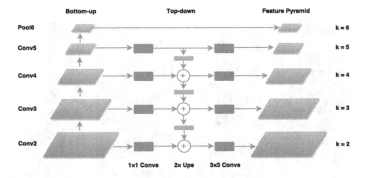

Fig. 3. Feature Pyramid. The bottom-up pathway generates feature maps in forward propagation. The feature pyramid is constructed by 5 level features after top-down pathway operations.

3.1   Multilevel Selective Learning

In the wild, person scales usually vary significantly, which motivates us to capture scale-aware multilevel features in pedestrian detection procedure. Besides, we are inspired by the fact that some lower-level features such as clothing color play an essential role in the re-identification task. Therefore, instead of extracting features from a specified layer, we introduce an FPN module to assemble features from different levels and select the most suitable ones according to box scales. In this case, it can produce more accurate pedestrian proposals and learn discriminative representations selectively.

As shown in Fig. 3, Conv2 ~ Conv5 and Pool6 in the bottom-up pathway are generated by forward propagation. As for the top-down pathway, semantically stronger but spatially coarser feature maps are upsampled and then combined with lower-level ones to form a feature pyramid.

Once feature pyramid is generated, we assign a specific pyramid level k to a RoI based on its height h and width w

$$k = \lfloor k_0 + log_2(\sqrt{wh}/224) \rfloor, \; where \; k_0 = 4 \; and \; k \in [2, \; 6] \tag{1}$$

As a result, a smaller pedestrian box is associated with a spatially finer representation to emphasize lower-level features.

3.2   Collaborative Attention Learning

In our network, we design a collaborative attention learning mechanism consisting of hard regional attention and soft pixel-wise attention modules. The attention mechanism alleviates the influence of box misalignment, background noise and occlusion problems, and results in more discriminative representations for people.

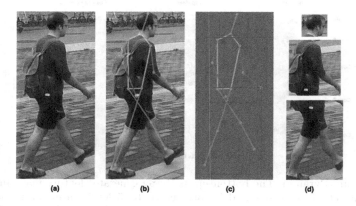

Fig. 4. In hard attention module, keypoints are used to locate upper, middle and lower regions precisely. (a) is the original image. (b) is added with 14 keypoints based on (a). (c) shows upper, middle and lower regions partitioned by keypoints. (d) illustrates cropped results.

Hard Regional Attention. The hard regional attention module aims to locate body parts precisely. In this case, redundant background noise is wiped out and negative influence of misalignment is palliated. Once pedestrian proposals are available, corresponding features obtained by RoIAlign go through a series of

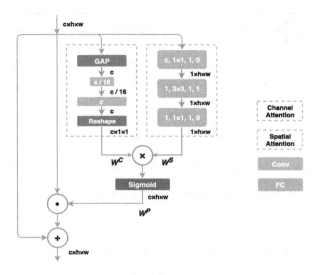

Fig. 5. Soft Pixel-wise Attention Module. It consists of spatial and channel attention modules. And the Conv layer is denoted as (channel, kernel, stride, padding).

convolutions leading to 14 heatmaps. And the location of a keypoint is determined by the pixel with highest response on corresponding heatmap. We take advantage of these 14 keypoints to determine upper, middle and lower three regions as Fig. 4. More details of keypoints and partition procedure are described in the supplementary material. Finally, we can obtain three corresponding local features produced by RoIAlign from the feature pyramid.

Soft Pixel-Wise Attention. Considering that the quality of different person regions may vary largely due to occlusion or other reasons, applying a pixel-wise attention learning to acquire the most discriminative regions is fairly reasonable. We design a parameter-economic but efficient, joint spatial [32] and channel [11] residual attention module (Fig. 5). Note that all convolution layers are followed by Batch Normalization [12] and ReLU operations in spatial attention modules.

Given feature maps, the channel attention module takes advantage of a global average pooling layer and two fully connected (FC) layers to generate $W^C \in R^{c \times 1 \times 1}$. To reduce the amount of parameters, the size of first FC layer is one-sixteenth of the second. And the spatial attention module produces a spatial quality aware weight map $W^S \in R^{1 \times h \times w}$ after several convolutions.

The soft pixel-wise attention operator $W^P \in R^{c \times h \times w}$ is acquired by

$$W^P = W^C \times W^S \tag{2}$$

and a sigmoid operation. By implementing the soft attention module in a separate spatial and channel manner, the network significantly reduce the parameters and computation. We draw on the successful experience of ResNet [10] and

Algorithm 1. $OHMRSS(c, C, K, H, N)$

1: **for** $i = 1$ to N **do**
2:     **if** c_i is 0 **then**
3:         $s_1 = 0$
4:         set $s_2,...,s_K$ from $\{1, 2, ..., C\}$ randomly
5:     **else**
6:         $s_1 = 0$
7:         $s_2 = c_i$
8:         set $s_3,...,s_K$ from $\{1, 2, ..., C\}\backslash c_i$ randomly
9:     **end if**
10:     $S = \{s_1, s_2, ..., s_K\}$, and calculate the Softmax loss of sample i based on S
11: **end for**
12: Find H largest losses out of N samples, and calculate their average L
13: **return** L

---

implement soft attention module with a residual addition as

$$X = X + W^P \odot X \tag{3}$$

where $\odot$ denotes the element-wise multiplication. In this way, the soft pixel-wise attention module is well applied in a deep network to avoid gradient vanishing. Additionally, it makes the features of some discriminative regions more competitive and accelerates the optimization of network.

3.3 Online Hard Mined Random Sampling Softmax

In view of the limited number of pedestrians on an image, a mass of candidate RoIs usually seem redundant. Besides, only a few classes appearing in one mini-batch will lead to a severe suppression to others. Therefore, inspired by Online Hard Example Mined strategy [27] and Random Sampling Softmax (RSS) [35], we adopt an effective Online Hard Mined Random Sampling Softmax (OHMRSS) defined as Algorithm 1 to select a set of hard examples and optimize a small part of class discriminant functions at a time. As for parameters in Algorithm 1, c is the label class, K means selecting K different classes out of $C + 1$ classes (C identities denoted from 1 to C and background denoted 0), H means picking H hardest samples out of all N samples in a batch.

In this case, our designed OHMRSS loss function is defined as

$$L = -\frac{1}{H} \sum_{i=1}^{H} \sum_{j=1}^{K} t_{is_j} \log y_{is_j}, \; where \; y_{is_j} = \frac{e^{x_{is_j}}}{\sum_{l=1}^{K} e^{x_{is_l}}} \tag{4}$$

where s_j is the j-th element of S, x_{is_j} is the j-th sampled classifier score of the i-th selected example and t_{is_j} is its associated binary label (0 or 1). We use $H = 64$ and $K = 100$ in our experiments.

The designed loss has double benefits. Firstly, it improves the performance of our network in person search task. Secondly, it accelerates the convergence of network and eliminates several hyperparameters in common use.

4   Experiments

In this section, we conduct extensive experiments to compare the performance of our framework and state-of-the-art methods for person search task on two benchmarks. Our method works especially well on the PRW [43] dataset, showing a significant performance gain over the state-of-the-art methods.

4.1   Datasets and Evaluation Metrics

PRW. PRW [43] dataset provides 10-h videos captured by 6 cameras in Tsinghua university. And 11,816 frames are selected and 43,110 pedestrian bounding boxes are manually annotated, among which 34,304 boxes related to 932 pedestrians are labeled from 1 to 932. The other unidentified persons are annotated with -2. The dataset is split into training and testing two parts. Training set contains 5,704 frames and 482 identities. As for testing set, it provides a query set of 2,057 persons and a gallery set of 6,112 images covering the other 450 labeled pedestrians.

CUHK-SYSU. CUHK-SYSU [36] is a large-scale person search dataset collected from street and movie snapshots. There are totally 18,184 images and 96,143 bounding boxes. Among them, 8,432 pedestrians are annotated with identities and others with -1 label. It is ensured that there are no overlapped images and labeled pedestrians between training and testing subsets. As for training set, it covers 11,206 images and 5,532 identities. The test 2,900 queries have different gallery images with size ranging from 50 to 4,000, and jointly cover all 6,978 test images.

Evaluation Protocols. To compare with previous works, we adopt both top-K matching rate and mean averaged precision (mAP) in our experiments. The top-K matching rate treats person search as a matching problem, and a matching is counted if there is at least one of predicted top-K pedestrian boxes overlapping with the ground truth with intersection-over-union (IoU) larger than 0.5. As for mAP, it is borrowed from the detection task and is the mean of all queries' APs based on precision-recall curve.

4.2   Implementation Details

Eight GTX 1080Ti GPUs are utilized, and the mini-batch size per GPU is 1. There are total 50k iterations during training. The learning rate is set to 0.01 for first 20k iterations, and drops to 0.001 and 0.0001 at the 30k and the 40k iterations, respectively. Additionally, there is a warm-up during first 4k iterations. An SGD optimizer with momentum 0.9 and weight decay 0.0001 is applied. As for testing, all query and gallery features are normalized, then cosine distances are calculated for inference.

4.3   Comparison with State-of-the-art Methods

To evaluate the performance of MCAN, we compare it with many other existing person search frameworks. Generally speaking, previous works can be categorized into two types. One kind is end-to-end person search network, and the other is combination framework of pedestrian detector and identification recognizer.

Table 1. Comparison of MCAN's performance with state-of-the-art methods on PRW.

Method	top-1(%)	mAP(%)
DPM-Alex + LOMO + XQDA	34.1	13.0
DPM-Alex + IDE$_{det}$	47.4	20.3
DPM-Alex + IDE$_{det}$ + CWS	48.3	20.5
ACF-Alex + LOMO +XQDA	30.6	10.3
ACF-Alex + IDE$_{det}$	43.6	17.5
ACF-Alex + IDE$_{det}$ + CWS	45.2	17.8
LDCF + LOMO + XQDA	31.1	11.0
LDCF + IDE$_{det}$	44.6	18.3
LDCF + IDE$_{det}$ + CWS	45.5	18.3
OIM	49.9	21.3
NPSM	53.1	24.2
MCAN	**60.1**	**29.1**

Results on PRW. In PRW dataset, the large size of gallery brings a fairly great challenge. We compare MCAN performance with 9 combination results of 3 detectors (DPM [5], ACF [3], LDCF [23]) and 3 recognizers (LOMO+XQDA [18], IDE [43], IDE+CWS [43]) as well as two from end-to-end systems (OIM [36] and NPSM [21]). All results are given in Table 1.

On the basis of [43], we are aware that person search performance is jointly determined by detector and recognizer for an ensemble system. From Table 1, it is obvious that DPM outperforms other two detectors because it is more robust to pedestrian deformations. Besides, as for recognizer, CNN-based IDE leads to a significant improvement than hand-crafted LOMO. Confidence Weighted Similarity (CWS) further enlarges the gap. Compared with these ensemble systems, the end-to-end OIM and NPSM have better performance indicating that additional gains come from joint optimization of detection and re-identification tasks. Finally, we can see that our method achieves the best result with large margins, outperforming NPSM by **7.0%** and **4.9%** on top-1 accuracy and mAP respectively. All of these results strongly demonstrate that our network accomplishes pedestrian detection successfully and learns more refined discriminative representations for detected persons.

Results on CUHK-SYSU. We report MCAN performance on CUHK-SYSU dataset in Table 2. All results are obtained in the condition of gallery size equal to 100. Apart from recognizers and metrics mentioned above, we also introduce BoW [42], DSIFT [41], KISSME [14] and IDNet [36] (re-identification part of OIM). It is clear that our method achieves a fairly good performance in comparison with other methods. Although OIM achieves a better result than ours, we hold an opinion that the gain is mainly from a specially designed loss on this benchmark rather than its architecture. We notice that nearly 70% pedestrian boxes in training set on this benchmark are unidentified. In this case, a specific loss designing for these unlabeled pedestrians is more likely to improve the re-identification performance. To verify our viewpoint, we replace the loss of our network with OIM loss and obtain obviously improved results (**79.2%** on top-1 accuracy and **76.3%** on mAP) which is better than OIM.

Table 2. Comparison of MCAN's performance with existing methods on CUHK-SYSU.

Method	top-1(%)	mAP(%)
CCF + DSIFT + Euclidean	11.7	11.3
CCF + DSIFT + KISSME	13.9	13.4
CCF + BOW + Cosine	29.4	26.9
CCF + LOMO + XQDA	46.4	41.2
CCF + IDNet	57.1	50.9
ACF + DSIFT + Euclidean	25.9	21.7
ACF + DSIFT + KISSME	38.1	32.3
ACF + BOW + Cosine	48.4	42.4
ACF + LOMO + XQDA	63.1	55.5
ACF + IDNet	63.0	56.5
Faster R-CNN + RSS	62.7	55.7
OIM	78.7	75.5
MCAN	71.8	67.9
MCAN + OIM	**79.2**	**76.3**

4.4 Ablation Study

In this section, we conduct several ablation experiments on PRW dataset to investigate individual contributions of different components of our MCAN. The results in Table 3 demonstrate that all proposed components including multilevel selective learning, attention module and Online Hard Mined Random Sampling Softmax (OHMRSS) loss are beneficial to person search in our framework. To further investigate the influence of detection module, we replace detection proposals with ground truths and find that this only leads to a small gain for MCAN. It indicates that MCAN has an excellent detection performance and deals with misalignment situations successfully to some extent.

Table 3. Ablation study results on PRW.

Methods	Components					top-1(%)	mAP(%)
	Hard	Soft	KS	FP	OHMRSS		
MCAN	√	√	√	√	√	**60.1**	**29.1**
MCAN + GT	√	√	√	√	√	62.0	30.7
w/o Soft	√		√	√	√	57.4	27.8
w/o Att			√	√	√	51.8	23.4
w/o Att & KS				√	√	50.6	25.8
w/o Att & KS & FP					√	48.0	24.9
w/o OHMRSS	√	√	√	√		57.8	26.2

"GT" refers to ground truth bounding boxes.
The following five components are respectively hard attention ("Hard"), soft attention ("Soft"), keypoint supervison ("KS"), feature pyramid ("FP") and OHMRSS.

Multilevel Selective Learning. Compared with pure re-identification task, we can not scale detected pedestrian boxes to a defined standard in an end-to-end system for person search. In this case, we implement a multilevel selective learning to generate more accurate pedestrian proposals and discriminative features. Comparing "w/o Att & KS & FP" with "w/o Att & KS", we see that multilevel selective learning helps our framework acquire a 2.6% gain on top-1 accuracy and 0.9% on mAP. For a small bounding box, MCAN can pay more attention to its lower-level feature maps to extract a more discriminative representation. For a large box, the network will focus on higher-level features and lead to a more refined representation.

Attention Learning. In MCAN, both hard regional attention and soft pixel-wise attention are applied to extract more discriminative features. The result of "w/o Att & KS" with top-1 accuracy 50.6% and mAP 25.8% is obtained by only learning global features without any attention module and keypoint supervision. We define this setting as our base model. Compared with our state-of-the-art result of MCAN, it drops 9.5% on top-1 and 3.3% on mAP. Specially, we notice that MCAN has one additional keypoint supervision. To figure out whether this supervision contributes to the gain or not, we conduct an experiment with keypoint supervision but without attention learning, and the result "w/o Att" shows that gain of this supervision is relatively small. Therefore, we draw a conclusion that attention learning module in our framework plays a fairly important role in extracting discriminative representations.

It is obvious that network with hard attention module, which means learning part regional features, achieves a superior result denoted as "w/o Soft" in Table 3 than the base model. The improvements over top-1 accuracy and mAP are 6.8% and 2.0%, respectively. Therefore the hard attention module in MCAN effectively alleviates negative influence of misalignment and cluttered background noise.

Table 4. More Analysis on Soft Attention on PRW.

Method	top-1(%)	mAP(%)
MCAN	**60.1**	**29.1**
w/o Soft	57.4	27.8
w/o Channel in Soft	58.0	27.1
w/o Spatial in Soft	57.7	27.2
w/o Global Soft	58.2	27.9
w/o Local Soft	58.8	28.3
w/o Residual in Soft	59.6	28.8

Fig. 6. Images obtained after attention learning module. The left is original image and the right is combined with feature maps after soft attention processing.

Besides, from Table 3 we can see that MCAN without a soft attention learning drops 2.7% and 1.3% on two metrics respectively. Although the lifting effect is smaller than hard attention, we can still take a recognition that soft attention learning indeed helps the network learn more refined pixel-wise features.

More Analysis on Soft Attention. In this part, we conduct more analysis on soft attention module in terms of channel and spatial submodules, global and local components, and residual design. The results described in Table 4 are obtained from PRW benchmark. Given results of "w/o Spatial in Soft" and "w/o Channel in Soft", we find that each individual effect of these two parts is not so obvious compared with setting "w/o Soft". Once they are applied together in our framework, it can obtain a fairly good result. Additionally, it is clear that applying a soft attention module on either global or local part will lead to a relative large improvement on result. And we can see that the global part acquires a greater gain than the local one. It is quite reasonable since global feature generally contains more background noise and the pose-guided local branch has already eliminated much noise. Finally, the entry "w/o Residual in Soft" shows that gain of the residual component on a single layer is limited compared with other designs above, and it inspires us that the residual design may be more efficient when applying it on more layers.

To show the effectiveness of this module, we visualize some pedestrians' feature maps (Fig. 6) after attention processing extracted from global branch. We can see that our framework highlights the most discriminative regions, and handles cluttered background as well as occlusion situations successfully.

Online Hard Mined Random Sampling Softmax. The OHMRSS loss is designed to optimize the network more efficiently. Replacing it with Softmax loss, we obtain a result ("w/o OHMRSS" in Table 3) that top-1 accuracy drops from 60.1% to 57.8%, and mAP from 29.1% to 26.2%. This demonstrates that person search task can benefit from this loss in our framework. Furthermore, we find that network with OHMRSS converges more quickly during training especially in early periods demonstrated in Fig. 7.

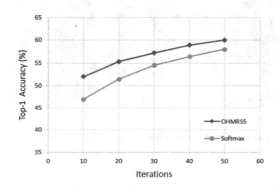

Fig. 7. Top-1 accuracy of MCAN on PRW benchmark during training.

5   Conclusions

In this work, we present a novel Multilevel Collaborative Attention Network (MCAN) for fulfilling person search task in an end-to-end manner. In contrast to previous methods, our framework efficiently deals with large scale variation of detected pedestrian boxes and misalignment, overwhelming background noise as well as occlusion. Extensive experiments demonstrate its superiority over current state-of-the-art methods.

Acknowledgement. This paper is supported by NSFC (No. 61772330, 61533012, 61876109, 61472075, 61876085), the Basic Research Project of Shanghai "Innovation Action Plan" (16JC1402800) and the interdisciplinary Program of Shanghai Jiao Tong University (YG2015MS43).

References

1. Cai, Z., Saberian, M., Vasconcelos, N.: Learning complexity-aware cascades for deep pedestrian detection. In: ICCV, pp. 3361–3369 (2015)
2. Ding, S., Lin, L., Wang, G., Chao, H.: Deep feature learning with relative distance comparison for person re-identification. Pattern Recognit. **48**(10), 2993–3003 (2015)
3. Dollár, P., Appel, R., Belongie, S., Perona, P.: Fast feature pyramids for object detection. PAMI **36**(8), 1532–1545 (2014)
4. Farenzena, M., Bazzani, L., Perina, A., Murino, V., Cristani, M.: Person re-identification by symmetry-driven accumulation of local features. In: CVPR, pp. 2360–2367. IEEE (2010)
5. Felzenszwalb, P.F., Girshick, R.B., McAllester, D., Ramanan, D.: Object detection with discriminatively trained part-based models. PAMI **32**(9), 1627–1645 (2010)
6. Gheissari, N., Sebastian, T.B., Hartley, R.: Person reidentification using spatiotemporal appearance. In: CVPR, vol. 2, pp. 1528–1535. IEEE (2006)
7. Gray, D., Tao, H.: Viewpoint invariant pedestrian recognition with an ensemble of localized features. In: Forsyth, D., Torr, P., Zisserman, A. (eds.) ECCV 2008. LNCS, vol. 5302, pp. 262–275. Springer, Heidelberg (2008). https://doi.org/10.1007/978-3-540-88682-2_21
8. Hamdoun, O., Moutarde, F., Stanciulescu, B., Steux, B.: Person re-identification in multi-camera system by signature based on interest point descriptors collected on short video sequences. In: ICDSC, pp. 1–6. IEEE (2008)
9. He, K., Gkioxari, G., Dollár, P., Girshick, R.: Mask R-CNN. In: ICCV, pp. 2980–2988. IEEE (2017)
10. He, K., Zhang, X., Ren, S., Sun, J.: Deep residual learning for image recognition. In: CVPR, pp. 770–778 (2016)
11. Hu, J., Shen, L., Sun, G.: Squeeze-and-excitation networks. arXiv preprint arXiv:1709.01507 (2017)
12. Ioffe, S., Szegedy, C.: Batch normalization: accelerating deep network training by reducing internal covariate shift. arXiv preprint arXiv:1502.03167 (2015)
13. Jaderberg, M., Simonyan, K., Zisserman, A., et al.: Spatial transformer networks. In: Advances in Neural Information Processing Systems, pp. 2017–2025 (2015)
14. Koestinger, M., Hirzer, M., Wohlhart, P., Roth, P.M., Bischof, H.: Large scale metric learning from equivalence constraints. In: CVPR, pp. 2288–2295. IEEE (2012)
15. Li, S., Xiao, T., Li, H., Zhou, B., Yue, D., Wang, X.: Person search with natural language description
16. Li, W., Zhao, R., Xiao, T., Wang, X.: DeepReID: deep filter pairing neural network for person re-identification. In: CVPR, pp. 152–159 (2014)
17. Li, W., Zhu, X., Gong, S.: Harmonious attention network for person re-identification. arXiv preprint arXiv:1802.08122 (2018)
18. Liao, S., Hu, Y., Zhu, X., Li, S.Z.: Person re-identification by local maximal occurrence representation and metric learning. In: CVPR, pp. 2197–2206 (2015)
19. Liao, S., Li, S.Z.: Efficient PSD constrained asymmetric metric learning for person re-identification. In: ICCV, pp. 3685–3693 (2015)
20. Lin, T.Y., Dollár, P., Girshick, R., He, K., Hariharan, B., Belongie, S.: Feature pyramid networks for object detection. In: CVPR, vol. 1, p. 4 (2017)
21. Liu, H., et al.: Neural person search machines. In: ICCV (2017)
22. Ma, L., Sun, Q., Georgoulis, S., Van Gool, L., Schiele, B., Fritz, M.: Disentangled person image generation. arXiv preprint arXiv:1712.02621 (2017)

23. Nam, W., Dollár, P., Han, J.H.: Local decorrelation for improved pedestrian detection. In: Advances in Neural Information Processing Systems, pp. 424–432 (2014)
24. Paisitkriangkrai, S., Shen, C., van den Hengel, A.: Learning to rank in person re-identification with metric ensembles. In: CVPR, pp. 1846–1855 (2015)
25. Pumarola, A., Agudo, A., Sanfeliu, A., Moreno-Noguer, F.: Unsupervised person image synthesis in arbitrary poses
26. Ren, S., He, K., Girshick, R., Sun, J.: Faster R-CNN: towards real-time object detection with region proposal networks. In: Advances in Neural Information Processing Systems, pp. 91–99 (2015)
27. Shrivastava, A., Gupta, A., Girshick, R.: Training region-based object detectors with online hard example mining. In: Proceedings of the IEEE Conference on Computer Vision and Pattern Recognition, pp. 761–769 (2016)
28. Song, G., Leng, B., Liu, Y., Hetang, C., Cai, S.: Region-based quality estimation network for large-scale person re-identification. arXiv preprint arXiv:1711.08766 (2017)
29. Su, C., Li, J., Zhang, S., Xing, J., Gao, W., Tian, Q.: Pose-driven deep convolutional model for person re-identification. In: ICCV, pp. 3980–3989. IEEE (2017)
30. Tao, D., Guo, Y., Song, M., Li, Y., Yu, Z., Tang, Y.Y.: Person re-identification by dual-regularized kiss metric learning. IEEE Trans. Image Process. 25(6), 2726–2738 (2016)
31. Tian, Y., Luo, P., Wang, X., Tang, X.: Deep learning strong parts for pedestrian detection. In: ICCV, pp. 1904–1912 (2015)
32. Wang, F., et al.: Residual attention network for image classification. arXiv preprint arXiv:1704.06904 (2017)
33. Wang, X., Doretto, G., Sebastian, T., Rittscher, J., Tu, P.: Shape and appearance context modeling. In: ICCV, pp. 1–8. IEEE (2007)
34. Wei, L., Zhang, S., Yao, H., Gao, W., Tian, Q.: GLAD: global-local-alignment descriptor for pedestrian retrieval. In: ACMMM, pp. 420–428. ACM (2017)
35. Xiao, T., Li, S., Wang, B., Lin, L., Wang, X.: End-to-end deep learning for person search. arXiv preprint (2016)
36. Xiao, T., Li, S., Wang, B., Lin, L., Wang, X.: Joint detection and identification feature learning for person search. In: CVPR, pp. 3376–3385. IEEE (2017)
37. Xu, Y., Ma, B., Huang, R., Lin, L.: Person search in a scene by jointly modeling people commonness and person uniqueness. In: ACMMM, pp. 937–940. ACM (2014)
38. Yang, B., Yan, J., Lei, Z., Li, S.Z.: Convolutional channel features. In: ICCV, pp. 82–90. IEEE (2015)
39. Zajdel, W., Zivkovic, Z., Krose, B.: Keeping track of humans: have I seen this person before? In: ICRA, pp. 2081–2086. IEEE (2005)
40. Zhang, L., Lin, L., Liang, X., He, K.: Is faster R-CNN doing well for pedestrian detection? In: Leibe, B., Matas, J., Sebe, N., Welling, M. (eds.) ECCV 2016. LNCS, vol. 9906, pp. 443–457. Springer, Cham (2016). https://doi.org/10.1007/978-3-319-46475-6_28
41. Zhao, R., Ouyang, W., Wang, X.: Unsupervised salience learning for person re-identification. In: CVPR, pp. 3586–3593. IEEE (2013)
42. Zheng, L., Shen, L., Tian, L., Wang, S., Wang, J., Tian, Q.: Scalable person re-identification: a benchmark. In: ICCV, pp. 1116–1124 (2015)
43. Zheng, L., Zhang, H., Sun, S., Chandraker, M., Tian, Q.: Person re-identification in the wild. arXiv preprint (2017)
44. Zhou, Z., Huang, Y., Wang, W., Wang, L., Tan, T.: See the forest for the trees: joint spatial and temporal recurrent neural networks for video-based person re-identification. In: CVPR, pp. 6776–6785. IEEE (2017)

Enhancing Perceptual Attributes
with Bayesian Style Generation

Aliaksandr Siarohin[1]([✉])[iD], Gloria Zen[1][iD], Nicu Sebe[1][iD], and Elisa Ricci[1,2][iD]

[1] DISI, University of Trento, Trento, Italy
{aliaksandr.siarohin,gloria.zen,niculae.sebe,e.ricci}@unitn.it
[2] Fondazione Bruno Kessler (FBK), Trento, Italy

Abstract. Deep learning has brought an unprecedented progress in computer vision and significant advances have been made in predicting subjective properties inherent to visual data (e.g., memorability, aesthetic quality, evoked emotions, etc.). Recently, some research works have even proposed deep learning approaches to modify images such as to appropriately alter these properties. Following this research line, this paper introduces a novel deep learning framework for synthesizing images in order to enhance a predefined perceptual attribute. Our approach takes as input a natural image and exploits recent models for deep style transfer and generative adversarial networks to change its style in order to modify a specific high-level attribute. Differently from previous works focusing on enhancing a specific property of a visual content, we propose a general framework and demonstrate its effectiveness in two use cases, *i.e.* increasing image memorability and generating scary pictures. We evaluate the proposed approach on publicly available benchmarks, demonstrating its advantages over state of the art methods.

Keywords: Style transfer · GANs · Memorability · Scariness

1 Introduction

The recent advances in predicting and understanding subjective properties of visual data (*e.g.* beauty, memorability, interestingness, etc.) enabled by deep learning models [3,9,13,17,18,22] have motivated researchers in computer vision to take a step forward and investigate automatic techniques to manipulate images in order to modify these properties. For instance, recent works have proposed methods to edit images in order to increase their memorability [25], to improve their aesthetic quality [30] or to evoke specific emotional reactions into users [21]. Recently, deep style transfer methods [6,11,15,29] which allow the users to modify pictures by blending them with style images have gained popularity. These methods have significantly widened the set of editing operations available in traditional image enhancement tools, fostering the diffusion of novel software for turning user pictures into artworks. While earlier methods for neural style transfer [6,29] considered a fixed set of styles and relied on slow

C. V. Jawahar et al. (Eds.): ACCV 2018, LNCS 11361, pp. 483–498, 2019.
https://doi.org/10.1007/978-3-030-20887-5_30

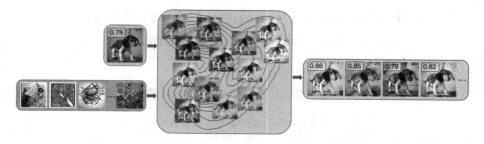

Fig. 1. Idea behind our approach. Given a generic input image (yellow box) our framework provides as output a set of stylized images (green box) obtained by applying the styles which maximally enhance a given perceptual attribute. The attribute value (shown on top left corners of input and output images) is automatically assessed by a deep network. The style selection process is achieved by modeling the style space (light blue box) as a probability distribution automatically learned from a given training set of style images (orange box) using a generative adversarial network. (Color figure online)

optimization processes, more recent approaches [11,15] are highly flexible, enable the generation of arbitrary styles and have close to realtime performance.

Motivated by these recent advances, in this paper we propose a novel approach for generating stylized images in order to enhance a given perceptual attribute. Similarly to previous deep style transfer methods [11,15], the stylized images are obtained by training a feed-forward neural network which receives as input the original images and the style pictures. Opposite to previous works, the style choice is not made by a user but it is automatic and is driven by a specific criterion, *i.e.* increasing the value of the given perceptual attribute. At the core of our style selection process there is a novel probabilistic framework which exploits recent Generative Adversarial Networks (GANs) to learn a probability distribution modeling the style space and Markov Chain Monte Carlo (MCMC) methods to sample from the learned distribution and compute the best styles. We named the proposed approach BAE (Bayesian Attribute Enhancement). While our framework is generic and can be used for different types of perceptual attributes, in this work we focus on two applications, *i.e.* increasing memorability, defined as the probability of an image to be remembered [13] and generating scary pictures. We quantitatively and qualitatively evaluate the proposed approach on publicly available datasets, demonstrating superior performance over state of the art methods. Figure 1 illustrates the intuition behind our method.

Contributions. To summarize, the contribution of this work is threefold. (i) We propose a novel framework to automatically modify an input image in order to alter its inherent perceptual attributes. To preserve the semantic content of the original image, our approach relies on a neural style transfer method. In this way, the problem of perceptual attribute enhancement naturally translates to that of retrieving the best styles to apply to the given image. Opposite to previous works which focus on modifying a specific subjective property [21,25,30], our method can be applied to any arbitrary attribute. While we tested it on two scenarios, we expect the method to be useful in other applications, *e.g.* for enhancing the aesthetic quality of images or for increasing their virality score. (ii) By exploiting state of the art deep style transfer techniques [11] within a novel probabilistic framework for modelling the style space, our approach does not simply select the best styles from a small predefined set but also allows to generate arbitrary new styles. Thus, a higher increase of the attribute score can be obtained with respect to previous approaches [25]. (iii) Our framework is highly flexible and allows not only to automatically select the best styles but also the degree of stylization. Furthermore, by resorting on MCMC sampling methods, it can be used to compute multiple styles. In this way we keep the users in the loop, suggesting the best styles for attribute increase but still allowing the users to choose among multiple stylized images according to their personal preferences.

2   Related Works

Our work lies at the intersection between two main research lines. The first line focuses on the problem of understanding and predicting subjective properties from visual data, the second one includes works proposing novel deep models for automatic image editing.

Predicting Perceptual Attributes from Visual Data. In the last decade several works in computer vision and multimedia have addressed the problem of modelling and predicting perceptual attributes from images and videos. These studies have focused on the automatic assessment of aesthetic value [17,18], interestingness [9], memorability [13], virality [3], symmetry [5], etc. In some cases, typically where a large amount of training data is available, automatic systems can even reach human-level performances. For instance, Khosla *et al.* [13] showed that a deep learning model trained on LaMem, the largest memorability dataset so far, can predict image memorability with an accuracy close to that of human annotators. Similarly, recent methods for computing automatically the aesthetic value of images are quite precise, achieving an accuracy superior to 75% on the AVA dataset [17,18,20]. In this work we focus not only on predicting subjective attributes but we also address the more challenging task of image enhancement.

Deep Models for Automatic Image Manipulation. Deep learning models and, in particular, neural style transfer methods [6,11,15,29] and deep generative networks [7,12] have enabled significant advances for automatic image editing and generation. In the wake of these progresses, recent works have taken a step beyond perceptual attributes prediction and have proposed methods to manipulate images in order to modify these intrinsic attributes [25,28,30]. For instance, Wang *et al.* [30] addressed the task of increasing the aesthetic value of an image by finding the best crop. Tsai *et al.* [28] proposed a deep model for image harmonization which adjusts the appearance of the image foreground in order to better adapt it to the background. Liao *et al.* [16] introduced a method to alter intrinsic image properties like color, texture or style based on deep analogy and visual property transfer. However, these works simply propose strategies to modify a specific property of images but do not provide a general framework to systematically enhance an arbitrary perceptual attribute and quantitatively assess its value increase. Recently, Siarohin et al. [25] moved a step forward in this direction, by proposing an approach which selects the best styles for a given image in order to increase its memorability. Still, their method relies on a pre-defined set of styles and the degree of stylization is also fixed a priori. In this work, we overcome these limitations by introducing a more general and flexible approach which operates on a large set of styles and where the trade-off between style and content is regulated by a user-defined hyper-parameter α.

3   Enhancing Perceptual Attributes with BAE

As stated in Sect. 1, the proposed approach deals with the problem of automatically modifying an arbitrary input image in order to enhance a specific perceptual attribute, e.g. its memorability, the likelihood to evoke specific emotional reactions from users, etc. This task is addressed within a novel Bayesian framework and by resorting on a state of the art neural style transfer method [11]. In fact, our approach aims to modify the given image increasing its perceptual attribute score by changing its style while retaining the semantic content. In the following we briefly describe the neural style transfer method in [11] and then introduce the proposed approach providing some details on our implementation.

3.1   Arbitrary Style Transfer

Given an input image I and a style image S, let us denote with $\hat{I} = T(I, S)$ the modified image obtained by applying the style transfer model T. In this work we consider the style transfer approach in [11] as, oppositely to earlier methods [6,29], (i) it is not tied to a fixed set of styles, allowing to generate arbitrary new styles, (ii) it performs style transfer in realtime, and (iii) it is very flexible, enabling to control the degree of stylization also at test time.

The deep architecture T proposed in [11] has a simple encoder-decoder structure. The encoder f_E is used to compute the feature maps $f_E(I)$ and $f_E(S)$ associated respectively to the input and to the style images. The computed feature maps are then fed to a specific feature alignment layer, the Adaptive Instance Normalization (AdaIN) layer. This layer aligns the mean μ and variance σ of the image features to those of the style features, producing the target feature maps:

$$t = \text{AdaIN}(f_E(I), f_E(S)) = \sigma(f_E(S))\nu(f_E(I)) + \mu(f_E(S)) \qquad (1)$$

where $\nu(x) = \frac{x - \mu(x)}{\sigma(x)}$. The decoder $f_D(t)$ is trained to map the target feature maps t back to the image space, generating the stylized image $\hat{I} = T(I, S) = f_D(t)$. As typically done in neural style transfer methods, the network T is trained by optimizing a loss function which is the weighted sum of two terms, i.e. $\mathcal{L} = \mathcal{L}_c + \gamma \mathcal{L}_S$, where $\mathcal{L}_c$ and $\mathcal{L}_S$ are the content and the style loss respectively and γ is a user defined parameter regulating at training time the trade-off between semantic content and stylization. We refer the reader to the original paper [11] for details on the definition of the loss functions.

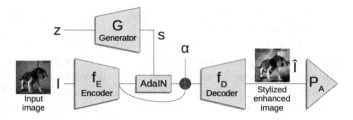

Fig. 2. Overview of our approach: given an input image I our method generates through G the style s which maximizes the perceptual attribute score, computed by P_A, of the stylized image $\hat{I}$.

A prominent feature of the neural style transfer method in [11] is the possibility to control the degree of stylization not only at training time by changing γ but also at test time. In particular, a parameter α is introduced and the stylized image is computed as:

$$\hat{I} = T(I, S, \alpha) = f_D(\alpha f_E(I) + (1 - \alpha)t) \qquad (2)$$

Here, $\alpha = 1$ corresponds to the case where no style transfer is performed, while $\alpha = 0$ corresponds to full stylization.

3.2 Bayesian Attribute Enhancement

The main idea behind the proposed approach is to construct a model which, given an arbitrary image I and a set of style images $\Omega_S = \{S_i\}_{i=1}^{K}$, is able to automatically compute a novel set of M styles that, applied to the input image, fully enhance a specific perceptual attribute, $e.g.$ increase its memorability. To build this model, inspired by [11], we first introduce a compact representation for styles in terms of mean and standard deviation of activations. Formally, given a style image S and a pre-trained encoder network f_E, we define a style $s = \{\mu_s, \sigma_s\}$ where $\mu_s = \mu(f_E(S))$ and $\sigma_s = \sigma(f_E(S))$.

Given the set of style images $\Omega_S = \{S_i\}_{i=1}^{K}$ and the associated representations $\mathcal{S} = \{s_i\}_{i=1}^{K}$, we propose to learn a probability density function $P_{\mathcal{S}}(s)$ modelling the style space. While different methods can be used for this purpose, motivated by the recent successes of deep generative models [7], in this paper we consider a Generative Adversarial Network (GAN). A GAN consists of two networks, a generator G and a discriminator D. These two networks play a minimax game in which the task of D is to distinguish the samples generated by G from the real samples and the task of G is to increase the chances of D producing a high probability for a synthetic example. In [7] it is shown that the equilibrium in this game is achieved when the probability density of the generated samples is equal to the probability density of the real ones. For our application we use $\mathcal{S}$ as real samples, and learn the generator G in order to produce styles $s = G(z)$, where $z \sim P(z)$. The input z to the generator is sampled from some simple noise distribution $P(z)$ such as a Gaussian distribution $N(0, \mathbb{1})$. For training the

GAN model we use an efficient version of Wasserstein GANs [4], and specifically WGAN-GP, recently proposed in [8].

In addition to the GAN model, we propose to learn two additional deep networks. The first network implements the neural style transfer approach described in Sect. 3.1. In the following, given a style s we denote as $\hat{I} = Z(I, s) = f_D\left(\sigma_s \nu(f_E(I)) + \mu_s\right)$ the stylized image. The second network is used to learn a function $P_A(I)$ which, given an input image I, outputs a probability score reflecting the strength of a given perceptual attribute. The design of this network and its training strategy, described in Sect. 3.3, is at the core of our method and depends on the chosen subjective attribute. In particular in this paper we consider two attributes, memorability and scariness, $i.e.$ we propose two different criteria for modifying pictures: increasing their memorability and maximizing their likelihood to evoke scary reactions into users. Given the above definitions, we propose to build a probability density for the joint model:

$$P(\hat{I}, z) = P(\hat{I}|z)P(z) = P_A(Z(I, G(z)))N(z; 0, \mathbb{1}) \tag{3}$$

where the last term is derived considering the learned models G, Z and P_A. In this way, we obtain a probability over z which, in our work, can be seen as a latent representation of a style s. We propose to exploit $P(\hat{I}, z)$ in order to find the styles which better enhance a given perceptual attribute. Specifically, to obtain a diverse set of styles corresponding to high values of the target attribute, we propose to sample from $P(\hat{I}, s, z)$ using MCMC methods. The best styles, in fact, correspond to the modes of the distribution.

We also extend the proposed Bayesian framework in order to compute automatically not only the best styles but also the degree of stylization. To this aim, we consider Eq. 2 and define $\hat{Z}(I, S, \alpha) = f_D(\alpha f_E(I) + (1 - \alpha)t)$. However, instead of setting α as a constant, we assume that α is a random variable. In this case, similarly to Eq. 3, we define the joint probability:

$$P(\hat{I}, z, \alpha) = P(\hat{I}|z, \alpha)P(z)P(\alpha) \tag{4}$$

where $P(\hat{I}|z, \alpha) = P_A(\hat{Z}(I, G(z), \alpha))$ and $P(\alpha)$ is a prior probability. In this case with MCMC sampling we obtain a set of latent style representations $\mathcal{S}' = \{z_1, z_2, ...z_M\}$, as well as a set of stylization coefficients $\mathcal{A}' = \{\alpha_1, \alpha_2, ...\alpha_M\}$. In the following, we refer to our method as Bayesian Attribute Enhancer (BAE), while its adaptive version where we also automatically compute α value is called ABAE. An overview of the proposed framework is illustrated in Fig. 2.

3.3 Implementation

In this Section we report additional details on the implementation of the proposed method. In particular, we describe the adopted deep network architectures and provide further details on the considered MCMC sampling strategies.

Network Architectures. The neural style transfer network is implemented following the original paper [11]. The encoder f_E is built from the first four convolutional layers of a pre-trained VGG-19 [26]. The decoder f_D is implemented

---

Algorithm 1. Langevin MCMC

Data: $\mathbb{O}$: energy function, M: number of samples, τ: learning rate
Result: Set $\mathcal{S}' = \{z_0, z_1, ..., z_M\}, z_i \sim exp(\mathbb{O})$
// Initialization
1   $z \sim N(0, \mathbb{1})$, $i := 0$, $\mathcal{S}' := \emptyset$;
2 **while** $i \lnot M$ **do**
     // Generate candidate point
3       $\hat{z} := z + l\mathbb{O}_g(z) + \sqrt{2\tau}\epsilon$;
     // Calculate acceptance ratio
4       $r := \exp\left(\mathbb{O}(\hat{z}) - \mathbb{O}(z) + \frac{\|\hat{z}-z+\tau\nabla\mathbb{O}(z)\|}{4\tau} - \frac{\|z-\hat{z}+\tau\nabla\mathbb{O}(\hat{z})\|}{4\tau}\right)$;
5   **if** $\mathbb{U}(0,1) \leq \min(1, r)$ **then**
        // Accept candidate point
6          $\mathcal{S}' \leftarrow \hat{z}$, $i := i + 1$, $z := \hat{z}$;
7   **end**
8 **end**

---

with a structure mirroring the encoder, with all pooling layers replaced with up-sampling layers. In the case of ABAE we limit the range of the coefficient α between $[0, 1]$ introducing a clipping function $\text{clip}(\alpha) = \min(1, \max(0, \alpha))$. It is worth noting that, while we consider the method in [11], our framework allows using different style transfer approaches such as the one proposed in [15]. In this case, the only difference would be the representation of style s, which in [15] is modelled in terms of mean and covariance matrix.

The implementation of the perceptual attribute predictor P_A depends on the considered attribute. For memorability we resort on the Memnet model introduced in [13] to allow fair comparison with [25]. As suggested in [13], we consider the HybridCNN network [32] and finetune it on LaMem dataset [13]. Following this protocol, the resulting model $\hat{P}_A(I)$ implements a regressor, *i.e.* $\hat{P}_A(I) \in [-\infty, +\infty]$. To normalize the output scores of the memorability predictor we compute $P_A(I) = \Sigma(\hat{P}_A(I))^\lambda$, where Σ is a sigmoid function and λ a user defined parameter. We follow a similar approach for deriving $P_A(I)$ in the case of scariness. We use InceptionV3 network as one of the best general purpose models [27]. We trained this model on images with their binary labels from the BAM dataset [31] to derive $\hat{P}_A(I)$ and then compute $P_A(I) = \hat{P}_A(I)^\lambda$.

In the proposed GAN model the generator G is implemented as a neural network with the following structure: FC_{128}^R - FC_{512}^R - FC_{1024}, where FC_P^R denotes a fully-connected layer with P output units and Relu activation, while FC_P indicates a fully-connected layer with P output units without activation. Similarly, the architecture of the discriminator D is defined as: FC_{512}^R - FC_{256}^R - FC_{128}^R - FC_1.

Style Sampling Methods. In this work we used MCMC sampling in order to find the best styles. MCMC is a general method for sampling from a multivariate probability distribution. We define the energy function $\mathbb{O}(z) = \log\left(P_A(Z(I, G(z)))N(z; 0, \mathbb{1})\right)$ and we chose Langevin MCMC [23] as our sampling method (see Algorithm 1). For simplicity here we report the formulas only for BAE. The algorithm is similar for ABAE. We also experiment with the two other popular MCMC methods: Metropolis Hastings [10] and Hamiltonian [19]. The effect of using different MCMC methods for creating new styles is discussed in Sect. 4.2. We also introduced two modifications to the traditional methods to help increasing the acceptance rate (line 5 - Algorithm 1):

- *Adaptive gradient:* Instead of using $\nabla \mathbb{O}(z)$ we consider it adaptive version, in analogy to Adam [14]. We found this strategy especially helpful for ABAE, because the gradient for α can be several order of magnitude higher than the gradient for z.
- *Adaptive learning rate:* At step 5 in Algorithm 1 upon rejection we decrease the learning rate τ (*e.g.* $\tau := 0.9\tau$) while upon acceptance we set τ to the initial value. This strategy eliminates the need of tuning the learning rate τ for each image.

4   Experimental Validation

In this Section we report the results of our experimental evaluation. First, we provide some details on the used datasets and our experimental setup (Sect. 4.1). Then, in Sect. 4.2 we quantitatively evaluate the performance of our method in enhancing two different perceptual attributes: memorability and scariness. In the case of memorability, we also discuss the effect of using different sampling methods. Finally, we report qualitative results comparing our method with baselines. Our code is available online [1].

4.1   Experimental Setup and Datasets

Datasets. We considered three datasets in our work.

The *DevianArt* dataset [24] is a collection of 500 abstract art paintings collected from an online social network site, deviantart.com, devoted to user-generated art. The dataset was used in [25] to define the style set.

LaMem [13] is a collection of 58,741 images annotated with the corresponding memorability score. The scores were collected through an efficient version of the memorability game. We encourage the reader to refer at [13] for further details. This dataset was also considered in [25].

The *Behance-Artistic-Media (BAM)* dataset [31] is a very large dataset with automatically labeled binary attribute scores. It comprises about 20 attributes, including emotional attributes like scary, gloomy, happy and peaceful. It contains 14,585 images (with positive or negative labels) originally crowdsourced from human annotators for the scary attribute. We were able to download a subset of 11,698 images from this dataset. We use this set to train our scariness predictor.

Experimental Setup. We now provide further details on our experimental setting and implementation. We follow an experimental protocol similar to [25] in order to allow a fair comparison with their work. Note that [25] only focuses on memorability, while our approach deals with arbitrary attributes.

Styles Set. For the style set Ω_S we considered 500 abstract art images from the DeviantArt [24] dataset. While Siarohin *et al.* [25] considered a pre-defined set of styles selecting 100 images from this dataset, our approach by learning a style probability density function can potentially learn from and generate an infinite number of styles. As described in Sect. 3.2, we use a GAN model to represent the probability density over the set of styles. The GAN was trained with batch size equal to 64 and for about 100k iterations. All the other hyper-parameters were set as indicated in [8].

Baseline Methods. In the case of memorability, we compare the performance of our method with [25]. The code from [25] is available online [2]. This is the closest work to ours, where a set of only 100 styles is considered for increasing memorability. We also consider an additional baseline method $\mathcal{B}$ which uses the same set of 500 styles of our approach. Specifically, this baseline consists in applying the style transfer method in [11] to the given image considering all the style pictures in the style set and then compare the obtained stylized images with those we obtained with our method setting $M = 500$. In the case of scariness, we simply compare with $\mathcal{B}$.

Perceptual Attribute Predictors. For the target attributes, memorability and scariness, we trained two predictors from two independent set of images. The first predictor, which we denote as the internal predictor $\mathbb{M}$, is used for generating the stylized images and corresponds to $\hat{P}_A$, while the second is employed only for assessing the performance of our method and we call it the external predictor $\mathbb{E}$, indicated as $\hat{P}_A^{\mathbb{E}}$. Specifically we use the second predictor to compute the score increases between the original image and the stylized images obtained with our method. In the case of memorability, we split LaMem into two sets of 22,500 images each and use these sets to learn the two predictors. This is exactly the same setup used in [25]. We also used the same training parameters. To verify that these predictors are valid and have performance close to human annotators, following [13], we compute the rank correlation and we obtain a value of 0.63 for both models. As reported in [13], this is close to human performance (0.68). In the case of scariness, we finetuned InceptionV3 [27] (considering only the two top inception blocks) originally trained on ImageNet using labels from BAM dataset [31]. We split the BAM labeled set into two disjoint sets of 5,849 images each and trained two scariness prediction models.

Style Transfer. As stated above for the style transfer network, we used the recent approach from [11]. We considered the pretrained network released by [11]. For the baseline $\mathcal{B}$ and our method BAE we used $\alpha = 0.5$. In the case of adaptive alpha, we used a Gaussian distribution as prior $P(\alpha) = N(\alpha; 0.5, 0.25)$ (see Eq. 2).

Hyper-Parameters. For the experiment on memorability we set $\tau = 1e - 1$ and $\lambda = 100$, while for scariness we consider $\tau = 1e - 2$ and $\lambda = 10$. In general, we found that the higher is λ the higher is the attribute increase, but when λ is too high nearly all the candidates points $\hat{z}$ are rejected at step 5 in Algorithm 1. So, we set λ to the highest value (we try values on a log scale, i.e. $\lambda \in \{1, 10, 100, 1000\}$) for which this effect is not observed. A similar trend is observed for the learning rate τ. If τ is high, we obtain more diverse styles. Still, when the learning rate is too high almost all candidate points are rejected. Similarly to λ, we set the initial learning rate τ to the highest possible value for which we do not observe this effect (we also try values on log scale $\tau \in \{1e-3, 1e-2, 1e-1, 1\}$). The problem of choosing the optimal learning rate is partially overcome with the adaptive learning rate strategy described in Sect. 3.3. Still, choosing an optimal initial learning rate can greatly improve the overall method speed.

Image Test Set. We evaluate the performance of our approach on the same test set as in [25], which we call $\mathcal{V}$, consisting of 1,001 generic images. We used this test set also for the experiment on scariness.

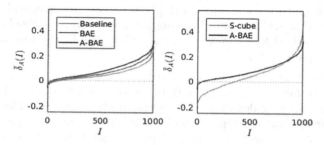

Fig. 3. Sorted memorability differences $\bar{\delta}_A(I)$ for the images $I \in \mathcal{V}$ obtained by averaging over the top 10 results retrieved with each method. Comparison of our methods with (left) the baseline $\mathcal{B}$ and (right) the competing work S-cube [25].

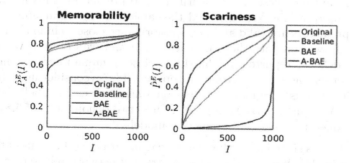

Fig. 4. Perceptual attributes scores. Sorted scores for the original images $I \in \mathcal{V}$ and the top results retrieved in the case of (left) memorability and (right) scariness: original image scores $\hat{P}_A^{\mathbb{E}}(I)$ and comparison with the top results $\hat{P}_A^{\mathbb{E}}(\hat{I})$ obtained with the baseline $\mathcal{B}$ and our method.

4.2   Results

Evaluation Metrics. Similarly to [25], we use the *Top N* results and compute the average *score difference* $\bar{\delta}_A(I)$ as evaluation measure. Specifically, for each method computing M stylized images, we rank these images based on the attribute scores calculated with the internal predictor $\hat{P}_A(\hat{I})$. The *Top N* results corresponds to the subset of N images which rank the highest according to these scores. Then, given a generic image I and a corresponding stylized image $T(I,s)$, we define $\delta_A(I,s)$ as the difference between the attribute scores of these two images, based on the external predictor, *i.e.* $\delta_A(I,s) = \hat{P}_A^{\mathbb{E}}(T(I,s)) - \hat{P}_A^{\mathbb{E}}(I)$. Finally, given the Top N results, we compute for each image I the corresponding average score difference $\bar{\delta}_A(I)$ by averaging over $\delta_A(I,s)$ from the Top N set.

Quantitative Results. We first perform some experiments in order to compare our approach with baseline methods on the memorability enhancement task. Figure 3 reports the average memorability differences obtained for all the images $I \in \mathcal{V}$. In the plot on the left we compare our approach in the case of fixed α (BAE) and adaptive α (ABAE) with the baseline. It is straightforward to see that our approach performs better than the baseline $\mathcal{B}$ and that the adaptive method ABAE guarantees a higher memorability gain with respect to BAE using a fixed α. This indicates that the possibility to automatically set the degree of stylization is beneficial in terms of performance. In the plot on the right we compare our best performing method ABAE with the competing work [25]. It can be noted that in the case of [25] the average memorability differences are negative for a large set of test images. This difference may be explained by the fact that in the case of [25] the top N styles are retrieved from a pool of only 100 art images, while our method learns the style space from an initial set of 500 styles. This result highlights the importance of considering a wide set of styles, in order to find those which better suit a given image. In this respect, our method is very powerful, being able to interpolate between the styles of a given style set, thus achieving a higher memorability increase.

Figure 4 reports the results of a similar analysis. Specifically, it depicts the sorted scores of the original image set $\mathcal{V}$ and of the corresponding sorted scores of stylized images obtained with our methods and with the baseline $\mathcal{B}$ in the

Table 1. Increasing (a) memorability and (b) scariness. Performance of our method with fixed α (BAE) and adaptive α (ABAE) compared to the baseline $\mathcal{B}$ and, in the case of memorability, also to [25]. Performances are measured in terms of memorability score differences averaged over the top N results $\bar{\delta}_A(I)$.

N	S-cube [25]	$\mathcal{B}$	BAE	ABAE
1	0.0792	0.0677	0.0812	**0.1067**
5	0.0594	0.0590	0.0762	**0.0976**
10	0.0488	0.0544	0.0723	**0.0911**

(a) memorability

N	$\mathcal{B}$	BAE	ABAE
1	0.4151	0.5362	**0.6960**
5	0.3500	0.5194	**0.6775**
10	0.3153	0.5075	**0.6631**

(b) scariness

case of (left) memorability and (right) scariness enhancement. For each image I and all the methods we consider only the best stylized image, *i.e.* the one for which we measure the highest attribute score increase. From the figure we can observe that both plots exhibits a similar trend: our method outperforms $\mathcal{B}$ and the adaptive α version ABAE outperforms the fixed α version BAE. It can also be observed that the score increases are more significant in the case of scary images. This result may be partially due to characteristics of the considered dataset: in the case of memorability, most of the images in the dataset exhibit an initial memorability score higher than 0.5, while in the case of scariness the original score is lower than 0.2 for the large majority of the images.

A comparison between our approach and the competing methods is also provided in Table 1(a), where we report the average memorability increases over the Top N results for test set $\mathcal{V}$ in the cases of $N = 1$, $N = 5$ and $N = 10$. In all cases our method performs better than the baseline $\mathcal{B}$ and the competing approach [25] and the highest performance is obtained with the adaptive version of our approach. It is also interesting to note that the performances of the baseline $\mathcal{B}$ are sometimes comparable or inferior to those of method in [25].

Table 2. Performance of our methods (top) BAE and (bottom) A-BAE considering different sampling strategies. Performances are measured in terms of memorability score differences averaged over the top N results $\bar{\delta}_A(I)$.

	Top N	M.H.	Langevin	Hamiltonian
BAE	1	0.0766	**0.0812**	0.0780
	5	0.0725	**0.0762**	0.0734
	10	0.0692	**0.0723**	0.0700
A-BAE	1	0.1070	0.1067	**0.1094**
	5	0.0995	0.0976	**0.1012**
	10	0.0939	0.0911	**0.0955**

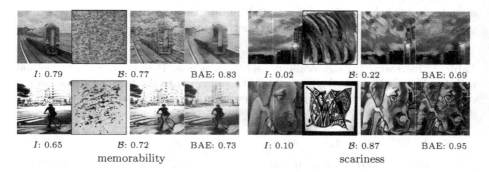

| I: 0.79 | $\mathcal{B}$: 0.77 | BAE: 0.83 | I: 0.02 | $\mathcal{B}$: 0.22 | BAE: 0.69 |
| I: 0.65 | $\mathcal{B}$: 0.72 | BAE: 0.73 | I: 0.10 | $\mathcal{B}$: 0.87 | BAE: 0.95 |

memorability   scariness

Fig. 5. Qualitative results: (left) original image and corresponding top result obtained with (center) the baseline $\mathcal{B}$ and (right) our method BAE. The predicted memorability and scariness scores $\hat{P}_A^E(\cdot)$ are reported below each image.

Indeed, $\mathcal{B}$ and [25] are based on two different style transfer approaches, and this may explain the small performance gaps, especially in the case of small N. A similar trend is observed in the experiments on the scariness scenario (Table 1(b)).

Comparison Between Different Sampling Methods. Table 2 reports the results of our approach using different sampling methods in the case of memorability enhancement. As expected, Metropolis-Hastings MCMC corresponds to the worst performance, while the other two methods are comparable. Qualitatively, we did not observe significant differences between the three methods. In light of these results, in all our experiments we use Langevin MCMC as sampling strategy as it represents the best trade-off between performance and computational speed.

Running Time. The running times for the Langevin MCMC of one image on Nvidia Titan X are, respectively, 1 m 41 s for the baseline (500 style images) and 7 m 20 s for A-BAE (500 MCMC iterations). However, by decreasing the number of iterations in A-BAE to 100 the running time can be reduced to 1 m 28 s, while the top1 average difference is still higher than the baseline (0.55 vs 0.41).

User Study. We run a user study to show the advantage of using our method for increasing image attribute in the case of scariness. The user study consisted in showing pairs of images to a user who was asked to indicate, for each pair, the image which looked more scary. We randomly selected 100 images out of our test set, and considered for each image the corresponding top results obtained respectively with the baseline $\mathcal{B}$ and our method ABAE. We run the study with 11 people (6 male, 5 female); viewers voted for the image modified with ABAE in the 72.36% of the cases in average. The inter user agreement for this user study, measured with cronbach alpha coefficient is 0.78, thus validating the study.

Qualitative Results. Finally, we report some qualitative results. Figure 5 depicts sample stylized images obtained with our method and with the baseline $\mathcal{B}$ in the case of (left) memorability and (right) scariness enhancement. Given

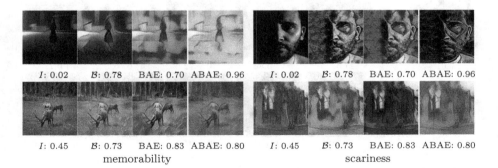

I: 0.02 $\mathcal{B}$: 0.78   BAE: 0.70   ABAE: 0.96 I: 0.02 $\mathcal{B}$: 0.78   BAE: 0.70   ABAE: 0.96

I: 0.45 $\mathcal{B}$: 0.73   BAE: 0.83   ABAE: 0.80 I: 0.45 $\mathcal{B}$: 0.73   BAE: 0.83   ABAE: 0.80

memorability   scariness

Fig. 6. Qualitative results: original input image and top result obtained with the baseline $\mathcal{B}$ and our method with fixed α (BAE) and adaptive α (ABAE). The corresponding memorability and scariness scores $\hat{P}_A^{\mathbb{E}}(\cdot)$ are reported below each image.

an input image $I_\mathcal{V}$ of the test set $\mathcal{V}$, we report the top stylized image computed by $\mathcal{B}$ and BAE. In both cases, the coefficient α is set to 0.5. In the case of $\mathcal{B}$, we also display the corresponding selected style. For the figure it is interesting to observe that, by generating new styles, our method allows to better customize the style to a given image and to achieve an higher increase in terms of attribute score. In Fig. 6 we report additional results to show the effects of further adapting the style to the given image by computing the optimal stylization coefficient α. For each image, we report the top result obtained with the baseline $\mathcal{B}$ and with our methods BAE and ABAE. (Due to space limitations, we do not report the original style image for $\mathcal{B}$). Our methods produce a significant increase in terms of perceptual score with respect to $\mathcal{B}$ and generally creates a style which better suit the input image. Furthermore, the performance of ABAE are always close or significantly better than those of BAE.

So far, we compared the Top 1 results obtained with different methods. In Fig. 7 instead we report the top 5 results corresponding to some sample images on the two considered scenarios. Specifically, we show a comparison between (left) $\mathcal{B}$ and BAE in the case of memorability and (right) $\mathcal{B}$ and ABAE in the case of scariness. The result images obtained with our method usually obtain higher score increases. As a counterpart, these increases come with a small loss

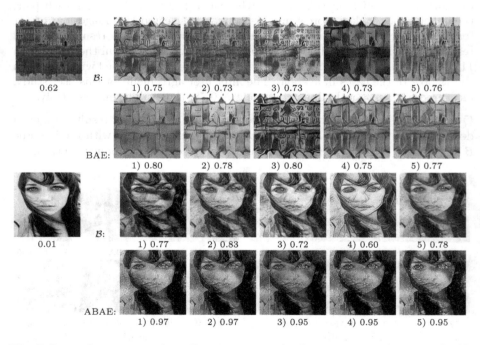

Fig. 7. Increasing perceptual attributes: top 5 results for a given sample image. (Left) Original image and (right) comparison (top) in the memorability scenario between $\mathcal{B}$ and BAE and (bottom) in the scariness scenario between $\mathcal{B}$ and ABAE.

in terms of diversity for the top stylized images with respect to $\mathcal{B}$. In Fig. 8 we report a few cases where our method does not perform as expected. We report sample results in the scariness scenario. In one case (left) our method performs poorly with respect to the baseline. In the other case (right) neither the baseline nor our method can find a suitable solution to create a scary picture.

I_V: 0.02 $\mathcal{B}$: 0.66   BAE: 0.45 ABAE: 0.51 I_V: 0.06 $\mathcal{B}$: 0.24   BAE: 0.21 ABAE: 0.29

Fig. 8. Increasing scariness: sample images where our method is not effective.

5   Conclusions

We presented BAE, a novel framework for generating stylized images in order to enhance a predefined perceptual attribute. By exploiting recent advances on neural style transfer and generative adversarial models, we showed that it is possible to edit images such as to increase their memorability and scariness. Future work will be devoted to exploit different style transfer approaches and consider other subjective properties.

Acknowledgments. We gratefully acknowledge Fondazione Caritro for supporting SMARTourism project and NVIDIA Corporation for the donation of the TitanX GPUs.

References

1. https://github.com/aliaksandrsiarohin/bae
2. https://github.com/aliaksandrsiarohin/mem-transfer
3. Alameda-Pineda, X., Pilzer, A., Xu, D., Sebe, N., Ricci, E.: Viraliency: pooling local virality. In: CVPR (2017)
4. Arjovsky, M., Chintala, S., Bottou, L.: Wasserstein GAN. arXiv preprint arXiv:1701.07875 (2017)
5. Funk, C., Liu, Y.: Beyond planar symmetry: modeling human perception of reflection and rotation symmetries in the wild. In: ICCV (2017)
6. Gatys, L.A., Ecker, A.S., Bethge, M.: Image style transfer using convolutional neural networks. In: CVPR (2016)
7. Goodfellow, I., et al.: Generative adversarial nets. In: NIPS (2014)
8. Gulrajani, I., Ahmed, F., Arjovsky, M., Dumoulin, V., Courville, A.C.: Improved training of wasserstein GANs. In: NIPS (2017)
9. Gygli, M., Grabner, H., Riemenschneider, H., Nater, F., Van Gool, L.: The interestingness of images. In: ICCV (2013)
10. Hastings, W.K.: Monte Carlo sampling methods using Markov chains and their applications. Biometrika **57**(1), 97–109 (1970)

11. Huang, X., Belongie, S.: Arbitrary style transfer in real-time with adaptive instance normalization. In: ICCV (2017)
12. Isola, P., Zhu, J.Y., Zhou, T., Efros, A.A.: Image-to-image translation with conditional adversarial networks (2017)
13. Khosla, A., Raju, A.S., Torralba, A., Oliva, A.: Understanding and predicting image memorability at a large scale. In: ICCV (2015)
14. Kingma, D.P., Ba, J.: Adam: a method for stochastic optimization. arXiv preprint arXiv:1412.6980 (2014)
15. Li, Y., Fang, C., Yang, J., Wang, Z., Lu, X., Yang, M.H.: Universal style transfer via feature transforms. In: NIPS (2017)
16. Liao, J., Yao, Y., Yuan, L., Hua, G., Kang, S.B.: Visual attribute transfer through deep image analogy. ACM Trans. Graph. (TOG) **36**(4), 120 (2017)
17. Lu, X., Lin, Z., Shen, X., Mech, R., Wang, J.Z.: Deep multi-patch aggregation network for image style, aesthetics, and quality estimation. In: ICCV (2015)
18. Mai, L., Jin, H., Liu, F.: Composition-preserving deep photo aesthetics assessment. In: CVPR (2016)
19. Neal, R.M., et al.: MCMC using Hamiltonian dynamics. In: Handbook of Markov Chain Monte Carlo, vol. 2, no. 11 (2011)
20. Peng, K.C., Chen, T.: Toward correlating and solving abstract tasks using convolutional neural networks. In: IEEE Winter Conference on Applications of Computer Vision (WACV) (2016)
21. Peng, K.C., Chen, T., Sadovnik, A., Gallagher, A.C.: A mixed bag of emotions: model, predict, and transfer emotion distributions. In: CVPR (2015)
22. Porzi, L., Rota Bulò, S., Lepri, B., Ricci, E.: Predicting and understanding urban perception with convolutional neural networks. In: ACM Multimedia (2015)
23. Rossky, P., Doll, J., Friedman, H.: Brownian dynamics as smart Monte Carlo simulation. J. Chem. Phys. **69**(10), 4628–4633 (1978)
24. Sartori, A., Yanulevskaya, V., Salah, A.A., Uijlings, J., Bruni, E., Sebe, N.: Affective analysis of professional and amateur abstract paintings using statistical analysis and art theory. ACM Trans. Interact. Intell. Syst. **5**(2), 8 (2015)
25. Siarohin, A., Zen, G., Majtanovic, C., Alameda-Pineda, X., Ricci, E., Sebe, N.: How to make an image more memorable?: a deep style transfer approach. In: ACM ICMR (2017)
26. Simonyan, K., Zisserman, A.: Very deep convolutional networks for large-scale image recognition. arXiv preprint arXiv:1409.1556 (2014)
27. Szegedy, C., Vanhoucke, V., Ioffe, S., Shlens, J., Wojna, Z.: Rethinking the inception architecture for computer vision. In: CVPR (2016)
28. Tsai, Y.H., Shen, X., Lin, Z., Sunkavalli, K., Lu, X., Yang, M.H.: Deep image harmonization. In: CVPR (2017)
29. Ulyanov, D., Lebedev, V., Vedaldi, A., Lempitsky, V.S.: Texture networks: feedforward synthesis of textures and stylized images. In: ICML (2016)
30. Wang, W., Shen, J.: Deep cropping via attention box prediction and aesthetics assessment. In: ICCV (2017)
31. Wilber, M.J., Fang, C., Jin, H., Hertzmann, A., Collomosse, J., Belongie, S.: Bam! the behance artistic media dataset for recognition beyond photography. In: ICCV (2017)
32. Zhou, B., Lapedriza, A., Xiao, J., Torralba, A., Oliva, A.: Learning deep features for scene recognition using places database. In: NIPS (2014)

Deep Convolutional Compressed Sensing for LiDAR Depth Completion

Nathaniel Chodosh[✉], Chaoyang Wang[✉], and Simon Lucey[✉]

Carnegie-Mellon University, Pittsburgh, PA 15213, USA
{nchodosh,chaoyanw,slucey}@andrew.cmu.edu

Abstract. In this paper we consider the problem of estimating a dense depth map from a set of sparse LiDAR points. We use techniques from compressed sensing and the recently developed Alternating Direction Neural Networks (ADNNs) to create a deep network which performs multi-layer convolutional compressed sensing. Our architecture internally performs the optimization for extracting convolutional sparse codes from the input which are then used to make a prediction. Our results demonstrate that with only three layers and 1800 parameters we achieve performance which is competitive with the state of the art, including deep networks with orders of magnitude more parameters and layers.

Keywords: Depth completion · Super LiDAR ·
Convolutional sparse coding

1   Introduction

In recent years 3D information has become an important component of robotic sensing. Usually, this information is presented in 2.5D as a depth map, either measured directly using LiDAR or computed using stereo correspondence. Since LiDAR and stereo techniques yield few samples relative to modern image sensors, it has become desirable to convert sparse depth measurements into high resolution depth maps as shown in Fig. 1.

Recent works [13,21] have directly applied deep networks to depth completion from sparse measurements. However, common network architectures have two drawbacks when applied to this task: (1) They implicitly pose depth completion as finding a mapping from sparse depth maps to dense ones, instead of as finding a depth map that is consistent with the sparse input. This essentially throws away information and we observe that feed forward networks do not learn to propagate the input points through to the output. Qualitative evidence of this can be seen in Fig. 1. (2) Common networks are sensitive to the sparsity of the input since they treat all pixels equally, regardless of whether or not they represent samples or missing input. Special CNN networks have been designed to address this problem, but they still do not express the constraints given by the input [21]. In this paper we address both of these issues by using Alternating Direction Neural Networks to create a novel deep architecture capable of internally optimizing its prediction with respect to a compressed sensing (CS) objective.

© Springer Nature Switzerland AG 2019
C. V. Jawahar et al. (Eds.): ACCV 2018, LNCS 11361, pp. 499–513, 2019.
https://doi.org/10.1007/978-3-030-20887-5_31

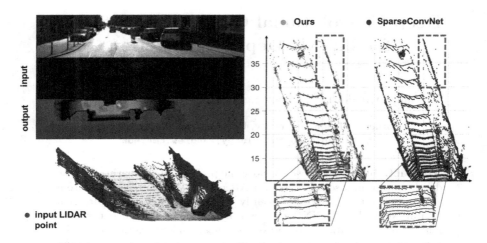

Fig. 1. The bicyclist and bollards can barely be seen in the input map but are clearly represented in the output. Our method also accurately reconstructs very thin objects such as the sign post. On the right it can be seen that our method enforces that its prediction should match the input points while the SparseConvNet systematically underestimates the depth.

Formally, CS is concerned with recovering signals from incomplete measurements by enforcing that signals be sparse when measured in an appropriate over-complete basis. CS provides a strong theoretical foundation for this task but has fallen out of fashion since the existing algorithms have difficult to interpret hyper-parameters, and often do not achieve good performance without careful tuning of these parameters. Our method avoids these issues by learning the parameters jointly with the overcomplete-basis.

The choice of dictionary is crucial for recovering the signal efficiently, especially when the dimensionality is high. For high resolution imagery data, such as depth maps, multi-layer convolutional sparse coding (CSC) [17] is effective as it explicitly models local interactions through the convolution operator with tractable computational and model complexity. However, none of the existing multi-layer convolutional sparse coding algorithms are designed for learning from sparse ground truth data. This is reflected by the fact that recent works [7,12] applying CS to depth completion are restricted to using single-level, hand crafted dictionaries.

Modern work in the formal analysis of deep learning has shown that convolutional neural networks and convolutional sparse coding are closely related. Specifically it has been shown that CNNs with ReLU activation functions are carrying out a specific form of the layered thresholding algorithm for CSC. Layered thresholding is a simple algorithm for solving multi-layered convolutional sparse coding (ML-CSC) problems, which can be effective when there is little noise and the coherence of the dictionary is high. Motivated by the work of Murdock *et al.* [15], in this paper we propose a network architecture which encodes a more sophisticated algorithm for ML-CSC. Encoding the ML-CSC objective in a deep network allows us to learn the dictionaries and parameters together in an end to end fashion. We show that by better approximating this objective, we can achieve competitive performance on the KITTI depth completion benchmark while using far fewer parameters and layers. Furthermore, this work builds on the Alternating Direction Neural Network (ADNN) framework of Murdock *et al.* which gives theoretical insight into deep learning, allows for encoding complex constraints on predictions and we believe is a promising new area of research.

To summarize, the main contributions of this paper are:

1. We frame an end-to-end multi-layer dictionary learning algorithm as a neural network. This allows us to effectively learn dictionaries and hyper-parameters from a large dataset. In comparison, existing CS algorithms either use hand crafted dictionaries, separately learned multi-level dictionaries, or are inapplicable to incomplete training data [20], as is our case.
2. Our method allows for explicit encoding of the constraints from the input sparse depth. Current deep learning approaches [13] simply feed in a sparse depth map and rely solely on data to teach the network to identify which inputs represent missing data. Some recent models [21] explicitly include masks to achieve sparsity invariance, but none have a guaranteed way of encoding that the input is a noise corrupted subset of the desired output. In contrast our method directly optimizes the predicted map with respect to the input.
3. Our method demonstrates performance comparable to the best performing models with much fewer parameters compared to deep networks. In fact, using only two layers of dictionaries and 1600 parameters, our method already substantially outperforms some modern deep networks which use more than 20 layers and over 3 million parameters [13]. As a result of having fewer parameters, our approach trains faster and requires less data.

Even though the experimental portion of this paper focuses on the problem of LiDAR depth completion - we believe our paper is of broader theoretical appeal to the vision community. Specifically, we demonstrate how the classical problem of sparse signal reconstruction can be re-interpreted through the modern lens of deep learning. In particular, we demonstrate how much of the traditional guess work associated with compressed learning - can be removed - while still leaving the firm theoretical foundations upon which compressed sensing has been built upon over the last few decades.

2 Related Work

Since our proposed method is a fusion of deep learning and compressed sensing, we will review in this section previous work that has used either technique for depth estimation.

2.1 Compressed Sensing

Compressed sensing is a technique in signal processing for recovering signals from a small set of measurements. Naturally it has been applied to depth completion in previous work, but has been limited to single-level hand-crafted dictionaries. The earliest is Hawe *et al.* [7], who show that disparity maps can be represented sparsely using the wavelet basis. Liu *et al.* [12] built on that by combining wavelets with contourlets and investigated the effect of different sampling patterns. Both methods were out performed by Ma & Karaman [14] who exploit the simple structures of man-made indoor scenes to achieve full depth reconstruction. In contrast to all of these works, our approach learns multi-level convolutional dictionaries from a large dataset of incomplete ground truth depth maps.

2.2 Deep Learning

Previous work such as [23] has formulated an specialized optimization algorithm as a deep neural network, but none have been applied to the task of depth estimation. Depth estimation using deep learning has largely been restricted to single-shot, RGB to depth prediction. This line of inquiry started with Eigen *et al.* [3] who showed that a deep network could reasonably estimate depth using only an RGB image. Many variants of this method have since been explored [6,9,11]. Lania *et al.* [10] introduced up-projection blocks which allowed for very deep networks and several other works have proposed variants of their architecture. The most relevant of these variants is the Sparse-to-Dense network of Ma and Karaman [13], which they also apply to depth completion from LiDAR points. Uhrig *et al.* [21] introduced the KITTI depth completion dataset, and showed that CNNs which explicitly encode the sparsity of the input achieve much better performance. Riegler *et al.* [18] designed ATGV-Net, a deep network for depth map super resolution, but they assume a rectangular grid of inputs so it is not applicable to LiDAR completion. At the time of submission there are many deep learning models listed on the KITTI depth completion which outperform our method, however at the time of writing none of them have been published. Therefore we will use the published methods of Ma & Karaman and Urhig *et al.* as our baseline comparisons.

Notations. We define our notations throughout as follows: lowercase boldface symbols (*e.g.* $\mathbf{x}$) denote vectors, uppercase boldface symbols (*e.g.* $\mathbf{W}$) denote matrices;

3 Preliminary

3.1 Compressed Sensing

Compressed sensing concerns the problem of recovering a signal from a small set of measurements. In our case, we're interested in reconstructing the depth map $\mathbf{d}$ with full resolution from the sparse depth map $\mathbf{d}_s$ produced by LiDAR. To achieve this, certain prior knowledge of the signal is required. The most widely used prior assumption is that the signal can be reconstructed with a sparse linear combination of basis elements from an over-complete dictionary $\mathbf{W}$. This gives an optimization problem similar to sparse coding:

$$\min_{\mathbf{z}} \|\mathbf{MWz} - \mathbf{d}_s\| + b\,\|\mathbf{z}\|_0, \tag{1}$$

where $\mathbf{z}$ is the code, $\mathbf{Wz}$ produces our predicted depth map, and $\mathbf{M}$ is a diagonal matrix with 0 and 1s on its diagonal. It's used to mask out the unmeasured portions of the signal, such that the reconstruction error is only applied to the pixels which have been measured.

The key question to apply CS in Eq. 1 is: (1) For high dimensional signals such as the depth map, how to design the dictionary such that it encourages uniqueness of the code while still being computationally feasible; (2) How to learn the dictionary to get best reconstruction accuracy. In Sect. 4, we are going to show that the dictionary can be factored into a structure equivalent to performing multi-layer convolution, and that we can unroll the optimization of Eq. 1 into a network similar to a deep recurrent neural network. This allows us to learn the dictionary together with other hyper-parameters (*e.g.* b) through end-to-end training.

3.2 Deep Component Analysis

Equation (1) can be generalized to multi-layered sparse coding in which one seeks a very high level sparse representation $\mathbf{z}_\ell$ such that $\mathbf{d} = \mathbf{W}_1\mathbf{W}_2\ldots\mathbf{W}_{\ell-1}\mathbf{z}_\ell$ and each intermediate product $\mathbf{z}_i = \mathbf{W}_i\mathbf{W}_{i+1}\ldots\mathbf{W}_{\ell-1}\mathbf{z}_\ell$ is also sparse. This formulation makes using a large effective dictionary computationally tractable, and when the dictionaries have a convolutional structure it allows for increased receptive fields while keeping the number of parameters manageable. This is further generalized to Deep Component Analysis (DeepCA) by the recent work of Murdock *et al.* which replaces the ℓ_0 loss with arbitrary sparsity-encouraging penalties. The DeepCA objective function is stated in [15] as:

$$\min_{\{\mathbf{z}_i\}} \sum_{i=1}^{\ell} \frac{1}{2} \|\mathbf{z}_{i-1} - \mathbf{W}_i\mathbf{z}_i\|_2^2 + \Phi_i(\mathbf{z}_i), \tag{2}$$

where the Φ_j are sparsity encouraging regularizers. Previous work has shown that the specific choice of $\Phi(\mathbf{x}) = I(\mathbf{x} > 0) + b\,\|\mathbf{x}\|_1$ yields optimization algorithms very similar to a feed-forward neural network with Relu activation functions. By using the ADMM algorithm to solve Eq. 2, Murdock *et al.* create Alternating Direction Neural Networks, a generalization of feed forward neural networks which internally solve optimization problems with the form of (2). Alternating Direction Neural Networks (ADNNs) perform the optimization in a fully differentiable manner and cast the activation functions of each layer as the proximal operators of penalty function Φ_i of that layer. This allows for learning the dictionaries W_i and parameters b through gradient descent and back propagation with respect to an arbitrary loss function on the sparse codes. To mirror neural networks, Murdock *et al.* apply various loss functions to the highest level of codes, which take the place of the output layer in traditional NNs. In the following sections we will show how ADNNs can be adapted to the depth completion problem within the framework of compressed sensing.

4 Deep Convolutional Compressed Sensing

4.1 Inference

Directly applying compressed sensing to the DeepCA objective gives

$$\min_{\{\mathbf{z}_i\}} \frac{1}{2} \|\mathbf{d}_s - \mathbf{M}\mathbf{W}_1\mathbf{z}_1\|_2^2 + \sum_{i=2}^{\ell} \frac{1}{2} \|\mathbf{z}_{i-1} - \mathbf{W}_i\mathbf{z}_i\|_2^2 + \sum_{i=1}^{\ell} \Phi_i(\mathbf{z}_i), \qquad (3)$$

where $\mathbf{d}_s$ is the input sparse depth map. However, if we take the $\mathbf{W}_i$ to have a convolutional structure then an element $\mathbf{z}_1$ will not be recovered if its spatial support contains no valid depth samples. Thus, extracting the higher level codes is itself a missing data problem and can be written the same way. This gives the full Deep Convolutional Compressed Sensing objective:

$$\min_{\{\mathbf{z}_i | i > 0\}} \sum_{i=1}^{\ell} \frac{1}{2} \|\mathbf{M}_{i-1}\mathbf{z}_{i-1} - \mathbf{M}_{i-1}\mathbf{W}_i\mathbf{z}_i\|_2^2 + \Phi_i(\mathbf{z}_i). \qquad (4)$$

Here, to simplify notation, we merge the depth reconstruction cost (left term in Eq. 3) and the reconstruction cost of the codes together, with $\mathbf{z}_0 = \mathbf{d}_s$ and $\mathbf{M}_0$ denotes the mask $\mathbf{M}$ used in (3). Each $\mathbf{M}_i$ is a mask encoding which elements of $\mathbf{z}_i$ had any valid inputs in their spatial support. In practice computing $\mathbf{M}_i$ is done with a maxpooling operation with the same stride and kernel size as the convolution represented by $\mathbf{W}_{i+1}^T$.

We solve (4) using the ADMM algorithm, which introduces auxiliary variables $\mathbf{y}_i$ that we constrain to be equal to the codes $\mathbf{z}_i$ as below:

$$\min_{\{\mathbf{y}_i, \mathbf{z}_i | i > 0\}} \sum_{i=1}^{\ell} \frac{1}{2} \|\mathbf{M}_{i-1}\mathbf{y}_{i-1} - \mathbf{M}_{i-1}\mathbf{W}_i\mathbf{z}_i\|_2^2 + \Phi_i(\mathbf{y}_i) \qquad (5)$$
$$\text{s.t.} \quad \mathbf{z}_i = \mathbf{y}_i.$$

Here, we again refer the input sparse depth $\mathbf{d}_s$ as $\mathbf{y}_0$. With this, the augmented Lagrangian of (5) with dual variables $\boldsymbol{\lambda}$ and a quadratic penalty weight ρ is:

$$L_\rho(\mathbf{z}, \mathbf{y}, \boldsymbol{\lambda}) = \sum_{i=1}^{\ell} \frac{1}{2} \|\mathbf{M}_{i-1}\mathbf{y}_{i-1} - \mathbf{M}_{i-1}\mathbf{W}_i\mathbf{z}_i\|_2^2 + \Phi_i(\mathbf{y}_i) + \boldsymbol{\lambda}_i^T(\mathbf{z}_i - \mathbf{y}_i) + \frac{\rho}{2} \|\mathbf{z}_i - \mathbf{y}_i\|_2^2. \quad (6)$$

The ADMM algorithm then minimizes L_ρ over each variable in turn, while keeping all others fixed. Following Murdock *et al.* we will incrementally update each layer instead of first solving for all $\mathbf{z}_i$ followed by all $\mathbf{y}_i$. They show this order leads to faster convergence. The ADMM updates for each variable are as follows:

1. At each iteration $t+1$, $\mathbf{z}_i$ is first updated by minimizing L_ρ with the associated auxiliary variable $\mathbf{y}_i$ from the previous iteration, and $\mathbf{z}_{i-1}$ from the current iteration fixed:

$$\mathbf{z}_i^{[t+1]} = \operatorname*{argmin}_{\mathbf{z}_i} L_\rho(\mathbf{z}_i, \mathbf{y}_{i-1}^{[t+1]}, \mathbf{y}_i^{[t]}, \boldsymbol{\lambda}_i^{[t]})$$

$$= (\mathbf{W}_i^T\mathbf{M}_{i-1}^T\mathbf{M}_{i-1}\mathbf{W}_i + \rho\mathbf{I})^{-1}(\mathbf{W}_i^T\mathbf{M}_{i-1}^T\mathbf{M}_{i-1}\mathbf{W}\mathbf{y}_{i-1}^{[t+1]} + \rho\mathbf{y}_i^{[t]} - \boldsymbol{\lambda}_i^{[t]}). \qquad (7)$$

This gives a fully differentiable update of $\mathbf{z}_i$ but the matrix inversion is computationally expensive, especially since W_i is in practice very large. To deal with this problem we make the approximation that $\mathbf{W}_i$ is a Parseval tight frame [15], that is we assume $\mathbf{W}_i \mathbf{W}_i^T = \mathbf{I}$. In addition to being common practice in autoencoders with tied weights, this assumption is also made by Murdock $et\ al.$ and has previously been explicitly enforced in deep neural networks [2]. We can then use the binomial matrix identity to rewrite the $\mathbf{z}_i$ update as:

$$\mathbf{z}^{[t+1]} = \tilde{\mathbf{y}}_i^{[t]} + \frac{1}{1+\rho} \mathbf{W}_i^T \mathbf{M}_{i-1}^T (\mathbf{M}_{i-1} \mathbf{y}_{i-1}^{[t+1]} - \mathbf{M}_{i-1} \mathbf{W}_i \tilde{\mathbf{y}}_i^{[t]}), \tag{8}$$

where $\tilde{\mathbf{y}}_i^{[t]} \triangleq \mathbf{y}_i^{[t]} - \frac{1}{\rho}$.

2. Similarly, the update rule for the auxiliary variables $\mathbf{y}_i$ is:

$$\mathbf{y}_i^{[t+1]} = \underset{\mathbf{y}_i}{\text{argmin}}\ L_\rho(\mathbf{z}_i^{[t+1]}, \mathbf{z}_{i+1}^{[t]}, \mathbf{y}_i, \boldsymbol{\lambda}_i^{[t]})$$

$$= \phi_i \left(\frac{1}{1+\rho} \mathbf{W}_{i+1} \mathbf{z}_{i+1}^{[t]} + \frac{\rho}{1+\rho} (\mathbf{z}_i^{[t+1]} + \frac{\boldsymbol{\lambda}_i^{[t]}}{\rho}) \right) \tag{9}$$

$$\mathbf{y}_\ell = \phi_i \left(\mathbf{z}_i^{[t]} + \frac{\boldsymbol{\lambda}_i^{[t]}}{\rho} \right).$$

Here ϕ_i is the proximal operator associated with the penalty function Φ_i. For appropriate choices of Φ_i, ϕ_i is differentiable and can be computed efficiently. With this in mind, we choose $\Phi_i(\mathbf{x}) = I(\mathbf{x} > 0) + b\|\mathbf{x}\|_1$ so that $\phi_i(\mathbf{x}) = \text{ReLU}(\mathbf{x} - \frac{b}{\rho})$.

3. Finally the dual variable $\boldsymbol{\lambda}_i$ is updated by:

$$\boldsymbol{\lambda}_i^{[t+1]} = \boldsymbol{\lambda}_i^{[t]} + \rho(\mathbf{z}_i^{[t+1]} - \mathbf{y}_i^{[t+1]}). \tag{10}$$

The full procedure is detailed in algorithm (1). As shown in above, all the operations used in the ADMM iteration are differentiable, and can be implemented with deep learning layers $e.g.$ convolution, convolution transpose, and ReLU. We unroll the ADMM iteration for a constant number of iterations T, and output our optimized code $\mathbf{z}_\ell$ for the last layer. We can then extract our prediction of the depth map by applying the effective dictionary to the high level code z_ℓ as shown in equation (11). This is different from the standard decoder portion of a deep autoencoder, where the nonlinear activations are applied in between each convolution. Our approach does not require this since the internal optimization of z_ℓ enforces equality constraints between layers, which is not the case for conventional autoencoders. We choose to reconstruct the depth from z_ℓ instead of a lower layer because its elements have the largest receptive field and therefore z_ℓ will have the fewest number of missing entries.

$$\mathbf{d}_{\text{pred}} = \mathbf{W}_1 \mathbf{W}_2 \ldots \mathbf{W}_\ell \mathbf{z}_\ell \tag{11}$$

Algorithm 1. Deep Convolutional Compressed Sensing

Input : model parameters $\mathbf{W}_i, b_i$, iterations T, sparse depth $\mathbf{y}_0 = \mathbf{d}_s$, mask $\mathbf{M}$

Output: sparse codes $\mathbf{z}_i^{[T]}$, predicted depth $\mathbf{d}_{\text{pred}}$

for $i \leftarrow 1$ **to** ℓ **do**

> $\mathbf{z}_i^{[0]} \leftarrow \mathbf{W}_i^T \mathbf{M}_{i-1}^T \mathbf{y}_{i-1}$;
>
> $\mathbf{y}_i^{[0]} \leftarrow ReLU(\mathbf{z}_i^{[0]} - b/\rho)$;
>
> $\boldsymbol{\lambda}_i^{[0]} \leftarrow 0$;

for $t \leftarrow 1$ **to** T **do**

> **for** $i \leftarrow 1$ **to** ℓ **do**
>
> > Update $\mathbf{z}_i^{[t]}$ using equation (8);
> >
> > Update $\mathbf{y}_i^{[t]}$ using equation (9);
> >
> > Update $\boldsymbol{\lambda}_i^{[t]}$ using equation (10);
>
> Predict $\mathbf{d}_{\text{pred}}$ using equation (11);

4.2   Learning

With the ADMM update unrolled to T iterations as described above, the entire inference procedure can be thought of as a single differentiable function:

$$\mathbf{d}_{\text{pred}} = f_{\text{DCCS}}^{[T]}(\mathbf{M}, \mathbf{d}_s; \{\mathbf{W}_i, b_i\}) \tag{12}$$

Thus the dictionaries $\mathbf{W}_i$ and the bias term b_i which are the parameters for $f_{\text{DCCS}}^{[T]}$ can be learned through stochastic gradient descent over a suitable loss function. Using the standard sum of squared loss error, dictionary learning is formed as minimizing the depth reconstruction error $\mathcal{L}_{\text{reconstruct}}$:

$$\min_{\{\mathbf{W}_i, b_i\}} \sum_{n=1}^{N} \left\| \mathbf{d}_{\text{gt}}^{(n)} - \mathbf{M}'^{(n)} \mathbf{d}_{\text{pred}}^{(n)}) \right\| \tag{13}$$

Where $\mathbf{d}_{\text{gt}}^{(n)}$ is the ground truth depth map of the n_{th} training example. We allow the ground truth depth map to have missing value by using mask $\mathbf{M}'^{(n)}$ to segment out the invalid pixels in the ground truth depth map.

In practice we found that due to the sparsity of the training data, the depth maps our method predicted were rather noisy. To fix this issue we included the well known anisotropic total variation loss (TV-L1) when training to encourage smoothness of the predicted depth map. The TV loss is given by

$$\mathcal{L}_{\text{TV-L1}} = \sum_{x,y} |I(x+1, y) - I(x, y)| + |I(x, y+1) - I(x, y)|$$

where I is the input depth image. Note that this change has no significant impact on the quantitative error metrics, but produces more visually pleasing outputs. The total loss is then given by summation of the depth reconstruction loss and the TV-L1 smoothness loss, with hyper-parameter α to control the weighting for the smoothness penalty:

$$\mathcal{L} = \mathcal{L}_{reconstruct} + \alpha \mathcal{L}_{\text{TV-L1}}. \tag{14}$$

We empirically determined that $\alpha = 0.1$ produces the best results.

5   Experiments

5.1   Implementation Details

We implemented three variants of algorithm (1) for the cases $\ell = 1, 2, 3$. For the single layer case we let $\mathbf{W}_1^T$ be a 11×11 convolution with striding of 2 and 8 filters. For $\ell = 2$ we let $\mathbf{W}_1^T$ be an 11×11 convolution with 8 filters and W_1^T be a 7×7 convolution with 16 filters. Finally for the $\ell = 3$ case: $\mathbf{W}_1^T$ is an 11×11 convolution with 8 filters, $\mathbf{W}_2^T$ is a 5×5 convolution with 16 filters, and $\mathbf{W}_3^T$ is a 3×3 convolution with 32 filters. For both $\ell = 2$ and $\ell = 3$, all convolutions have striding of 2. For the single layer case we learned the dictionaries with the number of iterations set to 5 and then at test time increased the number of iterations to 20. For the two and three layer cases the number of iterations was fixed at train and test time to 10 except in Sect. 5.4 where the number of test and training iterations is varied. All training was done with the ADAM optimizer with the standard parameters: learning rate $= 0.001$, $\beta_1 = 0.9$, $\beta_2 = 0.999$, $\epsilon = 10^{-8}$.

Error Metrics. For evaluation on the KITTI benchmark we use the conventional error metrics [5, 21], e.g. root mean square error (RMSE), mean absolute error (MAE), mean absolute relative error (MRE). We also use the percentage of inliers metric, δ_i which counts the percent of predictions whose relative error is within a threshold raised to the power i. Here, we use smaller thresholds (1.01^i) compared to the more widely used ones (1.5^i) in order to compare differences in performance under tighter metrics.

5.2   KITTI Depth Completion Benchmark

We evaluate our method on the new KITTI Depth Completion Benchmark [21] instead of directly comparing against the LiDAR measurements from the raw KITTI dataset. The raw LiDAR points given in KITTI are corrupted by noise, motion of the vehicle during sampling, image rectification artifacts, and accounts to only 4% of the total number of pixels in the image. Thus it's not ideal for evaluating depth completion systems. Instead, the benchmark proposed in [21] resolved these issues by accumulating LiDAR measurements from nearby frames in the video sequences, and automatically removing accumulated LiDAR points that deviate too far from the points reconstructed by semi-global matching. This provides quality ground truth and effectively simulates the main application of interest: recovering dense depth from a single LiDAR sweep.

Table 1. Validation error of various methods on the KITTI Depth Completion benchmark. All results except for SparseConvNet and Ma's are taken as reported from [21]. Our method outperforms all previous depth only completion methods (Middle) as well as those that use RGB images for guidance (Top). Test set RMSE and MAE were also obtained from the benchmark website: 1.32 and 0.44 for our 3 layer method, vs 1.60 and 0.48 for the SparseConvNet.

	RMSE (m)	MAE (m)	MRE	$\delta_1 < 1.01$	$\delta_2 < 1.01^2$	$\delta_3 < 1.01^3$
Bilateral NN [8]	4.19	1.09	-	-	-	-
SGDU [19]	2.5	0.72	-	-	-	-
Fast Bilateral Solver [1]	1.98	0.65	-	-	-	-
TGVL [4]	4.85	0.59	-	-	-	-
Closest Depth Pooling	2.77	0.94	-	-	-	-
Nadaraya Watson [16,22]	2.99	0.74	-	-	-	-
ConvNet	2.97	0.78	-	-	-	-
ConvNet + mask	2.24	0.79	-	-	-	-
SparseConvNet [21]	1.82	0.58	0.035	0.33	0.65	0.82
Ma and Karaman [13]	1.68	0.70	0.039	0.21	0.41	0.59
Ours 1 Layer	2.77	0.83	0.054	0.3	0.47	0.59
Ours 2 Layers	1.45	0.47	0.028	0.41	0.68	0.8
Ours 3 Layers	**1.35**	**0.43**	**0.024**	**0.48**	**0.73**	**0.83**

In Table 1, we form a close comparison against the very deep Sparse-to-Dense network (Ma & Karaman [13]) and the Sparsity Invariant CNN (SparseConvNet [21]) which are the best published methods for this task.

The Sparse-to-Dense network uses a similar deep network architecture as those used for single shot depth prediction – with Resnet-18 as the encoder and up-projection blocks for the decoder. While the Sparse-to-Dense network is able to achieve good RMSE, it falls behind the SparseConvNet on MAE. We believe that this is because the deeper network can better estimate the average depth of a region but is unable to predict fine detail, leading to a higher MAE. By comparison, our method is able to both estimate the correct average depth and reconstruct fine detail due to its ability to directly optimize the prediction with respect to the input. Most notably our method outperforms all of the existing methods by a wide margin, including those that use RGB images and those that use orders of magnitude more parameters than our method.

Varying Sparsity Levels. Uhrig *et al.* show that their Sparsity Invariant CNNs are very robust to a mismatch between the training and testing levels of sparsity. While we do not see a practical use for disparities as large as those tested in [21], we do believe that depth completion systems should perform well under reasonable sparsity changes. To this end we adjusted the level of input sparsity in the KITTI benchmark by dropping input samples with probability p, for various values of p. The results of this experiment are shown in Fig. 2.

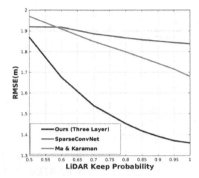 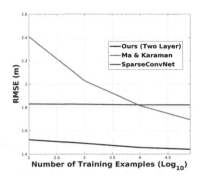

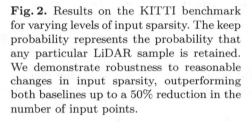

Fig. 2. Results on the KITTI benchmark for varying levels of input sparsity. The keep probability represents the probability that any particular LiDAR sample is retained. We demonstrate robustness to reasonable changes in input sparsity, outperforming both baselines up to a 50% reduction in the number of input points.

Fig. 3. Results of selected methods on the KITTI benchmarks for varying training set sizes. Our method performs well with training sizes ranging from 100-86k but still benefits from larger training sizes.

While it is clear that our method does not achieve the level of sparsity invariance of the SparseConvNet, it still outperforms both baseline results even when the only 50% of the input samples are kept.

5.3 Effect of Amount of Training Data

Modern deep learning models typically have tens of thousands to millions of parameters and therefore require enormous training sets to achieve good performance. This is in fact the motivation for the KITTI depth completion dataset, since previous benchmarks did not have enough data to train deep networks. In this section we investigate the dependence on the amount of training data on the performance of our method in comparison with a standard deep network and the sparsity invariant variety.

Figure 3 shows the results of evaluating these models on the 1k manually selected validation depth maps after training on varying subsets of the 86k training maps. Our method outperforms both baselines for all training sizes. As expected Ma & Karaman's method fails to generalize well when trained on a small dataset since the model has 3.4M parameters but performs well once trained on the full dataset. It is interesting to observe that the method of Uhrig *et al.* does not gain any performance from training on more data. As a result it is ultimately out performed by the deep network which does not take sparsity into account. Our method is able to perform comparably to the sparsity invariant network with only 100 training examples but does increase in performance when

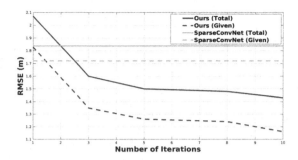

Fig. 4. Results on the depth completion benchmark for different numbers of ADMM iterations. The total error is shown in blue while the red line shows the error on just those points given as input. The dotted lines show the same metrics but for the SparseConvNet of Uhrig *et al.* [21].

given more data, validating the need for learning layered sparse coding dictionaries from large training sets. Additionally we find that even when training on the full dataset our method converges in fewer epochs than the Sparse-to-Dense Model (9 and 16 for our 2 and 3 layer models, vs 20 for the deep network).

5.4   Effect of Iterative Optimization

In this section we demonstrate that the success of our approach comes from its ability to refine depth estimates over multiple iterations. Applying a feed forward neural network to this problem frames it as finding a mapping from sparse LiDAR points to true depth maps. This is a reasonable approach but it doesn't utilize all of the available information, specifically it doesn't encode the relationship that input samples are a subset of the output that has been corrupted by noise. In contrast, our approach of phrasing depth completion as a compressed sensing missing data problem directly expresses that relationship. By solving this problem in an iterative fashion our network that is able to find depth maps that are both consistent with the input constraints and have sparse representations.

The importance of iterative optimization is shown in Fig. 4 where we examine the performance of our method as a function of the number of ADMM iterations it uses. It is clear that with few iterations our network fails to enforce the constraints and performs comparably to the SparseConvNet. This is also consistent with Murdock *et al.* observation that a feed forward network resembles a single iteration of an ADNN. As we increase the number of iterations our method is able to better optimize its prediction and gains a substantial performance boost.

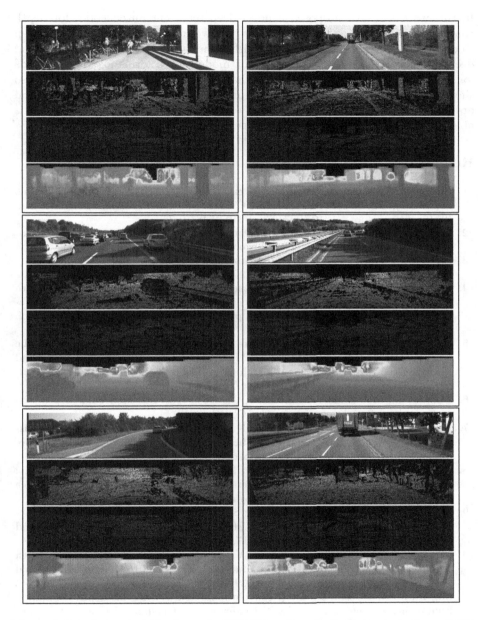

Fig. 5. Selected visual results form the KITTI benchmark. From top to bottom: RGB Image, Ground truth, input LiDAR points, Predicted depth.

6   Conclusion

In this work we have proposed a novel deep recurrent autoencoder for depth completion. Our architecture builds on the work of Murdock *et al.* on Deep Component Analysis and further establishes the link between sparse coding and deep learning. We demonstrate that our model outperforms existing methods for depth completion, including those that leverage RGB information. We also show that the success of our method is fundamentally a product of the internal optimization it performs, and that due to its small number of parameters it is able to perform well even without a large training set (Fig. 5).

References

1. Barron, J.T., Poole, B.: The fast bilateral solver. In: Leibe, B., Matas, J., Sebe, N., Welling, M. (eds.) ECCV 2016. LNCS, vol. 9907, pp. 617–632. Springer, Cham (2016). https://doi.org/10.1007/978-3-319-46487-9_38
2. Cissé, M., Bojanowski, P., Grave, E., Dauphin, Y., Usunier, N.: Parseval networks: improving robustness to adversarial examples. In: ICML (2017)
3. Eigen, D., Puhrsch, C., Fergus, R.: Depth map prediction from a single image using a multi-scale deep network. In: NIPS (2014)
4. Ferstl, D., Reinbacher, C., Ranftl, R., Ruether, M., Bischof, H.: Image guided depth upsampling using anisotropic total generalized variation. In: 2013 IEEE International Conference on Computer Vision, pp. 993–1000, December 2013. https://doi.org/10.1109/ICCV.2013.127
5. Geiger, A., Lenz, P., Urtasun, R.: Are we ready for autonomous driving? The KITTI vision benchmark suite. In: 2012 IEEE Conference on Computer Vision and Pattern Recognition, pp. 3354–3361, June 2012. https://doi.org/10.1109/CVPR.2012.6248074
6. Godard, C., Mac Aodha, O., Brostow, G.J.: Unsupervised monocular depth estimation with left-right consistency. In: CVPR (2017)
7. Hawe, S., Kleinsteuber, M., Diepold, K.: Dense disparity maps from sparse disparity measurements. In: 2011 International Conference on Computer Vision, pp. 2126–2133, November 2011. https://doi.org/10.1109/ICCV.2011.6126488
8. Jampani, V., Kiefel, M., Gehler, P.V.: Learning sparse high dimensional filters: image filtering, dense CRFs and bilateral neural networks. In: IEEE Conference on Computer Vision and Pattern Recognition (CVPR), June 2016
9. Kuznietsov, Y., Stuckler, J., Leibe, B.: Semi-supervised deep learning for monocular depth map prediction. In: 2017 IEEE Conference on Computer Vision and Pattern Recognition (CVPR), July 2017. https://doi.org/10.1109/cvpr.2017.238
10. Laina, I., Rupprecht, C., Belagiannis, V., Tombari, F., Navab, N.: Deeper depth prediction with fully convolutional residual networks. In: 2016 Fourth International Conference on 3D Vision (3DV), pp. 239–248. IEEE (2016)
11. Liu, F., Shen, C., Lin, G.: Deep convolutional neural fields for depth estimation from a single image. In: 2015 IEEE Conference on Computer Vision and Pattern Recognition (CVPR), pp. 5162–5170, June 2015. https://doi.org/10.1109/CVPR.2015.7299152
12. Liu, L., Chan, S.H., Nguyen, T.Q.: Depth reconstruction from sparse samples: representation, algorithm, and sampling. IEEE Trans. Image Process. **24**(6), 1983–1996 (2015). https://doi.org/10.1109/TIP.2015.2409551

13. Ma, F., Karaman, S.: Sparse-to-dense: depth prediction from sparse depth samples and a single image. In: 2018 IEEE International Conference on Robotics and Automation (ICRA), pp. 1–8, May 2018. https://doi.org/10.1109/ICRA.2018.8460184

14. Ma, F., Carlone, L., Ayaz, U., Karaman, S.: Sparse sensing for resource-constrained depth reconstruction. In: 2016 IEEE/RSJ International Conference on Intelligent Robots and Systems (IROS), pp. 96–103. IEEE (2016)

15. Murdock, C., Chang, M.-F., Lucey, S.: Deep component analysis via alternating direction neural networks. In: Ferrari, V., Hebert, M., Sminchisescu, C., Weiss, Y. (eds.) ECCV 2018. LNCS, vol. 11219, pp. 851–867. Springer, Cham (2018). https://doi.org/10.1007/978-3-030-01267-0_50

16. Nadaraya, E.: On estimating regression. Theory Probab. Appl. **9**(1), 141–142 (1964). https://doi.org/10.1137/1109020

17. Papyan, V., Romano, Y., Sulam, J., Elad, M.: Theoretical foundations of deep learning via sparse representations: a multilayer sparse model and its connection to convolutional neural networks. IEEE Signal Process. Mag. **35**(4), 72–89 (2018). https://doi.org/10.1109/MSP.2018.2820224

18. Riegler, G., Rüther, M., Bischof, H.: ATGV-Net: accurate depth super-resolution. In: Leibe, B., Matas, J., Sebe, N., Welling, M. (eds.) ECCV 2016. LNCS, vol. 9907, pp. 268–284. Springer, Cham (2016). https://doi.org/10.1007/978-3-319-46487-9_17

19. Schneider, N., Schneider, L., Pinggera, P., Franke, U., Pollefeys, M., Stiller, C.: Semantically guided depth upsampling. In: Rosenhahn, B., Andres, B. (eds.) GCPR 2016. LNCS, vol. 9796, pp. 37–48. Springer, Cham (2016). https://doi.org/10.1007/978-3-319-45886-1_4

20. Sulam, J., Papyan, V., Romano, Y., Elad, M.: Multilayer convolutional sparse modeling: pursuit and dictionary learning. IEEE Trans. Signal Process. **66**(15), 4090–4104 (2018). https://doi.org/10.1109/TSP.2018.2846226

21. Uhrig, J., Schneider, N., Schneider, L., Franke, U., Brox, T., Geiger, A.: Sparsity invariant CNNs. CoRR (2017). http://arxiv.org/abs/1708.06500v2

22. Watson, G.S.: Smooth regression analysis. Sankhyā: Indian J. Stat. Ser. A (1961–2002) **26**(4), 359–372 (1964). http://www.jstor.org/stable/25049340

23. Zheng, S., et al.: Conditional random fields as recurrent neural networks. In: 2015 IEEE International Conference on Computer Vision (ICCV), pp. 1529–1537, December 2015. https://doi.org/10.1109/ICCV.2015.179

Learning for Video Super-Resolution Through HR Optical Flow Estimation

Longguang Wang, Yulan Guo$^{(\boxtimes)}$, Zaiping Lin, Xinpu Deng, and Wei An

National University of Defense Technology, Changsha, China
{wanglongguang15,yulan.guo,linzaiping,dengxinpu,anwei}@nudt.edu.cn

Abstract. Video super-resolution (SR) aims to generate a sequence of high-resolution (HR) frames with plausible and temporally consistent details from their low-resolution (LR) counterparts. The generation of accurate correspondence plays a significant role in video SR. It is demonstrated by traditional video SR methods that simultaneous SR of both images and optical flows can provide accurate correspondences and better SR results. However, LR optical flows are used in existing deep learning based methods for correspondence generation. In this paper, we propose an end-to-end trainable video SR framework to super-resolve both images and optical flows. Specifically, we first propose an optical flow reconstruction network (OFRnet) to infer HR optical flows in a coarse-to-fine manner. Then, motion compensation is performed according to the HR optical flows. Finally, compensated LR inputs are fed to a super-resolution network (SRnet) to generate the SR results. Extensive experiments demonstrate that HR optical flows provide more accurate correspondences than their LR counterparts and improve both accuracy and consistency performance. Comparative results on the Vid4 and DAVIS-10 datasets show that our framework achieves the state-of-the-art performance.

Keywords: Video super-resolution ·
Optical flow reconstruction network · Temporal consistency

1 Introduction

Super-resolution (SR) aims to generate high-resolution (HR) images or videos from their low-resolution (LR) counterparts. As a typical low-level computer vision problem, SR has been widely investigated for decades [1–3]. Recently, the prevalence of high-definition display further advances the development of SR. For single image SR, image details are recovered using the spatial correlation in a single frame. In contrast, inter-frame temporal correlation can further be exploited for video SR.

Since temporal correlation is crucial to video SR, the key to success lies in accurate correspondence generation. Numerous methods [6–8] have demonstrated that the correspondence generation and SR problems are closely interrelated and can boost each other's accuracy. Therefore, these methods integrate

© Springer Nature Switzerland AG 2019
C. V. Jawahar et al. (Eds.): ACCV 2018, LNCS 11361, pp. 514–529, 2019.
https://doi.org/10.1007/978-3-030-20887-5_32

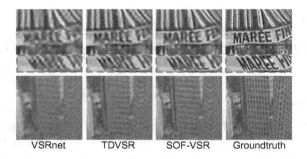

VSRnet   TDVSR   SOF-VSR   Groundtruth

Fig. 1. Temporal profiles under ×4 configuration for VSRnet [4], TDVSR [5] and our SOF-VSR on *Calendar* and *City*. Purple boxes represent corresponding temporal profiles. Our SOF-VSR produces finer details in temporal profiles, which are more consistent with the groundtruth. (Color figure online)

the SR of both images and optical flows in a unified framework. However, current deep learning based methods [4,5,9–12] mainly focus on the SR of images, and use LR optical flows to provide correspondences. Although LR optical flows can provide sub-pixel correspondences in LR images, their limited accuracy hinders the performance improvement for video SR, especially for scenarios with large upscaling factors.

In this paper, we propose an end-to-end trainable video SR framework to generate both HR images and optical flows. The SR of optical flows provides accurate correspondences, which not only improves the accuracy of each HR image, but also achieves better temporal consistency. We first introduce an optical flow reconstruction net (OFRnet) to reconstruct HR optical flows in a coarse-to-fine manner. These HR optical flows are then used to perform motion compensation on LR frames. A space-to-depth transformation is therefore used to bridge the resolution gap between HR optical flows and LR frames. Finally, the compensated LR frames are fed to a super-resolution net (SRnet) to generate each HR frame. Extensive evaluation is conducted to test our framework. Comparison to existing video SR methods shows that our framework achieves the state-of-the-art performance in terms of peak signal-to-noise ratio (PSNR) and structural similarity index (SSIM). Moreover, our framework achieves better temporal consistency for visual perception (as shown in Fig. 1).

Our main contributions can be summarized as follows: (1) We integrate the SR of both images and optical flows into a single SOF-VSR (super-resolving optical flow for video SR) network. The SR of optical flows provides accurate correspondences and improves the overall performance; (2) We propose an OFRnet to infer HR optical flows in a coarse-to-fine manner; (3) Extensive experiments have demonstrated the effectiveness of our framework. It is shown that our framework achieves the state-of-the-art performance.

2   Related Work

In this section, we briefly review some major methods for single image SR and video SR.

2.1   Single Image SR

Dong *et al.* [13] proposed the pioneering work to use deep learning for single image SR. They used a three-layer convolutional neural network (CNN) to approximate the non-linear mapping from the LR image to the HR image. Recently, deeper and more complex network architectures have been proposed [14–16]. Kim *et al.* [14] proposed a very deep super-resolution network (VDSR) with 20 convolutional layers. Tai *et al.* [15] developed a deep recursive residual network (DRRN) and used recursive learning to control the model parameters while increasing the depth. Hui *et al.* [16] proposed an information distillation network to reduce computational complexity and memory consumption.

2.2   Video SR

Traditional Video SR. To handle complex motion patterns in video sequences, Protter *et al.* [17] generalized the non-local means framework to address the SR problem. They used patch-wise spatio-temporal similarity to perform adaptive fusion of multiple frames. Takeda *et al.* [18] further introduced 3D kernel regression to exploit patch-wise spatio-temporal neighboring relationship. However, the resulting HR images of these two methods are over-smoothed. To exploit pixel-wise correspondences, optical flow is used in [6–8]. Since the accuracy of correspondences provided by optical flows in LR images is usually low [19], an iterative framework is used in these methods [6–8] to estimate both HR images and optical flows.

Deep Video SR with Separated Motion Compensation. Recently, deep learning has been investigated for video SR. Liao *et al.* [9] performed motion compensation under different parameter settings to generate an ensemble of SR-drafts, and then employed a CNN to recover high-frequency details from the ensemble. Kappelar *et al.* [4] also performed image alignment through optical flow estimation, and then passed the concatenation of compensated LR inputs to a CNN to reconstruct each HR frame. In these methods, motion compensation is separated from CNN. Therefore, it is difficult for them to obtain the overall optimal solution.

Deep Video SR with Integrated Motion Compensation. More recently, Caballero *et al.* [11] proposed the first end-to-end CNN framework (namely, VESPCN) for video SR. It comprises a motion compensation module and a sub-pixel convolutional layer used in [20]. Since that, end-to-end framework with motion compensation dominates the research of video SR. Tao *et al.* [10] used

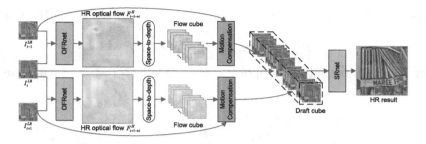

Fig. 2. Overview of the proposed framework. Our framework is fully convolutional and can be trained in an end-to-end manner.

the motion estimation module in VESPCN, and proposed an encode-decoder network based on LSTM. This architecture facilitates the extraction of temporal context. Liu et al. [5] customized ESPCN [20] to simultaneously process different numbers of LR frames. They then introduced a temporal adaptive network to aggregate multiple HR estimates with learned dynamic weights. Sajjadi et al. [21] proposed a frame-recurrent architecture to use previously inferred HR estimates for the SR of subsequent frames. The recurrent architecture can assimilate previous inferred HR frames without increase in computational demands.

It is already demonstrated by traditional video SR methods [6–8] that simultaneous SR of images and optical flows produces better result. However, current CNN-based methods only focus on the SR of images. Different from previous works, we propose an end-to-end video SR framework to super-resolve both images and optical flows. It is demonstrated that the SR of optical flows facilitates our framework to achieve the state-of-the-art performance.

3   Network Architecture

Our framework takes N consecutive LR frames as inputs and super-resolves the central frame. The LR inputs are first divided into pairs and fed to OFRnet to infer an HR optical flow. Then, a space-to-depth transformation [21] is employed to shuffle the HR optical flow into LR grids. Afterwards, motion compensation is performed to generate an LR draft cube. Finally, the draft cube is fed to SRnet to infer the HR frame. The overview of our framework is shown in Fig. 2.

3.1   Optical Flow Reconstruction Net (OFRnet)

It is demonstrated that CNN has the capability to learn the non-linear mapping between LR and HR images for the SR problem [13]. Recent CNN-based works [22,23] have also shown the potential for motion estimation. In this paper, we incorporate these two tasks into a unified network to infer HR optical flows from LR images. Specifically, our OFRnet takes a pair of LR frames I_i^L and I_j^L as

inputs, and reconstruct an optical flow between their corresponding HR frames
I_i^H and I_j^H:

$$F_{i \to j}^H = \mathbf{Net}_{OFR}(I_i^L, I_j^L; \Theta_{OFR}), \qquad (1)$$

where $F_{i \to j}^H$ represents the HR optical flow and Θ_{OFR} is the set of parameters.

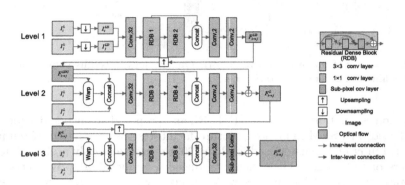

Fig. 3. Architecture of our OFRnet. Our OFRnet works in a coarse-to-fine manner. At each level, the output of its previous level is used to compute a residual optical flow.

Motivated by the pyramid optical flow estimation method in [24], we use a coarse-to-fine approach to handle complex motion patterns (especially large displacements). As illustrated in Fig. 3, a 3-level pyramid is employed in our OFRnet.

Level 1: The pair of LR images I_i^L and I_j^L are downsampled by a factor of 2 to produce I_i^{LD} and I_j^{LD}, which are further concatenated and fed to a feature extraction layer. Then, two residual dense blocks (RDB) [25] with 5 layers and a growth rate of 32 are customized. Within each residual dense block, the first 4 layers are followed by a leaky ReLU using a leakage factor of 0.1, while the last layer performs feature fusion. The residual dense block works in a local residual learning manner with a local skip connection at the end. Once dense features are extracted by the residual dense blocks, they are concatenated and fed to a feature fusion layer. Then, the optical flow $F_{i \to j}^{LD}$ at this level is inferred by the subsequent flow estimation layer.

Level 2: Once the raw optical flow $F_{i \to j}^{LD}$ is obtained from level 1, it is upscaled by a factor of 2. The upscaled flow $F_{i \to j}^{LDU}$ is then used to warp I_i^L, resulting in $\hat{I}_{i \to j}^L$. Next, $\hat{I}_{i \to j}^L$, I_j^L and $F_{i \to j}^{LDU}$ are concatenated and fed to a network module. Note that, this module at level 2 is similar to that at level 1, except that residual learning is used.

Level 3: The module at level 2 generates an optical flow $F_{i \to j}^L$ with the same size as the LR input I_j^L. Therefore, the module at level 3 works as an SR part to infer the HR optical flow. The architecture at level 3 is similar to level 2 except that the flow estimation layer is replaced by a sub-pixel convolutional layer [20] for resolution enhancement.

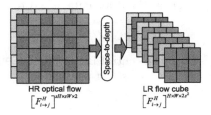

Fig. 4. Illustration of space-to-depth transformation. The space-to-depth transformation folds an HR optical flow in LR space to generate an LR flow cube.

Although numerous networks for SR [15,16,26,27] and optical flow estimation [28–30] can be found in literature, our OFRnet is, to the best of our knowledge, the first unified network to integrate these two tasks. Specifically, our OFRnet learns to infer HR flows between latent HR images from LR inputs. Though some existing approaches can also obtain optical flows of full resolution by performing pre-interpolation on LR inputs [31] or post-interpolation on LR flows [21], the flow estimation is still performed in LR space since interpolation does not introduce additional information for SR problem [20]. Note that, inferring HR optical flow from LR images is quite challenging, our OFRnet has demonstrated the potential of CNN to address this challenge. Our OFRnet is compact, with only 0.6M parameters. It is further demonstrated in Sect. 4.3 that the resulting HR optical flows benefit our video SR framework in both accuracy and consistency performance.

3.2 Motion Compensation

Once HR optical flows are produced by OFRnet, space-to-depth transformation is used to bridge the resolution gap between HR optical flows and LR frames. As illustrated in Fig. 4, regular LR grids are extracted from the HR flow and placed into the channel dimension to derive a flow cube with the same resolution as LR frames:

$$\left[F_{i \to j}^H\right]^{sH \times sW \times 2} \to \left[F_{i \to j}^H\right]^{H \times W \times 2s^2}, \tag{2}$$

where H and W represent the size of the LR frame, s is the upscaling factor. Note that, the magnitude of optical flow is divided by a scalar s during the transformation to match the spatial resolution of LR frames.

Then, slices are extracted from the flow cube to warp the LR frame I_i^{LR}, resulting in multiple warped drafts:

$$C_{i \to j}^L = \mathrm{W}(I_i^L, \left[F_{i \to j}^H\right]^{H \times W \times 2s^2}), \tag{3}$$

where $\mathrm{W}(\cdot)$ denotes warping operation using bilinear interpolation and $C_{i \to j}^L \in R^{H \times W \times s^2}$ represents the warped drafts after concatenation, namely draft cube.

3.3   Super-Resolution Net (SRnet)

After motion compensation, all the drafts are concatenated with the central LR frame, as shown in Fig. 2. Then, the draft cube is fed to SRnet to infer the HR frame:

$$I_0^{SR} = \mathbf{Net}_{SR}(C^L; \Theta_{SR}), \tag{4}$$

where I_0^{SR} is the SR result of the central LR frame, $C^L \in R^{H \times W \times (2s^2+1)}$ represents the draft cube and Θ_{SR} is the set of parameters.

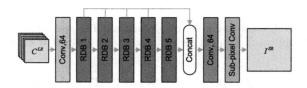

Fig. 5. Architecture of our SRnet.

As illustrated in Fig. 5, the draft cube is first passed to a feature extraction layer with 64 kernels, and then the output features are fed to 5 residual dense blocks (which are similar to our OFRnet). Here, we increase the number of layers to 6 for each residual dense block. Afterwards, we concatenate all the outputs of residual dense blocks and use a feature fusion layer to distillate the dense features. Finally, a sub-pixel layer is used to generate the HR frame.

The combination of densely connected layers and residual learning in residual dense blocks has been demonstrated to have a contiguous memory mechanism [25,32]. Therefore, we employ residual dense blocks in our SRnet to facilitate effective feature learning from preceding and current local features. Furthermore, feature reuse in the residual dense blocks improves the model compactness and stabilizes the training process.

3.4   Loss Function

We design two loss terms $\mathcal{L}_{\text{OFR}}$ and $\mathcal{L}_{\text{SR}}$ for OFRnet and SRnet, respectively. For the training of OFRnet, intermediate supervision is used at each level of the pyramid:

$$\mathcal{L}_{\text{OFR}} = \sum_{i \in [-T, T], i \neq 0} \frac{\mathcal{L}_{level3,i} + \lambda_2 \mathcal{L}_{level2,i} + \lambda_1 \mathcal{L}_{level1,i}}{2T}, \tag{5}$$

where

$$\begin{cases} \mathcal{L}_{level3,i} = \left\| W(I_i^H, F_{i \to 0}^H) - I_0^H \right\|_2^2 + \lambda_3 \left\| \nabla F_{i \to 0}^H \right\|_1 \\ \mathcal{L}_{level2,i} = \left\| W(I_i^L, F_{i \to 0}^L) - I_0^L \right\|_2^2 + \lambda_3 \left\| \nabla F_{i \to 0}^L \right\|_1 \\ \mathcal{L}_{level1,i} = \left\| W(I_i^{LD}, F_{i \to 0}^{LD}) - I_0^{LD} \right\|_2^2 + \lambda_3 \left\| \nabla F_{i \to 0}^{LD} \right\|_1 \end{cases} \tag{6}$$

T denotes the temporal window size and $\left\| \nabla F_{i \to 0}^{H} \right\|_1$ is the regularization term to constrain the smoothness of the optical flow. We empirically set $\lambda_2 = 0.25$ and $\lambda_1 = 0.125$ to make our OFRnet focus on the last level. We also set $\lambda_3 = 0.01$ as the regularization coefficient.

For the training of SRnet, we use the widely applied mean square error (MSE) loss:

$$\mathcal{L}_{\text{SR}} = \left\| I_0^{SR} - I_0^H \right\|_2^2. \tag{7}$$

Finally, the total loss used for joint training is $\mathcal{L} = \mathcal{L}_{\text{SR}} + \lambda_4 \mathcal{L}_{\text{OFR}}$, where λ_4 is empirically set to 0.01 to balance the two loss terms.

4   Experiments

In this section, we first conduct ablation experiments to evaluate our framework. Then, we further compare our framework to several existing video SR methods.

4.1   Datasets

We collected 152 1080P HD video clips from the CDVL Database[1] and the Ultra Video Group Database[2]. The collected videos cover diverse natural and urban scenes. We used 145 videos from the CDVL Database as the training set, and 7 videos from the Ultra Video Group Database as the validation set. Following the configuration in [7, 9, 10], we downsampled the video clips to the size of 540×960 as the HR groundtruth using Matlab $imresize$ function. In this paper, we only focus on the upscaling factor of 4 since it is the most challenging case. Therefore, the HR video clips were further downsampled to produce LR inputs of size 135×240.

For fair comparison to the state-of-the-arts, we chose the widely used Vid4 benchmark dataset. We also used another 10 video clips from the DAVIS dataset [33] for further comparison, which we refer to as DAVIS-10.

4.2   Implementation Details

Following [5, 13], we converted input LR frames into YCbCR color space and only fed the luminance channel to our network. All metrics in this section are computed in the luminance channel. During the training phase, we randomly extracted 3 consecutive frames from an LR video clip, and randomly cropped a 32×32 patch as the input. Meanwhile, its corresponding patch in HR video clip was cropped as groundtruth. Data augmentation was performed through rotation and reflection to improve the generalization ability of our network.

We implemented our framework in PyTorch. We applied the Adam solver [34] with $\beta_1 = 0.9$, $\beta_2 = 0.999$ and batch size of 16. The initial learning rate was set to 10^{-4} and reduced to half after every 50K iterations. We trained our network from scratch for 300K iterations. All experiments were conducted on a PC with an Nvidia GTX 970 GPU.

---

[1] www.cdvl.org.
[2] ultravideo.cs.tut.fi.

4.3 Ablation Study

In this section, we present ablation experiments on the Vid4 dataset to justify our design choices.

Network Variants. We proposed several variants of our SOF-VSR to perform ablation study. All the variants were re-trained for 300K iterations on the training data.

SOF-VSR w/o OFRnet. To handle complex motion patterns in video sequences, optical flow is used for motion compensation in our framework. To test the effectiveness of motion compensation for video SR, we removed the whole OFRnet and fed LR frames directly to our SRnet. Note that, replicated LR frames were used to match the dimension of the draft cube C^L.

SOF-VSR w/o OFRnet$_{level3}$. The SR of optical flows provides accurate correspondences for video SR and improves the overall performance. To validate the effectiveness of HR optical flows, we removed the module at level 3 in our OFRnet. Specifically, the LR optical flows at level 2 were directly used for motion compensation and subsequent processing. To match the dimension of the draft cube, compensated LR frames were also replicated before feeding to SRnet.

SOF-VSR w/o OFRnet$_{level3}$+ upsampling. Super-resolving the optical flow can also be simply achieved using interpolation-based methods. However, our OFRnet can recover more reliable optical flow details. To demonstrate this, we removed the module at level 3 in our OFRnet, and upsampled the LR optical flows at level 2 using bilinear interpolation. Then, we used the modules in our original framework for subsequent processing.

Experimental Analyses. To test the accuracy of individual output image, we used PSNR/SSIM as metrics. To further test the consistency performance, we used the temporal motion-based video integrity evaluation index (T-MOVIE) [35]. Besides, we used MOVIE [35] and video quality measure with variable frame delay (VQM-VFD) [36] for overall evaluation. The MOVIE and VQM-VFD metrics are correlated with human perception and widely applied in video quality assessment. Evaluation results of our original framework and the 3 variants achieved on the Vid4 dataset are shown in Table 1.

Table 1. Comparative results achieved by our framework and its variants on the Vid4 dataset under ×4 configuration. Best results are shown in boldface.

	PSNR(↑)	SSIM(↑)	T-MOVIE(↓) ($\times 10^{-3}$)	MOVIE(↓) ($\times 10^{-3}$)	VQM-VFD(↓)
SOF-VSR w/o OFRnet	25.80	0.760	20.08	4.54	0.240
SOF-VSR w/o OFRnet$_{level3}$	25.88	0.764	19.95	4.48	0.235
SOF-VSR w/o OFRnet$_{level3}$ + upsampling	25.86	0.763	19.92	4.50	0.231
SOF-VSR	**26.01**	**0.771**	**19.78**	**4.32**	**0.227**

Motion Compensation. It can be observed from Table 1 that motion compensation plays a significant role in performance improvement. If OFRnet is removed, the PSNR/SSIM values are decreased from 26.01/0.771 to 25.80/0.760. Besides, the consistency performance is also degraded, with T-MOVIE value being increased to 20.08. That is because, it is difficult for SRnet to learn the non-linear mapping between LR and HR images under complex motion patterns.

Table 2. Average EPE results achieved on the Vid4 dataset under ×4 configuration. Best results are shown in boldface.

	Upsampled optical flow	Super-resolved optical flow
Calendar	0.43	**0.39**
City	0.59	**0.49**
Foliage	0.59	**0.36**
Walk	0.63	**0.55**
Average	0.56	**0.45**

HR Optical Flow. If modules at levels 1 and 2 are introduced to generate LR optical flows for motion compensation, the PSNR/SSIM values are increased to 25.88/0.764. However, the performance is still inferior to our SOF-VSR method using HR optical flows. That is because, HR optical flows provide more accurate correspondences for performance improvement. If bilinear interpolation is used to upsample LR optical flows, no consistent improvement can be observed. That is because, upsampling operation cannot recover reliable correspondence details as the module at level 3. To demonstrate this, we further compared the super-resolved optical flow (output at level 3), upsampled optical flow (upsampling result of the output at level 2) to the groundtruth. Since no groundtruth optical flow is available for the Vid4 dataset, we used the method proposed by Hu et al. [37] to compute the groundtruth optical flow. We used the average endpoint error (EPE) for quantitative comparison, and present the results in Table 2.

It can be seen from Table 2 that the super-resolved optical flow significantly outperforms the upsampled optical flow, with an average EPE being reduced from 0.56 to 0.45. It demonstrates that the module at level 3 effectively recovers the correspondence details. Figure 6 further illustrates the qualitative comparison on *City* and *Walk*. In the upsampled optical flow, we can roughly distinguish the outlines of the building and the pedestrian. In contrast, more distinct edges can be observed in the super-resolved optical flow, with finer details being recovered. Although some checkboard artifacts generated by the sub-pixel layer can also be observed [38], the resulting HR optical flow provides highly accurate correspondences for the video SR task.

4.4 Comparisons to the State-of-the-Art

We first compared our framework to IDNnet [16] (the latest state-of-the-art single image SR method) and several video SR methods including VSRnet [4], VESCPN [11], DRVSR [10], TDVSR [5] and FRVSR [21] on the Vid4 dataset. Then, we conducted comparative experiments on the DAVIS-10 dataset.

For IDNnet and VSRnet, we used the codes provided by the authors to produce the results. For DRVSR and TDVSR, we used the output images provided by the authors. For VESCPN and FRVSR, the results reported in their papers [11,21] are used. Here, we report the performance of FRVSR-3-64 since its network size is comparable to our SOF-VSR. Following [39], we crop borders of $6 + s$ for fair comparison.

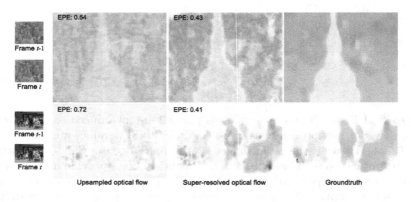

Fig. 6. Visual comparison of optical flow estimation results achieved on *City* and *Walk* under ×4 configuration. The super-resolved optical flow recovers fine correspondences, which are consistent with the groundtruth.

Note that, DRVSR and FRVSR are trained on a degradation model different from other networks. Specifically, the degradation model used in IDNnet, VSRnet, VESCPN and TDVSR is bicubic downsampling with Matlab *imresize* function (denoted as BI). However, in DRVSR and FRVSR, the HR images are first blurred using Gaussian kernel and then downsampled by selecting every s^{th} pixel (denoted as BD). Consequently, we re-trained our framework on the BD degradation model (denoted as SOF-VSR-BD) for fair comparison to DRVSR and FRVSR.

Without optimization of the implementation, our SOF-VSR network takes about 250 ms to generate an HR image of size 720×576 under ×4 configuration on an Nvidia GTX 970 GPU.

Quantitative Evaluation. Quantitative results achieved on the Vid4 dataset and the DAVIS-10 dataset are shown in Tables 3 and 4.

Table 3. Comparison of accuracy and consistency performance achieved on the Vid4 dataset under ×4 configuration. Note that, the first and last two frames are not used in our evaluation since VSRnet and TDVSR do not produce outputs for these frames. Results marked with * are directly copied from the corresponding papers. Best results are shown in boldface.

	BI degradation model					BD degradation model		
	IDNnet [16]	VSRnet [4]	VESCPN [11]	TDVSR [5]	SOF-VSR	DRVSR [10]	FRVSR-3-64 [21]	SOF-VSR-BD
PSNR(↑)	25.06	24.81	25.35*	25.49	**26.01**	25.99	26.17*	**26.19**
SSIM(↑)	0.715	0.702	0.756*	0.746	**0.771**	0.773	**0.798***	0.785
T-MOVIE(↓) (×10⁻³)	23.98	26.05	-	23.23	**19.78**	18.28	-	**17.63**
MOVIE(↓) (×10⁻³)	5.99	6.01	5.82*	4.92	**4.32**	4.00	-	**4.00**
VQM-VFD(↓)	0.268	0.273	-	0.238	**0.227**	0.217	-	**0.215**

Evaluation on the Vid4 Dataset. It can be observed from Table 3 that our SOF-VSR achieves the best performance for the BI degradation model in terms of all metrics. Specifically, the PSNR and SSIM values achieved by our framework are better than other methods by over 0.5 and 0.15. That is because, more accurate correspondences can be provided by HR optical flows and therefore more reliable spatial details and temporal consistency can be well recovered.

For the BD degradation model, although the FRVSR-3-64 method achieves higher SSIM, our SOF-VSR-BD method outperforms FRVSR-3-64 in terms of PSNR. Compared to the DRVSR method, PSNR, SSIM and T-MOVIE values achieved by our SOF-VSRBD method are improved by a notable margin, while a comparable performance is achieved in terms of MOVIE and VQM-VFD.

We further show the trade-off between consistency and accuracy in Fig. 7. It can be seen that our SOF-VSR and SOF-VSR-BD methods achieve the highest PSNR values, while maintaining superior T-MOVIE performance.

Evaluation on the DAIVIS-10 Dataset. It is clear in Table 4 that our SOF-VSR and SOF-VSR-BD methods surpass the state-of-the-arts for both the BI and BD degradation models in terms of all metrics. Since the DAVIS-10 dataset comprises scenes with fast moving objects, complex motion patterns

Table 4. Comparative results achieved on the DAVIS-10 dataset under ×4 configuration. Best results are shown in boldface.

	BI degradation model			BD degradation model	
	IDNnet [16]	VSRnet [4]	SOF-VSR	DRVSR [10]	SOF-VSR-BD
PSNR(↑)	33.74	32.63	**34.32**	33.02	**34.27**
SSIM(↑)	0.915	0.897	**0.925**	0.911	**0.925**
T-MOVIE(×10⁻³)(↓)	12.16	14.60	**11.77**	14.06	**10.93**
MOVIE(×10⁻³)(↓)	2.19	2.85	**1.96**	3.15	**1.90**
VQM-VFD(↓)	0.146	0.163	**0.119**	0.142	**0.127**

(especially large displacements) lead to deterioration of existing video SR methods. In contrast, more accurate correspondences are provided by HR optical flows in our framework. Therefore, complex motion patterns can be handled more robustly and better performance can be achieved.

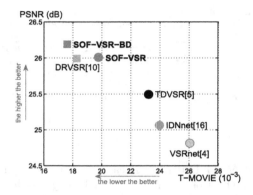

Fig. 7. Consistency and accuracy performance achieved on the Vid4 dataset under ×4 configuration. Dots and squares represent performance for BI and BD degradation models, respectively. Our framework achieves the best performance in terms of both PSNR and T-MOVIE.

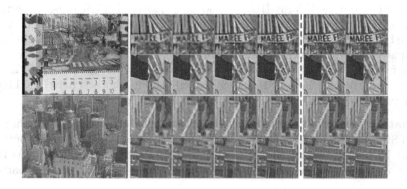

Fig. 8. Visual comparison of ×4 SR results on *Calendar* and *City*. Zoom-in regions from left to right: IDNnet [16], VSRnet [4], TDVSR [5], our SOF-VSR, DRVSR [10] and our SOF-VSR-BD. IDNnet, VSRnet, TDVSR and SOF-VSR are based on the BI degradation model, while DRVSR and SOF-VSR-BD are based on the BD degradation model.

Qualitative Evaluation. Figure 8 illustrates the qualitative results on two scenarios of the Vid4 dataset. It can be observed from the zoom-in regions that our framework recovers finer and more reliable details, such as the word "MAREE" and the stripes of the building. The qualitative comparison on the DAVIS-10 dataset (as shown in Fig. 9) also demonstrates the superior visual

Fig. 9. Visual comparison of ×4 SR results on *Boxing* and *Demolition*. Zoom-in regions from left to right: IDNnet [16], VSRnet [4], our SOF-VSR, DRVSR [10] and our SOF-VSR-BD. IDNnet, VSRnet and SOF-VSR are based on the BI degradation model, while DRVSR and SOF-VSR-BD are based on the BD degradation model.

quality achieved by our framework. The pattern on the shorts, the word "PEUA" and the logo "CAT" are better recovered by our SOF-VSR and SOF-VSR-BD methods.

Figure 1 further shows the temporal profiles achieved on *Calendar* and *City*. It can be observed that the word "MAREE" can hardly be recognized by VSRnet in both image space and temporal profile. Although finer results are achieved by TDVSR, the building is still obviously distorted. In contrast, smooth and reliable patterns with fewer artifacts can be observed in temporal profiles of our results. In summary, our framework produces temporally more consistent results and better perceptual quality.

5   Conclusions and Future Work

In this paper, we propose a deep end-to-end trainable video SR framework to super-resolve both images and optical flows. Our OFRnet first super-resolves the optical flows to provide accurate correspondences. Motion compensation is then performed based on HR optical flows and SRnet is used to infer the final results. Extensive experiments have demonstrated that our OFRnet can recover reliable correspondence details for the improvement of both accuracy and consistency performance. Comparison to existing video SR methods has shown that our framework achieves the state-of-the-art performance.

Since groundtruth optical flows are unavailable for our training dataset, we trained our OFRnet in an unsupervised manner. Introducing another state-of-the-art supervised flow estimation network (*e.g.*, [28–30]) to provide supervision may bring further performance improvement, and this would be our future work.

Acknowledgments. This work is supported by the National Natural Science Foundation of China (No. 61605242, No. 61602499 and No. 61471371).

References

1. Nguyen, N., Milanfar, P., Golub, G.: A computationally efficient superresolution image reconstruction algorithm. IEEE Trans. Image Process. **10**(4), 573–583 (2001)
2. Fattal, R.: Image upsampling via imposed edge statistics. ACM Trans. Graph. **26**(3), 95 (2007)
3. Freedman, G., Fattal, R.: Image and video upscaling from local self-examples. ACM Trans. Graph. **30**(2), 12:1–12:11 (2011)
4. Kappeler, A., Yoo, S., Dai, Q., Katsaggelos, A.K.: Video super-resolution with convolutional neural networks. IEEE Trans. Comput. Imaging **2**(2), 109–122 (2016)
5. Liu, D., Wang, Z., Fan, Y., Liu, X.: Robust video super-resolution with learned temporal dynamics. In: ICCV, pp. 2526–2534 (2017)
6. Fransens, R., Strecha, C., Van Gool, L.J.: Optical flow based super-resolution: a probabilistic approach. Comput. Vis. Image Underst. **106**(1), 106–115 (2007)
7. Liu, C., Sun, D.: On bayesian adaptive video super resolution. IEEE Trans. Pattern Anal. Mach. Intell. **36**(2), 346–360 (2014)
8. Ma, Z., Liao, R., Tao, X., Xu, L., Jia, J., Wu, E.: Handling motion blur in multi-frame super-resolution. In: CVPR, pp. 5224–5232 (2015)
9. Liao, R., Tao, X., Li, R., Ma, Z., Jia, J.: Video super-resolution via deep draft-ensemble learning. In: ICCV, pp. 531–539 (2015)
10. Tao, X., Gao, H., Liao, R., Wang, J., Jia, J.: Detail-revealing deep video super-resolution. In: ICCV, pp. 4482–4490 (2017)
11. Caballero, J., et al.: Real-time video super-resolution with spatio-temporal networks and motion compensation. In: CVPR, pp. 2848–2857 (2017)
12. Liu, D., et al.: Learning temporal dynamics for video super-resolution: a deep learning approach. IEEE Trans. Image Process. **27**(7), 3432–3445 (2018)
13. Dong, C., Loy, C.C., He, K., Tang, X.: Learning a deep convolutional network for image super-resolution. In: Fleet, D., Pajdla, T., Schiele, B., Tuytelaars, T. (eds.) ECCV 2014. LNCS, vol. 8692, pp. 184–199. Springer, Cham (2014). https://doi.org/10.1007/978-3-319-10593-2_13
14. Kim, J., Kwon Lee, J., Mu Lee, K.: Accurate image super-resolution using very deep convolutional networks. In: CVPR, pp. 1646–1654 (2016)
15. Tai, Y., Yang, J., Liu, X.: Image super-resolution via deep recursive residual network. In: CVPR, pp. 2790–2798 (2017)
16. Hui, Z., Wang, X., Gao, X.: Fast and accurate single image super-resolution via information distillation network. In: CVPR (2018)
17. Protter, M., Elad, M., Takeda, H., Milanfar, P.: Generalizing the nonlocal-means to super-resolution reconstruction. IEEE Trans. Image Process. **18**, 36–51 (2008)
18. Takeda, H., Milanfar, P., Protter, M., Elad, M.: Super-resolution without explicit subpixel motion estimation. IEEE Trans. Image Process. **18**(9), 1958–1975 (2009)
19. Seok Lee, H., Mu Lee, K.: Simultaneous super-resolution of depth and images using a single camera. In: CVPR, pp. 281–288 (2013)
20. Shi, W., et al.: Real-time single image and video super-resolution using an efficient sub-pixel convolutional neural network. In: CVPR, pp. 1874–1883 (2016)
21. Sajjadi, M.S.M., Vemulapalli, R., Brown, M.: Frame-recurrent video super-resolution. In: CVPR (2018)
22. Dosovitskiy, A., et al.: Flownet: learning optical flow with convolutional networks. In: CVPR, pp. 2758–2766 (2015)
23. Ilg, E., Mayer, N., Saikia, T., Keuper, M., Dosovitskiy, A., Brox, T.: Flownet 2.0: evolution of optical flow estimation with deep networks. In: CVPR, pp. 1647–1655 (2017)

24. Bouguet, J.-Y.: Pyramidal implementation of the lucas kanade feature tracker: description of the algorithm. Technical report, Intel Corporation (1999)
25. Zhang, Y., Tian, Y., Kong, Y., Zhong, B., Fu, Y.: Residual dense network for image super-resolution. In: CVPR (2018)
26. Sajjadi, M.S.M., Schölkopf, B., Hirsch, M.: Enhancenet: single image super-resolution through automated texture synthesis. In: ICCV, pp. 4501–4510 (2017)
27. Lai, W.-S., Huang, J.-B., Ahuja, N., Yang, M.-H.: Deep laplacian pyramid networks for fast and accurate super-resolution. In: CVPR, pp. 5835–5843 (2017)
28. Sun, D., Yang, X., Liu, M.-Y., Kautz, J.: PWC-Net: CNNs for optical flow using pyramid, warping, and cost volume. In: CVPR (2018)
29. Ranjan, A., Black, M.J.: Optical flow estimation using a spatial pyramid network. In: CVPR, pp. 2720–2729 (2017)
30. Hui, T.-W., Tang, X., Change Loy, C.: LiteFlowNet: a lightweight convolutional neural network for optical flow estimation. In: CVPR (2018)
31. Makansi, O., Ilg, E., Brox, T.: End-to-end learning of video super-resolution with motion compensation. In: GCPR, pp. 203–214 (2017)
32. Huang, G., Liu, Z., van der Maaten, L., Weinberger, K.Q.: Densely connected convolutional networks. In: CVPR, pp. 2261–2269 (2017)
33. Pont-Tuset, J., Perazzi, F., Caelles, S., Arbelaez, P., Sorkine-Hornung, A., Van Gool, L.: The 2017 DAVIS challenge on video object segmentation. arXiv:1704.00675, pp. 1–9 (2017)
34. Kingma, D.P., Ba, J.: Adam: A method for stochastic optimization. In: ICLR (2015)
35. Seshadrinathan, K., Bovik, A.C.: Motion tuned spatio-temporal quality assessment of natural videos. IEEE Trans. Image Process. **19**(2), 335–350 (2010)
36. Wolf, S., Pinson, M.H.: Video quality model for variable frame delay (VQM-VFD). US Dept. Commer., Nat. Telecommun. Inf. Admin., Boulder, CO, USA, Tech. Memo TM-11-482 (2011)
37. Hu, Y., Li, Y., Song, R.: Robust interpolation of correspondences for large displacement optical flow. In: CVPR, pp. 4791–4799 (2017)
38. Odena, A., Dumoulin, V., Olah, C.: Deconvolution and checkerboard artifacts. Distill **1**(10), e3 (2016)
39. Timofte, R., Agustsson, E., Van Gool, L., Yang, M.-H., Zhang, L., et al.: NTIRE 2017 challenge on single image super-resolution: methods and results. In: CVPR, pp. 1110–1121 (2017)

Deep Multiple Instance Learning
for Zero-Shot Image Tagging

Shafin Rahman[1,2(✉)] and Salman Khan[1,2,3]

[1] Australian National University, Canberra, ACT 2601, Australia
{shafin.rahman,salman.khan}@anu.edu.au
[2] Data61, CSIRO, Canberra, ACT 2601, Australia
[3] Inception Institute of AI, Abu Dhabi, UAE

Abstract. In-line with the success of deep learning on traditional recognition problem, several end-to-end deep models for zero-shot recognition have been proposed in the literature. These models are successful to predict a single unseen label given an input image, but does not scale to cases where multiple unseen objects are present. In this paper, we model this problem within the framework of Multiple Instance Learning (MIL). To the best of our knowledge, we propose the first end-to-end trainable deep MIL framework for the multi-label zero-shot tagging problem. Due to its novel design, the proposed framework has several interesting features: (1) Unlike previous deep MIL models, it does not use any off-line procedure (e.g., Selective Search or EdgeBoxes) for bag generation. (2) During test time, it can process any number of unseen labels given their semantic embedding vectors. (3) Using only seen labels per image as weak annotation, it can produce a bounding box for each predicted label. We experiment with large-scale NUS-WIDE dataset and achieve superior performance across conventional, zero-shot and generalized zero-shot tagging tasks.

Keywords: Zero-shot learning · Zero-shot tagging · Object detection

1 Introduction

In recent years, numerous single label zero-shot classification methods have been proposed aiming to recognize novel concepts with no training examples [2,8,17,21,40,41]. However in real life settings, images often come with multiple objects or concepts that may or may not be observed during training. For example, enormous growth in on-line photo collections require automatic image tagging algorithms that can provide both seen and unseen labels to the images. Despite the importance and practical nature of this problem, there are very few existing methods with the capability to address the zero-shot image tagging task [10,16,19,42]. This is primarily due to the challenging nature of the problem. Notably, any object or concept can either be present at a localized region or be inferred from the holistic scene information (e.g., *'sun'* vs *'sunset'*).

© Springer Nature Switzerland AG 2019
C. V. Jawahar et al. (Eds.): ACCV 2018, LNCS 11361, pp. 530–546, 2019.
https://doi.org/10.1007/978-3-030-20887-5_33

The available label space is also significantly larger comprising of thousands of possible tags. Moreover, objects and concepts are often occluded in natural scenes. Therefore, assigning multiple tags to an image requires searching for both local and global details present at different scales, orientations, poses and illumination conditions.

The existing attempts on zero-shot image tagging (e.g., [10, 16, 19, 42]) mostly used deep CNN features of the whole images. But, as objects or concepts are more likely to be present at the localized regions, the extracted features from the whole image cannot represent all possible labels. In this paper, we propose a deep Multiple Instance Learning (MIL) framework that operates on a bag of instances generated from each image. MIL assumes that each bag contains at least one instance of the true labels defined by the ground-truth. A distinguishing feature of our approach is that the bag of instances not only encodes the global information about a scene, but also have a rich representation for localized object-level features. This is different from the existing literature in image tagging, where MIL based strategies do not consider local image details. For example, [9] used image-level features at different scale of the network whereas [43] used whole image features from different video frames as the bag of instance. These techniques therefore work only for the most prominent objects but often fail for non-salient concepts due to the lack of localized information.

In addition to the use of localized features, we integrate the instance localization within a unified system that can be trained jointly in an end-to-end manner. We note that the previous approaches [33, 36–38] applied off-line procedures like Selective Search [34]/EdgeBoxes [44]/BING [5] for object proposal or patch generation which served as a bag-of-instance. Patches obtained from any external procedure were used in three ways: (1) to extract features for each patch from a pre-trained deep network to create bag-of-instance [32, 36], (2) to use a set of patches as a bag and feed the bag as an input to a deep network to perform MIL such that the image features can be fine-tined [37, 38], (3) to feed an input image with all patch locations and later perform a ROIPooling to generate bag of features [11, 33]. One common issue with these approaches is that none of the methods are end-to-end trainable because of the dependency on the external non-differentiable procedure in either feature extraction or bag of instance generation. Furthermore, the above methods do not address zero-shot tagging problem and cannot relate the visual and semantic concepts. A recent work from Ren et al. [29] proposed a solution for zero-shot tagging in-line with the third approach above, that cannot dynamically adapt the patch locations.

Finally, our proposed network maps local and global information inside the bag to a semantic embedding space so that the correspondences with both seen and unseen classes can be found. With respect to zero-shot tagging, [29, 42] are closest to our work as they handle a visual-semantic mapping to bridge seen and unseen classes. However, [42] uses off the shelf features and does not consider local details. To extract local image information, our approach finds localized patches similar to [29]. However, instead of using an external off-line procedure as in [29] for bag generation, our network can generate a bag as well. Moreover, it can simultaneously fine-tune the feature extraction and bag

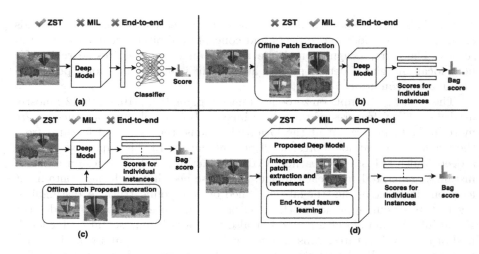

Fig. 1. Overview of different multi-label image annotation architecture. (a) [32,36,42] extract deep features separately, then feed those features to a neural network for classification (b) [37,38] use external procedure to extract image patches for bag generation then feed those patches to a deep model to get final bag score for seen classes (c) [11,29,33] process the whole image as well as some patches obtained using external process together where all patches from one image share the same CNN layers to get bag score for seen classes (d) Our proposed MIL model simply takes an image as an input and produces bag score for both seen and unseen classes.

generation process instead of just using pre-trained patch features. In Fig. 1, we illustrate a block diagram of different kinds of image tagging frameworks. Our framework encompasses all different modules of MIL into only a single integrated network, which results in state-of-the-art performances for conventional and zero-shot image tagging.

The key contributions of our paper are:

- We propose the first end-to-end deep MIL framework for multi-label image tagging that can work in both conventional and zero-shot settings.
- The proposed method does not require any off-line procedure to generate bag of instances for MIL, rather incorporates this step in a joint framework.
- The proposed method is extendable to any number of novel tags (open vocabulary) as it does not use any prior information about the unseen concepts.
- The proposed method can annotate a bounding box for both seen and unseen tags without even using any ground-truth bounding box during training.

2   Related Work

Zero-Shot Learning: In recent years, we have seen some exciting papers on Zero Shot Learning (ZSL). The overall goal of those efforts is basically to classify an image to an unseen class for which no training is performed. Investigations are focused on domain adaptation [8], class attribute association [7], unsupervised semantics [8], hubness effect [41] and generalized setting [40] of inductive [25] or transductive ZSL learning [17]. The major shortcoming of these approaches is their inability to assign multiple labels to an image, which is a major limitation in real-world settings. In line with the general consideration of only a singe label per image, traditional ZSL methods use recognition datasets which mostly contain only one prominent concept per image. Here, we present an end-to-end deep zero shot tagging method that can assign multiple tags per image.

Object Detection: Image tags can corresponds to either whole image or a specific location inside it. To model the relations between tag and its corresponding locations we intend to localize objects in a scene. To do so, we are interested in end-to-end object detection framework. Popular examples of such frameworks are Faster R-CNN [28], R-FCN [15], SSD [18] and YOLO [27]. The main distinction among these models is the object localization process. R-CNN [28] and R-FCN [15] used a Region Proposal Network (RPN) to generate object proposals whereas SSD [18] and YOLO [27] propose bounding box and classify it in a single step. The later group of models usually runs faster but are relatively less accurate. Due to the focus on highly accurate object detection, we built on Faster R-CNN [26, 28] as the backbone architecture in the current work.

Zero-Shot Image Tagging: Instead of assigning one unseen label to an image during recognition task, zero-shot tagging assigns multiple unseen tags to an image and/or ranking the array of unseen tags. Although interesting, this problem is not well-addressed in the zero-shot learning literature. An early attempt extended a zero-shot recognition work [22] to perform zero-shot tagging by proposing a hierarchical semantic embedding to make the label embedding more reliable [16]. [10] proposed a transductive multi-label version of the problem where a predefined and relatively small set of unseen tags were considered. In a recent work, [42] proposed a fast zero-shot tagging approach that can be trained using only seen tags and tested using both seen and unseen tags for test images. [29] proposed a multi-instance visual-semantic embedding approach that can extract localized image features. The main drawback of these early efforts is the dependence on pre-trained CNN features (in [10,16,42]) or fast-RCNN [11] features (in [29]) and therefore not end-to-end trainable. In this work, we propose a fully end-to-end solution for both conventional and zero-shot tagging.

3   Our Method

3.1   Problem Formulation

Suppose, we have a set of 'seen' tags denoted by $\mathcal{S} = \{1, \ldots, S\}$ and another set of 'unseen' tags $\mathcal{U} = \{S + 1, \ldots, S + U\}$, such that $\mathcal{S} \cap \mathcal{U} = \phi$ where, S and U represents the total number of seen and unseen tags respectively. We also denote $\mathcal{C} = \mathcal{S} \cup \mathcal{U}$, such that C = S + U is the cardinality of the tag-label space. For each of the tag $c \in \mathcal{C}$, we can obtain a 'd' dimensional word vector $\mathbf{v}_c$ (word2vec or GloVe) as semantic embedding. The training examples can be defined as a set

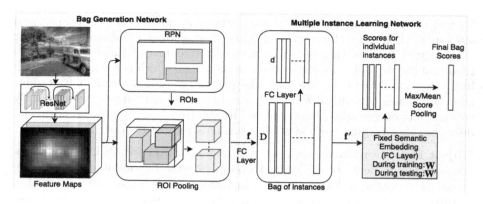

Fig. 2. Proposed network architecture for MIL based zero-shot image tagging.

of tuples, $\{(\mathbf{X}_s, \mathbf{y}_s) : s \in [1, M]\}$, where $\mathbf{X}_s$ is the sth input image and $\mathbf{y}_s \subset \mathcal{S}$ is the set of relevant seen tags. We represent u^{th} testing image as $\mathbf{X}_u$ which corresponds to a relevant seen and/or unseen tag $\mathbf{y}_u \subset \mathcal{C}$. Note that, $\mathbf{X}_u, \mathbf{y}_u, \mathcal{U}$ and its corresponding word vectors are not observed during training. Now, we define the following problems:

- *Conventional tagging:* Given $\mathbf{X}_u$ as input, assign relevant seen tags $\mathbf{y}_u \subset \mathcal{S}$.
- *Zero-shot tagging (ZST):* Given $\mathbf{X}_u$, assign relevant unseen tags $\mathbf{y}_u \subset \mathcal{U}$.
- *Generalized zero-shot tagging (GZST):* Given $\mathbf{X}_u$ as input, assign relevant tags from both seen and unseen $\mathbf{y}_u \subset \mathcal{C}$.

MIL Formulation: We formulate the above mentioned problems in Multiple Instance Learning (MIL) framework. Let us represent the sth training image with a bag of $n + 1$ instances $\mathbf{X}_s = \{\mathbf{x}_{s,0} \ldots \mathbf{x}_{s,n}\}$, where i^{th} instance $\mathbf{x}_{si}$ represents either an image patch (for $i > 0$) or the complete image itself (for $i = 0$). We assume that each instance $\mathbf{x}_{s,i}$ has an individual label $\ell_{s,i}$ which is unknown during training. As $\mathbf{y}_s$ represents relevant seen tags of $\mathbf{X}_s$, according to MIL assumption, the bag has at least one instance for each tag in the set $\mathbf{y}_s$ and no instance for $\mathcal{S} \setminus \mathbf{y}_s$ tags. The bag- and instance-level labels are related by:

$$y \in \mathbf{y}_s \quad \text{iff} \quad \sum_{i=0}^{n} \llbracket \ell_{s,i} = y \rrbracket > 0, \quad \forall y \in \mathcal{S}. \tag{1}$$

Thus, instances in $\mathbf{X}_s$ can work as a positive example for $y \in \mathbf{y}_s$ and negative example for $y' \in \{\mathcal{S} \setminus \mathbf{y}_s\}$. This formulation does not use instance level tag annotation which makes it a weakly supervised problem. Our aim is to design and learn an end-to-end deep learning model that can itself generate the appropriate bag of instances and simultaneously assign relevant tags to the bag.

3.2   Network Architecture

The proposed network architecture is illustrated in Fig. 2. It is composed of two parts: bag generation (*left*) and multiple instance learning (MIL) network (*right*). The bag generation network generates the bag of instances as well as their visual features, and the MIL network processes the resulting bag of instance features to find the final multi-label prediction which is calculated by a global pooling of the prediction scores of individual instances. In this manner, bag generation and zero-shot prediction steps are combined in a unified framework that effectively transfers learning from seen to unseen tags.

Bag Generation: In our proposed method, the bag contains image patches which are assumed to cover all objects and concepts presented inside the image. Many closely related traditional methods [29,33,37,38] apply some external procedure like Selective Search [34], EdgeBoxes [44], BING [5] etc. for this purpose. Such a strategy creates three problems: **(1)** the off-line external process does not allow an end-to-end learnable framework, **(2)** patch generation process is prone to more frequent errors because it can not get fine-tuned on the target dataset, and **(3)** the MIL framework need to process patches rather than the image itself. In this paper, we propose to solve these problems by generating useful bag of patches by the network itself. The recent achievements of object detection framework, such as the Faster-RCNN [28], allows us to generate object proposals and later perform detection within a single network. We adopt this strategy to generate bag of image patches for MIL. Remarkably, the original Faster-RCNN model is designed for supervised learning, while our MIL framework extends it to weakly supervised setting.

A Faster-RCNN model [28] with Region Proposal Network (RPN) is learned using the ILSVRC-2017 detection dataset. This architecture uses a base network ResNet-50 [14] which is shared between RPN and classification/localization network. As practiced, the base network is initialized with the pre-trained weights. Though not investigated in this paper, other popular CNN models e.g., VGG [30] and GoogLeNet [31] can also be used as the shared base network. Now, given a training image $\mathbf{X}_s$, the RPN can produce a fixed number (n) of region of interest (ROI) proposals $\{\mathbf{x}_{s,1} \ldots \mathbf{x}_{s,n}\}$ with a high recall rate. For image tagging, all tags may not represent an object. Rather, tags can be concepts that describe the whole image e.g., nature and landscape. To address this issue, we add a global image ROI (denoted by $\mathbf{x}_{s,0}$) comprising of complete image to the ROI proposal set generated by the RPN. Afterwards, ROIs are fed to ROI-Pooling and subsequent densely connected layers to calculate a D-dimensional feature $\mathcal{F}_s = [\mathbf{f}_{s,0} \ldots \mathbf{f}_{s,n}] \in \mathbb{R}^{D \times (n+1)}$ for each ROI where $\mathbf{f}_{s,0}$ is the feature representation of the whole image. This bag is forwarded to MIL network for prediction.

Training: We first initialize a Faster-RCNN model with the pre-trained weights for ILSVRC-2017 object detection dataset. After that, the last two layers (i.e. the classification and localization head) are popped out to produce bag generation network. Our network design then comprises of three fully connected (FC) layers and a global pooling layer (either max or mean pooling) with the resulting network to predict scores for S number of seen tags. The network is then fine-tuned on target tagging dataset i.e. NUS-WIDE [6]. Note that, at this stage, the resulting network cannot train the RPN further because no localization branch is left in the network. However, the shared base network (ResNet-50) gets fine-tuned for better proposal generation. Similarly, the weights connecting the newly added layers remains trainable along with the other parts of the network. The only exception is the last FC layer, where we use a **fixed** semantic embedding $\mathbf{W} = [\mathbf{v}_1 \ldots \mathbf{v}_S] \in \mathbb{R}^{d \times S}$ containing word vector of seen tags as non-trainable weights for that layer. Notably, a non-linear activation ReLU is applied between the first two FC layer but not considered between second and third layers. The role of the first two layers is to process bag of features by calculating $\mathcal{F}'_s = [\mathbf{f}'_{s,0} \ldots \mathbf{f}'_{s,n}] \in \mathbb{R}^{d \times (n+1)}$ for the projection onto the fixed embedding. The third FC layer projects $\mathcal{F}'_s$ in the semantic tag space and produce scores for individual instances. Given, a bag of instance feature $\mathcal{F}'_s$, we compute the prediction scores of individual instances, $\mathbf{P}_s = [\mathbf{p}_{s,0} \ldots \mathbf{p}_{s,n}] \in \mathbb{R}^{S \times (n+1)}$ as follows:

$$\mathbf{P}_s = \mathbf{W}^T \mathcal{F}'. \tag{2}$$

Since the supervision is only available for bag-level predictions (i.e., image tags), we require an aggregation mechanism to combine predictions scores $\mathbf{P}_s$ for individual instances in a bag. To this end, we introduce the following global pooling (max or mean) layer to calculate final multi-label prediction score for each bag:

$$\mathbf{o}_s = \max \left\{ \mathbf{p}_{s,0}, \mathbf{p}_{s,1}, \ldots, \mathbf{p}_{s,n} \right\} \quad or \quad \mathbf{o}_s = \frac{1}{n+1} \sum_{j=0}^{n} \mathbf{p}_{s,j}. \tag{3}$$

Theoretical Analysis: It can be proved that the proposed MIL framework can approximate any general function defined on the bag $\mathbf{X}_s$. Since our formulation is based on object proposals $\mathbf{x}_{s,i}$, we can approximate the general function $\mathcal{G}(\cdot)$ on the bag with the following object level decomposition:

$$\mathcal{G}(\mathbf{X}_s = \{\mathbf{x}_{s,i}\}_0^n) \approx h(\{g(\mathbf{x}_{s,i})\}_0^n), \quad s.t., \ g(\mathbf{x}_{s,i}) = \mathbf{p}_{s,i}, \tag{4}$$

where $g(\cdot)$ is the transformation function defined using a deep network and $h(\cdot)$ is a symmetric function that is invariant to permutations of the object proposals. Such a decomposition is intuitive because the proposal set is unordered and its cardinality can vary, neither of these two factors should effect the bag level predictions. The function $\mathcal{G}(\cdot)$ can be approximated adequately using the symmetric transformations in Eq. 3 according to the following corollaries:

Corollary 1: Max Pooling - From the Theorem of Universal approximation for continuous set functions [24], a Hausdorff symmetric function $\mathcal{G}(\cdot)$ can be approximated with in the bounds $\epsilon \in \mathbb{R}^+$ if $g(\cdot)$ is a continuous function and $h(\cdot)$ is a element-wise vector maximum operator (denoted as 'max'), i.e.,

$$\| \mathcal{G}(\mathbf{X}_s) - \max\{g(\mathbf{x}_{s,i})\} \| < \epsilon. \tag{5}$$

Corollary 2: Mean Pooling - From the the Chevalley-Shephard-Todd (CST) theorem [3], a permutation invariant continuous function $\mathcal{G}(\cdot)$ operating on the set $\mathbf{X}_s$ can be arbitrarily approximated if $g(\cdot)$ is a transformation function implemented as a neural network, where neural networks are universal approximators [13], and $h(\cdot)$ is an element wise mean operator (denoted as 'mean'), i.e.,

$$\mathcal{G}(\mathbf{X}_s) \approx \mathrm{mean}\{g(\mathbf{x}_{s,i})\}. \tag{6}$$

Therefore, our MIL formulation can learn a permutation invariant function on bags of visual instances consisting of both object and concept representations.

Loss Formulation: Suppose, for s^{th} training image, $\mathbf{o}_s = [o_1 \ldots o_S]$ contains final multi-label prediction of a bag for seen classes. This bag is a positive example for each tag $y \in \mathbf{y}_s$ and negative examples for each tag $y' \in \{\mathcal{S} \setminus \mathbf{y}_s\}$. Thus, for each pair of y and y', the difference $o_{y'} - o_y$ represents the disparity between predictions for positive and negative tags. Our goal is to minimize these differences in each iteration. We formalize the loss of a bag considering it to contain both positive and negative examples for different tags:

$$L_{tag}(\mathbf{o}_s, \mathbf{y}_s) = \frac{1}{|\mathbf{y}_s||\mathcal{S} \setminus \mathbf{y}_s|} \sum_{y' \in \{\mathcal{S} \setminus \mathbf{y}_s\}} \sum_{y \in \mathbf{y}_s} \log\left(1 + \exp(o_{y'} - o_y)\right).$$

We minimize the overall loss on all training images as follows:

$$L = \underset{\Theta}{\arg\min} \frac{1}{M} \sum_{s=1}^{M} \left(L_{tag}(\mathbf{o}_s, y_s)\right).$$

Here, Θ denote the network parameters and M is the number of training images.

Prediction: During testing, we modify the fixed embedding $\mathbf{W}$ with both seen and unseen word vectors instead of only seen word vectors. Suppose, after modification $\mathbf{W}$ becomes $\mathbf{W}' = [\mathbf{v}_1 \ldots \mathbf{v}_S, \mathbf{v}_{S+1} \ldots \mathbf{v}_{S+U}] \in \mathbb{R}^{d \times C}$. With the use of $\mathbf{W}'$ in Eq. 2, we get prediction score of both seen and unseen tags for each individual instance in the bag. Then, after the global pooling (mean or max), we get the final prediction score for each seen and unseen tags. Finally, based on the tagging task (conventional/zero-shot/generalized zero-shot), we assign top K target tags (from the set $\mathcal{S}, \mathcal{U}$ or $\mathcal{C}$) with higher scores.

4   Experiment

4.1   Setup

Dataset: We perform our experiments using a real-world web image dataset namely NUS-WIDE [6]. It contains 269,648 images with three sets of per image tags from Flickr. The first, second and third set contain 81, 1000 and 5018 tags respectively. The tags inside the first set are carefully chosen, therefore less noisy whereas third set has the highest noise in annotations. Following the previous work [42], we use 81 first set tags as unseen. We notice that the tag 'interesting' comes twice within the second set. After removing this inconsistency and selecting 81 unseen tags from the second set results in 924 tags which we use as seen for our experiment. The dataset provides 161,789 training and 107,859 testing images. We use this recommended setting ignoring the non-tagged images.

Visual and Semantic Embedding: Unlike previous attempts in zero-shot tagging [16,42], our model works in an end-to-end manner using ResNet-50 [14] as a base network. It means the visual feature are originating from ResNet-50, but they are updated during iterations. As the semantic embedding, we use ℓ_2 normalized 300 dimensional GloVe vectors [23]. We are unable to use word2vec vectors [20] because the pre-trained word-vector model cannot provide vectors for all of 1005 (924 seen + 81 unseen) tags.

Evaluation Metric: Following [42], we calculate precision (P), recall (R) and F1 score of the top K predicted tags ($K = 3$ and 5 is used) and Mean image Average Precision (MiAP) as evaluation metric. MiAP of an input image I is:

$$MiAP(I) = \frac{1}{R} \sum_{j=1}^{|\mathcal{G}|} \frac{r_j}{j} \delta(I, t_j),$$

where, R = total number of relevant tags, $|\mathcal{G}|$ = total number of ground truth tags, r_j = number of relevant tags of jth rank and $\delta(I, t_j) = 1$ if j^{th} tag t_j is associated with the input image I, otherwise $\delta(I, t_j) = 0$.

Implementation Details: We used the following settings during Faster-RCNN training, following the proposed settings of [28]: rescaling shorter size of image as 600 pixels, RPN stride = 16, three anchor box scale 128, 256 and 512 pixels, three aspect ratios 1:1, 1:2 and 2:1, non-maximum suppression (NMS) with IoU threshold = 0.7 with maximum proposal = 300 for faster-RCNN. During training our MIL framework, we generated one bag of instances at each iteration from an image to feed our network. We chose a fixed n number of RoIs proposed by RPN which archives best objectness score. We carried out 774k training iterations using Adam optimizer with a learning rate of 10^{-5}, $\beta_1 = 0.9$ and $\beta_2 = 0.999$. We implemented our model in *Keras* library.

4.2   Tagging Performance

We evaluate the performance of our framework on three problem variants, namely conventional, zero-shot and generalized zero-shot tagging (Sect. 3.1).

Table 1. Results for conventional tagging. K denotes the number of assigned tags.

Method	MiAP	$K = 3$			$K = 5$		
		P	R	F1	P	R	F1
Fast0Tag [42]	35.73	20.24	34.48	25.51	16.16	45.87	23.90
Baseline	40.45	22.95	39.09	28.92	17.99	51.09	26.61
Ours (Bag: 32)	**53.97**	30.17	51.41	38.03	22.66	64.35	33.52
Ours (Bag: 64)	53.94	**30.23**	**51.61**	**38.18**	**22.73**	64.54	33.62

Table 2. Results for zero-shot and generalize zero-shot tagging tasks.

Method	Zero-shot tagging							Generalized zero-shot tagging						
	MiAP	$K = 3$			$K = 5$			MiAP	$K = 3$			$K = 5$		
		P	R	F1	P	R	F1		P	R	F1	P	R	F1
ConSE [22]	18.91	8.39	14.30	10.58	7.16	20.33	10.59	7.27	2.11	3.59	2.65	8.82	5.69	6.92
Fast0Tag [42]	24.73	13.21	22.51	16.65	11.00	31.23	16.27	10.36	5.21	8.88	6.57	12.41	8.00	9.73
Baseline	29.75	16.64	28.34	20.97	13.49	38.32	19.96	12.07	5.99	10.20	7.54	14.28	9.21	11.20
Ours (Bag: 32)	37.50	21.16	36.06	26.67	16.62	47.20	24.59	**20.55**	**27.66**	**10.70**	**15.43**	**23.67**	**15.27**	**18.56**
Ours (Bag: 64)	**39.01**	**22.05**	**37.56**	**27.79**	**17.26**	**49.01**	**25.53**	20.32	27.09	10.48	15.12	23.27	15.01	18.25

Compared Methods: To compare our results, we have reimplemented two similar published methods (ConSE [22] and Fast0Tag [42]) and one simple baseline based on ResNet-50. We choose these methods for comparison because of their suitability to perform zero-shot tasks.

ConSE [22] was originally introduced for zero-shot learning. This approach first learns a classifier for seen tags and generates a semantic embedding for unseen input by linearly combining word vectors of seen classes using seen prediction scores. In this way, it can rank unseen tags based on the distance of generated semantic embedding and the embedding of unseen tags.

Fast0Tag [42] is the main competitor of our work. This is a deep feature based approach, where features are calculated from a pre-trained VGG-19 [30]. Afterwards, a neural network is trained on these features to classify seen and unseen input. This approach outperforms many established methods (e.g., [4,12]) on conventional tagging task. Therefore, we do not compare against these low-performing methods. Similarly, we do not compare with [29] as they are limited to single-label ZST while ours is a multi-label framework. The performance reported in this paper using Fast0Tag method is relatively different from the published results because of few reasons: (1) We use the recent ResNet-50 whereas [42] reported results on VGG-19, (2) Although [42] experimented on NUS-WIDE, they only used a subset 223,821 images in total, (3) The implementation for [42] did not consider the repetition of the seen tag 'interesting'.

The baseline method is a special case of our proposed method which uses the whole image as a single instance inside the bag. It breaks the multiple instance learning consideration but does not affect the end-to-end nature of the solution.

Results: As mentioned earlier, we perform our training with 924 seen tags and testing with 81 unseen tags for zero-shot settings. However, in conventional tagging case, all tags are considered as seen. Therefore, we use the 81 tag set in both training and testing. Note that, in all of our experiments the same test images are used. Thus, the basic difference between conventional vs. zero-shot tagging is whether those 81 tags were used during training or not. For generalized zero-shot tagging case same testing image set is used, but instead of predicting tags from 81 tag set, our method predicts tags from seen 924 + unseen 81 = 1005 tag set. The performances of ours and compared methods on the tagging tasks are reported in Tables 1 and 2 for the case of NUS-WIDE dataset. Our method outperforms all competitor methods by a significant margin. Notably, the following observations can be developed from the results: (1) The performance of conventional tagging is much better than zero-shot cases because unseen tags and associated images are not present during training for zero-shot tasks. One can consider that the performance of conventional case is an upper-bound for zero-shot tagging case. (2) Similar to previous work [40], the performance for the generalized zero-shot tagging task is even poorer than the zero-shot tagging task. This can be explained by the fact that the network gets biased towards the seen tags and scores low on unseen categories. This subsequently leads to a decrease in performance. (3) Similar to the observation from [42], Fast0Tag beats ConSE [22] in zero-shot cases. The main reason is that the ConSE [22] does not use semantic word vectors during its training which is crucial to find a bridge between seen and unseen tags. No results are reported with ConSE for the conventional tagging case because it is only designed for zero-shot scenarios. (4) The baseline beats other two compared methods across all tasks because of the end-to-end training considering word vectors in the learning phase. This approach is benefited by the appropriate adaptation of feature representations for the tagging task. (5) Our approach outperforms all other competitors because it utilizes localized image features based on MIL, perform end-to-end training and integrate semantic vectors of seen tags within the network. We also illustrate some qualitative comparisons in Fig. 3.

Ablation Study: The proposed framework can work for different number of instances in the bag. Moreover, the global pooling before the last layer can be based on either mean or max pooling. In Table 3, we perform an ablation study for zero-shot tagging based on different combinations of network settings. We observe that with only a few number of instances (e.g., 4) in the bag, our method can beat state-of-the-art approaches [22,42]. The required bag size is actually depended on the dataset and pooling type. We notice that a large bag-size improves tagging performance for mean pooling and vise versa for max pooling case. This variation is related to the noise inside the tag annotation of the ground truth. Many previous deep MIL networks [33,37] recommended max-pooling for MIL where they experimented on object detection dataset containing the ground-truth annotation without any label noise. In contrast, other than the 81-tag set, NUS-WIDE contains significant noise in the tag annotations.

Table 3. Ablation study: Impact of pooling type and bag size on zero-shot tagging

Bag size $(n+1)$	Mean pooling							Max pooling						
	MiAP	K = 3			K = 5			MiAP	K = 3			K = 5		
		P	R	F1	P	R	F1		P	R	F1	P	R	F1
4	32.85	18.28	31.15	23.04	14.79	42.00	21.88	**35.75**	**20.20**	**34.42**	**25.46**	**16.13**	**45.79**	**23.85**
8	36.33	20.68	35.23	26.06	16.38	46.51	24.23	34.89	19.61	33.42	24.72	15.67	44.49	23.18
16	37.20	21.08	35.92	26.57	16.58	47.08	24.53	32.03	18.01	30.68	22.69	14.85	42.17	21.97
32	37.50	21.16	36.06	26.67	16.62	47.20	24.59	29.29	16.57	28.23	20.88	14.00	39.75	20.70
64	**39.01**	**22.05**	**37.56**	**27.79**	**17.26**	**49.01**	**25.53**	32.05	17.68	30.13	22.29	14.38	40.82	21.27

Therefore, mean-pooling with large bag size achieves a balance in the noisy tags, outperforming max-pooling in general for NUS-WIDE. Notably, the bag size of our framework is far less compared to other MIL approaches [33,37]. Being dependent on external bag generator [5,34,44] previous methods lose control inside the generation process. Thus, a large bag size helps them to get enough proposals to choose the best score in last max-pooling layer. Conversely, our method controls the bag generation network by fine-tuning shared ResNet-50 layers which eventually can relax the requirement of large bag sizes.

Tagging in the Wild: Since our method does not use any information about unseen tags in zero-shot settings, it can process an infinite number of unseen tags from an open vocabulary. We test such setting using 5018 tag set of NUS-WIDE. We remove 924 seen tags and ten other tags (hand-sewn, interestingness, manganite, marruecos, mixs, monochromia, shopwindow, skys, topv and uncropped for which no GloVe vectors were found) to produce a large set of 4084 unseen tags. After training with 924 seen tags, the performance of zero-shot tagging with this set is shown in Table 4. Because of extreme noise in the annotations, the results are very poor in general, but our method still outperforms other competitors [22,42].

Table 4. Results for zero-shot tagging task with 4,084 unseen tags.

Method	MiAP	K = 3			K = 5		
		P	R	F1	P	R	F1
ConSE [22]	0.36	0.08	0.06	0.07	0.10	0.13	0.11
Fast0Tag [42]	3.26	3.15	2.40	2.72	2.51	3.18	2.81
Baseline	3.61	3.51	2.67	3.04	2.83	3.59	3.16
Ours (Bag: 32)	**5.85**	**5.42**	**4.12**	**4.68**	**4.42**	**5.60**	**4.94**
Ours (Bag: 64)	5.52	5.01	3.81	4.33	4.10	5.20	4.59

4.3 Zero Shot Recognition (ZSR)

Our proposed framework is designed to handle zero-shot multi-label problem. Therefore, it can also be used for single label ZSR problem. To evaluate the performance of ZSR, we experiment with the Caltech-UCSD Birds-200-2011 (CUB) dataset [35]. Although the size of this dataset is relatively small containing 11,788 images belonging to 200 classes, it is popular for fine-grain recognition tasks. In ZSR literature [39,40], the standard train/test split allows fixed 150 seen and 50 unseen classes for experiments. We follow this traditional setting without using

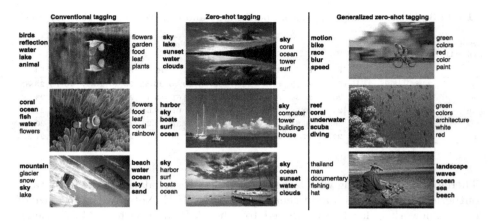

Fig. 3. Example of top 5 predicted tags across different tasks by our method (left/blue) and fast0tag [42] (right/black). **Bold** text represents the correct ground-truth tags. Top two rows illustrate successful examples and third row shows negative cases. (Color figure online)

bounding boxes annotation, per image part annotation (like [2]) and descriptions (like [41]). To be consistent with the rest of the paper, we consider 400-d unsupervised GloVe (glo) and word2vec (w2v) vectors used in [39].

For a test image, our network predicts unseen class scores and an image is classified to the unseen class which gets the maximum score. As per standard practice, we report the mean Top1 accuracy of unseen classes in Table 5. Our method achieves superior results in comparison to state-of-the-art methods using the same settings. Note that, all other methods are deep feature based approaches but do not train a joint framework in an end-to-end manner. In contrast, our method is end-to-end learnable based on ResNet-50 and additionally generates bounding boxes without using any box annotations.

Table 5. Zero-shot recognition on CUB using mean pooling based MIL. For fairness, we only compared with inductive setting of other methods without per image part annotation.

Top1 Accuracy	w2v	glo
Akata'16 [2]	33.90	-
DMaP-I'17 [17]	26.38	30.34
SCoRe'17 [21]	31.51	-
Akata'15 [1]	28.40	24.20
LATEM'16 [39]	31.80	32.50
DMaP-I'17 [17]	26.28	23.69
Ours (Bag size: 32)	31.77	29.56
Ours (Bag size: 64)	**36.55**	**33.00**

4.4 Discussion

How Does MIL Help in the Multi-label Zero-Shot Annotation? We explain this aspect using the illustration in Fig. 3. One can observe that several tags pertain to localized information in a scene that is represented by only a small subset of the whole image, e.g., fish, coral, bike and bird. This demonstrates that a multi-label tagging method should consider localized regions in

Fig. 4. Tag discovery. Bounding boxes are shown for Top 2 tags.

conjunction with the whole image. Our proposed method incorporates such consideration using MIL. Therefore, it can annotate those localized tags where previous method, fast0tag, [42] usually fails (see rows 1–2 in Fig. 3). However, tags like beach, sunset, landscape in the third row of the figure are related to the global information in an image which does not depend on the localized features. Therefore, in this respect, our method sometimes fail in compared to fast0tag [42] (see row 3 in Fig. 3). However, as illustrated in Fig. 3 (the non-bold tags in blue and black colors), the predicted tags of our method in those failure cases are still meaningful and relevant compared to the failure cases of fast0tag [42].

Image Location and Tag Correspondence: As a byproduct, our approach can generate a bounding box for each assigned tag. In Fig. 4, we illustrate some boxes (for top 2 tags) to indicate the correspondence between image locations and associated tags. Notably, our method often selects the whole image as one bounding box because we consider whole image as an instance inside the bag. This consideration is particularly helpful for NUS-WIDE dataset because it contains many tags which are not only related to objects but are relevant to the overall scene e.g., natural concept (sky, water, sunset), aesthetic style (reflection, tattoo) or action (protest, earthquake, sports). Any quantitative analysis for this weakly supervised box detection task was not possible because the NUS-WIDE dataset does not provide any localization ground-truth for tags in an image.

5   Conclusion

While traditional zero-shot learning methods only handle single unseen label per image, this paper attempts to assign multiple unseen tags. For the first time, we propose an end-to-end, deep MIL framework to tackle multi-label zero-shot tagging. Unlike previous models for traditional image tagging, our MIL framework does not depend on off-line bag generator. Building on recent advancements in object detection, our model automatically generates the bag of instances in an efficient manner and can assign both seen and unseen labels to input images. Moreover, any number of unseen tags from an open vocabulary could be employed during test time. In addition, our method can be viewed as a weakly supervised learning approach because of its ability to find a bounding box for each tag without requiring any box annotation during training. We validate our framework by achieving state-of-the-art performance on a large-scale tagging dataset outperforming established methods in the literature.

References

1. Akata, Z., Reed, S., Walter, D., Lee, H., Schiele, B.: Evaluation of output embeddings for fine-grained image classification. In: CVPR, 07–12 June 2015, pp. 2927–2936 (2015)
2. Akata, Z., Malinowski, M., Fritz, M., Schiele, B.: Multi-cue zero-shot learning with strong supervision. In: CVPR, June 2016
3. Bourbaki, N.: Eléments de mathématiques: théorie des ensembles, chapitres 1 à 4, vol. 1. Masson (1990)
4. Chen, M., Zheng, A., Weinberger, K.Q.: Fast image tagging. In: ICML, January 2013
5. Cheng, M.M., Zhang, Z., Lin, W.Y., Torr, P.: Bing: binarized normed gradients for objectness estimation at 300fps. In: CVPR, pp. 3286–3293 (2014)
6. Chua, T.S., Tang, J., Hong, R., Li, H., Luo, Z., Zheng, Y.T.: NUS-WIDE: a real-world web image database from National University of Singapore. In: CIVR, Santorini, Greece, 8–10 July 2009
7. Demirel, B., Gokberk Cinbis, R., Ikizler-Cinbis, N.: Attributes2classname: a discriminative model for attribute-based unsupervised zero-shot learning. In: ICCV, October 2017
8. Deutsch, S., Kolouri, S., Kim, K., Owechko, Y., Soatto, S.: Zero shot learning via multi-scale manifold regularization. In: CVPR, July 2017
9. Feng, J., Zhou, Z.H.: Deep MIML network. In: AAAI, pp. 1884–1890 (2017)
10. Fu, Y., Yang, Y., Hospedales, T., Xiang, T., Gong, S.: Transductive multi-label zero-shot learning. arXiv preprint arXiv:1503.07790 (2015)
11. Girshick, R.: Fast R-CNN. In: ICCV, December 2015
12. Gong, Y., Jia, Y., Leung, T., Toshev, A., Ioffe, S.: Deep convolutional ranking for multilabel image annotation. arXiv preprint arXiv:1312.4894 (2013)
13. Hassoun, M.H.: Fundamentals of Artificial Neural Networks. MIT Press, Cambridge (1995)
14. He, K., Zhang, X., Ren, S., Sun, J.: Deep residual learning for image recognition, vol. 2016, pp. 770–778, January 2016. Cited by 107
15. Dai, J., Li, Y., He, K., Sun, J.: R-FCN: object detection via region-based fully convolutional networks. arXiv preprint arXiv:1605.06409 (2016)
16. Li, X., Liao, S., Lan, W., Du, X., Yang, G.: Zero-shot image tagging by hierarchical semantic embedding. In: RDIR, pp. 879–882. ACM (2015)
17. Li, Y., Wang, D., Hu, H., Lin, Y., Zhuang, Y.: Zero-shot recognition using dual visual-semantic mapping paths. In: CVPR, July 2017
18. Liu, W., et al.: SSD: single shot multibox detector. In: Leibe, B., Matas, J., Sebe, N., Welling, M. (eds.) ECCV 2016. LNCS, vol. 9905, pp. 21–37. Springer, Cham (2016). https://doi.org/10.1007/978-3-319-46448-0_2
19. Mensink, T., Gavves, E., Snoek, C.G.: COSTA: co-occurrence statistics for zero-shot classification. In: CVPR, pp. 2441–2448 (2014)

20. Mikolov, T., Sutskever, I., Chen, K., Corrado, G.S., Dean, J.: Distributed representations of words and phrases and their compositionality. In: NIPS, pp. 3111–3119 (2013)
21. Morgado, P., Vasconcelos, N.: Semantically consistent regularization for zero-shot recognition. In: CVPR, July 2017
22. Norouzi, M., et al.: Zero-shot learning by convex combination of semantic embeddings. In: ICLR (2014)
23. Pennington, J., Socher, R., Manning, C.D.: GloVe: global vectors for word representation. In: EMNLP, pp. 1532–1543 (2014)
24. Qi, C.R., Su, H., Mo, K., Guibas, L.J.: PointNet: deep learning on point sets for 3D classification and segmentation. In: CVPR, vol. 1, no. 2, p. 4 (2017)
25. Rahman, S., Khan, S., Porikli, F.: A unified approach for conventional zero-shot, generalized zero-shot, and few-shot learning. IEEE Trans. Image Process. 27(11), 5652–5667 (2018)
26. Rahman, S., Khan, S., Porikli, F.: Zero-shot object detection: learning to simultaneously recognize and localize novel concepts. In: Asian Conference on Computer Vision (ACCV). Springer, December 2018
27. Redmon, J., Farhadi, A.: Yolo9000: Better, faster, stronger. arXiv preprint arXiv:1612.08242 (2016)
28. Ren, S., He, K., Girshick, R., Sun, J.: Faster R-CNN: towards real-time object detection with region proposal networks. IEEE TPAMI 39(6), 1137–1149 (2017)
29. Ren, Z., Jin, H., Lin, Z., Fang, C., Yuille, A.: Multiple instance visual-semantic embedding. In: BMVC (2017)
30. Simonyan, K., Zisserman, A.: Very deep convolutional networks for large-scale image recognition. arXiv preprint arXiv:1409.1556 (2014)
31. Szegedy, C., et al.: Going deeper with convolutions. In: CVPR, 07–12 June 2015, pp. 1–9 (2015)
32. Tang, P., Wang, X., Feng, B., Liu, W.: Learning multi-instance deep discriminative patterns for image classification. IEEE TIP 26(7), 3385–3396 (2017)
33. Tang, P., Wang, X., Huang, Z., Bai, X., Liu, W.: Deep patch learning for weakly supervised object classification and discovery. Pattern Recogn. 71, 446–459 (2017)
34. Uijlings, J.R.R., van de Sande, K.E.A., Gevers, T., Smeulders, A.W.M.: Selective search for object recognition. IJCV 104(2), 154–171 (2013)
35. Wah, C., Branson, S., Welinder, P., Perona, P., Belongie, S.: The Caltech-UCSD Birds-200-2011 dataset. Technical report, CNS-TR-2011-001, California Institute of Technology (2011)
36. Wang, X., Zhu, Z., Yao, C., Bai, X.: Relaxed multiple-instance SVM with application to object discovery, pp. 1224–1232 (2015)
37. Wei, Y., et al.: HCP: a flexible CNN framework for multi-label image classification. IEEE TPAMI 38(9), 1901–1907 (2016)
38. Wu, J., Yu, Y., Huang, C., Yu, K.: Deep multiple instance learning for image classification and auto-annotation. In: CVPR, pp. 3460–3469, June 2015
39. Xian, Y., Akata, Z., Sharma, G., Nguyen, Q., Hein, M., Schiele, B.: Latent embeddings for zero-shot classification. In: CVPR, June 2016
40. Xian, Y., Schiele, B., Akata, Z.: Zero-shot learning - the good, the bad and the ugly. In: CVPR (2017)
41. Zhang, L., Xiang, T., Gong, S.: Learning a deep embedding model for zero-shot learning. In: CVPR, July 2017
42. Zhang, Y., Gong, B., Shah, M.: Fast zero-shot image tagging. In: CVPR, June 2016

43. Zhou, Y., Sun, X., Liu, D., Zha, Z., Zeng, W.: Adaptive pooling in multi-instance learning for web video annotation. In: ICCV, October 2017
44. Zitnick, C.L., Dollár, P.: Edge boxes: locating object proposals from edges. In: Fleet, D., Pajdla, T., Schiele, B., Tuytelaars, T. (eds.) ECCV 2014. LNCS, vol. 8693, pp. 391–405. Springer, Cham (2014). https://doi.org/10.1007/978-3-319-10602-1_26

Zero-Shot Object Detection: Learning to Simultaneously Recognize and Localize Novel Concepts

Shafin Rahman[1,2]([✉])[iD], Salman Khan[1,2,3][iD], and Fatih Porikli[1,4][iD]

[1] Australian National University, Canberra, ACT 2601, Australia
{shafin.rahman,salman.khan,fatih.porikli}@anu.edu.au
[2] Data61, CSIRO, Canberra, ACT 2601, Australia
[3] Inception Institute of AI, Abu Dhabi, UAE
[4] Huawei, Santa Clara, CA 95050, USA

Abstract. Current Zero-Shot Learning (ZSL) approaches are restricted to recognition of a single dominant unseen object category in a test image. We hypothesize that this setting is ill-suited for real-world applications where unseen objects appear only as a part of a complex scene, warranting both 'recognition' and 'localization' of an unseen category. To address this limitation, we introduce a new *'Zero-Shot Detection'* (ZSD) problem setting, which aims at simultaneously recognizing and locating object instances belonging to novel categories without any training examples. We also propose a new experimental protocol for ZSD based on the highly challenging ILSVRC dataset, adhering to practical issues, e.g., the rarity of unseen objects. To the best of our knowledge, this is the first end-to-end deep network for ZSD that jointly models the interplay between visual and semantic domain information. To overcome the noise in the automatically derived semantic descriptions, we utilize the concept of meta-classes to design an original loss function that achieves synergy between max-margin class separation and semantic space clustering. Furthermore, we present a baseline approach extended from recognition to ZSD setting. Our extensive experiments show significant performance boost over the baseline on the imperative yet difficult ZSD problem.

Keywords: Zero-shot learning · Zero-shot detection · Object detection

1 Introduction

Since its inception, zero-shot learning research has been dominated by the object classification problem [3,4,8,13,15,17,20,24,29,34,36–38]. Although it still remains as a challenging task, the zero-shot recognition has a number of

Electronic supplementary material The online version of this chapter (https://doi.org/10.1007/978-3-030-20887-5_34) contains supplementary material, which is available to authorized users.

© Springer Nature Switzerland AG 2019
C. V. Jawahar et al. (Eds.): ACCV 2018, LNCS 11361, pp. 547–563, 2019.
https://doi.org/10.1007/978-3-030-20887-5_34

Fig. 1. ZSD deals with a more complex label space (object labels and locations) with considerably less supervision (i.e., no examples of unseen classes). (**a**) Traditional recognition task only predicts seen class labels. (**b**) Traditional detection task predicts both seen class labels and bounding boxes. (**c**) Traditional zero-shot recognition task only predicts unseen class labels. (**d**) The proposed ZSD predicts both seen and unseen classes and their bounding boxes.

limitations that render it unusable in real-life scenarios. *First*, it is destined to work for simpler cases where only a single dominant object is present in an image. *Second*, the attributes and semantic descriptions are relevant to individual objects instead of the entire scene composition. *Third*, zero-shot recognition provides an answer to unseen categories in elementary tasks, e.g., classification and retrieval, yet it is unable to scale to advanced tasks such as scene interpretation and contextual modeling, which require a fundamental reasoning about all salient objects in the scene. *Fourth*, global attributes are more susceptible to background variations, viewpoint, appearance and scale changes and practical factors such as occlusions and clutter. As a result, image-level ZSL fails for the case of complex scenes where a diverse set of competing attributes that do not belong to a single image-level category would exist.

To address these challenges, we introduce a new problem setting called the *zero-shot object detection*. As illustrated in Fig. 1, instead of merely classifying images, our goal is to simultaneously detect and localize each individual instance of new object classes, even in the absence of any visual examples of those classes during the training phase. In this regard, we propose a new zero-shot detection protocol built on top of the ILSVRC - Object Detection Challenge [30]. The resulting dataset is very demanding due to its large scale, diversity, and unconstrained nature, and also unique due to its leveraging on WordNet semantic hierarchy [22]. Taking advantage of semantic relationships between object classes, we use the concept of '*meta-classes*'[1] and introduce a novel approach to update the semantic embeddings automatically. Raw semantic embeddings are learned in an unsupervised manner using text mining and therefore they have considerable noise. Our optimization of the class embeddings proves to be an effective way to reduce this noise and learn robust semantic representations.

ZSD has numerous applications in novel object localization, retrieval, tracking, and reasoning about object's relationships with its environment using only

---

[1] Meta-classes are obtained by clustering semantically similar classes.

available semantics, e.g., an object name or a natural language description. Although a critical problem, ZSD is remarkably difficult compared to its classification counterpart. While the zero-shot recognition problem assumes only a single primary object in an image and attempts to predict its category, the ZSD task has to predict both the multi-class category label and precise location of each instance in the given image. Since there can be a prohibitively huge number of possible locations for each object in an image and because the semantic class descriptions are noisy, a detection approach is much more susceptible to incorrect predictions compared to classification. Therefore, it would be expected that a ZSD method predicts a class label that might be incorrect but visually and semantically similar to the corresponding true class. For example, wrongly predicting a 'spider' as 'scorpion' where both are semantically similar because of being invertebrates. To address this issue, we relax the original detection problem to independently study the confusions emanating from the visual and semantic resemblance between closely linked classes. For this purpose, alongside the ZSD, we evaluate on zero-shot meta-class detection, zero-shot tagging, and zero-shot meta class tagging. Notably, the proposed network is trained only 'once' for ZSD task and the additional tasks are used during evaluations only.

Although deep network based solutions have been proposed for zero-shot recognition [8,17,36], to the best of our knowledge, we propose the first end-to-end trainable network for the ZSD problem that concurrently relates visual image features with the semantic label information. This network considers semantic embedding vector of classes within the network to produce prediction scores for both seen and unseen classes. We propose a novel loss formulation that incorporates max-margin learning [38] and a semantic clustering loss based on class-scores of different meta-classes. While the max-margin loss tries to separate individual classes, semantic clustering loss tries to reduce the noise in semantic vectors by positioning similar classes together and dissimilar classes far apart. Notably, our proposed formulation assumes predefined unseen classes to explore the semantic relationships during model learning phase. This assumption is consistent with recent efforts in the literature which consider class semantics to solve the domain shift problem in ZSL [5,10] and does not a constitute transductive setting [6,9,13]. Based on the premise that unseen class semantics may be unknown during training in several practical zero-shot scenarios, we also propose a variant of our approach that can be trained without predefined unseen classes. Finally, we propose a comparison method for ZSD by extending a popular zero-shot recognition framework named ConSE [24] using Faster-RCNN [28].

In summary, this paper reports the following advances:

- We introduce a new problem for zero-shot learning, which aims to jointly recognize and localize novel objects in complex scenes.
- We present a new experimental protocol and design a novel baseline solution extended from conventional recognition to the detection task.
- We propose an end-to-end trainable deep architecture that simultaneously considers both visual and semantic information.

– We design a novel loss function that achieves synergistic effects for max-margin class separation and semantic clustering based on meta-classes. Beside that, our approach can automatically tune noisy semantic embeddings.

2   Problem Description

Given a set of images for seen object categories, ZSD aims at the *recognition* and *localization* of previously unseen object categories. In this section, we formally describe the ZSD problem and its associated challenges. We also introduce variants of the detection task, which are natural extensions of the original problem. First, we describe the notations used in the following discussion.

Preliminaries: Consider a set of 'seen' classes denoted by $\mathcal{S} = \{1, \ldots, S\}$, whose examples are available during the training stage and S represents their total number. There exists another set of 'unseen' classes $\mathcal{U} = \{S+1, \ldots, S+U\}$, whose instances are only available during the test phase. We denote the set of all object classes by $\mathcal{C} = \mathcal{S} \cup \mathcal{U}$, such that $C = S + U$ denote the cardinality of the label space.

We define a set of meta (or super) classes by grouping similar object classes into a single meta category. These meta-classes are denoted by $\mathcal{M} = \{z_m : m \in [1, M]\}$, where M denote the total number of meta-classes and $z_m = \{k \in \mathcal{C} \ s.t., g(k) = m\}$. Here, $g(k)$ is a mapping function which maps each class k to its corresponding meta-class $z_{g(k)}$. Note that the meta-classes are mutually exclusive i.e., $\cap_{m=1}^{M} z_m = \phi$ and $\cup_{m=1}^{M} z_m = \mathcal{C}$.

The set of all training images is denoted by $\mathcal{X}^s$, which contains examples of all seen object classes. The set of all test images containing samples of unseen object classes is denoted by $\mathcal{X}^u$. Each test image $\mathbf{x} \in \mathcal{X}^u$ contains at least one instance of an unseen class. Notably, no unseen class object is present in $\mathcal{X}^s$, but $\mathcal{X}^u$ may contain seen objects.

We define a d dimensional word vector $\mathbf{v_c}$ (word2vec or GloVe) for every class $c \in \mathcal{C}$. The ground-truth label for an ith bounding box is denoted by y_i. The object detection task also involves identifying the background class for negative object proposals, we introduce the extended label sets: $\mathcal{S}' = \mathcal{S} \cup y_{bg}$, $\mathcal{C}' = \mathcal{C} \cup y_{bg}$ and $\mathcal{M}' = \mathcal{M} \cup y_{bg}$, where $y_{bg} = \{C+1\}$ is a singleton set denoting the background label.

Task Definitions: Given the observed space of images $\mathcal{X} = \mathcal{X}^s \cup \mathcal{X}^u$ and the output label space $\mathcal{C}'$, our goal is to learn a mapping function $f : \mathcal{X} \mapsto \mathcal{C}'$ which gives the minimum regularized empirical risk ($\hat{\mathcal{R}}$) as follows:

$$\arg\min_{\Theta} \hat{\mathcal{R}}(f(\mathbf{x}; \Theta)) + \Omega(\Theta), \tag{1}$$

where, $\mathbf{x} \in \mathcal{X}^s$ during training, Θ denotes the set of parameters and $\Omega(\Theta)$ denotes the regularization on the learned weights. The mapping function has the following form:

$$f(\mathbf{x}; \Theta) = \arg\max_{y \in \mathcal{C}} \max_{b \in \mathcal{B}(\mathbf{x})} \mathcal{F}(\mathbf{x}, y, b; \Theta), \tag{2}$$

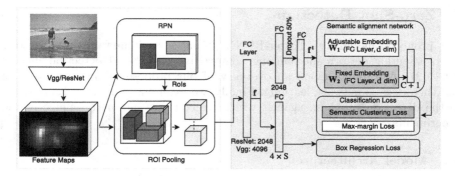

Fig. 2. Network Architecture - *Left:* Image level feature maps are used to propose candidate object boxes and their corresponding features. *Right:* The features are used for classification and localization of new classes by utilizing their semantic concepts.

where $\mathcal{F}(\cdot)$ is a compatibility function, $\mathcal{B}(\mathbf{x})$ is the set of all bounding box proposals in a given image $\mathbf{x}$. Intuitively, Eq. 2 finds the best scoring bounding boxes for each object category and assigns them the maximum scoring object category. Next, we define the zero-shot learning tasks which go beyond a single unseen category recognition in images. Notably, the training is framed as the challenging ZSD problem, however the remaining task descriptions are used during evaluation to relax the original problem:

T1 *Zero-shot detection (ZSD):* Given a test image $\mathbf{x} \in \mathcal{X}^u$, the goal is to categorize and localize each instance of an unseen object class $u \in \mathcal{U}$.

T2 *Zero-shot meta-class detection (ZSMD):* Given a test image $\mathbf{x} \in \mathcal{X}^u$, the goal is to localize each instance of an unseen object class $u \in \mathcal{U}$ and categorize it into one of the super-classes $m \in \mathcal{M}$.

T3 *Zero-shot tagging (ZST):* To recognize one or more unseen classes in a test image $\mathbf{x} \in \mathcal{X}^u$, without identifying their location.

T4 *Zero-shot meta-class tagging (ZSMT):* To recognize one or more meta-classes in a test image $\mathbf{x} \in \mathcal{X}^u$, without identifying their location.

Among the above mentioned tasks, the ZSD is the most difficult problem and difficulty level decreases as we go down the list. The goal of the later tasks is to distill the main challenges in ZSD by investigating two ways to relax the original problem: **(a)** The effect of reducing the unseen object classes by clustering similar unseen classes into a single super-class (T2 and T4). **(b)** The effect of removing the localization constraint. To this end we investigate the zero-shot tagging problem, where the goal is to only recognize all object categories in an image (T3 and T4).

The state-of-the-art in zero-shot learning deals with only recognition/tagging. The proposed problem settings add the missing detection task which indirectly encapsulates traditional recognition and tagging tasks.

3   Zero-Shot Detection

Our proposed model uses Faster-RCNN [28] as a backbone architecture, due to its superior performance among competitive end-to-end detection models [12,19,27]. We first provide an overview of our proposed model architecture and then discuss network learning. Finally, we extend a popular ZSL approach to the detection problem, against which we compare our performance in the experiments.

3.1   Model Architecture

The overall architecture is illustrated in Fig. 2. It has two main components marked in color: the first provides object-level feature descriptions and the second integrates visual information with the semantic embeddings to perform zero-shot detection. We explain these in detail next.

Object-Level Feature Encoding: For an input image $\mathbf{x}$, a deep network (VGG/ResNet) is used to obtain the intermediate convolutional activations. These activations are treated as feature maps, which are forwarded to a Region Proposal Network (RPN). The RPN generates a set of candidate object proposals by automatically ranking the anchor boxes at each sliding window location. The high-scoring proposals can be of different sizes, which are mapped to fixed sized representation using a RoI pooling layer that operates on the initial feature maps and the proposals generated by the RPN. The resulting object level features for each candidate are denoted as '$\mathbf{f}$'. Note that the RPN generates object proposal based on the objectness measure. Thus, a trained RPN on seen objects can generate proposals for unseen objects also. In the second block of our architecture, these feature representations are used alongside the semantic embeddings to learn useful representations for both the seen and unseen object-categories.

Integrating Visual and Semantic Contexts: The object-level feature $\mathbf{f}$ is forwarded to two branches in the second module. The **top branch** is trained to predict the object category for each candidate box. Note that this can assign a class $c \in \mathcal{C}'$, which can be a seen, unseen or background category. The branch consists of two main sub-networks, which are key to learning the semantic relationships between seen and unseen object classes.

The first component is the '*Semantic Alignment Network*' (SAN), which consist of an adjustable FC layer, whose parameters are denoted as $\mathbf{W}_1 \in \mathbb{R}^{d \times d}$, that projects the input visual feature vectors to a semantic space with d dimensions. The resulting feature maps are then projected onto the **fixed** semantic embeddings, denoted by $\mathbf{W}_2 \in \mathbb{R}^{d \times (C+1)}$, which are obtained in an unsupervised manner by text mining (e.g., Word2vec and GloVe embeddings). Note that, here we consider both seen and unseen semantic vectors which require unseen classes to be predefined. This consideration is inline with a very recent effort [10] which adopt this setting to explore the cluster manifold structure of the semantic embedding space and address domain shift issue. Given a feature representation input to SAN in the top branch, $\mathbf{f}^t$, the overall operation can be represented as:

$$\mathbf{o} = (\mathbf{W_1}\mathbf{W_2})^T \mathbf{f}^t. \tag{3}$$

Here, $\mathbf{o}$ is the output prediction score. The $\mathbf{W_2}$ is formed by stacking semantic vectors for all classes, including the background class. For background class, we use the mean word vectors $\mathbf{v}_b = \frac{1}{C}\sum_{c=1}^{C}\mathbf{v}_c$ as its embedding in $\mathbf{W_2}$.

Notably, a non-linear activation function is not applied between the adjustable and fixed semantic embeddings in the SAN. Therefore, the two projections can be understood as a single learnable projection on to the semantic embeddings of object classes. This helps in automatically updating the semantic embeddings to make them compatible with the visual feature domain. It is highly valuable because the original semantic embeddings are often noisy due to the ambiguous nature of closely related semantic concepts and the unsupervised procedure used for their calculation. In Fig. 3, we visualize modified embedding space when different loss functions are applied during training.

The **bottom branch** is for bounding box regression to add suitable offsets to the proposals to align them with the ground-truths such that precise location of objects can be predicted. This branch is set up similar to Faster-RCNN [28].

3.2 Training and Inference

We follow a two step training approach to learn the model parameters. The **first** part involves training the backbone Faster-RCNN for only seen classes using the training set $\mathcal{X}^s$. This training involves initializing weights of shared layers with a pre-trained Vgg/ResNet model, followed by learning the RPN, classification and detection networks. In the **second** step, we modify the Faster-RCNN model by replacing the last layer of Faster-RCNN classification branch with the proposed semantic alignment network and an updated loss function (see Fig. 2). While rest of the network weights are used from the first step, the weights $\mathbf{W_1}$ are randomly initialized and the $\mathbf{W_2}$ are fixed to semantic vectors of the object classes and not updated during training.

While training in second step, we keep the shared layers trainable but fix the layers specific to RPN since the object proposals requirements are not changed from the previous step. The same seen class images $\mathcal{X}^s$ are again used for training. For each given image, we obtain the output of RPN which consists of a total of 'R' ROIs belonging to both positive and negative object proposals.

Each proposal has a corresponding ground-truth label given by $y_i \in \mathcal{S}'$. Positive proposals belong to any of the seen class $\mathcal{S}$ and negative proposals contain only background. In our implementation, we use an equal number of positive and negative proposals. Now, when object proposals are passed through ROI-Pooling and subsequent dense layers, a feature representation $\mathbf{f}_i$ is calculated for each ROI. This feature is forwarded to two branches, the classification branch and regression branch. The overall loss is the summation of the respective losses in these two branches, i.e., classification loss and bounding box regression loss.

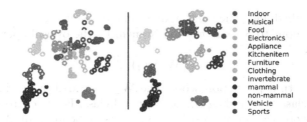

Fig. 3. The 2D tSNE embedding of modified word vectors $\mathbf{W_1W_2}$ using only max-margin loss, L_{mm} (left) and with clustering loss, $L_{mm} + L_{mc}$ (right). Semantically similar classes are embedded more closely in cluster based loss.

$$L(\mathbf{o}_i, b_i, y_i, b_i^*) = \arg\min_{\Theta} \frac{1}{T} \sum_i \left(L_{cls}(\mathbf{o}_i, y_i) + L_{reg}(b_i, b_i^*) \right)$$

where Θ denotes the parameters of the network, $\mathbf{o}_i$ is the classification branch output, $T = N \times R$ represents the total number of ROIs in the training set with N images. b_i and b_i^* are parameterized coordinates of predicted and ground-truth bounding boxes respectively and y_i represents the true class label of the ith object proposal.

Classification Loss: This loss deals with both seen and unseen classes. It comprises of a max-margin loss (L_{mm}) and a meta-class clustering loss (L_{mc}):

$$L_{cls}(\mathbf{o}_i, y_i) = \lambda L_{mm}(\mathbf{o}_i, y_i) + (1 - \lambda)L_{mc}(\mathbf{o}_i, g(y_i)), \tag{4}$$

where, hyper-parameter λ controls the trade-off between two losses. We define,

$$L_{mm}(\mathbf{o}_i, y_i) = \frac{1}{|\mathcal{C}' \setminus y_i|} \sum_{c \in \mathcal{C}' \setminus y_i} \log\left(1 + \exp(o_c - o_{y_i})\right), \text{ and}$$

$$L_{mc}(\mathbf{o}_i, g(y_i)) = \frac{1}{|\mathcal{M}' \setminus z_{g(y_i)}||z_{g(y_i)}|} \sum_{c \in \mathcal{M}' \setminus z_{g(y_i)}} \sum_{j \in z_{g(y_i)}} \log\left(1 + \exp(o_c - o_j)\right)$$

where, o_k represents the prediction response of class $k \in \mathcal{S}$. L_{mm} tries to separate the prediction response of true class from rest of the classes. In contrast, L_{mc} tries to cluster together the members of each super-class and pulls further apart the classes belonging to different meta-classes.

We illustrate the effect of clustering loss on the learned embeddings in Fig. 3. The use of L_{mc} enables us to cluster semantically similar classes together which results in improved embeddings in the semantic space. For example, all animals related meta-classes are in close position whereas food and vehicle are far apart. Such a clear separation in semantic space helps in obtaining a better ZSD performance. Moreover, meta-class based clustering loss does not harm fine-grained detection because the hype-parameter λ is used to put more emphasis on the max-margin loss (L_{mm}) as compared to the clustering part (L_{mc}) of the overall loss (L_{cls}). Still, the clustering loss provides enough guidance to the noisy semantic embeddings (e.g., unsupervised w2v/glove) such that similar classes are clustered together as illustrated in Fig. 3. Note that w2v/glove try to place similar words nearby with respect to millions of text corpus, it is therefore not fine-tuned for just 200 class recognition setting.

Regression Loss: This part of the loss fine-tunes the bounding box for each seen class ROI. For each $\mathbf{f}_i$, we get $4 \times S$ values representing 4 parameterized co-ordinates of the bounding box of each object instance. The regression loss is calculated based on these co-ordinates and parameterized ground truth co-ordinates. During training, no bounding box prediction is done for background and unseen classes due to unavailability of visual examples. As an alternate approach, we approximate the bounding box for an unseen object through the box proposal for a closely related seen object that achieves maximum response. This is a reasonable approximation because visual features of unseen classes are related to that of similar seen classes.

Prediction: We normalize each output prediction value of classification branch using $\hat{o}_c = \frac{o_c}{\|\mathbf{v}_c\|_2 \|\mathbf{f}^t\|_2}$. It basically calculates the cosine similarity between modified word vectors and image features. This normalization maps the prediction values within 0 to 1 range. We classify an object proposal as background if maximum responds among $\hat{o}_c$ where $c \in \mathcal{C}'$ belongs to y_{bg}. Otherwise, we detect an object proposal as unseen object if its maximum prediction response among $\hat{o}_u$ where $u \in \mathcal{U}$ is above a threshold α.

$$y_u = \arg \max_{u \in \mathcal{U}} \hat{o}_u \quad s.t., \ \hat{o}_u > \alpha. \tag{5}$$

The other detection branch finds b_i which is the set of parameterized co-ordinates of bounding boxes for S seen classes. Among them, we choose a bounding box corresponding to the class having the maximum prediction response in $\hat{o}_s$ where $s \in \mathcal{S}$ for the classified unseen class y_u. For the tagging tasks, we simply use the mapping function $g(.)$ to assign a meta-class for any unseen label.

3.3 ZSD Without Pre-defined Unseen

While applying clustering loss in Sect. 3.2, the meta-class assignment adds high-level supervision in the semantic space. While doing this assignment, we consider both seen and unseen classes. Similarly, the max-margin loss considers the set $\mathcal{C}'$ that has both seen and unseen classes. This problem setting helps to identify the clustering structure of the semantic embeddings to address domain adaptation for zero-shot detection. However, in several practical scenarios, unseen classes may not be known during training. Here, we report a simplified variant of our approach to train the proposed network without pre-defined unseen classes.

For this problem setting, we use only seen+bg word vectors (instead of seen+unseen+bg vectors) as the fixed embedding $\mathbf{W}_2 \in \mathbb{R}^{d \times (S+1)}$ to train the whole framework with only the max-margin loss, L'_{mm}, defined as follows: $L'_{mm}(\mathbf{o}_i, y_i) = \frac{1}{|\mathcal{S}' \backslash y_i|} \sum_{c \in \mathcal{S}' \backslash y_i} \log \left(1 + \exp(o_c - o_{y_i})\right)$. Since the output classification layer cannot make predictions for unseen classes, we apply a procedure similar to ConSE during the testing phase [24]. Here, the choice of [24] is made due to two main reasons: **(a)** In contrast to other ZSL methods which train separate models for each class [4,26], ConSE can work on the prediction score of a single model. **(b)** It is straight-forward to extend a single network to ZSD using ConSE, since [24] uses semantic embeddings only during the test phase.

Suppose, for an object proposal, vector $\mathbf{o} \in \mathbb{R}^{S+1}$ contains final probability values of only seen classes and background. As described earlier, we ignore an object proposal if the background gets highest score. For other cases, we sort the vector $\mathbf{o}$ in descending order to compute a list of indices l and the sorted list $\hat{\mathbf{o}}$:

$$\hat{\mathbf{o}}, l = \text{sort}(\mathbf{o}) \quad s.t., \quad o_j = \hat{o}_{l_j}. \tag{6}$$

Then, top K score values (s.t., $K \leq S$) from $\hat{\mathbf{o}}$ are combined with their corresponding word vectors using the equation: $\mathbf{e}_i = \sum_{k=1}^{K} \hat{o}_k \mathbf{v}_{l_k}$. We consider $\mathbf{e}_i$ as a semantic space projection of an object proposal which is a combination of word vectors weighted by top K seen class probabilities. The final prediction is made by finding the maximum cosine similarity among $\mathbf{e}_i$ and all unseen word vectors,

$$y_u = \arg\max_{u \in \mathcal{U}} \cos(\mathbf{e}_i, \mathbf{v}_u).$$

In this paper, we use $K = 10$ as proposed in [24]. For bounding box detection, we choose the box for which corresponding seen class gets maximum score.

4   Experiments

4.1   Dataset and Experiment Protocol

Dataset: We evaluate our approach on the standard ILSVRC-2017 detection dataset [30]. This dataset contains 200 object categories. For training, it includes 456,567 images and 478,807 bounding box annotations around object instances. The validation dataset contains 20,121 images fully annotated with the 200 object categories which include 55,502 object instances. A category hierarchy has been defined in [30], where some objects have multiple parents. Since, we also evaluate our approach on meta-class detection and tagging, we define a single parent for each category (see supplementary material for detail).

Seen/Unseen Split: Due to lack of an existing ZSD protocol, we propose a challenging seen/unseen split for ILSVRC-2017 detection dataset. Among 200 object categories, we randomly select 23 categories as unseen and rest of the 177 categories are considered as seen. This split is designed to follows the following practical considerations: *(a)* unseen classes are rare, *(b)* test categories should be diverse, *(c)* the unseen classes should be semantically similar with at least some of the seen classes. The details of split are provided in supplementary material.

Train/Test Set: A zero-shot setting does not allow any visual example of an unseen class during training. Therefore, we customize the training set of ILSVRC such that images containing any unseen instance are removed. This results in

Table 1. mAP of the unseen classes. Ours (with L'_{mm}) and Ours (with L_{cls}) denote the performance without predefined unseen and with cluster loss respectively (Sects. 3.3 and 3.2). For cluster case, $\lambda = 0.8$.

Network	ZSD			ZSMD			ZST			ZSMT		
	Baseline	Ours (L'_{mm})	Ours (L_{cls})	Baseline	Ours (L'_{mm})	Ours (L_{cls})	Baseline	Ours (L'_{mm})	Ours (L_{cls})	Baseline	Ours (L'_{mm})	Ours (L_{cls})
R+w2v	12.7	15.0	**16.0**	13.7	15.4	**15.4**	23.3	27.5	**30.0**	28.8	33.4	**39.3**
R+glo	12.0	12.3	**14.6**	12.9	14.1	**16.1**	22.3	24.5	**26.2**	29.2	31.5	**36.3**
V+w2v	10.2	**12.7**	11.8	11.4	**12.5**	11.8	23.3	25.6	**26.2**	29.0	31.3	**36.0**
V+glo	9.0	10.8	**11.6**	9.7	11.3	**11.8**	20.3	22.9	**23.9**	27.3	29.2	**34.2**

a total of 315,731 training images with 449,469 annotated bounding boxes. For testing, the traditional zero-shot recognition setting is used which considers only unseen classes. As the test set annotations are not available to us, we cannot separate unseen classes for evaluation. Therefore, our test set is composed of the left out data from ILSVRC training dataset plus validation images having at least one unseen bounding box. The resulting test set has 19,008 images and 19,931 bounding boxes.

Semantic Embedding: Traditionally ZSL methods report performance on both supervised attributes and unsupervised word2vec/glove as semantic embeddings. As manually labeled supervised attributes are hard to obtain, only small-scale datasets with these annotations are available [7,16]. ILSVRC-2017 detection dataset used in the current work is quite huge and does not provide attribute annotations. In this paper, we work on ℓ_2 normalized 500 and 300 dimensional unsupervised word2vec [21] and GloVe [25] vector respectively to describe the classes. These word vectors are obtained by training on several billion words from Wikipedia dump corpus.

Evaluation Metric: We report average precision (AP) of individual unseen classes and mean average precision (mAP) for the overall performance of unseen classes.

Implementation Details: Unlike Faster-RCNN, our first step is trained in one step: after initializing shared layer with pre-trained weights, RPN and detection network of Fast-RCNN layers are learned together. Some other settings includes rescaling shorter size of image as 600 pixels, RPN stride = 16, three anchor box scale 128, 256 and 512 pixels, three aspect ratios 1:1, 1:2 and 2:1, non-maximum suppression (NMS) on proposals class probability with IoU threshold = 0.7. Each mini-batch is obtained from a single image having 16 positive and 16 negative (background) proposals. Adam optimizer with learning rate 10^{-5}, $\beta_1 = 0.9$ and $\beta_2 = 0.999$ is used in both state training. First step is trained over 10 million mini-batches without any data augmentation, but data augmentation through repetition of object proposals is used in second step (details in supplementary material). During testing, the prediction score threshold was 0.1 for baseline and Ours (with L'_{mm}) and 0.2 for clustering method (Ours with L_{cls}). We implement our model in *Keras*.

Table 2. Average precision of individual unseen classes using ResNet+w2v and loss configurations L'_{mm} and L_{cls} (cluster based loss with $\lambda = 0.6$). We have grouped unseen classes into two groups based on whether visually similar classes present in the seen class set or not. Our proposed method achieve significant performance improvement for the group where similar classes are present in the seen set.

	OVERALL	p.box	syringe	harmonica	maraca	burrito	pineapple	bowtie	s.trunk	d.washer	canopener	p.rack	bench	e.fan	iPod	scorpion	snail	hamster	tiger	ray	train	unicycle	golfball	h.bar
		Similar classes NOT present ZSD Baseline = 6.3, Ours (L'_{mm}) = **6.5**, Ours (L_{cls}) = 4.4											Similar classes present ZSD Baseline = 18.6, Ours (L'_{mm}) = 22.7, Ours (L_{cls}) = **27.4**											
							Zero-Shot Detection (ZSD)																	
Baseline	12.7	0.0	3.9	**0.5**	0.0	36.3	**2.7**	1.8	1.7	**12.2**	2.7	**7.0**	1.0	0.6	22.0	19.0	1.9	40.9	**75.3**	0.3	28.4	**17.9**	12.0	4.0
Ours (L'_{mm})	15.0	0.0	**8.0**	0.2	0.2	**39.2**	2.3	**1.9**	**3.2**	11.7	**4.8**	0.0	0.0	**7.1**	23.3	25.7	**5.0**	**50.5**	**75.3**	0.0	44.8	7.8	**28.9**	**4.5**
Ours (L_{cls})	**16.4**	**5.6**	1.0	0.1	0.0	27.8	1.7	1.5	1.6	7.2	2.2	0.0	**4.1**	5.3	**26.7**	**65.6**	4.0	47.3	71.5	**21.5**	**51.1**	3.7	26.2	1.2
							Zero-Shot Tagging (ZST)																	
Baseline	23.3	2.9	13.4	9.6	3.1	61.7	20.7	16.3	7.5	29.4	8.6	**12.2**	8.5	4.9	46.2	30.7	11.0	51.8	77.6	9.0	46.1	**39.0**	12.7	12.6
Ours (L'_{mm})	27.5	2.9	**20.8**	10.5	3.3	**72.5**	**27.7**	16.7	7.9	22.9	**14.3**	2.8	6.7	**14.5**	46.8	42.6	**16.0**	**59.1**	**80.0**	12.9	67.3	34.1	**34.0**	**17.1**
Ours (L_{cls})	**30.6**	**12.6**	10.2	**11.9**	**4.9**	48.9	21.8	**17.9**	**29.1**	**32.2**	10.0	4.1	**20.7**	10.7	**52.2**	**82.6**	12.3	58.5	75.5	**48.9**	**72.2**	16.9	33.9	15.5

Meta-class	OVERALL	Indoor	Musical	Food	Clothing	Appli.	Kitchen	Furn.	Electronic	Invertebra.	Mammal	Fish	Vehicle	Sport
					Zero-Shot Meta Detection (ZSMD)									
Baseline	13.7	3.3	**0.3**	24.0	4.0	12.2	2.1	1.0	12.1	17.0	70.7	0.3	22.1	8.5
Ours (L'_{mm})	15.4	**8.1**	0.1	18.4	2.3	11.7	3.0	0.0	14.3	27.8	**73.6**	0.0	**32.1**	9.0
Ours (L_{cls})	**15.6**	3.5	0.1	10.0	1.9	7.2	1.2	**4.1**	**15.3**	**31.4**	66.8	**21.5**	31.2	**9.3**
					Zero-Shot Meta-class Tagging (ZSMT)									
Baseline	28.8	15.2	12.0	55.6	25.2	29.4	10.7	8.5	31.5	36.5	75.8	9.0	48.4	17.0
Ours (L'_{mm})	33.4	**24.1**	13.6	**55.9**	31.3	22.9	**14.7**	6.7	33.0	49.4	82.6	12.9	64.2	23.2
Ours (L_{cls})	**39.9**	19.2	**15.5**	45.6	**38.5**	**32.2**	12.4	**20.7**	**40.3**	**58.2**	**84.8**	**48.9**	**74.7**	**27.1**

4.2 ZSD Performance

We compare different versions of our method (with loss configurations L'_{mm} and L_{cls} respectively) to a baseline approach. Note that the baseline is a simple extension of Faster-RCNN [28] and ConSE [24]. We apply the inference strategy mentioned in Sect. 3.3 after first step training as we can still get a vector $\mathbf{o} \in \mathbb{R}^{S+1}$ on the classification layer of Faster-RCNN network. We use two different architectures i.e., VGG-16 (V) [31] and ResNet-50 (R) [11] as the backbone of the Faster-RCNN during the first training step. In second step, we experiment with both Word2vec and GloVe as the semantic embedding vectors used to define $\mathbf{W}_2$. Figure 5 illustrates some qualitative ZSD examples. More performance results of ZSD on other datasets is provided in the supplementary material.

Overall Results: Table 1 reports the mAP for all approaches on four tasks: ZSD, ZSMD, ZST, and ZSMT across different combinations of network architectures. We can make following observations: *(1)* Our cluster based method outperforms other competitors on all four tasks because its loss utilizes high-level semantic relationships from meta-class definitions which are not present in other methods. *(2)* Performances get improved from baseline to Ours (with L'_{mm}) across all zero-shot tasks. The reason is baseline method did not consider word vectors during the training. Thus, overall detection could not get enough supervision about the semantic embeddings of classes. In contrast, L'_{mm} loss formulation considers word vectors. *(3)* Performances get improved from ZST to ZSMT across all methods whereas similar improvement is not common from ZSD to ZSMD. It's not surprising because ZSMD can get some benefit if meta-class of the predicted class is same as the meta-class of true class. If this is violated

Table 3. Zero shot recognition on CUB using $\lambda = 1$ because no meta-class assignment is done here. For fairness, we only compared our result with the inductive setting of other methods without per image part annotation and description. We refer V = VGG, R = ResNet, G = GoogLeNet.

Top1 accuracy	Network	w2v	glo
Akata'16 [2]	V	33.90	-
DMaP-I'17 [18]	G+V	26.38	30.34
SCoRe'17 [23]	G	31.51	-
Akata'15 [1]	G	28.40	24.20
LATEM'16 [33]	G	31.80	32.50
DMaP-I'17 [18]	G	26.28	23.69
Ours	R	**36.77**	**36.82**

frequently, we cannot expect significant performance improvement in ZSMD. *(4)* In comparison of traditional object detection results, ZSD achieved significantly lower performance. Remarkably, even the state-of-the-art zero-shot classification approaches perform quite low e.g., a recent ZSL method [36] reported 11% hit@1 rate on ILSVRC 2010/12. This trend does not undermine to significance of ZSD, rather highlights the underlying challenges.

Individual Class Detection: Performances of individual unseen classes indicate the challenges for ZSD. In Table 2, we show performances of individual unseen classes across all tasks with our best (R+w2v) network. We observe that the unseen classes for which visually similar classes are present in their meta-classes achieve better detection performance (ZSD mAP 18.6, 22.7, 27.4) than those which do not have similar classes (ZSD mAP 6.3, 6.5, 4.4) for the all methods (baseline, our's with L'_{mm} and L_{cls}). Our proposed cluster method with loss L_{cls} outperforms the other versions significantly for the case when visually similar classes are present. For the all classes, our cluster method is still the best (mAP: cluster 16.4 vs. baseline 12.7). However, our's with L'_{mm} method performs better for when case similar classes are not present (mAP 6.5 vs 4.4).

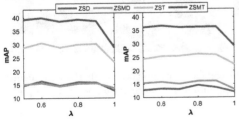

Fig. 4. Effect of varying λ in different zero-shot tasks for ResNet+w2v (left) and ResNet+glo (right).

For the easier tagging tasks (ZST and ZSMT), the cluster method gets superior performance in most of the cases. This indicates that one potential reason for the failure cases of our cluster method for ZSD might be confusions during localization of objects due to ambiguities in visual appearance of unseen classes.

Varying λ: The hyperparameter λ controls the weight between L_{mm} and L_{mc} in L_{cls}. In Fig. 4, we illustrate the effect of varying λ on four zero-shot tasks for R+w2v and R+glo. It shows that performances has less variation in the range of $\lambda = .5$ to $.9$ than $\lambda = .9$ to 1. For a larger λ, mAP starts dropping since the impact of L_{mc} decreases significantly.

4.3   Zero Shot Recognition (ZSR)

Being a detection model, the proposed network can also perform traditional ZSR. We evaluate ZSR performance on popular Caltech-UCSD Birds-200–2011 (CUB) dataset [32]. This dataset contains 11,788 images from 200 classes and provides single bounding boxes per image. Following standard train/test split [34], we use 150 seen and 50 unseen classes for experiments. For semantics embedding, we use 400-d word2vec (w2v) and GloVe (glo) vector [33]. Note that, we do not use per image part annotation (like [2]) and descriptions (like [36]) to enrich semantic embedding. For a given test image, our network predicts unseen class bounding boxes. We pick only one label with the highest prediction score per image. In this way, we report the mean Top1 accuracy of all unseen classes in Table 3. One can find our proposed solution achieve significant performance improvement in comparison with state-of-the-art methods.

Fig. 5. Selected examples of ZSD of our cluster ($\lambda = .6$) method with R+w2v, using the prediction score threshold $= 0.3$. (See supplementary material for more examples)

4.4   Challenges and New Directions

ZSD is Challenging: Our empirical evaluations show that ZSD needs to deal with the following challenges: *(1)* Unseen classes are rare compared to seen classes; *(2)* Small unseen objects are hard to detect and relate with their semantics; *(3)* The lack of similar seen classes leads to inadequate description of an unseen class; *(4)* As derived in an unsupervised manner, the noise of semantic space affects ZSD. These issues are discussed in detail in supplementary material.

Future Challenges: The ZSD problem warrants further investigation. *(1)* Instead of mapping image feature to the semantic space, the reverse mapping may help ZSD similar to ZSR [14,36]. *(2)* One can consider the fusion of different word vectors (word2vec and GloVe) to improve ZSD. *(3)* Like generalized ZSL [18,34,35], one can explore generalized ZSD setting. Moreover, weakly/semi-supervised version of ZSD/GZSD is also interesting.

5   Conclusion

While traditional ZSL research focuses on only object recognition, we propose to extend the problem to object detection (ZSD). To this end, we offer a new experimental protocol with ILSVRC-2017 dataset specifying the seen-unseen, train-test split. We also develop an end-to-end trainable CNN model to solve this problem. We show that our solution is better than a strong baseline.

Overall, this research throws some new challenges to ZSL community. To make a long-standing progress in ZSL, the community needs to move forward in the detection setting rather than merely recognition.

References

1. Akata, Z., Reed, S., Walter, D., Lee, H., Schiele, B.: Evaluation of output embeddings for fine-grained image classification. In: CVPR, 07–12 June 2015, pp. 2927–2936 (2015). https://doi.org/10.1109/CVPR.2015.7298911
2. Akata, Z., Malinowski, M., Fritz, M., Schiele, B.: Multi-cue zero-shot learning with strong supervision. In: CVPR, June 2016
3. Akata, Z., Perronnin, F., Harchaoui, Z., Schmid, C.: Label-embedding for image classification. IEEE Trans. Pattern Anal. Mach. Intell. **38**(7), 1425–1438 (2016). https://doi.org/10.1109/TPAMI.2015.2487986
4. Changpinyo, S., Chao, W.L., Gong, B., Sha, F.: Synthesized classifiers for zero-shot learning. In: CVPR, 2016 January, pp. 5327–5336 (2016)
5. Deng, J., et al.: Large-scale object classification using label relation graphs. In: Fleet, D., Pajdla, T., Schiele, B., Tuytelaars, T. (eds.) ECCV 2014. LNCS, vol. 8689, pp. 48–64. Springer, Cham (2014). https://doi.org/10.1007/978-3-319-10590-1_4
6. Deutsch, S., Kolouri, S., Kim, K., Owechko, Y., Soatto, S.: Zero shot learning via multi-scale manifold regularization. In: CVPR, July 2017
7. Farhadi, A., Endres, I., Hoiem, D., Forsyth, D.: Describing objects by their attributes. In: CVPR, pp. 1778–1785. IEEE (2009)
8. Frome, A., et al.: DeVISE: a deep visual-semantic embedding model. In: Burges, C.J.C., Bottou, L., Welling, M., Ghahramani, Z., Weinberger, K.Q. (eds.) NIPS, pp. 2121–2129. Curran Associates, Inc. (2013)

9. Fu, Y., Yang, Y., Hospedales, T., Xiang, T., Gong, S.: Transductive multi-label zero-shot learning. arXiv preprint arXiv:1503.07790 (2015)
10. Fu, Z., Xiang, T., Kodirov, E., Gong, S.: Zero-shot learning on semantic class prototype graph. IEEE Trans. Pattern Anal. Mach. Intell. **PP**(99), 1 (2017). https://doi.org/10.1109/TPAMI.2017.2737007
11. He, K., Zhang, X., Ren, S., Sun, J.: Deep residual learning for image recognition, 2016 January, pp. 770–778 (2016). Cited by 107
12. Dai, J., Li, Y., He, K., Sun, J.: R-FCN: object detection via region-based fully convolutional networks. arXiv preprint arXiv:1605.06409 (2016)
13. Kodirov, E., Xiang, T., Fu, Z., Gong, S.: Unsupervised domain adaptation for zero-shot learning. In: ICCV, December 2015
14. Kodirov, E., Xiang, T., Gong, S.: Semantic autoencoder for zero-shot learning. In: CVPR, July 2017
15. Lampert, C.H., Nickisch, H., Harmeling, S.: Attribute-based classification for zero-shot visual object categorization. IEEE Trans. Pattern Anal. Mach. Intell. **36**(3), 453–465 (2014). https://doi.org/10.1109/TPAMI.2013.140
16. Lampert, C., Nickisch, H., Harmeling, S.: Learning to detect unseen object classes by between-class attribute transfer. In: CVPR Workshops, pp. 951–958 (2009). https://doi.org/10.1109/CVPRW.2009.5206594
17. Lei Ba, J., Swersky, K., Fidler, S., et al.: Predicting deep zero-shot convolutional neural networks using textual descriptions. In: CVPR, pp. 4247–4255 (2015)
18. Li, Y., Wang, D., Hu, H., Lin, Y., Zhuang, Y.: Zero-shot recognition using dual visual-semantic mapping paths. In: CVPR, July 2017
19. Liu, W., et al.: SSD: single shot multibox detector. In: Leibe, B., Matas, J., Sebe, N., Welling, M. (eds.) ECCV 2016. LNCS, vol. 9905, pp. 21–37. Springer, Cham (2016). https://doi.org/10.1007/978-3-319-46448-0_2
20. Bucher, M., Herbin, S., Jurie, F.: Improving semantic embedding consistency by metric learning for zero-shot classification. In: Leibe, B., Matas, J., Sebe, N., Welling, M. (eds.) ECCV 2016. LNCS, vol. 9909, pp. 730–746. Springer, Cham (2016). https://doi.org/10.1007/978-3-319-46454-1_44
21. Mikolov, T., Sutskever, I., Chen, K., Corrado, G.S., Dean, J.: Distributed representations of words and phrases and their compositionality. In: Burges, C.J.C., Bottou, L., Welling, M., Ghahramani, Z., Weinberger, K.Q. (eds.) NIPS, pp. 3111–3119. Curran Associates, Inc. (2013)
22. Miller, G.A.: WordNet: a lexical database for english. Commun. ACM **38**(11), 39–41 (1995)
23. Morgado, P., Vasconcelos, N.: Semantically consistent regularization for zero-shot recognition. In: CVPR, July 2017
24. Norouzi, M., et al.: Zero-shot learning by convex combination of semantic embeddings. In: ICLR (2014)
25. Pennington, J., Socher, R., Manning, C.D.: GloVe: global vectors for word representation. In: EMNLP, pp. 1532–1543 (2014)
26. Rahman, S., Khan, S.H., Porikli, F.: A unified approach for conventional zero-shot, generalized zero-shot and few-shot learning. arXiv preprint arXiv:1706.08653 (2017)
27. Redmon, J., Farhadi, A.: Yolo9000: better, faster, stronger. In: The IEEE Conference on Computer Vision and Pattern Recognition (CVPR), July 2017
28. Ren, S., He, K., Girshick, R., Sun, J.: Faster R-CNN: towards real-time object detection with region proposal networks. IEEE Trans. Pattern Anal. Mach. Intell. **39**(6), 1137–1149 (2017). https://doi.org/10.1109/TPAMI.2016.2577031

29. Romera-Paredes, B., Torr, P.: An embarrassingly simple approach to zero-shot learning. In: ICML, pp. 2152–2161 (2015)
30. Russakovsky, O., et al.: ImageNet large scale visual recognition challenge. IJCV **115**(3), 211–252 (2015). https://doi.org/10.1007/s11263-015-0816-y
31. Simonyan, K., Zisserman, A.: Very deep convolutional networks for large-scale image recognition. arXiv preprint arXiv:1409.1556 (2014)
32. Wah, C., Branson, S., Welinder, P., Perona, P., Belongie, S.: The Caltech-UCSD Birds-200-2011 dataset. Technical report, CNS-TR-2011-001, California Institute of Technology (2011)
33. Xian, Y., Akata, Z., Sharma, G., Nguyen, Q., Hein, M., Schiele, B.: Latent embeddings for zero-shot classification. In: CVPR, June 2016
34. Xian, Y., Schiele, B., Akata, Z.: Zero-shot learning - the good, the bad and the ugly. In: CVPR (2017)
35. Xu, X., Shen, F., Yang, Y., Zhang, D., Shen, H.T., Song, J.: Matrix tri-factorization with manifold regularizations for zero-shot learning. In: CVPR (2017)
36. Zhang, L., Xiang, T., Gong, S.: Learning a deep embedding model for zero-shot learning. In: CVPR, July 2017
37. Zhang, Z., Saligrama, V.: Zero-shot learning via semantic similarity embedding. In: ICCV, December 2015
38. Zhang, Z., Saligrama, V.: Zero-shot learning via joint latent similarity embedding. In: CVPR, June 2016

Vision-Based Freezing of Gait Detection with Anatomic Patch Based Representation

Kun Hu[1](✉), Zhiyong Wang[1], Kaylena Ehgoetz Martens[2], and Simon Lewis[2]

[1] School of Information Technologies, The University of Sydney,
Sydney, Australia
hukun_sdu@hotmail.com
[2] Parkinson's Disease Research Clinic, Brain and Mind Centre,
The University of Sydney, Sydney, Australia

Abstract. Parkinson's disease (PD) impacts millions of people in the world. As freezing of gait (FoG) is a common symptom of PD patients that leads to falls and nursing home placement, it is very important to identify FoG event effectively and efficiently. Direct observation based assessment of FoG by doctors or trained experts is the *de facto* 'gold standard' for clinical diagnosis, which is time consuming. While several computer aided FoG event detection methods have been proposed, they were not particularly designed for video data collected during clinical diagnosis. In this paper, we treat video based FoG detection as a fine grained human action recognition problem and reduce the interference of visual content which is irrelevant to FoG, such as gait motion of supporting staff involved in clinical assessment. In order to effectively characterize FoG patterns, we propose to identify anatomic patches as the candidate regions which could be relevant to FoG events, and formulate FoG detection as a weakly-supervised learning task. The formulation will help identify the patches contributing to FoG events. To take both the global context of a clinical video and the local anatomic patches into account, several fusion strategies are investigated. Experimental results on videos collected from 45 subjects during clinical trials demonstrated promising results of our proposed method in terms of AUC of 0.869. To the best of our knowledge, this is one of the first studies on automatic FoG detection from clinical assessment videos.

Keywords: Freezing of Gait · Deep learning · Action recognition

1 Introduction

Parkinson's disease (PD) is a neurodegenerative disorder of the central nervous system for which there is no cure at this stage. It impacts more than 6 million people in the world [11]. In general, more than 70% of PD patients develop freezing of gait (FoG) which is a movement pause or discontinuity against a

© Springer Nature Switzerland AG 2019
C. V. Jawahar et al. (Eds.): ACCV 2018, LNCS 11361, pp. 564–576, 2019.
https://doi.org/10.1007/978-3-030-20887-5_35

patient's intention and happens daily in over 20% of PD patients [10, 15]. With the progression of the disease, FoG happens more frequently and often leads to falls [2, 14], which eventually affects the mobility, independence and quality of life of a PD patient [22]. While it is critical to detect FoG events [6] for accurate assessment, annotation of FoG events from video recordings is burdensome for trained experts. Hence, it is highly desirable to have computer aided solutions for automatic FoG detection method.

Many motion sensor based FoG detection methods have been proposed with some promising performance to date [1, 20, 21, 27]. However, there are two key limitations in relation to these methods. First, motion information obtained from motion sensors is noisy and may not be directly associated with FoG events due to the locations where sensors are attached. Second, since it is the *de facto* golden standard to identify FoG events through observing subjects directly or indirectly from recorded videos [16], preparing ground truth annotation requires trained experts to observe videos and label its corresponding motion data simultaneously, which is time consuming. Therefore, directly working on video data to automatically detect FoG event is a straightforward and compelling way. Moreover, vision-based methods are convenient to use due to their non-contact nature.

Although several video based Parkinsonian gait analysis methods have been proposed [12, 17, 18], they are not particularly designed for FoG event detection. In addition, in the clinical setting, support staff are often involved in walking closely to the patient to ensure the safety of the subject. Therefore, multiple people could appear in the recorded videos, whilst existing methods assume that a video only contains the patient walking independently. Second, existing datasets are too small to reach a convincing conclusion. Lastly, these methods mainly use hand-crafted features and have not leveraged the recent success of deep learning techniques (e.g., [26]).

To address FoG detection in this paper, we propose an anatomic patch based deep learning method. Since FoG events are generally observed at anatomic regions (e.g., the joints of knees and feet), we treat the videos as a set of anatomic patches. When a video is annotated as FoG, there is at least one patch contributing to the FoG event. On the contrary, a normal video should not contain any patches contributing to an FoG event. To reduce the cost of patch level annotations, we further formulate FoG detection as a weakly-supervised deep learning task which only requires label information at the temporal level. In addition, to take global context from the entire video into account, a fusion framework was developed and several fusion strategies were investigated.

In summary, the key contributions of this work are summarized as follows:

- We propose a method to take both global context and local features around human joints by treating FoG detection as a fine-grained human action recognition problem. To the best of our knowledge, this is one of the first studies devised for clinical videos. The anatomic patch based representation is also helpful for interpreting learning results.

- We formulate the anatomic patch based FoG detection as a weakly-supervised learning task.
- We evaluate our proposed method on a large video dataset collected during the clinical assessment of 45 subjects.

The remainder of this paper is organized as follows. In Sect. 2 we review related works for Parkinsonian gait analysis and human action recognition. In Sect. 3 we introduce the details of our proposed method. In Sect. 4 we present and discuss the experimental results. Finally, in Sect. 5, we draw the conclusions and discuss the future work.

2   Related Works

In this section, we review the related work from two aspects: vision based Parkinsonian gait analysis methods and deep learning based human action recognition. Note that we leave traditional human action recognition methods (e.g., hand-crafted feature based ones) as deep learning based methods achieve the state-of-the-art recognition performance for benchmark datasets.

2.1   Vision Based Parkinsonian Gait Analysis

Several vision based Parkinsonian gait analysis methods have been proposed [12, 17, 18]. In [12], human gait was recorded in the sagittal view by placing a camera laterally to the human subject. Posture lean and stride cycle related features were used to characterize gait motion pattern and cosine similarity was utilized for motion matching between normal and abnormal gait. Matching percentage was obtained as prediction score to identify abnormal/normal gaits. Only 7 subjects were involved in this study and it lacked the ability to temporally localize the abnormal patterns within a video. In [18], gait were characterized with various motion features such as average cycle time, stride length, and leg angles and binary classifiers were used to identify abnormal gait patterns. Note that feature extraction in this method followed a strict recording protocol. For example, each subject was requested to walk independently. However, this is not always practically feasible in patients, as those subjects could have mobility difficulties and may need external support to protect them from possible falls. In addition, the abnormal cases were simulated by healthy persons. Frontal-view videos have also been used for Parkinsonian gait analysis due to the convenience of its space saving setup, where a subject is required to walk towards and away from the recording camera [17]. Moreover, the setup is similar to clinical assessment and can avoid the issue that one leg is occluded by the other.

While the existing methods provide gait analysis of PD patients at a coarse level, they are not particularly designed for identifying FoG events. In addition, these methods follow a traditional pattern recognition pipeline: extract hand-crafted features and feed them into a machine learning model such as support vector machine (SVM). However, extracting hand-crafted features usually relies on strong assumptions which may not always be feasible in realistic scenarios. For example, these methods require that only the subjects are in the video and there is a need to ignore external support from health-care staff during the recording for clinical assessment. The ground-breaking success of deep learning techniques for many visual understanding tasks provide a great opportunity for developing deep learning based FoG detection methods.

2.2 Deep Learning Based Human Action Recognition

In general, there are three major types of deep learning based human action recognition methods: convolution neural network (CNN) based methods, recurrent neural network based methods and two-stream based methods. The first type in general extends the 2D CNN architecture to 3D counterpart to handle 3-dimensional video data, such as C3D [26], P3D [19] and I3D [30]. By considering a video as 2D image sequence, the second type aims to model the temporal structure with recurrent neural networks such as LSTM (Long Short Term Memory) and GRU (Gaited Recurrent Units) [5,23]. The last type aims to represent video content by two streams: appearance and motion parts, respectively. [7,25]. The two streams are processed via two separate CNN and finally fused to produce the prediction. However, these action recognition methods mainly deal with actions which involve significant variation of motion, object and background context. They are not well designed for detecting FoG events of which the intra-class variation is subtle. In this perspective, FoG detection can be viewed as a fine-grained action recognition task. There are methods proposed to handle such fine-grained classification problems. Pose-CNN was proposed to generate patch proposals with help of pose estimation for fine grained human action recognition and use the proposals for improving recognition performance [4]. However, a simple pooling and concatenation strategy was utilized to generate the descriptor for each patch proposal, which is not able to best represent the candidate FoG patches. Bilinear pooling were also used to deal with fine-grained classification. For example, a bilinear estimation based soft-attention model was proposed [8]. However, the potential issue is that it increases the model complexity for training. Therefore, in this paper, we aimed to leverage FoG related knowledge to design an effective fine-grained human action recognition method for FoG detection.

3 Proposed Method

As shown in Fig. 1, our proposed method consisted of four components, namely, candidate anatomic proposal identification, deep feature extraction, LSTM based temporal motion modeling, and fusion prediction. Firstly, key anatomic patches were estimated from an input video. Secondly, these patch proposals were fed into a pre-trained CNN - Res-Net 50 to extract spatial descriptors frame by frame. Note that, in order to consider the global context of a video, the entire video segment was also fed for deep feature extraction. Thirdly, spatial descriptors

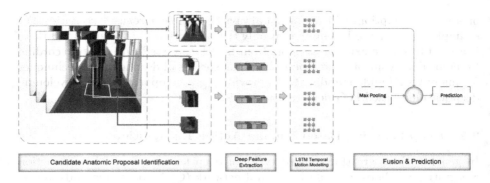

Fig. 1. Illustration of our proposed method which uses both global context and anatomic patch proposals to generate deep spatial features. These spatial features are fed to LSTM to model temporal motion patterns. In the end, a fusion step is utilized to take advantage of both global and local features for the classification of the input video segment.

of each patch and the entire video segment were further modelled by a LSTM network to characterize temporal motion patterns. Finally, a maximum pooling step for the output score of each patch was performed and the obtained score was further fused with the global output for final prediction. In this section, we provide the details of each part of the proposed method.

3.1 Anatomic Joint Patches

For FoG detection, it is reasonable to pay attention to the motion of anatomic joints. To locate anatomic joints, we apply convolutional pose machines (CPM) to accomplish pose estimation [3,28]. For a video segment V, we first choose the middle frame to compute the anatomic joints of poses with CPM. Next a square window centred at each detected pose joint is identified. The side length l_{pm} of the window can be customized as a hyper parameter. Although the square windows (i.e., patches) are generated at the middle frame, it is applied to other frames as well since the length of a video segment is short. Therefore, a patch video proposal is extracted for each anatomic joint. Denote these patch video proposals of V as $\mathbb{V}_p = \{v_i | i = 1...n\}$, where n refers to the total number of anatomic joints detected, and v_i is an anatomic joint patch.

Note that for each video, CPM finds anatomic joints of all the people appearing in the video. Thus, $\mathbb{V}_p$ contains all their joint patch proposals. Let

$$\mathbb{V}_p = \mathbb{V}_p^F \cup \mathbb{V}_p^{NF}, \tag{1}$$

where $\mathbb{V}_p^F$ denotes the key patch proposals contributing to FoG, and $\mathbb{V}_p^{NF}$ denotes the proposals not related to FoG event.

3.2   Weakly-Supervised Learning for Patch Proposals

In this work, only the label of an entire video segment V is available. There is no prior annotation for its related proposals $v_i \in \mathbb{V}_p$ and the key information is missing: it is unclear whether v_i contributes to FoG $(v_i \in \mathbb{V}_p^F)$ or not $(v_i \in \mathbb{V}_p^{NF})$. Therefore, we propose a weakly-supervised learning based solution to deal with such anatomic patch based action representation.

Generally, for a video v and a classifier f, the estimation or prediction score $\hat{y}$ can be computed as: $\hat{y} = f(v)$. In particular, for $v_i \in \mathbb{V}_p$ and a proposal prediction model f_p, we have:

$$\hat{y}_i = f_p(v_i). \tag{2}$$

As there are no FoG patterns related to $v_i \in \mathbb{V}_p^{NF}$, prediction score $\hat{y}_i$ of the elements in this set should be always close to 0 no matter what the label of V is; for $v_i \in \mathbb{V}_p^F$ contributing to FoG events, the prediction score $\hat{y}_i$ should be close to 1 respectively. Therefore, we have

$$max_{v_i \in \mathbb{V}_p^F}(\hat{y}_i) \geq max_{v_i \in \mathbb{V}_p^{NF}}(\hat{y}_i). \tag{3}$$

Clearly, the maximum of $\hat{y}_i = f_p(v_i)$, where $v_i \in \mathbb{V}_p$, can be viewed as an estimation of y. We denote this as $\hat{y}_{patch}$:

$$\hat{y}_{patch} = max_i(\hat{y}_i). \tag{4}$$

As a binary classification problem, cross-entropy loss function is used to optimize the model f_p. We add superscript n to the variables mentioned above to indicate the n-th training sample. The loss function can be written as:

$$J = -\frac{1}{N} \sum_{n=1}^{N} [y^{(n)} log(\hat{y}_{patch}^{(n)}) + (1 - y^{(n)}) log(1 - \hat{y}_{patch}^{(n)})]. \tag{5}$$

In addition to obtaining the prediction $\hat{y}_{patch}$, by ranking the scores $\hat{y}_i$, the top contributors to FoG events can be localized. It is very helpful to interpret the classification results for clinical assessment.

3.3   Patch and Global Feature Fusion

For anatomic patches, we use weakly-supervised method to obtain the final predictions. However, it lacks the information of the context out of these patches and relationship among them. Therefore, we further apply a global model f_g of which the input is the entire video segment V itself. A prediction $\hat{y}_{global} = f_g(V)$ is derived and further fused it with $\hat{y}_{patch}$.

Model f_p and f_g have similar structure except the input dimension since the spatial size of the entire video segment V and proposal $V_i \in \mathbb{V}_p$ are different. In order not to increase model complexity, we first leverage a pre-trained ResNet-50 to obtain spatial features frame by frame [9]. Next, we treat these frame features as a temporal sequence and feed them as the input to a LSTM network for modeling the temporal patterns of FoG. The output from the LSTM network is the probability that FoG event appears in the input segment V.

Without increasing the complexity of the model's objective function, the global and patch models are trained independently. Generally, there are several methods to fuse $\hat{y}_{global}$ and $\hat{y}_{patch}$ to get a comprehensive prediction, and we list three as follows:

$$\hat{y} = \alpha\hat{y}_{global} + (1 - \alpha)\hat{y}_{patch}, \alpha \in [0,1], \tag{6}$$

$$\hat{y} = \hat{y}_{global}\hat{y}_{patch}, \tag{7}$$

$$\hat{y} = max(\hat{y}_{global}, \hat{y}_{patch}). \tag{8}$$

Obviously the values of the three functions are in $[0,1]$, which indicate the probability of a FoG event happening in V. If $\hat{y}_{global}$ and $\hat{y}_{full}$ achieve high values simultaneously, fusing them obtains high value as well and V tends to contain a FoG event.

(a) Positive (i.e., FoG) sample frames.

(b) Negative (i.e., non-FoG) sample frames.

Fig. 2. Sample frames of FoG and non-FoG events.

4   Experimental Results and Discussions

4.1   Dataset

The dataset consisted of videos collected from 45 PD patients who were clinically assessed for FoG in their practically defined Off state having omitted their usually dopaminergic medications overnight. Note that this dataset has the largest number of subjects in the literature of video based Parkinsonian gait analysis. For example, 11 subjects were involved in the work of [12] and 30 subjects in [18]. Each subject follows instructions to participate in a Timed Up and Go (TUG) test which is a process to assess the mobility of a person [24]. Videos were recorded in a frontal view at 25 frames per second with frame resolution of 720×576. The FoG ground truth was annotated by trained experts at frame level. For FoG detection, we partition these videos into 1-s long non-overlapped video clips and aim to classify each video as either FoG or non-FOG class. A video clip is labelled as FoG class, if any frame of the clip is annotated as FoG frame according to the ground truth; otherwise it is labelled as non-FoG class. In total, 25.5-h video are acquired and 8.7% of the time includes FoG patterns. The FoG ratio of the dataset is 8.7%, which is highly imbalanced. Sample frames of FoG (i.e., positive) and non-FoG (i.e., negative) clips for two subjects are shown in Fig. 2.

4.2 Implementation Details

We set the size of each anatomic patch to 50×50 pixels. Hence the anatomic proposal at each joint is $25 \times 50 \times 50 \times 3$ for each 1-s video clip, where 3 denotes RGB channels and 25 is the frame rate. As the size of the input to Res-Net 50 is $224 \times 224 \times 3$, the size of the last pooling layer of Res-Net 50 for f_p is set to $(2, 2)$, and the size of the output feature for each proposal is 25×2048. To obtain the global feature, we first resize a frame with the same aspect ratio, and crop a 224×224 sub-image at the center as the input to the Res-Net 50 of f_g. Finally, a 25×2048 dimensional Res-Net 50 global feature vector is obtained. A two-layer LSTM is utilized to model temporal patterns for both f_p and f_g.

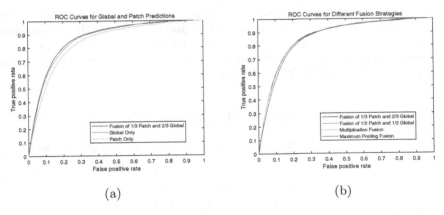

Fig. 3. Comparison of the detection results in terms of ROC Curves. (a) Comparison among independent methods and the best fusion strategy. (b) Comparison among different fusion strategies.

We train f_g and f_p independently so as not to increase model complexity, and fuse their outputs for final prediction. Multiple fusion strategies such as linear fusion, product fusion and maximum fusion, are utilized. For linear fusion, we choose $\alpha = 0.5$ and $\alpha = \frac{2}{3}$ (which achieved the best image classification performance in [29]). As the dataset is imbalanced, we use all of the positive samples and randomly select the same number of negative samples from the training set in each epoch for model training.

To comprehensively evaluate the performance of the proposed method, 5-fold cross validation was utilized. We randomly partitioned 45 subjects into 5 groups evenly. Each time the videos of 4 groups are chosen to train the model and the videos of the remaining group were used for testing, so that the video clips of each subject could only appear in either the training set or the testing set. After repeating the procedure 5 times, the average performance on testing datasets was obtained as the overall detection performance.

4.3   FoG Detection Results

As the FoG dataset was imbalanced, we use a number of performance metrics in addition to accuracy (i.e., the number of correctly classified samples over the total number of testing samples), including ROC (Receiver Operating Characteristic) curve which illustrates the diagnostic ability of a binary classification model, Area Under Curve (AUC), sensitivity, and specificity. Figure 3 shows ROC curves for different estimations $\hat{y}$. More detailed results are listed in in Table 1. It is noticed that the linear fusion method with $\alpha = \frac{2}{3}$ achieves the best overall performance 0.869 in terms of AUC, while the differences among different fusion strategies are marginal, such as 0.869 vs 0.868. In particular, patch and global based methods together enhance the performance, as they provide multi-scale information for the modelling.

To calculate other metrics shown in Table 1, we chose a cut-off which maximizes Youden's J statistics: $J = sensitivity + specificity - 1$ [31]. The statistic is designed to find a cut-off to balance sensitivity and specificity, and maximize the the sum of these two values. Sensitivity and specificity values of linear fusion with $\alpha = \frac{2}{3}$ are 84.5% and 76.5% respectively.

Table 1. Comparison of FoG detection performance

	AUC	Sensitivity	Specificity	Likelihood ratio positive	Likelihood ratio negative	Accuracy
$\hat{y}_{global}$	0.863	84.9%	74.8%	3.365	0.202	75.7%
$\hat{y}_{patch}$	0.841	81.8%	73.8%	3.120	0.247	74.5%
$\frac{2}{3}\hat{y}_{global} + \frac{1}{3}\hat{y}_{patch}$	**0.869**	**84.5%**	**76.5%**	**3.598**	**0.203**	**77.2%**
$0.5\hat{y}_{global} + 0.5\hat{y}_{patch}$	0.868	82.4%	78.3%	3.805	0.225	78.7%
$max(\hat{y}_{global}, \hat{y}_{patch})$	0.860	83.4%	76.2%	3.511	0.218	76.9%
$\hat{y}_{global} \times \hat{y}_{patch}$	0.868	82.1%	78.5%	3.814	0.228	78.8%

4.4   Key Patch Localization

In order to further understand how each anatomic joint patch contributes to detecting FoG event (i.e., computing $\hat{y}_{patch}$), we visualize the localization of key patches as shown in Fig. 4. Note that in our dataset there was no region or

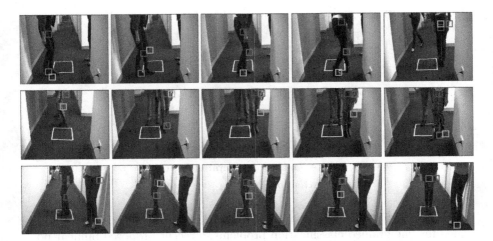

Fig. 4. Illustration of localizing key patches of FoG events. (Color figure online)

patch level annotation of FoG events, we were not able to provide quantitative evaluation. In general, several persons appeared in each video and produced patch proposals collectively. In a positive video clip, the key patches should only appear on the patient subject. Otherwise, the algorithm could have produced the prediction incorrectly.

The key patches of three subjects (one subject per row) are visualized in Fig. 4. We choose 5 consecutive seconds per subject in which FoG happened and use the middle frame of each 1-s video for visualization. Three square boxes for each sub-figure are drawn to indicate three key patches according to the top 3 scores of $\hat{y}_i$. The box with darker outline color indicates its higher score. It is noticed that most of the boxes are located on the patient subjects while some boxes occasionally appear on the supporting health care staff.

4.5 Comparison

2D CNN and LSTM was used in the proposed framework to formulate the spatial and temporal FoG patterns. To further demonstrate that it formulated the FoG patterns sufficiently, in this subsection, we present several additional experimental results of state-of-the-art algorithms for comparison. As shown in Table 2, the performance of C3D [26] and P3D [19] indicates that 3D convolution filters are not the best choice to adequately characterize FoG temporal pattern due to the higher model complexity. An alternative method to formulate the frame by frame spatial feature is the dilated temporal CNN, which is proposed to obtain long-term patterns for video classification [13]. However, it fails to obtain promising temporal pattern comparing with LSTM. This is because that dilated temporal convolution filter is good at capturing long-term fixed pattern whilst FoG shows a short term dynamic pattern. In addition, as bilinear methods are successfully

Table 2. Performance of the state-of-the-art methods

Method	AUC
P3D Global	0.8124
P3D Patch	0.6784
P3D Linear Fusion	0.7986
C3D Global	0.8526
C3D Patch	0.8388
C3D Linear Fusion	0.8589
2D CNN + Dilated Temporal CNN	0.8399
Bilinear Attention Pooling	0.8477

applied for fine-grained classification problems, a recent proposed bilinear pooling attention method was selected. For the global model, the linear computation of the output gate of f_g is changed into a bilinear pooling operation: $H^T a^T b H$, where H is the hidden state of the LSTM cell and $a^T b$ is the rank-1 estimation of bilinear transformation [8]. Although such bilinear operations can enhance the model performance for other applications, its model complexity degrades the FoG recognition results.

5   Conclusion

In this paper, we present a novel fine-grained human action recognition method for detecting FoG events from PD patients. To the best of our knowledge, our work is one of the first studies on FoG detection from clinical videos collected from PD patients. We propose an anatomic patch based representation and formulate the detection problem as a weakly supervised classification task. In order to take both global deep features and local patch based deep features into account, several fusion strategies are investigated. We evaluated our proposed method from our in-house dataset, which has the largest number of subjects in the literature of PD gait analysis. Experimental results demonstrated the superior performance of our proposed method over similar methods. In addition, the patch based representation provides an intuitive interpretation for the detection results by localizing key patches in a FoG video. In our future work, we will focus on eliminating the number of incorrect key patches while further improving the detection performance.

Acknowledgement. This research was partially supported by NHMRC-ARC Dementia Fellowship #1110414. We thank our patients who participated into the clinical assessment video collection. We would like to acknowledge and thank Moran Gilat, Julie Hall, Alana Muller, Jennifer Szeto and Courtney Walton for conducting and scoring the freezing of gait assessments. We would also like to acknowledge ForeFront, a large collaborative research group dedicated to the study of neurodegenerative diseases. It is funded by the National Health and Medical Research Council of Australia Program

Grant (#1037746), Dementia Research Team Grant (#1095127) and NeuroSleepCentre of Research Excellence (#1060992), as well as the Australian Research Council Centre of Excellence in Cognition and its Disorders MemoryProgram (#CE110001021), and the Sydney Research Excellence Initiative 2020.

References

1. Bachlin, M., et al.: Wearable assistant for Parkinson's disease patients with the freezing of gait symptom. IEEE Trans. Inf. Technol. Biomed. **14**(2), 436–446 (2010)
2. Bloem, B.R., Hausdorff, J.M., Visser, J.E., Giladi, N.: Falls and freezing of gait in Parkinson's disease: a review of two interconnected, episodic phenomena. Mov. Disord. **19**(8), 871–884 (2004)
3. Cao, Z., Simon, T., Wei, S.E., Sheikh, Y.: Realtime multi-person 2D pose estimation using part affinity fields. In: IEEE Conference on Computer Vision and Pattern Recognition (2017)
4. Cheron, G., Laptev, I., Schmid, C.: P-CNN: pose-based CNN features for action recognition. In: IEEE International Conference on Computer Vision, pp. 3218–3226 (2015)
5. Donahue, J., et al.: Long-term recurrent convolutional networks for visual recognition and description. In: IEEE Conference on Computer Vision and Pattern Recognition, pp. 2625–2634 (2015)
6. Donovan, S., et al.: Laserlight cues for gait freezing in Parkinson's disease: an open-label study. Park. Relat. Disord. **17**(4), 240–245 (2011)
7. Feichtenhofer, C., Pinz, A., Zisserman, A.: Convolutional two-stream network fusion for video action recognition. In: IEEE Conference on Computer Vision and Pattern Recognition, pp. 1933–1941 (2016)
8. Girdhar, R., Ramanan, D.: Attentional pooling for action recognition. In: Advances in Neural Information Processing Systems, pp. 33–44 (2017)
9. He, K., Zhang, X., Ren, S., Sun, J.: Deep residual learning for image recognition. In: Proceedings of the IEEE Conference on Computer Vision and Pattern Recognition, pp. 770–778 (2016)
10. Hely, M.A., Reid, W.G., Adena, M.A., Halliday, G.M., Morris, J.G.: The Sydney multicenter study of Parkinson's disease: the inevitability of dementia at 20 years. Mov. Disord. **23**(6), 837–844 (2008)
11. Jankovic, J.: Parkinson's disease: clinical features and diagnosis. J. Neurol. Neurosurg. Psychiatry **79**(4), 368–376 (2008)
12. Khan, T., Westin, J., Dougherty, M.: Motion cue analysis for Parkinsonian gait recognition. Open Biomed. Eng. J. **7**, 1 (2013)
13. Lea, C., Flynn, M.D., Vidal, R., Reiter, A., Hager, G.D.: Temporal convolutional networks for action segmentation and detection. In: IEEE Conference on Computer Vision and Pattern Recognition, pp. 1003–1012. IEEE (2017)
14. Lewis, S.J., Barker, R.A.: A pathophysiological model of freezing of gait in Parkinson's disease. Park. Relat. Disord. **15**(5), 333–338 (2009)
15. Macht, M., et al.: Predictors of freezing in Parkinson's disease: a survey of 6,620 patients. Mov. Disord. **22**(7), 953–956 (2007)
16. Morris, T.R., et al.: Clinical assessment of freezing of gait in Parkinson's disease from computer-generated animation. Gait & Posture **38**(2), 326–329 (2013)

17. Nieto-Hidalgo, M., Ferrández-Pastor, F.J., Valdivieso-Sarabia, R.J., Mora-Pascual, J., García-Chamizo, J.M.: Vision based gait analysis for frontal view gait sequences using RGB camera. In: García, C.R., Caballero-Gil, P., Burmester, M., Quesada-Arencibia, A. (eds.) UCAmI 2016, Part I. LNCS, vol. 10069, pp. 26–37. Springer, Cham (2016). https://doi.org/10.1007/978-3-319-48746-5_3

18. Nieto-Hidalgo, M., Ferrández-Pastor, F.J., Valdivieso-Sarabia, R.J., Mora-Pascual, J., García-Chamizo, J.M.: A vision based proposal for classification of normal and abnormal gait using RGB camera. J. Biomed. Inform. **63**, 82–89 (2016)

19. Qiu, Z., Yao, T., Mei, T.: Learning spatio-temporal representation with pseudo-3D residual networks. In: IEEE International Conference on Computer Vision, pp. 5534–5542. IEEE (2017)

20. Ravi, D., Wong, C., Lo, B., Yang, G.Z.: A deep learning approach to on-node sensor data analytics for mobile or wearable devices. IEEE J. Biomed. Health Inform. **21**(1), 56–64 (2017)

21. Rodríguez-Martín, D., et al.: Home detection of freezing of gait using support vector machines through a single waist-worn triaxial accelerometer. PloS One **12**(2), e0171764 (2017)

22. Schaafsma, J., Balash, Y., Gurevich, T., Bartels, A., Hausdorff, J.M., Giladi, N.: Characterization of freezing of gait subtypes and the response of each to levodopa in Parkinson's disease. Eur. J. Neurol. **10**(4), 391–398 (2003)

23. Shi, X., Chen, Z., Wang, H., Yeung, D.Y., Wong, W.K., Woo, W.C.: Convolutional LSTM network: a machine learning approach for precipitation nowcasting. In: Cortes, C., Lawrence, N.D., Lee, D.D., Sugiyama, M., Garnett, R. (eds.) Advances in Neural Information Processing Systems 28, pp. 802–810. Curran Associates, Inc. (2015)

24. Shumway-Cook, A., Brauer, S., Woollacott, M.: Predicting the probability for falls in community-dwelling older adults using the timed up & go test. Phys. Ther. **80**(9), 896–903 (2000)

25. Simonyan, K., Zisserman, A.: Two-stream convolutional networks for action recognition in videos. In: Advances in Neural Information Processing Systems, pp. 568–576 (2014)

26. Tran, D., Bourdev, L., Fergus, R., Torresani, L., Paluri, M.: Learning spatiotemporal features with 3D convolutional networks. In: IEEE International Conference on Computer Vision, pp. 4489–4497. IEEE (2015)

27. Tripoliti, E.E., et al.: Automatic detection of freezing of gait events in patients with Parkinson's disease. Comput. Methods Programs Biomed. **110**(1), 12–26 (2013)

28. Wei, S.E., Ramakrishna, V., Kanade, T., Sheikh, Y.: Convolutional pose machines. In: IEEE Conference on Computer Vision and Pattern Recognition, pp. 4724–4732 (2016)

29. Xiao, T., Xu, Y., Yang, K., Zhang, J., Peng, Y., Zhang, Z.: The application of two-level attention models in deep convolutional neural network for fine-grained image classification. In: IEEE Conference on Computer Vision and Pattern Recognition, pp. 842–850 (2015)

30. Xie, S., Sun, C., Huang, J., Tu, Z., Murphy, K.: Rethinking spatiotemporal feature learning for video understanding (2017). arXiv preprint: arXiv:1712.04851

31. Youden, W.J.: Index for rating diagnostic tests. Cancer **3**(1), 32–35 (1950)

Fast Video Shot Transition Localization with Deep Structured Models

Shitao Tang[✉], Litong Feng, Zhanghui Kuang, Yimin Chen, and Wei Zhang

Sensetime Research, Hong Kong, China
shitaot@gmail.com

Abstract. Detection of video shot transition is a crucial pre-processing step in video analysis. Previous studies are restricted on detecting sudden content changes between frames through similarity measurement and multi-scale operations are widely utilized to deal with transitions of various lengths. However, localization of gradual transitions are still under-explored due to the high visual similarity between adjacent frames. Cut shot transitions are abrupt semantic breaks while gradual shot transitions contain low-level spatial-temporal patterns caused by video effects, e.g. dissolve. In this paper, we propose a structured network aiming to detect these two shot transitions using targeted models separately. Considering speed performance trade-offs, we design the following framework. In the first stage, a light filtering module is utilized for collecting candidate transitions on multiple scales. Then, cut transitions and gradual transitions are selected from those candidates by separate detectors. To be more specific, the cut transition detector focus on measuring image similarity and the gradual transition detector is able to capture temporal pattern of consecutive frames, even locating the positions of gradual transitions. The light filtering module can rapidly exclude most of the video frames from further processing and maintain an almost perfect recall of both cut and gradual transitions. The targeted models in the second stage further process the candidates obtained in the first stage to achieve a high precision. With one TITAN GPU, the proposed method can achieve a 30× real-time speed. Experiments on public TRECVID07 and RAI databases show that our method outperforms the state-of-the-art methods. To train a high-performance shot transition detector, we contribute a new database ClipShots, which contains 128636 cut transitions and 38120 gradual transitions from 4039 online videos. ClipShots intentionally collect short videos for more hard cases caused by hand-held camera vibrations, large object motions, and occlusion. The database is avaliable at https://github.com/Tangshitao/ClipShots.

Keywords: Shot boundary detection · Deep structured model · ClipShots

1   Introduction

Shot transition detector is a necessary component in many video recognition tasks [10,23,29]. The goal of shot transition detection is to find semantic breaks

© Springer Nature Switzerland AG 2019
C. V. Jawahar et al. (Eds.): ACCV 2018, LNCS 11361, pp. 577–592, 2019.
https://doi.org/10.1007/978-3-030-20887-5_36

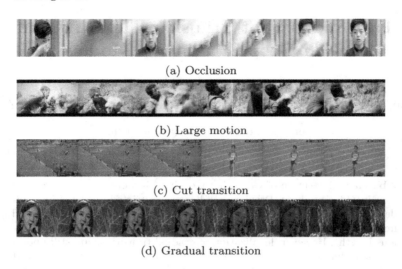

(a) Occlusion

(b) Large motion

(c) Cut transition

(d) Gradual transition

Fig. 1. Challenge of shot boundary detection

in videos. Cut transitions are defined as abrupt transitions from one sequence to another while gradual transitions are almost the same but in a gradual manner. They share one common attribute, the start of a transition and the end of a transition are semantically different. Previous methods focus on finding both cut transitions and gradual transitions with one similarity function [25,27]. Such methods have shown a great success in cut transition detection in the aspects of both speed and accuracy. However, when applied to gradual transition detection, it is not effective in the detection of gradual transitions. As Fig. 1 shows, it is widely recognized that many large motions or occlusion, e.g. camera movement, are detected as positive when only measuring similarity. In order to overcome this shortcoming, recent research [9,17] begins to explore the temporal pattern of gradual transitions. Therefore, in [9], the C3D ConvNet is adopted to classify segments into three classes (cut, gradual and background), which achieves state-of-the-art performance. Yet C3D ConvNet not only consumes too much computing resources, but is also not an effective architecture for handling both cut and gradual transitions, i.e. the lengths of gradual transitions are varying but C3D ConvNet is not designed for multi-scale detection. Inspired by this method and previous similarity measurement method, we present a cascade framework, consisting of a targeted cut transition detector and a targeted gradual transition detector. The cut transition detector, for measuring the image similarity, is fast and accurate while the gradual transition detector is capable of capturing the temporal pattern of gradual transitions in multi-scale level. In addition, compared to deepSBD, our framework can locate both cut transitions and gradual transitions accurately.

In this work, we present a new cascade framework, a fast and accurate approach for shot boundary detection. The first stage applies a ridiculously fast method to initially filter the whole video and selects the candidate segments. This stage is for accelerating the framework (up to 2 times faster than not) and facilitate the training for the cut/gradual detector. In the second stage, we use a well designed 2D ConvNet learning the similarity function between two images to locate the cut transitions. The third stage utilizes a novel C3D ConvNet model to locate positions of gradual transitions. Typically, we use the notation of default boxes introduced in [15] and propose a novel single shot boundary detector (SSBD).

In sum, our framework is fast and accurate for shot boundary detection and achieves state-of-the-art performance on many public databases running at 700 FPS without any bells and whistles.

Current datasets, i.e. TRECVID and RAI, are not sufficient for training deep neural net due to limited dataset size. Besides, the training set is various in different work when evaluating supervised methods on TRECVID and RAI databases. For training a high performance neural network and a fair comparison between different methods, we contribute a new large-scale video shot database ClipShots consisting of different types of videos collected from Youtube and Weibo. ClipShots is the first large-scale database for shot boundary detection and will be released.

Aspects of novelty of our work include:

- We separate cut transition detection and gradual transition detection, designing targeted network structures with different purposes.
- We design a cascade framework for accelerating the processing speed.
- We collect the first large-scale database for shot boundary detection training and evaluation.

2   Related Work

In this section, we introduce the work related to our proposed framework.

Unsupervised Shot Boundary Detection Method. In decades, many researchers explore to design similarity function finding transitions with hand-crafted features. In [27], average Intensity Measurement (AIM), Histogram Comparison (HC), Likelihood Ratio (LR) is used as the feature extractor. It is observed that similarities often vary gradually within a shot but abruptly in shot boundaries so the paper proposes an adaptive threshold should be applied when selecting positive samples. This method greatly improves the gradual transition performance compared to methods that only use static thresholds. Besides, another benefit is that it runs very fast so we integrate it in our framework to select potential shot boundaries. Yuan et al. [25] proposes a graph partition model to perform temporal data segmentation. It treats every frame as a node and calculate the similarity metrix and the scores of the cuts, selecting feasible cuts whose scores are the local minima of the corresponding neighborhoods. These two methods all rely on well designed hand-crafted features to calculate the similarity of two images.

Supervised Shot Boundary Detection Method. Due to the shortcoming of unsupervised methods, Yuan et al. [26] adopts a supervised way, a support vector machine trained to classify different shot boundaries with extracted features. In [16], shot boundaries are classified into 6 categories (cut, fast dissolve, fade in, fade out, dissolve, wipe). Different features are used to train different SVMs targeting at different shot boundaries. Researchers explore which features can most effectively classify the shot boundaries.

Shot Boundary Detection with Deep Learning. Hassanien et al. [9] introduces a simple C3D network that takes a segments of fixed length as input and classify it into 3 categories (cut, gradual, background). This method shows the effectiveness of ConvNet in this task. However, this method deals with gradual transitions of different scales in the same way and cannot locate the accurate' boundaries. Gygli [7] also adopts fully convolutional network. It takes the whole video sequence as input and assigns the positive label to the frames in transitions.

Image Similarity Comparison. Deep learning has been successful on image similarity comparison task. In [28], three architectures are proposed to compute image similarities, siamese net, image concatenation net, pseudo-siamese net. Empirical experiments show the image concatenation network and its variants obtain the best performance. In [22], a ranking model that employs deep learning techniques to learn similarity metric directly from images. We apply the similarity measurement only for the cut transition detection.

Object Detection. State-of-the-art methods for general object detection are mainly based on deep ConvNet to extract rich semantic features from images. Liu et al. [15] introduces single shot detector (SSD) using default boxes to match the feature to ground truth and achieve the speed of 19–46 fps. Our gradual detection model design share the same spirit with SSD.

Action Recognition. Carreira and Zisserman [13] has released the kinetics database for large-scale action classification. I3D [3] shows a good weights initialization is necessary to train the C3D network. Qiu et al. [19] proposes a fast network architecture based a spatial convolution kernel and temporal kernel to explore the temporal information. Action recognition is closely related to our work because we want to use temporal information to distinguish large motions and the gradual transitions.

Action Detection. This task focuses on learning how to detect action instances in untrimmed videos. Recently, many approaches adopt detection by classification' framework. Xu et al. [24] builds faster-RCNN style architecture for fast classifying and locating actions. It first selects potential segments with region proposal network and proposes the ROI 3D pooling layer to extract rich features for further classification. In [14], the single shot detector locates action on feature map extracted from well trained action classification ConvNets. Escorcia et al. [6] proposes to generate a set of proposals based on the RNN network. Zhao et al. [30] models the temporal structure of each action instance via a structured temporal pyramid. Although some of the methods can be applied to gradual

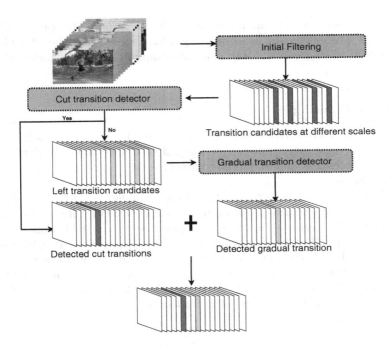

Fig. 2. An overview of our framework

detection directly, these methods rely on extracting rich spatial-temporal features from a heavy ConvNet body, so these methods are far slower than our proposed methods.

3 Our Approach

In this section, we will introduce our approach in details. The framework of our approach is shown in Fig. 2.

3.1 An Overview

The framework takes a video as input and predicts the locations of transitions. The proposed method, as shown in Fig. 2, is composed of three modules, including initial filtering, cut transition detector and gradual transition detector, implemented with three stages. (1) Adaptive thresholding produces a set of transition candidates. Each candidate comes with a center frame index indicating whether the content in frames has drastic changes. These positions may be transitions or caused by large motion, e.g. camera movement. (2) The candidate transitions are further feed into a strong cut transition detector to filter out false cut transitions. (3) For the remaining center frames which have negative responses to the cut detector, we expand them by x frames on both forward and backward temporal directions to form candidate segments. The gradual transition detector processes all these segments, locating the gradual transitions. The whole framework is designed in a cascade way and the computation of the earlier stage is lighter than the later.

3.2 Initial Filtering

As most of the consecutive video frames are highly similar to each other, a trivial unsupervised algorithm can be applied to reduce the candidate regions for further processing. A fast method, adaptive thresholding, is chosen as the initial filtering step.

Let I_n and I_{n+1} be the potential transition candidates and F_{n-a+1}, F_{n-a+2}, ..., F_{n+a} be a set of features extracted from consecutive video frames in a sliding window of length $2a$ centered at frame n. In practice, we use the feature extracted from SqueezeNet [11] trained on Imagenet [4]. The computation cost in this step is subtle. We calculate the similarity metric of each frame S_i, which is represented as the cosine distance between the current frame feature and its neighboring frame feature. Given the similarity metric of these frames as S_{n-a+1}, S_{n-a+2}, S_{n-a+3}, ..., S_{n+a-1}, the threshold of a window is calculated as

$$T = t + \frac{\sigma}{2\alpha} \sum_{i=n-a+1}^{n+a-1} (1 - S_i) \tag{1}$$

The hyper-parameter σ is the dynamic threshold ratio and t is the static threshold. In practice, we set σ to 0.05 and t to 0.5. The frame is selected as a candidate center if $1 - S_n$ is larger than T. Lengths of gradual transitions vary greatly. In order not to miss any gradual transition, we down-sample frames with multiple temporal scales. At scale ω, we sample one video frame every ω frames and do the above thresholding operations on these down-sampled frames. Finally, results of different scales are merged together. If two candidates on different scales are too close, i.e., within a distance of 5 frames. The candidate with a lower scale will be kept. In practice, we use scales of 1, 2, 4, 8, 16, and 32.

3.3 Cut Model

Some image pairs are semantically similar even when they are cut transitions, i.e. images containing the same object but the backgrounds are different. Therefore, a stronger cut transition detector is needed to filter out these negative cut candidates from the candidates selected by adaptive thresholding. Zagoruyko and Komodakis [28] show CNN can learn the similarity function directly from image pairs. We design a ConvNet to determine whether a image pair is a cut transition or not. In this paper, we compare four models, including siamese, image concatenation, feature concatenation and C3D ConvNet. In contrast to deepSBD, where the position of the cut transition is unknown in one segment, adaptive thresholding can find the cut transition position accurately since it selects the pair of adjacent frames with the largest dissimilarity as the center, facilitating the learning task for our cut detector.

Siamese. A siamese neural network consists of twin networks that accept distinct images and output their features. The parameters are shared between the twin networks and each network computes the same function. An energy loss function is added to the top for optimization. In our problem, we choose contrastive loss as the top energy function. The siamese net outputs a similarity score. At inference, we select the score above some threshold.

Feature Concatenation. This network can be seen as a variant of siamese network. More specifically, it has the structure of the siamese net described above, computing the feature using the same network architecture and weights. The loss energy function is not applied directly to the features. Instead, we concatenate features from both images and add cross entropy loss function to the top.

Image Concatenation. We simply consider the two patches of an RGB image pairs as a 6-channel image and feed it to a generic network. This network provides greater flexibility compared to the above models as it starts by processing the two patches jointly. It is fast to train and infer. Further more, it allows to concatenate multiple images as a input. We find the performance is much improved when using more images.

C3D ConvNet. Hassanien et al. [9] shows the C3D ConvNet is capable of classifying cut transitions. Therefore, we also test this structure for comparison. However, the C3D ConvNet is more complex than 2D ConvNet, which requires much computation resources.

3.4 Gradual Model

Inspired by region proposal network [20] and single shot detector [15], we propose a single shot boundary network, a novel network to locate gradual transitions in a continuous video stream. The network, illustrated in Fig. 3, consists of 2 components, a shared C3D ConvNet feature extractor and subnets for classification and localization.

Feature Hierarchies. Innovated by deepSBD, the C3D ConvNet shows impressive performance in this task. Therefore, we use a C3D ConvNet to extract rich temporal feature hierarchies from a given input video buffer. The input to our model is a sequence of RGB video frames with a dimension of $3 \times L \times H \times W$ and we use ResNet-18 proposed in [8] as the backbone network. We modify all the temporal strides to 1 in ResNet-18 so that the length of the final feature map is also L. The number of frames L can be arbitrary and is only limited by memory.

Subnets for Classification and Location. Since the lengths of gradual transitions are various, we use the same notion default boxes introduced in [15]. In our task, we call it default segments. Default segments are predefined multi-scale

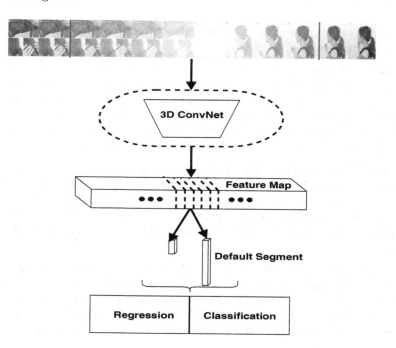

Fig. 3. An overview of gradual detector

windows centered at a location. we put one default segment every $l \times (1 - a)$ frames where l is the length of the default segment and a is the positive IOU threshold. Therefore, each ground truth whose length is between l/a and $l \times a$ can be matched to a default segment. The total number of default segments is $L/(l \times (1 - a))$. The default segments serve as reference segments for ground truth matching. To get features for predicting gradual transitions, we first apply a spatial global average pooling layer to reduce the spatial dimension to 1×1. At each location which has k default segments, we apply a $2k \times 3 \times 1 \times 1$ filter A for binary classification, and a $2k \times 3 \times 1 \times 1$ filter B for location refinement. For both A and B, 3 is the size of the temporal convolution kernel. For A, 2 corresponds to binary classification of a gradual transition or not. For B, 2 corresponds to two relative offsets of $\{\delta c_i, \delta l_i\}$ to the center location and the length of each default segment respectively, where the ground truth of $\{\delta c_i, \delta l_i\}$ is defined as

$$\delta c_i = (c - c_i)/l_i \tag{2}$$
$$\delta l_i = log(l/l_i) \tag{3}$$

The mark c_i and l_i are the center location and the length of default segments while c and l is the ground truth position and length.

Optimization Strategy. In training, positive/negative labels are assigned to default segments. Following the same protocol in object detection, positive labels are assigned if default segments are overlapped with some ground truth if intersection of union $IOU > a$ and negative labels are assigned for default segments if $IOU < b$. Segments with IOU between a and b are ignored during training. In practice, we set a to 0.5 and b to 0.1, which achieves the best performance. As the length of the gradual transitions in our training data ranges in 3 to 40, we use 2 default segments of length 6 and 20 to cover all true transitions. Similar to single shot detector, we implement hard negative example mining and dynamically balance the positive and negative examples with a ratio of 1:1 during training. To utilize the GPU efficiently, we fixed the length of each segment, consisting of L consecutive frames, i.e., L is 64 in our experiment.

We train the network by optimizing the classification and the regression losses jointly with a fixed learning rate of 0.001 for 5 epochs. We adopt softmax loss for classification and smooth L1 loss for regression. The loss function is given in (4). The hyper-parameter λ is set to 1 in practice. Y_i^1 is the predicted score and T_i^1 is the assigned label. $Y_i^2 = \{\delta c_i, \delta l_i\}$ is the predicted relative offset to the default segments and T_i^2 is the target location. The loss function is the same as [15], which is

$$Loss = \frac{1}{N_{cls}} \sum_i L_{cls}(Y_i^1, T_i^1) + \lambda \frac{1}{N_{loc}} \sum_i L_{loc}(Y_i^2, T_i^2) \qquad (4)$$

Inference. At inference, the framework processes input videos of varying lengths. However, in order not to exceed the limit of memory, a video will be divided into segments of length T_{seg} with a overlap of $\frac{1}{2}T_{seg}$ such that transitions won't be missed due to the division. After predicting one video, we apply non maximum suppression (NMS) to all the predictions. If two predicted gradual transitions are overlapped, we remove the one with lower classification score.

4   ClipShots

Current datasets, i.e. TRECVID and RAI, are not sufficient for training deep neural network due to a limited size. In addition, previous work utilized different training sets when evaluating their supervised methods on TRECVID and RAI. Therefore, a benchmark is made for comparing different methods fairly. ClipShots is the first large-scale dataset for shot boundary detection collected from Youtube and Weibo covering more than 20 categories, including sports, TV shows, animals, etc. In contrast to TRECVID2007 and RAI, which only consist of documentaries or talk shows where the frames are relatively static, we construct a database containing 4039 short videos from Youtube and Weibo. Many short videos are home-made, with more challenges, e.g. hand-held vibrations and large occlusion. The training set consists of 3539 videos, 122760 cut transitions, and 35698 gradual transitions while the evaluation set consists of 500 videos, 5876 cut transitions, and 2422 gradual transitions. The types of these videos are various, including movie spotlights, competition highlights, family videos recorded by mobile phones etc. Each video has a length of 1–20 min. The gradual transitions in our database include dissolve, fade in fade out, and sliding in sliding out. In order to annotate such a large dataset, we design an annotation tool allowing annotators to watch multiple frames on a single page and select the begin frame and the end frame of transitions. More details are given in the appendix.

5   Experiments

5.1   Databases and Evaluation Metrics

Training and Evaluation Set. The proposed framework is trained and tested on ClipShots. In order to illustrate the effectiveness of our approach and Clip-Shots, we also evaluated them on two public databases (TRECVID2007, RAI).

Evaluation Metrics. For all 3 databases, we use the standard TRECVID evaluation metrics: one-to-one match if the predicted boundary has at least 1 frame overlapped with the ground truth. For our testing set, we add an additional criterion using IOU to measure the localization performance. We assess performance quantitatively using precision (P), recall (R) and F-score (F).

5.2   Experiments Configuration

We adopt adaptive thresholding to find candidate segments and adjust the parameters to make sure it achieves nearly 100% recall for both cut and gradual transitions. For cut detector, 122760 positive examples and 224312 negative examples are used for training. For gradual detector, the training set contains 35698 ground truths. The potential segments filtered by adaptive thresholding are divided into subsegments of fixed length 64, with overlapped length of 32 between 2 consecutive segments. We choose ResNet-18 3D- ConvNet as the backbone, setting all the strides in the temporal dimension to 1 so that the temporal length of the output feature is identical with the input length. The weights of 3D ResNet-18 are initialized with model pretrained on kinetics database, as the inflated 3D-Conv [3]. For both cut and gradual model, the positive examples and negative examples are highly unbalanced so the positive and negative samples are dynamically balanced with ratio 1:1 in each mini-batch.

5.3   Experiments on ClipShots

Cut Detector Comparison. In this section, we choose four potential models introduced in Sect. 3.3 and test their performance. We use ResNet-50 as the backbone for all models and a fixed learning rate of 0.0001, We train each model for 5 epochs from scratch. For C3D, we adopt the same configuration as deepSBD. For image concatenation model, we evaluated it with different number of images. We expand x frames to the forward and backward in the temporal direction. As Table 1 shows, the image concatenation model obtains best performance among these four models when using 4 or more frames. Siamese net performs worse than image concatenation (2 frames) and C3D network. Given the fact that siamese

Table 1. Comparison of cut models. Image concat(6 frames) obtains the best performance.

Model	P	R	F
Image concat(2 frames)	0.771	0.793	0.782
Image concat(4 frames)	0.775	0.862	0.816
Image concat(6 frames)	0.776	0.934	0.848
Feature concat(2 frames)	0.231	0.574	0.329
Siamese(2 frames)	0.638	0.852	0.729
C3D(16 frames)	0.760	0.910	0.831

net cannot explore information on multiple frames and its computation cost is much larger than image concatenation, this architecture is not adopted in our framework. C3D network (16 frames) is a little better than image concatenation (2 frames), but much worse than image concatenation (4 frames or 6 frames). Feature concatenation is not a working architecture, but we still list it here. For image concatenation, we also study the relationship between the number of input images and performance. More input frames can improve performance. The model gains improvements when increasing the frame number from 2 to 6 and saturates around 6. Therefore, we use an input of 6 frames in our method considering both performance and the processing speed.

Table 2. All methods under a unified viewpoint. Different cut models and gradual models are compared.

Method	Initial filtering	Cut	Gradual
(1)	No	DeepSBD	DeepSBD
(2)	Yes	DeepSBD	DeepSBD
(3)	Yes	Image concat(6 frames)	DeepSBD
(4)	Yes	Image concat(6 frames)	SSBD

Ablation Study. We conduct ablation study with different options. The detailed setting is shown in Table 2. The difference is mainly at cut models, gradual models, and whether initial filtering is used. We also implement deepSBD but the post processing technology introduced in [9] is abandoned for a fair comparison. We adopt 3D ResNet-18 as the backbone for both deepSBD and our single shot boundary detector (Table 3).

Method (1). The model classifies segments directly into 3 categories (cut, gradual, and background).

Method (2). Compared to method (1), initial filtering is utilized to find candidate segments for deepSBD. As is shown, the performance of gradual transition is

Table 3. Performance of different methods. Our method (4) obtains the best performance in both cut transition detection and gradual transitions detection.

Methods	Cut			Gradual		
	P	R	F	P	R	F
(1)	0.765	0.910	0.831	0.770	0.622	0.688
(2)	0.757	0.902	0.823	0.699	0.810	0.750
(3)	0.776	0.934	0.848	0.711	0.830	0.766
(4)	0.776	0.934	0.848	0.840	0.904	0.870

higher than the original deepSBD. It is implied that the initial filtering can also improve performance of deepSBD.

Method (3). For gradual transitions, the deepSBD model only classifies the segments into 2 categories (gradual transition and background) so cut transitions are treated as negative samples. For cut detector, we use image concatenation model. The results show the single shot boundary detector is better than deepSBD by a large margin.

Method (4). The results reveals that our single shot boundary detector is far better than deepSBD. We attribute the performance gain to the following reasons: (1) The receptive field of our model is much bigger than deepSBD, hence the detector can exploit more temporal information. (2) Our default segment design is effective for dealing with gradual transitions of multi scales.

Benchmark in ClipShots. We implement [9] and evaluate them in ClipShots. Table 4 summaries performance of different methods. DeepSBD with 3D ResNet-18 is significantly better than the original network (3D Alexnet alike).

Table 4. Benchmark in ClipShots

Methods	Cut			Gradual		
	P	R	F	P	R	F
deepSBD (Original)	0.731	0.921	0.815	0.837	0.386	0.528
deepSBD (ResNet-18)	0.765	0.910	0.831	0.770	0.622	0.688
DSM (Ours)	0.776	0.934	0.848	0.840	0.904	0.870

Speed Comparison. In this section, we compare the speeds of different models as shown in Table 5. The code is implemented using PyTorch and tested with one TITAN XP GPU. Our method is nearly 2 times faster than the original deepSBD on account of adaptive thresholding based initial filter (Table 6).

Gradual Model Localization Performance. An accurate localization of gradual transitions is important in many video recognition task. Therefore, we

Table 5. Comparison of speed

Model	Speed (FPS)
deepSBD	382
Adaptive thresholding+deepSBD	680
Ours	700

Table 6. Localization performance. We calculate the F1-score at different IOU threshold.

IOU	P	R	F
0.1	0.813	0.865	0.839
0.5	0.726	0.772	0.748
0.75	0.599	0.637	0.618

also evaluate performance of the gradual transition localization using the proposed framework. F1 scores are measured at different IOU level $(0.1, 0.5, 0.75)$. A predicted gradual transition is considered as correct only if its $IOU > a$, otherwise it's considered wrong. When the IOU is 0.75, we can still obtain a F1 score of 0.618, indicating the proposed gradual detector is able to accurately locate gradual transitions.

Table 7. Trecvid07 top performers.

Methods	Cut			Gradual		
	P	R	F1	P	R	F1
ATT [16]	0.996	0.979	0.972	0.802	0.709	0.753
THU11 [26]	0.982	0.968	0.975	0.733	0.718	0.725
Marburg [18]	0.942	0.945	0.944	0.595	0.766	0.670
NHT [12]	0.975	0.816	0.945	0.768	0.578	0.66
Priya [5]	0.972	0.976	0.974	0.869	0.719	0.78
DeepSBD [9]	0.978	0.968	0.973	0.826	0.731	0.776
Ours	0.971	0.988	0.980	0.813	0.806	0.810
Ours (correct label)	0.981	0.997	0.989	0.838	0.845	0.841

5.4 Experiments on TRECVID07

TRECVID07 contains a total of 17 videos, including 2236 cut transitions and 225 gradual transitions. They are all color and black/white documentaries. The videos include cases such as global illumination variation, smoke, fire, and fast non-rigid motion. We take the ground truth from TRECVID07 SBD task. In addition, the experimental results of the proposed method over this database are compared to the top performers of TRECVID07 SBD task. We find some of the ground truths are wrong, so we correct these labels. Evaluation results using original labels and corrected labels are both reported. The cut and gradual models are trained with the same training setting described in Sect. 5.2.

In Table 7, we present a comparative evaluation of the shot boundary detection performance with existing state-of-the-art approaches in terms of F1-score and report the results using both the original ground truth and the corrected ground truth. We evaluate cut transitions and gradual transitions separately. Cut transitions are the most part of all transitions in a video so it plays a dominate role in the overall performance. For cut transitions, we improve the-state-of-art by 0.6%, which is a huge improvement considering there is no much space for improvement. In fact, the errors concentrate in black/white videos due to the lack of similar ones in the training set. Further improvement can be achieved through adding more black/white videos into the training set. For gradual transitions, we achieve 2.9% improvement comparing to the state-of-the-art when using the original ground truth and 6.4% improvement when using the corrected ground-truth (Table 8).

Table 8. RAI comparison

	F1 score
Apostolidis et al. [1]	0.84
Baraldi et al. [2]	0.84
Song et al. [21]	0.68
Michael et al. [7]	0.88
Hassanien et al. [9]	0.934
Ours	0.935

5.5 Experiments on RAI

RAI database is a collection of ten randomly selected broadcasting videos from the Rai Scuola video archive 1, which are mainly documentaries and talk shows. This database includes 722 cut transitions and 263 gradual transitions. Shots have been manually annotated by a set of human experts. The proposed method achieves a competitive results compared to deepSBD. It is noted that DeepSBD adopts posting-processing technology, i.e. filtering the segments whose HSV similarity under a threshold, which is not used in our methods. We perform evaluations on TRECVID and RAI using the same models, weights, and hyperparameters, which indicates the proposed framework are robust on different databases.

6 Conclusion

We propose a cascade shot transition detection framework and annotate the first large-scale shot boundary database. Adaptive thresholding is adopted to find candidate regions for acceleration. The cut and gradual transition detector are designed separately. The cut transition detector is for measuring similarity while the gradual transition detector is for capturing temporal patterns. Especially, the gradual detector is able to locate gradual transitions of multi-scales. We outperform state-of-the-art methods on both TRECVID and RAI databases. In addition, our framework is very fast, achieving a 30× real-time speed.

References

1. Apostolidis, E., Mezaris, V.: Fast shot segmentation combining global and local visual descriptors. In: 2014 IEEE International Conference on Acoustics, Speech and Signal Processing (ICASSP), pp. 6583–6587. IEEE (2014)
2. Baraldi, L., Grana, C., Cucchiara, R.: Shot and scene detection via hierarchical clustering for re-using broadcast video. In: Azzopardi, G., Petkov, N. (eds.) CAIP 2015, Part I. LNCS, vol. 9256, pp. 801–811. Springer, Cham (2015). https://doi.org/10.1007/978-3-319-23192-1_67
3. Carreira, J., Zisserman, A.: Quo vadis, action recognition? A new model and the kinetics dataset. In: 2017 IEEE Conference on Computer Vision and Pattern Recognition (CVPR), pp. 4724–4733. IEEE (2017)
4. Deng, J., Dong, W., Socher, R., Li, L.J., Li, K., Fei-Fei, L.: ImageNet: a large-scale hierarchical image database. In: IEEE Conference on Computer Vision and Pattern Recognition, CVPR 2009, pp. 248–255. IEEE (2009)
5. Domnic, S.: Walsh-Hadamard transform kernel-based feature vector for shot boundary detection. IEEE Trans. Image Process. **23**(12), 5187–5197 (2014)
6. Escorcia, V., Caba Heilbron, F., Niebles, J.C., Ghanem, B.: DAPs: deep action proposals for action understanding. In: Leibe, B., Matas, J., Sebe, N., Welling, M. (eds.) ECCV 2016, Part III. LNCS, vol. 9907, pp. 768–784. Springer, Cham (2016). https://doi.org/10.1007/978-3-319-46487-9_47
7. Gygli, M.: Ridiculously fast shot boundary detection with fully convolutional neural networks (2017). arXiv preprint: arXiv:1705.08214
8. Hara, K., Kataoka, H., Satoh, Y.: Can spatiotemporal 3D CNNs retrace the history of 2D CNNs and ImageNet? (2017). arXiv preprint: arXiv:1711.09577
9. Hassanien, A., Elgharib, M., Selim, A., Hefeeda, M., Matusik, W.: Large-scale, fast and accurate shot boundary detection through spatio-temporal convolutional neural networks (2017). arXiv preprint: arXiv:1705.03281
10. Huang, Q., Xiong, Y., Xiong, Y., Zhang, Y., Lin, D.: From trailers to storylines: an efficient way to learn from movies (2018). arXiv preprint: arXiv:1806.05341
11. Iandola, F.N., Han, S., Moskewicz, M.W., Ashraf, K., Dally, W.J., Keutzer, K.: SqueezeNet: AlexNet-level accuracy with 50x fewer parameters and <0.5 MB model size (2016). arXiv preprint: arXiv:1602.07360
12. Kawai, Y., Sumiyoshi, H., Yagi, N.: Shot boundary detection at TRECVID 2007. In: TRECVID (2007)
13. Kay, W., et al.: The kinetics human action video dataset (2017). arXiv preprint: arXiv:1705.06950
14. Lin, T., Zhao, X., Shou, Z.: Single shot temporal action detection. In: Proceedings of the 2017 ACM on Multimedia Conference, pp. 988–996. ACM (2017)
15. Liu, W., et al.: SSD: single shot MultiBox detector. In: Leibe, B., Matas, J., Sebe, N., Welling, M. (eds.) ECCV 2016, Part I. LNCS, vol. 9905, pp. 21–37. Springer, Cham (2016). https://doi.org/10.1007/978-3-319-46448-0_2

16. Liu, Z., Gibbon, D., Zavesky, E., Shahraray, B., Haffner, P.: At&t research at TRECVID 2007. In: Proceedings of TRECVID Workshop, pp. 19–26 (2007)
17. Lu, Z.M., Shi, Y.: Fast video shot boundary detection based on svd and pattern matching. IEEE Trans. Image Process. **22**(12), 5136–5145 (2013)
18. Mühling, M., Ewerth, R., Stadelmann, T., Zöfel, C., Shi, B., Freisleben, B.: University of Marburg at TRECVID 2007: shot boundary detection and high level feature extraction. In: TRECVID (2007)
19. Qiu, Z., Yao, T., Mei, T.: Learning spatio-temporal representation with pseudo-3D residual networks. In: 2017 IEEE International Conference on Computer Vision (ICCV), pp. 5534–5542. IEEE (2017)
20. Ren, S., He, K., Girshick, R., Sun, J.: Faster R-CNN: towards real-time object detection with region proposal networks. In: Advances in Neural Information Processing Systems (NIPS) (2015)
21. Song, Y., Redi, M., Vallmitjana, J., Jaimes, A.: To click or not to click: automatic selection of beautiful thumbnails from videos. In: Proceedings of the 25th ACM International on Conference on Information and Knowledge Management, pp. 659–668. ACM (2016)
22. Wang, J., et al.: Learning fine-grained image similarity with deep ranking (2014). arXiv preprint: arXiv:1404.4661
23. Wang, L., Xiong, Y., Lin, D., Van Gool, L.: UntrimmedNets for weakly supervised action recognition and detection. In: IEEE Conference on Computer Vision and Pattern Recognition, vol. 2 (2017)
24. Xu, H., Das, A., Saenko, K.: R-C3D: region convolutional 3D network for temporal activity detection. In: The IEEE International Conference on Computer Vision (ICCV), vol. 6, p. 8 (2017)
25. Yuan, J., Li, J., Lin, F., Zhang, B.: A unified shot boundary detection framework based on graph partition model. In: Proceedings of the 13th Annual ACM International Conference on Multimedia, pp. 539–542. ACM (2005)
26. Yuan, J., et al.: A formal study of shot boundary detection. IEEE Trans. Circ. Syst. Video Technol. **17**(2), 168–186 (2007)
27. Yusoff, Y., Christmas, W.J., Kittler, J.: Video shot cut detection using adaptive thresholding. In: BMVC, pp. 1–10 (2000)
28. Zagoruyko, S., Komodakis, N.: Learning to compare image patches via convolutional neural networks. In: 2015 IEEE Conference on Computer Vision and Pattern Recognition (CVPR), pp. 4353–4361. IEEE (2015)
29. Zhang, K., Chao, W.-L., Sha, F., Grauman, K.: Video summarization with long short-term memory. In: Leibe, B., Matas, J., Sebe, N., Welling, M. (eds.) ECCV 2016, Part VII. LNCS, vol. 9911, pp. 766–782. Springer, Cham (2016). https://doi.org/10.1007/978-3-319-46478-7_47
30. Zhao, Y., Xiong, Y., Wang, L., Wu, Z., Tang, X., Lin, D.: Temporal action detection with structured segment networks. In: The IEEE International Conference on Computer Vision (ICCV), vol. 8 (2017)

Traversing Latent Space Using Decision Ferns

Yan Zuo$^{(\boxtimes)}$, Gil Avraham$^{(\boxtimes)}$, and Tom Drummond$^{(\boxtimes)}$

ARC Centre of Excellence for Robotic Vision, Monash University, Melbourne,
Australia
{yan.zuo,gil.avraham,tom.drummond}@monash.edu

Abstract. The practice of transforming raw data to a feature space so
that inference can be performed in that space has been popular for many
years. Recently, rapid progress in deep neural networks has given both
researchers and practitioners enhanced methods that increase the rich-
ness of feature representations, be it from images, text or speech. In this
work we show how a constructed latent space can be explored in a controlled manner and argue that this complements well founded inference
methods. For constructing the latent space a Variational Autoencoder is
used. We present a novel controller module that allows for smooth traversal in the latent space and construct an end-to-end trainable framework.
We explore the applicability of our method for performing spatial transformations as well as kinematics for predicting future latent vectors of a
video sequence.

1 Introduction

A large part of human perception and understanding relies on using visual information from the surrounding environment and interpreting this information to
subsequently act upon these interpretations. However, these interpretations are
not applied directly to what is being seen; rather, there exists an abstraction
mechanism from the image space into a more informative space so that complex
inferences can be made there [1]. Similarly in machine learning, we would like
machines to inherit this ability to abstract as it is the key to understanding and
learning when it comes to real world data.

Recently, deep learning has demonstrated enormous success across various
vision-related tasks such as image classification, object detection and semantic
segmentation [2–6]. However, this level of success has yet to transition across
to more complicated tasks such as video prediction. Many of the popular deep
learning methods approach these more challenging tasks in a similar manner to
image classification or segmentation, choosing to learn directly from the image

---

This work was supported by the Australian Research Council Centre of Excellence for
Robotic Vision (project number CE1401000016).
Y. Zuo and G. Avraham are contributed equally.

© Springer Nature Switzerland AG 2019
C. V. Jawahar et al. (Eds.): ACCV 2018, LNCS 11361, pp. 593–608, 2019.
https://doi.org/10.1007/978-3-030-20887-5_37

space [7,8]. This presents a challenge because often the image space is high-dimensional and complex; there exists a large semantic gap between the input pixel representation of the data and the desired transformation of said data for complex inference tasks such as prediction.

To address this challenge, we leverage the compact encoding space provided by a Variational Autoencoder (VAE) [9] to learn complex, higher order tasks in a more feasible way. Inference in the latent space takes advantage of solving a far more tractable problem than performing the operation in the image space as it gives a more natural way of separating the task of image construction and inferring semantic information. Other works have similarly utilised a compact encoding space to learn complex functions [10–13]; however, even this encoding space can be strongly entangled and highly non-linear. As such, we construct a residual decision fern based architecture which serves as a controller module that provides the necessary non-linearity to properly disentangle encodings in the latent space. To this end, we introduce a novel framework for controlled traversal of the latent space called the Latent Space Traversal Network (LSTNet) and offer the following contributions:

- We discuss the benefits of operating in a latent space for complex inference tasks, introducing a novel decision fern based controller module which enables the use of control variables to traverse the latent space (Sects. 4.1, 4.2 and 4.3).
- We create a unified, end-to-end trainable framework which incorporates our controller module with a residual VAE framework, offering an encoding space for learning high order inference tasks on real world data. Additionally, this framework offers a key insight into separating the tasks of pixel reconstruction and the high order inference (Sects. 4.3 and 5.3).
- We demonstrate significant qualitative and quantitative improvements offered by LSTNet over popular models that impose geometrical and kinematic constraints on the prediction search space across the MNIST and KITTI datasets (Sects. 5.1, 5.2 and 5.3).

2   Related Work

2.1   Learning Representations for Complex Inference Tasks

Operating in the latent space for demonstrating certain properties has been shown in several works. Applying a convolution architecture to a GAN framework [14] showed that following the construction of the latent space and applying arithmetic operations between two latent vectors observes the semantic logic we would expect from such an operation. InfoGAN [15] added a regularisation term which maximises mutual information between image space and some latent variables. Both these works claim to yield a disentangled latent vector representation of the high dimensional input data, demonstrating this by choosing a specific latent variable and interpolating across two values and showing smooth image transformation. However, due to their unsupervised nature, there are no guarantees on what attributes they will learn and how this will distort the intrinsic properties of the underlying data.

The work of [12] divides the learned latent space in a VAE into extrinsic and intrinsic variables. The extrinsic variables are forced to represent controllable parameters of the image and the intrinsic parameters represent the appearance that is invariant to the extrinsic values. Although it is trained in a VAE setting, this method requires full supervision for preparing training batches. [11] introduces a fully supervised method, introducing a Gaussian Process Regression (GPR) in the constructed latent space of a VAE. However, using GPR imposes other limitations and assumptions that do not necessarily apply in training a VAE. [13] uses a semi-supervised approach for video prediction, training a VAE-GAN [16] instead of a standard VAE. The VAEGAN leads to sharper looking images but imposes a discriminator loss which complicates the training procedure drastically. This work is similar to ours in that control variables are used to guide learning in the latent space. However, this method is not end-to-end trainable and uses a simple framework for processing the latent space; [7] showed this framework underperformed in next frame prediction and next frame steering prediction.

[8, 10] offered a video prediction framework which combines an adversarial loss along with an image gradient difference/optical flow loss function to perform next frame prediction and also multi-step frame prediction. Similarly, [7] performed predictive coding by implementing a convolutional LSTM network that is able to predict the next frame and also multi-step frame prediction in a video sequence. In contrast to our work and [13], these works optimise learning in the image space. This approach suffers from poor semantic inference (especially over large time intervals) and will be a focus of investigation in our work.

2.2 Decision Forests and Ferns

Decision forests are known for their flexibility and robustness when dealing with data-driven learning problems [17]. Random decision forests were first developed by [18], where it was found that randomly choosing the subspace of features for training had a regularising effect which overcame variance and stability issues in binary decision trees. Following this, various methods which added more randomness quickly followed which helped further stabilise training of decision trees [19–21]. Further work extended decision trees into a related method of decision ferns, finding use in applications such as keypoint detection [22, 23]. Recently, an emerging trend has seen the incorporation of decision forests within deep learning frameworks [24–27], utilising the non-linear discriminating capabilities offered by deep decision trees.

3 Background

3.1 Variational Autoencoders

There are many works that approximate probability distributions. For a comprehensive review, refer to [28] and more recently [29]. [9] proposed a unified encoder-decoder framework for minimising the variational bound:

$$\log P(X) - \mathcal{D}[\mathcal{Q}(z|X)||P(z|X)]$$
$$= E_{z \sim \mathcal{Q}}[log P(X|z)] - \mathcal{D}[\mathcal{Q}(z|X)||P(z)] \tag{1}$$

It states that minimising the term $\mathcal{D}[\mathcal{Q}(z|X)||P(z)]$ by choosing $z \sim \mathcal{N}(\mu, \sigma)$ and imposing a normal distribution on the output of the encoder will result in approximating $P(X)$ up to an error that depends on the encoder-decoder reconstruction error. In practice, the encoder and decoder are constructed to have sufficient capacity to encode and decode an image from the given distribution properly, and also the choice of latent space to be in a high enough dimensionality to have the descriptive capacity to encode the image information needed for reconstruction.

3.2   Decision Forests and Ferns

Decision Trees. A decision tree is composed of a set of internal decision nodes and leaf nodes ℓ. The decision nodes, $\mathcal{D} = \{d_0, \cdots, d_{N-1}\}$, each contain a decision function $d(\boldsymbol{x}; \theta)$, which perform a routing of a corresponding input based on its decision parameters θ. The set of input samples $\boldsymbol{x}$ start at the root node of a decision tree and are mapped by the decision nodes to one of the terminating leaf nodes $\ell = \mathcal{D}(\boldsymbol{x}, \Theta)$, which holds a real value $\boldsymbol{q}$ formed from the training data (Θ denotes the collected decision parameters of the decision tree). Hence, a real value q in a leaf node will be denote by:

$$q(\ell) = \frac{\sum_i \omega(\mathcal{D}(x_i))v_i}{n_\ell} \tag{2}$$

where n_ℓ is the total count of training samples mapped to leaf node ℓ, ω specifies the decision tree mapping of the sample to leaf ℓ and v_i specifies the value of sample i.

Decision Ferns. A decision fern is related to a decision forest. Similarly to decision trees, decision ferns route samples from the root node to a leaf node ℓ. The key difference is that all decisions are made simultaneously rather than in sequence. Thus, a decision fern contains a single root node which holds the parameters Θ of the entire decision fern to map from the root to a leaf node.

4   Operating in Latent Space

To operate in latent space, there must exist a mechanism which allows for transition between the latent space and image space. Hence, there is an inherent trade off between obtaining a good semantic representation of the image space via its corresponding latent space, and the reconstruction error introduced when transitioning back to the image space. In this work, we show that the benefits of working in a compact latent space far outweigh the loss introduced by image reconstruction. The latent space emphasises learning the underlying changes in semantics related to an inference task which is paramount when learning high order inference tasks.

4.1   Constructing a Latent Space

When constructing a latent space, several viable options exist as candidates. The most naive method would be to perform Principal Component Analysis (PCA) directly on images [30], selecting the N most dominant components. Clearly, this method results in a large loss in information and is less than ideal. Generative Adversarial Networks (GANs) [14,31,32] can produce realistic looking images from a latent vector but lack an encoder for performing inference and are difficult to train. Techniques which combine GANs with Variational Autoencoders [16, 29,33–35] offer an inference framework for encoding images but prove to be cumbersome to train due to many moving parts and instability accompanying their adversarial training schemes.

Hence, to construct our latent space, we use the relatively straightforward framework of a Variational Autoencoder (VAE) [9]. VAEs contain both a decoder as well as an encoder; the latter of which is used to infer a latent vector given an image. This encoding space offers a low dimensional representation of the input and has many appealing attributes: it is semi-smooth and encapsulates the intrinsic properties of image data. It is separable by construction; it maintains an encoding space that embeds similar objects in the image space near each other in the latent space. Furthermore, a VAE can be trained in a stable manner and in an unsupervised way, making it an ideal candidate to learn complex higher order inference tasks.

4.2   Traversing the Latent Space

Recent works attempted to learn the latent space transformation using various models, such as RNNs [13] and a Gaussian Process Regression [11]. In our work we recognise that although the original input data is reduced to a lower dimension encoding space, inference on this space is a complex operation over a space which if constructed correctly, has no redundant dimensions; hence all latent variables should be utilised for the inference task. Under the assumption of a smooth constructed manifold and a transformation that traverses this smooth manifold under a narrow constraint (in the form of a control variable or side information), a reasonable model for the controller module is:

$$z_{t+h} = z_t + F(z_t, z_{t-1}, z_{t-2}, ..., \theta) \qquad (3)$$

where $\{z_t, z_{t-1}, z_{t-2}...\}$ are the latent vectors corresponding to input data $\{x_t, x_{t-1}, x_{t-2}...\}$, θ is the control variable and z_{t+h} is the output of the model corresponding to given the inputs. The operator $F(\mathbf{z}, \theta)$ can be interpreted as:

$$\frac{z_{t+h} - z_t}{h} = \frac{1}{h} F(z_t, z_{t-1}, z_{t-2}, ..., \theta)$$
$$\frac{\partial z_t}{\partial h} = F(z_t, z_{t-1}, z_{t-2}, ..., \theta) \qquad (4)$$

where $\frac{1}{h}$ can be absorbed into $F(\mathbf{z}, \theta)$ and by doing so, we can interpret it as a residual term that is added for smoothly traversing from input z_t to z_{t+h}, given side information θ and the history $\{z_t, z_{t-1}, z_{t-2}...\}$. This construction allows us to implement Eq. 4 using a neural network, which we denote as the *Transformer Network*. Equation 3 encapsulates the complete controller, which is a residual framework that delivers the final transformed latent vector (denoted as the *Controller Module*). The Transformer Network, will be doing the heavy lifting of inferring the correct step to take for obtaining the desired result $\hat{z}_{t+h}$ and as such should be carefully modelled. In the following section, we discuss our chosen implementation and the considerations that were taken when constructing the Transformer Network.

4.3   Latent Space Traversal Network

Our Latent Space Traversal Network (LSTNet) consists of two main components:

1. A VAE with an encoder and decoder. The encoder learns a latent representation to encode the real set of training images into this space. The decoder learns a mapping from latent space back to the image space.
2. A Controller Module (C) with a Transformer Network (TN), that applies an operation in the latent space offered by the VAE.

An overview of our model is shown in Fig. 1. In the rest of this section, we detail and justify the choice of architecture for the components of our model. To construct our latent space, we adopt the approach in [36] and construct a residual encoder and decoder to form our VAE. This residual VAE offers a low-dimensional, dense representation of input image data, allowing for a latent space which makes higher order inference tasks easier to learn in. LSTNet is trained on a loss function composed of three terms:

$$\mathbb{L}_{VAE} = \frac{1}{N} \sum_{i \in \mathcal{B}} \left\| \hat{I}_i - I_i \right\|^2 + \frac{1}{N} \sum_{i \in \mathcal{B}} KL(z_i, \mathcal{N}(0, U)), \tag{5}$$

$$z \sim \mathcal{N}(\mu_{enc}(I), \sigma_{enc}^2(I))$$

$$\mathbb{L}_z = \frac{1}{N} \sum_{i \in \mathcal{B}} \| z_{target} - \hat{z}_{target} \|^2, \hat{z}_{target} = C(z_{t-n}, ..., z_{t-1}, \theta), \tag{6}$$

$$z_{target} = \mu_{enc}(I_t)$$

$$\mathbb{L}_I = \frac{1}{N} \sum_{i \in \mathcal{B}} \left\| I_{target} - \hat{I}_{target} \right\|^2, \hat{I}_{target} = \mathcal{P}(I_{t-n}, ..., I_{t-1}, \theta), \tag{7}$$

$$I_{target} = I_t$$

where $\mathbb{L}_{VAE}$ is the loss for the VAE which updates the encoder and decoder parameters, $\mathbb{L}_z$ is the controller loss which updates the controller network's parameters and $\mathbb{L}_I$ is the predicted image loss which updates the controller and decoder of the VAE. In this case, I is an image, z is a latent vector in the encoding space, $\mathcal{B}$ is a minibatch, U is an identity matrix, μ_{enc} and σ_{enc}^2 are the respective mean and variance of the encoder's output, C is the controller network and $\mathcal{P}$ denotes the LSTNet (passing an input image(s) through the encoder, controller and decoder to generate a transformation/prediction).

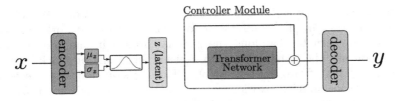

Fig. 1. Overview of our LSTNet model architecture. x denotes input data and y denotes the output with a transformation applied in the latent space.

Fern-Based Transformer Network. Even in the latent space, learning complex tasks such as video prediction can be difficult. In the experiments section, we motivate the use of the fern-based controller over a linear variant composed of stacked fully connected layers with ReLU non-linearities.

Our transformer network employs an ensemble of soft decision ferns as a core component. The use of soft decision ferns allows them to be differentiable such that they can be integrated and trained within an end-to-end framework. One way to achieve this is construct decision functions which apply a sigmoid to each input activation biased with a threshold value, yielding a soft value between $[0, 1]$:

$$d_n(\boldsymbol{x}, \boldsymbol{t}) = \sigma((x_n - t_n)) \tag{8}$$

where $\sigma(x)$ is a sigmoid function. x_n and t_n are the respective input activation and corresponding threshold values assigned towards the decision. To illustrate this, for a depth two fern using two activations which create the soft routes to its corresponding four leaves, its output Q is:

$$\begin{aligned} Q = q_0 \times p_0 \times p_1 + q_1 \times p_0 \times (1 - p_1) \\ + q_2 \times (1 - p_0) \times p_1 + q_3 \times (1 - p_0) \times (1 - p_1) \end{aligned} \tag{9}$$

where p_0 and p_1 are the respective probability outputs of the decision functions of the decision ferns. q_0, q_1, q_2 and q_3 are the corresponding leaf nodes of the decision fern (illustrated in Fig. 2a). Equation 9 can be reparameterised as:

$$Q = b + d_0 \times x + d_1 \times y + d_0 \times d_1 \times z \tag{10}$$

where:

$$d_0 = \tanh(x_0), d_1 = \tanh(x_1)$$

$$b = \frac{1}{2^h} \times (q_0 + q_1 + q_2 + q_3), x = \frac{1}{2^h} \times (q_0 - q_1 + q_2 - q_3) \tag{11}$$

$$y = \frac{1}{2^h} \times (q_0 + q_1 - q_2 - q_3), z = \frac{1}{2^h} \times (q_0 - q_1 - q_2 + q_3)$$

x_0 and x_1 are the assigned activations to the decision fern and h is the fern depth. b, x, y, z can be represented by fully connected linear layers which encapsulates all decision ferns in the ensemble.

Figure 2b shows the architecture of our transformer network. We adopt the residual framework in [4], modifying it for a feedforward network (FNN) and adding decision fern blocks along the residual branch in the architecture. In Fig. 2c, we outline the construction of a decision fern building block in the TN, consisting of ferns of two levels in depth. The decision nodes of the fern are reshaped from incoming activations, to which Batch Normalisation [37] and a Hyperbolic Tangent function is applied. This compresses the activations between the range of $[-1, 1]$ and changes their role to that of making decisions on routing to the leaf nodes. A split and multiply creates the conditioned depth two decisions of the fern, which is concatenated with the depth one decisions. Finally, a FC linear layer serves to interpret the decisions made by the decision fern and form the leaf nodes which are free to take any range of values.

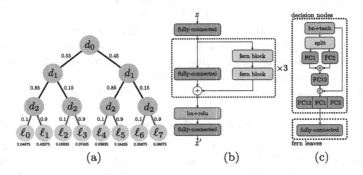

Fig. 2. (a) A decision fern with soft routing. (b) The architecture of our proposed transformer network. (c) The structure of a fern block used in our transformer network.

Hence, our final controller network is expressed as:

$$C(z_{t-n}, ..., z_{t-1}, \theta) = z_{t-1} + TN(z_{t-n}, ..., z_{t-1}, \theta) \tag{12}$$

5   Experiments

In our experiments, we explore operation sets that can be achieved by working in the latent space. We choose two applications to focus on - the first application looks towards imposing a spatial transformation constraint on the latent vector; the second application is a more complex one, looking at video prediction. For the first application, we use the MNIST dataset [38] as a toy example and investigate rotating and dilation operations on the dataset. For the second application, we use the KITTI dataset [39] and perform video prediction and steering prediction. For all experiments, we trained using the ADAM Optimiser [40] with learning rate of 0.0001, first and second moment values of 0.9 and 0.999 respectively, using a batch size of 64.

5.1   Imposing Spatial Transformation

For imposing spatial transformations, we present rotating and dilation operations and show how to constrain the direction a latent vector will traverse by using a spatial constraint. This constrained version of LSTNet applies a transformation to a single image which either rotates or erodes/dilates the given image. For both rotation and dilation experiments, we use a small residual VAE architecture along with our specified controller module with 1 residual layer with decision fern blocks (refer to Fig. 2b and c for details). We choose a latent vector size of 100 dimensions. The encoder consists of 2 residual downsampling layers (refer to [36] for details on the residual layers), Batch Normalisation and ReLU activations in between, ending with 2 fully connected linear layers for emitting the mean and variance of the latent vector. The decoder also consists of 2 residual upsampling layers, Batch Normalisation and ReLU activations in between, ending with a Hyperbolic Tangent function. To compare our method, we use two baselines. The first method is the most obvious comparison; we implement a baseline CNN which learns a target transformation, given an input image and corresponding control variable θ (CNN-baseline). This CNN-baseline is composed of 2 strided 3×3 convolution layers, 2 FC linear layers and 2 strided 3×3 deconvolution layers with ReLU non-linearities used for activation. The number of output channels in the hidden layers was kept at a constant 128. Additionally, we implement the Deep Convolutional Inverse Graphics Network [12] as specified in their paper for comparison. For each of the three methods compare under the same conditions by providing the control variable θ as an input during training and testing.

Rotation. We create augmented, rotated samples from the MNIST dataset. Specifically, we randomly choose 600 samples from the data, ensuring a even distribution of 60 samples per class label are chosen. For each sample, we generate 45 rotation augmentations by rotating the sample in the range of $-45° < \theta < 45°$. We add this augmented set to the original MNIST data and train the VAE with controller module end-to-end for 20k iterations. To inject the control variable into the input, we concatenate it to the encoded latent vector before feeding it to the controller module.

To train our controller module, we note that there are $\binom{45}{2}$ possible pairs in each example giving us much more training data than needed. Hence, for every iteration of training, we randomly choose a batch of triplets (I_i, I_j, θ), where θ is a rotation control variable specifying the rotation in radians. For inference, we randomly sample images from the MNIST dataset that were not selected to be augmented and perform a rotation parameter sweep. This results in smooth rotation of the image, while preserving the shape (see Fig. 3a). Note that other works (*i.e* [15]) have shown that by altering a variable in the latent space, a rotated image can be retrieved, but fail to preserve image shape, leading to distortion. This indicates that the specific variable for rotation is not only responsible for rotation, but has influence over other attributes of the image. In contrast, we observed this difference across several variables between z_i and z_j. This gives the insight that in order to perform a rotation (or similar spatial transform operation) and preserve the original image shape, several variables in the latent vector need to change which justifies the use of a highly non-linear network to approximate this operation.

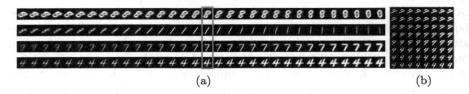

(a)   (b)

Fig. 3. (a) Rotation and dilation operations performed by LSTNet on samples from MNIST. The top two rows show rotation, whilst the bottom two rows show dilation. The original samples are shown in the middle highlighted in blue (b) LSTNet applying a combination of rotation and dilation operations to a sample from MNIST. (Color figure online)

Thickening. For learning the dilation operation we created augmentations in a similar manner to rotation operations. We randomly choose 5000 samples (evenly distributed across the 10 class labels) from the original dataset, and augmented every sample 4 times with 2 steps of dilation (thickening) and 2 steps of erosion (thinning). We train our VAE and controller module in a similar way to the rotation operation for 20k iterations, specifying batches of triplets (I_i, I_j, θ); θ changes its role to specifying a dilation factor that takes one of 5 discrete values in the range of $[-2, 2]$. Note that although the network was trained on 5 discrete levels of dilation, it manages to learn to smoothly interpolate when performing the operation sweep during inference (as shown in Fig. 3a).

Combining Operations. One immediate extension that LSTNet offers when performing spatial transformation operations is the ease in which multiple spatial transformation operations can be combined together into a single framework. In Fig. 3b, we show the samples produced by an LSTNet with 2 controller modules, sharing a single latent space offered by the VAE. It is important to note here that neither the rotation or dilation controller modules saw the other's training data. Hence, both operations are applied consecutively in the latent space and decoded back for visualisation.

In Table 1, we show the mean squared error (MSE) of LSTNet, comparing against the two baseline methods across rotation, dilation and combined rotation plus dilation operations. We can see that LSTNet outperforms in Rotation and Dilation MSE and handily outperforms in the combined operation. This indicates a generality in learning in the latent space; LSTNet has learned the semantics behind rotation and dilation operations and thus can seamlessly combine these two operations.

Table 1. Mean squared error values on MNIST for rotation, dilation and rotation+dilation operations across CNN-baseline, DCIGN and LSTNet. Note that significant improvement LSTNet offers for the combined operation of rotation and dilation, indicating modularity.

Model	Rotation (MSE)	Dilation (MSE)	Rotation+Dilation (MSE)
DCIGN [12]	0.07373484	0.02599725	0.08349568
CNN-baseline	0.02819574	0.00841950	0.06508310
LSTNet	**0.02380177**	**0.00835836**	**0.04410466**

5.2 Imposing Kinematics

We now move towards the more complex inference task of video prediction. Similar to imposing spatial transformation, we use a larger, residual VAE architecture along with a larger controller module to account for the more complex inference task. We increase the base dimension of our latent vector to 256 dimensions. The VAE's encoder consists of 3 residual downsampling layers, Batch Normalisation and ReLU activations in between, ending with a fully connected linear layer. The decoder consists of 3 residual upsampling layers, Batch Normalisation and ReLU activations in between, ending with a Hyperbolic Tangent function. For the controller network we test out two variants. The first is the fern-based controller as described in Sect. 4.3. For motivating the use of the fern-based controller, a linear variant controller network is also used: this is a feedforward linear network consisting of 4 FC linear layers with ReLU activations, matching the fern-based controller in terms of model capacity.

Similarly to imposing spatial transformations, we randomly create batches of triplets (I_i, I_j, θ), where I_i and I_j are the respective current frame and target future frame (randomly chosen to be within 5 time steps of the current frame) and θ is the corresponding time step from the current frame to the target frame. Our controller module receives the latent vectors of the current frame as well as a sequence of latent vectors belonging to the previous 5 frames to the current frame. We train our framework end-to-end for 150k iterations; the VAE is trained to minimise reconstruction loss along with KL regularisation, the controller is minimised using the latent error between the target latent and predicted latent vectors and both decoder and controller are jointly optimised using the error between target frames and predicted frames (refer to Eqs. 5, 6 and 7). The VAE is trained at a ratio of 5:1 against the controller module to ensure a proper representation of the input in its encoding space.

Figure 6 shows qualitative results comparing our model with the PredNet model in [7]. We can see that the further the prediction is over time, the less accurate PredNet becomes, whilst LSTNet remains much more robust to changes in the scene over time. Observing the samples of PredNet, a recurring phenomena is that in areas of the frame where object movement should occur, moving objects are instead smeared and blurred. In the case of LSTNet, predicted movement is better observed (particularly over large time steps), whilst vehicles and

Model	Avg. MSE	Avg. SSIM
Copy Last Frame	0.03829	0.615
PredNet [7]	0.02436	0.604
PredNet (Finetuned) [7]	0.01524	0.679
Linear Controller Variant	0.02083	0.631
LSTNet (Ours)	**0.01316**	**0.694**

(a)

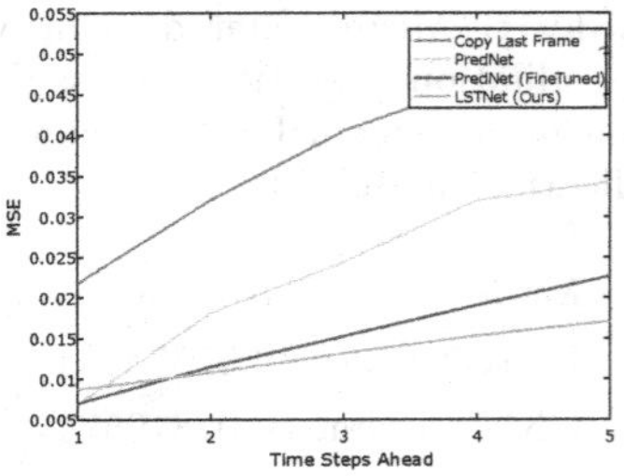

(b)

Fig. 4. (a) Averaged MSE and SSIM over 5 timesteps of frame prediction (b) Individual MSE for each predicted time step plotted for Copy Last Frame, PredNet and LSTNet.

street furniture (*i.e* road signs and markings) are better placed. This gives an indication that LSTNet is more reliable for prediction mechanism within the context of the task. PredNet implicitly learns to predict on a fixed timestep and hence predicts over a large time interval by rolling out over its predicted images. On the other hand, LSTNet offers a more general approach via the control variable θ. It learns a transformation which allows it to directly predict the target video frame without the computational overhead of rolling out fixed timesteps to reach the target. In Fig. 4a, we depict the average MSE and SSIM over 5 future frames (500 ms time lapse) for the KITTI test set. We can see that LST-Net outperforms the compared methods of Copy Last Frame, PredNet and a Linear Controller Variant (as discussed in Sect. 4.3). Looking at Fig. 4b, we can see that LSTNet achieves lowest MSE for all timesteps, except $\hat{I}_{t+1}$, where Pred-Net has slightly lower MSE. However, despite slightly higher initial MSE, the performance of LSTNet quickly exceeds both methods as inferring good prediction begins playing a larger factor over time (see Sect. 5.3 for a full discussion). These quantitative results correlate well with our qualitative results and again indicate that LSTNet is able to outperform on the task of prediction rather than on image reconstruction.

5.3 Latent Space for Prediction

In addition to the computational and memory footprint benefits, we show that projecting an image onto the latent space, operating on the latent vector using our controller module and reprojecting back into the pixel space has on average a lower MSE in the pixel space over operations that are increasingly harder to perform; for example: chaining spatial transformations and predicting more than 1 time step into the future.

Moving vs. Non-moving Objects. An attribute that does not favour a latent space framework is the inference of the fine grained detail in images with a lot of static, non-moving components. Figure 5 shows an example of a $\hat{I}_{t+1}$ prediction for PredNet (Finetuned) [7], our LSTNet and the ground truth for comparison. Across these three images, red and green highlighted patches show texture and movements of objects respectively. Across the patches, it is visually apparent that LSTNet does not generate a fine detailed texture of the tree leaves, although it captures the movement of a car well as the camera viewpoint changes. Conversely, PredNet is able to capture the fine grain texture of the tree leaves, but fails to capture the movement of the stationary car from a camera viewpoint change.

We perform a simple patch-based test to illustrate the importance of emphasising prediction on moving objects versus non-moving objects. For $\hat{I}_{t+1}$, we randomly choose 20 patches of 20×20 pixels that we identify where movement occurs; computing the average MSE yielded values for LSTNet = **0.0020** and PredNet = 0.0134 showing we significantly improve on predicting movement. Similarly, we sample 20 patches identified as being static with no moving objects which yielded an MSE for LSTNet = 0.0024 and PredNet = **0.0021**, with the results favouring PredNet. This result correlates well with the competitive results shown on $\hat{I}_{t+1}$ in Fig. 4b; between two consecutive frames, the majority of parts in a scene are static with little to no movement. Hence, prediction plays a smaller role in pixel space MSE over such a short time frame. However, as the time between predicted and current frame increases, getting better predictions on movements plays a larger role, which accounts for the results shown in Fig. 4b.

Auxiliary Parameter Predictions. Furthermore, inferring auxiliary tasks such as steering angle does not require the fine detailed knowledge contained in the pixel space of a scene. For performing such inference tasks, the main requirement is the semantic information contained in the scene. For this task, we used a pretrained LSTNet and added a FC layer to the controller to output a single value and finetuned to learn the steering angle. This is where LSTNet shines; it considerably outperforms PredNet [7] and copying steering angles from the last seen frame as shown in Table 2. This further correlates with our MSE prediction results that LSTNet is inherently able to distill the semantics from a scene for complex inference tasks.

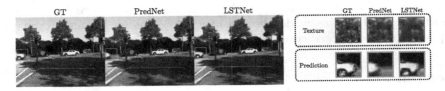

Fig. 5. On the left hand side we present the next frame prediction $(t+1)$ for the Ground-Truth, PredNet (Finetuned) and our LSTNet. On the right hand side are patches that match the rectangular markings on the images with corresponding labels. Our LSTNet excels at predicting the semantic changes that are important for maintaining the correct structure of a scene; and at times may fail (as shown) at outputting the fine-grained details of the scene objects, due to reconstruction.

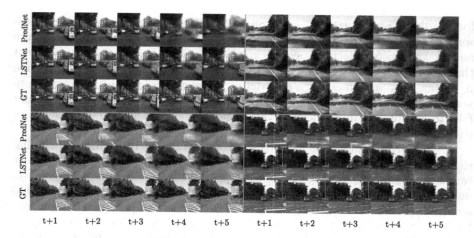

Fig. 6. Multi-frame predictions. This figure depicts of 4 sequences selected from the KITTI test set where $\{I_{t+1}, I_{t+2}, I_{t+3}, I_{t+4}, I_{t+5}\}$ are predicted using a past sequence of 5 frames.

Table 2. Steering angle prediction MSE on the KITTI test data.

Model	Steering angle MSE (Degrees^2)
Copy Last Frame	1.3723866
PredNet (FineTuned) [7]	2.4465750
LSTNet (Ours)	**0.5267124**

6   Conclusion

In this work, we present a novel, end-to-end trainable framework for operating on a latent space constructed using a VAE. We explore the power of learning in latent space on two operations: spatial transformations and video prediction, and show semantic gains for increasingly harder inference tasks which subsequently translates to a more meaningful result in the pixel space. Furthermore, as a direct extension to this work, the use of a VAE presents an opportunity to explore multi-model predictions for further robustness in predictive tasks.

References

1. Utgoff, P.E., Stracuzzi, D.J.: Many-layered learning. Neural Comput. **8**, 2497–2529 (2002)
2. Krizhevsky, A., Sutskever, I., Hinton, G.E.: Imagenet classification with deep convolutional neural networks. Adv. Neural Inf. Process. Syst. **25**, 1097–1105 (2012)
3. Simonyan, K., Zisserman, A.: Very deep convolutional networks for large-scale image recognition. arXiv preprint arXiv:1409.1556 (2014)

4. He, K., Zhang, X., Ren, S., Sun, J.: Deep residual learning for image recognition. In: Proceedings of the IEEE Conference on Computer Vision and Pattern Recognition, pp. 770–778 (2016)
5. Girshick, R., Donahue, J., Darrell, T., Malik, J.: Rich feature hierarchies for accurate object detection and semantic segmentation. In: Computer Vision and Pattern Recognition (2014)
6. Long, J., Shelhamer, E., Darrell, T.: Fully convolutional models for semantic segmentation. In: Computer Vision and Pattern Recognition (2015)
7. Lotter, W., Kreiman, G., Cox, D.: Deep predictive coding networks for video prediction and unsupervised learning. arXiv preprint arXiv:1605.08104 (2016)
8. Mathieu, M., Couprie, C., LeCun, Y.: Deep multi-scale video prediction beyond mean square error. arXiv preprint arXiv:1511.05440 (2015)
9. Kingma, D.P., Welling, M.: Auto-encoding variational bayes. arXiv preprint arXiv:1312.6114 (2013)
10. Liang, X., Lee, L., Dai, W., Xing, E.P.: Dual motion GAN for future-flow embedded video prediction. arXiv preprint (2017)
11. Yoo, Y., Yun, S., Chang, H.J., Demiris, Y., Choi, J.Y.: Variational autoencoded regression: high dimensional regression of visual data on complex manifold. In: Proceedings of the IEEE Conference on Computer Vision and Pattern Recognition, 3674–3683 (2017)
12. Kulkarni, T.D., Whitney, W.F., Kohli, P., Tenenbaum, J.: Deep convolutional inverse graphics network. In: Advances in Neural Information Processing Systems, pp. 2539–2547 (2015)
13. Santana, E., Hotz, G.: Learning a driving simulator. arXiv preprint arXiv:1608.01230 (2016)
14. Radford, A., Metz, L., Chintala, S.: Unsupervised representation learning with deep convolutional generative adversarial networks. arXiv preprint arXiv:1511.06434 (2015)
15. Chen, X., Duan, Y., Houthooft, R., Schulman, J., Sutskever, I., Abbeel, P.: Infogan: interpretable representation learning by information maximizing generative adversarial nets. In: Advances in Neural Information Processing Systems, 2172–2180 (2016)
16. Larsen, A.B.L., Sønderby, S.K., Larochelle, H., Winther, O.: Autoencoding beyond pixels using a learned similarity metric. arXiv preprint arXiv:1512.09300 (2015)
17. Caruana, R., Niculescu-Mizil, A.: An empirical comparison of supervised learning algorithms. In: Proceedings of the 23rd International Conference on Machine Learning, pp. 161–168. ACM (2006)
18. Ho, T.K.: The random subspace method for constructing decision forests. IEEE Trans. Pattern Anal. Mach. Intell. **20**, 832–844 (1998)
19. Breiman, L.: Bagging predictors. Mach. Learn. **24**, 123–140 (1996)
20. Breiman, L.: Random forests. Mach. Learn. **45**, 5–32 (2001)
21. Geurts, P., Ernst, D., Wehenkel, L.: Extremely randomized trees. Mach. Learn. **63**, 3–42 (2006)
22. Ozuysal, M., Calonder, M., Lepetit, V., Fua, P.: Fast keypoint recognition using random ferns. IEEE Trans. Pattern Anal. Mach. Intell. **32**, 448–461 (2010)
23. Kursa, M.B.: rFerns: an implementation of the random ferns method for general-purpose machine learning. arXiv preprint arXiv:1202.1121 (2012)
24. Bulo, S.R., Kontschieder, P.: Neural decision forests for semantic image labelling. In: CVPR, vol. 5, pp. 81–88 (2014)

25. Kontschieder, P., Fiterau, M., Criminisi, A., Bulo, S.R.: Deep neural decision forests. In: 2015 IEEE International Conference on Computer Vision (ICCV), pp. 1467–1475. IEEE (2015)
26. Zuo, Y., Drummond, T.: Fast residual forests: rapid ensemble learning for semantic segmentation. In: Conference on Robot Learning, pp. 27–36 (2017)
27. Zuo, Y., Avraham, G., Drummond, T.: Generative adversarial forests for better conditioned adversarial learning. arXiv preprint arXiv:1805.05185 (2018)
28. Bengio, Y., Courville, A., Vincent, P.: Representation learning: a review and new perspectives. IEEE Trans. Pattern Anal. Mach. Intell. **35**, 1798–1828 (2013)
29. Rosca, M., Lakshminarayanan, B., Warde-Farley, D., Mohamed, S.: Variational approaches for auto-encoding generative adversarial networks. arXiv preprint arXiv:1706.04987 (2017)
30. Turk, M.A., Pentland, A.P.: Face recognition using eigenfaces. In: IEEE Computer Society Conference on Computer Vision and Pattern Recognition, Proceedings CVPR 1991, pp. 586–591. IEEE (1991)
31. Goodfellow, I., et al.: Generative adversarial nets. In: Advances in Neural Information Processing Systems, pp. 2672–2680 (2014)
32. Arjovsky, M., Chintala, S., Bottou, L.: Wasserstein GAN. arXiv preprint arXiv:1701.07875 (2017)
33. Dumoulin, V., et al.: Adversarially learned inference. arXiv preprint arXiv:1606.00704 (2016)
34. Donahue, J., Krähenbühl, P., Darrell, T.: Adversarial feature learning. arXiv preprint arXiv:1605.09782 (2016)
35. Mescheder, L., Nowozin, S., Geiger, A.: Adversarial variational bayes: unifying variational autoencoders and generative adversarial networks. arXiv preprint arXiv:1701.04722 (2017)
36. Kingma, D.P., Salimans, T., Jozefowicz, R., Chen, X., Sutskever, I., Welling, M.: Improved variational inference with inverse autoregressive flow. In: Advances in Neural Information Processing Systems, pp. 4743–4751 (2016)
37. Ioffe, S., Szegedy, C.: Batch normalization: accelerating deep network training by reducing internal covariate shift. In: International Conference on Machine Learning, pp. 448–456 (2015)
38. LeCun, Y.: The MNIST database of handwritten digits. http://yann.lecun.com/exdb/mnist/
39. Geiger, A., Lenz, P., Stiller, C., Urtasun, R.: Vision meets robotics: the kitti dataset. Int. J. Robot. Res. **32**, 1231–1237 (2013)
40. Kingma, D.P., Ba, J.: Adam: a method for stochastic optimization. arXiv preprint arXiv:1412.6980 (2014)

Parallel Convolutional Networks
for Image Recognition via a Discriminator

Shiqi Yang and Gang Peng[✉]

School of Automation, Huazhong University of Science and Technology,
Wuhan, China
{albert_yang,penggang}@hust.edu.cn

Abstract. In this paper, we introduce a simple but quite effective recognition framework dubbed D-PCN, aiming at enhancing feature extracting ability of CNN. The framework consists of two parallel CNNs, a discriminator and an extra classifier which takes integrated features from parallel networks and gives final prediction. The discriminator is core which drives parallel networks to focus on different regions and learn different representations. The corresponding training strategy is introduced to ensures utilization of discriminator. We validate D-PCN with several CNN models on benchmark datasets: CIFAR-100, and ImageNet, D-PCN enhances all models. In particular it yields state of the art performance on CIFAR-100 compared with related works. We also conduct visualization experiment on fine-grained Stanford Dogs dataset to verify our motivation. Additionally, we apply D-PCN for segmentation on PASCAL VOC 2012 and also find promotion.

1 Introduction

Since the AlexNet [1] sparked off the passion for research on convolutional neural networks (CNNs), CNNs have been improving the performance of image classification continuously. And heterogeneous successive brilliant CNN models lead this wave with compelling results, besides, state of the art of various vision tasks, such as detection [2–4] and segmentation [5,6], is advancing rapidly leveraging the power of CNN.

A number of recent papers [7–11] have tried to lend insights on interpretability of CNN. These methods focus on understanding CNN by visualizing learned representations. An interesting conclusion [8,9] has been drawn that CNN has ability to localize objects without any supervision in classification task. As shown in Fig. 1, we visualize the VGG16 using Grad-CAM [11]. We posit that single network may not notice all informative regions or details which leads to misclassification as exhibited in Fig. 1, meaning focusing on specific areas since some different categories may have these regions in similar. Based on this point, in

Electronic supplementary material The online version of this chapter (https://doi.org/10.1007/978-3-030-20887-5_38) contains supplementary material, which is available to authorized users.

this paper we propose a parallel networks architecture dubbed D-PCN, which coordinates parallel networks to achieve diverse representations under the guide of a discriminator. The final prediction is reported by the extra classifier. We adopt a training method which is modified from adversarial learning to achieve our goal.

We implement D-PCN on CIFAR-100 [12], ImageNet [13,14] datasets with NIN [15], ResNet [16], WRN [17], ResNeXt [18] and DenseNet [19]. In experiments, the performance of D-PCN ascends greatly compared with single base CNN, and it's proved that performance improvement is not merely from more parameters. In particular, our method has outperformed all advanced related approaches which use multiple subnetworks on CIFAR-100. We also apply Grad-CAM [11] to visualize D-PCN with VGG16 [20] on a fine-grained classification dataset, Stanford Dogs [21], and the result verifies our motivation. In addition, we introduce D-PCN into FCN [5] on PASCAL VOC 2012 segmentation task, and experiment result demonstrates that D-PCN enhances the network.

Fig. 1. Grad-CAM visualization of VGG16, which aims to distinguish diverse categories of dogs. The second row shows the class-discriminative regions localized by CNN. The cls result means whether network predicts correctly.

We summarize our contributions as follows:

- We propose the D-PCN (**P**arallel **C**onvolutional **N**etwork via **D**iscriminator), a simple but quite effective framework to enhance feature extracting ability of CNN, which outperforms other related methods.
- Two parallel networks in D-PCN focus on different regions of input respectively, that leads to more discriminative representations after features fusion.
- We propose a novel training method, and it's of high efficiency to be applied for D-PCN.

2 Related Work

CNN based models occupy advanced performance in almost all computer vision areas, including classification, detection and segmentation. Many attempts have been made to design efficient CNN architecture. In early stage, the networks from AlexNet [1] to VGGnet [20] tend to get deeper, and thanks to skip connection, ResNet [16] can contain extreme deeper layers. WRN [17] demonstrates the fact that increasing width can improve performance too. And there also exist other innovative designed models, such as Inception [22], ResNeXt [18], DenseNet [19], which improve capability of CNN further. Besides, many new modules have been constructed. For example, serials of activation functions have been introduced, like PReLU [23] which accelerates convergence, and the interesting SeLU [24]. Additionally, batch normalization [25] is used to normalize input of layers and improve performance. Some other works achieve enhancement by employing regularizer such as dropout [26] and maxout [27].

Although all these methods turn out to be very helpful, but sometimes designing new network models or activation units is of high complexity. Speculating on how to strengthen ability of existed CNN models is a feasible approach. Several works have paid attention to that, including Bilinear CNN [28], HD-CNN [29], DDN [30] and DualNet [31], all of which resort to multiple networks. HD-CNN [29] embeds CNN into two-level category hierarchy, it separates easy classes with a coarse category classifier while distinguishing different classes using a fine classifier. And DDN [30] automatically builds a network that splits the data into disjoint clusters of classes which would be handled by the subsequent expert networks.

Bilinear CNN [28] and DualNet [31] are more related to our work which all use parallel networks. But our work is distinctive from them. In Bilinear CNN [28] the parallel networks have different parameters numbers and receptive fields, and in fact Bilinear CNN is eventually implemented with a single CNN using weights sharing, while D-PCN has two identical networks. DualNet [31] is the first to focus on the cooperation of multiple CNNs. Although it shares same philosophy with us which means deploying identical parallel networks, it puts an extra classifier in the end to participate in joint training with parallel networks. The extra classifier guarantees divergence of two networks in DualNet, and final prediction is a weighted average over three classifiers. While D-PCN uses a discriminator to drive two networks to learn different representations, and the added extra classifier doesn't take part in training with parallel. Moreover, the final prediction in D-PCN is reported by extra classifier alone. Besides, the motivation is different, two subnetworks in D-PCN are expected to localize distinctive regions of input. A novel training strategy adapted from adversarial learning is proposed to achieve it in D-PCN. Additionally, D-PCN is much easy to implement compared with related works.

3 D-PCN

3.1 Motivation

Nowadays, neural networks are still trained with back propagation to optimize the loss function, the process is driven by losses generated at higher layers. Consequently, as demonstrated in [31]: In the optimization of single network, some distinctive details of the objects, which are low-level but essential to discriminate the classes of strong similarity, are likely to be dropped in the middle layers or overwhelmed by massive useless information, since the loss signals received by shallow layers for parameter update have been filtered by multiple upper layers. And these may happen constantly in whole propagation process when loss signals flow from higher to lower layers. All in all, it is tough for a single network to extract all details of input.

As mentioned in Sect. 1, there are various intriguing works for visualization on CNN lately, which point out that CNN can localize related target object in a spontaneous way. We conjecture that the missing of some information in the optimizing process [31] may lead to inaccurate localization and misclassification as shown in Fig. 1. We want to utilize this characteristic of CNN to improve performance of vision tasks. Therefore we hope to find a way to compel multiple networks to focus on different regions or details, which implies one network can learn features omitted by others.

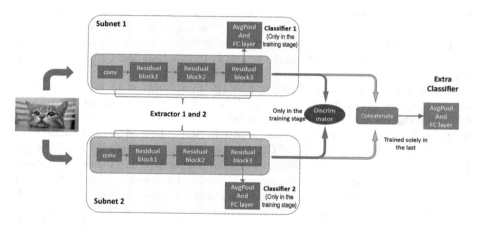

Fig. 2. The architecture of D-PCN based on ResNet-20. Noted that classifier of two networks and the discriminator just appear in training process.

Recently, Generative Adversarial Nets (GAN) [32] are prevalent. In a GAN, the discriminator and generator are playing a max-min game which is realized by adversarial learning. This competition between them can drive both teams to improve their methods until the spurious are indistinguishable from the genuine ones. The adversarial learning is depicted as below:

$$\min_{G} \max_{D} V(D,G) = \mathbb{E}_{x \sim p_{data}(\mathbf{x})}[log D(\mathbf{x})] + \mathbb{E}_{z \sim p_{\mathbf{z}}(\mathbf{z})}[log(1 - D(G(\mathbf{z})))]$$

D here represents discriminator and E means generator.

Arguably, the discriminator can be regarded as a relay through which generator acquires information from real input, meanwhile it differentiates generated one and real one. [32] reformulates $\max_{D} V(D,G)$ as:

$$\max_{D} V(D,G) = -log4 + 2 \cdot JSD(p_{data}||p_g)$$

JSD means Jensen-Shannon divergence. The generator is to minimize the divergence so as to generate indistinguishable object.

Inspired by it, we propose a parallel networks framework named D-PCN by transforming adversarial learning to a maximization optimization problem. In D-PCN, there are two identical parallel networks, a discriminator, and an extra classifier giving final prediction. The key component of D-PCN is the discriminator which can coordinate parallel networks to learn features from different regions or aspects. It's achieved by a training strategy as below:

$$\max_{E_1,E_2} \max_{D} V(D,E_1,E_2) = \mathbb{E}_{x\sim input}[logD(E_1(x))] + \mathbb{E}_{x\sim input}[log(1-D(E_2(x)))] \quad (1)$$

where the E_1, E_2 symbolize extractors of subnetwork 1 and 2 respectively, equal to generator in GAN. And $E_1(x), E_2(x)$ means features learned by networks. The equation 1 is to enforce two subnetworks to learn two different features spaces. Unlike GAN, we want to maximize $\max_{D} V(D,E_1,E_2) = -log4 + 2 \cdot JSD(p_{E_1(x)}||p_{E_2(x)})$ to expand distribution distance between two extractors, by which means subnetworks can learn different features. Though there maybe a solution to Eq. 1 where the ordering of features are perturbed between $E_1(x)$ and $E_2(x)$, we posit that since we force subnets to achieve good performance on classification meanwhile, the perturbation (even very minor) between $E_1(x)$ and $E_2(x)$ will always exist, which will result in extra information learned. This will be proved in Sect. 4.3.

3.2 Architecture

In D-PCN, there are two subnetworks with same architecture. Two subnetworks can be replaced with any present CNN model.

In order to articulate the framework, we take a D-PCN based on ResNet-20 [16] for example, which is presented in Fig. 2. We separate a single network into an extractor and a classifier, corresponding to the figure. The classifier merely contains a pool layer and a fully connected (fc) layer, the other lower layers belong to extractor. The discriminator is comprised by several convolutional layers, with batch normalization and Leaky ReLU [33]. Noted that in experiments sigmoid activation is added in the end of discriminator specially for D-PCN based on NIN [15].

The reasons why we choose features of higher layer to be sent to discriminator lie in two aspects. First, CNN extracts hierarchical representations from edges to almost entire object with encoded features [7], so the features in high layers are much discriminative. Discriminator which takes in these features can acquire enough information to guide training. Second, the feature size in higher layers is much small, and this will reduce computational cost. All other D-PCNs with

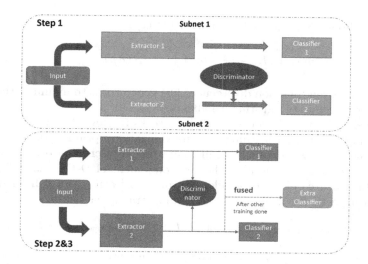

Fig. 3. The training strategy of D-PCN. At first, parallel networks start being trained, the loss of discriminator is attached to one of them to realize different parameters of two networks. Then two subnetworks are trained jointed in accompany with discriminator. In completion of joint training, extra classifier will be trained with fused features from two extractors.

various CNN models adopt similar position as division of extractor and classifier. Just as shown in Fig. 2, during training procedure, extractors along with their own classifiers and discriminator will get trained in the beginning. Then features will be integrated and input to extra classifier, which will be trained solely. In inference stage, there are no discriminator and classifier in both subnetworks, and final prediction is reported by extra classifier.

For simplicity, we adopt concatenating as fusing method. After features integrated, the whole framework can obtain more discriminative representations. The extra classifier keeps same architecture as classifiers in subnetworks, but with width doubling because of concatenating. By the way, since NIN just has a pool layer in the end for prediction, we add a *fc* layer in extra classifier to maintain proper structure.

3.3 Training Method

The proposed training method is crucial to coordinate parallel networks to localize diversely and learn different features. The process contains three steps. The discriminator in D-PCN works like a binary classifier, and it tells which network the features are from, and the loss signal it spreads to parallel networks will encourage one network to learn features omitted by another. Since there is no constraint to make feature of subnets complementary totally, duplications will exist in representations of subnets, but it's important that subnets catch different details, which will be proved in Sect. 4.3. Unlike using iterative training in DualNet [31], we deploy a 3-step training, illustrated in Fig. 3.

Training Step 1. Because it's tough for parallel networks with same values of parameters to converge by our joint training method, we need to initialize subnetworks with different weights. Since we have a discriminator in D-PCN, we opt to make full use of it. In this stage, discriminator is initialized and fixed, and the loss value from discriminator is added to one of the subnetwork, by which means features learned by parallel networks can be discriminative and distinctive simultaneously. For subnetwork 1, the loss function is defined as:

$$L_1 = L_{cls_1} \tag{2}$$

while loss function of another is defined as:

$$L_2 = L_{cls_2} + \lambda L_{D_2} \tag{3}$$

$$L_{D_2} = \frac{1}{n} \sum_{}^{n} [D(E_2(input))]^2 \tag{4}$$

The L_{cls} means cross entropy loss for classification, and L_{D_2} is a L2 loss from discriminator. After a few epoches, Step 1 is finished.

Training Step 2. Joint training based on Eq. 1 starts. Loss function of subnetwork 1 is defined as:

$$L_1 = L_{cls_1} + \lambda L_{D_1} \tag{5}$$

$$L_{D_1} = \frac{1}{n} \sum_{}^{n} [1 - D(E_1(input))]^2 \tag{6}$$

In the meantime the corresponding one of subnetwork 2 is defined as:

$$L_2 = L_{cls_2} + \lambda L_{D_2} \tag{7}$$

$$L_{D_2} = \frac{1}{n} \sum_{}^{n} [D(E_2(input))]^2 \tag{8}$$

As for discriminator, it follows the paradigm of GAN:

$$L_D = L_{D_1} + L_{D_2} \tag{9}$$

Extractors in parallel networks can be regarded as counterparts of generator in GAN. In above training, L_{cls} ensures that learned features are discriminative, meanwhile L_{D_1}, L_{D_2} make sure that features from subnetworks are different. By the way, L_{D_1} and L_{D_2} both can be seen as regularization to some extend.

Training Step 3. In this stage, we remove classifiers in two-stream networks, so does discriminator. Features from two networks get integrated and are sent to extra classifier. All extractors are fixed, and we only train extra classifier.

In training, λ is set to 1, and we find it's sufficient to promote performance of CNN significantly. We emphasize that discriminator receives discriminative features from subnetwork 1 and can instruct the training of subnetwork 2 and vice verse, although the process is operated in class level. In addition, our method is much easy to implement compared with related works.

Table 1. Compared with recent related works on CIFAR-100 without data augmentation. The accuracy means the top-1 accuracy on CIFAR-100 test datasets. *-with data augmentation and 10 view testing.

Method	Test accuracy
Maxout Network [27]	61.43%
Tree based priors [34]	63.15%
Network in Network [15]	64.32%
DSN [35]	65.43%
NIN+LA units [36]	65.60%
HD-CNN* [29]	67.38%
DDN [30]	68.35%
DNI, DualNet [31]	69.76%
D-PCN (ours)	**71.10%**

4   Experiments

In this section, we empirically demonstrate the effectiveness of D-PCN with several CNN models on various benchmark datasets and compare it with related state of the art methods. Additional experiments for visualization and segmentation are also conducted. All experiments are implemented with PyTorch[1] on a TITAN Xp GPU.

4.1   Classification Results on CIFAR-100

The CIFAR-100 [12] dataset consists of 100 classes and total 60000 images with 32×32 pixels each, in which there are 50000 for training and 10000 for testing. We simply apply normalization for images using means and standard deviations in three channels.

For convenience of making comparison with related works, which use NIN [15] as base model, such as HD-CNN [29], DDN [30], DualNet [31], we build a D-PCN

---

[1] http://pytorch.org/.

Table 2. Comparison between DualNet and our D-PCN on CIFAR-100. Here we report predictions from all classifiers in two frameworks. The top half shows the results of DualNet, in which the *classifier average* means the weighted average of all three classifiers in DualNet. The bottle half shows the results of our D-PCN. * means that DualNet uses changing contrast, brightness and color shift as additional data augmentation ways, while + means our D-PCN only adopts cropping randomly with padding.

Method				
DualNet [31]	NIN	ResNet-20*	ResNet-34*	ResNet-56*
base network	66.91%	*69.09%*	*69.72%*	*72.81%*
Iter training (extra classifier)	69.01%	71.93%	73.06%	75.24%
Iter training (classifier average)	69.51%	72.29%	73.31%	75.53%
Joint finetuning (classifier average)	**69.76%**	**72.43%**	**73.51%**	**75.57%**
D-PCN	NIN	ResNet-20+	ResNet-34+	ResNet-56+
base network	66.63%	67.89%	68.70%	71.98%
Classifier of subnet1	68.03%	68.15%	69.76%	73.44%
Classifier of subnet2	67.96%	68.69%	69.94%	74.01%
extra classifier	**71.10%**	**72.39%**	**74.07%**	**76.39%**

based on NIN. We follow the setting of NIN in [30, 31], and D-PCN is trained without data augmentation. Table 1 shows performance comparisons between several works. Noted directly compared to D-PCN are [15, 29–31], which are all built on NIN and deploying multiple networks. And HD-CNN actually uses cropping and 10 view testing [1] as data augmentation, but it's a representative work using multiple subnetworks, for which reason it's listed here. To the best of our knowledge, DualNet reports highest accuracy on CIFAR-100 without augmentation before D-PCN. Our work surpasses DualNet by 1.34%.

Furthermore, we make several elaborate comparisons between D-PCN and DualNet in order to prove the effectiveness of our work. As shown in Table 2, we list all prediction results in DualNet and D-PCN. DualNet consists of two parallel networks and an extra classifier, final prediction is given by a weighted average of all three classifiers, while ours is provided by extra classifier alone. For ResNet, DualNet takes additional data augmentation approaches, which may explain why accuracy of base network in DualNet overtakes ours. However, D-PCN still outperforms DualNet except for ResNet-20. The accuracy of two subnetworks in D-PCN already exceeds base network. It's worth nothing that the extra classifiers achieve significant boost over base single network after features integration. These promising results may signify representations learned by parallel networks in D-PCN are indeed different.

Moreover, we also evaluate other celebrated CNN models, including WRN [17], DenseNet [19] and ResNeXt [18]. The results are shown in Table 3. We can find D-PCN improve the accuracy of all these models.

Table 3. Accuracy of D-PCNs based on several models on CIFAR-100. All results are run by ourselves and produced with cropping randomly implemented as the only data augmentation method.

D-PCN	DenseNet-40	WRN-16-4	ResNeXt-29,8x64d
base network	70.03%	76.72%	81.77%
classifier of subnet1	70.33%	77.63%	81.98%
classifier of subnet2	70.10%	77.76%	82.41%
extra classifier	**71.43%**	**80.19%**	**84.59%**

Analysis of D-PCN on ResNet and DenseNet. Just as discussed in DPN [37], ResNet tends to reuse the feature and fails to explore new ones while DenseNet is able to extract new features. In short, DenseNet can learn comprehensive features as much as possible. And results in Tables 2 and 3 reflect these characteristics, where D-PCN can bring more promotions for ResNet than DenseNet.

Compared with Model Ensemble. For sake of further verifying effectiveness of D-PCN, we also report results of model ensemble. Specifically, we train two CNN models independently, initialized with normal distribution or using Glorot [38] and He [23] initialization, and final prediction is obtained with their predictions averaged. As illustrated in Table 4, ensemble is still inferior to D-PCN. More importantly, D-PCN is orthogonal to model ensemble just like HD-CNN [29] and DaulNet [31], ensemble of D-PCNs can further improve the performance.

Table 4. Comparison between model ensemble and D-PCN on CIFAR-100. Model ensemble deploys two identical CNN models with different initialization and takes averaged prediction as final result.

	NIN	ResNet-20	ResNeXt-29,8x64d
base network	66.63%	67.89%	81.77%
model ensemble	68.94%	70.01%	83.03%
D-PCN	**71.10%**	**72.39%**	**84.59%**

Compared with Doubling Width. We take NIN for this experiment. After doubling number of channels directly, accuracy is improved to 68.89%, still lower than 71.10% of D-PCN.

Experiments on ensemble and doubling width demonstrate that the improvement of D-PCN is not merely from more parameters.

Where to Feed the Discriminator. We take ResNet-20 for this experiment. Original D-PCN sends features from block3 to discriminator as shown in Fig. 2, here we choose block2 and block1 instead. And it just gets 71.50% and 68.78% accuracy respectively, lower than 72.39% of original one. Furthermore, we try to bring features from both block2 and block3 to discriminator through a convolutional layer, and we achieve 72.89% accuracy, a little higher than 72.39% of original D-PCN.

How to Aggregate Features. Here we experiment on NIN and WRN-16-4. We replace concatenating with sum, and it achieves 70.27%, 78.84% for NIN and WRN respective, lower than 71.10% and 80.19% using concatenating.

More Subnetworks. We take NIN for this experiment. By adjusting the loss function in Sect. 3.3, which is clarified in supplementary material, we can deploy three parallel networks in D-PCN. We get 71.97% accuracy, a little higher than 71.10% of original one.

4.2   Classification Results on ImageNet

In this experiment, we investigate D-PCN with ResNet-18 [16] on ImageNet32 × 32 [13], NIN-ImageNet on ImageNet [14], to prove that D-PCN can generalize to more complex dataset. ImageNet32 × 32 is a downsampled version of ImageNet with 32 × 32 pixel per image. The applied ResNet-18 has same structure as ResNet for CIFAR, but numbers of channels keep pace with ResNet for ImageNet. The structure of NIN-ImageNet stays the same as in supplementary material of DualNet [31]. For ImageNet32 × 32, we only shift dataset to range from 0 to 1 and then zero-center the datasets. The results are shown in Table 5. D-PCN attains 4.394% and 9.45% promotion in Top1 and Top5 accuracy on ImageNet32 × 32 respectively.

Table 5. Accuracy of D-PCN on ImageNet without data augmentation. No data augmentation method is adopted except zero-centering preprocessing.

D-PCN	Top1 accuracy	Top5 accuracy
base ResNet-18	45.738%	59.78%
classifier of subnet1	45.732%	
classifier of subnet2	45.884%	
extra classifier	**50.132%**	**69.23%**

For original ImageNet, the structure of NIN-ImageNet stays the same as in supplementary material of DualNet [31]. As presented in Table 6, D-PCN surpasses DualNet, and gains 2.33% improvement versus base NIN-ImageNet.

For all above experiments, since philosophy of D-PCN is to coordinate two-stream networks to learn different representations, it's natural to use single CNN

Table 6. Top 1 accuracy on ILSVRC-2012 ImageNet. Both DualNet and D-PCN are based on NIN-ImageNet. The overall accuracy means accuracy of final prediction in DualNet and D-PCN on validation set.

	DualNet [31]	D-PCN
base NIN-ImageNet	59.15%	58.94%
Overall Accuracy	60.44%	**61.27%**

as baseline. And we can draw a conclusion that our work makes sense by introducing the novel framework and training strategy to compel two CNNs to learn distinctive features.

4.3 Visualization

In experiments, loss from discriminator is getting quite small. We conjecture that in adversarial learning discriminator always wins, *i.e.*, loss of the discriminator goes to very low fast. And the training rule in Sect. 3.3 will make optimization of discriminator even easy.

To confirm our point of view, we apply Grad-CAM [11] to give visual explanations for base VGG16 and two networks in D-PCN on Stanford Dogs [21] dataset, which possesses larger resolution better for visualization. Grad-CAM is extended from CAM [9] and is applicable to a wide variety of CNN models. Stanford Dogs [21] is a fine grained classification dataset, which has 120 categories of dogs and total 20580 images, where 12000 are for training. It's quite suitable for proving effectiveness of D-PCN since some different categories of dogs have very similar traits and are hard to be distinguished. D-PCN based on VGG16 is trained like other CNN models above. Here we select some representative pictures for visualization, as shown in Fig. 4, where red zone represents class-discriminative regions for network while blue zone is on the contrary. Like picture 4 in Fig. 4, original VGG only focuses on mouth, which maybe the reason of misclassification. Subnetworks localize more related areas and final result of D-PCN is correct.

And we select two images consisting of a cat, and a dog belonging to one category in Stanford Dogs, to test the models. Images are from internet. Visualization is shown in Fig. 5. We list category predictions of networks. That manifests the reason why CNNs misclassify some categories of dogs maybe some features vital to distinguish similar object get omitted, since network fails to observe some aspects. An extreme arresting discovery is that two networks even localize cat with no supervision information (Please pay attention to blue zone).

From these experiments, we can see that two subnetworks not only focus on different regions, but also localize more accurately than base network, which means discriminator does play a part in D-PCN even though loss from discriminator will be very small in training. We conjecture that there are always minor perturbations between subnetworks which make them diverse. Sometimes one of parallel networks predicts wrong or both misclassify objects, but extra classifier in D-PCN gives right answer eventually. Pleased noted like mentioned in Sect. 3.3 there are duplications among representations of subnets, but difference exists. This certifies the diversity (with redundance) of features from two-stream networks in D-PCN. By the way, accuracy of base VGG16 is 72.88%, and accuracy of subnetworks and extra classifier are 73.36%,73.53% and 75.86% respectively, which also signifies generalization of D-PCN on large dataset. Visualizations of different types can be found in supplementary material which further verifies our motivation.

4.4 Segmentation on PASCAL VOC 2012

Since D-PCN can coordinate parallel networks to learn different features, we think it can improve performance of other vision tasks. Here we put D-PCN into FCN-8s [5] on PASCAL VOC 2012 semantic segmentation task. ResNet-18 and ResNet-34 are chosen as base model. In training we set all λ in Sect. 3.3 to 0.2, and extra classifier turns into convolutional layers corresponding to FCN. Two networks are initialized with pre-trained model on ImageNet. Experiment results are shown in Table 7. D-PCN achieves 1.458% and 1.304% mIoU improvement respectively. Although improvements of testing mIoU are quite small, we found that training mIoU increases greatly. The results imply that convergence of networks rises significantly[2].

Table 7. Segmentation results on PASCAL VOC 2012. We take FCN as base segmentation model. Training part of dataset is for training while validation part is for testing.

	ResNet-18	ResNet-34
base model		
training mIoU of base model	**69.139%**	**74.259%**
testing mIoU of base model	50.352%	55.335%
D-PCN		
training mIoU of subnetwork1	**85.436%**	**87.557%**
training mIoU of subnetwork2	**85.341%**	**87.647%**
testing mIoU of subnetwork1	50.536%	56.026%
testing mIoU of subnetwork2	50.601%	56.101%
testing mIoU of D-PCN	51.810%	56.639%

---

[2] We think it may explain why parallel networks can localize cat in Sect. 4.3, because subnetworks catch enough information to know what's dog.

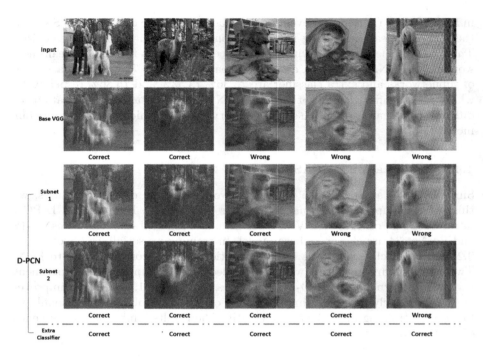

Fig. 4. Grad-CAM visualization of VGG16 on Stanford Dogs. The correct or wrong means whether input is classified correctly by its own classifier. Last row represents results of extra classifier in D-PCN. There is no visualization for extra classifier since it takes in fused features. (Color figure online)

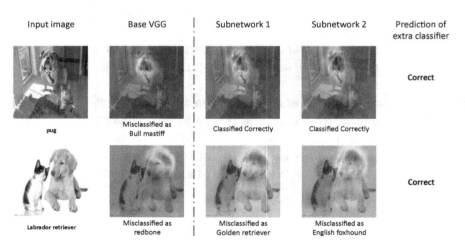

Fig. 5. Grad-CAM visualization of VGG16 trained on Stanford Dogs. On the right of dotted line is visualization of D-PCN. Below pictures are predictions, correctness of final prediction from D-PCN is in the rightmost. (Color figure online)

5 Conclusion

In this paper, we propose a novel framework named D-PCN to boost the performance of CNN. The parallel networks in D-PCN can learn discriminative and distinctive features via a discriminator. The fused features are more discriminative. An effective training method inspired by adversarial learning is introduced. D-PCNs based on various CNN models are investigated on CIFAR-100 and ImageNet datasets, and achieve promotion. In particular, it gets state-of-the-art performance on CIFAR-100 compared with related works. Additional experiments are conducted for visualization and segmentation. In the future, we will deploy D-PCN in other tasks efficiently, such as detection and segmentation.

References

1. Krizhevsky, A., Sutskever, I., Hinton, G.E.: ImageNet classification with deep convolutional neural networks. In: NIPS (2012)
2. Ren, S., He, K., Girshick, R., Sun, J.: Faster R-CNN: towards real-time object detection with region proposal networks. In: NIPS (2015)
3. Redmon, J., Divvala, S., Girshick, R., Farhadi, A.: You only look once: unified, real-time object detection. In: CVPR (2016)
4. Liu, W., et al.: SSD: single shot multibox detector. In: Leibe, B., Matas, J., Sebe, N., Welling, M. (eds.) ECCV 2016. LNCS, vol. 9905, pp. 21–37. Springer, Cham (2016). https://doi.org/10.1007/978-3-319-46448-0_2
5. Long, J., Shelhamer, E., Darrell, T.: Fully convolutional networks for semantic segmentation. In: CVPR (2015)
6. Chen, L.C., Papandreou, G., Kokkinos, I., Murphy, K., Yuille, A.L.: DeepLab: semantic image segmentation with deep convolutional nets, atrous convolution, and fully connected CRFs. PAMI **40**, 834 (2017)
7. Zeiler, M.D., Fergus, R.: Visualizing and understanding convolutional networks. In: Fleet, D., Pajdla, T., Schiele, B., Tuytelaars, T. (eds.) ECCV 2014. LNCS, vol. 8689, pp. 818–833. Springer, Cham (2014). https://doi.org/10.1007/978-3-319-10590-1_53
8. Zhou, B., Khosla, A., Lapedriza, A., Oliva, A., Torralba, A.: Object detectors emerge in deep scene CNNs. In: ICLR (2015)
9. Zhou, B., Khosla, A., Lapedriza, A., Oliva, A., Torralba, A.: Learning deep features for discriminative localization. In: CVPR (2016)
10. Bau, D., Zhou, B., Khosla, A., Oliva, A., Torralba, A.: Network dissection: quantifying interpretability of deep visual representations. In: CVPR (2017)
11. Selvaraju, R.R., Cogswell, M., Das, A., Vedantam, R., Parikh, D., Batra, D.: Grad-CAM: visual explanations from deep networks via gradient-based localization. In: ICCV (2017)
12. Krizhevsky, A., Hinton, G.: Learning multiple layers of features from tiny images (2009)
13. Chrabaszcz, P., Loshchilov, I., Hutter, F.: A downsampled variant of imageNet as an alternative to the CIFAR datasets. arXiv preprint arXiv:1707.08819 (2017)
14. Deng, J., Dong, W., Socher, R., Li, L.J., Li, K., Fei-Fei, L.: ImageNet: a large-scale hierarchical image database. In: CVPR (2009)
15. Lin, M., Chen, Q., Yan, S.: Network in network. In: ICLR (2014)

16. He, K., Zhang, X., Ren, S., Sun, J.: Deep residual learning for image recognition. In: CVPR (2016)
17. Zagoruyko, S., Komodakis, N.: Wide residual networks. In: BMVC (2016)
18. Xie, S., Girshick, R., Dollár, P., Tu, Z., He, K.: Aggregated residual transformations for deep neural networks. In: CVPR (2017)
19. Huang, G., Liu, Z., van der Maaten, L., Weinberger, K.Q.: Densely connected convolutional networks. In: CVPR (2017)
20. Simonyan, K., Zisserman, A.: Very deep convolutional networks for large-scale image recognition. arXiv preprint arXiv:1409.1556 (2014)
21. Khosla, A., Jayadevaprakash, N., Yao, B., Li, F.F.: Novel dataset for fine-grained image categorization: Stanford dogs. In: CVPR Workshop (2011)
22. Szegedy, C., et al.: Going deeper with convolutions. In: CVPR (2015)
23. He, K., Zhang, X., Ren, S., Sun, J.: Delving deep into rectifiers: surpassing human-level performance on imagenet classification. In: ICCV (2015)
24. Klambauer, G., Unterthiner, T., Mayr, A., Hochreiter, S.: Self-normalizing neural networks. In: NIPS (2017)
25. Ioffe, S., Szegedy, C.: Batch normalization: accelerating deep network training by reducing internal covariate shift. In: ICML (2015)
26. Srivastava, N., Hinton, G., Krizhevsky, A., Sutskever, I., Salakhutdinov, R.: Dropout: a simple way to prevent neural networks from overfitting. J. Mach. Learn. Res. **15**, 1929–1958 (2014)
27. Goodfellow, I.J., Warde-Farley, D., Mirza, M., Courville, A., Bengio, Y.: Maxout networks. In: ICML (2013)
28. Lin, T.Y., RoyChowdhury, A., Maji, S.: Bilinear CNN models for fine-grained visual recognition. In: ICCV (2015)
29. Yan, Z., et al.: HD-CNN: hierarchical deep convolutional neural networks for large scale visual recognition. In: ICCV (2015)
30. Murthy, V.N., Singh, V., Chen, T., Manmatha, R., Comaniciu, D.: Deep decision network for multi-class image classification. In: CVPR (2016)
31. Saihui Hou, X.L., Wang, Z.: DualNet: learn complementary features for image recognition. In: ICCV (2017)
32. Goodfellow, I., et al.: Generative adversarial nets. In: NIPS (2014)
33. Maas, A.L., Hannun, A.Y., Ng, A.Y.: Rectifier nonlinearities improve neural network acoustic models. In: ICML Workshop (2013)
34. Srivastava, N., Salakhutdinov, R.R.: Discriminative transfer learning with tree-based priors. In: NIPS (2013)
35. Lee, C.Y., Xie, S., Gallagher, P., Zhang, Z., Tu, Z.: Deeply-supervised nets. In: Artificial Intelligence and Statistics, pp. 562–570 (2015)
36. Agostinelli, F., Hoffman, M., Sadowski, P., Baldi, P.: Learning activation functions to improve deep neural networks. arXiv preprint arXiv:1412.6830 (2014)
37. Chen, Y., Li, J., Xiao, H., Jin, X., Yan, S., Feng, J.: Dual path networks. In: NIPS (2017)
38. Glorot, X., Bengio, Y.: Understanding the difficulty of training deep feedforward neural networks. In: AISTATS (2010)

ENG: End-to-End Neural Geometry for Robust Depth and Pose Estimation Using CNNs

Thanuja Dharmasiri$^{(\boxtimes)}$, Andrew Spek, and Tom Drummond

Monash University, Melbourne, Australia
{thanuja.dharmasiri,andrew.spek,tom.drummond}@monash.edu

Abstract. Recovering structure and motion parameters given a image pair or a sequence of images is a well studied problem in computer vision. This is often achieved by employing Structure from Motion (SfM) or Simultaneous Localization and Mapping (SLAM) algorithms based on the real-time requirements. Recently, with the advent of Convolutional Neural Networks (CNNs) researchers have explored the possibility of using machine learning techniques to reconstruct the 3D structure of a scene and jointly predict the camera pose. In this work, we present a framework that achieves state-of-the-art performance on single image depth prediction for both indoor and outdoor scenes. The depth prediction system is then extended to predict optical flow and ultimately the camera pose and trained end-to-end. Our framework outperforms previous deep-learning based motion prediction approaches, and we also demonstrate that the state-of-the-art metric depths can be further improved using the knowledge of pose.

Keywords: Depth · Optical flow · Pose prediction ·
Indoor and outdoor datasets

1 Introduction

The importance of navigation and mapping to the fields of robotics and computer vision has only increased since its inception. Vision based navigation in particular is an extremely interesting field of research due to its discernible resemblance to human navigation and the wealth of information an image contains. Although creating a machine that understands structure and motion purely from

T. Dharmasiri and A. Spek—These authors contributed equally. This research was supported by the Australian Research Council Centre of Excellence for Robotic Vision (project number CE14010006).

Electronic supplementary material The online version of this chapter (https://doi.org/10.1007/978-3-030-20887-5_39) contains supplementary material, which is available to authorized users.

C. V. Jawahar et al. (Eds.): ACCV 2018, LNCS 11361, pp. 625–642, 2019.
https://doi.org/10.1007/978-3-030-20887-5_39

RGB images is challenging, the computer vision community has developed a plethora of algorithms to replicate useful aspects of human vision using a computer. Tracking and mapping remains an unsolved problem, with many popular approaches. Photometric based techniques rely on establishing correspondences across different viewpoints of the same scene and the matching points are then used to perform triangulation. Based on the density of the map, the field can be divided into dense [27], semi-dense [6] and sparse [25] approaches, each comes with advantages and disadvantages.

Applying machine learning techniques to solve vision problems has been another popular area of research. Great advances have been made in the fields of image classification [11,12,19] and semantic segmentation [24,40] and this has led geometry based machine learning methods to follow suit. The massive growth in neural network driven research has largely been facilitated by the increased availability of low-cost high performance GPUs as well as the relative accessibility of machine learning frameworks such as Tensorflow and Caffe.

In this work we draw from both machine learning approaches as well as SfM techniques to create a unified framework which is capable of predicting the depth of a scene and the motion parameters governing the camera motion between an image pair. We construct our framework incrementally where the network is first trained to predict depths given a single color image. Then a color image pair as well as their associated depth predictions are provided to a flow estimation network which produces an optical flow map along with an estimated measure of confidence in x and y motion. Finally, the pose estimation block utilises the outputs of the previous networks to estimate a motion vector corresponding to the logarithm of the Special Euclidean Transformation $\mathbb{SE}(3)$ in $\mathbb{R}^3$, which describes the relative camera motion from the first image to the second.

We summarise the contributions made in this paper as follows:

- We achieve state-of-the-art results for single image depth prediction on both NYUv2 (indoor) and KITTI (outdoor) datasets. [Sect. 4: Tables 1 and 2].
- We outperform previous camera motion prediction frameworks on both TUM and KITTI datasets. [Sect. 4: Tables 5 and 4].
- We also present the first approach to use a neural network to predict the full information matrix which represents the confidence of our optical flow estimate.

2   Related Work

Estimating motion and structure from two or more views is a well studied vision problem. In order to reconstruct the world and estimate camera motion, sparse feature based systems [18,25] compute correspondences through feature matching while the denser approaches [6,27] rely on brightness constancy across multiple viewpoints. In this work, we leverage CNNs to solve the aforementioned tasks and we summarize the existing works in the literature that are related to the ideas presented in this paper.

2.1 Single Image Depth Prediction

Predicting depth from a single RGB image using learning based approaches has been explored even prior to the resurgence of CNNs. In [32], Saxena *et al.* employed a Markov Random Field (MRF) to combine global and local image features. Similar to our approach Eigen *et al.* [5] introduced a common CNN architecture capable of predicting depth maps for both indoor and outdoor environments. This concept was later extended to a multi-stage coarse to fine network by Eigen *et al.* in [4]. Advances were made in the form of combining graphical models with CNNs [23] to further improve the accuracy of depth maps, through the use of related geometric tasks [3] and by making architectural improvements specifically designed for depth prediction [21]. Kendall *et al.* demonstrated that predicting depths and uncertainties improve the overall accuracy in [16]. While most of these methods demonstrated impressive results, explicit notion of geometry was not used during any stage of the pipeline which opened the way for geometry based depth prediction approaches.

In one of the earliest works to predict depth using geometry in an unsupervised fashion, Garg *et al.* used the photometric difference between a stereo image pair, where the target image was synthesized using the predicted disparity and the known baseline [8]. Left-right consistency was explicitly enforced in the unsupervised framework of Goddard *et al.* [10] as well as in the semi-supervised framework of Kuznietsov *et al.* [20], which is a technique we also found to be beneficial during training on sparse ground truth data.

2.2 Optical Flow Prediction

An early work in optical flow prediction using CNNs was [7]. This was later extended by Ilg *et al.* to FlowNet 2.0 [13] which included stacked FlowNets [7] as well as warping layers. Ranjan and Black proposed a spatial pyramid based optical flow prediction network [29]. More recently, Sun *et al.* proposed a framework which uses the principles from geometry based flow estimation techniques such as image pyramid, warping and cost volumes in [35]. As our end goal revolves around predicting camera pose, it becomes necessary to isolate the flow that was caused purely from camera motion, in order to achieve this we extend upon these previous works to predict both the optical flow and the associated information matrix of the flow. Although not in a CNN context [38] showed the usefulness of estimating flow and uncertainty.

2.3 Pose Estimation

CNNs have been successfully used to estimate various components of a Structure from Motion pipeline. Earlier works focused on learning discriminative image based features suitable for ego-motion estimation [2, 14]. Yi *et al.* [39] showed a full feature detection framework can be implemented using deep neural networks. Rad and Lepetit in BB8 [28] showed the pose of objects can be predicted even under partial occlusion and highlighted the increased difficulty of predicting 3D quantities over 2D quantities. Kendall and Cipolla demonstrated that camera pose prediction from a single image catered for relocalization scenarios [15].

However, each of the above works lack a representation of structure as they do not explicitly predict depths. Our work is more closely related to that of Zhou *et al.* [41] and Ummenhofer *et al.* [37] and their frameworks SfM-Learner and DeMoN. Both of these approaches also predict a single confidence map in contrast to ours which estimates the confidence in x and y directions separately. Since our framework predicts metric depths in comparison to theirs we are able produce far more accurate visual odometry and combat against scale drift. CNN SLAM by Tateno *et al.* [36] incorporated depth predictions of [21] into a SLAM framework. Our method performs competitively with CNN-SLAM as well as ORB-SLAM [26] and LSD-SLAM [6] which have the added advantage of performing loop closures and local/global bundle adjustments despite solely computing sequential frame-to-frame alignments.

3   Method

3.1   Network Architecture

The overall architecture consists of 3 main subsystems in the form of a depth, flow and camera pose network. A large percentage of the model capacity is invested in to the depth prediction component for two reasons. Firstly, the output of the depth network also serves as an additional input to the other subsystems. Secondly, we wanted to achieve superior depths for indoor and outdoor environments using a common architecture[1]. In order to preserve space and to provide an overall understanding of the data flow a high level diagram of the network is shown in Fig. 1. An expanded architecture with layer definitions for each of the subsystems is included in the supplementary materials.

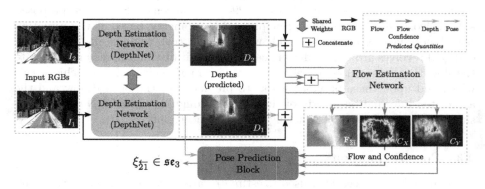

Fig. 1. Overview of our system full pipeline. Please note that we use the notation $\overleftarrow{ij}$ to indicate from j to i

---

[1] Although there are separate models for indoor and outdoor scenes the underlying architecture is common.

3.2   Depth Prediction

The depth prediction network consists of an encoder and a decoder module. The encoder network is largely based on the DenseNet161 architecture described in [12]. In particular we use the variant pre-trained on ImageNet [31] and slightly increase the receptive field of the pooling layers. As the original input is down-sampled 4 times by the encoder, during the decoding stage the feature maps are up-sampled back 4 times to make the model fully convolutional. We employ skip connections in order to re-introduce the finer details lost during pooling. Since the first down-sampling operation is done at a very early stage of the pipeline and closely resemble the image features, these activations are not reused inside the decoder. Up-project blocks are used to perform up-sampling in our network, which provide better depth maps compared to de-convolutional layers as shown in [21].

Due to the availability of dense ground truth data for indoor datasets (e.g NYUv2 [33], RGB-D [34]) this network can be directly utilised to perform supervised learning. Unfortunately, the ground truth data for the outdoor datasets (KITTI) are much sparser and meant we had to incorporate a semi-supervised learning approach in order to provide a strong training signal. Therefore, during training on KITTI, we use a Siamese version of the depth network with complete weight-sharing, and enforce photometric consistency between the left-right image pairs through an additional loss function. This is similar to the previous approaches [8, 20] and is only required during the training stage, during inference only a single input image is required to perform depth estimation using our network.

3.3   Flow Prediction

The flow network provides an estimation of the optical flow along with the associated confidences given an image pair. These outputs combined with predicted depths allow us to predict the camera pose. As part of our ablation studies we integrated the flow predictions of [13] with our depths, however, the main limitation of this approach was the lack of a mechanism to filter out the dynamic objects which are abundant in outdoor environments. This was solved by estimating confidence, specifically the information matrix in addition to the optical flow. More concretely, for each pixel our flow network predicts 5 quantities, the optical flow $\mathbf{F} = [\Delta u, \Delta v]^T$ in the x and y direction, and the quantities $\hat{\alpha}$, $\hat{\gamma}$ and $\hat{\beta}$, which are required to compute the information matrix ($\mathcal{I}$) or the inverse of the covariance matrix as shown below.

$$\mathcal{I} = \begin{bmatrix} C_x & C_{xy} \\ C_{xy} & C_y \end{bmatrix}, \qquad C_x = e^{\hat{\alpha}}, C_y = e^{\hat{\gamma}}, C_{xy} = e^{\frac{\hat{\gamma}+\hat{\alpha}}{2}} \tanh(\hat{\beta}). \qquad (1)$$

This parametrisation guarantees $\mathcal{I}$ is positive-definite and can be used to parametrise any 2×2 information matrix. We found that the gradients are much more stable compared to predicting the information matrix directly as the determinant of the matrix is always greater than zero since $\tanh(\hat{\beta}) = \pm 1$ only when $\hat{\beta} \to \pm\infty$.

With respect to the architecture we borrow elements from FlowNet [7] as well as FlowNet 2.0 [13]. As mentioned in [13], FlowNet 2.0 was unable to reliably estimate small motions, which we address with two key changes. Firstly, our flow network takes the predicted depth map as an input, allowing the network to learn the relationship between depth and flow explicitly, including that closer objects appear to move more compared to the objects that are further away from the camera. Secondly, we use "warp-concatenation", where coarse flow estimates are used to warp the CNN features during the decoder stage. This appears to resolve small motions more effectively particularly on the TUM [34] dataset.

3.4 Pose Estimation

We take two approaches to pose estimation, shown in Fig. 2, an iterative and a fully-connected (FC). This contrasts the ability of a neural network to estimate using the available information, and the simplicity of a standard computer vision approach using the available predicted quantities. We use FC layers to provide the network with as wide a receptive field as possible, to compare more equivalently against using the inferred quantities in the iterative approach.

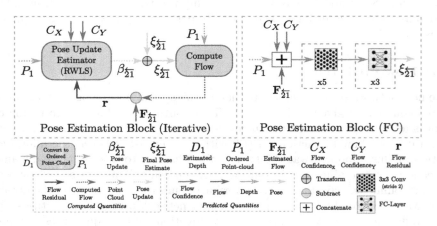

Fig. 2. We detail the two approaches we took to estimating the relative pose alignment between adjacent frames (best viewed in colour). *Left* shows the iterative approach we took, that incorporates a re-weighted least-squares solver (RWLS) into a pose estimation loop. *Right* shows our fully-connected (FC) approach, which incorporates a succession of 3×3 strided convolutions, followed by several FC layers.

Iterative. This approach uses a more conventional method for computing relative pose estimates. We use a standard re-weighted least squares solver based on the residual flow, given an estimate of the relative transformation. More concretely we attempt to minimise the following error function with respect to the relative transformation parameters ($\xi_{\overline{21}} \in \mathbb{R}^6$)

$$\mathbf{e} = \sum_{i=1}^{N} ||(\mathbf{x}_i - \mathrm{T}_{\overline{21}}\mathbf{x}_i)_{[u,v]} - \mathbf{F}_{\overline{21}}(\mathbf{u}_i)||_2 = \sum_{i=1}^{N} ||\mathbf{F}_{\overline{21}}^+(\mathbf{u}_i) - \mathbf{F}_{\overline{21}}(\mathbf{u}_i)||_2 = \sum_{i=1}^{N} \mathbf{r}_i^2, \quad (2)$$

where $\mathbf{e} \in \mathbb{R}^2$ is the total residual flow in normalised camera coordinates, the subscript $_{[u,v]}$ indicates only the first two dimensions of the vector are used, $\mathbf{x} \in P_1$ is the ith inverse depth coordinate $\mathbf{x} = \begin{bmatrix} u & v & 1 & q \end{bmatrix}^T$ of an ordered point cloud (P_1), $\mathbf{F}_{\overline{21}}(\mathbf{u}_i)$ and $\mathbf{F}_{\overline{21}}^+(\mathbf{u}_i)$ are the ith predicted flow and estimated flow respectively, and $\mathbf{u}_i = \begin{bmatrix} x & y \end{bmatrix}_i^T$ is the ith pixel coordinate. $\mathrm{T}_{\overline{21}} \in \mathbb{SE}(3)$ is the current transformation estimate, and can be expressed by the matrix exponential as $\mathrm{T}_{\overline{21}} = e^{\sum_{j=0}^{6} \alpha_j \mathbf{G}_j}$, where $\alpha_j \in \xi_{\overline{21}}$ is the jth component of the motion vector $\xi_{\overline{21}} \in \mathbb{R}^6$, which is a member of the Lie-algebra $\mathfrak{se}_3$, and $\mathbf{G}_j$ is the generator matrix corresponding to the relevant motion parameter. As This pipeline is implemented in Tensorflow [1] it allows us to train the network end to end. Please see the supplementary material for a more detailed explanation.

Fully-Connected. Similar to Zhou *et al.* [41] and Ummenhofer *et al.* [37] we also constructed a fully connected layer based pose estimation network. This network utilises 3 stacked fully connected layers and uses the same inputs as our iterative method. While we outperform the pose estimation benchmarks of [37,41] using this network the iterative network is our recommended approach due to its close resemblance to conventional geometry based techniques.

3.5   Loss Functions

Depth Losses

For supervised training on indoor and outdoor datasets we use a reverse Huber loss function [21] defined by

$$\mathcal{L}_{\mathrm{B}}(D_i, D_i^*) = \begin{cases} |D_i - D_i^*| & |D_i - D_i^*| < c, \\ ((D_i - D_i^*)^2 + c^2)/2c & |D_i - D_i^*| > c, \end{cases} \quad (3)$$

where $c = \frac{1}{5}max(D_i - D_i^*)$, and $D_i = D(\mathbf{u}_i)$ and $D_i^* = D^*(\mathbf{u}_i)$ represent the ith predicted and the ground truth depth respectively. For the KITTI dataset we employed an additional photometric loss during training as the ground truth is highly sparse. This unsupervised loss term enforces left-right consistency between stereo pairs, defined by

$$
\begin{aligned}
\mathcal{L}_{\mathrm{C}} = \frac{1}{n} \sum_{i=1}^{n} &|I_L(\mathbf{u}_i) - I_R(\pi(\mathrm{KT}_{\overline{RL}}\pi^{-1}(D_i^L, \mathbf{u}_i)))| \\
+ \frac{1}{n} \sum_{i=1}^{n} &|I_R(\mathbf{u}_i) - I_L(\pi(\mathrm{KT}_{\overline{LR}}\pi^{-1}(D_i^R, \mathbf{u}_i)))|,
\end{aligned}
\tag{4}
$$

where I_L and I_R are the left and right images and D_i^L and D_i^R are their corresponding depth maps, $\pi(x) = ((x_0/x_2), (x_1/x_2))^T$ is a normalisation function where $x \in \mathbb{R}^3$, K is the camera intrinsic matrix, $\pi^{-1}(D, \mathbf{u}) = DK^{-1}(\mathbf{u})$ is the transformation from pixel to camera coordinates, and $\mathrm{T}_{\overline{RL}} \in \mathbb{SE}(3)$ and $\mathrm{T}_{\overline{LR}} \in \mathbb{SE}(3)$ define the relative transformation matrices from left-to-right and right-to-left respectively. In this case the rotation is assumed to be the identity and the matrices purely translate in the x-direction. Additionally, we use a smoothness term defined by

$$
\mathcal{L}_{\mathrm{S}} = \frac{1}{n} \sum_{i=1}^{n} (|\nabla_x D_i| + |\nabla_y D_i|),
\tag{5}
$$

where ∇_x and ∇_y are the horizontal and vertical gradients of the predicted depth. This provides qualitatively better depths as well as faster convergence. The final loss function used to train KITTI depths is given by

$$
\mathcal{L}_{ss} = \lambda_1 \mathcal{L}_{\mathrm{B}} + \lambda_2 \mathcal{L}_{\mathrm{C}} + \lambda_3 \mathcal{L}_{\mathrm{S}}, \qquad \lambda_1 = 2, \ \lambda_2 = 1, \ \lambda_3 = e^{-4},
\tag{6}
$$

where $\mathcal{L}_{\mathrm{B}}$ and $\mathcal{L}_{\mathrm{S}}$ are computed on both left and right images separately.

Flow Loss

The probability distribution of multivariate Gaussian in 2D can be defined as follows.

$$
p(\mathbf{x}|\mu, \boldsymbol{\mathcal{I}}) = ((|\boldsymbol{\mathcal{I}}|^{\frac{1}{2}})/(2\pi))e^{-\frac{1}{2}(\mathbf{x}-\mu)^T \boldsymbol{\mathcal{I}}(\mathbf{x}-\mu)},
\tag{7}
$$

where $\boldsymbol{\mathcal{I}} = \Sigma^{-1}$ is the information matrix or inverse covariance matrix Σ^{-1}. The flow loss $\mathcal{L}_F$ criterion can now be defined by

$$
\mathcal{L}_F = \frac{1}{2}((\mathbf{F}_{\overline{21}} - \mathbf{F}_{\overline{21}}^*)^T \boldsymbol{\mathcal{I}}(\mathbf{F}_{\overline{21}} - \mathbf{F}_{\overline{21}}^*) - \log(|\boldsymbol{\mathcal{I}}|)),
\tag{8}
$$

where $\mathbf{F}_{\overline{21}}$ is the predicted flow, and $\mathbf{F}_{\overline{21}}^*$ is the ground truth flow. This optimises by maximising the log-likelihood of the probability distribution over the residual flow error.

Pose Loss

Given two input images I_1, I_2, the predicted depth map D_1 of I_1 and the predicted relative pose $\xi_{\overline{21}} \in \mathbb{R}^6$ the unsupervised loss $\mathcal{L}_{\mathrm{U}}$ and pose loss $\mathcal{L}_{\mathrm{P}}$ can be defined as

$$\mathcal{L}_U = \frac{1}{n} \sum_{i=1}^{n} |I_1(\mathbf{u}_i) - I_2(\pi(KT_{\overleftarrow{21}}\pi^{-1}(D_1, \mathbf{u}_i)))|, \text{ and} \tag{9}$$

$$\mathcal{L}_P = ||\xi_{\overleftarrow{21}} - \log_e(T_{\overleftarrow{21}}^*)||_2 = ||\xi_{\overleftarrow{21}} - \xi_{\overleftarrow{21}}^*)||_2, \tag{10}$$

where $\log_e(T)$ maps a transformation T from the Lie-group $\mathbb{SE}(3)$ to the Lie-algebra $\mathfrak{se}_3$, such that $\log_e(T) \in \mathbb{R}^6$ can be represented by its constituent motion parameters, and $\xi_{\overleftarrow{21}}^*$ is the ground truth relative pose parameters.

3.6 Training Regime

We train our network end-to-end on NYUv2 [33], TUM [34] and KITTI [9] datasets. We use the standard test/train split for NYUv2 and KITTI and define our scene split for TUM. It is worth mentioning that the amount of training data we used is radically reduced compared to [37, 41]. More concretely, for NYUv2 we use $\approx 3\%$ of the full dataset, for KITTI $\approx 25\%$. We use the Adam optimiser [17] with an initial learning rate of 1e-4 for all experiments and chose Tensorflow [1] as the learning framework and train using an NVIDIA-DGX1. We provide a detailed training schedule and breakdown in the supplementary material.

4 Results

In this section we summarise the single-image depth prediction and relative pose estimation performance of our system on several popular machine learning and SLAM datasets. We also investigate the effect of using alternative optical flow estimates from [13, 30] in our pose estimation pipeline as an ablation study. The entire model contains $\approx 130M$ parameters. Our depth estimator runs at 5 fps on an NVIDIA GTX 1080Ti, while other sub-networks run at ≈ 30fps.

4.1 Depth Estimation

We summarise the results of evaluating our single-image depth estimation of the datasets NYUv2 [33], RGB-D [34] and KITTI [9] in Tables 1, 2 and 3 respectively using the established metrics of [5].

   We train *Ours(baseline)* model to showcase the improvement we get by purely using the depth loss. This is then extended to use the full end-to-end training loss (depth + flow + pose losses) in the *Ours(full)* model which demonstrates a consistent improvement across all datasets. Most notably in Tables 2 and 3 for which ground truth pose data was available for training. This validates our approach for improving single image depth estimation performance, and demonstrates a network can be improved by enforcing more geometric priors on the loss functions. We would like to mention that the improvement we gain from

Table 1. The performance of several approaches evaluated on single-image depth estimation using the standard testset of NYUv2 [33] proposed in [4].

Method	Lower better			Higher better		
	RMS_{lin}	RMS_{ln}	Rel_{abs}	δ	δ^2	δ^3
Eigen$_{vgg}$ [4]	0.641	0.214	0.16	76.9%	95.0%	98.8%
Laina *et al.* [21]	0.573	0.195	0.13	81.1%	95.3%	98.8%
Kendall *et al.* [16]	0.506	-	**0.110**	81.7%	95.9%	98.9%
Ours (baseline)	0.487	0.164	0.113	86.7%	97.7%	99.4%
Ours (full)	**0.478**	**0.161**	0.111	**87.2%**	**97.8%**	**99.5%**

Ours(baseline) to *Ours(full)* is purely due to the novel combined loss terms as the flow and pose sub networks do not increase the model capacity of the depth subnet itself.

Additionally we include qualitative results for NYUv2 [33] and KITTI [9] in Figs. 3 and 4 respectively. Each of which illustrates a noticeable improvement over previous methods. We also demonstrate that the improvement is beyond the numbers, as our approach generates more convincing depths even when the RMSE may be higher, as is the case in the second row of Fig. 3, where [21] computes a lower RMSE. More impressive still are the results in Fig. 4, where we compare against previous approaches that are both trained on much larger training sets than our own and still show noticeable qualitative and quantitative improvements.

4.2 Pose Estimation

To demonstrate the ability of our approach to perform accurate relative pose estimation, we compare our approach on several unseen sequences from the datasets for which ground-truth poses were available. To quantitatively evaluate the trajectories we use the absolute trajectory error (ATE) and the relative pose error (RPE) as proposed in [34]. To mitigate the effect of scale-drift on these quantities we scale all poses to the groundtruth associated poses during evaluation. By using both metrics it provides an estimate of the consistency of each pose estimation approach. We summarise the results of this quantitative analysis for KITTI [9] in Table 4 and for RGB-D [34] in Table 5. We include comparisons of the performance against other state-of-the-art pose estimation networks namely SFM-Learner [41] and DeMoN [37]. Additionally we include results from current state-of-the-art SLAM systems also, namely ORB-SLAM2 [25] and LSD-SLAM [6].

In Table 4 we show the most comparable performance of our approach to state-of-the-art SLAM systems. We demonstrate a noticeable improvement over SfM-Learner on both sequences in all metrics. We evaluate SfM-Learner on its frame-to-frame tracking performance for adjacent frames (*SFM-Learner(1)*) and

Table 2. The performance of previous state-of-the-art approaches evaluated on the standard testset of the KITTI dataset [9].

Cap	Method	Lower better			Higher better		
		RMS_{lin}	RMS_{ln}	Rel_{abs}	δ	δ^2	δ^3
0–80 m	Zhou et al. [41]	6.856	0.283	0.208	67.8%	88.5%	95.7%
	Godard et al. [10]	4.935	0.206	0.141	86.1%	94.9%	97.6%
	Kuznietsov et al. [20]	4.621	0.189	0.113	86.2%	96.0%	98.6%
	Ours (baseline)	4.394	0.178	0.095	89.4%	96.6%	98.6%
	Ours (full)	**4.301**	**0.173**	**0.096**	**89.5%**	**96.8%**	**98.7%**
0–50 m	Zhou et al. [41]	5.181	0.264	0.201	69.6%	90.0%	96.6%
	Garg et al. [8]	5.104	0.273	0.169	74.0%	90.4%	96.2%
	Godard et al. [10]	3.729	0.194	0.108	87.3%	95.4%	97.9%
	Kuznietsov et al. [20]	3.518	0.179	0.108	87.5%	96.4%	98.8%
	Ours (baseline)	3.359	0.168	0.092	90.5%	97.0%	98.8%
	Ours (full)	**3.284**	**0.164**	**0.092**	**90.6%**	**97.1%**	**98.9%**

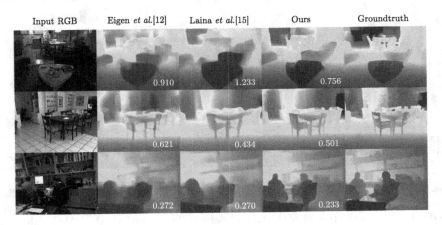

Fig. 3. Resulting single image depth estimation for several approaches and ours against the ground truth on the dataset NYUv2 [33]. The RMSE for each prediction is included

separations of 5 frames (*SFM-Learner(5)*), as they train their approach to estimate this size frame gap. Even with the massive reduction in accumulation error expected by taking larger frame gaps (demonstrated in reduced ATE) our system still produces more accurate pose estimates (Fig. 6).

We show the resulting scaled trajectories of sequence 09 in Fig. 5, as well as the relative scaling of each trajectories poses in a box-plot. The spread of scales present for SFM-Learner indicates scale is essentially ignored by their system, with scale drifts ranging across a full log scale, while ORB-SLAM and our approach are barely visible at this scale. Another thing to note is that our

Table 3. The performance of previous state-of-the-art approaches on a randomly selected subset of the frames from the RGB-D dataset [34]. We post separate entries for DeMoN(est) and DeMoN(gt), former is scaled by the estimated scale of their system while the latter is scaled by the median groundtruth depth.

Method	Lower better				Higher better		
	RMS_{lin}	RMS_{log}	Rel_{abs}	Rel_{sqr}	δ	δ^2	δ^3
Laina *et al.* [21]	1.275	0.481	0.189	0.371	75.3%	89.1%	91.8%
DeMoN(est)[37]	2.980	0.910	1.413	5.109	21.0%	36.6%	48.9%
DeMoN(gt)[37]	1.584	0.555	0.301	0.581	52.7%	70.7%	80.7%
Ours(baseline)	1.068	0.353	0.128	0.236	86.9%	92.2%	93.5%
Ours(full)	**0.996**	**0.329**	**0.108**	**0.194**	**90.3%**	**93.6%**	**94.5%**

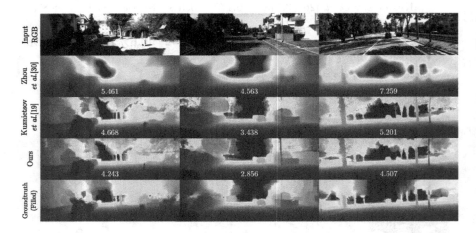

Fig. 4. The resulting single image depth estimation for several approaches including Zhou *et al.* [41](SfM-Learner), Kuznietsov *et al.* [20] and Ours against a ground truth filled using [22] on the testset of the KITTI dataset [9]. We include the RMSE values for each methods prediction. Filled depths are included for visualisation purposes during evaluation the predictions are evaluated against the sparse velodyne ground truth data

scale is centered around 1.0, as we estimate scale directly by estimating metric depths. This seems to provide a strong benefit in terms of reducing scale-drift and we believe makes our system more usable in practice.

In Table 5 we show a significant improvement in performance against existing machine learning approaches across several sequences from the RGB-D dataset [34]. We evaluate against DeMoN [37] in two ways, frame-to-frame (*DeMoN(1)*) and we again try to provide the same advantage to DeMoN as SfM-Learner by using wider baselines, which they claim improves their depth estimations [37], using a frame gap of 10 (*DeMoN(10)*). It can be observed that even with the massive reduction in accumulation error over our frame-to-frame approach, we still manage to significantly out-perform their approach in ATE, even surpassing

LSD-SLAM on the sequence *fr1-xyz*. ORB-SLAM is still the clear winner, as they massively benefit from the ability to perform local bundle-adjustments on the sequences used, which are short trajectories of small scenes. We include an example of a frame from the sequence *fr3-walk-xyz* in Fig. 7, which shows this scene is not static, but our system has the ability to deal with this through the flow confidence estimates, discussed in Sect. 4.3.

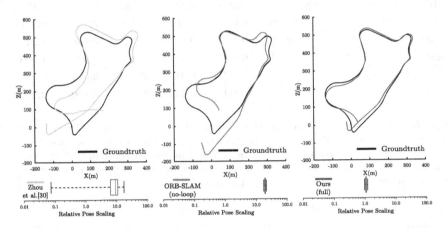

Fig. 5. *Top* the scaled and aligned trajectories for Zhou *et al.* [41](SfM-Learner), ORB-SLAM2 [25] (without loop-closure enabled) and Ours respectively. *Bottom* box-plots of the relative pose scaling required to bring the predicted translation to the same magnitude as the ground-truth pose

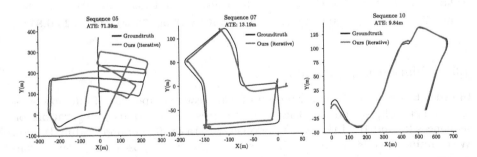

Fig. 6. Resulting trajectories using our iterative approach on 3 additional sequences of KITTI [9]. Sequence 05 shows a failure for our approach, where accumulated drift causes the trajectory to not be well aligned. These sequences are not used for training

Table 4. Performance of several approaches evaluated on two sequences of the KITTI dataset [9]. SfM-Learner(1) and SfM-Learner(5) indicates the different frame gaps used to construct the trajectories. The results are separated by SLAM and machine learning approaches. We highlight the strongest results in bold for each type of approach.

Sequence	09			10		
Method	ATE(m)	RPE(m)	RPE(°)	ATE(m)	RPE(m)	RPE(°)
ORB-SLAM (no-loop) [25]	57.57	**0.040**	0.103	8.090	0.033	0.105
ORB-SLAM (full) [25]	**9.104**	0.056	**0.084**	**7.349**	**0.031**	**0.100**
SfM-learner (5) [41]	58.31	0.077	0.803	31.75	0.069	1.242
SfM-learner (1) [41]	81.09	0.050	0.976	75.89	0.045	1.599
Ours (fully connected)	41.50	0.087	0.387	29.29	0.081	0.486
Ours (full)	**16.55**	**0.047**	**0.128**	**9.846**	**0.039**	**0.138**

Table 5. Performance of pose estimation on several sequences from the RGB-D dataset [34]. DeMoN(1) and DeMoN(10) indicates the trajectories were constructed with a frame gap of 1 and 10 respectively. Both [25] and [6] fail to track on *fr2-360-hs*. The results are separated by SLAM and machine learning approaches. We highlight the strongest results in bold for each type of approach.

Sequence	fr1-xyz			fr2-360-hs			fr3-walk-xyz		
Method	ATE (m)	RPE (m)	RPE (°)	ATE (m)	RPE (m)	RPE (°)	ATE (m)	RPE (m)	RPE (°)
LSD-SLAM [6]	0.090	-	-	-	-	-	0.124	-	-
ORB-SLAM [25]	**0.009**	**0.007**	**0.645**	-	-	-	**0.012**	**0.013**	**0.694**
DeMoN(10)[37]	0.178	**0.021**	**1.193**	0.601	0.035	2.243	0.265	0.049	1.447
DeMoN(1)[37]	0.183	0.037	3.612	0.669	0.032	3.233	0.279	0.040	3.174
Ours (fully connected)	0.169	0.028	1.887	0.883	0.030	1.799	0.268	0.044	1.698
Ours (iterative)	**0.071**	0.024	1.237	**0.461**	**0.020**	**0.736**	**0.240**	**0.026**	**0.811**

4.3 Ablation Experiments

In order to examine the contribution of using each component of our pose estimation network, we compare the pose estimates under various configurations on sequences 09 and 10 of the KITTI odometry dataset [9], summarised in Table 6. We examine the relative improvement of iterating on our pose estimation till convergence, against a single weighted-least-squares iteration, which demonstrates iterating has a significantly positive effect. We demonstrate the improved utility of our flows by replacing our flow estimates with other state-of-the-art flow estimation methods from [13,30] in our pose estimation pipeline, and consistently demonstrate an improvement using our approach. We show the result of optimising with and without our estimated confidences, demonstrating quantitatively how important they are to pose estimation accuracy, with significant reductions across all metrics.

Table 6. Results of pose estimation on KITTI [9] with various components of the network removed or replaced. We highlight the strongest results in bold.

Sequence	09			10		
Method	ATE(m)	RPE(m)	RPE(°)	ATE(m)	RPE(m)	RPE(°)
Ours (noconf)	53.40	0.356	0.931	58.50	0.308	1.058
Ours (noconf,iterative)	33.18	0.248	0.421	35.87	0.280	0.803
Flownet2.0 [13]	29.64	0.349	0.838	51.90	0.222	0.954
Flownet2.0 (iterative) [13]	24.61	0.185	0.400	22.61	0.142	0.484
EpicFlow [30]	119.0	0.566	0.931	20.98	0.199	0.853
EpicFlow (iterative) [30]	59.79	0.379	0.459	14.80	0.154	0.581
Ours (full-single iteration)	31.20	0.089	0.324	24.10	0.095	0.389
Ours (full-til convergence)	**16.55**	**0.047**	**0.128**	**9.846**	**0.039**	**0.138**

We also demonstrate qualitatively one of the ways in which estimating confidence improves our pose estimation in Fig. 7. This shows that our system has learned the confidence on moving objects is lower than its surroundings and the confidences of edges are higher, helping our system focus on salient information during optimisation in an approach similar to [6].

Fig. 7. For a frame pair (I_i and I_j) from the sequence *fr3-walk-xyz*, $\mathbf{F}_{\overleftarrow{ji}}$ is the estimated optical flow from I_i to I_j, and C_x and C_y are the estimated flow confidences in the x and y direction respectively

5   Conclusion and Further Work

We present the first piece of work that performs least squares based pose estimation inside a neural network. Instead of replacing every component of the SLAM pipeline with CNNs, we argue it's better to use CNNs for tasks that greatly benefit from feature extraction (depth and flow prediction) and use geometry for tasks its proven to work well (motion estimation given the depths and flow). Our formulation is fully differentiable and is trained end-to-end. We achieve state-of-the-art performances on single image depth prediction for both NYUv2 [33] and KITTI [9] datasets. We demonstrate both qualitatively and quantitatively that our system is capable of producing better visual odometry that considerably reduces scale-drift by predicting metric depths.

References

1. Abadi, M., et al.: Tensorflow: large-scale machine learning on heterogeneous distributed systems. arXiv preprint arXiv:1603.04467 (2016)
2. Agrawal, P., Carreira, J., Malik, J.: Learning to see by moving. In: IEEE International Conference on Computer Vision (ICCV) (2015)
3. Dharmasiri, T., Spek, A., Drummond, T.: Joint prediction of depths, normals and surface curvature from RGB images using CNNs. arXiv preprint arXiv:1706.07593 (2017)
4. Eigen, D., Fergus, R.: Predicting depth, surface normals and semantic labels with a common multi-scale convolutional architecture. In: IEEE International Conference on Computer Vision (ICCV) (2015)
5. Eigen, D., Puhrsch, C., Fergus, R.: Depth map prediction from a single image using a multi-scale deep network. In: Advances in Neural Information Processing Systems (NIPS) (2014)
6. Engel, J., Schöps, T., Cremers, D.: LSD-SLAM: large-scale direct monocular SLAM. In: Fleet, D., Pajdla, T., Schiele, B., Tuytelaars, T. (eds.) ECCV 2014. LNCS, vol. 8690, pp. 834–849. Springer, Cham (2014). https://doi.org/10.1007/978-3-319-10605-2_54
7. Fischer, P., et al.: FlowNet: learning optical flow with convolutional networks. In: IEEE International Conference on Computer Vision (ICCV) (2015)
8. Garg, R., B.G., V.K., Carneiro, G., Reid, I.: Unsupervised CNN for single view depth estimation: geometry to the rescue. In: Leibe, B., Matas, J., Sebe, N., Welling, M. (eds.) ECCV 2016. LNCS, vol. 9912, pp. 740–756. Springer, Cham (2016). https://doi.org/10.1007/978-3-319-46484-8_45
9. Geiger, A., Lenz, P., Stiller, C., Urtasun, R.: Vision meets robotics: the KITTI dataset. Int. J. Robot. Res. (IJRR) **32**, 1231–1237 (2013)
10. Godard, C., Mac Aodha, O., Brostow, G.J.: Unsupervised monocular depth estimation with left-right consistency. In: IEEE Conference on Computer Vision and Pattern Recognition (CVPR) (2017)
11. He, K., Zhang, X., Ren, S., Sun, J.: Deep residual learning for image recognition. In: IEEE Conference on Computer Vision and Pattern Recognition (CVPR) (2016)
12. Huang, G., Liu, Z., van der Maaten, L., Weinberger, K.Q.: Densely connected convolutional networks. In: IEEE Conference on Computer Vision and Pattern Recognition (CVPR) (2017)
13. Ilg, E., Mayer, N., Saikia, T., Keuper, M., Dosovitskiy, A., Brox, T.: Flownet 2.0: evolution of optical flow estimation with deep networks. In: IEEE Conference on Computer Vision and Pattern Recognition (CVPR) (2017)
14. Jayaraman, D., Grauman, K.: Learning image representations tied to ego-motion. In: IEEE International Conference on Computer Vision (ICCV) (2015)
15. Kendall, A., Cipolla, R.: Modelling uncertainty in deep learning for camera relocalization. In: IEEE International Conference on Robotics and Automation (ICRA), pp. 4762–4769 (2016)
16. Kendall, A., Gal, Y.: What uncertainties do we need in Bayesian deep learning for computer vision? In: Advances in Neural Information Processing Systems (NIPS) (2017)
17. Kingma, D.P., Ba, J.: Adam: a method for stochastic optimization. arXiv preprint arXiv:1412.6980 (2014)
18. Klein, G., Murray, D.: Parallel tracking and mapping for small AR workspaces. In: IEEE and ACM International Symposium on Mixed and Augmented Reality (ISMAR) (2007)

19. Krizhevsky, A., Sutskever, I., Hinton, G.E.: ImageNet classification with deep convolutional neural networks. In: Advances in Neural Information Processing Systems (NIPS) (2012)
20. Kuznietsov, Y., Stückler, J., Leibe, B.: Semi-supervised deep learning for monocular depth map prediction. In: IEEE Conference on Computer Vision and Pattern Recognition (CVPR) (2017)
21. Laina, I., Rupprecht, C., Belagiannis, V., Tombari, F., Navab, N.: Deeper depth prediction with fully convolutional residual networks. In: International Conference on 3D Vision (3DV) (2016)
22. Levin, A., Lischinski, D., Weiss, Y.: Colorization using optimization. In: ACM Transactions on Graphics (2004)
23. Liu, F., Shen, C., Lin, G.: Deep convolutional neural fields for depth estimation from a single image. In: IEEE Conference on Computer Vision and Pattern Recognition (CVPR) (2015)
24. Long, J., Shelhamer, E., Darrell, T.: Fully convolutional networks for semantic segmentation. In: IEEE Conference on Computer Vision and Pattern Recognition (CVPR) (2015)
25. Mur-Artal, R., Montiel, J.M.M., Tardos, J.D.: ORB-SLAM: a versatile and accurate monocular slam system. IEEE Trans. Rob. **31**(5), 1147–1163 (2015)
26. Mur-Artal, R., Tardós, J.D.: ORB-SLAM2: an open-source SLAM system for monocular, stereo and RGB-D cameras. IEEE Trans. Rob. **33**(5), 1255–1262 (2017)
27. Newcombe, R.A., Lovegrove, S.J., Davison, A.J.: DTAM: dense tracking and mapping in real-time. In: IEEE International Conference on Computer Vision (ICCV) (2011)
28. Rad, M., Lepetit, V.: BB8: A scalable, accurate, robust to partial occlusion method for predicting the 3D poses of challenging objects without using depth. In: IEEE International Conference on Computer Vision (ICCV) (2017)
29. Ranjan, A., Black, M.J.: Optical flow estimation using a spatial pyramid network. In: IEEE Conference on Computer Vision and Pattern Recognition (2017)
30. Revaud, J., Weinzaepfel, P., Harchaoui, Z., Schmid, C.: Epicflow: edge-preserving interpolation of correspondences for optical flow. In: IEEE Conference on Computer Vision and Pattern Recognition (CVPR) (2015)
31. Russakovsky, O., et al.: Imagenet large scale visual recognition challenge. Int. J. Comput. Vis. (IJCV) **115**(3), 211–252 (2015). https://doi.org/10.1007/s11263-015-0816-y
32. Saxena, A., Chung, S.H., Ng, A.Y.: Learning depth from single monocular images. In: Advances in Neural Information Processing Systems (2006)
33. Silberman, N., Hoiem, D., Kohli, P., Fergus, R.: Indoor segmentation and support inference from RGBD images. In: Fitzgibbon, A., Lazebnik, S., Perona, P., Sato, Y., Schmid, C. (eds.) ECCV 2012. LNCS, vol. 7576, pp. 746–760. Springer, Heidelberg (2012). https://doi.org/10.1007/978-3-642-33715-4_54
34. Sturm, J., Engelhard, N., Endres, F., Burgard, W., Cremers, D.: A benchmark for the evaluation of RGB-D SLAM systems. In: International Conference on Intelligent Robot Systems (IROS) (2012)
35. Sun, D., Yang, X., Liu, M.Y., Kautz, J.: PWC-Net: CNNs for optical flow using pyramid, warping, and cost volume. arXiv preprint arXiv:1709.02371 (2017)
36. Tateno, K., Tombari, F., Laina, I., Navab, N.: CNN-SLAM: real-time dense monocular slam with learned depth prediction. In: IEEE Conference on Computer Vision and Pattern Recognition (CVPR) (2017)

37. Ummenhofer, B., et al.: Demon: depth and motion network for learning monocular stereo. In: IEEE Conference on Computer Vision and Pattern Recognition (CVPR) (2017)

38. Wannenwetsch, A.S., Keuper, M., Roth, S.: Probflow: joint optical flow and uncertainty estimation. In: IEEE International Conference on Computer Vision (ICCV) (2017)

39. Yi, K.M., Trulls, E., Lepetit, V., Fua, P.: LIFT: learned invariant feature transform. In: Leibe, B., Matas, J., Sebe, N., Welling, M. (eds.) ECCV 2016. LNCS, vol. 9910, pp. 467–483. Springer, Cham (2016). https://doi.org/10.1007/978-3-319-46466-4_28

40. Zhao, H., Shi, J., Qi, X., Wang, X., Jia, J.: Pyramid scene parsing network. In: IEEE Computer Vision and Pattern Recognition (CVPR) (2017)

41. Zhou, T., Brown, M., Snavely, N., Lowe, D.G.: Unsupervised learning of depth and ego-motion from video. In: IEEE Conference on Computer Vision and Pattern Recognition (CVPR) (2017)

Multi-scale Adaptive Structure Network for Human Pose Estimation from Color Images

Wenlin Zhuang[1], Cong Peng[2], Siyu Xia[1], and Yangang Wang[1(✉)]

[1] School of Automation, Southeast University, Nanjing, China
ygwangthu@gmail.com
[2] College of Automation Engineering, Nanjing University of Aeronautics and
Astronautics, Nanjing, China

Abstract. Human pose estimation is formulated as a joint heatmap regression problem in the deep learning based methods. Existing convolutional neural networks usually adopt fixed kernel size for generating joint heatmaps without regard to the size of human shapes. In this paper, we propose a novel method to address this issue by adapting the kernel size of joint heatmaps to the human scale of the input images in the training stage. We present a normalization strategy of how to perform the adaption between the kernel size and human scale. Beyond that, we introduce a novel limb region representation to learn the human pose structural information. Both the adaptive joint heatmaps as well as the limb region representation are combined together to construct a novel neural network, which is named **Multi-scale Adaptive Structure Network (MASN)**. The effectiveness of the proposed network is evaluated on two widely used human pose estimation benchmarks. The experiments demonstrate that our approach could obtain the state-of-the-art results and outperform the most existing methods over all the body parts.

Keywords: Multi-scale · Adaptive heatmaps · Human pose estimation

1 Introduction

Human pose estimation, which is defined as the problem of localization of human skeleton joints, is a vital yet challenging task in the field of computer vision. It has many important applications such as human-machine interactions, intelligent manufacturing, autonomous driving and etc. In recent years, human pose estimation from single RGB image has gained significant improvements by Convolutional Neural Networks (ConvNets) [1–6]. However, achieving accurate human pose from color images is still very difficult due to variant appearances, strong articulations, heavy occlusions and etc.

The main stream of work for human pose estimation from color images with deep learning has been motivated by formulating it as a joint regression problem,

© Springer Nature Switzerland AG 2019
C. V. Jawahar et al. (Eds.): ACCV 2018, LNCS 11361, pp. 643–658, 2019.
https://doi.org/10.1007/978-3-030-20887-5_40

where joint coordinates [1] or joint heatmaps [2] are the most widely used human pose representation. Since joint heatmaps are robust to the data noise and have better performance [2], they attract more and more attentions from researchers in this field. Typically, a joint heatmap is described as a 2D map with a circle at the position of the joint. The values in the circle are computed from a 2D Gaussian probability density function. The variance of the density function determines the influence range of the circle in the joint heatmap. In this paper, the variance is named as the **kernel size** of the joint heatmap.

Existing methods [7,8] usually adopt fixed kernel size without considering the the size of human shapes in the input image, which is named as **human scale** in this paper. However, we argue that the human scale is very important for regressing the accurate joint heatmaps. Suppose there is a very small person in a given input image, fixed kernel size might generate joint heatmaps which cover the whole pixels of the human. This scenario is not what we expect and we want the neural network would have available receptive fields according to the image content. In other words, the kernel size of joint heatmaps should be appropriate to the human scale in the given input images. We want the designed neural network could have the abilities that large person has large kernel size and small person has small kernel size. The 'large' and 'small' are relative to the coverage ratio of human in the same size of rescaled input image. More importantly, the joint heatmaps for different human scales should be normalized into a unified framework while performing the end-to-end training with the convolutional neural networks.

It is noted that several previous works attempt to address the human scales in the methods of deep learning based human pose estimation. Data augmentation with rescaling the input images [9] is the widely used strategy. Nevertheless, none of previous data augmentation with scaling strategies construct a concrete normalization between the human scale and human pose. Feature pyramid network [6] explicitly consider several different sizes of human shapes by constructing the pyramid structures of image features. However, they can only address limited scales (always 3 or 4) in the network structure. And their method does not consider the image normalization among the input dataset. Our key idea is to adapt the kernel size of joint heatmaps to the human scale of input images in the training stage. We deduct a mathematical normalization strategy (Sect. 3) to perform the adaption between the kernel size and human scale. We demonstrate that the normalized human scale as well as human pose could benefit with each other in an end-to-end neural network framework and has superior performance as described in Sect. 4.

In order to further improve the performance of human pose estimation, we also encode the pose structural information by referring the idea of part affinity [7]. Different from their directional representation of limb, we propose a simple yet effective limb region representation to capture the pose structural information. The limb region is composed of the pixels covered by the connecting line between two adjacent joints. We generate the limb region maps, where only the pixels in the limb region are filled with 1. Similar as the proposed adaptive joint heatmaps, the limb region maps are also associated with the human scales. Both the adaptive joint heatmaps as well as the limb region maps are combined together to construct a novel neural network, which is named as **Multi-scale Adaptive Structure Network (MASN)**. For exploiting the abilities of the proposed method, we use the Hourglass module [5] to construct our network architecture. The whole network structure is visualized in Fig. 1. We evaluate the proposed method on the MPII Human Pose dataset [9], and there is a clear average accuracy improvement over the original Hourglass network [5]. In addition, our model has fewer parameters, as shown in Table 1. In the following sections, we will introduce each individual components of the proposed method in details.

2 Related Works

Human Pose Estimation. Early human pose estimation usually adopt pictorial structure [10–14]. Due to the influence of human flexbility and occlusion, the application conditions are too harsh and these approachs are not robust enough. In recent years, pose estimation has been greatly developed under deep learning [1–3,15,16]. DeepPose [1] uses ConvNets to regress the coordinates of body joints. However, ConvNets cannot learn the mapping from image to location faultlessly. Therefore, the method of directly detecting heatmap appears, and its effect is better and more robust. Convolutional Pose Machine [4] is one of first adpot heatmap, which show ConvNets can learn image features and implicitly model long-range dependencies for different body parts. Hourglass [5], the most famous method in recent years, differs from other networks mainly in its more symmetric distribution of capacity between bottom-up processing (from high resolutions to low resolutions) and top-down processing (from low resolutions to high resolutions). Based on this, chu et al. [17] added attention to different resolutions of Hourglass. Yang et al. [6] made improvements to the basic module (Pyramid Residual Module) in Hourglass, and can obtain richer features at each scale. The detecting heatmap methods used above use the same kernel size of joint heatmaps. However, the shape of human body in two-dimensional images is different. For different human body shape, different kernel size should be selected. Our method uses an adaptive heatmap generation method to generate heatmaps of different kernel size according to different human body shape in color image.

Human Pose Structure. The human skeleton structure is the most critical information of human pose, which is crucial for inferring joints that are invisible under the influence of occlusion, illumination and etc. The pictorial structure [13] is based on human body structure using a visual description method to build a model of the human body structure. Chu et al. [18] used geometrical transform kernels in ConvNets to establish the dependencies between joints, and use a bi-directional tree to model the human upper body structure. Ke et al. [8] use Hourglass to learn body structure information to characterize body structure information using multiple joint heatmaps. These methods have different defects, and the structure of the human body is incomplete. Openpose [7] models the areas between the joints of the two connections, and uses the human skeleton connection relationship to perform multi-person pose estimation. In this work, we adopt the method of establishing a model of the area between the two joints that have a connection relationship, and can completely learn the human skeleton structural information.

Multi-scale Structure Information. In the past two years, multi-scale network architecture has had a significant role in multiple tasks of vision [19–21]. In human pose estimation, the champion of COCO2017 [22] cascaded the RefineNet after FPN to better merge different scale feature information. Sun et al. [23] used multi-scale fully concolutional network to detect joints. Hourglass [5] has acquired information on multi-scale but has not conducted supervised learning at multi-scale. Ke et al. [8] added supervised learning of heatmaps at multi-scale of hourglass, but lacked more important body structural information. We supervise structural information at multi-scale of the network and describe human information in images more comprehensively.

3   Method

An overview of our framework is illustrated in Fig. 1. Our method achieves end-to-end learning which inputs RGB images and outputs joint heatmaps. We adopt the highly modularied Hourglass module [5] as the basic network structure to investigate the effect of Multi-scale Adaptive Structure Network (MASN). MASN add multi-scale adaptive structural information supervision on the basis

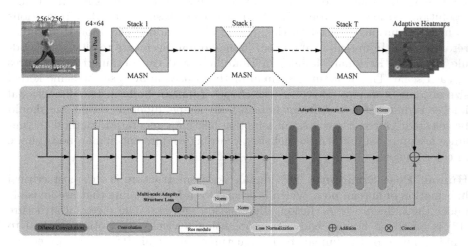

Fig. 1. Overview of our framework. **Top**: the Multi-scale Adaptive Structure Network (MASN), stacked T MASN sequentially. **Bottom**: details of each stack of MASN. In each stack of MASN, we generate multi-scale adaptive structure maps.

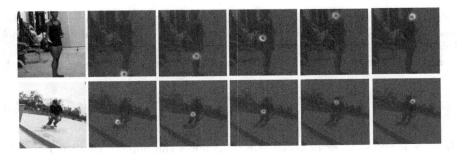

Fig. 2. Adaptive heatmaps. **Top**: large-scale human body uses large heatmaps. **Bottom**: small-scale human body uses small heatmaps.

of Hourglass module, and output adaptive heatmaps at the end. The ground-truth multi-scale adaptive structural information is represented by limb region, which is the area between the two joints that have a connection relationship. The ground-truth adaptive heatmaps are generated using 2D Gaussian.

In this section, we specifically elaborate our approach in three areas: adaptive heatmaps, limb region (structural information), and multi-scale adaptive structure supervision.

3.1 Adaptive Heatmaps

In many pose estimation methods [2,7,22] including Hourglass, the ground-truth heatmap h_j^* of joint j is generated by 2D Gaussian:

$$h_j^*(p) = exp(-\frac{\|p - y_j\|^2}{2\sigma^2}) \tag{1}$$

where p is heatmap location, $j = 1, 2, ..., J$ is the index of each joint and J is the number of joints, y_j is the ground-truth at location p. σ controls the spread of the peak.

All existing methods use the same σ to generate heatmaps of the same kernel size, but obviously this is flawed. Although it is necessary to crop images according to the size of the human body area in the single-person pose estimation, the scale of the human body in the cropped image region is also inconsistent. In the far left side of Fig. 2, these are two training images of the same size ($w \times h$), their human body scales are different. It is obviously not suitable to use the heatmaps of same kernel size to represent joints positions. We adpot adaptive heatmaps, which means that different σ is generated according to the size of the different body scale to control the kernel size of heatmaps. We chose the simplest and most appropriate method to generate adaptive heatmaps. Given an image ($w \times h$), the maximum scale m of the human body shape is found by traversing according to the positions of whole body joints. Then the appropriate σ produced by linear interpolation,

$$\sigma = \frac{m}{max(w, h)}(\sigma_{max} - \sigma_{min}) + \sigma_{min} \tag{2}$$

where σ_{max} is the upper limit of σ, σ_{min} is the lower limit. In our experiments, $\sigma_{max} = 2.5$, $\sigma_{min} = 1.0$. Hence the ground-truth is $h^*_{j,\sigma}$. In Fig. 2, different kernel sizes of heatmaps are generated according to different body scale.

For each stack Hourglass i, the loss function L^i_h in [5] for training the model,

$$L^i_h = \frac{1}{N} \sum_{n=1}^{N} \frac{1}{J} \sum_{j=1}^{J} \frac{1}{P} \sum_p \left\| \widehat{y}^i_j(p) - h^*_j(p) \right\|_2^2 \tag{3}$$

where N is the number of batch size, P denotes the number of all point p, $\widehat{y}^i_j$ denotes the predicted heatmap for joint j at stack Hourglass i. However, in our method, the large ground-truth heatmaps produce large loss function, the small ground-truth heatmaps produce small loss function. Therefore, it is necessary to normalize the loss function. For Gaussian functions,

$$\int exp(-\frac{x^2}{2\sigma^2}) = \sigma\sqrt{2\pi} \tag{4}$$

$$\sum (exp(-\frac{x^2}{2\sigma^2}))^2 = \sum exp(-\frac{x^2}{2(\frac{\sqrt{2}}{2}\sigma)^2}) \approx \int exp(-\frac{x^2}{2(\frac{\sqrt{2}}{2}\sigma)^2}) = \sigma\sqrt{\pi} \tag{5}$$

so the loss of every training sample $L^i_{h,n}$ is proportional to σ,

$$L^i_{h,n} \propto \sigma \tag{6}$$

In a batch, normalize each training sample. The final loss function is

$$L^i_h = \frac{1}{N} \sum_{n=1}^{N} \frac{1}{J \times \sigma_n} \sum_{j=1}^{J} \frac{1}{P} \sum_p \left\| \widehat{y}^i_j(p) - h^*_j(p, \sigma_n) \right\|_2^2 \tag{7}$$

where σ_n is generated by Eq. (2) for training sample n, $\widehat{y}^i_j$ denotes the predicted heatmap at stack i.

The adaptive heatmaps help to solve the problem that the heatmaps of joints detected is not accurate enough when the human body area is too large or the human body area is too small.

3.2   Limb Region

The human body structural information is the skeleton connection relationship of the human body. Obviously, the human body structure is one of the most important information for human pose estimation. Directly detecting joint heatmaps does not learn structural information primely, and how to characterize structural information is greatly important. Hence we propose a non-parametric representation called limb region to learn structural information, which denotes location information across the region of limb. In addition, our method can also be understood as an attention model, which is different from the previous

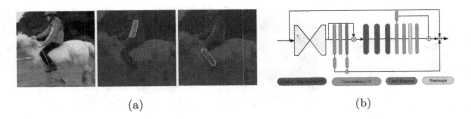

Fig. 3. (a): Structural information is represented by limb region, e.g. left arm, left thigh. **(b)**: An illustration of the structural network. It consists of two parts: limb region detection and heatmaps detection.

method [17] focusing on the entire human body area. It breaks the attention of whole human body into the attention of limbs and makes it more accurate for the region of human body.

The limb region is a scalar region between the two joints that have a connection relationship. In the known human pose estimation datasets, the ground-truth of limb region is not given. It is necessary to split different limbs according to the ground-truth positions of joints. Consider a single limb l (as show in the middle/right of Fig. 3(a)), y_{j_a} and y_{j_b} are the ground-truth positions of joints adjacent to the limb l. The limb l is characterized by a rectangular box, the length is the distance $length$ between y_{j_a} and y_{j_b}, and the width is the set value $width$. If a point p lies on the rectangular box, the value of ground-truth limb region $r_l^*(p)$ is 1; for all other points, the value is 0. The ground-truth limb region at a map as,

$$r_l^*(p) = \begin{cases} 1, & if\ p\ on\ limb\ l \\ 0, & otheriwise \end{cases} \tag{8}$$

We first used Hourglass for the detection of the limb region, followed by the joint heatmaps detection network, as shown in Fig. 3(b). Hourglass outputs 64×64 features, the channel is 256, after two 1×1 convolution to get the output of limb regions $L \times 64 \times 64$, where L is the number of limbs. Then concat Hourglass features with limb regions to get $(256 + L)$-d features. Sequentially stacked three 3×3 dilated convolution [24, 25] and two 1×1 convolution, joint heatmaps $J \times 64 \times 64$ are obtained. It is worth noting that after the first 3×3 dilated convolution, the feature dimension is reduced to 256. Similar to convolutional Pose Machine [4], the purpose of stacking three dilated convolution is to expand the receptive field so that the receptive field at the position of each joint can at least cover the neighboring limb region and make full use of the structural information of the limb region. However, convolutional Pose Machine uses large convolutional kernels($9 \times 9, 11 \times 11$) and generates a large amount of parameters. Significantly, we use the dilated convolution, which can both guarantee the receptive field and reduce the amount of parameters.

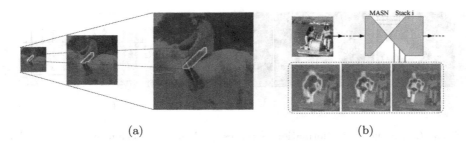

(a)   (b)

Fig. 4. Multi-scale adaptive structure supervision. **(a)**: Take the right thigh as an example to show the representation of the limb region on three scales. **(b)**: An example of Multi-scale limb regions output within a MASN. These limb regions are summed into a sigle map.

In the training phase of the limb region, we used the same loss function as heatmaps, the Mean Square Error loss. For each stack i, the loss function L_r^i is

$$L_r^i = \frac{1}{N}\sum_{n=1}^{N}\frac{1}{L}\sum_{l=1}^{L}\frac{1}{P}\sum_p \left\|\widehat{r}_l^i(p) - r_l^*(p)\right\|_2^2 \tag{9}$$

where $\widehat{r}_l^i$ denotes the predicted limb region at stack i.

3.3   Multi-scale Adaptive Structure Supervision

In Sect. 3.2, we specifically state that human body structural information can be characterized by the limb region, and this section will expound human structural information at multi-scale. The first question, how do we generate multi-scale of the limb region? One possible way is downsampling the ground-truth limb region produced at the original scale. However, when the limb region is small, downsampling is extremely easy to distort, resulting in a discontinuous limb region. To address the limitation, we generate the ground-truth limb region on the corresponding scales by scaled the positions of joints. In our experiments, we generate three scales limb regions, as shown in Fig. 4(a). We enlarge the different scale of the limb region of the network to the original scale, which can be clearly found that the smaller scale, the more comprehensive the information described; the larger scale, more accurate, as shown in Fig. 4(b). For different scales of the limb region, the length is the distance between adjacent joints, but the width is inconsistent. In addition, adaptive width c is also needed for body scale, similar to adaptive σ in Eq. (2),

$$c_n^s = \frac{m_n}{max(w, h)}(c_{max}^s - c_{min}^s) + c_{min}^s \tag{10}$$

where c_n^s denotes the adaptive width of nth training sample at scale s, m_n denotes the maximum scale of body shape of nth training sample, c_{max}^s is the upper limit of limb region width at scale s, c_{min}^s is the lower limit of limb region width. In our experiments, $s = 1, \frac{1}{2}, \frac{1}{4}$, $c_{max}^1 = 2.5$, $c_{min}^1 = 1.0$, $c_{max}^{\frac{1}{2}} = 1.5$, $c_{min}^{\frac{1}{2}} = 0.75$, $c_{max}^{\frac{1}{4}} = 1.0$, $c_{min}^{\frac{1}{4}} = 0.5$.

Similar to the adaptive sigma, the loss function of the adaptive limb region needs to be normalized, and the loss functions of different scales are also normalized. The multi-scale adaptive limb region loss function as,

$$L_r^i = \sum_s \frac{s^2}{N} \sum_{n=1}^{N} \frac{1}{L \times c_n^s} \sum_{l=1}^{L} \frac{1}{P^s} \sum_{p^s} \left\| \widehat{r_l^i}(p^s) - r_l^*(p^s, c_n^s) \right\|_2^2 \tag{11}$$

where $\widehat{r_l^i}(p^s)$ denotes the predicted limb region position p^s of scale s, $r_l^*(p^s, c_n^s)$ denotes the ground-truth limb region which generates according to adaptive width c_n^s, P^s denotes the number of all point at scale s. It should be noted that s^2 is a value set to balance the loss of limb region between multi-scale. In addition to the learning of the limb region, each complete MASN also needs to detect the joint heatmaps, so the complete loss function is

$$L = \sum_i L^i = \sum_i \alpha L_h^i + \beta L_r^i \tag{12}$$

where α and β is to balance the two loss functions. In our experiments, $\alpha = 1$, $\beta = 0.1$.

4   Experiments

4.1   Implementation Details

Dataset. We evaluate our method on two widely benchmark datasets, MPII Human Pose [9] and Leeds Sports Poses (LSP) [26] and its extended training dataset. MPII Human Pose dataset includes around 25 K images containing over 40 K people with annotated body joints, which covers a wide range of human activities. LSP dataset is composed of 12 K images with challenging poses in sports activities.

Data Augmentation. We crop the images with the annotated body center and scale, and resize to $256 * 256$. Then we augment the images by performing random scaling factor between 0.75 and 1.25, horizontal flipping, and rotating across $\pm 30°$. In addition, we add color noise to make the model more robust. During testing, we crop the images with approximate location and scale of each person for MPII dataset. Since the LSP dataset has no location and scale, we use the image center and image size. All our experimental results is conducted on 6-scale image pyramids with flipping.

Experiment Settings. Due to the shortage of hardware devices, we only stack 4 MASNs as our complete network structure. Our method is implemented using PyTorch open-source framework and we use RMSprop [27] algorithm to optimize the network on 1 NVIDIA GTX 1080Ti GPU with a mini-batch size of 6 for 200 epochs. The initial learning rate is 1.5×10^{-4} and is dropped by 10 at 110th and the 160th epoch.

(a) Examples of predicted pose on MPII test set.

(b) Examples of predicted pose on LSP test set.

Fig. 5. Qualitative results on two datasets.

4.2   Results

Evaluation Measure. We use the Percentage Correct Keypoints (PCK) [33] measure on the LSP dataset, which reports the percentage of predicted joint position within a normalized distance of the ground-truth. For MPII dataset, the PCKh [9], as an improved PCK, uses the head size for normalization, making the evaluation more accurate.

MPII Dataset. Our results are reported in Table 1 using PCKh@0.5 measure, 50% of the head size for normalization. Our model is trained by dividing the MPII complete training data set into a training set and a validation set as [28]. We reach a test score of 91.7%, which is 0.8% higher than 8-stack Hourglass [5]. Obviously, it is a 0.2% improvement over the improved attention mechanic [17] based on Hourglass, which is also a stack of 8 modules. The results of Yang et al. [6] are basically the same as ours, but comparing the model parameters, our model has fewer parameters. In particular, for the most challenging joints, our method achieves 1.0% and 1.0% improvements on wrist and ankle compared with 8-stack Hourglass. Examples of quantitative results are shown in Fig. 5(a).

LSP Dataset. We add MPII training data to the LSP dataset and its extended training dataset in a similar way as before [4,17]. Table 2 summarizes the PCK

Table 1. Comparison MPII dataset results (PCKh@0.5 score) and model parameters

Method	Head	Sho.	Elb.	Wri.	Hip	Knee	Ank.	Mean	Para.(M)
Carreira et al. [3]	95.7	91.7	81.7	72.4	82.8	73.2	66.4	81.3	–
Tompson et al. [28]	96.1	91.9	83.9	77.8	80.9	72.3	64.8	82.0	–
Gkioxary et al. [29]	96.2	93.1	86.7	82.1	85.2	81.4	74.1	86.1	–
Wei et al. [4]	97.8	95.0	88.7	84.0	88.4	82.8	79.4	88.5	–
Bulat et al. [30]	97.9	95.1	89.9	85.3	89.4	85.7	81.7	89.7	–
Newell et al. [5]	98.2	96.3	91.2	87.1	90.1	87.4	83.6	90.9	23.7
Chu et al. [17]	98.5	96.3	91.9	88.1	90.6	88.0	85.0	91.5	>23.7
Yang et al. [6]	98.4	96.5	91.9	**88.2**	**91.1**	**88.6**	**85.3**	**91.8**	26.9
Ours	**98.5**	**96.7**	**92.1**	88.1	90.9	88.2	84.6	91.7	**21.6**

Table 2. Comparison LSP dataset results (PCK@0.2 score) and model parameters

Method	Head	Sho.	Elb.	Wri.	Hip	Knee	Ank.	Mean	Para.(M)
Belagiannis et al. [31]	95.2	89.0	81.5	77.0	83.7	87.0	82.8	85.2	–
Lifshitz et al. [32]	96.8	89.0	82.7	79.1	90.9	86.0	82.5	86.7	–
Wei et al. [4]	97.8	92.5	87.0	83.9	91.5	90.8	89.9	90.5	–
Bulat et al. [30]	97.2	92.1	88.1	85.2	92.2	91.4	88.7	90.7	–
Chu et al. [17]	98.1	93.7	89.3	86.9	93.4	94.0	92.5	92.6	>23.7
Ours	**98.1**	**94.6**	**91.5**	**89.0**	**94.2**	**94.4**	**93.2**	**93.6**	**21.6**

scores at threshold of 0.2 with PC (Person-Centric) annotations. Our method increased by 1.0% compared to the previous method, and in particular, there was a 2.2% and 2.1% increase in elbow and wrist. In Fig. 6, we show that our method performs significantly better than before when the normalized distance is greater than 0.08. Examples of predicted pose are demonstrated in Fig. 5(b).

4.3 Ablation Study

In order to study whether the method mentioned in Sect. 3 is really effective for pose estimation. We conducted a series of experiments on the MPII validation set [28]. First, we use 1-stack Hourglass as our baseline for comparison. We need to analyze each part of our approach, including: Adaptive Heatmaps, the Limb Region (structural information), and Multi-scale Adaptive Structure (1-stack MASN). We use the most challenging joints and the overall score as the basis for evaluation to compare with 1-stack Hourglass score of 86.2% at PCKh@0.5, as shown in Fig. 7(a).

Adaptive Heatmaps. In order to explore whether the size of the heatmap affect the model effect, we first evaluate the adaptive heatmaps method. By comparing

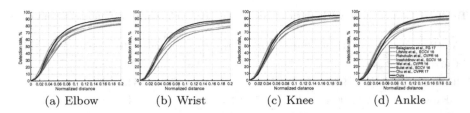

(a) Elbow   (b) Wrist   (c) Knee   (d) Ankle

Fig. 6. PCK curves on the LSP dataset on some challenging body joints.

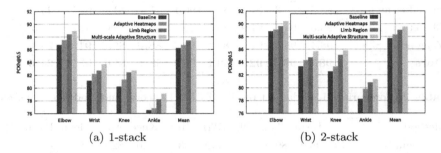

(a) 1-stack   (b) 2-stack

Fig. 7. Ablation Study. Comparison PCKh@0.5 score on MPII validation set. **(a)**: 1-stack Hourglass as baseline. **(b)**: 2-stack Hourglass as baseline.

with baseline, we got a PCKh@0.5 score of 86.7% with a 0.5% improvement. The most significant increase in knees is 1.1%.

Limb Region. The Limb Region, as an expression of the human body structure, performs very well in our experiments. Compared to baseline, this method achieves a PCKh@0.5 score of 87.4%, with significant effects at several challenging joints.

Multi-scale Adaptive Structure. We supervise the limb region at $\frac{1}{4}$, $\frac{1}{2}$, and original scale. Due to the inconsistency of the human body size in the image, we adopt an adaptive approach for the limb region at each scale and adaptive heatmaps for joints. At the same time, according to the method in Sect. 3, the loss function is normalized. In the end, we get a 87.9% PCKh@0.5 score.

1-stack Hourglass vs. 1-stack MASN. We obtain a PCKh@0.5 score of 87.9% by stacking one MASN, which is a 1.7% increase over 1-stack Hourglass. We visualize some of the results and found that it has a significant effect on the illumination, occlusion and other complex situations.

The comparison of 2-stack. Similarly, we use 2-stack Hourglass as our baseline for comparison. Analyzing Adaptive Heatmaps, the Limb Region (structural information), and Multi-scale Adaptive Structure (1-stack MASN), as shown in Fig. 7(b). Same as 1-stack, each part of our approach has a better score than baseline, and 2-stack MASN achieves a PCKh@0.5 score of 89.5%, a 1.8% increase over 2-stack Hourglass.

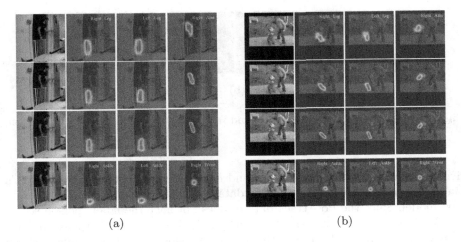

(a)   (b)

Fig. 8. Two examples of qualitative analysis. 1st row to 3rd row: $\frac{1}{4}$, $\frac{1}{2}$, and original scale of adaptive limb region. 4th row:adaptive heatmaps of joints. (**a**): Large body size, illumination and occlusion. (**b**): Small body size, occlusion.

4.4   Discussion

Some of the visualized examples, as shown in Fig. 5, illustrate that our method has significant results for occlusion, illumination, crowding, complex backgrounds, rare pose, and so on. In this section, we first conduct a specific analysis about the performance of MASN.

Qualitative Analysis. The MASN module detects the adaptive limb region at $\frac{1}{4}$, $\frac{1}{2}$, and the original scale, and then detects the adaptive heatmaps of joints, as shown in Fig. 8. The limb region in small-scale can provide context information for the limb region refinement in large-scale. The limb region fully learns the structure information of the human body, which plays a key role in the prediction of invisible joints under the conditions of occlusion, illumination and so on. Moreover, the self-adaptive method of the limb regions and the heatmaps proposed by us is also crucial for different sizes of human body joints. It can help to predict appropriate results for human bodies of different sizes, thereby improving accuracy.

Model Parameters. In Sects. 4.2 and 4.3, we specifically illustrate the performance advantages of our approach over Hourglass, and we also compare the number of parameters. The number of parameters for stacking 4 MASN modules is 21.6M, and the number of parameters for stacking 8 Hourglass modules is 23.7M, as shown in Table 1. We improve performance while reducing the amount of parameters.

PoseTrack Dataset [34]. Our approach may fail under some extreme conditions, as shown in Fig. 9. For example, too abnormal human pose, heavy occlusion, and abnormal human body. Further analysis, we find that there are rela-

656   W. Zhuang et al.

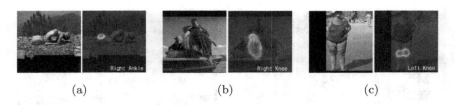

(a)   (b)   (c)

Fig. 9. Failure caused by (a) occlusion and rare pose, (b) overlapping and occlusion, (c) abnormal body (overweight).

tively few such data in the dataset, and our model has not yet perfected this kind of situation. So we use a larger dataset, PoseTrack dataset [34], to train ours model. Since the PoseTrack dataset does not have the label "person scale" like MPII dataset, we get the "person scale" directly from the marked joint points. Due to the limitations of the device, our model is not pre-trained on the COCO dataset [35] like other methods [36], and does not take advantage of the relationship between frames, but directly train on the labeled images. Although we don't get result as good as [36], the focus of our future work will be here, that is, combines our methods with human detection, end-to-end multi-person pose estimation, and the ability to use the continuous information of the video to estimate a smooth human body.

5 Conclusions

In this paper, we discuss the importance of variant sizes of the joint heatmaps for human pose estimation. We propose a novel method to address this issue by adapting the kernel size of joint heatmaps to the human scale of the input images in the training stage. A normalization strategy of how to perform the adaption is deduced. Besides, we introduce a novel limb region representation to learn the human pose structural information. Both the adaptive joint heatmaps as well as the limb region representation are combined together to construct a novel neural network, which is named **Multi-scale Adaptive Structure Network (MASN)**. We demonstrate that our method could obtain the state-of-the-art human pose estimation results and outperform the most existing methods.

Acknowledgment. This work is supported by the National Natural Science Foundation of China (No. 61806054, 6170320, 61728103 and 61671151), Jiangsu Province Science Foundation for Youths (No. BK20180355 and BK20170812) and Foundation of Southeast University (No. 3208008410 and 1108007121).

References

1. Toshev, A., Szegedy, C.: Deeppose: human pose estimation via deep neural networks. In: Proceedings of the IEEE Conference on Computer Vision and Pattern Recognition, pp. 1653–1660 (2014)
2. Pfister, T., Charles, J., Zisserman, A.: Flowing convnets for human pose estimation in videos. In: Proceedings of the IEEE International Conference on Computer Vision, pp. 1913–1921 (2015)
3. Carreira, J., Agrawal, P., Fragkiadaki, K., Malik, J.: Human pose estimation with iterative error feedback. In: Proceedings of the IEEE Conference on Computer Vision and Pattern Recognition, pp. 4733–4742 (2016)
4. Wei, S.E., Ramakrishna, V., Kanade, T., Sheikh, Y.: Convolutional pose machines. In: Proceedings of the IEEE Conference on Computer Vision and Pattern Recognition, pp. 4724–4732 (2016)
5. Newell, A., Yang, K., Deng, J.: Stacked hourglass networks for human pose estimation. In: Leibe, B., Matas, J., Sebe, N., Welling, M. (eds.) ECCV 2016. LNCS, vol. 9912, pp. 483–499. Springer, Cham (2016). https://doi.org/10.1007/978-3-319-46484-8_29
6. Yang, W., Li, S., Ouyang, W., Li, H., Wang, X.: Learning feature pyramids for human pose estimation. In: The IEEE International Conference on Computer Vision (ICCV), vol. 2 (2017)
7. Cao, Z., Simon, T., Wei, S.E., Sheikh, Y.: Realtime multi-person 2D pose estimation using part affinity fields. In: CVPR, vol. 1, p. 7 (2017)
8. Ke, L., Chang, M.C., Qi, H., Lyu, S.: Multi-scale structure-aware network for human pose estimation. arXiv preprint arXiv:1803.09894 (2018)
9. Andriluka, M., Pishchulin, L., Gehler, P., Schiele, B.: 2D human pose estimation: new benchmark and state of the art analysis. In: Proceedings of the IEEE Conference on Computer Vision and Pattern Recognition, pp. 3686–3693 (2014)
10. Felzenszwalb, P.F., Huttenlocher, D.P.: Pictorial structures for object recognition. Int. J. Comput. Vis. **61**, 55–79 (2005)
11. Ramanan, D.: Learning to parse images of articulated bodies. In: Advances in Neural Information Processing Systems, pp. 1129–1136 (2007)
12. Buehler, P., Everingham, M., Huttenlocher, D.P., Zisserman, A.: Long term arm and hand tracking for continuous sign language TV broadcasts. In: Proceedings of the 19th British Machine Vision Conference, pp. 1105–1114. BMVA Press (2008)
13. Andriluka, M., Roth, S., Schiele, B.: Pictorial structures revisited: people detection and articulated pose estimation. In: IEEE Conference on Computer Vision and Pattern Recognition, 2009. CVPR 2009, pp. 1014–1021. IEEE (2009)
14. Pishchulin, L., Andriluka, M., Gehler, P., Schiele, B.: Strong appearance and expressive spatial models for human pose estimation. In: 2013 IEEE International Conference on Computer Vision (ICCV), pp. 3487–3494. IEEE (2013)
15. Taylor, G.W., Fergus, R., Williams, G., Spiro, I., Bregler, C.: Pose-sensitive embedding by nonlinear NCA regression. In: Advances in Neural Information Processing Systems, pp. 2280–2288 (2010)
16. Chen, X., Yuille, A.L.: Articulated pose estimation by a graphical model with image dependent pairwise relations. In: Advances in Neural Information Processing Systems, pp. 1736–1744 (2014)
17. Chu, X., Yang, W., Ouyang, W., Ma, C., Yuille, A.L., Wang, X.: Multi-context attention for human pose estimation. arXiv preprint arXiv:1702.07432, vol. 1 (2017)

18. Chu, X., Ouyang, W., Li, H., Wang, X.: Structured feature learning for pose estimation. In: Proceedings of the IEEE Conference on Computer Vision and Pattern Recognition, pp. 4715–4723 (2016)
19. Lin, T.Y., Dollár, P., Girshick, R., He, K., Hariharan, B., Belongie, S.: Feature pyramid networks for object detection. In: CVPR, vol. 1, p. 4 (2017)
20. Gidaris, S., Komodakis, N.: Object detection via a multi-region and semantic segmentation-aware CNN model. In: Proceedings of the IEEE International Conference on Computer Vision, pp. 1134–1142 (2015)
21. Girshick, R., Iandola, F., Darrell, T., Malik, J.: Deformable part models are convolutional neural networks. In: Proceedings of the IEEE Conference on Computer Vision and Pattern Recognition, pp. 437–446 (2015)
22. Chen, Y., Wang, Z., Peng, Y., Zhang, Z., Yu, G., Sun, J.: Cascaded pyramid network for multi-person pose estimation. arXiv preprint arXiv:1711.07319 (2017)
23. Sun, K., Lan, C., Xing, J., Zeng, W., Liu, D., Wang, J.: Human pose estimation using global and local normalization. arXiv preprint arXiv:1709.07220 (2017)
24. Yu, F., Koltun, V.: Multi-scale context aggregation by dilated convolutions. arXiv preprint arXiv:1511.07122 (2015)
25. Yang, Z., Hu, Z., Salakhutdinov, R., Berg-Kirkpatrick, T.: Improved variational autoencoders for text modeling using dilated convolutions. arXiv preprint arXiv:1702.08139 (2017)
26. Johnson, S., Everingham, M.: Clustered pose and nonlinear appearance models for human pose estimation (2010)
27. Tieleman, T., Hinton, G.: Lecture 6.5-rmsprop: divide the gradient by a running average of its recent magnitude. COURSERA: Neural Netw. Mach. Learn. **4**, 26–31 (2012)
28. Tompson, J., Goroshin, R., Jain, A., LeCun, Y., Bregler, C.: Efficient object localization using convolutional networks. In: Proceedings of the IEEE Conference on Computer Vision and Pattern Recognition, pp. 648–656 (2015)
29. Gkioxari, G., Toshev, A., Jaitly, N.: Chained predictions using convolutional neural networks. In: Leibe, B., Matas, J., Sebe, N., Welling, M. (eds.) ECCV 2016. LNCS, vol. 9908, pp. 728–743. Springer, Cham (2016). https://doi.org/10.1007/978-3-319-46493-0_44
30. Bulat, A., Tzimiropoulos, G.: Human pose estimation via convolutional part heatmap regression. In: Leibe, B., Matas, J., Sebe, N., Welling, M. (eds.) ECCV 2016. LNCS, vol. 9911, pp. 717–732. Springer, Cham (2016). https://doi.org/10.1007/978-3-319-46478-7_44
31. Belagiannis, V., Zisserman, A.: Recurrent human pose estimation. In: 2017 12th IEEE International Conference on Automatic Face & Gesture Recognition (FG 2017), pp. 468–475. IEEE (2017)
32. Lifshitz, I., Fetaya, E., Ullman, S.: Human pose estimation using deep consensus voting. In: Leibe, B., Matas, J., Sebe, N., Welling, M. (eds.) ECCV 2016. LNCS, vol. 9906, pp. 246–260. Springer, Cham (2016). https://doi.org/10.1007/978-3-319-46475-6_16
33. Yang, Y., Ramanan, D.: Articulated human detection with flexible mixtures of parts. IEEE Trans. Pattern Anal. Mach. Intell. **35**, 2878–2890 (2013)
34. Andriluka, M., et al.: A benchmark for human pose estimation and tracking. In: Proceedings of the IEEE Conference on Computer Vision and Pattern Recognition, pp. 5167–5176 (2018)
35. Lin, T.-Y., et al.: Microsoft coco: common objects in context. In: Fleet, D., Pajdla, T., Schiele, B., Tuytelaars, T. (eds.) ECCV 2014. LNCS, vol. 8693, pp. 740–755. Springer, Cham (2014). https://doi.org/10.1007/978-3-319-10602-1_48
36. Xiao, B., Wu, H., Wei, Y.: Simple baselines for human pose estimation and tracking. arXiv preprint arXiv:1804.06208 (2018)

Full Explicit Consistency Constraints in Uncalibrated Multiple Homography Estimation

Wojciech Chojnacki[ID] and Zygmunt L. Szpak[(✉)][ID]

Australian Institute for Machine Learning, The University of Adelaide,
Adelaide, SA 5005, Australia
{wojciech.chojnacki,zygmunt.szpak}@adelaide.edu.au

Abstract. We reveal a complete set of constraints that need to be imposed on a set of 3×3 matrices to ensure that the matrices represent genuine homographies associated with multiple planes between two views. We also show how to exploit the constraints to obtain more accurate estimates of homography matrices between two views. Our study resolves a long-standing research question and provides a fresh perspective and a more in-depth understanding of the multiple homography estimation task.

Keywords: Multiple homographies · Consistency constraints ·
Latent variables · Parameter estimation · Scale invariance ·
Maximum likelihood

1 Introduction

Two images of the same planar surface in space are related by a homography—a transformation which can be described, to within a scale factor, by an invertible 3×3 matrix. This basic fact is what makes estimating a *single* homography from image measurements one of the primary tasks in computer vision. Three-dimensional reconstruction, mosaicing, camera calibration, and metric rectification are examples of the applications making use of a single homography [19]. A recent addition to this list is the problem of color transfer [14,17]. Various methods for estimating a single homography are available [19] and new techniques emerge on a regular basis [3,18,27,31,41].

A task closely related to estimating a single homography is that of estimating *multiple* homographies. Multiple planar surfaces are ubiquitous in urban environments, and, as a result, estimating multiple homographies between two views from image measurements is an important step in many applications such as non-rigid motion detection [21,40], enhanced image warping [16], multiview 3D reconstruction [22], augmented reality [30], indoor navigation [32], multi-camera calibration [37], camera-projector calibration [28], or ground-plane recognition

This research was supported by the Australian Research Council.

© Springer Nature Switzerland AG 2019
C. V. Jawahar et al. (Eds.): ACCV 2018, LNCS 11361, pp. 659–675, 2019.
https://doi.org/10.1007/978-3-030-20887-5_41

for object detection and tracking [1]. Surprising as it may seem, a vast array of techniques for estimating multiple homographies, including many robust multi-structure estimation methods [5, 6, 12, 15, 20, 25, 29, 38, 39, 42] applicable to the task of estimating multiple homographies, are deficient in a fundamental way—they fail to recognise that a set of homography matrices does not represent a set of genuine homographies between two views of the same scene unless appropriate *consistency constraints* are satisfied. These constraints reflect the rigidity of the motion and the scene. If the constraints are not deliberately enforced, they do not hold in typical scenarios. Hence, one of the fundamental problems in estimating multiple homography matrices is to find a way to enforce the consistency constraints—a task reminiscent of that of enforcing the rank-two constraint in the case of the fundamental matrix estimation [19, Sect. 11.1.1].

Being unable to specify explicit formulae for all relevant constraints, various researchers have managed over the years to identify and enforce various reduced sets of constraints. As pioneers in this regard, Shashua and Avidan [33] found that homography matrices induced by four or more planes in a 3D scene appearing in two views span a four-dimensional linear subspace. Chen and Suter [4] derived a set of strengthened constraints for the case of three or more homographies in two views. Zelnik-Manor and Irani [40] have shown that another rank-four constraint applies to a set of so-called relative homographies generated by two planes in four or more views. These latter authors also derived constraints for larger sets of homographies and views. Finally, in recent work [36] Szpak et al. introduced what they dubbed the multiplicity and singularity constraints that apply to two or more, and three or more, homographies between two views, respectively.

Once isolated, the available constraints are typically put to use in a procedure whereby first individual homography matrices are estimated from image data, and then the resulting estimates are upgraded to matrices satisfying the constraints. Following this pattern, Shashua and Avidan as well as Zelnik-Manor and Irani used low-rank approximation under the Frobenius norm to enforce the rank-four constraint. Chen and Suter enforced their set of constraints also via low-rank approximation, but then employed the Mahalanobis norm with covariances of the input homographies. All of these estimation procedures produce matrices that satisfy only incomplete constraints so their true consistency cannot be guaranteed.

Without knowledge of *explicit* formulae for all of the constraints, it is still possible to *implicitly* enforce full consistency by exploiting a natural parametrisation of the family of all fixed-size sets of compatible homography matrices (see Sect. 2). Following this path, Chojnacki et al. [10, 11] employed this parametrisation and a distinct cost function to develop an upgrade procedure based on unconstrained optimisation. Szpak et al. [35] used the same parametrisation and the Sampson distance to develop an alternative estimation technique with a sound statistical basis. The parameters encoding compatible homographies constitute the *latent variables* in the model explaining the dependencies between the homographies involved. While the use of latent variables guarantees the enforcement of all of the underlying consistency constraints, it also has some notable drawbacks. Specifically, the latent variable based method does not provide a means to directly *measure* the extent to which a collection of homography matrices are compatible. Furthermore, finding suitable initial values for the latent variables is a non-trivial task. The initialisation methods utilised by Chojnacki et al. [10, 11] and Szpak et al. [35] are based on factorising a collection of homography matrices. The factorisation procedure is described in detail in [11, Sect. 6.2] and summarised in [11, Algorithm 1]. It involves a series of algebraic manipulations and a singular value decomposition. Each of these steps is sensitive to noise, and so when some of the given homographies have substantial uncertainty, the resulting initial latent variables will correspond to compatible homographies with high reprojection error. The high reprojection means that the subsequent optimisation process could converge to a sub-optimal local minimum. This predicament is explained and illustrated in Fig. 4 of [35].

In this paper we exhibit a full set of explicit constraints for multiple homographies between two views. This constitutes a theoretical contribution and also has practical ramifications. We use the deduced set of constraints to define a quantifiable measure to assess the extent to which separately estimated homographies are mutually incompatible. Based on this measure, we demonstrate experimentally that unless the consistency constraints are explicitly enforced, estimates of multiple homographies cannot be treated as *bona fide* homographies between two views. The palpable advantage of our constrained homography estimation procedure is evident in Fig. 1. By imposing consistency constraints, one improves not only the accuracy of the homographies but also ensures that any derived quantities, e.g. camera projection matrices, will be more accurate.

2 Path to Constraints

As already pointed out in the introduction, when estimating a set of homographies associated with multiple planes from image correspondences between two views, one must recognise that the homographies involved are interdependent. To reveal the nature of the underlying dependencies, consider two fixed uncalibrated cameras giving rise to two camera matrices $\mathbf{P}_1 = \mathbf{K}_1\mathbf{R}_1[\mathbf{I}_3, -\mathbf{t}_1]$ and $\mathbf{P}_2 = \mathbf{K}_2\mathbf{R}_2[\mathbf{I}_3, -\mathbf{t}_2]$. Here, the length-3 translation vector $\mathbf{t}_k$ and the 3×3 rotation matrix $\mathbf{R}_k$ represent the Euclidean transformation between the k-th ($k = 1, 2$) camera and the world coordinate system, $\mathbf{K}_k$ is a 3×3 upper triangular calibration matrix encoding the internal parameters of the k-th camera, and $\mathbf{I}_3$ denotes the 3×3 identity matrix. Suppose, moreover, that a set of I planes in a 3D scene have been selected. Given $i = 1, \ldots, I$, let the i-th plane from the collection have a unit outward normal $\mathbf{n}_i$ and be situated at a distance d_i from the origin of the world coordinate system. Then, for each $i = 1, \ldots, I$, the i-th plane gives rise to a planar homography between the first and second views described by the 3×3 matrix

$$\mathbf{H}_i = w_i\mathbf{A} + \mathbf{b}\mathbf{v}_i^\top, \tag{1}$$

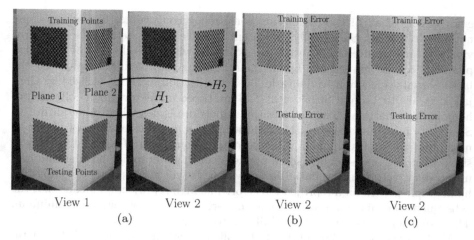

View 1   View 2   View 2   View 2
(a)   (b)   (c)

Fig. 1. Enforcing full consistency constraints improves the quality of homography estimates. Panel (a) illustrates red feature points on two planes which were used to estimate two homography matrices (H_1 and H_2). The quality of the estimated homographies can be evaluated by how accurately they map the green feature points into the second view. Panel (b) shows the result of mapping points in the first view to the second view using homographies estimated using the *gold standard* bundle-adjustment method which does not enforce homography consistency constraints. Panel (c) illustrates the result of mapping points in the first view to the second view using homographies estimated using bundle-adjustment while simultaneously enforcing the specific constraints proposed in this paper. Note how the gold standard method fails to map the green points associated with the second plane accurately; this is indicated by the red arrow. In contrast, our proposed solution produces a substantially more accurate result. (Color figure online)

where

$$\mathbf{A} = \mathbf{K}_2\mathbf{R}_2\mathbf{R}_1^{-1}\mathbf{K}_1^{-1}, \quad w_i = d_i - \mathbf{n}_i^\top\mathbf{t}_1,$$
$$\mathbf{b} = \mathbf{K}_2\mathbf{R}_2(\mathbf{t}_1 - \mathbf{t}_2), \quad \mathbf{v}_i = \mathbf{K}_1^{-\top}\mathbf{R}_1\mathbf{n}_i \tag{2}$$

(cf. [13,34]). In the case of calibrated cameras when one may assume that $\mathbf{K}_1 = \mathbf{K}_2 = \mathbf{I}_3$, $\mathbf{t}_1 = \mathbf{0}$, $\mathbf{R}_1 = \mathbf{I}_3$, $\mathbf{R}_2 = \mathbf{R}$, system (2) reduces to

$$\mathbf{A} = \mathbf{R}, \quad w_i = d_i,$$
$$\mathbf{b} = \mathbf{t}, \quad \mathbf{v}_i = \mathbf{n}_i, \tag{3}$$

with $\mathbf{t} = -\mathbf{R}\mathbf{t}_2$, and equality (1) becomes the familiar *direct nRt* representation $\mathbf{H}_i = d_i\mathbf{R} + \mathbf{t}\mathbf{n}_i^\top$ (cf. [2], [24, Sect. 5.3.1]). We stress that all of our subsequent analysis concerns the general uncalibrated case, with $\mathbf{A}$, $\mathbf{b}$, w_i's and $\mathbf{v}_i$'s to be interpreted according to (2) rather than (3).

An entity naturally associated with the matrices $\mathbf{H}_i$ is the $3 \times 3I$ (horizontal) concatenation matrix $\mathbf{H}_{\mathrm{cat}} = [\mathbf{H}_1, \dots, \mathbf{H}_I]$. With vec denoting column-wise vectorisation [23], if we let

$$\boldsymbol{\eta} = [(\mathrm{vec}\,\mathbf{A})^\top, \mathbf{b}^\top, \mathbf{v}_1^\top, \dots, \mathbf{v}_I^\top, w_1, \dots, w_I]^\top$$

and

$$\boldsymbol{\Pi}(\boldsymbol{\eta}) = [\boldsymbol{\Pi}_1(\boldsymbol{\eta}), \dots, \boldsymbol{\Pi}_I(\boldsymbol{\eta})],$$

where

$$\boldsymbol{\Pi}_i(\boldsymbol{\eta}) = w_i \mathbf{A} + \mathbf{b} \mathbf{v}_i^\top \quad (i = 1, \dots, I), \tag{4}$$

then $\mathbf{H}_{\mathrm{cat}}$ can be written as

$$\mathbf{H}_{\mathrm{cat}} = \boldsymbol{\Pi}(\boldsymbol{\eta}).$$

Here $\boldsymbol{\eta}$ represents a vector of latent variables that link all the constituent matrices together and provide a natural parametrisation of the set $\mathcal{H}$ of all $\mathbf{H}_{\mathrm{cat}}$'s. Since $\boldsymbol{\eta}$ has a total of $4I + 12$ entries, the aggregate of all matrices of the form $\boldsymbol{\Pi}(\boldsymbol{\eta})$ has dimension no greater than $4I + 12$, with the relevant notion of dimension being here that of dimension of a semi-algebraic set [8,9]. By employing a rather subtle argument, one can calculate exactly the dimension of the set of all $\boldsymbol{\Pi}(\boldsymbol{\eta})$'s (which is the same as the dimension of $\mathcal{H}$ given that the set of all $\boldsymbol{\Pi}(\boldsymbol{\eta})$'s is identical with $\mathcal{H}$) and this turns out to be equal to $4I + 7$ [8,9]. The difference between the dimension of the set of all $\boldsymbol{\eta}$'s and the dimension of $\mathcal{H}$ is indicative of five degrees of the internal gauge freedom present in the parametrisation $\boldsymbol{\Pi}$; this occurrence will be crucially exploited in what follows—see Sect. 3. Since $4I + 7 < 9I\ (= 3 \times 3I)$ whenever $I \geq 2$, it follows that $\mathcal{H}$ is a proper subset of the set of all $3 \times 3I$ matrices for $I \geq 2$. It is now clear that the requirement that $\mathbf{H}_{\mathrm{cat}}$ take the form as per (4) whenever $I \geq 2$ can be seen as an implicit constraint on $\mathbf{H}_{\mathrm{cat}}$, with the consequence that the $\mathbf{H}_i$'s are all interdependent. This further begs the question as to how to turn the implicit constraint into a system of explicit constraints (not involving the latent variables) that has to be put upon a set of matrices $\mathbf{H}_i\ (i = 1, \dots, I)$ in order that the $\mathbf{H}_i$'s represent genuine homographies between two views. We shall subsequently answer this question in steps, with the first step being taken in the next section.

3   Problem

Let $\mathbb{R}$ denote the set of real numbers and let $\mathbb{R}^{m \times n}$ denote the set of $m \times n$ matrices with entries in $\mathbb{R}$. We formally formulate the main purpose of this paper as an answer to the following problem:

Problem 1. Given I invertible matrices $\mathbf{H}_1, \dots, \mathbf{H}_I \in \mathbb{R}^{3 \times 3}$, find a system of equations that the $\mathbf{H}_i$'s have to satisfy in order to be representable in the form

$$\mathbf{H}_i = w_i \mathbf{A} + \mathbf{b} \mathbf{v}_i^\top \quad (i = 1, \dots, I) \tag{5}$$

for some matrix $\mathbf{A} \in \mathbb{R}^{3 \times 3}$, some vectors $\mathbf{b}, \mathbf{v}_1, \dots, \mathbf{v}_I \in \mathbb{R}^3$, and some scalars $w_1, \dots, w_I \in \mathbb{R}$.

We start with two observations that will greatly facilitate solving the problem. First we note that, for each i, if $\mathbf{H}_i$ can be represented as $w_i\mathbf{A} + \mathbf{bv}_i^\top$, then necessarily $w_i \neq 0$. Indeed, if $w_i = 0$ held for some i, then $\mathbf{H}_i$ would be equal to $\mathbf{bv}_i^\top$ and hence would be of rank one, contravening the assumption that all the $\mathbf{H}_i$'s are invertible. Next we observe that if (5) holds for a set of $\mathbf{A}$, $\mathbf{b}$, $\mathbf{v}_i$'s, and w_i's, then it also holds for various other sets of $\mathbf{A}$, $\mathbf{b}$, $\mathbf{v}_i$'s, and w_i's. Indeed, if $\mathbf{H}_i = w_i\mathbf{A} + \mathbf{bv}_i^\top$ for each i, then also $\mathbf{H}_i = w_i'\mathbf{A}' + \mathbf{b}'\mathbf{v}_i'^\top$ for each i, where $\mathbf{A}' = \beta\mathbf{A} + \mathbf{bc}^\top$, $\mathbf{b}' = \alpha\mathbf{b}$, $\mathbf{v}_i' = \alpha^{-1}\mathbf{v}_i - \alpha^{-1}\beta^{-1}w_i\mathbf{c}$, and $w_i' = \beta^{-1}w_i$, with α and β being non-zero scalars and $\mathbf{c}$ being a length-3 vector. We now exploit this last observation by letting $\alpha = 1$, $\beta = w_1$, and $\mathbf{c} = \mathbf{v}_1$; critically, by our first observation, β is non-zero. Then $\mathbf{A}'$ becomes $\mathbf{H}_1$ and we further have $\mathbf{H}_i = w_i'\mathbf{H}_1 + \mathbf{bv}_i'^\top$ with $w_i' = w_1^{-1}w_i$ and $\mathbf{v}_i' = \mathbf{v}_i - w_1^{-1}w_i\mathbf{v}_1$ for each i. In light of this, we see that Problem 1 can equivalently be restated as follows:

Problem 2. Given I invertible matrices $\mathbf{H}_1, \ldots, \mathbf{H}_I \in \mathbb{R}^{3\times3}$, find a system of equations that the $\mathbf{H}_i$'s have to satisfy in order that

$$\mathbf{H}_i = w_i\mathbf{H}_1 + \mathbf{bv}_i^\top \quad (i = 2, \ldots, I) \tag{6}$$

hold for some vectors $\mathbf{b}$, $\mathbf{v}_2, \ldots, \mathbf{v}_I \in \mathbb{R}^3$ and some scalars $w_2, \ldots, w_I \in \mathbb{R}$.

In what follows we reveal a solution to Problem 2. This will give us a sought-after set of explicit homography constraints.

4   Algebraic Prerequisites

To make the derivation of a constraint set more accessible, we start in this section with some necessary technical prerequisites.

4.1   The Characteristic Polynomial

Let $\mathbf{A}, \mathbf{B} \in \mathbb{R}^{3\times3}$. The *linear matrix pencil* of the matrix pair $(\mathbf{A}, \mathbf{B})$ is the matrix function $\lambda \mapsto \mathbf{A} - \lambda\mathbf{B}$. The *characteristic polynomial* of $(\mathbf{A}, \mathbf{B})$, $p_{\mathbf{A},\mathbf{B}}$, is defined by $p_{\mathbf{A},\mathbf{B}}(\lambda) = \det(\mathbf{A} - \lambda\mathbf{B})$. Adopting MATLAB's notation to let $\mathbf{M}_{:i}$ represent the ith column of the matrix $\mathbf{M}$, one verifies directly that $p_{\mathbf{A},\mathbf{B}}$ can be explicitly written as $p_{\mathbf{A},\mathbf{B}}(\lambda) = \sum_{n=0}^{3}(-1)^n c_n \lambda^n$, where

$$
\begin{aligned}
c_0 &= \det\mathbf{A}, \\
c_1 &= \det[\mathbf{B}_{:1}, \mathbf{A}_{:2}, \mathbf{A}_{:3}] + \det[\mathbf{A}_{:1}, \mathbf{B}_{:2}, \mathbf{A}_{:3}] + \det[\mathbf{A}_{:1}, \mathbf{A}_{:2}, \mathbf{B}_{:3}], \\
c_2 &= \det[\mathbf{A}_{:1}, \mathbf{B}_{:2}, \mathbf{B}_{:3}] + \det[\mathbf{B}_{:1}, \mathbf{A}_{:2}, \mathbf{B}_{:3}] + \det[\mathbf{B}_{:1}, \mathbf{B}_{:2}, \mathbf{A}_{:3}], \\
c_3 &= \det\mathbf{B}.
\end{aligned}
\tag{7}
$$

The characteristic polynomial arises in connection with the generalised eigenvalue problem

$$\mathbf{Ax} = \lambda\mathbf{Bx}. \tag{8}$$

As with the standard eigenvalue problem, eigenvalues for the problem (8) occur precisely where the matrix pencil $\lambda \to \mathbf{A} - \lambda\mathbf{B}$ is singular. In other words, the eigenvalues for the pair $(\mathbf{A}, \mathbf{B})$ are the roots of $p_{\mathbf{A},\mathbf{B}}$.

A fact that will be of significance in what follows is that if the generalised eigenvalue problem (8) has a double eigenvalue, then this eigenvalue is a double root of $p_{\mathbf{A},\mathbf{B}}$. For the sake of completeness, we recall the argument presented in [36] which validates this fact and correct a misprint that has slipped into the original proof.

Suppose that the generalised eigenvalue problem (8) has a double eigenvalue μ, which means that there exist linearly independent length-3 vectors $\mathbf{v}_1$ and $\mathbf{v}_2$ such that $\mathbf{A}\mathbf{v}_i = \mu\mathbf{B}\mathbf{v}_i$ for $i = 1, 2$. With a view to showing that μ is a double root of $p_{\mathbf{A},\mathbf{B}}$, select arbitrarily a length-3 vector $\mathbf{v}_3$ that does not belong to the linear span of $\mathbf{v}_1$ and $\mathbf{v}_2$; for example, we may assume that $\mathbf{v}_3 = \mathbf{v}_1 \times \mathbf{v}_2$. Then $\mathbf{v}_1, \mathbf{v}_2$, and $\mathbf{v}_3$ form a basis for $\mathbb{R}^3$, and hence the matrix $\mathbf{S} = [\mathbf{v}_1, \mathbf{v}_2, \mathbf{v}_3]$ is non-singular. Let $\tilde{\mathbf{A}} = \mathbf{S}^{-1}\mathbf{A}\mathbf{S}$ and $\tilde{\mathbf{B}} = \mathbf{S}^{-1}\mathbf{B}\mathbf{S}$. For $i = 1, 2, 3$, let $\mathbf{j}_i$ denote the i-th standard unit vector in $\mathbb{R}^3$, with 1 in the i-th position and 0 in all others. Then, clearly, $\mathbf{v}_i = \mathbf{S}\mathbf{j}_i$ for $i = 1, 2, 3$. It is immediate that, for $i = 1, 2$, $\tilde{\mathbf{A}}\mathbf{j}_i = \mu\tilde{\mathbf{B}}\mathbf{j}_i$ and so $(\tilde{\mathbf{A}} - \lambda\tilde{\mathbf{B}})\mathbf{j}_i = (\mu - \lambda)\tilde{\mathbf{B}}\mathbf{j}_i$. Hence the pencil $\tilde{\mathbf{A}} - \lambda\tilde{\mathbf{B}}$ takes the form

$$\tilde{\mathbf{A}} - \lambda\tilde{\mathbf{B}} = \begin{bmatrix} (\mu - \lambda)\tilde{b}_{11} & (\mu - \lambda)\tilde{b}_{12} & \tilde{a}_{13} - \lambda\tilde{b}_{13} \\ (\mu - \lambda)\tilde{b}_{21} & (\mu - \lambda)\tilde{b}_{22} & \tilde{a}_{23} - \lambda\tilde{b}_{23} \\ (\mu - \lambda)\tilde{b}_{31} & (\mu - \lambda)\tilde{b}_{32} & \tilde{a}_{33} - \lambda\tilde{b}_{33} \end{bmatrix}$$

and we have

$$p_{\tilde{\mathbf{A}},\tilde{\mathbf{B}}}(\lambda) = \begin{vmatrix} (\mu - \lambda)\tilde{b}_{11} & (\mu - \lambda)\tilde{b}_{12} & \tilde{a}_{13} - \lambda\tilde{b}_{13} \\ (\mu - \lambda)\tilde{b}_{21} & (\mu - \lambda)\tilde{b}_{22} & \tilde{a}_{23} - \lambda\tilde{b}_{23} \\ (\mu - \lambda)\tilde{b}_{31} & (\mu - \lambda)\tilde{b}_{32} & \tilde{a}_{33} - \lambda\tilde{b}_{33} \end{vmatrix} = (\mu - \lambda)^2 \begin{vmatrix} \tilde{b}_{11} & \tilde{b}_{12} & \tilde{a}_{13} - \lambda\tilde{b}_{13} \\ \tilde{b}_{21} & \tilde{b}_{22} & \tilde{a}_{23} - \lambda\tilde{b}_{23} \\ \tilde{b}_{31} & \tilde{b}_{32} & \tilde{a}_{33} - \lambda\tilde{b}_{33} \end{vmatrix},$$

which shows that μ is a double root of $p_{\tilde{\mathbf{A}},\tilde{\mathbf{B}}}$. But $p_{\tilde{\mathbf{A}},\tilde{\mathbf{B}}}$ coincides with $p_{\mathbf{A},\mathbf{B}}$, given that

$$\begin{aligned} p_{\tilde{\mathbf{A}},\tilde{\mathbf{B}}}(\lambda) &= \det(\mathbf{S}^{-1}(\mathbf{A} - \lambda\mathbf{B}_3)\mathbf{S}) \\ &= \det\mathbf{S}^{-1} \det(\mathbf{A} - \lambda\mathbf{B}_3) \det\mathbf{S} \\ &= (\det\mathbf{S})^{-1} \det(\mathbf{A} - \lambda\mathbf{B}_3) \det\mathbf{S} = p_{\mathbf{A},\mathbf{B}}(\lambda). \end{aligned}$$

Therefore μ is *a fortiori* a double root of $p_{\mathbf{A},\mathbf{B}}$.

4.2 A Double Root of the Characteristic Polynomial of a Cubic Polynomial

Let $\mathbf{A}$ and $\mathbf{B}$ be two 3×3 matrices such that $p_{\mathbf{A},\mathbf{B}}$ has a double root which is not a triple root. Then, as it turns out, the root is uniquely determined and is given by an explicit formula. This is a consequence of a more general result that we present next.

Let $p(\lambda) = a\lambda^3 + b\lambda^2 + c\lambda + d$ be a cubic polynomial with $a \neq 0$. Suppose that μ is a double root of p,

$$p(\mu) = p'(\mu) = 0, \qquad (9)$$

but not a triple root, $p''(\mu) \neq 0$; we shall term such a double root non-degenerate. Then equations (9) can explicitly be written as

$$a\mu^3 + b\mu^2 + c\mu + d = 0, \qquad (10)$$
$$3a\mu^2 + 2b\mu + c = 0. \qquad (11)$$

When we multiply the first of these equations by 3 and the second by μ and next subtract the second equation from the first, we get

$$b\mu^2 + 2c\mu + 3d = 0. \qquad (12)$$

Restating (10) and (12) as

$$3a\mu^2 + 2b\mu = -c,$$
$$b\mu^2 + 2c\mu = -3d,$$

we obtain a system of linear equations in μ^2 and μ. Solving for μ and μ^2 gives

$$\mu = \frac{9ad - bc}{2(b^2 - 3ac)} \quad \text{and} \quad \mu^2 = \frac{c^2 - 3bd}{b^2 - 3ac}. \qquad (13)$$

Here $b^2 \neq 3ac$ for otherwise Eq. (11) would have its quadratic discriminant $4(b^2 - 3ac)$ equal to zero, with the consequence that μ would be a repeated root for p' and hence a triple root for p. Now, the first equation in (13) provides a formula for a non-degenerate double root of a cubic polynomial. We see in particular that if a cubic polynomial has a non-degenerate double root, then this root is uniquely determined.

In light of the above discussion it is clear that if $p_{\mathbf{A},\mathbf{B}}$ has a non-degenerate double root, then the root is unique, and when we denote this root by $\mu_{\mathbf{A},\mathbf{B}}$, we have $\mu_{\mathbf{A},\mathbf{B}} = \omega_{\mathbf{A},\mathbf{B}}$, where, with the notation from (7),

$$\omega_{\mathbf{A},\mathbf{B}} = \frac{c_1 c_2 - 9c_0 c_3}{2(c_2^2 - 3c_1 c_3)}. \qquad (14)$$

5   Full Constraints

Here we finally present a solution to Problem 2 (and hence also to Problem 1).

Let $\mathbf{H}_1, \ldots, \mathbf{H}_I \in \mathbb{R}^{3 \times 3}$ be such that (6) holds for some $\mathbf{b}, \mathbf{v}_2, \ldots, \mathbf{v}_I \in \mathbb{R}^3$ and $w_2, \ldots, w_I \in \mathbb{R}$. Fix $i \in \{2, \ldots, I\}$ arbitrarily. If $\mathbf{c}$ is a length-3 vector orthogonal to $\mathbf{v}_i$, then

$$\mathbf{H}_i \mathbf{c} = w_i \mathbf{H}_1 \mathbf{c} + \mathbf{b}\mathbf{v}_i^\top \mathbf{c} = w_i \mathbf{H}_1 \mathbf{c},$$

showing that $(w_i, \mathbf{c})$ is an eigenpair for the pair $(\mathbf{H}_i, \mathbf{H}_1)$. Since length-3 vectors orthogonal to $\mathbf{v}_i$ form a two-dimensional linear space, it follows that w_i is in fact a double eigenvalue for $(\mathbf{H}_i, \mathbf{H}_1)$. Using the material from Sect. 4 and assuming that all double roots of intervening characteristic polynomials are non-degenerate (which is generically true), we conclude that, for each $i = 2, \ldots, I$, w_i is uniquely defined, namely $w_i = \omega_{\mathbf{H}_i, \mathbf{H}_1}$ (recall the definition given in (14)). For each $i = 2, \ldots, I$, let

$$\mathbf{J}_i = \mathbf{H}_i - \omega_{\mathbf{H}_i, \mathbf{H}_1} \mathbf{H}_1, \qquad (15)$$

and let

$$\mathbf{J} = [\mathbf{J}_2, \ldots, \mathbf{J}_I]. \qquad (16)$$

In view of (6), $\mathbf{J}_i = \mathbf{b}\mathbf{v}_i^\top$ for each $i = 2, \ldots, I$. Hence, letting $\mathbf{w} = [\mathbf{v}_2^\top, \ldots, \mathbf{v}_I^\top]^\top$, we have

$$\mathbf{J} = [\mathbf{b}\mathbf{v}_2^\top, \ldots, \mathbf{b}\mathbf{v}_I^\top] = \mathbf{b}[\mathbf{v}_2^\top, \ldots, \mathbf{v}_I^\top] = \mathbf{b}\mathbf{w}^\top,$$

which implies that $\mathbf{J}$ has rank one.

Conversely, if $\mathbf{J}$ has rank one, then $\mathbf{J} = \mathbf{b}\mathbf{w}^\top$ for some length-3 vector $\mathbf{b}$ and some length-$3(I-1)$ vector $\mathbf{w}$ which, as any vector of this length, can be represented as $\mathbf{w} = [\mathbf{v}_2^\top, \ldots, \mathbf{v}_I^\top]^\top$ for some length-3 vectors $\mathbf{v}_2, \ldots, \mathbf{v}_I$. This, in conjunction with the definitions (15) and (16), leads to $\mathbf{H}_i = \omega_{\mathbf{H}_i, \mathbf{H}_1} \mathbf{H}_1 + \mathbf{b}\mathbf{v}_i^\top$ for each $i = 2, \ldots, I$, which is a representation of the form required in Problem 2.

In light of the above, we see that Problem 2 reduces to finding the requirement in algebraic form that $\mathbf{J}$ have rank one. As is well known, the relevant condition is that all 2×2 minors of $\mathbf{J}$ should vanish [26, Sect. V.2.2, Theorem 3]. To express this condition explicitly, we introduce some notation. Given an $m \times n$ matrix $\mathbf{A}$ and positive integers $a_1, \ldots, a_k$ with $1 \leq a_1 < a_2 < \ldots < a_{k-1} < a_k \leq m$ and positive integers $b_1, \ldots, b_l$ with $1 \leq b_1 < b_2 < \ldots < b_{l-1} < b_l \leq n$, let $\mathbf{A}(a_1, a_2, \ldots, a_{k-1}, a_k; b_1, b_2, \ldots, b_{l-1}, b_l)$ denote the submatrix of $\mathbf{A}$ contained in the rows indexed by $a_1, a_2, \ldots, a_{k-1}, a_k$ and the columns indexed by $b_1, b_2, \ldots, b_{l-1}, b_l$. With this notation, the condition that all the 2×2 minors of $\mathbf{J}$ should vanish can be stated as

$$\det \mathbf{J}(a, b; c, d) = 0 \quad (a, b \in \{1, 2, 3\}, a < b; c, d \in \{1, \ldots, 3I-3\}, c < d). \qquad (17)$$

It is directly verified that if $\lambda_1, \ldots \lambda_I$ are non-zero scalars, then

$$\mathbf{J}_i(\lambda_1 \mathbf{H}_1, \lambda_i \mathbf{H}_i) = \lambda_i \mathbf{J}_i(\mathbf{H}_1, \mathbf{H}_i)$$

for each $i = 2, \ldots, I$. This implies that

$$[\det \mathbf{J}(a, b; c, d)](\lambda_1 \mathbf{H}_1, \ldots, \lambda_I \mathbf{H}_I) = \lambda_{i_c} \lambda_{i_d}[\det \mathbf{J}(a, b; c, d)](\mathbf{H}_1, \ldots, \mathbf{H}_I), \qquad (18)$$

where the indices i_c and i_d are such that the c-th column of $\mathbf{J}$ belongs to $\mathbf{J}_{i_c}$ and the d-th column of $\mathbf{J}$ belongs to $\mathbf{J}_{i_d}$ ($i_c, i_d \in \{2, \ldots, I\}$), respectively. The above identity reveals that the vanishing of $[\det \mathbf{J}(a, b; c, d)](\lambda_1 \mathbf{H}_1, \ldots, \lambda_i \mathbf{H}_I)$ is equivalent to the vanishing of $[\det \mathbf{J}(a, b; c, d)](\mathbf{H}_1, \ldots, \mathbf{H}_I)$. Thus equations in (17) are genuine constraints on the homographies represented by the matrices $\mathbf{H}_i$.

Another consequence of (18) is that, being scale dependent, the functions $(\mathbf{H}_1, \ldots, \mathbf{H}_I) \mapsto [\det \mathbf{J}(a, b; c, d)](\mathbf{H}_1, \ldots, \mathbf{H}_I)$ cannot be directly used as building blocks for a measure qualifying the extent to which members of a given set of I homographies are mutually incompatible. Instead, these functions have to be replaced by their scale-invariant counterparts given by

$$\phi_{abcd}(\mathbf{H}_1, \ldots, \mathbf{H}_I) = \|\mathbf{H}_{i_c}\|_{\mathrm{F}}^{-1} \|\mathbf{H}_{i_d}\|_{\mathrm{F}}^{-1} [\det \mathbf{J}(a, b; c, d)](\mathbf{H}_1, \ldots, \mathbf{H}_I). \qquad (19)$$

Here, for a given matrix $\mathbf{A}$, $\|\mathbf{A}\|_{\mathrm{F}}$ denotes the Frobenius norm of $\mathbf{A}$. Strictly speaking, the functions ϕ_{abcd} are positive scale independent—their sign may still change with a change of the scales of the homography matrices. However, the squares of these functions are genuinely scale invariant. With this in mind, a natural measure for assessing the amount of incompatibility amongst a set of I homographies can be defined by the expression

$$\psi := \sum_{\substack{a, b \in \{1,2,3\}, a<b; \\ c, d \in \{1, \ldots, 3I-3\}, c<d}} \phi_{abcd}^2. \qquad (20)$$

It is obvious that the constraints given in (17), or, equivalently, all the constraints of the form $\phi_{abcd} = 0$, are satisfied if and only if

$$\psi = 0. \qquad (21)$$

6   Maximum Likelihood Estimation

Let $\{\{\tilde{\mathbf{m}}_{ij}, \tilde{\mathbf{m}}'_{ij}\}_{j=1}^{J_i}\}_{i=1}^I$ be a collection of I sets of pairs of corresponding inhomogeneous points in two images, arising from I planar surfaces in the 3D scene. Suppose that homography estimates $\hat{\mathbf{H}}_1, \ldots, \hat{\mathbf{H}}_I$ are to be evolved based on $\{\{\tilde{\mathbf{m}}_{ij}, \tilde{\mathbf{m}}'_{ij}\}_{j=1}^{J_i}\}_{i=1}^I$ in such a way that the aggregate $\hat{\mathbf{H}}_1, \ldots, \hat{\mathbf{H}}_I$ satisfies constraints (17). One statistically meaningful approach to this estimation problem involves the *maximum likelihood* cost function, also called the *reprojection error* cost function,

$$J_{\mathrm{ML}}(\{\mathbf{H}_i\}_{i=1}^I, \{\{\underline{\tilde{\mathbf{m}}}_{ij}\}_{j=1}^{J_i}\}_{i=1}^I)$$
$$= \sum_{i=1}^I \sum_{j=1}^{J_i} \left(\left\| \tilde{\mathbf{m}}_{ij} - \underline{\tilde{\mathbf{m}}}_{ij} \right\|^2 + \left\| \tilde{\mathbf{m}}'_{ij} - \mathrm{hom}^{-1}\left(\mathbf{H}_i \, \mathrm{hom}\left(\underline{\tilde{\mathbf{m}}}_{ij} \right) \right) \right\|^2 \right) (22)$$

that has for its collective argument the homography matrices $\mathbf{H}_i$ and the corrections $\underline{\tilde{\mathbf{m}}}_{ij}$ of the points $\tilde{\mathbf{m}}_{ij}$ in the first image [19, Sect. 4.5]. Here, hom and hom^{-1} denote the operators of homogenisation and dehomogenisation given by

$$\mathrm{hom}\,(\tilde{\mathbf{x}}) = [x_1, x_2, 1]^{\top} \quad \text{for } \tilde{\mathbf{x}} = [x_1, x_2]^{\top}$$

and

$$\text{hom}^{-1}\left(\mathbf{y}\right)=\left[y_1/y_3,y_2/y_3\right]^{\top}\quad\text{for }\mathbf{y}=\left[y_1,y_2,y_3\right]^{\top},$$

respectively; these operators convert between the Cartesian and homogeneous coordinate representation of a given 2D point. Minimisation of J_{ML} subject to the constraints $\phi_{abcd}=0$ (recall the definition given in (19)) yields estimates $\widehat{\mathbf{H}}_i$ and $\widehat{\underline{\mathbf{m}}}_{ij}$. The $\widehat{\underline{\mathbf{m}}}_{ij}$ may be discarded and then the remaining $\widehat{\mathbf{H}}_i$ constitute the *gold standard* maximum likelihood homography estimates. We remark that, with a view to easing implementation, an alternative optimisation approach can be adopted—and, in fact, we use this approach in our experiments—whereby J_{ML} is minimised subject to constraints (17) and the additional constraints $\|\mathbf{H}_i\|_{\text{F}}=1$ $(i=2,\ldots,I)$.

7   Experiments

We investigated the stability and accuracy of our method by conducting experiments on both synthetic and real data. We compared our results with the gold standard bundle-adjustment method which does not enforce homography constraints [19, Sect. 4.5], as well as bundle-adjustment which imposes all constraints implicitly using the parametrisation outlined in [35]. For all of our experiments, we ensured that there were no mismatched corresponding points.

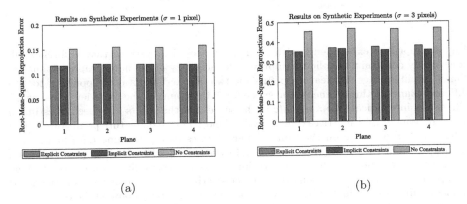

(a)   (b)

Fig. 2. Experiments with simulated data evidence that enforcing full consistency constraints explicitly or implicitly improves the accuracy of the homography estimates. For each experimental trial we generated a random scene with four planar surfaces and sampled 50 corresponding points within arbitrarily sized rectangular regions. We subsequently added zero-mean isotropic Gaussian noise to the correspondences. The accuracy of the estimates was evaluated by computing the root-mean-square reprojection error. Panels (a) and (b) show the results for Gaussian noise with a standard deviation of one and three pixels, respectively. The results are based on a thousand trials. (Color figure online)

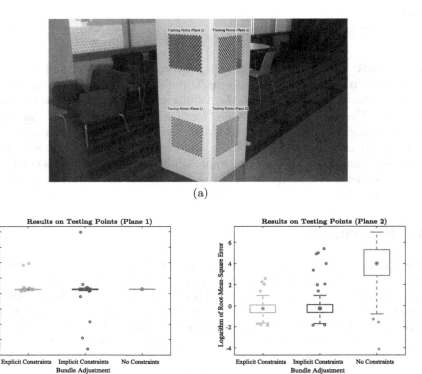

(a)

(b)   (c)

Fig. 3. Repeated experiments with realistic images demonstrate that enforcing full consistency constraints explicitly improves the accuracy of the homography estimates. Panel (a) illustrates red feature points on two checkerboards corresponding to two different planes in three-dimensional space (only the points in the first view are shown). The red points indicate *training data* in the sense that the points were used to estimate a pair of homography matrices (one for each plane). The green feature points on the two other checkerboards lie on the same planar surfaces as the training data and served as *testing data*. Note that the training data associated with the first plane spanned the entire checkerboard, whereas the training data corresponding to the second plane occupied only a small 4 × 4 square on the checkerboard. The quality of the estimated homographies was evaluated by analysing how accurately they mapped the testing data (green points) into the second view. We conducted numerous experiments in which we fixed the training data for the first plane, and exhaustively varied the small 4 × 4 square from which the training data on the second plane were sampled. Panels (b) and (c) show the logarithm of the root-mean-square reprojection error for the testing data in the first and second plane respectively. Because of the abundance of training data the estimators produced almost identical results for the first plane. However, when operating with fewer training data points, bundle adjustment without constraints produced results for the second plane which are orders of magnitude worse than variants of bundle adjustment which enforced explicit or implicit constraints. (Color figure online)

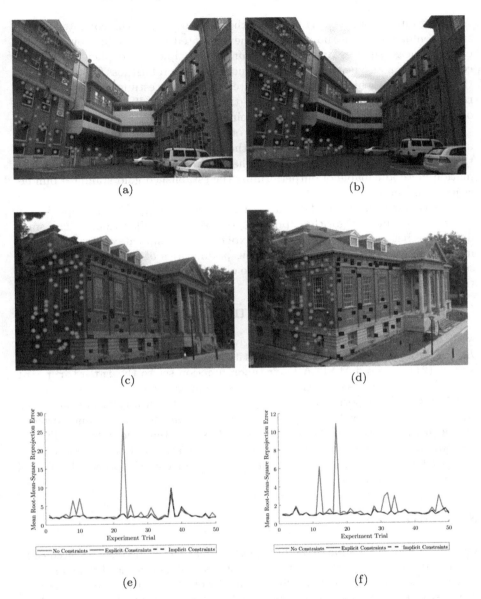

Fig. 4. Repeated experiments with images taken from the AdelaideRMF dataset [39]. Panels (a)–(d) illustrate manually matched SIFT feature points on two planar surfaces of the *nese* and *library* buildings. For each building, we conducted 50 experiments in which we estimated a pair of homographies from ten randomly selected corresponding points. We quantified the quality of the homography estimates by evaluating the reprojection errors on the remaining set of corresponding points. Panels (e) and (f) illustrate the average root-mean-square reprojection error for the feature points in the *nese* and *library* buildings, respectively. The results of the implicit and explicit constraint enforcement algorithms are indistinguishable, and both are superior to standard bundle-adjustment. (Color figure online)

Avoiding outliers allowed us to assess the contribution of the consistency constraint enforcement on the quality of the estimated homographies by using the canonical least-squares reprojection error. In principle, our explicit consistency constraints can also be enforced in conjunction with a robust loss function such as the Huber norm which can accommodate outliers. We estimated initial homography matrices using the direct linear transform and all estimation methods operated on Hartley-normalised data points [7].

Details on the design and outcome of our experiments with simulated and authentic image data are presented in the captions of Figs. 2, 3, and 4, respectively. The results demonstrate that we have formulated a new homography estimation method capable of outperforming the established gold standard bundle-adjustment method.

The conclusions on simulated data suggest that the explicit and implicit constraint enforcement algorithms produce, on average, similar results. The small differences between the performance of explicit versus implicit constraint enforcement in Fig. 2b can be attributed to the peculiarities of different optimisation schemes and the non-linear nature of the objective function. The objective function with implicit constraints

$$(\boldsymbol{\eta}, \{\{\underline{\tilde{\mathbf{m}}}_{ij}\}_{j=1}^{J_i}\}_{i=1}^{I}) \mapsto J_{\mathrm{ML}}(\{\boldsymbol{\Pi}_i(\boldsymbol{\eta})\}_{i=1}^{I}, \{\{\underline{\tilde{\mathbf{m}}}_{ij}\}_{j=1}^{J_i}\}_{i=1}^{I}),$$

with J_{ML} given in (22) and $\boldsymbol{\Pi}_i(\boldsymbol{\eta})$ given in (4), was optimised using the *Levenberg-Marquardt* algorithm. To optimise J_{ML} subject to the explicit constraint (21), we used MATLAB's `fmincon` function set to the `interior point` algorithm. The results on authentic images are in agreement with the simulated conclusions. The experiments with real data stress the utility of imposing constraints when very few feature points are observed on one of the planar surfaces. The logarithmic scale of the y-axis in Fig. 3c shows that the homography corresponding to the second planar surface was estimated with superior accuracy. The results presented in Figs. 4e and f further underscore the practical utility of the proposed algorithm.

8   Conclusion

Our paper addressed a long-standing question that has evaded the research community. We have identified a complete set of constraints that need to be imposed on a set of homography matrices linking images of planar surfaces between a pair of views to ensure consistency between all the matrices. Furthermore, we have demonstrated how the constraints can be incorporated into a non-linear constrained optimisation method. Our experiments with simulated and real images illustrated the benefits of imposing constraints in practical scenarios.

References

1. Arrospide, J., Salgado, L., Nieto, M., Mohedano, R.: Homography-based ground plane detection using a single on-board camera. IET Intell. Transp. Syst. **4**(2), 149–160 (2010)
2. Baker, S., Datta, A., Kanade, T.: Parameterizing homographies. Technical report. CMU-RI-TR-06-11, Robotics Institute, Carnegie Mellon University, Pittsburgh, PA (2006)
3. Barath, D., Hajder, L.: A theory of point-wise homography estimation. Pattern Recognit. Lett. **94**, 7–14 (2017)
4. Chen, P., Suter, D.: Rank constraints for homographies over two views: revisiting the rank four constraint. Int. J. Comput. Vis. **81**(2), 205–225 (2009)
5. Chin, T.J., Wang, H., Suter, D.: The ordered residual kernel for robust motion subspace clustering. Adv. Neural Inf. Process. Syst. **22**, 333–341 (2009)
6. Chin, T.J., Wang, H., Suter, D.: Robust fitting of multiple structures: the statistical learning approach. In: Proceedings of 12th International Conference on Computer Vision, pp. 413–420 (2009)
7. Chojnacki, W., Brooks, M.J., van den Hengel, A., Gawley, D.: Revisiting Hartley's normalized eight-point algorithm. IEEE Trans. Pattern Anal. Mach. Intell. **25**(9), 1172–1177 (2003)
8. Chojnacki, W., van den Hengel, A.: A dimensionality result for multiple homography matrices. In: Proceedings of 13th International Conference on Computer Vision, pp. 2104–2109 (2011)
9. Chojnacki, W., van den Hengel, A.: On the dimension of the set of two-view multi-homography matrices. Complex Anal. Oper. Theory **7**(2), 465–484 (2013)
10. Chojnacki, W., Szpak, Z., Brooks, M.J., van den Hengel, A.: Multiple homography estimation with full consistency constraints. In: Proceedings of International Conference on Digital Image Computing: Techniques and Applications, pp. 480–485 (2010)
11. Chojnacki, W., Szpak, Z.L., Brooks, M.J., van den Hengel, A.: Enforcing consistency constraints in uncalibrated multiple homography estimation using latent variables. Mach. Vis. Appl. **26**(2), 401–422 (2015)
12. Decrouez, M., Dupont, R., Gaspard, F., Crowley, J.L.: Extracting planar structures efficiently with revisited BetaSAC. In: Proceedings of 21st International Conference on Pattern Recognition, pp. 2100–2103 (2012)
13. Faugeras, O.D., Lustman, F.: Motion and structure from motion in a piecewise planar environment. Int. J. Pattern Recognit. Artif. Intell. **2**(3), 485–508 (1988)
14. Finlayson, G.D., Gong, H., Fisher, R.: Color homography: theory and applications. IEEE Trans. Pattern Anal. Mach. Intell. **41**(1), 20–33 (2019)
15. Fouhey, D.F., Scharstein, D., Briggs, A.J.: Multiple plane detection in image pairs using J-linkage. In: Proceedings of 20th International Conference on Pattern Recognition, pp. 336–339 (2010)
16. Gao, J., Kim, S.J., Brown, M.S.: Constructing image panoramas using dual-homography warping. In: Proceedings of IEEE Conference on Computer Vision and Pattern Recognition, pp. 49–56 (2011)
17. Gong, H., Finlayson, G.D., Fisher, R.B.: Recoding color transfer as a color homography. In: Proceedings of 27th British Machine Vision Conference (2016)
18. Guo, J., Cai, S., Wu, Z., Liu, Y.: A versatile homography computation method based on two real points. Image Vis. Comput. **64**, 23–33 (2017)

19. Hartley, R.I., Zisserman, A.: Multiple View Geometry in Computer Vision, 2nd edn. Cambridge University Press, Cambridge (2004)
20. Isack, H., Boykov, Y.: Energy-based geometric multi-model fitting. Int. J. Comput. Vis. **97**(2), 123–147 (2012)
21. Kähler, O., Denzler, J.: Rigid motion constraints for tracking planar objects. In: Hamprecht, F.A., Schnörr, C., Jähne, B. (eds.) DAGM 2007. LNCS, vol. 4713, pp. 102–111. Springer, Heidelberg (2007). https://doi.org/10.1007/978-3-540-74936-3_11
22. Karami, M., Afrouzian, R., Kasaei, S., Seyedarabi, H.: Multiview 3D reconstruction based on vanishing points and homography. In: Proceedings of 7th International Symposium on Telecommunications, pp. 367–370 (2014)
23. Lütkepol, H.: Handbook of Matrices. Wiley, Chichester (1996)
24. Ma, Y., Soatto, S., Košecká, J., Sastry, S.: An Invitation to 3-D Vision: From Images to Geometric Models, 2nd edn. Springer, New York (2005). https://doi.org/10.1007/978-0-387-21779-6
25. Mittal, S., Anand, S., Meer, P.: Generalized projection-based M-estimator. IEEE Trans. Pattern Anal. Mach. Intell. **34**(12), 2351–2364 (2012)
26. Mostowski, A., Stark, M.: Introduction to Higher Algebra. Pergamon Press/PWN-Polish Scientific Publishers, Oxford/Warszawa (1964)
27. Osuna-Enciso, V., Cuevas, E., Oliva, D., Zúñiga, V., Pérez-Cisneros, M., Zaldívar, D.: A multiobjective approach to homography estimation. Comput. Intell. Neurosci. **2016**, 1–12 (2016)
28. Park, S.Y., Park, G.G.: Active calibration of camera-projector systems based on planar homography. In: Proceedings of 20th International Conference on Pattern Recognition, pp. 320–323 (2010)
29. Pham, T.T., Chin, T.J., Yu, J., Suter, D.: The random cluster model for robust geometric fitting. IEEE Trans. Pattern Anal. Mach. Intell. **36**, 1658–1671 (2014)
30. Prince, S.J.D., Xu, K., Cheok, A.D.: Augmented reality camera tracking with homographies. IEEE Comput. Graph. Appl. **22**(6), 39–45 (2002)
31. Qi, N., Zhang, S., Cao, L., Yang, X., Li, C., He, C.: Fast and robust homography estimation method with algebraic outlier rejection. IET Image Process. **12**(4), 552–562 (2018)
32. Rodrigo, R., Zouqi, M., Chen, Z., Samarabandu, J.: Robust and efficient feature tracking for indoor navigation. IEEE Trans. Syst. Man Cybern.—Part B **39**(3), 658–671 (2009)
33. Shashua, A., Avidan, S.: The rank 4 constraint in multiple ($\geq$ 3) view geometry. In: Buxton, B., Cipolla, R. (eds.) ECCV 1996. LNCS, vol. 1065, pp. 196–206. Springer, Heidelberg (1996). https://doi.org/10.1007/3-540-61123-1_139
34. Szpak, Z.L.: Constrained parameter estimation in multiple view geometry. Ph.D. thesis, School of Computer Science, University of Adelaide (2013). http://hdl.handle.net/2440/82702
35. Szpak, Z.L., Chojnacki, W., Eriksson, A., van den Hengel, A.: Sampson distance based joint estimation of multiple homographies with uncalibrated cameras. Comput. Vis. Image Underst. **125**, 200–213 (2014)
36. Szpak, Z.L., Chojnacki, W., van den Hengel, A.: Robust multiple homography estimation: an ill-solved problem. In: Proceedings of the IEEE Conference on Computer Vision and Pattern Recognition, pp. 2132–2141 (2015)
37. Ueshiba, T., Tomita, F.: Plane-based calibration algorithm for multi-camera systems via factorization of homography matrices. In: Proceedings of 9th International Conference Computer Vision, vol. 2, pp. 966–973 (2003)

38. Wang, H., Chin, T.J., Suter, D.: Simultaneously fitting and segmenting multiple-structure data with outliers. IEEE Trans. Pattern Anal. Mach. Intell. **34**(6), 1177–1192 (2012)
39. Wong, H.S., Chin, T.J., Yu, J., Suter, D.: Dynamic and hierarchical multi-structure geometric model fitting. In: Proceedings of 13th International Conference on Computer Vision, pp. 1044–1051 (2011)
40. Zelnik-Manor, L., Irani, M.: Multiview constraints on homographies. IEEE Trans. Pattern Anal. Mach. Intell. **24**(2), 214–223 (2002)
41. Zhao, C., Zhao, H.: Accurate and robust feature-based homography estimation using half-sift and feature localization error weighting. J. Vis. Commun. Image R **40**, 288–299 (2016)
42. Zuliani, M., Kenney, C.S., Manjunath, B.S.: The multiRANSAC algorithm and its application to detect planar homographies. In: Proceedings of International Conference on Image Processing. **3**, 153–156 (2005)

A New Method for Computing
the Principal Point of an Optical Sensor
by Means of Sphere Images

Rudi Penne[1]([✉]), Bart Ribbens[1], and Steven Puttemans[2]

[1] Faculty of Applied Engineering, University of Antwerp,
Groenenborgerlaan 171, 2020 Antwerpen, Belgium
rudi.penne@uantwerpen.be
[2] EAVISE Research Group, KU Leuven,
Jan Pieter De Nayerlaan 5, 2860 Sint-Katelijne-Waver, Belgium
steven.puttemans@kuleuven.be

Abstract. For some applications it can be preferable to use images of
spheres in order to calibrate a 2D camera. All published sphere-based
algorithms need the complete knowledge of the elliptic sphere image,
i.e. 5 geometric parameters, in particular the ellipse orientation. Because
sphere images tend to be close to circular shapes, this orientation is often
very noise-sensitive. For example, it is common to compute the principal
point as the intersection of the lines through the major axes of the elliptic
images, but this procedure is quite unstable. We present a new method
for computing the principal point by means of three sphere images, without making use of the ellipse orientation. By mean of simulations and real
experiments we demonstrate that the proposed method is more accurate
and stable in finding the principal point as compared to sphere-based
calibration algorithms that use the complete ellipse geometry.

Keywords: 2D camera · Sphere-based calibration · Principal point

1 Introduction

A spherical object is easy to recognize in an image, and its silhouette is visible independent of the angle of view from which the image has been taken.
This has motivated various authors to use images of one or more spheres in
camera calibration, both intrinsic and extrinsic (e.g. [2,6,8,10–15,18]). In [14]
the authors claim that their sphere image based calibration improves the 2%
accuracy tollerance of the standard method of Zhang [19].

In a perfect pinhole model the image of a sphere is an ellipse [6], that is
only equal to a circle in a frontal view with the focal axis throught the sphere
center. Furthermore, if we assume square pixels, then the principal point of the
sensor can be proven to be collineair with the major axis of this elliptic image
(Proposition 1, [6,13]). Many intrinsic calibration algorithms based on sphere

© Springer Nature Switzerland AG 2019
C. V. Jawahar et al. (Eds.): ACCV 2018, LNCS 11361, pp. 676–690, 2019.
https://doi.org/10.1007/978-3-030-20887-5_42

images rely on this property for determining the principal point of the sensor as the intersection of the lines that carry the major axes of two sphere images [2,6,11,15,18]. Even for sensors with an aspect ratio not equal to one and with significant radial distortion, this *major-axes-method* is used to obtain an initial value for the principal point in a recursive optimization [14].

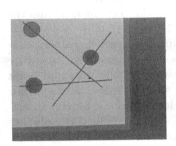 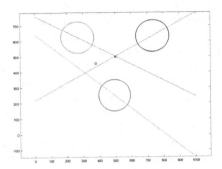

Fig. 1. Left, a real image with three spheres, and the lines through the major axes of the corresponding ellipses. In a perfect pinhole model, the principal point (small circle) should belong to each of these lines. Right, a synthetic discretized image of three spheres with principal point at $(500, 500)$, causing inaccuracies for the major axes of the fitting ellipses, even when no radial distortion or Gaussian noise has been added.

In Fig. 1 we show a real image of three spheres (biljard balls) and a synthetic image. For the real image, the ground truth for the principal point has been provided by means of Bouguet's toolbox [4] (indicated by a small circle). The lines extending the major axes of the elliptic sphere images should all contain the principal point. The observed inaccuracies are caused by sensor noise, but especially by nonlinear lens distortion, as demonstrated in Sect. 5. The error is most apparent for the major axis of the sphere image that is closest to the image center. Indeed, in this case the elliptic sphere image is almost a circle, causing the computation of the ellipse orientation to become strongly unstable, although many efficient ellipse detectors are available [7,9,16,17]. Another interesting approach to sphere-based camera calibration computes the image of the absolute conic (IAC), or its dual, by means of the equations of the elliptic sphere images, and obtains the calibration matrix K by means of Cholevsky decomposition (e.g. [6,10,15,18]). We will refer to these algorithms as IAC-methods. Although the IAC-methods do not seem to use the orientation of the major axes, they certainly need the complete symmetric matrix for each ellipse in the image. Consequently, these authors implicitly use the ellipse orientation, and we will show that this may cause serious inaccuracies for the computed principal point in "circular situations" for the used sphere images. Furthermore, in most cases the IAC-methods crash when the sphere images overlap (spheres occluding each other), because this may cause the matrix of the IAC to be not positive definite.

In this paper we propose an alternative method for determining the principal point from given sphere images that does not need the orientation of the major ellipse axes. Our method only uses the ellipse center $M = (M_x, M_y)$ and the sizes a and b of the half major axis and minor axis. These four parameters can be retrieved sufficiently correct for acquired sphere images, even when partially occluded [7]. By several simulations (Sect. 5) and real experiments (Sect. 6) we validate the accuracy and robustness of the proposed method, even in situations where the sphere images are small or close to the image center, whence $a \approx b$. We show that the proposed method outperforms the major-axes-method and the IAC-methods in these circular situations.

This paper does not present a complete procedure for intrinsic camera calibration based on sphere images, but it provides a sound argument to substitute the major-axes-method for computing (an intial value of) the principal point (wherever applied) by the proposed method. This method is also very suitable to recover the IAC-method in case they fail in circular situations. Indeed, inserting correct values for the principal point in the IAC-matrix reduces the number of unknowns and enables to restrict to the more appropriate sphere images (e.g., see Appendix B of [15]).

In Sect. 2 we briefly explain the geometric foundation that is needed to understand the relation between the shape of an elliptic sphere image and the intrinsic parameters of a pinhole camera with square pixels. The objective of that section is to present Eq. 1 in Proposition 2, a formula that has appeared in [3] and in [13]. This formula involves the principal point as well as the focal length of the camera, using the center M and the dimensions a and b of an elliptic sphere image. The key observation of this paper is presented in Sect. 3, where we perform a simple but novel trick for eliminating the focal length when Eq. 1 is applied to two sphere images. We observe that two sphere images yield a circular locus for the principal point in the same spirit as the major axis of one sphere image, while avoiding the unstable computation of the orientation of circle-like ellipses. In Sect. 4 we conclude the theoretical elaboration by describing the proposed algorithm that calibrates for the principal point independent of the focal length, by means of three sphere images. Finally, synthetic and real experiments validate the accuracy and stability of the proposed method (Sects. 5 and 6).

2   How the Perspective Image of a Sphere Is Related to the Camera Intrinsics

The following proposition appeared in [6] and has been proven in [13].

Proposition 1. *Let the ellipse $\mathcal{E}$ be the perspective image of a sphere, projected from a center O on a plane Π. Let C be the orthogonal projection of O on the projection plane Π. Then C is collinear with the major axis of $\mathcal{E}$.*

Notice that this is a Euclidean property. The concept of 'major axis of an ellipse' is even not an affine invariant. Therefore, in order to apply Proposition 1 for a pinhole camera, the elliptic sphere image must be captured with respect to (undistorted) square pixels.

Theoretically, Proposition 1 can be used to detect the principal point for an optical sensor with square pixels. As a matter of fact this is done in most intrinsic calibration algorithms based on sphere images [2,6,11,15,18]. Even for sensors with an aspect ratio not equal to one and with significant radial distortion, this *major-axes-method* is used to obtain an initial value for the principal point in a recursive optimization [14]. In order to apply this major-axes-method we need to carefully segment the elliptic boundaries of at least two ball images, and intersect (the lines through) their major axes. However, in spite of the existence of many efficient ellipse detectors, e.g. in [7,9,16,17], the detection of the orientation of the major ellipse axis is unstable, especially in case of a small difference between the lengths of the major and the minor axes (Fig. 1). On the other hand, the ellipse center $M = (M_x, M_y)$, and to a certain extent also the half axes a and b, can be retrieved sufficiently correct for acquired sphere images, even when partially occluded (e.g. by [7]). Consequently, our method for calibrating the principal point (Sect. 4) will be supported by the ellipse parameters (M_x, M_y, a, b) only, not using the ellipse orientation.

Our method for computing the principal point is based on Eq. 1, provided in Proposition 2, This formula appeared in [3], and has been proven in full detail in [13] using the Theorem of Dandelin-Quételet [5].

Proposition 2. *Assume an elliptic image $\mathcal{E}$ of a ball by a perfect pinhole camera with square pixels. Let δ_M denotes the distance between the principal point C and the center M of $\mathcal{E}$, let a denote the half major axis and b denote the half minor axis of $\mathcal{E}$ (all in pixel units). Then we can relate the focal length f of the camera to a, b and δ_M by means of the following formula:*

$$f^2(a^2 - b^2) = b^2(\delta_M^2 - a^2 + b^2) \tag{1}$$

Equation 1 can be used to compute the focal length f from the knowledge of (or a good guess for) the location of the principal point. Notice that the obtained value for f is not reliable if a is not sufficiently larger than b (less than 3% according to [13]).

Suppose we are given the image of a sphere $\mathcal{E}$ (with half major axes $a > b$). Motivated by Eq. 1, we define a circle $\gamma(f)$ in the sensor plane centered at the center M of the ellipse $\mathcal{E}$, with radius

$$R(f) = \frac{\sqrt{(a^2 - b^2)(f^2 + b^2)}}{b} \tag{2}$$

With fixed M, a and b, this radius $R(f)$ can be considered as a monotone increasing function in f. As an immediate consequence of Proposition 2 we can state:

Corollary 1. *If f equals the focal length of a pinhole camera with square pixels that captures the sphere image $\mathcal{E}$, then the circle $\gamma(f)$ with center M and radius $R(f)$ contains the principal point C.*

3 The Locus of the Principal Point Determined by Two Sphere Images

Recall that we assume a sensor with square pixels and with removed non-linear lens distortions. Assume moreover the availability of two elliptic ball images $\mathcal{E}_1, \mathcal{E}_2$, with centers $M_i = (p_i, q_i)$ and half major axes $a_i > b_i$ ($i = 1, 2$). Notice that we can assume that $a_1 \neq b_1$ and that $a_2 \neq b_2$, because otherwise one of the sphere images would be a perfect circle, implying that the focal axis would contain the sphere center (frontal view) and that the principal point could be recovered as the centre of this circular sphere image. For each possible value of the focal length f we can define two circles $\gamma_1(f)$ and $\gamma_2(f)$ associated with these respective elliptic images, centered around M_1 and M_2, and with radii given by Eq. 2. According to Corollary 1 the principal point must be a point of intersection of these two circles if f equals the exact focal length (Fig. 2).

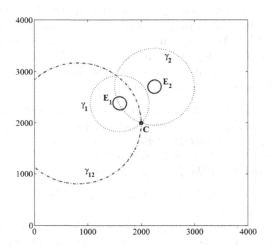

Fig. 2. Two sphere images and a fixed value for f always determine two circles γ_1 and γ_2. If f equals the correct focal length then the principal point C is a point of intersection of these circles.

$$\gamma_1(f) \leftrightarrow x^2 + y^2 - 2p_1x - 2q_1y = \frac{(a_1^2 - b_1^2)(f^2 + b_1^2)}{b_1^2} - p_1^2 - q_1^2 \qquad (3)$$

$$\gamma_2(f) \leftrightarrow x^2 + y^2 - 2p_2x - 2q_2y = \frac{(a_2^2 - b_2^2)(f^2 + b_2^2)}{b_2^2} - p_2^2 - q_2^2 \qquad (4)$$

The key observation of this paper is that we can eliminate f in these equations by the combination (assuming $a_1 \neq b_1$ and that $a_2 \neq b_2$)

$$(3) \cdot b_1^2/(a_1^2 - b_1^2) - (4) \cdot b_2^2/(a_2^2 - b_2^2) \qquad (5)$$

yielding a circle γ_{12} that necessarily contains the principal point C (Fig. 2):

$$\gamma_{12} \leftrightarrow x^2 + y^2 - 2p_{12}x - 2q_{12}y + k_{12} = 0 \tag{6}$$

$$p_{12} = (\frac{b_1^2 p_1}{a_1^2 - b_1^2} - \frac{b_2^2 p_2}{a_2^2 - b_2^2})/w_{12} \tag{7}$$

$$q_{12} = (\frac{b_1^2 q_1}{a_1^2 - b_1^2} - \frac{b_2^2 q_2}{a_2^2 - b_2^2})/w_{12} \tag{8}$$

$$k_{12} = (\frac{b_1^2(p_1^2 + q_1^2)}{a_1^2 - b_1^2} - \frac{b_2^2(p_2^2 + q_2^2)}{a_2^2 - b_2^2} - b_1^2 + b_2^2)/w_{12} \tag{9}$$

$$w_{12} = \frac{b_1^2}{a_1^2 - b_1^2} - \frac{b_2^2}{a_2^2 - b_2^2} \tag{10}$$

The expressions for the coefficients of the equation of γ_{12} can be reformulated in a more stable way (where we put $r_1 = a_1/b_1$ and $r_2 = a_2/b_2$):

$$p_{12} = \frac{p_1(r_2^2 - 1) - p_2(r_1^2 - 1)}{r_2^2 - r_1^2} \tag{11}$$

$$q_{12} = \frac{q_1(r_2^2 - 1) - q_2(r_1^2 - 1)}{r_2^2 - r_1^2} \tag{12}$$

$$k_{12} = \frac{(p_1^2 + q_1^2)(r_2^2 - 1) - (p_2^2 + q_2^2)(r_1^2 - 1) + (b_2^2 - b_1^2)(r_1^2 - 1)(r_2^2 - 1)}{r_2^2 - r_1^2} \tag{13}$$

Remark About Singular Case $r_1 \approx r_2$: A special but common case for this singularity happens when the sphere images have almost identical shape, hence $a_1 \approx a_2$ and $b_1 \approx b_2$. But in this case, we can eliminate the focal length f in Eqs. 3 and 4 by a simple subtracting, yielding a line for γ_{12}:

$$2(p_1 - p_2)x + 2(q_1 - q_2)y - p_1^2 - q_1^2 + p_2^2 + q_2^2 = 0 \tag{14}$$

Another special case for the singularity $r_1 = r_2$ is when $r_1 \approx 1 \approx r_2$. We meet this case when the observed spheres have small images or appear close to the image center. Here we can take the limit in the expressions for p_{12}, q_{12} and k_{12}, yielding

$$p_{12} \approx p_1 - p_2 \tag{15}$$

$$q_{12} \approx q_1 - q_2 \tag{16}$$

$$k_{12} \approx p_1^2 + q_1^2 - p_2^2 - q_2^2 \tag{17}$$

4   Determination of the Principal Point by Three Sphere Images

Next, suppose that we have (at least) three elliptic ball images $\mathcal{E}_1, \mathcal{E}_2, \mathcal{E}_3, \ldots$ available, either in one image or in several images by the same camera (with

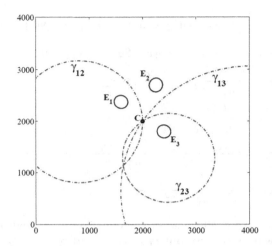

Fig. 3. For each pair out of three given sphere images we can consider a f-eliminating circle γ_{ij}. The circles γ_{12}, γ_{13} and γ_{23} have two points in common, one of which equals the principal point.

identical settings), still assuming square pixels. So, we know the center M_i and the half sizes of the major axes a_i and b_i $(a_i > b_i)$ for each of these ellipses $\mathcal{E}_i$.

Each pair $\{i, j\}$ of ellipses yields a locus for the principal point, namely the circle γ_{ij} given by Eq. 6. So the principal point can be located as the "most likely" of the two points of intersection of the circles γ_{12} and γ_{13} (Fig. 3). Notice that the third circle γ_{23} passes through the same points of intersections; hence it adds no new information. A suitable choice between both points of intersection might be guided by the minimal distance to the geometric midpoint of the image, or by another ball image $\mathcal{E}_4$, providing an additional locus γ_{14} (Fig. 4).

Remark. Since our method serves perfectly as a minimal solver using three sphere images within a RANSAC procedure that operates on a larger data set, the false second intersection will be identified as outlier.

Remark. Once the principal point C has been determined, the focal length f can be computed from Formula 1. However, as observed in [13], the computation of f by means of this formula might be unstable. Furthermore, this computation for f seems to be very sensitive to radial distortion. So, as suggested by the title, the contribution of this article is predominantly to propose a new method for principal point computation, that can be substituted in existing sphere-based calibration procedures [2,6,11,14,15,18]. Since the proposed method makes no use of the orientation of the major axis of the elliptic sphere image, we can expect that it is more robust against pixel noise and lens distortion. This expectation is validated in the next two sections.

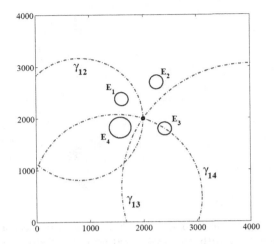

Fig. 4. We have plotted three out of six possible f-eliminating circles γ_{ij} in case of four given sphere images, The principal point is the unique point of intersection of these circles.

5   Synthetic Experiments

We simulated images of three spheres by positioning balls in the Euclidean 3-space and projecting them from the origin O on a plane parallel to the XY-plane ($z = f$). The z-axis coincides with the focal axis, such that the principal point coincides with the origin of the sensor plane $z = f$. The distance between O and the projection plane was chosen to realize a feasible angle of view. More precisely, with respect to the simulation unit, we considered spheres with a radius varying from 3 to 10, located at a depth equal to 200 (z-coordinate of the sphere center). The chosen focal length equals $f = 2$ and the position offset w.r.t. the focal axis varies from 5 to 105, such that the sphere images are situated in the square $[-1.5, 1.5] \times [-1.5, 1.5]$. The exact boundaries of the elliptic ball images have been disturbed by adding zero mean Gaussian noise to the (sub)pixel coordinates of these boundary points before being quantized to the simulated image grid. We used an increasing standard deviation up to $\sigma = 0.02$ (1% of the focal length). See Fig. 5 for an example of the image of two synthetic triples of spheres with radius 3, at distances 10 from the focal axis (images near the center), with Gaussian noise level $\sigma = 0.004$.

We will compare the proposed method with two alternative methods for obtaining the principal point. For each trial on a selected noise level, ellipses are fitted through three simulated ball images, that are needed in the three considered methods. For this ellipse fitting we chose the procedure of Fitzgibbon [7]. Providing the ellipse centers M_i, the size parameters a_i and b_i, and the ellipse orientation. This was done for a run of 50 trials at each noise level.

The proposed method computes the f-eliminant γ_{12} and γ_{13} for two pairs of sphere images, and then the most plausible intersection of both circles.

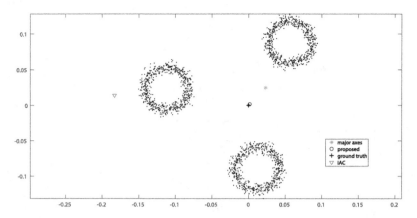

Fig. 5. Simulated image of three spheres near the center, with acceptable image noise (zero mean Gaussian noise with $\sigma = 0.004$) and no radial distortion. We present the means of the results of the three methods for a run of 50 trials.

This eliminant degenerates into a line in case two elliptic sphere images are congruent; other singular situations with $a_1/b_1 = a_2/b_2$ did not occur.

The major axis method actually only needs two sphere images, but it is fair to present the three circles to this algorithm, enabling us to select the point that minimizes the distance to the three major axes.

Another reference for comparison are the IAC-oriented calibration methods. We chose to implement the method of [10], because it is shown in this paper that this method has the same accuracy and precision as the known IAC-methods of [1] and [18]. The IAC-methods do not explicetely use the orientation of the ellipse, but since they use the complete symmetric matrix that represent the ellipse, the dependence on the ellipse orientation is implicit. Actually, the method of [10] computes the common self-polar triangle of two pairs of elliptic sphere images, sharing the image of sphere 1 say, yielding the computation of the vanishing line of the plane through the contour of the first sphere as it is viewed from the camera center. The complex points of intersection of this vanishing line and the image of the first sphere belong to the image of the absolute conic (IAC). The symmetric 3×3 matrix A representing IAC has 3 to 5 dof (depending on assumptions on skewness and aspect ratio). Finally, in the IAC-methods the calibration matrix K (including the coordinates of the principal point) is recovered by the Cholevsky decomposition $A = K^{-T} \cdot K^{-1}$ (we refer to [10] for more details.)

The first experiment is represented by Table 1, where we challenge the three methods to operate on three sphere images with a critical a/b-ratio (near to 1, approximately 1.0003).

In Table 1 we observe that the IAC-methods perform in a very unstable way when $r = a/b \approx 1$. Actually, the IAC-method as presented in [10] crashed in 40% to 50% of the cases where we simulated realistic image noise, due to

Table 1. The statistics for the computed principal point (u_0, v_0) computated by three simulated images of spheres with radius 3, depth 200, distance 5 to the focal axis, correct principal point $= (0,0)$ and focal distance $f = 2$ with added Gaussian noise ($\sigma = 0.004$), but no distortion.

	Proposed				Major axes				IAC			
σ	$\bar{u_0}$	$\bar{v_0}$	std(u_0)	std(v_0)	$\bar{u_0}$	$\bar{v_0}$	std(u_0)	std(v_0)	$\bar{u_0}$	$\bar{v_0}$	std(u_0)	std(v_0)
0	0	0	0	0	0	0	0	0	0	0	0	0
0.002	0.0002	0.0002	0.0216	0.0147	0.0032	−0.0068	0.0223	0.0288	−0.0812	−0.0318	0.5435	0.2455
0.004	0.0089	0.0005	0.0216	0.0217	0.0106	−0.0174	0.0251	0.0393	1.2318	0.8891	3.9930	3.0725
0.006	−0.0016	0.0029	0.0186	0.0162	−0.0052	0.0208	0.0411	0.0727	1.2318	0.8891	3.9930	3.0725
0.008	0.0039	−0.0033	0.0223	0.0215	0.0046	0.0022	0.0238	0.0304	12.5164	2.1061	58.3867	9.4463
0.010	−0.0044	−0.0013	0.0205	0.0188	0.0058	−0.0011	0.0251	0.0302	−0.4685	−0.4290	1.3611	1.2671
0.012	0.0005	0.0005	0.0181	0.0141	0.0017	−0.0042	0.0221	0.0253	0.2074	0.7046	0.9580	1.8791
0.014	0.0004	0.0019	0.0189	0.0139	0.0102	0.0106	0.0305	0.0311	−0.8798	−0.8897	4.2796	4.1437
0.016	0.0002	−0.0007	0.0151	0.0184	0.0033	0.0040	0.0238	0.0264	−0.3716	2.9294	1.7176	13.7421
0.018	0.0005	0.0005	0.0240	0.0158	0.0008	0.0009	0.0227	0.0196	−0.6755	1.6666	2.5598	6.3728
0.020	−0.0005	−0.0013	0.0174	0.0156	−0.0031	0.0001	0.0300	0.0277	0.3457	−0.1788	1.0477	0.5571

the fact that the computed matrix A (representing the IAC) was not positive definite (e.g., due to overkapping sphere inages), and hence not suitable for the Cholevsky decomposition. The proposed method is clearly more stable than the major axes method in these circumstances, because the proposed method does not use the ellipse orientation. The fact that the proposed method and the major-axes-method are both designed for the direct computation of the principal point seems to be an advantage in this aspect compared to the general calibration IAC-methods.

In Table 2 we created data samples consisting of sphere triples at different distances with respect to the focal axis Z, 50 trials at each distance. The radii of the spheres have been chosen to be equal to 10, and the boundary pixels of their elliptic images have been disturbed by limited Gaussian noise ($\sigma = 0.002 = f/1000$). As expected, the shortcomings of the major-axes-method and the IAC-method is most manifest for configurations of sphere that have a small offset with respect to the focal axis, since their images lie close to the image center and imply an unstable ratio $r = a/b \approx 1$ for recovering the ellipse orientation. Furthermore, the IAC-method crashed in ±50% of the sample at the smallest considered offset and in ±20% of the sample at the second smallest offset.

Next, in Table 3 we present the statistic results of a synthetic experiment with simulated radial distortion, still under increasing levels of image noise. More precisely, we used the polynomial model of radial distortion with coefficients $k_1 = -0.2507$ and $k_2 = 0.1345$ of r^2 and r^4 respectively, matching the calibrated distortion of the camera that was used in our real experiments. As can be concluded by Table 3 radial distortion combined with pixel noise causes the oriented axis of the fitting ellipse to deviate drastically from correct orientation of the pure perspective elliptic image. This implies that the major-axes-method is unreliable in the presence of radial distortion. This observation is already manifest in simulations with restricted radial distortion and no image noise (Fig. 1).

Table 2. The statistics for the computed principal point (u_0, v_0) computated by three simulated images of spheres with radius 10, depth 200, and with varying distances to the focal axis (from 5 to 105). The correct principal point lies at $(0, 0)$. The image noise is simulated by a zero-mean Gaussian with $\sigma = 0.002 = f/1000$.

	Proposed				Major axes				IAC			
Offset	$\overline{u_0}$	$\overline{v_0}$	std(u_0)	std(v_0)	$\overline{u_0}$	$\overline{v_0}$	std(u_0)	std(v_0)	$\overline{u_0}$	$\overline{v_0}$	std(u_0)	std(v_0)
5	0.0033	0.0051	0.0310	0.0233	−0.0061	0.0007	0.0377	0.0438	0.1590	0.8190	0.7308	2.9317
15	−0.0053	−0.0067	0.0525	0.0415	0.0038	−0.0219	0.0564	0.0744	0.0110	−0.0075	0.0965	0.0797
25	0.0101	0.0016	0.0380	0.0272	−0.0003	−0.0053	0.0306	0.0391	0.0022	−0.0002	0.0201	0.0200
35	−0.0020	−0.0016	0.0220	0.0214	0.0042	−0.0049	0.0217	0.0216	0.0022	−0.0037	0.0164	0.0163
45	−0.0006	−0.0031	0.0180	0.0149	0.0001	−0.0015	0.0150	0.0169	0.0000	−0.0025	0.0108	0.0124
55	−0.0005	−0.0007	0.0156	0.0120	0.0031	0.0048	0.0106	0.0125	0.0015	0.0010	0.0087	0.0093
65	−0.0002	0.0009	0.0140	0.0114	−0.0006	−0.0013	0.0110	0.0145	−0.0004	0.0002	0.0086	0.0096
75	0.0001	0.0020	0.0142	0.0100	0.0016	−0.0009	0.0074	0.0103	0.0012	0.0009	0.0055	0.0082
85	−0.0003	−0.0009	0.0107	0.0099	−0.0013	0.0013	0.0084	0.0101	−0.0010	0.0001	0.0064	0.0076
95	0.0011	−0.0014	0.0092	0.0092	−0.0006	−0.0022	0.0064	0.0116	0.0000	−0.0015	0.0054	0.0081
105	0.0020	0.0023	0.0090	0.0084	0.0002	−0.0023	0.0064	0.0065	0.0010	0.0008	0.0055	0.0063

On the other hand, these synthetic results indicate that radial distortion is less dramatic for the proposed method for computing the principal point. This is a crucial advantage in sphere-based calibration techniques that solve for the complete calibration matrix K as well as for a reduction model for radial distortion by means of an iterative optimization, requiring a reliable initial value for the principal point (e.g. [14]). Furthermore, the IAC-method of [10] that we used as reference method seems to fail completely in the presence of radial distortion, probably due to the fact that the IAC-methods strongly rely on properties of projective geometry, which are violated when the ellipses are distorted. Even for low or zero image noise, the IAC matrix ceases to be positive definite. Since the IAC-methods appear to require preprocessed images with removed radial distortion, they are not included in the results of Table 3.

Table 3. The statistics for the principal point (u_0, v_0) computated by three simulated sphere images, subject to radial distortion, disturbed by increasing image noise. In each trial we positioned spheres of radius 10 at depth 200 and offset 50 ($f = 2$ and correct principal point $= (0, 0)$.)

	Major axes				Proposed			
σ	$\overline{u_0}$	$\overline{v_0}$	std(u_0)	std(v_0)	$\overline{u_0}$	$\overline{v_0}$	std(u_0)	std(v_0)
0	−0.0003	0.0001	0.0012	0.0010	−0.0181	0.0153	0.0066	0.0056
0.002	0.0010	−0.0068	0.0193	0.0143	4.0524	0.7724	22.3606	4.1018
0.004	−0.0056	−0.0036	0.0374	0.0300	−0.9438	−0.1605	4.7358	0.8211
0.006	−0.0004	0.0019	0.0539	0.0538	−0.6330	−0.0481	2.5950	0.4592
0.008	0.0030	−0.0020	0.0645	0.0559	0.7729	0.1832	6.3221	1.1907
0.010	−0.0162	0.0001	0.0721	0.0718	−0.4937	−0.0207	3.2097	0.5734

6   Real Experiments

We used a set-up consisting of three biljard balls on a LED panel. In the first experiment we used a monochrome CCD camera with 1024×768 pixels and a 3.5 mm lens (Fig. 6). In a second set-up we used an optical sensor with 8 mm lens and a resolution of 1280×1024 pixels.

Fig. 6. The set-up for taking images of three biljard balls in our experiments.

To provide a ground truth for the computed principal points, we calibrated the cameras for this set-up by means of a state-of-the-art procedure by checkerboards. More precisely, we used Bouguet's toolbox [4], resulting into $(526.53,\ 359.43)$ for the principal point of the first sensor, and $(634.64,\ 499.85)$ for the second sensor. In each set-up the camera processed 30 sphere images, providing an abundance of sphere triples. Each of these triples have been used to compute the principal point, by the proposed method and by the ellipse-axes intersection method, yielding sufficient statistics for comparison. Figure 7 illustrates this for a random subset of 10 out of the 30 sphere images.

The results are given in Table 4. We observe that the calibration of the principal point by means of sphere images appears to be less stable and accurate than predicted by the simulations, probably due to nonlinear lens distortion. However, still the proposed method perfoms significantly better than the ellipse-axes method.

Table 4. The comparison of the proposed method to the ellipse-axes method in a performance on real data, implemented for two sensors. The ground truth has been provided by Bouguet's toolbox. Results have been rounded up to pixel unit.

Sensor	Major axes				Proposed				Ground truth	
	$\overline{u_0}$	$\overline{v_0}$	$\mathrm{std}(u_0)$	$\mathrm{std}(v_0)$	$\overline{u_0}$	$\overline{v_0}$	$\mathrm{std}(u_0)$	$\mathrm{std}(v_0)$	u_0	v_0
cam1	501	254	68	180	540	299	97	83	527	359
cam2	692	472	208	96	602	519	52	36	635	500

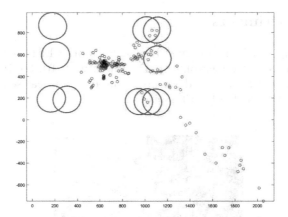

Fig. 7. For 10 segmented ellipses for 10 real sphere images we plot the computed pp for each possible triple, by the proposed method (black) and by the major axis method (blue). The pp found by the Matlab toolbox is given by the red star. (Color figure online)

7   Conclusions

The starting point of this article is the opinion that in some applications it can be useful to use images of spheres in order to calibrate a 2D camera. Many publications seem to confirm this, and we observe ongoing research for finetuning the sphere-based calibration procedures. Most of these procedures contain a separate module for computing the principal point, or an initial value of the principal point in case of recursive optimization. Existing sphere-based algorithms for computing the principal point rely on the orientation of the elliptic sphere images, either directly (the major-axes-method) or indirectly (the IAC-methods). Since the computation of the ellipse orientation becomes very unstable when $a \approx b$, as is often the case for sphere images, this article offers an alternative novel method for computing the principal point from sphere images, that does not make use of the major axes of the elliptic images. We proved that the proposed method is more accurate and robust than the major-axes-method and the IAC-methods, and that it holds up in the presence of radial distortion.

Finally, we report some warnings:

1. The major-axes-method only needs two sphere images as minimal solver, while the proposed method requires at least three sphere images. Furthermore, for three sphere images our method delivers two solutions for the principal point. The correct solution can be easily detected as the, most feasible one, or by using a fourth sphere image.
2. The correctness of the proposed method is proven for a perfect pinhole model with square pixels. Due to its robustness the computed principal point can serve as initial value for further numerical optimization of the calibration

parameters. The aspect ratio and the radial distortion parameters can also be determined by the sphere images [12, 14].

3. The singular positions of our method occurs when two spheres have the same shape ratio $r = a/b$. However, in practical situations this takes place when $a_1 = a_2$ and $b_1 = b_2$. We took care for this type of singularity by providing an alternative formula in this case.

4. The excellent performance of the proposed method in synthetic circumstances, simulating both pixel noise and radial distortion, was less clear in our real experiments, probably due to the negative impact of real lens distortions. However, also in real experiments our method surpasses the major-axes-method.

References

1. Agrawal, M., Davis, L.S.: Complete camera calibration using spheres: dual-space approach. IEEE **206**, 782–789 (2003)
2. Agrawal, M., Davis, L.S.: Camera calibration using spheres: a semi-definite programming approach. In: IEEE International Conference on Computer Vision, pp. 782–789 (2003)
3. Beardsley, P., Murray, D., Zisserman, A.: Camera calibration using multiple images. In: Sandini, G. (ed.) ECCV 1992. LNCS, vol. 588, pp. 312–320. Springer, Heidelberg (1992). https://doi.org/10.1007/3-540-55426-2_36
4. Bouguet, J.Y.: Camera calibration toolbox for Matlab. http://www.vision.caltech.edu/bouguetj/calib_doc/
5. Dandelin, G.P.: Mémoire sur quelques propriétés remarquables de la focale parabolique. Nouveaux mémoires de l'Académie Royale des Sciences et Belles-Lettres de Bruxelles T. **II**, 171–202 (1822)
6. Daucher, D., Dhome, M., Lapreste, J.: Camera calibration from spheres images. In: Proceedings European Conference Computer Vision, pp. 449–454 (1994)
7. Fitzgibbon, A., Pilu, M., Fisher, R.: Direct least-square fitting of ellipses. In: Proceedings International Conference on Pattern Recognition, pp. 253–257 (1996)
8. Guan, J., et al.: Extrinsic calibration of camera networks using a sphere. Sensors **15**(8), 18985–19005 (2015). https://doi.org/10.3390/s150818985, http://www.mdpi.com/1424-8220/15/8/18985
9. Ho, C., Chen, L.: A fast ellipse/circle detector using geometric symmetry. Pattern Recognit. **28**(1), 117–124 (1995)
10. Huang, H., Zhang, H., Cheung, Y.: Camera calibration based on the common self-polar triangle of sphere images. In: Cremers, D., Reid, I., Saito, H., Yang, M.-H. (eds.) ACCV 2014. LNCS, vol. 9004, pp. 19–29. Springer, Cham (2015). https://doi.org/10.1007/978-3-319-16808-1_2
11. Lu, Y., Payandeh, S.: On the sensitivity analysis of camera calibration from images of spheres. Comput. Vis. Image Underst. **114**(1), 8–20 (2010)
12. Penna, M.: Camera calibration: a quick and easy way to determine the scale factor. IEEE Trans. Pattern Anal. Mach. Intell. **13**(12), 1240–1245 (1991)
13. Penne, R., Ribbens, B., Mertens, L., Levrie, P.: What does one image of one ball tell us about the focal length? In: Battiato, S., Blanc-Talon, J., Gallo, G., Philips, W., Popescu, D., Scheunders, P. (eds.) ACIVS 2015. LNCS, vol. 9386, pp. 501–509. Springer, Cham (2015). https://doi.org/10.1007/978-3-319-25903-1_43

14. Sun, J., Chen, X., Gong, Z., Liu, Z., Zhao, Y.: Accurate camera calibration with distortion models using sphere images. Opt. Laser Technol. **65**, 83–87 (2015)
15. Teramoto, H., Xu, G.: Camera calibration by a single image of balls: from conics to the absolute conic. In: Asian Conference on Computer Vision, pp. 499–506 (2002)
16. Xie, Y., Ji, Q.: A new efficient ellipse detection method. In: International Conference on Pattern Recognition, pp. 957–960 (2002)
17. Yin, P., Chen, L.H.: A new method for ellipse detection using symmetry. J. Electron. Imaging **3**, 20–29 (1994)
18. Zhang, H., Wong, K., Zhang, G.: Camera calibration from images of spheres. IEEE Trans. Pattern Anal. Mach. Intell. **29**(3), 499–503 (2007)
19. Zhang, Z.: A flexible new technique for camera calibration. IEEE Trans. Pattern Anal. Mach. Intell. **22**(11), 1330–1334 (2000)

NightOwls: A Pedestrians at Night Dataset

Lukáš Neumann[1]([✉]), Michelle Karg[2], Shanshan Zhang[3],
Christian Scharfenberger[2], Eric Piegert[2], Sarah Mistr[2], Olga Prokofyeva[2],
Robert Thiel[2], Andrea Vedaldi[1], Andrew Zisserman[1], and Bernt Schiele[4]

[1] Department of Engineering Science, University of Oxford, Oxford, UK
lukas@robots.ox.ac.uk
[2] Continental Corporation, BU Advanced Driver Assistance Systems,
Hanover, Germany
[3] Nanjing University of Science and Technology, Nanjing, China
[4] Max Planck Institute for Informatics, Saarbrücken, Germany
http://www.nightowls-dataset.org/

Abstract. We introduce a comprehensive public dataset, *NightOwls*, for pedestrian detection at night. In comparison to daytime conditions, pedestrian detection at night is more challenging due to variable and low illumination, reflections, blur, and changing contrast.

NightOwls consists of 279k frames in 40 sequences recorded at night across 3 countries by an industry-standard camera, including different seasons and weather conditions. All the frames are fully annotated and contain additional object attributes such as occlusion, pose and difficulty, as well as tracking information to identify the same object across multiple frames. A large number of background frames for evaluating the robustness of detectors is included, a validation set for local hyperparameter tuning, as well as a testing set for central evaluation on a submission server is provided.

As a baseline for pedestrian detection at night time, we compare the performance of ACF, Checkerboards, Faster R-CNN, RPN+BF, and SDS-RCNN. In particular, we demonstrate that state-of-the-art pedestrian detectors do not perform well at night, even when specifically trained on night data, and we show there is a clear gap in accuracy between day and night detections. We believe that the availability of a comprehensive night dataset may further advance the research of pedestrian detection, as well as object detection and tracking at night in general.

L. Neumann, M. Karg and S. Zhang—Equal contribution.

Electronic supplementary material The online version of this chapter (https://doi.org/10.1007/978-3-030-20887-5_43) contains supplementary material, which is available to authorized users.

C. V. Jawahar et al. (Eds.): ACCV 2018, LNCS 11361, pp. 691–705, 2019.
https://doi.org/10.1007/978-3-030-20887-5_43

1   Introduction

Detecting and tracking people is one of the most important applied problems in computer vision. Significant applications such as entertainment, surveillance, robotics, and assisted and automated driving, are all centered around people. They thus require highly-reliable people detectors that can work in a variety of indoor and outdoor scenarios and are robust to challenging visual effects such as variable appearance, inhomogeneous illumination, low resolution, occlusions and limited field of view.

While recent progress in object detection has been substantial, current systems may still fail to measure up to the demands of such requirements, particularly when, as in autonomous driving and surveillance, detecting people with high reliability is paramount for safety. Unfortunately, current benchmarks are insufficient to assess such limitations in a reliable manner, let alone support further research to address them.

In order to fill this gap, we introduce *NightOwls*, a new dataset to assess the limitations of state-of-the-art pedestrian detectors when used in extreme but realistic conditions. We focus in particular on detection at nighttime, a problem largely underrepresented in the literature, but which is very important in many applications - in assistive driving, in surveillance monitoring, or in autonomous driving as a key input to the sensor fusion. Vision sensors also benefit from the advantage of high-quality shape and color information, human interpretability of the sensor output and low energy consumption (passive sensor), which is not the case for other sensor modalities.

Our work is inspired by datasets such as PASCAL VOC [8], ImageNet [13] and MS-COCO [14], whose introduction kickstarted new waves of fundamental research in classification and detection, moving the field from bag-of-visual-words, to deformable parts model, and finally to deep convolutional neural networks. Benchmarks such as Caltech pedestrians [4,5] had a similar impact in pedestrian detection.

For a dataset to be impactful, it must highlight important shortcomings in the current generation of algorithms. For pedestrian detection, the most frequently-used dataset, Caltech, is nearly saturated, with an average miss rate of 8.0% for state-of-the-art detectors [1,20] compared to 83.0% average miss rate at the time of introduction [5]. This tremendous improvement suggests that the Caltech benchmark is almost "solved", at least if we take human performance, estimated at 5.6% by [20], as an upper bound.

While Caltech and similar benchmarks may be saturated, we cannot conclude that pedestrian detection is "solved" in general. A limitation with most datasets is that they focus on detection during the daytime. While this requires to cope with challenges such as occlusions and variable appearance, scale, and pose, doing so in poor lighting, and at nighttime in particular, is more challenging still. Empirically, we will show that current detectors are far below human performance in such conditions.

In order to do so, *NightOwls* is designed to be representative of the following challenges:

Fig. 1. Sample images from the *NightOwls* dataset exhibiting the challenges of detection at night. Occlusion, low contrast, motion blur, image noise and inhomogeneous illumination (left) and different illumination and weather conditions (right).

1. **Motion blur and image noise:** Imaging at night requires a trade-off between long exposure times and sensor gain, resulting in significant motion blur or noise.
2. **Reflections and high dynamics:** The variations in light intensity in night scenes, caused by inhomogeneous light sources and their reflections, may exceed the dynamic range of a camera, resulting in inhomogeneous illumination with under- and over-saturated areas.
3. **Large variation in contrast, reduced color information:** Inhomogeneous illumination induces large contrast variations in images. Detection is difficult in low-contrast regions and may result in loss of color information and in confusing foreground and background regions.
4. **Weather and seasons:** Weather and seasons cause other visual variations impacting the performance of detectors. While snow has the potential to illuminate a scene more homogeneously, rain can reduce the contrast dramatically and add reflections to road surfaces.

In addition to addressing these challenges, *NightOwls* has a number of additional desirable properties: (i) images are captured by an industry-standard camera for automotive, whereas other datasets often use generic cameras, (ii) full annotations for each frame are provided, in the standard MS-COCO and Caltech formats, (iii) multiple European cities and countries are represented, (iv) track identity information when an object is detected in multiple frames is provided, (v) a central evaluation server for results submission and comparison is available, and (vi) additional classes (cyclists, motorbikes) and attributes (pose, difficult) are annotated.

Empirically, we demonstrate that state-of-the-art pedestrian detection methods do not perform well on this dataset, even when specifically trained on night data, and we show the gap in accuracy between day and night detections is quite significant. While we primarily focus on detecting pedestrians, we also believe that the availability of a comprehensive night dataset may initiate further research in other domains, such as general object detection or tracking.

2   Related Work

Existing Datasets. Over the last decade, several datasets have been created for pedestrian detection. Early efforts include INRIA [2], ETH [7], TUD-Brussels [17], and Daimler [6]. These datasets are now either too small (INRIA, ETH, TUD-Brussels) or only provide gray scale images (Daimler).

Table 1. Normalized histogram of pedestrian height in pixels (a), the number of objects per frame (b), the average image (c) and pedestrian patch (d) lightness of standard pedestrian datasets. Note that only datasets with an "occlusion" flag are shown for the patch statistics (d)

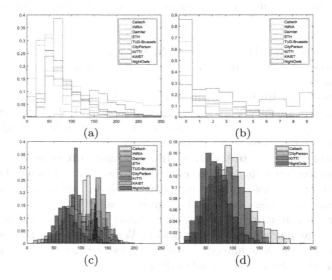

Recently, larger and richer datasets have been proposed and have become more popular, such as the Caltech [5], KITTI [10] and CityPersons [22] datasets. The Caltech dataset [4,5] has been widely used as it provides a large number of annotations, including around 250,000 frames and 185,000 pedestrian bounding boxes. Yet, the diversity of the annotations is limited as the video was recorded in only 11 sessions within a single city, the alignment quality of the annotations is poor due to the interpolation implemented between neighboring frames, and it is only recorded at daytime. The noisy annotations were then further improved in [20].

The focus of the KITTI dataset is to encourage research in the field of a multi-sensor setup consisting of cameras, a laser scanner and GPS/IMU localization providing data for multiple tasks such stereo matching, optical flow, visual odometry/SLAM, object detection and 3D estimation [9,10], but for pedestrian detection the dataset is relatively small.

The CityPersons dataset [22] consists of a large and diverse set of stereo video sequences recorded in streets from 27 cities in Germany and neighbouring countries. High quality bounding box annotations are provided for about 35k pedestrians in 5000 images. Additionally, fine pixel-level annotations of 30 visual classes are also available. The fine annotations include instance labels for persons and vehicles. However, the dataset does not have night or background images. Also, it does not have driving sequences (it consists of individual images), and consequently it does not have examples of the same objects across multiple frames.

To the best of our knowledge, the KAIST dataset [12] is currently the only public dataset that contains some night images for pedestrian detection (5 out of 10 recordings are at night). The data was captured in one city in one season, which limits the diversity, the camera used for recording is a consumer-grade camera, which resulted in poor recording quality and considerable additional image noise, and the dataset does not provide occlusion labels which severely limits the ability to train on this dataset (see Sect. 4). The focus of the KAIST dataset is multi-spectral pedestrian detection which considers the data fusion from a thermal sensor and a RGB camera, as an attempt to overcome the issues of pedestrian detection at night. We note, that using just the thermal sensor for object detection may not be feasible because of its low spatial and dynamic resolution, the limited thermal footprint of people in clothes and their currently prohibitive cost for production vehicles.

The number of images and annotations in different datasets is summarised in Table 2, key statistics are compared in Table 1.

Pedestrian Detection at Night. Apart from KAIST, all the above datasets and the vast majority of work on pedestrian detection [1,4,5,19,21,22] is focused on detection at daytime. Some early work attempted to solve the problem of object/pedestrian detection at night with the assistance from tracking methods [11,18] or stereo images [15].

However, to our knowledge pedestrian or object detection at night has not attracted much attention in the research community, despite its importance for robust vision applications. We suspect the main reason is the lack of publicly available data for such research.

3 Dataset

In this section, we describe the data capture procedure, the annotation protocol, our design choices and the statistics of the dataset.

Data Recording. The dataset has been recorded in several cities across Europe with a forward-looking industry-standard camera, using windshield mounting identical to professional mounts in production vehicles. The data was collected at dawn and nighttime throughout the whole year and under different weather conditions (see Fig. 1). In total, 40 individual recordings were captured and then split into the training, validation and test sets, maintaining uniform distribution of key parameters such as weather and pedestrian pose/height difficulty.

Table 2. Image and pedestrian annotations counts in pedestrian detection datasets.

Dataset	Training				Validation				Test				All	
	Images	Pedestrian Bboxes	Pedestrian Tracks	Background Images	Images	Pedestrian Bboxes	Pedestrian Tracks	Background Images	Images	Pedestrian Bboxes	Pedestrian Tracks	Background Images	Images	Objects / Frame
Caltech [5]	**128k**	**153k**	**1k**	67k	-	-	-	-	**121k**	**132k**	**869**	61k	250k	1.14
INRIA [2]	2k	1k	0	1k	-	-	-	-	288	589	0	0	2k	0.86
Daimler [6]	22k	14k	0	15k	-	-	-	-	-	-	-	-	22k	0.65
ETH [7]	2k	14k	0	5	-	-	-	-	-	-	-	-	2k	7.85
TUD [17]	508	1k	0	145	-	-	-	-	-	-	-	-	508	2.95
KITTI [10]	7k	4k	0	6k	-	-	-	-	-	-	-	-	7k	0.60
KAIST [12]	50k	41k	495	32k	-	-	-	-	45k	45k	675	26k	95k	0.90
night subset	*17k*	*17k*	*141*	*10k*	-	-	-	-	*16k*	*12k*	*156*	*10k*	*33k*	*0.86*
CityPersons	3k	17k	-	672	500	3k	-	102	1.5k	14k	-	249	5k	7.00
NightOwls	128k	38k	1657	**105k**	**51k**	**9k**	**262**	**45k**	103k	8k	196	**97k**	281k	0.20

Table 3. Pedestrian attributes statistics.

	Occlusion	Pose		Height			All
		Sideways	Frontal	Far	Medium	Near	
Train	5k [20%]	9k [35%]	18k [64%]	16k [58%]	7k [24%]	5k [18%]	27k[66%]
Vali	1k [13%]	2k [35%]	4k [64%]	3k [51%]	2k [29%]	1k [19%]	7k [16%]
Test	2k [25%]	2k [25%]	6k [75%]	5k [62%]	2k [26%]	1k [12%]	8k [18%]
All	8k [20%]	14k [33%]	28k [67%]	24k [58%]	11k [25%]	7k [17%]	**42k**

Image Quality and Size. Research datasets [5,12] are often recorded with consumer camera equipment, which results in high level of image noise and limited dynamic range. To provide a night dataset with realistic variations in contrast and blurriness, the dataset was captured by an industry-standard camera (image resolution 1024×640), very similar one to the ones used in production vehicles. The dataset includes both blurred and sharp images and the quality realistically depends on the scene illumination and the vehicle speed.

Annotation. The frame-rate is 15 fps and every frame was manually annotated. Every pedestrian, cyclists and motorcyclist (higher than 50px) is annotated with a bounding box, alongside with three attributes: occlusion, difficult (low contrast or unusual posture) and pose. People on posters, sculptures and groups where individuals are hard to separate are marked as "ignore". We note that compared to the existing datasets, the average number of objects per frame is lower, because naturally streets are less busy at night (see Table 1).

Fig. 2. Sample pedestrian, cyclist and motorcyclist annotations from the *NightOwls* dataset, including pose attribute.

As a result, the dataset contains $279k$ fully annotated frames with $42,273$ pedestrians, where $32k$ frames contain at least one annotated object and the remaining $247k$ are the background images. The annotations are provided in two standard MS-COCO [14] and Caltech (VBB) [4] formats, so that the new dataset can be plugged in to existing frameworks without any extra effort (Fig. 2).

Similarly to the Caltech dataset [4], the attributes are classified into several groups to allow more fine-grained evaluation using different data dimensions.

The pedestrian height is divided into *Far*, *Medium* and *Near* (see Table 3), based on the distance required to trigger automated breaking of a moving vehicle at different speeds. Using the pinhole camera model, the camera calibration

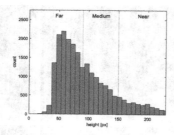 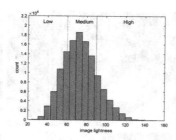

Fig. 3. Histogram of pedestrian scale (left) and image lightness (right) in the *NightOwls* dataset and the corresponding attribute categorisation

Table 4. Comparison of the annotation attributes.

Dataset	Year	Image Size	Occlusion	Difficult	Pose	Tracking Id	Night	Dusk/Dawn	# Cities	Different Seasons	Pedestrian	Cyclist	Motorcyclist
			Attributes				Images				Annotations		
Caltech [5]	2009	640x480	✓	x	x	✓	x	x	1	x	✓	x	x
Kitti [10]	2012	1392x512	✓	x	✓	x	x	x	1	x	✓	✓	✓
KAIST [12]	2015	640x480	x	✓	x	✓	✓	✓	1	x	✓	Driver	
CityPersons [23]	2017	2048x1024	✓	x	x	x	x	x	27	✓	✓	Rider	
NightOwls	2018	1024x640	✓	✓	✓	✓	✓	✓	7	✓	✓	✓	✓

parameters and average person height of (1.6m, 1.8m), the annotation height h is divided as

$h \leq 90$ *Far*   braking distance at $\sim 50\,\text{km/h}$

$90 \leq h \leq 150$ *Medium*   braking distance at $\sim 40\,\text{km/h}$

$h \geq 150$ *Near*   braking distance at $\sim 30\,\text{km/h}$

We note that a majority of the pedestrians are categorized as *Far* (see Fig. 3 left), which is due to the exhaustive labeling process of every frame. Similarly, we also categorize image lightness as *Low*, *Medium* and *High*, based on the histogram of mean image lightness (see Sect. 4 and Fig. 3 right).

The pose is annotated as left, right, front and back, but we refer to them as *Frontal* (front, back) and *Sideways* (left, right). We note that there is a bias towards the *Frontal* pose in the data (see Table 3), which is given by the fact how people/cyclists typically move on and alongside the road (Table 4).

Data Diversity. To achieve a high data diversity, which is desired for the generalization ability of detection algorithms, the recordings were collected in 7 cities across 3 countries (Germany, Netherlands, UK) during a period of five months. The dataset captures different weather conditions durin autumn, winter and spring, including rain and snow which change the lighting of the scene and add additional reflections.

Background Images. False positive rate is a major concern for real-world applications, because false alarms of safety-critical systems are not acceptable in driving scenarios. Moreover in these applications, the number of frames without any object of interest is significantly higher than the number of frames with it, which increases the chance of false positives even further.

In order to support the research of robust detectors with low false positive rates and to reliably estimate detector precision, $247k$ background images are included in the dataset. For night images, especially regions with low illumination or reflections are typically prone to such false positives.

Temporal Tracking. Most methods focus on detection from a single frame, which is inherently more prone to both false positive as well as false negative errors. In order to enable research of more robust multi-frame detection methods, the dataset includes temporal tracking annotations, so that the same object can be identified across different frames.

Validation and Testing Set. Similarly to the recent large-scale datasets such as MS-COCO [14] or CityPersons [22], we explicitly split the data for evaluation into a validation and a testing set. We publish images for both sets, but only the annotations for the validation set are published - the testing set annotations are then only to be used by the evaluation server (see below). Both sets have similar data statistics and the validation set is sufficiently large, so that it can be used by the researchers for local evaluation and hyper-parameter tuning. An additional benefit of a common validation set is that the hyper-parameter experiments become comparable between different methods.

Evaluation Server. A central submission server is provided for dataset download and evaluation. The submissions of detection results (JSON format) are automatically evaluated on the testing set and a leader board is presented, so that all detection methods are evaluated in a single place. The submissions are limited to one submission a day to reduce the possibility of over-fitting to the testing set. Additionally, because the testing set is sufficiently large, we only publish performance on one subset on the leader board, whilst the performance of the second sequestered subset will remain private - if there is a significant discrepancy in the accuracy on the both subsets, this points towards over-fitting or training on the testing data.

4   Experiments

Methods. We have evaluated 6 recently published pedestrian detection algorithms on the existing datasets, as well as the newly introduced dataset:

   ACF [3] Our experiments are based on the open source release of ACF[1]. One minor change we made is to use a larger model size (60×120 instead of 30×60 pixels), which shows to improve the vanilla ACF on several benchmarks, e.g. Caltech, KITTI. All other parameters are kept identical to the vanilla version.

---

[1] https://github.com/pdollar/toolbox.

Table 5. Comparison of state-of-the-art pedestrian detection methods trained and tested on the corresponding dataset. Average Miss Rate (MR) or mean Average Precision (mAP) shown, as per the dataset protocol, using Reasonable subset

Method *metric*	Caltech *MR*	KITTI *mAP*	CityPersons *MR*	**ours** *MR*
ACF [3]	27.63%	47.29%	33.10%	51.68%
Checkerboards [21]	18.50%	56.75%	31.10%	39.67%
Vanilla Faster R-CNN [16]	20.98%	65.91%	23.46%	20.00%
Adapted Faster R-CNN [22]	10.27%	66.72%	12.81%	18.81%
RPN+BF [19]	9.58%	61.29%		23.26%
SDS-RCNN [1]	7.36%	63.05%	13.26%	17.80%

Checkerboards [21] In contrast to ACF, the Checkerboards detector applies more filters with various sizes on top of the HOG+LUV channels in order to extract more representative features. We used the open source release[2] without tuning any parameters.

Vanilla Faster R-CNN [16] We reimplemented vanilla Faster R-CNN using the open source code[3]. We only changed the scales and aspect ratios in RPN network. We used a uniform scale step of 1.3, allowing the anchor boxes to cover the image height. Instead of the default multiple aspect ratios, we used only one (width/height = 0.41), which is consistent with our evaluation protocol. For training, we started with the VGG16 network pretrained on ImageNet and trained on our dataset for 100k iterations (LR = 10^{-3} for 60k and LR = 10^{-4} for another 40k).

Adapted Faster R-CNN [22] We followed the experimental findings from [22], and made corresponding modifications to vanilla Faster R-CNN for better performance. We started with the adapted Faster R-CNN model pretrained on the CityPersons dataset and then trained on our dataset for 100k iterations (LR = 10^{-3} for 60k and LR = 10^{-4} for another 40k).

RPN+BF [19] We followed the training procedure described by the authors, starting with the VGG16 network pretrained on ImageNet [13] and training the RPN network for 80k iterations (LR = 10^{-3}), followed by training the whole RPN+BF network for 80k iterations.

SDS-RCNN [1] Similarly to the previous method, we started with the VGG16 network pretrained on ImageNet and trained the RPN network for 120k iterations (LR = 10^{-3}), followed by 120k iterations for both RPN+BCN (full SDS-RCNN) network, using vanilla SGD.

Each method was trained on the *training* subset and evaluated on the validation subset (where available, otherwise the testing subset) of the dataset, keeping the training meta-parameters such as the learning rate or the number

---

[2] https://bitbucket.org/shanshanzhang/code_filteredchannelfeatures.

[3] https://github.com/rbgirshick/py-faster-rcnn.

of epochs identical for the given method between different datasets. We however calculated mean image color and subtracted it as a preprocessing step for each dataset individually - this value was same for all the methods. We followed the standard average Miss Rate (MR) metric [4,5] across all datasets, with the exception of the KITTI dataset, where the mean average precision (mAP) is typically used.

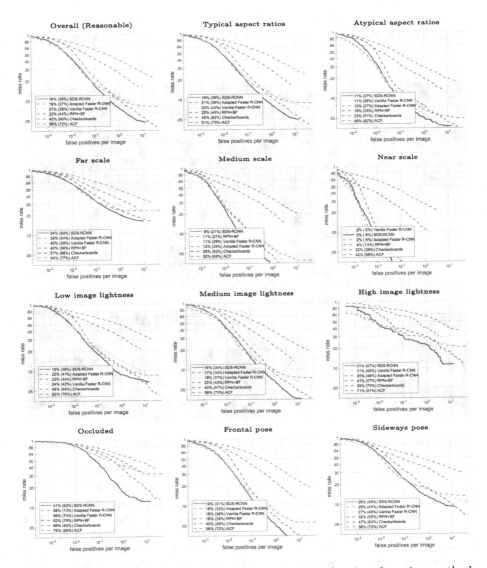

Fig. 4. Miss rates versus false positives of the recent pedestrian detection methods on the *NightOwls* dataset. Lower curve means better performance, the legend denotes average miss rate MR^{-2} [5] (MR^{-4} as in [20] in the parentheses)

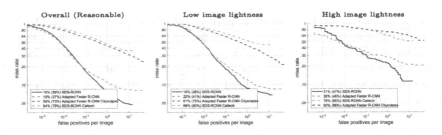

Fig. 5. Specifics of the night data. Comparing accuracy of methods trained on Caltech/CityPersons with methods trained directly on the *NightOwls* dataset

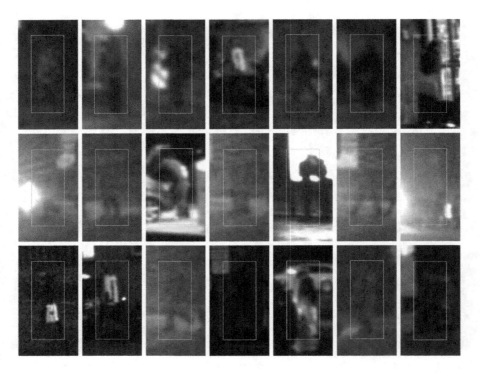

Fig. 6. Sample images of pedestrians missed by all the detection methods. *Far* scale (top row), *High image lightness* (middle row) and *Sideways pose* (bottom row)

Comparison with Other Datasets. Using the Reasonable subset [4], the SDS-RCNN [1] detector, which is the state-of-the-art method on the Caltech dataset, also achieved the lowest average miss rate for our dataset (see Table 5), however the error is still 2.5 times higher than for the Caltech and 50% higher than for the CityPersons dataset, which suggests that the proposed dataset is more challenging than the existing datasets. The gap in the miss rate between the vanilla Faster R-CNN and the improved version (SDS-RCNN) is also much smaller for our dataset, which suggests that the additional information brought by the instance segmentation in SDS-RCNN [1] is not as helpful for night scenarios.

We also train the Adapted Faster R-CNN detector [22] on the KAIST dataset [12], which is the only existing dataset with some nighttime images, and we compare the accuracy on both datasets (see Table 6). We show that on the KAIST test set, the model trained on the *NightOwls* dataset actually outperforms the model trained on the KAIST training set, which is most likely due to the problems with KAIST image and annotations quality (see Table 2). Note that because KAIST does not have an occlusion flag, we did not use the flag for either of the datasets in the above experiment, to make to comparison fair.

Aspect Ratio & Scale. We evaluated the performance of all the methods trained on the *NightOwls* dataset, depending on different ground truth attributes, in line with the standard evaluation introduced by Dollar et al. [4]. We show that the methods are not as sensitive to aspect ratios (Fig. 4 - top row), but they are very sensitive to the size of pedestrians (Fig. 4 - 2nd row). The deep-learning methods clearly benefit from the amount of training data and for the *Medium* and *Near* scales their error rate is comparable to daylight datasets, however for the small pedestrians in the *Far* scale ($h < 90$px), the miss rate rises dramatically and the accuracy of deep-learning methods is close to the traditional ones (see Fig. 6 - top row for sample images).

Illumination. We also compare the performance based on the average image lightness, where the lightness is luma L^4 in the HSL colour space (Fig. 4 - 2nd row). Perhaps counterintuitively, the error is higher for brighter images than the darker ones - this is caused by camera overexposure (see Fig. 6 - middle row), which makes the detection very challenging. Note that the evaluation based on pedestrian image patch lightness as opposed to whole image lightness and different lightness definitions give very similar results, hence we only include them in the supplementary material.

Pose. In contrast to the most commonly used datasets, we can also evaluate the detections based on the pedestrian pose - this clearly shows that all methods perform significantly better for people facing towards or away from the camera (*frontal pose*), than for pedestrians facing sideways (Fig. 4 - bottom row). We suggest this is due to higher ambiguity of the sideway pose, where there is a higher chance of confusion with other objects when a person is viewed from a side than from the front, but generally also due to lower number of pixels and therefore lower amount of information captured in the image for sideway poses (see Fig. 6 - bottom row).

Night Data Specifics. In order to evaluate how specific the night data is, we also run the state-of-the-art SDS-RCNN detector [1] trained on the Caltech dataset, which has a similar number of images, but it's exclusively captured in daytime. The model has an average miss rate of 7.36% on Caltech, but 63.99% on our dataset (the image mean subtracted as a pre-processing step was updated

---

^{4} $L = 0.299R + 0.587G + 0.114B$.

Table 6. Comparison of training and testing the Adapted Faster R-CNN detector on the KAIST-night and NightOwls datasets. The model trained on NightOwls performs better on both testing datasets. All numbers are MR on the Reasonable subset.

Test	Train	
	KAIST-night	NightOwls
KAIST-night	65%	**63%**
NightOwls	57%	**19%**

accordingly to make sure the image data is always centered around zero). Similarly, the Adapted Faster R-CNN model [22] trained on the CityPersons dataset has a miss rate of 59.05% (see Fig. 5). These results confirm the expectation that pedestrian detectors trained on daytime data do not work well at night and training specifically on night data as in the previous sections is required.

5 Conclusion

In this paper, we have introduced a novel comprehensive pedestrian dataset *NightOwls* to encourage research on night images. Recent benchmarks for pedestrian detection and - in general, for object detection in computer vision - have predominantly focused on images collected at daytime. Even though detection at nighttime is a more challenging task because of low illumination, changing contrast, and less color information, studies on nighttime data are underrepresented, rely on study-specific data, and are limited to individual case studies lacking official benchmarks. We believe that by introducing a comprehensive dataset and benchmark for pedestrian detection at night, cutting-edge research on the challenges of nighttime vision can be stimulated.

References

1. Brazil, G., Yin, X., Liu, X.: Illuminating pedestrians via simultaneous detection & segmentation. In: ICCV 2017 (2017)
2. Dalal, N., Triggs, B.: Histograms of oriented gradients for human detection. In: CVPR (2005)
3. Dollár, P., Appel, R., Belongie, S., Perona, P.: Fast feature pyramids for object detection. PAMI **36**, 1532–1545 (2014)
4. Dollár, P., Wojek, C., Schiele, B., Perona, P.: Pedestrian detection: a benchmark. In: CVPR (2009)
5. Dollár, P., Wojek, C., Schiele, B., Perona, P.: Pedestrian detection: an evaluation of the state of the art. PAMI **34**, 743–761 (2012)
6. Enzweiler, M., Gavrila, D.M.: Monocular pedestrian detection: survey and experiments. PAMI **31**, 2179–2195 (2009)
7. Ess, A., Leibe, B., Schindler, K., Van Gool, L.: A mobile vision system for robust multi-person tracking. In: CVPR (2008)

8. Everingham, M., Van Gool, L., Williams, C.K., Winn, J., Zisserman, A.: The Pascal Visual Object Classes (VOC) challenge. Int. J. Comput. Vis. **88**(2), 303–338 (2010)
9. Geiger, A., Lenz, P., Stiller, C., Urtasun, R.: Vision meets robotics: the kitti dataset. Int. J. Robot. Res. **32**(11), 1231–1237 (2013)
10. Geiger, A., Lenz, P., Urtasun, R.: Are we ready for autonomous driving? The kitti vision benchmark suite. In: 2012 IEEE Conference on Computer Vision and Pattern Recognition (CVPR), pp. 3354–3361. IEEE (2012)
11. Huang, K., Wang, L., Tan, T., Maybank, S.: A real-time object detecting and tracking system for outdoor night surveillance. Pattern Recognit. **41**(1), 432–444 (2008)
12. Hwang, S., Park, J., Kim, N., Choi, Y., So Kweon, I.: Multispectral pedestrian detection: benchmark dataset and baseline. In: Proceedings of the IEEE Conference on Computer Vision and Pattern Recognition, pp. 1037–1045 (2015)
13. Krizhevsky, A., Sutskever, I., Hinton, G.E.: Imagenet classification with deep convolutional neural networks. In: Advances in Neural Information Processing Systems, pp. 1097–1105 (2012)
14. Lin, T.-Y., et al.: Microsoft COCO: common objects in context. In: Fleet, D., Pajdla, T., Schiele, B., Tuytelaars, T. (eds.) ECCV 2014. LNCS, vol. 8693, pp. 740–755. Springer, Cham (2014). https://doi.org/10.1007/978-3-319-10602-1_48
15. Liu, X., Fujimura, K.: Pedestrian detection using stereo night vision. IEEE Trans. Veh. Technol. **53**(6), 1657–1665 (2004)
16. Ren, S., He, K., Girshick, R., Sun, J.: Faster R-CNN: towards real-time object detection with region proposal networks. In: Advances in Neural Information Processing Systems, pp. 91–99 (2015)
17. Wojek, C., Walk, S., Schiele, B.: Multi-cue onboard pedestrian detection. In: CVPR (2009)
18. Xu, F., Liu, X., Fujimura, K.: Pedestrian detection and tracking with night vision. IEEE Trans. Intell. Transp. Syst. **6**(1), 63–71 (2005)
19. Zhang, L., Lin, L., Liang, X., He, K.: Is faster R-CNN doing well for pedestrian detection? In: Leibe, B., Matas, J., Sebe, N., Welling, M. (eds.) ECCV 2016. LNCS, vol. 9906, pp. 443–457. Springer, Cham (2016). https://doi.org/10.1007/978-3-319-46475-6_28
20. Zhang, S., Benenson, R., Omran, M., Hosang, J., Schiele, B.: How far are we from solving pedestrian detection? In: CVPR, pp. 1259–1267 (2016)
21. Zhang, S., Benenson, R., Schiele, B.: Filtered channel features for pedestrian detection. In: CVPR (2015)
22. Zhang, S., Benenson, R., Schiele, B.: Citypersons: a diverse dataset for pedestrian detection. In: CVPR (2017)

Multi-view Consensus CNN for 3D Facial Landmark Placement

Rasmus R. Paulsen[1]([✉])(iD), Kristine Aavild Juhl[1](iD), Thilde Marie Haspang[1], Thomas Hansen[2](iD), Melanie Ganz[3,4](iD), and Gudmundur Einarsson[1](iD)

[1] Department of Applied Mathematics and Computer Science,
Technical University of Denmark, 2800 Kgs. Lyngby, Denmark
rapa@dtu.dk
[2] Institute of Biological Psychiatry,
Copenhagen University Hospital MHC Sct. Hans, Roskilde, Denmark
[3] Neurobiology Research Unit, Rigshospitalet, Copenhagen, Denmark
[4] Department of Computer Science, Copenhagen University, Copenhagen, Denmark
http://www.compute.dtu.dk

Abstract. The rapid increase in the availability of accurate 3D scanning devices has moved facial recognition and analysis into the 3D domain. 3D facial landmarks are often used as a simple measure of anatomy and it is crucial to have accurate algorithms for automatic landmark placement. The current state-of-the-art approaches have yet to gain from the dramatic increase in performance reported in human pose tracking and 2D facial landmark placement due to the use of deep convolutional neural networks (CNN). Development of deep learning approaches for 3D meshes has given rise to the new subfield called geometric deep learning, where one topic is the adaptation of meshes for the use of deep CNNs. In this work, we demonstrate how methods derived from geometric deep learning, namely multi-view CNNs, can be combined with recent advances in human pose tracking. The method finds 2D landmark estimates and propagates this information to 3D space, where a consensus method determines the accurate 3D face landmark position. We utilise the method on a standard 3D face dataset and show that it outperforms current methods by a large margin. Further, we demonstrate how models trained on 3D range scans can be used to accurately place anatomical landmarks in magnetic resonance images.

Keywords: 3D facial landmarks · Multi-view CNN · Geometric deep learning

1 Introduction

3D face recognition and analysis has a long history with important efforts dating back to work done in the early nineties [18] and with lots of work published in the early 2000s [4]. Initially, 3D scanning devices were expensive, complicated to use, and for laser scanning devices it required that the subject had to be still

© Springer Nature Switzerland AG 2019
C. V. Jawahar et al. (Eds.): ACCV 2018, LNCS 11361, pp. 706–719, 2019.
https://doi.org/10.1007/978-3-030-20887-5_44

and have closed eyes for a longer period of time. However, the availability of 3D scanning devices ranging from highly accurate clinical devices to consumer class products implemented in mobile devices has dramatically increased in the last decade. This means that human face recognition and analysis is moving from the 2D domain to the 3D domain. 3D morphometric analysis of human faces is an established research topic within human biology and medicine, where the applications range from 3D analysis of facial morphology [20] to plastic surgery planning and evaluation [9]. Broadly speaking, the analysis of facial 3D morphometry is based on either a sparse set of landmarks that serves as a simple measure of facial anatomy or the analysis of a dense set of points that are aligned to the 3D faces using a template matching approach. The approach of matching a dense template can then be used to solve the task of selecting a few anatomical landmarks as seen in for example [19,29].

While a substantial amount of work has been published on automated 3D landmark placement, only a limited number of publications have drawn on the recent drastic improvements in human pose estimation based on deep learning. In this paper, we describe a framework for automated 3D landmarking of facial surfaces using deep learning techniques from the field of human pose estimation.

A classic approach to finding facial landmarks and face parameterisations consists of fitting a 3D statistical shape model to either one or several views of the 3D surface or directly to the surface [2,24,42]. These approaches use a learned statistical deformation model based on both geometry and texture variations that is able to synthesise faces within a learned low-dimensional manifold. By computing the residuals between the actual views and the synthesised face rendering, the optimal parameterisation of the face model can be found. In the seminal paper [2] this is done in a standard penalised optimisation framework, while newer approaches cast the parameter optimisation into a deep learning framework [42]. These methods work well if the learned model is broad enough to fit all new faces, and can successfully recover the face pose. Unusual facial expressions or pathologies might confuse the methods, since they fall outside the learned appearance manifold.

Placing landmarks purely on the basis of surface geometry is described in [16], where landmarks are placed by computing the correspondence between a template face and a given face. The correspondence is computed by minimising bending energy between surface patches of the reference face and the target face. In [10] a machine learning approach is described where a set of local geometrical descriptors are extracted from facial scans and used to locate landmarks. The descriptors include surface curvature similar to what we propose. The concept of using local 3D shape descriptors to locate landmarks is also exploited in [30], where a facial landmark model is fitted to candidate locations found using curvature and local shape derivatives.

In this work, we use facial surfaces from a range scanner, where the data consist of surfaces with associated textures. We also use surfaces extracted as iso-surfaces from magnetic resonance (MR) images of the human head, where the face is only represented with its geometry. From a geometric point of view, the face is then a 2D surface embedded in 3D space and it is topologically equivalent to a disc. Recently, the analysis of this type of data using deep learning has seen a drastic increase under the term *geometric deep learning* [5], where one focus is the transformation of the representation of, for example, triangulated surfaces into a canonical representation that is suited for convolutional neural networks [17]. Spectral analysis of surfaces is one approach to the required domain adaptation as described in, for example, [3]. Another approach is to use a volumetric representation where the surface is embedded in a 3D volume [31] and the data is transformed into a volumetric occupancy grid. The goal of the method described in [31] is to classify entire objects into a set of predefined classes. Volumetric methods are still hampered by the drastic memory requirements for true 3D processing. In [31] the volume size is restricted to $30 \times 30 \times 30$, thus severely limiting the spatial resolution. An alternative approach is to render the surface or scene from multiple views and use a standard image based convolutional neural network (CNN) on the rendered views. This is described and analysed in detail in [31,37], the conclusion being that with the current memory limits and CNN architectures, multi-view approaches outperform volumetric methods. Multi-view approaches are conceptually similar to picking up an object and turning it around while looking (with one eye) for features or to identify the object. In [38], multi-view CNNs are used for silhouette prediction with convincing results. However, there is rapid development in all approaches such as the voxel based method in [33], which can also estimate the orientation of the object in question.

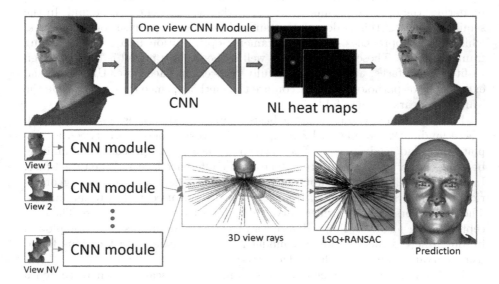

Fig. 1. Overall system overview. A 3D facial scan is rendered from several views that are fed to a CNN. After a 3D view ray voting process, the result is accurately placed 3D landmarks on the face surface.

In this paper, we propose a multi-view approach to identify feature points on facial surfaces. From one direction, a feature point can be seen as a ray in 3D space. By combining results from several views, a 3D landmark can be estimated as a consensus between *feature rays*, similarly to classic approaches from multi-view geometry [21]. A complete overview of the system can be seen in Fig. 1. A similar idea is presented in [15], where a depth image from a hand is synthesised into three orthogonal views. A CNN is then trained to generate 2D heatmaps of hand landmark locations for each of these three views. The final 3D landmark locations are found by fusing the heatmap probability distributions as a set of 3D Gaussian distributions and applying a learned parameterised hand-pose subspace. We propose using more 3D projections (in this paper, 100) and an outlier-robust method to fuse the individual heatmap results. Compared to the network used in [15], we propose a deeper network based on the stacked hourglass network that currently gives state-of-the-art results for human pose tracking and human face landmark detection [7, 8, 26, 39].

The inspiration for this work comes from recent advances in human pose tracking, where very deep convolutional networks have been trained to identify feature points on humans in a variety of poses and environments [26]. The method is based on heatmap regression, where each individual landmark is coded as a heatmap. The idea behind the used *hourglass network* is based on an aggregation of local evidence [6, 8], where heatmaps of individual landmarks are created in the first part of the network and fed into the next layer, thus enabling the network to refine its knowledge of the spatial coherence of landmark patterns. A recent paper [23], estimates 3D landmarks from 2D photos using the joint correspondence between frontal and profile landmarks and that method gets state-of-the-art results on standard test sets used for facial tasks. We propose a network architecture resembling the architecture described in [23] but our end goal is different.

In our work, the metric is the landmark localisation error measured in physical units (in this case, mm) since the end application is often related to physical estimation of facial morphology. However, in most state-of-the-art methods in landmark placement based on facial 2D photos, the accuracy is measured as Normalised Mean Error (NME), which is the average of landmark errors normalised by the bounding box size [23], which is not a particular good measure when working with 3D surfaces. In the work of [8, 23], 3D landmarks are estimated purely from 2D photos by regressing the unknown z-coordinate using a CNN network. The landmark accuracy in these cases is highly dependent on the heatmap resolution. They also do not utilise the availability of a true underlying 3D surface. In our work, we demonstrate a novel way to combine the output from state-of-the-art networks in a geometric based consensus to produce highly accurate 3D landmark predictions.

2   Methods

A summary of the method can be seen in Fig. 1. The overall idea is that the 3D surface is rendered from multiple views and for each view, landmark candidates are estimated. Each landmark estimate is now considered as a ray in 3D space and the final landmark 3D position is estimated using an outlier robust least squares estimate between *landmark rays*.

2.1 Multi-view Rendering

The scans are rendered using an OpenGL rendering pipeline. The scan is placed approximately at the origin of the coordinate system and the camera is placed in 100 random positions around the face, and with the focal point at the origin. The parameters of the camera are determined so the entire scan is in view. We use an orthographic projection. For each view, we render a view of the scan and store the 2D coordinates of the projected ground-truth 3D landmarks. A simple ambient white light source is used. All surfaces are rendered in a monochrome non-texture setup and surfaces with texture are also rendered with full RGB texture colours.

It is an ill-posed problem to pre-align a general surface scan to a canonical direction as also described in [37]. However, for facial surfaces it is common to start the pipeline by identifying feature points such as the eyes and the nose and using them to pre-align the scan or to crop an area of interest as in [19]. In this work, we do not rely on pre-alignment and simply have the loose assumption that the scanned surface contains a face among other information, such as shoulders, and that an approximate direction of the face is given by the scanner. It also means that we do not assume that a given camera position can be in any fixed position with relation to the facial anatomy. This makes the algorithm general and not specific to facial anatomies. The rendering pipeline is used in both generating 2D training images and computing the landmarks on an unseen 3D surface.

2.2 Geometric Derivatives

To enable the implicit use of geometry, a set of geometric representations are also rendered. The first is a distance map representation that is computed as being the OpenGL z-buffer, where the precision has been optimised by setting the near and far clipping planes as close as possible to the scans bounding box. Using a depth map for human feature recognition was popularised in the seminal articles on pose recognition using depth sensors [35] and later applied in, for example, 3D estimation of face geometry from 2D photos [34]. The standard geometry-only view is also rendered by disabling the texture. This is the representation used in the multi-view papers [31,37].

While curvature is implicitly represented in the depth map, we also render a view where the surface is grayscale coded according to the local mean curvature. Our aim is to use the method on surfaces containing surface noise, and therefore a robust curvature estimation is needed. Traditional methods that estimate curvature based on 1-ring neighbours of a vertex are too noisy. We use an approach where for each vertex P, the algorithm finds the curvature of the sphere that passes through P and that best approximates the set of neighbours P_N of P.

The curvature is estimated by first doing an inverse projection of neighbouring points, so they are lying on an approximate plane, and then performing an eigenvector analysis of this point cloud. In this work, neighbour points P_N are found by a mesh based region growing algorithm that includes points connected to P at a maximal distance empirically chosen to be 10 mm. Proofs and implementation details can be found in [12, 28].

2.3   Network Architecture and Loss Function

We use the two-stack hourglass model described in [26], which is based on the residual blocks described in [22]. We focus on having a higher resolution of the predicted heatmap than in previous work. After the heatmap prediction in the hourglass block, the heatmap is upsampled twice using nearest neighbours followed by a 3×3 learnable convolutional layers as suggested in [27]. The input to the network for a single view is 2D renderings of the 3D surface. The network is flexible with the number of input layers. Using renderings of the textured surface (RGB) adds three layers, while the depth rendering, the geometry rendering and the curvature rendering each add one layer. The input images are 256×256, the dimensions throughout the hourglass stacks are 128×128 and the heatmap is upsampled to 256×256. The entire network can be seen in Fig. 2. The ground truth is one heatmap per landmark, with a Gaussian kernel placed at the projected 2D position of the landmark. A cross entropy loss function is used. The heatmap estimates from the first and second hourglass modules are concatenated together to form a combined loss function, as demonstrated in previous work [26]. This ensures intermediate supervision of the network. Only one network is used and is able to recognise landmarks from all view directions.

The network is implemented with Tensorflow and trained using RMSPROP with an initial learning rate of 0.00025, a decay of 0.96 and a decay step of 2000. The batch size is between 4 and 8, the drop-out rate is 0.02. The network converged after around 60–100 epochs depending on the used rendering input. The network was trained and evaluated on one NVIDIA Titan X GPU card. The network is not optimised with respect to processing time and currently, it takes approximately 20 s to process 100 renderings, including overhead for reading and storing intermediate results.

2.4   Landmark Detection and Consensus Estimation

For a given input image, the output of the network are NL heatmaps, where NL is the number of landmarks. A 2D landmark is found as the position of the maximum value in the associated heatmap. This is illustrated in the top box of Fig. 1. However, to avoid using landmarks that are obviously not located correctly, only landmarks belonging to heatmap maxima over a certain threshold are considered. We have experimentally chosen a threshold of 0.5. When a 2D landmark candidate has been found, the parameters of the corresponding 3D ray are computed using the inverse camera matrix used (and stored) for that rendering. Details of camera geometry and view rays can be found in [21].

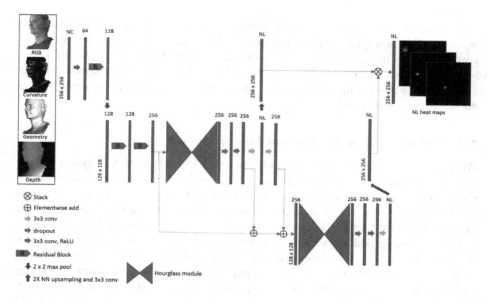

Fig. 2. Network architecture. The input is an image with a varying number of channels (NC) and the output is one heatmap per landmark. NL is the number of landmarks. Blue boxes are feature maps and the number on top is the number of feature maps. The spatial size of the layer is written in the lower left corner (where necessary). (Color figure online)

This is illustrated in the bottom part of Fig. 1 (3D view rays). For a given landmark, this results in up to NV rays in 3D space, where NV is the number of rendered views.

When the set of potential landmark rays for a given landmark has been computed, the landmark can be found as the crossing of these rays. In practice, the rays will not meet in a single point and some of the rays will be outliers due to incorrect 2D landmark detections. In order to robustly estimate a 3D point from several potentially noisy rays, we use a least squares (LSQ) fit combined with RANSAC selection [14]. When each ray is defined by an origin $\mathbf{a}_i$ and a unit direction vector $\mathbf{n}_i$, the sum of squared distances from a point $\mathbf{p}$ to all rays is:

$$\sum_i \mathbf{d}_i^2 = \sum_i \left[(\mathbf{p} - \mathbf{a}_i)^T (\mathbf{p} - \mathbf{a}_i) - \left[(\mathbf{p} - \mathbf{a}_i)^T \mathbf{n}_i \right]^2 \right] \tag{1}$$

Differentiating with respect to $\mathbf{p}$ results in a solution $\mathbf{p} = \mathbf{S}^+ \mathbf{C}$, where $\mathbf{S}^+$ is the pseudo-inverse [1] of $\mathbf{S}$. Here $\mathbf{S} = \sum_i (\mathbf{n}_i \mathbf{n}_i^T - \mathbf{I})$ and $\mathbf{C} = \sum_i (\mathbf{n}_i \mathbf{n}_i^T - \mathbf{I}) \mathbf{a}_i$

In the RANSAC procedure, the initial estimate of $\mathbf{p}$ is based on three random rays and the residual is computed as the sum of squared distances from the included rays to the estimated $\mathbf{p}$. The result is a robust estimate of a 3D point based on a consensus between rays where outlier rays are not included (Fig. 1,

LSQ+RANSAC). The final 3D landmark estimation is done by projecting the found point to the closest point on the target surface using an octree based space-division search algorithm. The result can be seen in Fig. 1(right).

3 Data

DTU-3D. The dataset consists of facial 3D scans of 601 subjects acquired using a `Canfield Vectra M3` surface scanner. The scanner accuracy is specified to be in the order of 1.2 mm (triangle edge length). Each face has been annotated with 73 landmarks using the scheme described in [13] and seen in Fig. 3. The faces in this dataset are all captured with a neutral expression. The data are used without being cropped, so both ears, neck and shoulders are partly present (see Fig. 1 for an example scan from the database).

BU-3DFE. The database contains 100 subjects (56 female, 44 male). Each subject performed seven expressions in front of the 3D face scanner, resulting in 2,500 3D textured facial expression models in the database [40]. Each scan is annotated with 83 landmarks and the faces have been cropped to only contain the facial region.

MR. One Magnetic Resonance (MR) volume of a human head acquired using a 3T Siemens Trio scanner with a T1-weighted MEMPRAGE sequence with an isotropic voxel size of 1 mm. We use the N3 algorithm [36] for bias field correction and the intensity normalisation approach from [11]. The outer skin surface is extracted using marching cubes [25] with an experimentally chosen iso-surface value of 20. The surface can be seen in Fig. 4(right); the skin surface is noisy due to inconsistences in MR values around the skin-air interface.

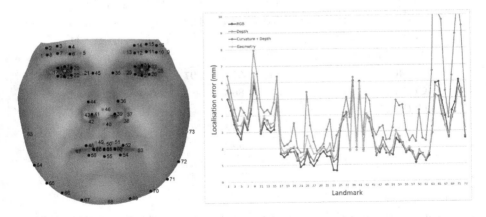

Fig. 3. (**Left**) The 73 landmark scheme used on the **DTU-3D** database [13]. (**Right**) landmark localisation errors on the **DTU-3D** database with different rendering configurations.

4 Results and Discussion

The **DTU-3D** data were divided into a training set with 541 faces and a validation set of 60 faces. For both the training and the validation set, the faces were rendered from 100 camera positions using RGB, geometry, curvature, and depth rendering. The network is trained and validated using different render combinations (as seen in Fig. 3). The **BU-3DFE** was divided into a training set with all scans belonging to 46 females and 34 males and a validation set with all scans belonging to 10 females and 10 males. This results in a training set of 2000 faces and a validation set of 500 faces. The network using **BU-3DFE** was trained and evaluated using RGB renderings.

For a given validation face, the 3D landmarks are computed using the proposed method and compared with the ground truth landmarks. The landmark localisation error is computed as the Euclidean distance between an annotated landmark and an estimated landmark. We report the mean and standard deviation of the landmark localisation error per landmark in millimeters as seen in Table 1 and Fig. 3.

Table 1. Results on the **BU-3DFE** database. Error is in mm. Improvements compared to previous work are in percentages. The number in parentheses after the landmark id is the corresponding landmark on Fig. 3

	Salazar [32]		Gilani [16]		Grewe [19]		This paper	
Images	350		2500		2500		500	
	Mean	Impr.	Mean	Impr.	Mean	Impr.	Mean	SD
Ex(L) (17)	9.63	73%	4.43	41%	2.95	11%	2.59	1.53
En(L) (21)	6.75	71%	4.75	64%	3.04	37%	1.89	0.98
Ex(R)(25)	8.49	66%	4.34	33%	3.22	11%	2.85	1.5
En(R) (29)	6.14	70%	3.29	33%	3.23	44%	1.8	0.89
Ac(L) (42)	6.47	59%	4.3	38%			2.61	1.41
Ac(R) (38)	7.17	58%	4.28	29%			2.96	1.56
Sn (40)			3.9	31%	1.97	−29%	2.52	1.69
Ch(L) (47)			6	86%			2.182	1.44
Ch(R) (53)			5.45	68%			2.42	1.44
Ls (50)			3.2	19%			2.33	1.31
Li (55)			6.9	99%			2.5	1.41
Mean	7.44		4.62		2.88		2.42	

As can be seen in Fig. 3, the method locates a large set of landmarks with a localisation error in the range of 2 mm. This is in the limit of what an experienced operator can achieve [13]. The landmarks that have high errors are also the landmarks that are typically very difficult for a human to place. Landmarks placed on the chin and on the eyebrows are very hard to place manually, due to the weak anatomical cues. An extensive analysis of landmark errors can be found in [13], where the inter-observer variance is reported for the landmark set used in the **DTU-3D**. It was found that the landmark around the chin has a manual inter-observer in the range of 6 mm. The errors in Fig. 3 for the **DTU-3D** base is probably more due to inconsistency in the manual annotation than from the presented method. The method handles uncropped 3D scans where both partial hair, ears, neck and shoulders are present as seen in Fig. 1, meaning that the only pre-processing step needed is a rough estimate of the overall direction of the head.

In Fig. 3 it can be seen that rendering using RGB textures generally performs very well. This is not surprising since textured surfaces contain many visual cues. The mode where depth is used in combination with the curvature also performs very well. This result enables the method to be used on data where RGB texture is not naturally present, such as iso-surfaces from modalities like computed tomography (CT) or MR imaging. Using the depth layer alone yields reasonable results on many landmarks, but in particular around the chin, where few visual cues are seen in the depth image, the errors are large. Using `geometry` rendering yields slightly worse results than the `RGB` and `depth+curvature` renderings, but still on par with the state-of-the-art algorithms. We also tested the `RGB + depth + curvature + geometry` configuration but did not achieve superior results.

The generalisability of the method was tested on the **MR** scan using `depth + curvature` rendering. The method successfully locates landmarks on the MR iso-surface despite significant noise. The landmarks around the eyes and nose have an error level in the range of 2–3 mm, while the errors on the chin are larger, as also seen in the results on the pure range scan. It can be seen that the estimation of the curvature yields reasonable visual results despite the very high noise levels on the surface. Placing landmarks in 3D volumes is notoriously difficult. While deep learning approaches are being used for true 3D landmark localisation [41], the methods are still limited by the prohibitive GPU memory requirements for true 3D processing. We believe that our approach offers an alternative way to handle complicated landmark placement in 3D volumes.

We have experimented with the number of views, using 25, 50, 75, and 100 view renderings. Going from 25 to 100 views decreases the landmark localisation error with less than a millimeter, meaning that in future applications the number of renderings could be significantly lowered. We have chosen to use random camera positions instead of selecting a fixed set of pre-defined positions. We believe that using random positions works as an extra data augmentation step. Finally, the results on the benchmark data set **BU-3DFE** as seen in Table 1 demonstrate that the proposed method outperforms state-of-the-art methods by a large margin.

Compared to the methods that are dominant for 2D photos faces-in-the-wild as for example [7,8,23], our metric is a landmark localisation error in mm, while they report the error as given as a percentage compared to either the eye-to-eye distance [7] or the bounding box diagonal length [8,23]. The average adult

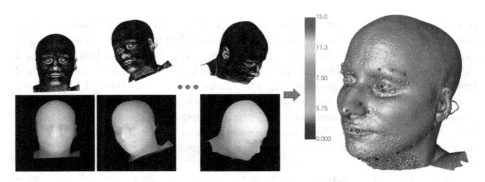

Fig. 4. Finding landmarks on iso-surfaces extracted from an MR image using combined curvature and depth rendering. (**Left**) examples of the rendered curvature and depth images. (**Right**) the 3D landmarks identified on the skin iso-surface. The colour coding is the localisation error compared to manually annotated landmarks. (Color figure online)

eye-to-eye distance is 62 mm. In [7] they report an error rate of 4.5% that is equivalent to 2.8 mm. This is a very impressive result but their measured error is still above what we report.

Since the 3D geometry of the face is available, it is possible to do a view-ray versus surface intersection test to for example determine if a landmark is placed on an occluded part of the surface. We have chosen not do that, since the method in practise is good at predicting the positions of landmarks on occluded parts of the face due to the high spatial correlation of landmark positions as for also demonstrated in [7,8,23]. Furthermore, the landmarks with low confidence are removed in the RANSAC procedure and this is often landmarks that from a given viewpoint are hard to see due to for example occlusion.

Future works includes improved maxima selection in the heatmap based on more rigorous statistical assumptions and using a fitting function to locate maxima with sub-pixel precision. In addition, different renderings and surface signature techniques to further enhance surface structure can be exploited. Furthermore, with increased GPU memory in sight, it will be interesting to design novel network architectures with higher spatial resolutions in mind and also test other blocks like the inception-resnet blocks used in [23].

5   Conclusion

In this paper, we demonstrated a multi-view pipeline to accurately locate 3D landmarks on facial surfaces. The method outperforms previous methods by a large margin with respect to landmark localisation error. Furthermore, the effect of using different rendering types was demonstrated, suggesting that a model trained on high-resolution 3D face scans could be used directly to accurately predict landmarks on surfaces from a completely different scanning modality.

It was demonstrated that using a combination of surface curvature rendering and depth maps performs on par with using RGB texture rendering, proving that implicit geometric information can be very valuable even when observed in 2D.

While geometric deep learning is a rapidly growing field and volumetric methods are gaining foothold, this paper shows that the concept of multi-view rendering of 3D surfaces currently produces state-of-the-art results with regard to feature point location.

Code, trained models and test data can be found here: http://ShapeML.compute.dtu.dk/

Acknowledgements. We gratefully acknowledge the support of NVIDIA Corporation with the donation of the Titan Xp GPU used for this research.

References

1. Ben-Israel, A., Greville, T.N.: Generalized Inverses: Theory and Applications. Springer, Heidelberg (2003). https://doi.org/10.1007/b97366
2. Blanz, V., Vetter, T.: A morphable model for the synthesis of 3D faces. In: Proceedings of Computer Graphics and Interactive Techniques, pp. 187–194 (1999)
3. Boscaini, D., Masci, J., Rodolà, E., Bronstein, M.: Learning shape correspondence with anisotropic convolutional neural networks. In: Proceedings of NIPS, pp. 3189–3197 (2016)
4. Bowyer, K.W., Chang, K., Flynn, P.: A survey of approaches and challenges in 3D and multi-modal 3D+ 2D face recognition. Comput. Vis. Image Underst. **101**(1), 1–15 (2006)
5. Bronstein, M.M., Bruna, J., LeCun, Y., Szlam, A., Vandergheynst, P.: Geometric deep learning: going beyond Euclidean data. IEEE Signal Process. Mag. **34**(4), 18–42 (2017)
6. Bulat, A., Tzimiropoulos, G.: Convolutional aggregation of local evidence for large pose face alignment. In: Proceedings of BMVC (2016)
7. Bulat, A., Tzimiropoulos, G.: Two-Stage convolutional part heatmap regression for the 1st 3D Face Alignment in the Wild (3DFAW) challenge. In: Hua, G., Jégou, H. (eds.) ECCV 2016. LNCS, vol. 9914, pp. 616–624. Springer, Cham (2016). https://doi.org/10.1007/978-3-319-48881-3_43
8. Bulat, A., Tzimiropoulos, G.: How far are we from solving the 2D & 3D face alignment problem? (and a dataset of 230,000 3D facial landmarks). arXiv preprint arXiv:1703.07332 (2017)
9. Chang, J.B., Small, K.H., Choi, M., Karp, N.S.: Three-dimensional surface imaging in plastic surgery: foundation, practical applications, and beyond. Plast. Reconstr. Surg. **135**(5), 1295–1304 (2015)
10. Creusot, C., Pears, N., Austin, J.: A machine-learning approach to keypoint detection and landmarking on 3D meshes. International journal of computer vision **102**(1–3), 146–179 (2013)
11. Dale, A.M., Fischl, B., Sereno, M.I.: Cortical surface-based analysis: I. segmentation and surface reconstruction. Neuroimage **9**(2), 179–194 (1999)
12. Delingette, H.: Modélisation, Déformation et Reconnaissance d'Objets Tridimensionnels á l'Aide de Maillages Simplexes. Ph.D. thesis, L'École Centrale de Paris (1994)

13. Fagertun, J., et al.: 3D facial landmarks: Inter-operator variability of manual annotation. BMC Med. Imaging **14**(1), 35 (2014)
14. Fischler, M.A., Bolles, R.C.: Random sample consensus: a paradigm for model fitting with applications to image analysis and automated cartography. In: Readings in Computer Vision, pp. 726–740. Elsevier (1987)
15. Ge, L., Liang, H., Yuan, J., Thalmann, D.: Robust 3D hand pose estimation in single depth images: from single-view CNN to multi-view CNNs. In: Proceedings of CVPR, pp. 3593–3601 (2016)
16. Gilani, S.Z., Shafait, F., Mian, A.: Shape-based automatic detection of a large number of 3D facial landmarks. In: Proceedings of CVPR, pp. 4639–4648. IEEE (2015)
17. Goodfellow, I., Bengio, Y., Courville, A., Bengio, Y.: Deep Learning. MIT press, Cambridge (2016)
18. Gordon, G.G.: Face recognition based on depth and curvature features. In: Proceedings 1992 IEEE Computer Society Conference on Computer Vision and Pattern Recognition, pp. 808–810. IEEE (1992)
19. Grewe, C.M., Zachow, S.: Fully automated and highly accurate dense correspondence for facial surfaces. In: Hua, G., Jégou, H. (eds.) ECCV 2016. LNCS, vol. 9914, pp. 552–568. Springer, Cham (2016). https://doi.org/10.1007/978-3-319-48881-3_38
20. Hammond, P., et al.: 3D analysis of facial morphology. Am. J. Med. Genet. Part A **126**(4), 339–348 (2004)
21. Hartley, R., Zisserman, A.: Multiple View Geometry in Computer Vision. Cambridge University Press, Cambridge (2003)
22. He, K., Zhang, X., Ren, S., Sun, J.: Deep residual learning for image recognition. In: Proceedings of CVPR, pp. 770–778 (2016)
23. Deng, J., Zhou, Y., Cheng, S., Zafeiriou, S.: Cascade multi-view hourglass model for robust 3D face alignment. In: FG (2018)
24. Jourabloo, A., Liu, X.: Pose-invariant 3D face alignment. In: Proceedings of ICCV, pp. 3694–3702 (2015)
25. Lorensen, W.E., Cline, H.E.: Marching cubes: a high resolution 3D surface construction algorithm. In: ACM Siggraph Computer Graphics, vol. 21, pp. 163–169. ACM (1987)
26. Newell, A., Yang, K., Deng, J.: Stacked hourglass networks for human pose estimation. In: Leibe, B., Matas, J., Sebe, N., Welling, M. (eds.) ECCV 2016. LNCS, vol. 9912, pp. 483–499. Springer, Cham (2016). https://doi.org/10.1007/978-3-319-46484-8_29
27. Odena, A., Dumoulin, V., Olah, C.: Deconvolution and Checkerboard Artifacts. Distill (2016) https://doi.org/10.23915/distill.00003
28. Paulsen, R.R.: Statistical shape analysis of the human ear canal with application to in-the-ear hearing aid design. Ph.D. thesis, Technical University of Denmark (2004)
29. Paulsen, R.R., Marstal, K.K., Laugesen, S., Harder, S.: Creating ultra dense point correspondence over the entire human head. In: Sharma, P., Bianchi, F.M. (eds.) SCIA 2017. LNCS, vol. 10270, pp. 438–447. Springer, Cham (2017). https://doi.org/10.1007/978-3-319-59129-2_37
30. Perakis, P., Passalis, G., Theoharis, T., Kakadiaris, I.A.: 3D facial landmark detection under large yaw and expression variations. IEEE Transact. Pattern Anal. Mach. Intell. **35**(7), 1552–1564 (2013)

31. Qi, C.R., Su, H., Nießner, M., Dai, A., Yan, M., Guibas, L.J.: Volumetric and multi-view cnns for object classification on 3D data. In: Proceedings of CVPR, pp. 5648–5656 (2016)
32. Salazar, A., Wuhrer, S., Shu, C., Prieto, F.: Fully automatic expression-invariant face correspondence. Mach. Vis. Appl. **25**(4), 859–879 (2014)
33. Sedaghat, N., Zolfaghari, M., Amiri, E., Brox, T.: Orientation-boosted Voxel nets for 3D object recognition. In: British Machine Vision Conference (BMVC) (2017)
34. Sela, M., Richardson, E., Kimmel, R.: Unrestricted facial geometry reconstruction using image-to-image translation. arXiv (2017)
35. Shotton, J., et al.: Real-time human pose recognition in parts from single depth images. In: Proceedings of CVPR, pp. 1297–1304 (2011)
36. Sled, J.G., Zijdenbos, A.P., Evans, A.C.: A nonparametric method for automatic correction of intensity nonuniformity in MRI data. IEEE Transact. Med. Imaging **17**(1), 87–97 (1998)
37. Su, H., Maji, S., Kalogerakis, E., Learned-Miller, E.: Multi-view convolutional neural networks for 3D shape recognition. In: Proceedings of the IEEE International Conference on Computer Vision, pp. 945–953 (2015)
38. Wiles, O., Zisserman, A.: SilNet: single-and multi-view reconstruction by learning from silhouettes. In: Proceedings of BMVC (2017)
39. Yang, J., Liu, Q., Zhang, K.: Stacked hourglass network for robust facial landmark localisation. In: Proceedings of CVPR, pp. 2025–2033. IEEE (2017)
40. Yin, L., Wei, X., Sun, Y., Wang, J., Rosato, M.J.: A 3D facial expression database for facial behavior research. In: Proceedings of FGR, pp. 211–216. IEEE (2006)
41. Zheng, Y., Liu, D., Georgescu, B., Nguyen, H., Comaniciu, D.: 3D deep learning for efficient and robust landmark detection in volumetric data. In: Navab, N., Hornegger, J., Wells, W.M., Frangi, A.F. (eds.) MICCAI 2015. LNCS, vol. 9349, pp. 565–572. Springer, Cham (2015). https://doi.org/10.1007/978-3-319-24553-9_69
42. Zhu, X., Lei, Z., Liu, X., Shi, H., Li, S.Z.: Face alignment across large poses: a 3D solution. In: Proceedings of CVPR, pp. 146–155 (2016)

Author Index